3rd International Symposium of Earth, Energy, Environmental Science, and Sustainable Development

IOP Conference Series: Earth and Environmental Science
Volume 1111

Depok, Indonesia
27 – 28 August 2022

Editor:

Herdis Herdiansyah

ISBN: 978-1-7138-6521-6
ISSN: 1755-1307

Printed from e-media with permission by:

Curran Associates, Inc.
57 Morehouse Lane
Red Hook, NY 12571

CURRAN ASSOCIATES INC.
proceedings
.com

Some format issues inherent in the e-media version may also appear in this print version.

Institute of Physics
Dirac House, Temple Back
Bristol BS1 6BE UK

Phone: 44 1 17 929 7481
Fax: 44 1 17 920 0979

techtracking@iop.org

Additional copies of this publication are available from:

Curran Associates, Inc.
57 Morehouse Lane
Red Hook, NY 12571 USA
Phone: 845-758-0400
Fax: 845-758-2633
Email: curran@proceedings.com
Web: www.proceedings.com

TABLE OF CONTENTS

IOP Conf. Series: Earth and Environmental Science 1111 (2022) 011001 doi:10.1088/1755-1315/1111/1/011001

Preface

This year, Journal of Environmental Science and Sustainable Development (JESSD) is proud to hold its second international virtual symposium, featuring world-class speakers and editors worldwide. We were incredibly honored to have invited RR. Ratih Dyah Kusumawati, Ph.D. from Center for Resilience and Sustainability Studies (CRESS), Department of Management, Faculty of Economics and Business, Universitas Indonesia, Assoc. Prof. Hui Nee Au Yong, Ph.D. from Faculty of Business and Finance, Universiti Tunku Abdul Rahman (UTAR), Mari Eko Mulyani, Ph.D. from School of Geography and The Environment, University of Oxford, Oxford, United Kingdom and School of Environmental Science, Universitas Indonesia, Indonesia, Assoc. Prof. Hironori Hamasaki from Graduate School of Fisheries and Environmental Sciences, Nagasaki University, Japan, Dian Novarina, M.Sc. from Deputy Director of Sustainability and Stakeholder Engagement, PT Riau Andalan Pulp and Paper (APRIL), Jessica Hanafi, Ph.D. from Executive Director PT. Life Cycle Indonesia, and Dr. Ir. Haruki Agustina M.Sc. from Director of Contaminated Land Recovery and Emergency Response Hazardous and Toxic Waste and Non-Hazardous and Toxic Materials, Ministry of Environment and Forestry.

The symposium was held in virtual format due to related restrictions during COVID-19 pandemic. I appreciate everyone's participation in the second JESSD Symposium: International Symposium of Earth, Energy, Environmental Science, and Sustainable Development, despite the restrictive situation caused by the global COVID-19 pandemic. Instead of being postponed, this symposium was conducted virtually in response to emerging issues mainly related to earth science, energy, environmental science, and sustainability. Moreover, this symposium also facilitates the publications of qualified research on COVID-19 to contribute to the global response towards the pandemic situation.

This symposium is slightly different from the previous symposium and conference held last year. By making the best use of development in telecommunication platforms, we are able to gather virtually to share our insights and progress covering numerous subjects ranging from: the broad earth science, energy, and environmental science, and the special issues of COVID-19 Research as well as Community Engagement for Better Environment. This symposium utilized Zoom and Youtube, a widely known cloud-based video conferencing tool, as its telecommunication platform.

The symposium was held on 27-28 August 2022. Committees are based in School of Environmental Science, Universitas Indonesia, Jakarta, Indonesia. We are aware that a virtual setup will not provide the same experience as an on-site symposium. However, it is in our nature as researchers to be undeterred by new challenges. The opportunity to host this event virtually allows us to facilitate cross-borders discussion and collaboration at lower cost and less impact on the environment through reducing carbon footprint. We have invited some prominent speakers from various regions to provide lectures on a wide range of environmental research. In addition, we have also invited world-class speakers and editors to help with your scientific publication process. The participants (presenters) are coming from 16 different countries all around the world, including: Malaysia, South Korea, Netherland, Ukraine, Thailand, Uzbekistan, Vietnam, Nigeria, Bangladesh, Russia, Brazil, Poland, China, Germany, Tunisia, and Indonesia.

We are glad that we received lots of submissions from the symposium and selected a bunch of high-quality papers and compiled them into the proceedings after reviewing them using the double-blind peer review procedure. The symposium is considered successful. Signal loss and some technical issues happened during parallel sessions but had been well and quickly addressed by technical assistants, without causing any significant delay. I express my deep gratitude to the editors: Editors committee: Dr. Ahyahudin Sodri, Dr. Herdis Herdiansyah, Mari E. Mulyani, D.Phil., Dr. Stefanie Steinebach, Prof. Dr. Ing. Ir. Misri Gozan, M.Tech., Dr. Djoni Hartono, Ratih Dyah Kusumastuti, Ph.D., Prof. Dr. Ivandini Tribidasari Anggraningrum, and committees Dr. Herdis Herdiansyah, Nova Amalia Sakina, Azhar Firdaus, Arty Dwi Januari, Hafizha Ilma Qadriina, Sinan Vidi Lazuardi, Salma Mar'atus Sholihah, and HIMPASILING UI Team for their time, hard work, and contribution to support this symposium during this challenging time.

I would like to take this opportunity also to express my gratitude for our sponsors, (Danone Indonesia, Induk Koperasi Kepolisian Negara Republik Indonesia (Inkopol), PT Arlindo Grafimedia, PT Riau Andalan Pulp and Paper (April Group), Bank Rakyat Indonesia, APP Sinar Mas, Harita Nickel, Greenera Consulting, Pertamina Patra Niaga, Starborn Chemical, PT. Prima Kana Energy, Waskita, Roatex, One Global Synergy, and Lintasarta) and also for Himpasiling (Himpunan Mahasiswa Pascasarjana Ilmu Lingkungan) Universitas Indonesia 2022 period for conducting the joint session. Last but not the least, we send our sincere thanks to the IOP publisher for publishing the proceedings.

Jakarta, 27 August 2022
Editor-in-Chief,

Dr. Herdis Herdiansyah

IOP Conf. Series: Earth and Environmental Science 1111 (2022) 011001 doi:10.1088/1755-1315/1111/1/011001

List of committees

Steering committee:

Chairman	: Dr. Ahyahudin Sodri
Editor-in-Chief	: Dr. Herdis Herdiansyah
Editorial Board	: Mari E. Mulyani, D.Phil.
	Dr. Stefanie Steinebach
	Prof. Dr. Ing. Ir. Misri Gozan, M.Tech.
	Dr. Djoni Hartono
	Ratih Dyah Kusumastuti, Ph.D.
	Prof. Dr. Ivandini Tribidasari Anggraningrum

Organizing committee:

Chairman	: Dr. Herdis Herdiansyah
Treasurer Section	: Nova Amalia Sakina
IT Section	: Azhar Firdaus
Program Section	: Arty Dwi Januari
	Hafizha Ilma Qadriina
	Sinan Vidi Lazuardi
	Salma Mar'atus Sholihah
Event Section	: HIMPASILING UI Team (https://sil.ui.ac.id/himpasiling-ui-2/)

IOP Conf. Series: Earth and Environmental Science 1111 (2022) 011002 doi:10.1088/1755-1315/1111/1/011002

Peer review declaration

All conference organizers/editors are required to declare details about their peer review. Therefore, please provide the following information:

- Type of peer review: Single-blind

Single-anonymous: authors' identities are known to the reviewers, reviewers' identities are hidden from authors

- Describe criteria used by Reviewers when accepting/declining papers. Was there the opportunity to resubmit articles after revisions? There are five main criteria: relevance to the conference topic, importance of the research/state-of-the-art, proper methodology, arguments and conclusions, and academic writing quality. All the criteria are included in the substantial review form (attached) provided for the reviewers. Organizers also prepared a technical review form (attached) to assess the compliance with IOP template and guidelines and to inform the result of plagiarism check. There was the opportunity to resubmit articles after revisions, up to a maximum of two (2) times based on substantial reviews.
- Conference submission management system: using Morressier Submission System, the complete submission guidelines are available in following link https://symposiumjessd.ui.ac.id/guideline/
- Number of submissions received: 193
- Number of submissions sent for review: 162
- Number of submissions accepted: 88
- Acceptance Rate (Number of Submissions Accepted / Number of Submissions Received X 100): 45,6%
- Average number of reviews per paper: 1-2
- Review process was done in a maximum of 14 days.
- Contact person for queries: Herdis Herdiansyah, School of Environmental Science, Universitas Indonesia, Jakarta, Indonesia, e-mail: herdis@ui.ac.id

Additional information

Review Period

- Review process will be done in a maximum of 14 days.

IOP Conf. Series: Earth and Environmental Science 1111 (2022) 011002 doi:10.1088/1755-1315/1111/1/011002

Substantial Review

1. Relevance to the Conference topic.
2. Importance of the research / State-of-the-art.
3. Proper methodology.
4. Arguments and conclusions.
5. Academic Writing quality.

Other Information

1. When the article is delivered to the publisher, every Author will receive Galley Proof. As an Author must agree with the editing version, except there is some correction from the Author.
2. Every Author will receive information and status of the article after submitted.

Submission Ethics

Ethics in JESSD Symposium is based on the Committee on Publication Ethics (COPE). Before submitting the article(s), completely represents the following statements.

1. The uploaded files have been carefully prepared based on the provided templates, both title page (title, author name(s), affiliation(s), abstract, and acknowledgements) and article (title, abstract, introduction, method, results and discussion, conclusion, and references)
2. The article represents a qualified scientific knowledge that might come from international collaborations
3. Author(s) has/have approved the article and agree(s) with the submission, therefore any withdrawal of the article is prohibited once it is submitted
4. Author(s) is/are willing to pay the symposium fee and publication fee charged by the organizers (symposium fee is mandatory charged for one author attendance virtually while publication fee is optionally charged before the symposium or after the article is listed in Scopus)
5. The article does not contain any form of plagiarism

6. Author(s) has/have made significant substantial contribution to the article, whether it is in the concept or design, acquisition of data, analysis and interpretation, or in all these areas
7. The article has not been published elsewhere and is not under consideration for publication elsewhere (either partly or wholly)
8. Author(s) has/have acknowledged any conflict of interest in the article, if necessary
9. Author(s) has/have acknowledged any source of funding in the article, if necessary
10. The article was written in English and carefully reviewed by a native English editor

Review Ethics

1. The author(s) is/are required to propose one reviewer and another one reviewer optionally. However, the organizers manage to prevent any conflict of interest that might happen during the review process. Therefore, please ensure that your suggested reviewer fits the following criteria.
2. He/she is an expert in the relevant field.
3. He/she has no competing interest with the author(s).
4. He/she has never been involved in any related work with the author(s).
5. He/she does not know the author(s) personally.
6. There is no certainty that the organizer will accept your suggested reviewer. The double-blind peer review will be conducted, therefore reviewers will not receive any author(s) information and vice versa.

Similarity Check

Each submitted article will be going to preliminary review automatically by the Morressier system. Moreover, the preliminary review will also consider return or direct rejection due to, but not limited to, following reasons.

1. Falling out of topic (Environmental Science)
2. Multiple submission or consideration for other publications.
3. Recommendation for rejection from reviewer.

Life Cycle Sustainability Assessment Review: Decision Making Category in Electrical-Vehicle Type

Salsa Bila Dian Islami[1], Faradhina Azzahra[1]*, and Diana Puspita Sari[1]
[1]Industrial Engineering Department, Faculty of Engineering, Diponegoro University, Semarang, 50275, Indonesia
[1]salsabdian@students.undip.ac.id, [2]*faradhinaazzahra@lecturer.undip.ac.id, [3]dianapuspitasari@lecturer.undip.ac.id

Abstract. The use of electric vehicles (EV) has begun to be widely used and is predicted to continue to increase. It has a positive impact on some aspects of the economy and society. Still, it has negative impacts, especially on the environmental aspect, such as hazardous and toxic waste, a huge water footprint, and exploitation of material for the production process, which cause damage to ecosystems. This study aims to figure out the advantages and disadvantages of EV types and provide recommendations on which one is the best EV for use. This study was done by the systematic review method that used the LCSA method, a combination of LCA, LCC, and S-LCA. This study assessed several types of electric vehicles: ICV, HEV, PHEV, and BEV. From an environmental aspect, ICV dominated the worst score than other types. From a social aspect, no specific EV contributed to a higher impact, but HEV had the highest reduction potential on human health. From an economic aspect, HEV were considered less profitable than the others. In the end, PHEV type was more potentially used due to giving a lower impact than the other types. Furthermore, this study could help public society decide the types of EV batteries based on their need.

1. Introduction

The demand for electric vehicles (EV) is increasing, which is predicted to reach 26.8% annual growth in 2030 starting in 2021 [1]. In 2030, it is predicted there will be 34.756 thousand EV units. The increasing demand is due to the wider access and availability of EV. In addition, eco-friendly products will be increasingly popular with awareness of the impact of emissions on the earth. Local emissions from gasoline-based vehicles also impact human health who live close to public transport infrastructure [2].

Based on the above explanation, this will lead to the demand for rechargeable batteries as EV fuel. Even if the emissions are low, there is still much to assess from an EV. This assessment is based on three pillars of Triple Bottom Line (TBL), which helps determine the long-term sustainability of electric vehicles. This should be assessed to achieve the target of Zero Carbon by 2050 in the environmental aspect. From a social aspect, it was found there were factories that produced rechargeable batteries for EV that employed minors [3], which is against the law and social norms. The economic aspect must also be considered to determine the long-term sustainability of EV.

The LCA method is a tool to assess the environmental effects of the production processes, transportation, use, and disposal phases, as well as the recovery and extraction of raw materials throughout the life cycle of products and activities [4, 5], especially this method is often used to evaluate the environmental impact of EV types. However, the LCA method is considered less effective in calculating uncertainties such as social aspects. To get the needs of other perspectives, this study uses the Montecarlo simulation technique to solve the uncertain nature of the LCSA of EV. The Life Cycle Sustainability Assessment (LCSA) framework is an approach that includes environmental, economic, and social dimensions. LCSA is a life cycle assessment technique or tool that integrates the three pillars of sustainability as an assessment dimension [6]. A life cycle assessment technique that focuses on environmental impact assessment is Life Cycle Assessment (LCA).

Meanwhile, Life Cycle Costing (LCC) is used if the assessment focuses on the economic impact, and Social Life Cycle Assessment (S-LCA) can be used to assess social impact. Many researchers from various fields have carried out research on LCSA. Research by [7] used LCSA to assess sustainability in wastewater management. Research by [8] used LCSA during the Spanish architect's Building Information Modelling (BIM) design phase.

The combination of LCA, LCC, and S-LCA can be used as an integrated technique to assess sustainability or make decisions according to the triple bottom line of sustainable development, namely planet, people, and profit [9]. In addition, input-output principles of the Triple Bottom Line assessment can also measure the three pillars of sustainability, which also takes a holistic aspect and makes it possible to evaluate a broad economic of the related systems [10]. This study aims to determine the advantages and disadvantages of these four types of EV from the TBL perspective and can provide recommendations for improvement from social, economic, and environmental perspectives.

There are various types of EV and batteries. The types of EV that the author examine further were the vehicles that are often used, namely: ICV, HEV, BEV, and PHEV. Figure 1 shows the visualization of each type of EV. ICV (Internal Combustion Engine) is a vehicle with a power supply from an ordinary internal combustion engine. ICV use fuel that combusts inside a combustion chamber with the help of an oxidizer (typical oxygen from the air). ICV most commonly use fuels derived from fossil fuels, including petrol, diesel, jet fuel, and compressed natural gas [11]. HEV (Hybrid Electric Vehicles) has a power supply from gas and electric motors as the driving force. This type of vehicle cannot be plugged in for recharging. The cars that use this technology are UMass Transit, Toyota Prius Japan, and Lohner-Porsche [12]. PHEV (Plug-in Hybrid Electric Vehicles) are vehicles powered by an engine and an electric motor as the driving force. Like regular hybrids, the PHEV can recharge its batteries via regenerative braking. In addition, power from a single gas tank can provide a PHEV of up to several hundred miles. The cars that use this technology are Audi A3 E-Tron, Mercedes S550e, and Toyota Prius [12]. BEV (Battery Electric Vehicle) has the full power of a gasoline engine and fully rechargeable batteries. BEV are zero-emission vehicles, with their energy fully coming from the rechargeable battery pack. The cars that use this technology are Tesla Model S and Nissan Leaf [12].

With this research, in the form of a literature review, it is hoped that it will be able to find out categories as a decision-making parameter with a Triple Bottom Line (TBL) perspective, helping industry and other researchers in their study. The most frequently used vehicle is the one called BEV. It is fully electrical and rechargeable.

2. Methods

This study is represented by exploratory and descriptive analysis with qualitative & quantitative passages to deconstruct existing research concerning LCSA for EV. The methodology of this exploration relies on [13].

2.1. Delimitations of the exploration

For the choice of the exploration, some delimitation benchmarks were considered. Just scientific papers published in journals were regarded. As for the times of publication, there was no delimitation since the topic of LCSA is relatively new. Still, this exploration comprises all the papers published until medial of 2021. The databases were demarcated, conforming to the program's benchmarks applied for data analysis. This study analyses the database Scopus. These databases contain the most significant number of peer-reviewed bibliographic citations and abstracts from multidisciplinary fields.

IOP Conf. Series: Earth and Environmental Science 1111 (2022) 012001 doi:10.1088/1755-1315/1111/1/012001

2.2. *Procedure for choosing the journals*

The choice of the journals commenced with some keywords explored in the databases. The keyword combinations were "Life Cycle Sustainability Assessment" or "Life Cycle Sustainability Analysis" or "LCSA"; "Electrical Vehicle"; "Sustainable Transportations". The additional benchmarks of the papers applied were (i) considering the environmental, economic, and social aspects and (ii) within the scope of the LCSA analysis, consisting of further evaluation using the LCA method.

2.3. *Systematic analysis*

The systematic analysis consisted of content analysis. Initially, a categorization of the papers corresponding to the type of study; sections of studies' operations; procedures for each life cycle analysis (LCA, LCC, and S-LCA) applied or elaborated; environmental, economic, and social indicators applied in the studies; and the final study results. The types of the study were categorized as the operation of procedures and elaboration of methodology and operation, review, or others that would be theoretical discussion papers on LCSA. The operation studies were detailed regarding the operation section in transportation. The final study results of the explorations were estimated as to how these results were offered, with or without data aggregation, if the individual results of each life cycle analysis were assessed and if some methodology was employed, or if there was the aggregation of the results of the analyses in a single sustainability score, and which system was used.

3. Results and Discussion

Each battery's environmental, social, and economic assessment is carried out according to their respective indicators in each aspect or assessment. With Life Cycle Sustainability Assessment (LCSA), research objectives are described. In this study, especially within the environmental assessment, the cradle-to-grave stage was used.

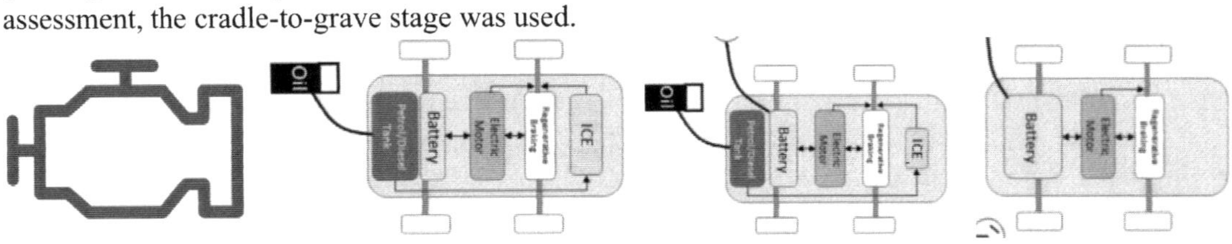

(a) (b) (c) (d)

Figure 1. Visualization of Electrical Vehicle Types: (a) ICV; (b) HEV; (c) PHEV; and (d) BEV *[11, 14]*.

Before starting the assessment of each aspect of EV, there are some characteristics of several EV types that can be advantages and disadvantages, as shown in table 1.

Table 1. Characteristics of Electric Vehicle Types.

Electric Vehicle Type	Battery Pack Capacity (kWh)	Actual Used Capacity (maximum %)	Power Required (kW)	Typical Voltage (v)	Chemistry and Energy Density (Wh/kg)	Weight of The Battery Pack (kg)	Operating Range (°C)	References
HEV	< 3	-	-	-	(Ni-MH) 40-50	60	-	[15]
PHEV	5.6 – 18	-	-	-	(Li-ion) 90-100	60-200	-	
	5 – 12	75	-	-	(Li-ion)	-	-30 to 52	[16, 17, 18]
	4 – 10	-	60	300-600	(Li-ion)	-	-	
BEV	35 – 54	-	-	-	(Li-ion) 90-100	450	-	[16, 17, 18]
	22 – 80	80	-	-	(Li-ion)	-	-40 to 50	
	15+	-	60	300-600	(Li-ion)	-	-	

The first aspect of the assessment is the environmental assessment, which makes a great deal. After identifying the study objectives, the next step is to look at the input-output of the EV battery process known as Life Cycle Inventory (LCI). In the sustainability of EV, the indicators looked into are CO_2 emissions, water withdrawal (l), energy consumption (MJ), hazardous waste (st), forestry (gha), CO_2 uptake land (gha), and cropland (gha) [10]. CO_2 emissions indicate the emissions of vehicles processed or used. Water and energy withdrawal is defined as the total of those resources used for the vehicle to be produced. Hazardous waste is defined as the amount of hazardous waste generated (by the U.S sector). Forestry refers to the amount of wood, timber, and firewood products sacrificed in the supply chain. Agricultural land refers to the most bio-productive land of all land used in the process. CO_2 absorption area is the amount of forest area as GHG absorption produced by each sector. [10]

For CO_2 emissions, it is crucial to see the future changes considering technological improvements in EV and fuel efficiency improvement to see whether it is relatable to make more improvements in action. From this indicator, PHEV has the most significant potential for maximum CO_2 reduction pollution in the long run, to 2050 [10]. For the second indicator, total energy, ICV has the most significant energy used with a total of 3.57 MJ, followed by BEV. BEV has the most significant water used (16.72l), followed by PHEV. From the production process, the waste created from ICV production is the biggest, with 0.0920 st, followed by HEV. Forestry, the indicator that looked into the amount of wood used in the supply chain, had the highest consumption of BEV. From CO_2 uptake land, ICV occupies the highest impact value, followed by BEV. Lastly, BEV took the highest impact value on the cropland indicator, followed by PHEV [10]. The value of total life cycle environmental impacts for several types of EV batteries can be seen in table 2.

Table 2. Environment Impacts of Electrical Vehicle. The measurement was based on the selected indicators with different units for each indicator. This table is equipped with a selection of the highest and lowest values.

Electric Vehicle	Environment Indicator						
	CO_2 Emissions (CO_2 eq)	Water Withdrawal (l)	Energy Consumption (MJ)	Hazardous waste (st)	Forestry (gha)	CO2 Uptake Land (gha)	Cropland (gha)
ICV	6.96	2.75	3.57	0.09	2.78E-07	6.52E-07	7.15E-07
HEV	7.52	2.63	2.39	0.07	2.79E-07	4.34E-07	7.02E-07
PHEV	11.1	9.137	2.15	0.048	2.82E-07	4.088E-07	6.975E-07
BEV	11.1	16.72	3.13	0.039	3.55E-07	4.46E-07	6.65E-07
Remark	ICV gives the lowest CO_2 emissions. PHEV and BEV have the highest CO_2 emissions.	HEV takes the least amount of water. PHEV takes the most significant amount of water.	PHEV consumes the least amount of energy. ICV consumes the most energy.	BEV gives the least amount of hazardous waste. ICV gives the most amount of hazardous waste.	ICV has the least impact on their supply chain. PHEV has the highest impact in their supply chain.	PHEV requires the least amount of land. ICV requires the most amount of land.	BEV consumes the least bio-productive area. ICV consumes the most bio-productive area.
References	[10]	[10]	[10]	[10]	[10]	[10]	[10]

The most used Life Cycle Impact Assessment (LCIA) method is ReCiPe, developed by Pre-Sustainability and collaborated with the Dutch National Institute for Public Health and the Environment, Radboud University Nijmegen, and Norwegian University of Science and Technology [10]. Of the reviewed, 77.2% of them used more than one impact category. For this paper's EV battery assessment review, the impact categories considered are climate change, energy use, and resource depletion, all calculated per kg of batteries. Climate change is calculated from cropland, CO_2 emissions, forestry, and CO_2 uptake. Energy use is calculated from energy consumption, and resource depletion comes from water withdrawal. Hazardous waste is included in other environmental information, waste type. [18]

Sustainability assessment indicators on social aspects consist of employment (emp-hr), income ($), injuries (#workers), government tax ($), and human health. The social indicators refer to the research of [18], which discusses the macro-level social, economic, and environmental impacts of alternative vehicle technologies, [18] which contains an alternative passenger car sustainability assessment with a decision-making framework that considers the environmental, social and economic impacts of each type of vehicle, and [10] which discusses new uncertainties related to environmental, economic, and social impacts as well as specific challenges for widespread adoption as a result of alternative technologies in vehicles. The employment indicator is assessed from full-time employees' work hours. The income indicator is defined as employee compensation, wages, and salaries. The injury indicator shows the number of employees or communities with non-fatal injuries. The government tax indicator is defined as government revenues derived from the imports and production of an electric vehicle. The human health indicator is the number of years lost due to employee illness, disability, or early death. Impacts on human health occur on humans, employees, and the community due to air pollution and climate change.

The employment indicator has the same total life cycle social impact value on all types of EV. The government tax indicator with the largest total life cycle social impact value is the ICV type, and the smallest is the BEV type. PHEV has the smallest value of the total life cycle social impact on the injury indicator. At the same time, the largest value on the injury indicator is BEV type. The income indicator has the smallest total life cycle social impact value for the PHEV type and the largest value for the BEV type. HEV has the highest reduction potential in human health indicators compared to others. Meanwhile, ICV has the highest value of human health impact compared to other types of EV. The value of total life cycle social impacts for several types of EV batteries can be seen in table 3.

Table 3. Social Impact of Electrical Vehicle. The measurement was based on the selected indicators with different units for each indicator. This table is equipped with a selection of the highest and lowest values.

Electric Vehicle	Social Indicator				
	Employment (emp-h/km)	Government Tax ($/km)	Injuries (#worker/km)	Income ($/km)	Human Health (DALY)
ICV	0.003	0.007	8.02E-08	0.056	6.80E-07
HEV	0.003	0.006	8.02E-08	0.056	4.70E-07
PHEV	0.003	-0,011	7.68E-08	0.054	4.95E-07
BEV	0.003	-0.024	8.76E-08	0.064	6.00E-07
Remark	All types have the same impact	BEV gives the lowest tax. ICV gives the highest tax.	BEV causes the most non-fatal injuries. PHEV causes the least non-fatal injuries.	PHEV provides the lowest income for employees. BEV provides the highest income for employees.	ICV has the greatest impact on human health. HEV has the least impact on human health.
Reference	[10, 18]	[10, 18]	[10, 18]	[10, 18]	[10, 18]

The third assessment aspect is economic. According to economic indicators, the types of EV are considered by the foreign purchase, national GDP contribution, business profit, and cost of air emissions refer to the research of [10]. Most of the imports happen during the vehicle manufacturing phases, accounting for 57%-87% of the total imports, while the imports share of the operation phase accounts for 13%-31%. On the other hand, savings from vehicle and battery end-of-life phases range from 1% to 3%. The contribution of EV manufacturing to imports is greatest for the BEV, accounting for 15% of total life cycle imports. Due to high imports from the purchase of solar modules for operation in solar charging stations, the BEV dominates. It is important to note that constructing solar charging stations significantly enlarged PHEV imports because of the imported solar modules to be used in constructing the solar charging stations [18].

EV total life cycle contribution has an impact on national GDP. The GDP contributions of the PHEV tend to be lower than those of the ICV and HEV, even though the ICV and HEV have consistently increasing trends due to the continually shifting balance between increases in Vehicle Miles Travelled (VMT) and advances in the energy economy. While the former stimulates energy consumption (gasoline and/or electricity), the ultimate (including battery effectiveness advancements) reduces gasoline and electricity consumption. PHEV initially contribute further to the GDP due to their higher operating and purchase costs when presented to the market. Still, as battery and EV prices decrease over time and energy efficiencies increase, the total quantity of money spent on batteries and/or electricity decreases, reducing the overall contribution to GDP to some extent [10]. Overall, BEV are projected to have the topmost contribution to the GDP by 2050, with ICV coming in a close alternative.

Indispensable vehicle technologies are more profitable and contribute further to GDP than ICV in terms of business profit and GDP impact. By constructing solar charging stations to power PHEV, the positive things of EV on GDP and business profit can be significantly raised. The operation phase is the most consequential in terms of position-specific costs of air emissions. The PHEV has the lowest impact, whereas the ICV causes the topmost profitable costs due to its air emissions. Using a solar charging station to power EV reduced the profitable costs of BEV air emissions by 45%. Solar charging stations are one of the cleanest electricity generation sources. However, upstream emissions from the manufacturing of solar panels and the construction of the stations significantly contribute to the costs of air pollution [18]. The value of total life cycle economic impacts for several types of EV batteries can be seen in table 4.

Table 4. Economic Impact of Electrical Vehicle. The measurement was based on the selected indicators with different units for each indicator. This table is equipped with a selection of the highest and lowest values.

Electric Vehicle	Economic Indicator			
	Foreign Purchase ($/km)	National GDP contribution ($/km)	Business profit ($/km)	Cost of air emission ($/km)
	0,177	0,102	0,105	0,027
BEV	0,235	0,123	0,241	0,024
HEV	0,167	0,100	0,099	0,021
PHEV	0,191	0,099	0,153	0,021
Remarks	BEV has the greatest value for importation. HEV hasthe lowest value for importation	PHEV makes the lowest contribution to national GDP. BEV makes the bigger contribution to the national GDP	HEV's considered less profitable than other vehicles. BEV's considered more profitable than the other electric vehicles	ICV has the highest air emission cost. PHEV has the lowest impact of air emission cost
Reference	[18]	[10]	[18]	[18]

From the environmental, social, and economic assessments, the better types of EV used for future development and application can be chosen. From the environmental aspect, ICV dominates the worst score because it requires a lot of resources during production. From the social aspects and indicators explained, there is no specific electric vehicle that contributes to a higher impact. On the notes of human health, HEV contributes more than other EV. But this does not make it to be the EV with the lowest score. On the social aspects, each indicator picks each type of EV, showing a contribution from the place and the manufacturing process. For the third aspect or category, economic, HEV has the worst value among the four indicators. Other EV besides HEV have a more significant potential to be more profitable. Each type of EV tends to have its pros and cons. From the discussion, the type of EV least mentioned tends to be more potential on having low impacts, which is BEV. On the contrary, ICV was often cited because it could potentially have greater impact.

The resulting impact of EV tends to differ in stages. There are EV that have an impact when used, such as ICV, and those that emit more emissions during the production process. Due to the increasing interest and market for electric vehicles, it is possible to provide lower prices for EV with less impact so that they are used more. This will be a worthless impact for years to come. Further research can compare the impact at each stage to provide more appropriate recommendations, such as the use of machines if the impact is greater during the production process.

This research provides a view of which electrical battery is suitable based on the abovementioned indicators. The background data coming from the US makes battery selection for other countries different. For example, in Indonesia, the resulting CO_2 emissions can be different due to various sources of electricity. California uses a source of electricity from natural gas sourced from methane.

Meanwhile, Indonesia uses most of its electricity source from coal. This will have an impact on the number of emissions produced. On the social aspect, government taxes differ according to the policies of the relevant state governments. This will also influence the price of the battery, which is a consideration for customer purchases. Another example is the economic aspect; each country's different GDPs influence battery prices.

4. Conclusion

This research contributes significantly to studies of LCSA on EV based on the elaboration of some articles and journals on this subject. This analysis provides information on which type of electric battery is best used in EV. In addition, the analysis of the leading environmental, economic, and social indicators showed that each type of EV battery has its advantages and disadvantages on each indicator.

ICV has advantages in environmental in the form of low levels of CO_2 emissions and minimal impact on the supply chain. However, ICV also disadvantages the environment, social, and economy. ICV consumes a large amount of energy, land, and bio-productive area, generates more hazardous waste, and has the greatest impact on human health, the highest taxes, and the highest air emissions cost. BEV has advantages in all aspects. BEV produce the least amount of hazardous waste, consume the least bio-productive area, have the lowest taxes, provide the highest income, and have the highest profit and national GDP.

Meanwhile, BEV causes high CO_2 emissions and non-fatal work accidents. HEV has the lowest impact on human health and low import value. On the other hand, HEV consumes the most water and provides the least profit.

From the discussion, the type of EV least mentioned tends to be more potential to have low impacts, which is BEV. On the contrary, ICV was often cited because it could potentially have greater impact. Finally, the results highlight that BEV has more potential to use on EV due to having a lower impact than the other types of EV.

References

[1] ResearchAndMarket, "Retrieved from Global Electric Vehicle Market (2021 to 2030) - Rising Demand for Electric Vehicles in the Automotive and Transportation Sectors Presents Opportunities," 2021. [Online]. [Diakses 2021].

[2] E. P. V. M. I. P. L. D. P. S. B. P. &. K. G. M. C. Dons, "Land use regression models as a tool for short, medium and long term exposure to traffic related air pollution," *Sci. Total Environ,* p. 476–477.

[3] Amnestry.org, "Exposed: Child labour behind smart phone and electric car batteries," 2019. [Online]. Available: amnesty.org/en/latest/news/2016/01/child-labour-behind-smart-phone-and-electric-car-batteries/.

[4] P. Ratna, Susanto, Adiaksa dan P. A, "Analysis of the eco-efficiency level in the dining table production process using life cycle assessment method to increase industry sustainability," *Analysis of the eco-efficiency level in the dining,* vol. 1072, 2021.

[5] P. Ratna, R. Zainal dan C. Simanjuntak, "Eco-Efficiency of Pencil Preduction Using Life Cycle Assessment to Increase the Manufacture Sustainability," *Jurnal Teknik Industri,* vol. 22, no. 1, pp. 47-54, 2020.

[6] D. Costa, P. Quinteiro dan A. Dias, "A systematic review of life cycle sustainability assessment: Current state, methodological challenges, and implementation issues," *Science of the Total Environment,* vol. 686, pp. 774-787, 2019.

[7] M. Furness, R. Bello-Mendoza, J. Dassonvalle dan R. Chamy-Maggi, "Building the 'Bio-factory': A bibliometric analysis of circular economies and Life Cycle Sustainability Assessment in wastewater treatment," *Journal of Cleaner Production,* vol. 23, 2021.

[8] C. Llatas, B. Soust-Verdaguer dan A. Passer, "Implementing Life Cycle Sustainability Assessment during design stages in Building Information Modelling: From systematic literature review to a methodological approach," *Building and Environment,* vol. 182, 2020.

[9] A. Ciroth, M. Finkbeiner, H. J, K. W, B. Mazijn, S. Prakash, S. G, M. Traverso, Ugaya dan Vickery-Niederman, "Towards Life Cycle Sustainability Assessment: making informed choices on products," United Nations Environment Programma (UNEP), Paris, France, 2011.

[10] N. C. Onat, M. Kucukvar dan O. Tatari, "Uncertainty-embedded dynamic life cycle sustainability assessment framework: An ex-ante perspective on the impacts of alternative vehicle options," *Energy,* vol. 112, pp. 715-728, 2016.

[11] Car Rental Gateway, "Internal combustion engine vehicle (ICEV)," 2020. [Online]. Available: https://www.carrentalgateway.com/glossary/internal-combustion-engine-vehicle/.

[12] EVgo, "Types of Electric Vehicles," 2020. [Online]. Available: https://www.evgo.com/ev-drivers/types-of-evs/.

[13] C. Visentin, A. W. d. S. Trentin, A. B. Braun dan A. Thome, "Life cycle sustainability assessment: A systematic literature review through the application perspective, indicators, and methodologies," vol. 270, 9 June 2020.

[14] K. Cheow, "Battery Electric Vehicle (BEV)," 2020. [Online]. Available: https://medium.com/@ackhor/battery-electric-vehicle-bev-d69f8683b141.

[15] J. Backes, A. D'Amico, N.Pauliks, S.Guarino, M. traverso dan V. L. Brano, "Life Cycle Sustainability Assessment of a dish-Stirling Concentrating Solar Power Plant in the Mediterranean area," *Sustainable Energy Technologies and Assessments,* vol. 47, 2021.

[16] V. Etacheri, R. Marom, R. Elazari, G. Salitra dan D. Aurbach, "Challenges in the development of advanced Li-ion batteries: a review," *Energy & Environmental Science,* vol. 4, no. 9, pp. 3243-3262, 2011.

[17] G. Hunt dan C. Motloch, "Freedom car battery test manual for power-assist hybrid electric vehicles," INEEL, Idaho Falls, 2003.

[18] N. C. Onat, M. Kucukvar dan O. Tatari, "Towards Life Cycle Sustainability Assessment of Alternative Passenger Vehicles," *Sustainability (Switzerland),* vol. 6, no. 12, pp. 9305-9342, 2014.

Environmental Challenges of The Green Economy: Case of Ukraine

Oleksandr Yermakov[1] and Iryna Kostetska[2*]
[1] Department of Economy, National University of Life and Environmental Science of Ukraine, 03041, Kyiv, Ukraine, 0000-0001-5070-6528
[2] Department of Economy, National University of Life and Environmental Sciences of Ukraine, 03041, Kyiv, Ukraine, 0000-0001-5340-0145
[1]ermakovou@ukr.net, [2*]irynakostetska@ukr.net

Abstract. The article analyses the essence of the "green economy." The historical context of the formation of the concept is considered. The postulates of the theory of "green economy" are studied; the essential postulate appears to be the impossibility of satisfying infinitely growing needs in limited resources. The directions of the "green economy" system are considered. Today, the known characteristics, features, and principles of the "green economy" and its multiplicative and anti-crisis potential are highlighted. The purpose of the study is to prepare proposals for a "green economy" model in Ukraine in the post-war period. An understanding of the essence and description of the goals of "green" technologies is proposed, which involves working not with the consequences but with the causes of environmental problems. The European aspect of the "green economy," ways of Poland's transformation, and the state of its implementation in Ukraine are analyzed. As well, normative-legal documents regulating the implementation of changes are given. Poland's experience in introducing the "green economy" is analyzed. An example of a "green" energy cluster in Konin (Poland) is given. Proposals for implementation have been formed, and the role and prospects of Ukraine's participation in the strategic development of the "green economy" have been determined.

1. Introduction

Each country's critical economic development task is to create conditions for achieving and maintaining sustainable economic growth. However, the global energy crisis precipitates the search for new models of economic growth and alternative energy sources to achieve sustainable development, i.e., balancing the rational consumption of material goods with minimal environmental risks. Contemporary issues related to the state of the environment are becoming more global. The promotion of the concept of a green economy has become relevant due to the many crises of recent years – climate, environmental, food, financial, and economic crises. In the first half of 2022, these issues have been exacerbated by the impact of the Russo-Ukrainian war. As a result, they can accelerate the advent of the global environmental crisis and, consequently, economic crisis [1].

This paper aims to study the European experience in implementing the green economy concept, identify successful cases for adaptation in Ukraine, and demonstrate the relationship between financial stability and the rational use of natural resources. The paper proposes a model for green economy development in Ukraine in the post-war period. The primary method used in the research is the empirical one. It is described in more.

2. Methods

The study is based on a systemic approach and upon the results of research efforts carried out over two years (2020-2021) during the project entitled "Transformation of Ukraine's Mono-Profile Territories." The project included, in particular, four study visits to Poland within the framework of the Poland-Ukraine Knowledge Exchange for Coal Regions in Transition Programme supported by the World Bank. The project aimed to study the experiences of just energy transition and the practical implementation of the "green economy" concept. The preparatory project activities included the studies of scholarly sources and the European Union's, Polish, and Ukrainian legal frameworks. The empirical research was conducted in the selected Polish energy cluster through observations and studies of specific phenomena. Appropriate generalizations were made, and the results were discussed in the results subsections of the article. The proposal for their practical implementation in Ukraine was presented in the conclusions.

The comparison method was employed for the research on the developments at the European level, Poland's efforts on the way to the introduction of the green economy, and Ukraine's steps toward European integration. The monographic method was used to study different experiences in the Polish energy cluster. The analytical method was used to focus on specific examples.

3. Results and Discussion

3.1. The emergence of the green economy

The concept of a green economy first appeared in 1989 in a report entitled Blueprint for a Green Economy prepared for the UK government by a group of British environmental economists, David Pierce, Anil Markandya, and Edward Barbier [2], who aimed to support the British government in its implementation of the concept of sustainable development [3–6]. The published document was revised in 1991 and 1994. The revisions expanded the perspective of global environmental issues such as climate change, ozone layer depletion, tropical deforestation, degradation, and loss of natural resources in developing countries [7]. The term "green economy" re-emerged in scientific circles in 2008 during a discussion initiated by the United Nations Environment Program (UNEP), which led to the Green Economy Initiative (GEI). It produced the Global Green New Deal report, published in 2009. The report did not deal exclusively with environmental aspects. It also looked for ways to combat the global economic crisis in the United States in 2007. The UNEP urged governments to promote green sectors and allocate significant resources to the accomplishment of three objectives: (a) economic recovery (economic objective); (b) poverty eradication (social objective); (c)reduced carbon emissions and ecosystem degradation (environmental objective).

The UNEP also published the Green Economy Report in 2011. It identified the green economy as one that results in improved human well-being and social equity while significantly reducing environmental risks and ecological scarcities. The UNEP published 30 reports on the green economy issues and 11 studies on the national strategies for its implementation in 2009–2012 [5].

The next step in implementing the green economy concept was made in 2015 when the UN General Assembly adopted the critical resolution entitled the 2030 Agenda for Sustainable Development. The document was signed by all 193 member states of the United Nations. It identified 17 Sustainable Development Goals. These goals will bring about global changes in environmental awareness, ensure environment-neutral production of energy and food, proper organization of transport, and mitigate the consequences of adverse climatic and other phenomena. These goals replaced the Millennium Development Goals adopted in 2000, which were to be achieved by 2015.

3.2. The European aspect of the green economy

The European Union has been actively involved in developing the concept of sustainable development. However, the concept of a green economy emerged relatively late in the EU documents. It was only on the eve of the Rio+20 Summit in 2012 that the European Commission identified the green economy as the "long-term survival of the planet" and a "low-carbon, resource-efficient, and socially inclusive economy" [8]. The first significant document concerning the green economy concept was the Europe 2020 Strategy. It focused on supporting sustainable development in the economy, environment, and society. As part of this Strategy, various documents were developed in 2011 to prepare for the practical implementation of the concept, including "Roadmap to a Resource Efficient Europe," "Roadmap for Moving to a Competitive Low Carbon Economy in 2050," and "EU Biodiversity Strategy to 2020" [9].

The green economy can be approached from conceptual and theoretical points of view, allowing for creating and implementing strategies and development programs. Thus, the green economy includes the following elements: green products and services, green investments, green sectors of the economy, green public procurement, and green jobs, as well as consumption that contributes to the greening of the economy [4]. The first element comprises products and services that have a low impact on the environment throughout their life cycle. This means that they are manufactured or provided using environment-neutral components and can be reused after processing in a so-called circular economy. Green products and services must meet specific environmental requirements. Green investments include the construction of infrastructural objects that protect the environment, purchases of equipment to use renewable energy sources, and other activities in the production or use of energy and raw materials. The green sectors of the economy have always been associated with agriculture and forestry, natural landscape conservation, renewable energy production, bio-economy development, and clean technologies.

On the other hand, green public procurement is an activity and process used by the authorities to purchase goods and services and implement projects that have a positive impact on the environment, as well as on economic and social growth and development. Green jobs help reduce energy and natural resource consumption, reduce waste and greenhouse gas emissions, protect ecosystems, and preserve biodiversity. The preservation of "green" sectors of the economy also depends on consumption that enables waste avoidance and rational waste management. Various actors are responsible for implementing the concept of a green economy: states, public administration entities, organizations, enterprises, local communities, households, and all citizens. That is the reason why the term "green economy" is also associated with the concepts of "green growth" and "sustainable development" proposed by the OECD [10].

The implementation of the green economy concept emphasized the use of the tax system that applied the polluter pays principle (high tax rates for polluting enterprises). This principle allowed Europe to prevent pollution growth and accumulate funds for environmental restoration. This management system was essential for natural habitats and industries that made intensive use of the elements of the natural environment. However, the concepts of sustainable development and green economy lead to the search for new tax solutions, new types of investment, and new resource management systems — especially energy sources that ensure environmental protection, economic stability, efficiency, and social progress. Efforts to increase the production of energy from renewable sources yielded visible results. In 2005–2013 alone, energy production from renewable sources in the EU increased from 3,847.3 to 17,066.6 MWh.

The European Commission unveiled the European Green Deal strategy [8] in its communication to other EU institutions in December 2019 to implement the 2030 Agenda for Sustainable Development. To implement this Strategy, the European Commission proposed to adopt Regulation Establishing the Framework for Achieving Climate Neutrality in the European Union (European Climate Law) in March 2020. The key types of activities provided by the Regulation are as follows: (a) setting emission reductions by 2050 to ensure predictability for businesses, stakeholders, and citizens; (b) development of a system of monitoring and reporting on the progress in achieving the goals; (c) ensuring economically efficient and socially just environmental transition. It came into force in July 2021.

3.3. Poland's green economy

Every year in Poland, the issue of creating a low-emission economy is increasingly discussed. These discussions resulted in developing and implementing the Energy Policy for Poland to 2030 strategy, as well as the document entitled Poland 2030 [11]. The Third Wave of Modernity. The

Long-Term Strategy of National Development was adopted in 2012 [12]. The latter provided for the following measures: (a) more efficient use of natural resources; (b) implementation of the program for the development of innovative environmental technologies;(c) support for leading research centers and enterprises in this field.

The Energy Policy of Poland until 2040 [13] sets the framework for energy transformation in the country. It includes strategic decisions on the technologies for building a low-emission energy system. PEP2040 promotes the implementation of the Paris Agreement concluded in December 2015 at the 21st Conference of the Parties to the United Nations Framework Convention on Climate Change (COP21), considering the need for transformation in a fair and solidary manner.

The planning and implementation of PEP2040 consider the scale of challenges related to the adaptation of the national economy to the EU regulatory framework arising from the climate and energy goals set to be achieved by 2030, the European Green Deal, the post-COVID-19 economic recovery plan, and the pursuit of climate neutrality under national obligations, as a contribution to the implementation of the Paris Agreement. PEP2040 is one of the nine integrated sectoral strategies resulting from the adopted Responsible Development Strategy. PEP2040 is in line with the National Energy and Climate Plan for 2021–2030. It is essential to demonstrate local approaches to PEP2040 implementation and the related funding sources.

The city of Konin (Greater Poland Voivodeship) offers a successful case of energy transformation: the Konin Green Energy Cluster [14]. The Konin City Council signed the agreement to establish the Konin Green Energy Cluster on June 21, 2018. The signatories were as follows: 1) Konin City; 2) municipal water supply company; 3) municipal heating company; 4) municipal waste treatment company; 5) Centrum Badań i Rozwoju Technologii DLA Przemysłu S.A (CBRTP, eng. Industrial Research and Development Center — commercial research and technology hub). Thus, the cluster participants represent self-government, municipal, and commercial entities. Moreover, the cluster includes a scientific institution that meets the criteria of the Triple Helix model [15]. The examples of the successfully implemented projects within the cluster are as follows:

a. The construction of a photovoltaic farm in the Left Bank neighborhood in Konin. The project goal: construction of a photovoltaic farm with a total capacity of 2.18 MW with associated infrastructure that will generate electricity using solar radiation (renewable energy source) and ensure environmental protection and energy security. Impact: reduction of greenhouse gas emissions by 1523 tons of CO_2 per year, reduction of dust emissions, and energy production with the capacity of 1890 MW/year using newly built generating units. Implementation period: from June 22, 2020, till May 31, 2022. Total cost: 15,042,991.69 zlotys, including co-financing of 8,307,919.24 zlotys.
b. The construction of a geothermal thermal power plant in Konin. The project goal: production of thermal energy based on a local geothermal (renewable energy) source. The project is in line with the EU's environmental policy, energy security strategy, and improving the conditions for regional development. Impact: construction of a geothermal thermal power plant with infrastructure and drilling of a Konin GT-3 geothermal well. The installed generating capacity will be 8.1 MW, and the planned annual heat production will be almost 150 thousand GJ. The estimated impact of this investment on environmental protection: annual reduction of CO_2 emissions by 6746 Mt. Implementation period: 2020-2022. Total cost: 56,088,436.04 zlotys, including co-financing of 24,141,131.00 zlotys.
c. The Pątnów-Adamów-Konin power plant complex (ZE PAK) will produce hydrogen from biomass. At the initial stage, it will produce at least 1 t of hydrogen per day. The goal is 2 t of hydrogen per day. This amount of hydrogen should be enough to refuel 100 buses with fuel cells. ZE PAK has signed a contract to supply HyLYZER 1000-30 electrolyzer to produce hydrogen. In addition to the electrolyzer, a 350-bar compressor station and gas distributors were also purchased. Today, electrolysis is the purest process of obtaining hydrogen — of course, if energy from clean sources is available — but it is also the most energy-intensive: the production of 1 kg of hydrogen requires about 50-60 kW/h of energy.

3.4. Ukraine's prospects concerning the green economy

International and European experience in implementing the green economy concept proves the effectiveness of several fiscal and financial instruments, including green tariffs, CO2 tax, environmental taxes, and green bonds. The development of renewable energy is directly related to the application of the "green" tariffs. The definition of this term appeared in Ukrainian legislation in 2003 [16]. It defined "green tariff" as a special price for energy produced from renewable energy sources — hydro, solar, wind, or biomass.

However, on July 21, 2020, the Verkhovna Rada adopted the Law "On Amendments to Certain Laws of Ukraine on Improving the Conditions for Supporting the Production of Electricity from Alternative Energy Sources." The law reduced the level of "green" tariffs for the already built facilities using solar and wind energy by setting reduction factors. In addition, the Rada proposed to cover 20% of the costs of the "green" tariff from the state budget. During the consideration of the draft law, an amendment was adopted to reduce the "green" tariff by 60% for solar power stations with a capacity of 1 to 75 MW that had been put into operation before April 1, 2021.

The critical strategic document of state policy in the environmental field is The Basic Principles (Strategy) of State Environmental Policy until 2030, approved by the Law of Ukraine on February 28, 2019. The current Strategy sets the goals of state environmental policy, in particular, to achieve a good state of the environment by introducing an ecosystem approach to all areas of socio-economic development of Ukraine to ensure the constitutional right of every citizen of Ukraine to a clean and safe environment, conserving and restoring natural ecosystems.

Increasing energy from renewable sources and alternative fuels is considered an essential part of Ukraine's Strategy to preserve traditional fuel and energy resources and reduce the associated negative impact on the environment. An important step was the adoption of the updated Nationally Determined Contribution to implement the Paris Agreement on July 30, 2021. It set the goal to reduce greenhouse gas emissions by 35% compared to the 1990 level and achieve carbon neutrality no later than 2060. Given the prospects for post-war development, we believe that the green economy will be paid critical attention to in Ukraine.

4. Conclusion

Implementation of the model of a "green" economy provides for the increased role of the state and intergovernmental bodies in economic Regulation, creating conditions for business development based on new environmental standards and technologies of cleaner production, and greening of industrial sectors. Given the resource and energy dependences in Ukraine, where environmentally harmful technologies are used in obsolete energy-inefficient enterprises, the gradual replacement of the "brown" industrial economy with a new "green" should be viewed as a strategic priority that gives a chance to strengthen the national security in the coming decades. According to the Green Deal strategy, the number of coal mines will decrease yearly. An additional factor is a Russia-Ukrainian war which makes it impossible to supply raw materials. Today, the need to develop a green economy is multiplying. As an alternative, it would be expedient to create an energy park on the Polish model in the western regions of Ukraine. This condition can help avoid energy shortages in both the war and post-war periods.

Green energy clusters can be a practical solution to the fuel shortages in Ukraine and the fiscal burden of the green tariffs. Since the western regions of Ukraine were much less affected by the Russo-Ukrainian war, developing green energy parks in those regions following the Polish model can proceed without insurmountable obstacles. The green tariffs, including the subsidies from the state budget, were suspended for the war period due to the fall in demand for electricity. However, the electricity produced from renewable sources (primarily photovoltaic and wind farms) could be used to operate electrolyzers producing hydrogen. This fuel can be exported to the EU or used to operate local hydrogen fuel cell buses if the local authorities support the transition to clean urban transport. The current scarcity of traditional fuels for urban transport in Ukraine makes the development of hydrogen energy clusters more attractive than ever for businesses and municipal or regional authorities. Thus, developing green energy clusters may help strengthen Ukraine's energy security and reduce greenhouse gas emissions.

The development of agriculture is the next factor that influences the implementation of the green economy, as agricultural activities affect the various areas related to sustainable development. The new economic Strategy proposed by the UN envisages a radical transformation of environmental management, industrial sectors, and environmental awareness, going beyond the post-industrial economy. Securing food supplies as well as improving their quality and environmental neutrality are included in European strategic plans. Nowadays, however, food security is undermined by such destructive factors as war. Fighting is taking place in large parts of Ukraine, where it is impossible to conduct a sowing campaign, and 25 million tons of grain last year were blocked in ports. This condition threatens the food security of Ukraine and other countries — buyers of Ukrainian food supplies. Therefore, it is essential to reconsider plans and create a new model of food security to prevent hunger.

References

[1] Waters J 1994 *Learning for a Sustainable Future Global Education Project*
[2] Anon 2021 *Zakon Ukrainy "Pro vnesennia zmin do deiakykh zakoniv Ukrainy shchodo udoskonalennia umov pidtrymky vyrobnytstva elektrychnoi enerhii z alternatyvnykh dzherel enerhii.*
[3] Bąk I and Cheba K 2020 *Zielona gospodarka jako narzędzie zrównoważonego rozwoju* (Warszawa: CeDeWu)
[4] Szyja P 2015 Pojęcie, tworzenie i pomiar zielonej gospodarki Gospodarka w Praktyce i Teorii 2 **2** 21–38
[5] Ryszawska B 2013 *Zielona gospodarka – teoretyczne podstawy koncepcji i pomiar jej wdrażania w Unii Europejskiej* ((Wrocław: Wydawnictwo uniwersytetu Ekonomicznego)
[6] Pearce D, Markandya A and Barbier E B 1989 *Blue Print for a Green Economy* (London: Earthscan)
[7] UNDESA 2012 Green economy, Green growth, and Low-carbon development–history, definitions, and a guide to recent publications. *Div. Sustain. Dev. Dep. Econ. Soc. Aff. United Nations, New York, August.* 1–63
[8] European Commision 2020 2050 Long Term Strategy *Eur. Commision*
[9] United Nations Conference on Sustainable Development Rio+20
[10] OECD 2019 *Declaration on Green Growth, Adopted at the Meeting of the Council at Ministerial Level on*
[11] Anon *Serwis Rzeczypospolitej Polskiej*
[12] Anon 2013 *Trzecia fala nowoczesności. Długookresowa Strategia Rozwoju Kraju* (Warszawa: Ministerstwo Administracji i Cyfryzacji)
[13] Anon *Polityka Energetyczna Polski do 2040* (PEP3040)
[14] Konin Konin
[15] Yermakov O and Kostetska I 2021 The role of industrial parks for strategic development of territories *Ekon. prostir* **174** 90–4
[16] Zakon 2003 *Ukrainy Pro alternatyvni dzherela enerhetyky*

Carbon Emission Assessment of Bioplastics from Durian Peel

Krittaphas Mongkoldhumrongkul[1*], Phatcharapron Sukkanta[2]
[1]Faculty of Science, Energy, and Environment, King Mongkut's University of Technology North Bangkok, Rayong, 21120, Thailand, 0000-0002-0930-5930
[2]Faculty of Science, Energy, and Environment, King Mongkut's University of Technology North Bangkok, Rayong, 21120, Thailand, 0000-0001-8739-0272
[1]*krittaphas.m@sciee.kmutnb.ac.th, [2]phatcharapron.sk@hotmail.com

Abstract. Plastic is a product widely used daily and contributes to huge waste, which can cause a potential environmental impact worldwide. The production of bioplastics can lessen plastic waste while also benefiting the environment. There are still few studies examining the influence of bioplastics in different contexts. This investigation assesses the cradle-to-gate life cycle of durian peel-derived bioplastics while considering their potential to contribute to global warming as measured by their carbon dioxide equivalent (CO2eq). The steps in the evaluation process were acquiring raw materials, extracting cellulose powder, synthesizing carboxymethyl cellulose, and making carboxymethyl cellulose film. Based on the results, the entire production of bioplastics emits 76.339 kgCO2eq, and the extraction process emits the maximum CO2eq. Bioplastic manufacturing can minimize global warming by reducing humidity before the extraction process.

1. Introduction

Plastics have become an essential and ubiquitous material in our economy and everyday activities, but it is also a primary environmental concern worldwide [1]. It is expected that 6.3 billion tons of plastic waste will be generated worldwide, 95% of plastics from fossil sources, which will have environmental impacts on terrestrial and aquatic organisms [2,3]. Therefore, supporting renewable materials and bioplastics to replace plastic while reducing waste and environmental impacts is essential.

Food waste or agricultural commodities are used to manufacture bioplastics [4]. Most bioplastics are created to replace petrochemical plastics, lessen the environmental effect, and reduce agricultural and plastic waste accumulation [5]. This study exploited the geography and climate of Thailand to examine the creation of bioplastics from agricultural waste [6]. Rayong Province in Thailand had a total output of durian to the market of 114,413 tons in 2020, according to reports from the Office of Agricultural Economics, Ministry of Agriculture, and Cooperatives [7]. This might be employed in the manufacture of bioplastics.

In earlier studies, [4] used SimaPro software's CML Baseline technique to perform a life cycle evaluation on whey plastics with a cradle-to-gate scope. According to the findings, 2,900 MJ of energy was needed to make 1,000 kilograms of whey plastics. In addition, compared to other typical plastics, 115.3 kgCO2eq had the lowest global warming potential on freshwater and marine aquatic toxicities. To that compare the results with high-density polyethylene (HDPE) and low-density polyethylene, [6] performed a life cycle analysis (LCA) of biodegradable polylactic acid (PLA) and bio-polyethylene (bio-PE) plastics from feedstock production through product end-of-life (EOL) (LDPE). The results revealed that bioplastics had the lowest GHG emissions, particularly bio-PE plastics. The life cycle assessment of bio-based polyethylene terephthalate (PET) bottles was reported by [8] using a cradle-to-grave perspective, and a comparison with petrochemical PET bottles was made in 13 categories of environmental effects. According to the findings, synthesizing terephthalic acid (PTA) and stretch blow molding were the processes most responsible for bio-based PET bottles' overall weighted environmental impacts. Using a life cycle assessment (LCA) - style business to consumer model, Boonniteewanich et al. [9] compared the carbon footprint of 1 kilogram of bioplastic straws and polypropylene (PP) straws (B2C). Due to additional production waste, the results indicated that bioplastic straws had a higher carbon footprint than PP straws.

While conventional bioplastics have been life cycle evaluated [4,5], bioplastics from durian peel have not been assessed. To support value addition to waste produced from agricultural products and propose guidelines for sustainable environmental conservation, this research aims to conduct a life cycle assessment of bioplastics made from durian peel with a focus on assessment of greenhouse gas emissions in the form of carbon dioxide equivalent emissions.

2. Methods

The Life Cycle Assessment (LCA) method evaluates the environmental impact of a product throughout its life cycle. There are two distinct forms of assessment: a complete LCA, known as a cradle-to-grave process, and a partial LCA, also called a cradle-to-gate process. Clarity on method and data sources, which can significantly affect outcomes, is a critical element of LCA investigations [5]. This research used a cradle-to-gate LCA, similar to LCA research on whey protein-based bioplastic synthesis [4].

2.1. Goal and scope definition

The cradle-to-grave LCA starts with the acquisition of raw materials and ends with the disposal of waste products. Even so, this study aims to perform the cradle-to-gate LCA of bioplastics made from durian peel by assessing the stage from which raw material acquisition to the stage at which product acquisition in order to analyze the environmental consequences of carbon dioxide emissions from those stages. The functional unit is one piece of bioplastics produced (0.005 kg). Regarding this unit of bioplastic production, the contributions of each process and its effects on the environment are evaluated. Figure 1 illustrates the inputs for the processes considered in this analysis, including energy and power consumption and the environmental impact of the chemicals employed in the production process.

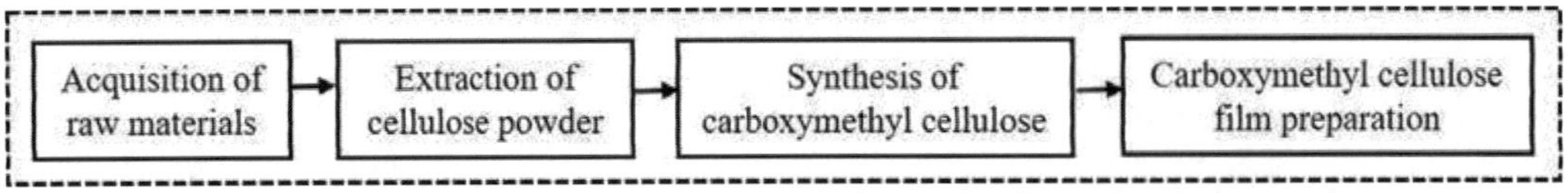

Figure 1. The conceptual boundary of bioplastics from the durian peel.

2.2. Life cycle inventory

The life cycle inventory considers a detailed description of raw materials, energy, and other inputs into the systems, as well as the emissions affecting the environment and waste outputs. Raw materials acquisition is based on data from the Department of Agriculture, considering the acquisition of 1 durian bark with a total yield of 70,000 grams.

2.3. Impact assessment

The total carbon dioxide emissions were calculated from carbon dioxide emissions according to the number of activities in each process by using equation (1) below.

$$CO2\ Emissions = Activity\ data \times Emission\ factor \quad \textbf{(1)}$$

Where
Activity data are the energy used in the process.
The emissions factor is found in the life cycle inventory database in Thailand.

2.4. Interpretation
Interpretation identifies the significant issue and evaluates a study based on its completeness, consistency, and sensitivity. The results of data analysis were obtained and suggested improvements to reduce global environmental impacts and improve resource efficiency.

3. Results and Discussion
3.1. Life cycle assessment scope of bioplastics
The findings of the life cycle assessment are discussed in terms of total carbon dioxide emissions. The outcomes comprise the total raw materials, cellulose powder extraction, carboxymethyl cellulose synthesis, preparation of carboxymethyl cellulose films, and process contributions. The functional unit is one piece of bioplastics produced, with dimensions for bioplastic comprising 2 cm × 3 cm × 0.3 cm. The weight of bioplastic is 0.005 kilograms.

3.2. Life cycle inventory of bioplastics
3.2.1. Acquisition of raw materials process
In the first week, 7.5 kilograms of organic fertilizer, 1.5 kilograms of chemical fertilizer (15-15-15), and 1,050 liters of water were used. In the 2nd week, chemical fertilizers (16-16-16) were used, containing.

0.4 grams each of nitrogen, phosphorus, and potassium. Weeds and pests included 0.05 kilograms of glyphosate, 0.05 kilograms of paraquat, and 1,450 liters of water. In the 3rd week, chemical fertilizers (16-16-16) were used, containing 0.4 grams each of nitrogen, phosphorus, and potassium. For herbicides and pesticides, 0.1 kilograms of glyphosate and 0.1 kilograms of paraquat were used for 1,450 liters of water. From the 4th to the 6th week, herbicides and pesticides were used as follows: 0.05 kilograms of glyphosate and 1,850 liters of water.

From the 7th to the 8th week, chemical fertilizers (16-16-16) were used, containing 0.2 grams each of nitrogen, phosphorus, and potassium. There were 1,600 liters of water and 0.05 kilograms of glyphosate utilized in the herbicide and pesticide doses. One thousand fifty liters of water were used in the 9th week, and 1,050 liters of water were utilized in the tenth. Chemical fertilizer (8-24-24) comprising 0.1 kilograms of nitrogen, 0.2 kilograms of phosphorus, and 0.2 kilograms of potassium was applied from the 11th to the 13th week.

One thousand two hundred liters of water and 0.05 kilograms of glyphosate were used to treat weeds and pests. Between the 14th and 15th weeks, 0.5 kilograms of chemical fertilizer (12-11-18) and 0.2 kilograms of nitrogen, 0.2 kilograms of phosphorus, and 0.3 kilograms of potassium were applied (15-15-15). 0.05 kilograms of glyphosate and 3,000 liters of water were used as herbicides and pesticides. Fertilizer was applied from the sixteenth to the eighteenth weeks. 0.2 kilograms of nitrogen, 0.2 kilograms of phosphorus, 0.3 kilograms of potassium, and 5,250 liters of water were utilized in the chemical formula 12-11-18. Based on research results from [9], 1.75 kilograms of chemical fertilizer (15-15-15) was applied from the 19th to the 20th week, and 4,200 liters of water were used with 2,250 litters in terms of yield and durian rinds to obtain durian peel, the primary raw material for cellulose extraction. 0.95 liters of diesel fuel were used during transportation.

The information above illustrates the input and output amounts of planting one durian tree, which yields 70 kilograms of durian and weighs 2.830 kilograms with 1.900 kilograms of peel each. To determine information on the acquisition of 80 grams of durian peel, the proportions were divided, and this was discovered that 0.012 kilograms of organic fertilizer were utilized overall. 0.006 kilograms of chemical fertilizer (15-15-15) were used. It was accomplished using fertilizers with 0.002 kilograms of nitrogen, 0.002 kilograms of phosphorus, and 0.003 kilograms of potassium. The herbicides and insecticides utilized totaled 0.0006 kilograms of glyphosate and 0.0003 kilograms of paraquat, and 41.395 liters of water were used in total, as demonstrated in Table 1.

Table 1. Inventory data for the acquisition of raw materials process.

Input/output	Amount	Unit
Water	41.395	L
Organic fertilizer	0.012	kg
Gunnysack	0.0017	kg
Diesel oil	0.95	L
Durian peel	80	g
Chemical fertilizer (15-15-15)	0.006	kg
Nitrogen	0.002	kg
Phosphorus	0.002	kg
Potassium	0.003	kg
Glyphosate	0.0006	kg
Paraquat	0.0003	kg

3.2.2. Extraction of cellulose powder process

The little amount of durian peel (80 grams) was baked using a Binder Model FD56 heater that consumed 1,100 watts of electricity for 48 hours at 55 degrees Celsius. The durian peel was then placed in a heated stirrer of the IKA brand, Model C-MAG HS 7, which required 1,020 watts of power to function, and boiled for an hour at 80 degrees Celsius with 50 milliliters of sodium hydroxide. The durian peel was filtered, cleaned with 800 milliliters of purified water, bleached for 30 minutes with 100 milliliters of 30% hydrogen peroxide, rinsed with 800 milliliters of distilled water, and then baked for 24 hours at 55 degrees Celsius using a Binder by WTB, model FD56. After baking, it was ground for 3 minutes with a sharp multi-purpose blender (Model EM-14), using 350 watts of energy, and then sieve shaken for 30 minutes with a RETSCH Model AS200 DIGIT, using 460 watts of electricity. 7.5 grams of cellulose were obtained by separating particles measuring 125 millimeters, as stated in Table 2.

Table 2. Inventory data for the extraction of cellulose powder process.

Input/output	Amount	Unit
Durian peel	80.000	g
Electricity	75.667	kWh
Sodium hydroxide	0.050	kg
Hydrogen peroxide (30%)	0.100	kg
Distilled water	0.001	m^3
Cellulose power	7.500	g

3.2.3. Synthesis of carboxymethyl cellulose process

An IKA brand, Model C-MAG HS 7 heated stirrer, was used to first bring 7.5 grams of cellulose powder to a boil at 40 degrees Celsius for 30 minutes with 40 milliliters of sodium hydroxide in a volume of 50 milliliters. Eighteen grams of chloroacetic acid and 350 milliliters of isopropyl alcohol were added. Using a WTB brand Binder, model FD56, cellulose powder was baked for 2.5 hours at 55 degrees Celsius. After baking, one hundred milliliters of methanol with a 70% concentration were added. Ninety percent acetic acid was used to correct the pH, then left alone for 10 minutes. By changing the pH, 100 milliliters of 70% ethanol concentration were used for washing, followed by 10 minutes of waiting, filtering out only the solid portion to wash, and 200 milliliters of 95% ethanol. To obtain 6.5 grams of cellulose, it was baked for 24 hours at 55 degrees Celsius using a WTB brand Binder Model FD56, as shown in Table 3.

Table 3. Inventory data for the synthesis of carboxymethyl cellulose process.

Input/output	Amount	Unit
Cellulose powder	7.5	g
Electricity	29.66	kWh
Isopropyl alcohol	350	ml
Sodium hydroxide	0.05	kg
Chloroacetic acid	0.018	kg
Methanol (40%)	0.1	kg
Ethanol (70%)	0.7	kg

3.2.4. *Carboxymethyl cellulose film preparation*

Preparation was started by boiling 6.5 grams of cellulose powder at 80 degrees Celsius for 10 minutes using an IKA stirrer, Model C-MAG HS 7, followed by baking at 55 degrees Celsius for 1 hour using a WTB brand Binder, Model FD56, as displayed in Table 4.

Table 4. Inventory data for the carboxymethyl cellulose film preparation.

Input/output	Amount	Unit
Cellulose powder	6.5	g
Electricity	1.27	kWh

3.3. *Impact assessment of bioplastics*

According to the evaluation's findings, 76.239 kgCO2eq of carbon dioxide was released during the bioplastics production from durian peel. The process that released the most carbon dioxide was the extraction of cellulose powder, the synthesis of carboxymethyl cellulose, the processing of raw materials, and the manufacture of carboxymethyl cellulose films, respectively, as revealed in Figure 2.

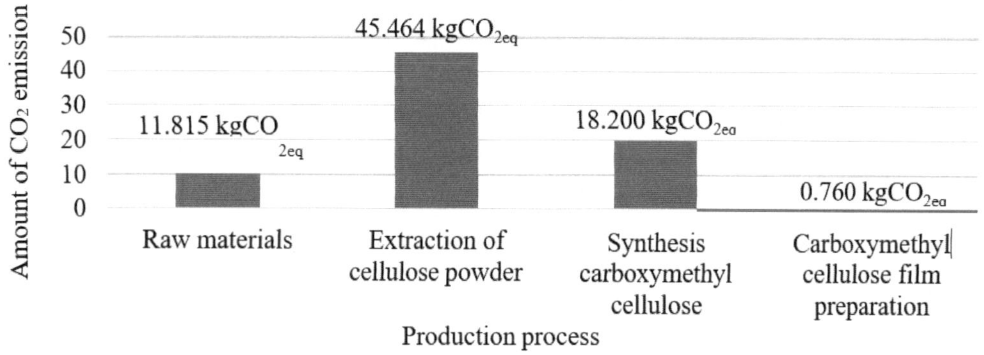

Figure 2. Carbon dioxide emissions of bioplastics from the durian peel.

The extraction of cellulose powder resulted in the most significant emissions, totaling 45.464 kgCO2eq, or 59.63 percent of all emissions. Most of the impact, or 45.294 kgCO2eq, came from the consumption of 97.55 percent power. The amount of 0.37 percent, or 0.170 kgCO2eq, is used by the chemicals. The carboxymethyl cellulose synthesis process has carbon dioxide emissions of 18.200 kgCO2eq, accounting for 23.87 percent, with most of the impact from electricity consumption 97.55 percent, or equal to 17.754 kgCO2eq, and from the chemicals use 2.45 percent, equal to 0.445 kgCO2eq.

The raw material process came next, with carbon dioxide emissions equivalent to 11.815 kgCO2eq, accounting for 15.50 percent of the overall carbon footprint. Most impact comes from water use at 99.74 percent or 11.784 kgCO2eq. The amount of fertilizer used is 0.16 percent or 0.018 kgCO2eq. The number of pesticides used, or 0.09 percent, is 0.011 kgCO2eq. The usage rate for organic fertilizers is 0.001 percent or 0.011 kgCO2eq. The fuel uses 0.01 percent, or 0.001 kgCO2eq, whereas the hemp bags utilize 0.01 percent, or 0.001 kgCO2eq.

The preparation of carboxymethyl cellulose films resulted in the lowest carbon dioxide emissions, 0.760 kgCO2eq, or 1 percent, with a resultant impact from 100 percent electricity use. The evaluation findings revealed that the production phase was where the most carbon dioxide was released. According to studies on bioplastics-straw-LCA, the production process generates more carbon dioxide emissions than the procurement and transportation of raw materials. Therefore, the carbon dioxide emissions from the disposal process are the largest [10]. The amount of carbon dioxide emissions from durian peel relative to the functional unit of bioplastics have a considerable influence.

Conversely, research has revealed that compared to plastics made from petrochemicals, carbon emissions from bioplastics are still less significant. For instance, the manufacturing of 1,000 kg of whey plastics results in the release of 1,939 kgCO2eq, less than that of PP, LDPE, and HDPE [4] and comparable to that of [11]. For both environmental effects and societal costs, polyhydroxyalkanoates (PHA) derived from urban biowaste can outperform PHA derived from first-generation biomass (sugarcane and maize) and fossil polyurethane [12]. By switching the sources used to generate electricity, the manufacture of bioplastic can be made more environmentally friendly [13].

3.4. Interpretation
The findings corroborate the previous study on the utilization of agricultural waste in the production of bioplastics. They indicate that the trend of carbon dioxide emissions from bioplastics generated from durian peel is high. On the other hand, one of the best solutions for disposing of agricultural waste is the production of bioplastics from durian peel. The development procedures must be improved and changed to decrease electricity usage and, consequently, carbon emissions produced throughout the manufacturing process. Additionally, a laboratory-scale study has been used for this investigation to calculate the number of raw materials used in the experiment. It would be able to maximize resources and have promising futures in terms of economic and environmental responsiveness with the increase of production in the industrial sector.

4. Conclusion
The acquisition of raw materials, followed by the extraction of cellulose powder, the synthesis of carboxymethyl cellulose, and the formation of carboxymethyl cellulose film, are the first steps in the life cycle assessment of carbon dioxide emissions of bioplastics made from durian peel. The outcomes revealed that the maximum carbon dioxide emissions were produced during the extraction of cellulose powder, followed by the production of carboxymethyl cellulose, the acquisition of raw materials, and the preparation of carboxymethyl cellulose films, respectively. The use of electricity has had the most significant influence since it is essential in the drying process to reduce the moisture content of cellulose.

This might be achieved by employing natural light to lower the durian peel's moisture content before commencing the process of extracting cellulose powder to create bioplastics, which could then be examined in more detail or molded into a selection of items. Because this assessment was conducted in a laboratory, the amount of carbon dioxide measured was somewhat higher than the functional unit. Further investigation, undertaken at the industrial level, will illustrate the environmental effects of each process. Planning management for sustainable production will benefit from this.

Acknowledgments
This research was funded by the National Science, Research and Innovation Fund (NSRF) and King Mongkut's University of Technology North Bangkok with Contract no. KMUTNB-FF-65-43.

References

[1] Geyer R, Jambeck J R and Law K L 2017 Production, use, and fate of all plastics ever made [Producción, uso y destino de todos los plásticos jamás fabricados] *Sci. Adv.* **3** e1700782
[2] Vince J and Stoett P 2018 From problem to crisis to interdisciplinary solutions: Plastic marine debris *Mar. Policy* **96** 200–3
[3] Schmaltz E, Melvin E C, Diana Z, Gunady E F, Rittschof D, Somarelli J A, Virdin J and Dunphy-Daly M M 2020 Plastic pollution solutions: emerging technologies to prevent and collect marine plastic pollution *Environ. Int.* **144**
[4] Chalermthai B, Giwa A, Schmidt J E and Taher H A B 2021 Life cycle assessment of bioplastic production from whey protein obtained from dairy residues *Bioresour. Technol. Reports* **15** 100695
[5] Bishop G, Styles D and Lens P N L 2021 Environmental performance comparison of bioplastics and petrochemical plastics: A review of life cycle assessment (LCA) methodological decisions *Resour. Conserv. Recycl.* **168** 105451
[6] Benavides P T, Lee U and Zarè-Mehrjerdi O 2020 Life cycle greenhouse gas emissions and energy use of polylactic acid, bio-derived polyethylene, and fossil-derived polyethylene *J. Clean. Prod.* **277**
[7] Office of Agricultural Economics 2022 Agricultural Production Information *Off. Agric. Econ.*
[8] Gursel I V, Moretti C, Hamelin L, Jakobsen L G, Steingrimsdottir M M, Junginger M, Høibye L and Shen L 2021 Comparative cradle-to-grave life cycle assessment of bio-based and petrochemical PET bottles *Sci. Total Environ.* **793** 148642
[9] Janthasri R, Pasuwityakoul S, Janthasri K and Boonraksa R 2006 Study on Growth and Development of Durian cv. Mon Tong in Ubon Ratchathani Province *UBU Eng. J.* **3** 15–25
[10] Boonniteewanich J, Pitivut S, Tongjoy S, Lapnonkawow S and Suttiruengwong S 2014 Evaluation of carbon footprint of bioplastic straw compared to petroleum based straw products *Energy Procedia* **56** 518–24
[11] Harding K G, Dennis J S, von Blottnitz H and Harrison S T L 2007 Environmental analysis of plastic production processes: Comparing petroleum-based polypropylene and polyethylene with biologically-based poly-β-hydroxybutyric acid using life cycle analysis *J. Biotechnol.* **130** 57–66
[12] Bassi S A, Boldrin A, Frenna G and Astrup T F 2021 An environmental and economic assessment of bioplastic from urban biowaste. The example of polyhydroxyalkanoate *Bioresour. Technol.* **327** 124813
[13] Kim H, Lee S, Ahn Y, Lee J and Won W 2020 Sustainable Production of Bioplastics from Lignocellulosic Biomass: Technoeconomic Analysis and Life-Cycle Assessment *ACS Sustain. Chem. Eng.* **8** 12419–29

Creating Peatland Restoration Policy for Supporting in Indonesian Economic in a Sustainable Way

Irfandi Pratama[1]*, Eko Priyono Purnomo[1,2], Dyah Mutiaran[2], Maisarah Mitra Adrian[1,2], Cici Sundari[1,2]
[1] Department of Government Affairs and Administration, Jusuf Kalla School of Government, Universitas Muhammadiyah Yogyakarta, Indonesia
[2] E-Governance and Sustainability Institute ESI, Yogyakarta, Indonesia
[1*]irfandipratama1998@gmail.com, [2]eko@umy.ac.id, [3]dyahmutiarin@umy.ac.id, [4]maisarahmitra@gmail.com, [5]cicisundari06@gmail.com

Abstract. As large tracts of degraded peatlands are caused by conversion and fires, peat restoration is carried out to restore peat ecosystems to their natural condition and support the socioeconomics of communities around peatlands that are affected because they depend on the resources provided by peatlands as livelihoods. This study aims to see the implementation of peat restoration in the Rewetting, Revegetation, and Revitalization (3R) approach in Jambi Province. The research focuses on the economic revitalization of people's livelihoods. This research method is exploratory qualitative, while data collection is through interviews, observations, documentation, and literature: data analysis techniques, data reduction, data presentation, and conclusion drawing. The results of research on the implementation of peat restoration in Jambi Province have been carried out since 2018; the implementation of the 3R has not been optimal, which is still far from the target set. In particular, assistance for the economic revitalization of people's livelihoods lacks help, so the program's sustainability is stopped halfway. In addition, the collaboration and coordination of the Regional Peat Restoration Team (TRGD) agencies did not go well.

1. Introduction

The world's peatlands are estimated at 397-423 million ha or equivalent to 3% of the land area; some of the most extensive peatlands are in Canada, the Soviet Union, the USA, and Indonesia [1]. Peatlands are in the world's spotlight as the largest absorbent and store of terrestrial carbon and regulators of water absorption up to 90% of their volume [2]. However, most of the world's peatlands, particularly in southeast Asia, are degraded and converted to oil palm plantations and industrial plantations [3]. Meanwhile, Indonesia's tropical peatlands reached 13,430,517 ha or 47%, representing the low global peatlands [1].

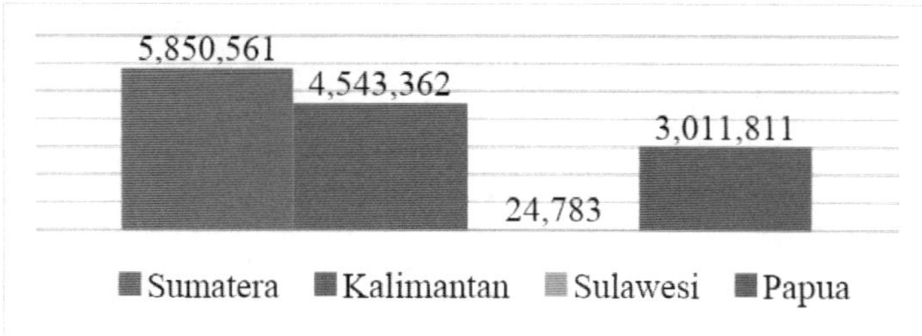

Figure 1. Peatland distribution in Indonesia

In figure 1, peatlands in Indonesia are spread across the islands of Sumatra, Kalimantan, Sulawesi, and Papua. Sumatra has the largest area of peatlands, reaching 5,850,561 ha, spread across Riau, South Sumatra, and Jambi Provinces. The biggest failure of peatland management in Indonesia occurred in the era of the new order, which was converted into rice fields [4]. Then the fires peaked in 2015 and burned 2.61 million ha of forest and land; there were 33% or 869,754 ha of peatland. All 130,000 ha burned in Jambi Province caused economic losses of 12 trillion. Even in 2019, 114,900.2 ha of forest and peatland fires were repeated [5,6]. The acceleration of restoration is legalized in Presidential Regulation Number 1 of 2016 concerning the Peatland Restoration Agency (BRG) to deal with damage to the peat ecosystem due to fires. Peatland restoration policy measures are carried out through the 3R, Rewetting, Revegetation, and Revitalization approach [7]. Rewetting is an activity of wetting peatlands through the construction of canal bulkhead infrastructure, drilled wells, and canal stockpiling [8].

After wetting, to restore vegetation cover, the Revegetation approach is carried out through the planting of native peat plants [9]. Meanwhile, revitalization provides economic carrying capacity to local communities [10]. Meanwhile, the implementation of peat restoration is carried out with multi-actor collaboration as a form of collaborative governance [11]. The collaboration is done to restore the peat ecosystem to its natural condition, on the other hand, to increase community participation in maintaining and operating and maintaining infrastructure in peatlands and helping to improve the economy [7]. The researchers highlighted that approximately 75% of non-concessionary forest and peatland areas are home to and livelihoods for 16.3 million people. Then the lack of alternative livelihoods triggers poverty due to the interests of people exploiting peatland resources [12].

The lack of expansion of land clearing by communities and the private sector for oil palm and rubber plantations impacts the vegetation cover of forests and peatlands [13]. Then the scholars suggested agroforestry to improve the economy and maintain peatland ecology. However, the success rate and low literacy accompanied by orientation economic benefits without environmental concerns are challenging [8]. In addition, the revitalization of the economy through livelihoods is faced with institutional capacity. Earlier scholars have emphasized key issues related to community participation, lack of monitoring, and evaluation of revitalization programs [3]. This research in Jambi Province, the third largest peatland in Sumatra, reached 621. 089 ha. The peat ecosystem in Jambi Province is in a critical condition of 8.07%, very essential at 18.05%, medium critical at 38.93%, and only 34.95% is in a relatively good state. In addition to being caused by fires, the destruction of peat ecosystems in Province Jambi due to the conversion of plantations and industrial forests also occurs in limited cultivated land [14,15].

In addition to the decrease in the total income of 269 million per hectare of land burned, socio-economical households are a vulnerable group that experiences a reduction in income [16]. Previous research conducted by scholars related to peat restoration still focuses a lot on the ecological success [3]. in addition to financing sustainable restoration and peat restoration stakeholder studies [9]. This research suggests further looking at the socio-economic aspects of peatland restoration policies. For this reason, the study fills the gap in the elements of peat restoration implementation in encouraging or creating alternative livelihoods for the community. This study aims to evaluate how successful the 3R approach to peat restoration has been in Jambi Province, with a particular emphasis on reviving the community's economy in Tanjung Regency Jabung Timur. This is because this region contains the largest peatland in Jambi Province, with most peatlands located in concession areas [17]. Hopefully, this research can contribute to the government as a form of monitoring and evaluating the implementation of peat restoration, especially in economic revitalization.

2. Methods

This paper analyzes the implementation of peat restoration of the 3R Approach in Jambi Province, focusing on the economic revitalization of communities in the Tanjung Jabung Timur Regency. Jambi Province is located at 0.45° o-2.45° South Latitude and between 101.10°-104.55° East Longitude. The fire in 2015 was in a peat area converted for agriculture and plantations in Tanjung Jabung Timur Regency, which has 46% peatland. In addition, this study is also based on the presentation of poverty, reaching the highest 11.39% in Jambi Province [18]. This research method is qualitative with an exploratory approach because it can analyze and examine more deeply a problem related to peat restoration; this method has been used in previous research [19]. This study describes and analyzes the field findings of the implementation of 3R peat restoration in Jambi Province when the research was carried out from June-July 2022.

This study's collection included primary data from observation, documentation, and semi-structured interviews with each source. We interviewed two speakers from each village represented by the community and the village government. The resource persons came from villages in Tanjung Jabung Timur Regency, namely Merbau, Pematang Rahim, Pandan Sejahtera, and Rantau Rasau villages. The interview questions we asked related to the implementation, impacts, and benefits of 3R peat restoration and the sustainability of economic revitalization. Meanwhile, the interview we conducted with OPD Province Jambi on the task of peat restoration assistance came from the Forestry Service, The Environment Agency, and the Planning Agency. The development of the Jambi Province was initiated by two speakers from each institution who received a direct task related to peat restoration. The interview questions focused on the implementation mechanisms, obstacles, and drivers of peat restoration, economic revitalization efforts, and the question related to stakeholders involved in peat restoration. Then we also conducted interviews with NGOs Walhi, Warsi, and KIFC to find information on the findings and success rates of implementing 3R peat restoration, as well as related to economic resilience and sustainability of economic revitalization packages. In the interviews we conducted to find research results based on the objectives of this study, secondary data were used to confirm research findings and build research literature. Secondary data for this study were obtained through journal literature, reports on restoration activities, and books or other documents related to peat restoration activities in Jambi Province. Data analysis techniques for data collection, data reduction, data presentation, and conclusion drawing can thus present the findings related to peat restoration [20].

3. Results and Discussion

We divide the flow of discussion and discussion in this study into the first step to implement peatland restoration policies, then the state of peatlands, followed by the implementation of restoration peat approach 3R Rewetting, Revegetation, and Economic Revitalization, and continued with mapping peat restoration stakeholders. Peatland restoration is a tangible manifestation of the government in protecting and restoring peat ecosystems through various schemes directly assigned by the President to the Peatland Restoration Agency (BRG) through Presidential Regulation No. 1 of 2016. Then it was changed through Presidential Regulation No. 120 of 2020 concerning the Peat and Mangrove Restoration Agency (BRGM). BRG's task of coordinating and facilitating peat restoration in 7 provinces; Riau, Jambi, South Sumatra, West Kalimantan, Central Kalimantan, South Kalimantan, and Papua.

The peatland of Jambi Province reaches 621,089 ha, the largest distributor of which is located in 3 districts; Tanjung Jabung Timur (46%), Tanjung Jabung Barat 20%, and Muaro Jambi 30% [14]. The MoU of BRG and the Government of Jambi Province produced the Surate of governor's decree No mor.383/KEP.GUB/BAPPEDA-4.3/2017 concerning establishing a Regional Peat Restoration Team (TRGD), which carries out peat restoration assistance tasks in other use areas, production forest areas that are not burdened with permits, and protected forests. The peat restoration plan prepared by BRG and the Provincial Government will be implemented in 13 Peat Hydrological Units (KHG) in peatland restoration units (SLRG) covering an area of 483,301.91 ha for protected functions. As for the cultivation function, it covers an area of 133,587.96 ha of function. However, the provincial government of Jambi targets peat restoration priorities for the restoration of peatlands affected by forest and land fires in 2015-2016, covering an area of 67,175 ha, while in cultivated and protected areas that already have a canal network covering an area of 324,651 ha [16].

Meanwhile, based on the Indicative Map of Peat Restoration, the area of land to be restored reaches 200,772 ha damaged by forest and land fires, which are divided; into protected areas of 46,415 ha, unlicensed cultivation areas (non-concessions) of 25,885 ha, and licensed cultivation areas (concessions) of 128,472 ha. The Indicative Map of Peat Restoration shows that there is a 63.9% obligation for concession holders of areas approved by oil palm plantation companies, Industrial Planting Forests (HTI), and Forest Tenure Rights (HPH) [21,22]. The achievement of peat restoration facilitated by the State Budget and partners in 2018 had an impact of 67,670 ha, of which the realization of rewetting included 301 units of drilled wells and 297 units of canal bulkheads and revitalization of 32 packages. Meanwhile, peat restoration in licensed or concession areas in 2019 reached 11,950 ha [17]. The absence of legal strictness can ensnare or cause a deterrent effect on companies that do not carry out peat restoration. In addition, a canal bulkhead was made only with a pile of mineral soil and wood that did not meet the specified specifications [21].

Peat restoration through the rewetting approach is carried out by building canal bulkheads; this aims to raise the water level in peatlands, in the implementation of which canal bulkheads are made at canal junction locations with a distance of 200 meters per canal bulkhead [12]. In addition, peat wetting is carried out by building drilled wells at areas burning 200 meters from water sources (rivers/canals) or between other well-drilled points [16]. In figure 2, we explain the implementation of the peat restoration rewetting approach through the construction of canal bulkheads and drilled wells carried out by TRGD.

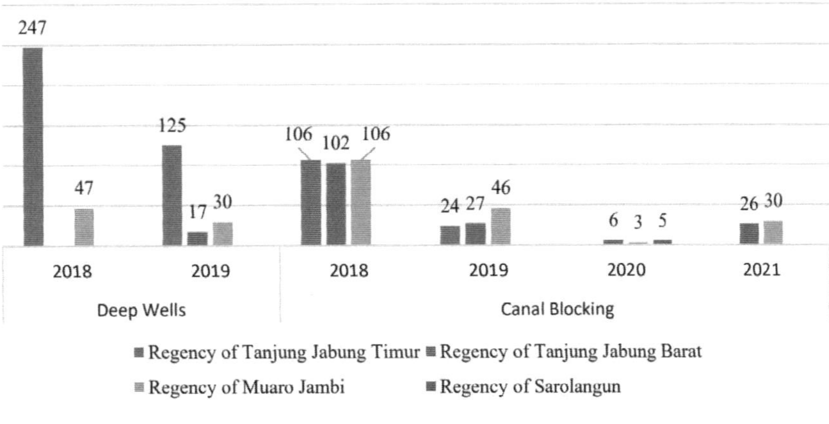

Figure 2. Number of drilled wells and canal blocking in Jambi Province
Source: Jambi provincial forestry service report

In figure 2, peat wetting infrastructure (rewetting) was constructed from 2018 to 2021; the drilled wells built reached 466 points, while the canal bulkheads were 481. The canal bulkhead was created independently by the surrounding community in the cultivation area and other uses of production forests that are not burdened with permits [23]. In our study, we observed several canal bulkheads made of unkempt and weathered wood, even in the grass. In addition, several canal bulkhead placement points located on the path of water and land transportation are prone to be damaged by mobilization activities. In addition to several canal bulkheads, it causes flooding, especially during the rainy season and high tide. According to the community and the village government, transferring locations and modification of canal partitions are not allowed. However, referring to the technical rules of canal bulkheads that are on the transportation route or not by the specifications and conditions of the canal or river can be modified. Then based on community reports and our observations directly, some of the drilled wells were not working or were deadlocked [24].

Revegetation or replanting of degraded peat areas is carried out by planting local types of crops. Degradation of peatlands is of several types; first, the area that has been cleared in forest areas and private and corporate areas, which are overgrown with shrubs, and degradation is caused by fire. First, revegetation is carried out employing natural succession in real with dense land cover, secondly through enrichment of plants in areas with slightly open peatland cover or sparse vegetation. Third, through planting maximum patterns on areas with available land cover. The selected plants must be adjusted if in the area of rewetting or protected function planted with forestry plants such as Jelutung (Dyer *Costulata*), Pulai Rawa (*Alstonia Sholaris L*), Bengal (Neonauclea Sp). Meanwhile, in the cultivation or non-concession function, production or plasticulture crops are planted, such as Betel Nut (*Areca Sp*), Coconut Dalam (*Cocos Nucifera*), Coffee (*Coffea Sp*), Chocolate (*Theobroma Cacao*), and others [25]. The following is peatland revegetation carried out by the TRGD of Jambi Province.

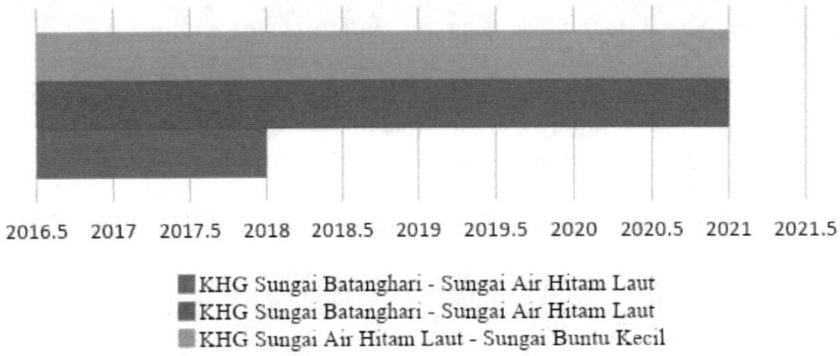

Figure 3. Area of land revegetated in Jambi Province

After the rewetting approach, apart from the construction of canal blocking and drilled wells as well as canal filling, the next step is revegetation. Because the land peat has been in an almost natural condition with a sufficient water supply, it is hoped that rewet peatlands will lower the fire threat or limit the death of planted tree seedlings. We know that degraded peatlands have a scarcity of water during the dry season and abundant water during the rainy season [26]. Figure 4 shows that the TRGD's revegetation efforts are still focused on Muaro Jambi Regency, the primary fire hotspot in 2015, reaching 45.95%. This demonstrates that, compared to other places, Muaro Jambi regency is nearly finished with rewetting implementation because it is the top priority of revegetation carried out on post-burn land. It also explains why peatlands are deteriorating so rapidly. Climate change, pest assaults, and fires are among the challenges that revegetation encounters.

Indirectly, peat restoration activities through rewetting and revegetation approaches have a socio-economic impact on the community. The development of peat wetting infrastructure is carried out by self-management, meaning that the community carries out the construction of canal bulkheads and the manufacture of boreholes. The result is that the funds received will be back to society. Not only that, but the community will also get technical guidance on how the development of peat wetting infrastructure will improve skills. In addition, the community will benefit from canal bulkheads and boreholes, namely the ease of accessing water for their plantation needs. The same is also done in the revegetation approach; planting is carried out independently by the community; in the function of cultivated land, the plant type has an economic value, which can be an alternative livelihood for people other than the main income.

Peatland restoration considers not only the environment but also the socio-economic situations of the community in a sustainable manner. This is because people's livelihoods on peatlands are still based on natural resources, particularly forestry, agriculture, plantation, and fisheries. This is a driving force behind land and forest cover changes caused by peatland conversion. As a result, peat restoration using an economic revitalization approach generates alternative livelihoods for local communities via paladikutur and agrosylbofishery [23]. Revitalization is also an endeavor to empower and expand the community's involvement in the repair process. Community empowerment is achieved by improving community capacity through developing alternative productive economies through restoration aims. Meanwhile, community participation is increasing in peatland restoration initiatives [27].

Community empowerment and participation in peat restoration is a type of incentive or compensation used to increase community awareness, willingness, active participation, and independence in peat restoration activities. Revitalization is also a kind of compensation for peat-soaking infrastructure construction initiatives that disturb local populations' livelihoods. Land-based, water-based, and environmental services-based revitalization operations exist [28]. The revitalization of Tanjung Jabung Timur Regency is explained here.

Table 1. Revitalization in Tanjung Jabung Timur Regency

Villages / Villages	Year	Revitalization Assistance
Fresh River Village	2018	1 goat cultivation package
Rice River Village	2018	1 goat cultivation package
Pandan Lagan Village	2018	1 package of cattle cultivation
Pandan Sejahtera Village	2018-2020	1 package of breeding and fattening cows - Breeding of bali cattle 7 heads and planting of gaja grass 0.5 ha
Parit Culum Village 1	2018-2020	1 package of cattle cultivation - 1 package of fattening of cows and planting of elephant grass 0.5 ha
Rahayu Chess Village	2018-2019-2020	1 liberica coffee cultivation package - 1 pepper cultivation package - breeding and fattening of 7 cows and planting of elephant grass 0.5 ha and cultivation of betel nut 8 ha
Teluk Dawan Village	2018	1 package of cattle cultivation
Parit Culum Village 2	2021	Catfish farming (biofloc) diameter 3m (5 ponds) and watermelon 1 ha
Sinar Wajo Village	2021	Goat livestock 25 heads, planting durian, avocado, and young jackfruit 1 ha
Pematang Rahim Village	2021	Pineapple cultivation 1.5 ha, red ginger 0.7 ha, pineapple derivative products, and compost house 0.5 ha
Koto Kandis Dendang Village	2021	Catfish Farming (biofloc 8x12 m), ducks 1200 heads, and cicada livestock
Merbau Village	2021	Goat livestock 34 heads of catfish cultivation 3000 heads and tilapia 3000 heads

Source: Jambi Provincial Forestry Service Report

In Table 1, the economic revitalization of community livelihoods in eastern Tanjung Jabung Regency is located in 12 villages and sub-districts in the form of land and water-based assistance. Revitalization assistance aims to develop the independence and welfare of communities on peatlands and their surroundings to support the restoration and sustainability of peat ecosystems. This is closely related to the habit of people who open land by burning because it is cost-effective, neutralizes peat acidity, and increases fertility [21]. Based on the results of interviews with the community and village governments as well as NGOs and TRGDs, the success and sustainability rate of economic revitalization assistance is still low. Among the contributing factors is that the socio-economic analysis of people's needs is not studied contingently. In addition, the village assistants are apathetic and do not understand the social conditions of the community they are accompanying because the village companions are not initially from the area. We found that in several villages, the focus of their livelihood was on plantations, such as those that produce oil palm, betel nut, and coconut, which are economically secure and have sufficient employment opportunities. But still have to accept economic revitalization assistance with the assumption that if they do not receive the aid, then government assistance from other programs that meet the community's needs will not be given again. We found this to be the case in some of the villages. Because of this, the service provided is not to the needs of the community; for example, villages with most farmers apply for water-based assistance, so the assistance is not well absorbed and sustainable.

On the contrary, we also found some villages with unstable economic conditions and livelihoods; economic revitalization assistance could not be adequately absorbed, which was influenced by the capacity of local institutions to politicize such aid. For example, community groups (Pokmas) are filled by relatives of the village government, so people who need or can manage revitalization assistance cannot reach them. Therefore, economic revitalization cannot be sustainable. Revitalization is also carried out by increasing community capacity through empowerment by providing training on peatland management without burning, paladiculture, and creative economy through processing plant products on peatlands such as processed products derived from pineapple, sago, purun, and others. However, the reality is that people are less interested because of limitations on product marketing. Become a record of the government in peat restoration to assess the involvement of village governments, the availability and capabilities of community groups, assistance, monitoring and evaluation, and accountability [29]. In addition, strengthening the role of district governments is crucial in peat restoration because they can certainly know precisely the situation, environment, and socioeconomics of its people. Next, we present the actors in peat restoration in auxiliary tasks.

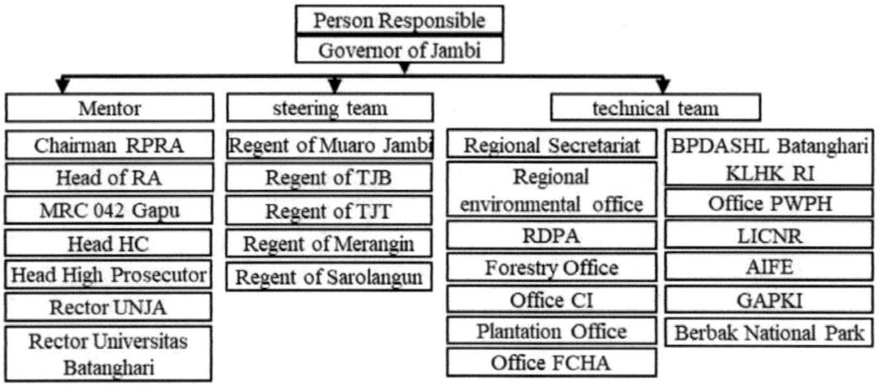

Figure 4. Peat restoration actors in Jambi Province

Notes: RPRA: Regional People's Representative Assembly; RA: Regional Police; MRC: Military Resort; Commander; Gapu: Garuda Putih; HC: High Court; UNJA: Universitas Negeri Jambi; TJB: Tanjung Jabung Barat; TJT: Tanjung Jabung Timur; RDPA: Regional Development Planning Agency; CI: Communications and Information; FCHA: Food Crops, Horticulture, and Agriculture Office; BPDASHL: Center for Watershed Management and Protected Forests; PWPH: Public Works and People's Housing; LICNR: The Local Institute for Conservation of Natural Resources; AIFE: Association of Indonesian forest entrepreneur; GAPKI: Indonesian Palm Oil Association;

In figure 5, we explain the TRGD actors involved in peat restoration assistance tasks formed by the Governor of Jambi through Decree No.394/KEP. GUB/DLH-3.2/2021 concerning establishing the TRGD of Jambi Province. The TRGD that was formed was filled out by several Jambi Province agencies, with the presence of a secretariat at the Provincial Environmental Service who carried out the task of monitoring and evaluating peat restoration. Meanwhile, the Forestry Service carries out the spearhead of the restoration's technical implementation, then the procurement of nurseries involving the Plantation Service and the Food Crops and Horticulture and Animal Husbandry Service (TPHP). The results of our interviews in three provincial government agencies explained that the collaboration and coordination of peat restaurant assistance tasks did not run optimally. The collaboration is made clear when we ask questions about peat restoration's primary duties and functions; some relevant agencies cannot answer or show their performance report. On the other side, it was slow to negotiate the administrative affairs of performance reporting. The district government only accepts the news of the program's implementation, as evidenced by the absence of a binding MoU. Collaboration in peat restoration with NGOs can only be carried out at the central level and implemented by BRGM and the ministry, one of which is a collaboration with KIFC. Meanwhile, at the TRGD level, it cannot be done through collaboration with NGOs based on interviews with the Forest Service and NGO Walhi and Warsi.

4. Conclusion

Peatland conditions in Jambi Province are degraded due to conversion, drainage, and fires. The aftermath of the 2015 fires was the beginning of a severe government to improve peat ecosystems. Peatland restoration in Jambi Province has been carried out since 2018, with the form of TRGD; restoration is carried out through the 3R Approach of Rewetting, Revegetation, and Revitalization of the eye economy in the search for society. Peatland wetting in Jambi Province is inseparable from the pros and cons of the community. Based on the Jambi Province Peat Ecosystem Protection and Management Plan (RPPEG), the peat restoration target set is still far from the implementation achievements. All speakers agreed that peat restoration is essential, but the success rate of peat restoration remains questionable. Then the economic revitalization of people's livelihoods is mainly unsustainable and has not been able to become a reliable alternative livelihood. Coordination and collaboration are still facing a deadlock. We are aware of the limitations of the object of study, so we recommend that subsequent research involve a more comprehensive range of things from the point of view of implementation at the village level. Furthermore, researchers can also explore stakeholder involvement in encouraging peat restoration in sustainable development studies.

References

[1] Anda M, Ritung S, Suryani E, Sukarman, Hikmat M, Yatno E, Mulyani A, Subandiono R E, Suratman and Husnain 2021 Revisiting tropical peatlands in Indonesia: Semi-detailed mapping, extent and depth distribution assessment *Geoderma* **402** 115235

[2] Xu J, Morris P J, Liu J and Holden J 2018 PEATMAP: Refining estimates of global peatland distribution based on a meta-analysis *Catena* **160** 134–40

[3] Mishra S, Page S E, Cobb A R, Lee J S H, Jovani-Sancho A J, Sjögersten S, Jaya A, Aswandi and Wardle D A 2021 Degradation of Southeast Asian tropical peatlands and integrated strategies for their better management and restoration *J. Appl. Ecol.* **58** 1370–87

[4] Sulaiman A A, Sulaeman Y and Minasny B 2019 A framework for the development of wetland for agricultural use in Indonesia *Resources* **8** 1–16

[5] Walhi Jambi 2020 Rilis Media Hasil Pemantauan Restorasi Gambut di Wilayah Konsesi Di Provinsi Jambi *Walhi Jambi*

[6] Agustiyara, Purnomo E P and Ramdani R 2021 Using Artificial Intelligence Technique in Estimating Fire Hotspots of Forest Fires *IOP Conf. Ser. Earth Environ. Sci.* **717**

[7] Dohong A 2019 Restoring Degraded Peatland in Indonesia: the 3R approach. *RSPO Man. Best Manag. Pract. Manag. Rehabil. Peatlands* 2016–7

[8] Yuwati T W, Rachmanadi D, Pratiwi, Turjaman M, Indrajaya Y, Nugroho H Y S H, Qirom M A, Narendra B H, Winarno B, Lestari S, Santosa P B, Adi R N, Savitri E, Putra P B, Wahyuningtyas R S, Prayudyaningsih R, Halwany W, Nasrul B, Bastoni and Mendham D 2021 Restoration of degraded tropical peatland in indonesia: A review *Land* **10** 1–31

[9] Sari A P, Dohong A and Wardhana B 2021 *Innovative financing for peatland restoration in indonesia* (Springer International Publishing)

[10] Pratiwi S W, Hastuti K P, Alviawasti E and Rahman A M 2021 Peatland Economic Revitalization Program Based on Community Participation in Jarenang Village, Kuripan District, Barito Kuala Regency, South Kalimantan *Proc. 2nd Int. Conf. Soc. Sci. Educ. (ICSSE 2020)* **525** 394–8

[11] Suharti S, Tata H L and Prameswari D 2022 Stakeholders mapping of peatlands restoration in Jambi : A case of Tanjung Jabung Barat Regency Stakeholders *IOP Conf. Ser. Earth Environ. Sci.*

[12] Dohong A, Abdul Aziz A and Dargusch P 2018 A Review of Techniques for Effective Tropical Peatland Restoration *Wetlands* **38** 275–92

[13] Medrilzam M, Smith C, Aziz A A, Herbohn J and Dargusch P 2017 Smallholder Farmers and the Dynamics of Degradation of Peatland Ecosystems in Central Kalimantan, Indonesia *Ecol. Econ.* **136** 101–13

[14] Dinas Lingkungan Hidup 2021 *Laporan Kegiatan Monitoring dan Evaluasi Tim Restorasi Gambut Daerah* (Jambi)

[15] Umayah D, Purnomo E P, Fadhlurrohman M I, Fathani A T and Salsabila L 2022 The Implementation of Indonesian Sustainable Palm Oil (ISPO) Policy in Managing Oil Palm Plantation in Indonesia *IOP Conf. Ser. Earth Environ. Sci.* **943**

[16] Badan Restorasi Gambut 2017 *Rencana Restorasi Ekosistem Gambut Provinsi Jambi Tahun 2017-2020* (Jakarta: Badan Restorasi Gambut Republik Indonesia)

[17] Tim ED Walhi Jambi 2020 *Laporan Hasil Temuan Pemantauan Restorasi Gambut di Provinsi Jambi* (Jambi)

[18] BPS Provinsi Jambi 2022 *Provinsi Jambi Dalam Angka 2022* (Jambi: BPS Provinsi Jambi)

[19] Priyo Purnomo E, Aditya F, Teguh Fathani A, Salsabila L and Rachmawatie D 2021 Penta-Helix Approach as a Strategy to Recovery Tourism in Bali Due to Covid-19 Pandemic *2021 The 11th International Conference on Information Communication and Management* ICICM 2021 (New York, NY, USA: Association for Computing Machinery) pp 122–127

[20] Brotosuilo A, Anita Rouly V and Pricilia C C 2021 Community involvement in peatland management: Social intervention approach and environmental law *IOP Conf. Ser. Earth Environ. Sci.* **716**

[21] Suprapto Y 2021 Perusahaan Biang Kebakaran Hutan Dihukum Denda Miliaran *Pantau Gambut*

[22] PRIMS 2020 Kubah Gambut di Konsesi HTI, Riwayatnya Kini *PRIMS*

[23] Dinas Kehutanan 2022 Laporan Rewweting TRGD Provinsi Jambi

[24] Dohong A, Cassiophea L, Sutikno S, Triadi B L, Wirada F, Rengganis P and Sigalingging L 2017 *Modul Pelatihan Pembangunan Infrastruktur Pembasahan Gambut Sekat Kanal Berbasis Masyarakat*

[25] Budiman I, Bastoni, Sari E N, Hadi E E, Asmaliyah, Siahaan H, Januar R and Hapsari R D 2020 Progress of paludiculture projects in supporting peatland ecosystem restoration in Indonesia *Glob. Ecol. Conserv.* **23** e01084

[26] Purnomo E P, Rahmasari F V, Trisnawati D W, Agustiyara and Erviana R 2021 Observed Data of Forest Fire Hotspots effects on Respiratory Disorder by Arc-GIS in Riau Province, Indonesia *IOP Conf. Ser. Earth Environ. Sci.* **717**

[27] Handoko T, Tiyas Tinov, Achmad Fajri Febrian, Risky Arya Putri, Frini Karina Andini and Ogi Rifansyah 2020 Peatlands Restoration As A Potential Solution To Resolve Peatlands Damage Based On Sustainable Development Goals (SDGS) In Sungai Tohor, Indonesia *J. Niara* **13** 123–31

[28] Kementerian LHK 2020 Permen LHK No.P.8/Menlhk/Setjen/Kum.1/2/2020 tentang Penugasan Sebagian Urusan Pemerintahan Bidang Lingkungan Hidup Dan Kehutanan Kepada 7 Gubernur Untuk Kegiatan Restorasi Gambut Tahun 2020

[29] Purnomo E P, Ramdani R, Agustiyara, Nurmandi A, Trisnawati D W and Fathani A T 2021 Bureaucratic inertia in dealing with annual forest fires in Indonesia *Int. J. Wildl. Fire* **30** 733–44

IOP Conf. Series: Earth and Environmental Science 1111 (2022) 012005 doi:10.1088/1755-1315/1111/1/012005

Sustainable Forest Governance: A New Policy Strategy in Handling Forest Fires in Jambi Province

Cici Sundari [1*,2], Eko Priyo Purnomo[1,2], Dyah Mutiarin[1], Maisarah Mitra Adrian[1,2], Cindy Fabrizia Suling[1,2], Irfandi Pratama[1,2]

[1] Department of Government Affairs and Administration, Jusuf Kalla School of Government, Universitas Muhammadiyah Yogyakarta, Indonesia

[2] E-Governance and Sustainability Institute, Indonesia

[1*]cicisundari06@gmail.com, [2]eko@umy.ac.id, [3]maisarahmitra@gmail.com, [4]cindyfabriziasuling@gmail.com, [5]irfandipratama1998@gmail.com

Abstract. Jambi Province is one of the areas most prone to forest destruction in Indonesia. The case of forest fires is an annual disaster that regularly occurs in Indonesia, including one in Jambi province. The problem of forest fires often occurs in Indonesia, whether human activities cause it or are caused by a long dry season. This study analyzes Sustainable Forest Governance from the point of view of the New Policy Strategy in Handling Forest Fires in Jambi Province. This research focuses on Jambi Province, one of Indonesia's regions with the largest forest. This research method is qualitative exploratory with secondary data carried out to describe and describe the research. Data were collected from field interviews, online media, and related literature related to the research topic. The analysis technique uses the NVivo 12 plus data processing application, a qualitative document analysis tool with the help of a computer. Equipment. They are easy to use and can word-process and explore word frequency, attributes, and cases from big data. They also generate factor or sub-factor categories in journalistic and research applications related to the research topic. The study results show that the government's program in making new policy strategies for handling forest fires in the province has several indicators, then the cooperation of actors in forest fire prevention in the region.

1. Introduction

Forests are one of the most important natural resources, providing rich biodiversity that contributes directly and indirectly to economic and social life [1]. Jambi Province is one of the areas most prone to forest destruction in Indonesia. The case of forest fires is an annual disaster that regularly occurs in Indonesia, including one in Jambi province. The problem of forest fires often occurs in Indonesia, whether human activities cause it or are caused by a long dry season. So that it has a negative impact caused by forest fires, such as ecological damage, decreased economic value of forests and land productivity, changes in micro and global climate, and also an impact on human health [2].

In modern times like today, the use of forests in tourism is increasingly being carried out by the community and local government. This forest is used to support the economy in the region, as well as a tangible manifestation of the synergy between the community and the local government, which always maintains and preserves the forest for the sake of survival in the future. Forests are a part of the environment. Forests are also a national development capital that has real benefits for the lives and livelihoods of the Indonesian people, both ecologically, socio-culturally, and economically. Forests play an essential role in human life, proven in the protection, shade, and products humans need for survival [3].

Figure 1. Indonesia Fire Statistics Data [2]

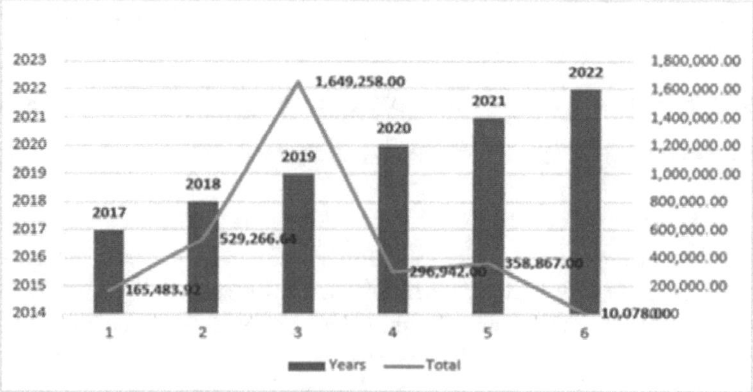

Figure 1. Describes the data on forest fires that occurred in the last five years in Indonesia and also in 2022, which shows that information on forest fires has increased and decreased in forest fires and areas where forest fires occurred in Indonesia [4]. Forest supervision and protection aim to prevent and minimize forest damage, maintain state rights to forests and their products, and have strategic value in people's and state's lives where forests function as biological natural resources, support life, and are regional assets that have ecological and economic benefits. Data from the Indonesian Central Statistics Agency shows that the decline in the number of forest and land fires every year from 2017-2022 will have a tremendous impact on the lives of Indonesian people in the future. This study uses an exploratory qualitative method, whereas, in qualitative research, the technique is used to explore the phenomenon of the object studied in the field [5].

Qualitative research has the responsibility to answer questions related to the complexity of the object of study so that the use of this method can explain the most fundamental problems and can provide construction recommendations. Utilization of this forest is used to support the economy in the region, as well as a tangible form the community and local government in synergy always to protect and preserve forests for the sake of future survival. The aim of the research is a new policy strategy for dealing with forest fires in Jambi province so that law enforcement and supervision from the central and local governments can help forest protection efforts for the balance of the surrounding environment and the sustainability of living things in the future. Based on this, it is essential to study and analyze new policy strategies for handling forest fires in Indonesia, especially in Jambi Province [6].

2. Methods

This type of research method is explorative qualitative, namely research that aims to explore the phenomenon of the object studied in the field. Secondary data is used to describe the research. Data were collected from field interviews, observations, questionnaires, and document studies [7]. Furthermore, data were also collected from online media and related literature related to the research topic. This research was conducted from June 2021 to May 2022 in Jambi Province, covering all regions in Jambi Province to see new policy strategies for handling forest fires in Jambi Province. The analysis technique uses the NVivo 12 plus data processing application, a qualitative document analysis tool with computer assistance [8].

Equipment. They are easy to use and can word-process and explore word frequency, attributes, and cases from big data. They also generate factor or sub-factor categories in journalistic and research applications related to the research topic. The focus of the problem in this research is Sustainable forest governance: a new policy strategy for dealing with forest fires in Jambi Province. This research was conducted because forest fires are one of the crucial environmental problems and have become a local and global concern [9].

Table 1. Informant data

No	Informant
1	Jambi Provincial Forestry Service Agency
2	Jambi Province Regional Disaster Management Agency
3	Indonesian Forum for the Environment (Walhi) Jambi Province
4	Jambi Provincial Fire Service
5	Forest Village Community

3. Results and Discussion

Identification of Forest and Land Fires in Jambi Province In the last five years, various districts in Jambi Province have experienced an increase and decrease in forest fires and the Area where the fires occurred. In 2017 the fire area was around 109.00 Ha, 2018 was around 1578.00 Ha, 2019, there was an increase in forest fire area around 56,592.00 Ha, 2020 was 1.002 Ha, and 2021 forest fires were around 540.00 Ha in Jambi Province. In solving forestry problems, it is necessary to involve stakeholders, namely the government, the private sector, and non-government institutions [10]. Jambi Province has registered protected forests in seven areas spread across Jambi Province, starting from Muara Jambi, Tanjung Jabung Barat, Tanjung Jabung Timur, Bungo, Tebo, Sarolangun, and Merangin areas, as well as four other areas that have not been recorded, namely protected forests. In Jambi City, Batanghari, Kerinci and Full River [11].

Jambi Province has five types of forest, namely (1) community forest is a form of social forestry scheme provided by the government to the community to manage and utilize forest areas to improve welfare, environmental sustainability, and community empowerment. (2) a village forest is a state forest managed by village institutions to help the community's economy and prosper the village, (3) a private forest is a forest located on land encumbered with certain land rights outside the state forest area, such as state forest. Forest. The private forest is evidenced by certificates of ownership rights, business rights, and usufructuary rights. (4) Customarily forest is a forest located within the territory of the customary law community, which aims to improve their welfare; (5) an industrial plantation forest is managed and managed based on the principle of optimal utilization by considering environmental sustainability. Natural resources use economic principles to obtain the maximum benefit [12].

One of the strategies that the Jambi Provincial government has prepared for the future is to form a task force in areas where forest and land fires often occur by cooperating with local communities. The government also makes efforts to prevent and control forest and land fires by following Law No. 41/1999 on Forestry, Law No. 18/2004 on Plantations, and Law No. 32/2009 on Environmental Guidelines and Management. Based on several indicators of forest and land fire prevention and control. Forest and land fires are currently seen as a form of disruption to environmental Management and sustainable development. Therefore, early prevention and the role of the community in tackling forest and land fires are essential through fire and land prevention activities to create a clean and smoke-free environment. The government needs to implement a new strategy for preventing forest fires by approaching areas where forest and land fires often occur, using a Co-management approach, namely an arrangement in which the responsibility for managing resources lies between the government and the community [13].

Table 2. Government Programs in Handling Fores Fires [14]

No	Indicators of forest and land fire management strategies
1.	Conducting an inventory of farmers' land ownership, types, soils, and plants
2.	Conducting fire prevention through outreach activities or socialization by several related technical agencies
3.	Assisting agricultural infrastructure facilities and appropriate technology and compensation so that farmers do not burn
4.	Creating a model or demonstration plot for land management without burning
5.	In a land fire, mobilization efforts to overcome land fires must be carried out.
6.	Installing a digital application of the analysis system for controlling forest and land fires (Digital Smoke)

Table 2. Explain the new strategy for handling forest fires in Jambi Province, which is included in the government program for preventing and controlling forest fires in Jambi Province. Programs one to five have been implemented for a long time, and the sixth is a new strategy prepared by the Jambi Provincial Government [15]. (1) conducting an inventory of farmers' land ownership, types, soils, and plants (2) conducting fire prevention through outreach activities or socialization by several related technical agencies (3) assisting agricultural infrastructure facilities and appropriate technology and compensation so that farmers do not burn (4) creating a model or demonstration plot for land management without burning (5) in the event of a land fire, mobilization efforts to overcome land fires must be carried out (6) installing a digital application of the analysis system for controlling forest and land fires (Digital Smoke). According to the authority, some of these steps have been responded to by the relevant agencies or agencies in the form of programs and activities. Each appropriate agency or agency has implemented them [16].

Figure 2. Actor Collaboration Scheme in Handling Forest Fire [3]

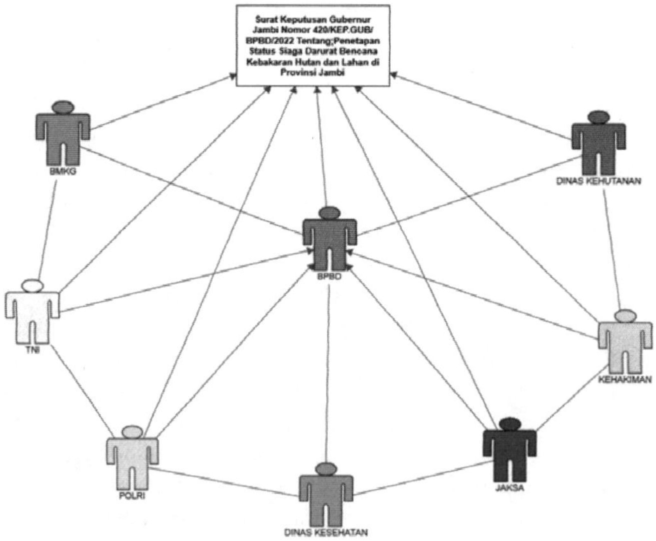

Figure 2. Shows the collaboration scheme of Penta helix actors in handling forest fires in Jambi Province based on the Jambi Governor's Decree Number 420/KEP/GUB/BPBD/2022 concerning the

establishment of an emergency alert status for forest and land fires in Jambi Province. The collaboration of the emergency task force on forest fire prevention in Jambi Province shows a network of Pentahelix actors consisting of the Regional Disaster Management Agency (BPBD), Meteorology Climatology and Geophysics Council (BMKG), Forestry Service, Indonesian National Army (TNI), Indonesia National Policy (Police), Health Service, Prosecutors, and Justice. This program is carried out by strengthening the model of excellent and sustainable institutional Management through the collaboration of relevant actors/stakeholders. The design of this program in a line of fire includes (1) a forest and land fire control task force, (2) a forest and land fire unit, and (3) a forest and land fire law enforcement unit [17].

Figure 3 . WordCloud Fire Forest [12]

Figure 3. Is the WordCloud result the researcher produced based on an analysis conducted using Nvivo 12 Plus software? The results from word cloud describe that forest and land fires are one the sectors with a significant impact, causing a lot of huge losses both materially and non-materially, such as fires. This impacts environmental, social, and economic aspects. Second, the resilience indicator is the attitude and behavior of each actor that is consistent in dealing with forest and land fires in Indonesia so that forests are maintained and preserved for all people in Indonesia, especially people in Jambi Province. The government has worked hard to solve this problem by issuing various policies and providing institutional and funding support; this is done with the hope that the number of forests and land fires can be minimized [18].

Table 3. Handling factors after forest and land fires [13]

No	Post-fire indicator
1	Information collection
2	Identification
3	Monitoring and evaluation
4	Rehabilitation
5	Law enforcement

Table 3. Describes handling activities after forest and land fires in Jambi Province, all efforts, actions, or activities that include inventory, monitoring and evaluation, and coordination to manage locations after the forest and land fires occur handling after the forest and land fires are carried out through activities (1) collecting information material which is carried out through field checks on burned areas using monitored hotspot data and collecting soil samples, plants, and hills in the fire area. (2) Identification is made to determine the cause, the fire area, the type of vegetation burned, and the effect on the environment and ecosystem. (3) Monitoring and evaluation are carried out to monitor the fire control activities that have been carried out and the development of the former fire area. (4) Rehabilitation is carried out in the context of rehabilitating ex-fire areas by considering recommendations or inputs based on data and information obtained from identification results. (5) Law enforcement is carried out in the context of efforts to take legal action in the field of forest fires by starting with the collection of materials and information relating to the occurrence of violations as investigation material [7].

4. Conclusion

Research results on new strategies in handling forest fires in Jambi Province, which are included in the government program in preventing and overcoming forest fires in Jambi Province. To take forest and land fires, several indicators prepare the government, then the Penta helix actor collaboration scheme in handling forest fires in Jambi Jambi Province based on the Decree of the Governor of Jambi Number 420/KEP/GUB/BPBD/2022 concerning the establishment of an emergency alert status for forest and land fires in Jambi Province. The collaboration of the forest fire emergency unit in Jambi Province and seeing community participation is no longer a matter of whether they want to participate but rather the extent to which they will benefit through this participation in their socio-economic life. The success of forest and land fire prevention and control (extinguishing) activities is very dependent on the success of humans and local communities in the emotions, feelings, and enthusiasm to maintain forest sustainability, which requires a forest and land management approach that understands the multiple aspects. The informal reaction of the community can be seen through the government's policy of making the Fire Care Group program based on community aspirations for forest and land fire incidents. The government created the Fire Care Group to suppress forest and land fires.

References

[1] Holuša J, Koreň M, Berčák R, Resnerová K, Trombik J, Vaněk J, Szczygieł R, and Chrome I 2021 A simple model indicates that there are sufficient water supply points for fighting forest fires in the Czech Republic *Int. J. Wildl. Fire* **30** 428–39

[2] Purnomo E P, Ramdani R, Agustiyara, Nurmandi A, Trisnawati D W and Fathani A T 2021 Bureaucratic inertia in dealing with annual forest fires in Indonesia *Int. J. Wildl. Fire* **30** 733–44

[3] Abdulsahib G M and Khalaf O I 2018 An improved algorithm to fire detection in a forest by using wireless sensor networks *Int. J. Civ. Eng. Technol.* **9** 369–77

[4] Costafreda-Aumedes S, Comas C, and Vega-Garcia C 2017 Human-caused fire occurrence modelling in perspective: A review *Int. J. Wildl. Fire* **26** 983–98

[5] Cidell J 2010 Content clouds as exploratory qualitative data analysis *Area* **42** 514–23

[6] Musri I, Ainuddin A N, Hyrul M H I, Hazandy A H, Azani A M and Mitra U 2020 Post forest fire management at tropical peat swamp forest: A review of Malaysian experience on rehabilitation and risk mitigation *IOP Conf. Ser. Earth Environ. Sci.* **504**

[7] Sundari C and Suswanta 2022 Artificial Intelligence Innovation in Capsule Bus in Jambi City *Iconpo 2021* **209** 104–10

[8] Sundari C, Nurmandi A, Muallidin I, Kurniawan D and Salahudin 2022 Analysis of Secondary Education Services During the COVID-19 Pandemic *Springer Int. Publ.* **319**

[9] Purnomo E P, Zahra A A, Malawani A D and Anand P 2021 The kalimantan forest fires: An actor analysis based on supreme court documents in Indonesia *Sustainability* **13** 1–12

IOP Conf. Series: Earth and Environmental Science 1111 (2022) 012005 doi:10.1088/1755-1315/1111/1/012005

[10] Sumoked G S J, Soepeno M H and Mamahit G N 2020 Tanggung Jawab Pemerintah Terhadap Kebakaran Hutan Dan Lahan (Karhutla) *Lex Adm.* **8** 157–68
[11] Syarifah H, Poli D T, Ali M, Rahmat H K and Widana I D K K 2020 Kapabilitas Badan Penanggulangan Bencana Daerah Kota Balikpapan Dalam Penanggulangan Bencana Kebakaran Hutan Dan Lahan *Nusant. J. Ilmu Pengetah. Sos.* **7** 408–20
[12] Suhendri S and Priyo Purnomo E 2017 Penguatan Kelembagaan Dalam Pencegahan dan Pengendalian Kebakaran Hutan dan Lahan di Kabupaten Muaro Jambi Provinsi Jambi *J. Gov. Public Policy* **4** 174–204
[13] Bouakkaz F, Ali W and Derdour M 2021 Forest fire detection using wireless multimedia sensor networks and image compression *Instrum. Mes. Metrol.* **20** 57–63
[14] Jufrida J, Basuki F R and Rahma S 2018 Potensi Kearifan Lokal Geopark Merangin Sebagai Sumber Belajar Sains Di SMP *EduFisika* **3** 1–16
[15] Laila N and Alhadi Z 2020 Upaya Pemerintah Dalam Penanggulangan Bencana Kebakaran Hutan dan Lahan (Karhutla) di Kabupaten Sarolangun (Studi Manajemen Bencana) Universitas Negeri Padang Email : najmilaila66@gmail.com zikrialhadi@fis.unp.ac.id Pendahuluan Daerah Kabupaten Sarola *J. Perspekt.* **3** 379–86
[16] Dani S and Wibawa A 2018 Challanges and Policy for Biomass Energy in Indoensia *Int. J. Business, Econ. Law* **15** 41047
[17] Arafah W, Nugroho L, Takaya R and Soekapdjo S 2018 Marketing strategy for renewable energy development in Indonesia context today *Int. J. Energy Econ. Policy* **8** 181–6
[18] Rauf R, Zainal R, Prayuda K, Rahman and Yuza A F 2020 Civil Society's Participatory Models: a Policy of Preventing Land and Forest Fire in Indonesia *Int. J. Innov. Creat. Chang* **14**

Carbon Monoxide (CO) and Particulate Matter (PM2.5) Concentration at Central Kalimantan, Indonesia

Kitso Kusin[1*,2], Yustinus Sulistiyanto[3], Aswin Usup[3] and Ardianor[3]
[1] Graduate Program of Environmental Science, Postgraduate Program, University of Palangka Raya. Indonesia, 73112, ORCID ID: 0000-0001-6019-0828
[2] Centre for the International Cooperation in Sustainable Management of Tropical Peatlands, University of Palangka Raya, Palangka Raya, Indonesia, 73112.
[3] Faculty of Agriculture, University of Palangka Raya, Palangka Raya, Indonesia, 73112.
[1*]kitso@cimtrop.upr.ac.id, [2]sulistiyanto_y@yahoo.co.uk, [3]aswin@agr.upr.ac.id, [4]ardianor@fish.upr.ac.id

Abstract. Peatland forest fires cause high concentrations of gases and particles in the air, especially carbon monoxide (CO) and PM2.5 which are very dangerous for health. Continuous monitoring of gas and particulate concentrations must be carried out during dry and rainy seasons so that the impact of hazards can be immediately identified and actions can be taken to overcome them. Monitoring equipment was placed from south to north at fire-prone locations to see the effect of wind and vegetation on the CO and PM2,5 concentrations. The concentration of Carbon Monoxide (CO) and Particulate Matter (PM2.5) at Central Kalimantan in the dry season is higher than in the rainy season. In the dry season of 2019, CO concentration in four locations was between 7104.1 – 22,127.5 μg/m3, while in 2021, it was the rainy season, so the CO concentration in six locations was lower, ranging from 38.7 – 1371.5 μg/m3. In the dry season of 2019, PM2,5 concentration in four locations was 196.8 – 360.5 μg/m3, while in 2021, at six locations was 5.0 – 40.0 μg/m3. The high concentration of CO and PM2.5 in the dry season needs to be anticipated with a serious handling program to prevent forest and peat fires

1. Introduction

Central Kalimantan, with an area of 15.36 million ha, is one of the provinces with an extensive tropical peat swamp forest, which is 2.55 million ha [1] or about 16.14% of the province's area. The rate of destruction of peat swamp forests continues to increase yearly, one of which is because of deforestation. From 2000 – 2008 the overall forest destruction rate reached 900 thousand hectares [2], and from 2013-2019 it was 511 thousand hectares [3]. The rate of destruction of peat swamp forests positively correlates with human needs in the economic, social, institutional, and infrastructure fields. Another forest damage is triggered by peatland forest fires that occur almost yearly because of improper forest management.

The history of peatland forest fires in Central Kalimantan is quite long, approximately 12 times in each dry season (El Nino) starting from 1972 – 2006 [4], then continuing in 2009, 2014, 2015, and 2019, emitting vast amounts of smoke and carbon. Peatland forest fires in humid conditions emit thick smoke, contain greenhouse gases [5] and cause particulate matter pollution [6-11]. In Peatland forest fires in 2019 at Palangka Raya, in one day, CO concentrations increased to 14,000 μg/m3 and PM2.5 concentrations up to 800 μg/m3 [12]. This condition is hazardous, exceeding the threshold of 8.000

μg/m3 for CO and 55.4 μg/m3 for PM2.5 [6]. PM2.5 is relatively small, enters the lungs, enters the bloodstream, and is transported to other tissues in the human body [7]; PM2.5 is much more dangerous than gaseous compounds [8], affecting the health and life of the people who live around it [9–11]. PM2.5 presents a high risk of deposition in the lungs' alveoli and is associated with a greater general health risk than coarse particles [12]. Each 1 mg/m3 increase in particulates results in approximately 4-8 infant deaths per 100,000 live births [13]. [14], inhaling air with high levels of particulate matter can impair breathing, resulting in inflammation of the airways and conditions such as coughing, bronchitis, respiratory difficulties, reduced lung function, and ultimately more severe obstructive respiratory distress.

Continuous monitoring of CO and PM2.5 concentrations is essential. The impact is enormous on healthy, especially for vulnerable people such as the elderly, pregnant women, and children. Monitoring is not only carried out in the dry season but in the rainy season by placing monitoring devices at specific points adjusting to the wind direction (North-South) to find out how significant the danger is. The tools used to measure CO and PM2.5 concentrations are portable, simple, and easy to implement.

2. Methods

2.1. Study area

The research was carried out for 12 months, starting from July – December 2019 and 2021, by using EL-USB-CO (Figure 1) devices for measuring concentration and P-Sensor (Figure 2) devices for measuring the concentration of PM2.5. Six locations for measuring at Saka Mangkahai Village; Buntoi Village, Tumbang Nusa Village and Taruna Jaya Village; Perum Pemda KM 10 Palangka Raya and Sei Gohong-Tangkiling (Figure 3). The selection of these locations is to see the impact of wind (south-north) and the effect of vegetation on CO and PM2.5 concentrations. In the dry season, the wind blows south, bringing the smog to the north. The data is downloaded every month to avoid memory on the device complete.

Figure 1. EL-USB-CO

Figure 2. P-Sensor

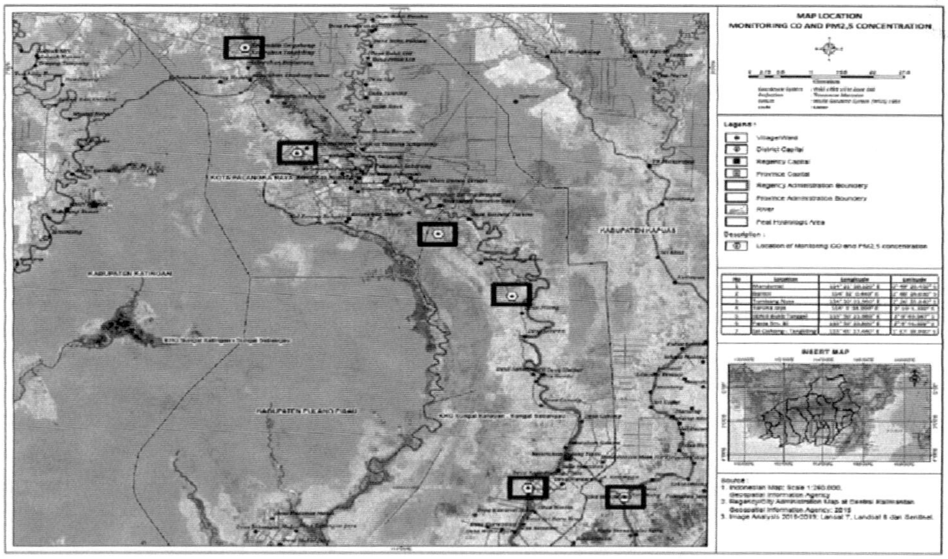

Figure 3. Location of CO and PM2.5 gauges are placed

2.2. Research design

EL-USB-CO measuring device to measure CO concentration (Figure 2) and P-Sensor to measure PM2.5 concentration (Figure 3), installed in a row from south to north, namely Saka Mangkahai Village, Kapuas Regency, Buntoi Village, Tumbang Nusa Village, Tanjung Taruna Village in Pulang Pisau Regency, Perum Pemda Km 10 Cilik Riwut and Sei Gohong-Tangkiling in Palangka Raya City. Sensors installed in July-December 2021, downloaded monthly to avoid memory full. Prepitation as supporting data from Center for International Cooperation in Sustainble Management of Tropical Peatland (CIMTROP) at Natural Laboratory of Peat Swamp Forest (NLPSF) Sabangau station in 2019, 2020 and 2021 (Figure 4).

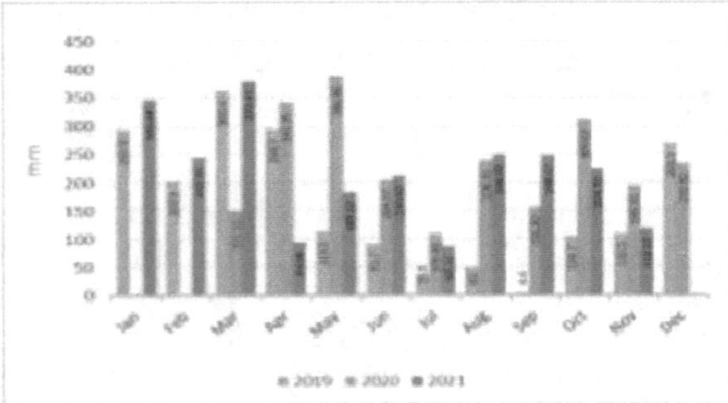

Figure 4. Prepitation in 2019, 2020 and 2021

3. Results and Discussion

3.1. CO concentration

Table 1. CO concentration

Location	2019			2021		
	Range (µg/cm³)	(Mean+SD) (µg/cm³)	Category	Range (µg/cm³)	(Mean+SD) (µg/cm³)	Category
Saka Mangkahai	-	-	-	0 – 137,2	16,2 + 19,2	Good
Buntoi	0 – 7104,1	1454,5 + 961,5	Medium (August)	0 – 795,1	42,1 + 82,2	Good
Tumbang Nusa	0 – 22.127,5	4654,1 – 3793,1	Dangerous (September)	0 – 38,6	12,5 + 3,3	Good
Taruna Jaya	0 – 19.065,5	5844,4 – 3326,1	Dangerous (September)	0 – 1371,5	85,1 + 101,3	Good
Pemda Km 10	0 – 8793,2	3345,2 – 1648,2	Dangerous (July)	0 – 213,5	36,8 + 29,6	Good
Sei Gohong	-	-	-	0 – 1371,5	11,7 – 103,4	Good

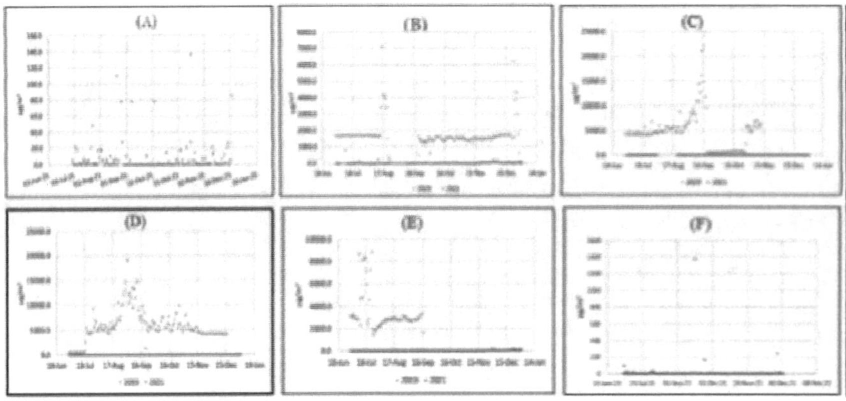

Figure 5. CO concentration; (A) Saka Mangkahai village; (B) Buntoi village; (C) Tumbang Nusa village; (D) Taruna Jaya village; (E) Perum Pemda Km 10 Palangka Raya city; and (F) Sei Gohong-Tangkiling.

3.2. PM2.5 concentration

Table 2. PM2,5 concentration

Location	2019			2021		
	Range (µg/cm³)	(Mean+SD) (µg/cm³)	Category	Range (µg/cm³)	(Mean+SD) (µg/cm³)	Category
Saka Mangkahai	-	-	-	0 – 11,3	3,5 + 1,6	Good
Buntoi	0 – 196,8	32,0 + 31,7	Dangerous (August)	0 – 9,2	4,9 + 3,0	Good
Tumbang Nusa	0 – 248,2	43,8 – 50,6	Dangerous (September)	0 – 50,0	2,6 + 1,3	Good
Taruna Jaya	-	-	-	-	-	-
Pemda Km 10	0 – 360,5	48,1 – 55,5	Very dangerous (September)	0 – 7,4	1,6 + 2,1	Good
Sei Gohong	-	-	-	0 – 40,0	6,8 – 5,4	Good

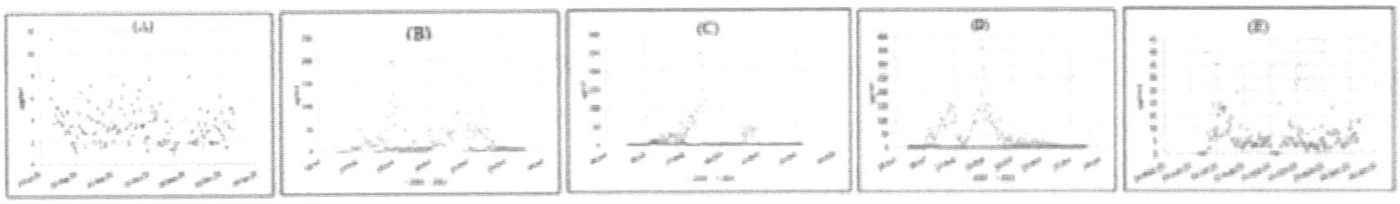

Figure 6. PM2,5 concentration;(A) Saka Mangkahai village; (B) Buntoi village; (C) Tumbang Nusa village; (D) Perum Pemda Km 10 Palangka Raya city; and (E) Sei Gohong-Tangkiling.

Overall, the CO and PM2.5 concentrations from July-December 2021 were in a suitable category (Table 1 and Table 2) because there was no dry season and peatland forest fires. High rainfall and green vegetation around the CO and PM2.5 concentration sensors are placed, to absorb harmful gases and particles, where this year being dominated by motor vehicle fumes. [15], heavy traffic is the cause of the increase in CO concentration, in addition to industrial activities, domestic fuel burning, and various agricultural activities, the highest particle concentrations tend to be in the dry season, but in some places with strong winds and at low temperatures, the concentration of gases and particles will decrease. Plants are air filters that are quite effective for cleaning the air and reducing pollution levels by absorbing, detoxifying, accumulating, and regulating metabolism in the air so that air quality increases through the release of oxygen by plants [16]. [17], plants can reduce air pollutants through oxygenation so that the air becomes clean. [18], pollutants are absorbed by plant tissues, especially leaves, without showing any damaging effects. [19], biochemical leaf metabolism processes can change air pollutants by absorbing them [20], both pollutants caused by humans and nature. The process of reducing PM concentration depends on plant morphology, climatic conditions, such as rain and wind, and the composition of PM itself. Plants absorb PM through leaves and then accumulate it [21], significantly correlated with leaf structure, size, and surface roughness. Then, they stabilize PM in the wax layer, which is influenced by the thickness and composition of wax in leaves [22].

The CO and PM2.5 concentrations in July - September 2019 was in the medium-dangerous category, because peatland forest fire occurs (Table 1 and Table 2). The process of oxygenation by vegetation does not work well because it is not supported by full sun lighting, which is blocked by smog. The oxygenation process is related to the opening and closing of the stomata leaf. In bright conditions, the stomata will open entirely so that more gas and pollutant particles are absorbed [23,24]. [20], in the dark conditions, the concentrations of O3 and NO2 gases were smaller than in the light conditions, reflecting that plant stomata open in response to light and close in the dark. [19], thereby reducing the capacity of plants to act as effective pollutant absorbers, both absorption capacity and susceptibility to pollutants. [25], air pollutants can also damage leaf tissue and cause changes in stomatal openings. The maximum CO concentration in Tumbang Nusa Village and Tanjung Taruna Village was in the dangerous category, while in Buntoi Village, the CO concentration was in the medium category. Possibly due to the distance of the fire locations and the influence of the wind that brought the smog to both locations. The distance of fire location to Taruna Jaya Village is about 0.5-kilometer, Tumbang Nusa Village is about 13-kilometer, Perum Pemda Km 10 Palangka Raya City about 0.1 km, and Buntoi Village is about 6-kilometer. Buntoi Village is also affected by smog, but its location in the southern part of the research location can quickly be carried by the wind to the north (Figure 7). In contrast to the case with PM2.5 concentrations, the highest occurred at Perum Pemda Km 10 Palangka Raya City, possibly because the location of the fire was very close to the monitoring site, and dust particles were carried by the smog, compared to Tumbang Nusa Village and Buntoi Village which were quite far away. The distance from the fire location has reduced the concentration due to the influence of wind, temperature, humidity, and vegetation. Smog occurs when dust and smoke particles are formed in relatively dry air, subsequently increasing in humidity resulting in hygroscopic growth and decreased visibility, thus becoming a threat to health, especially respiratory [26]. In addition, at Buntoi Village, monitoring devices were placed under a tree canopy, with an average of 15-20 meters tree high so that the leaf stomata first absorbed the smog. Meanwhile, the placement of measuring instruments at Tumbang Nusa Village, Taruna Jaya Village, and Perum Pemda Km 10 Palangka Raya are in open areas. Although a lot of vegetation is growing, the height is between 5-10 meters.

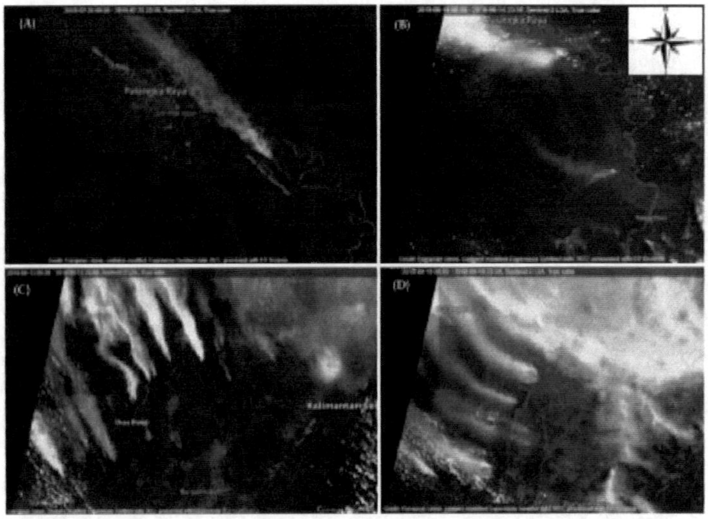

Figure 7. Map of Sentinel-2 L2A; (A) Fire in Tumbang Nusa village on July 25, 2019; (B) The smoke spread to Palangka Raya City on August 14, 2019; (C) Buntoi Village on September 13, 2019; (D) Central Kalimantan was covered by smoke from the fires on September 18, 2019.

Based on monitoring data on CO and PM2.5 concentrations, the highest concentrations are when peatland forest fires occur, which of course will have a huge impact on health, especially for vulnerable people such as the elderly, pregnant women, and children. appropriate efforts are needed to prevent forest and peatland fires to reduce their impact. School-age children are closed when forest and peatland fires occur, to prevent them from contracting respiratory inflammatory diseases, so it is necessary to find a solution to keep them in school. [27], one way to mitigate the impact of peatland forest fires so that children stay in school is to make smoke-free room model.

The coefficient correlation was significant at Tumbang Nusa village ($r = 0.947$), at Buntoi Village and Perum Pemda Km 10 Palangka Raya city there was no significant relationship between CO and PM2.5 concentrations (Figure 8). When fire occurred in July 2019 at Perum Pemda Km 10 Palangka Raya and in August 2019 in Buntoi Village, the CO concentration was quite high, then it fell to form a flat and stable graph, it is possible that the decrease in CO concentration was due to differences in fuel, burned area, vegetation, and wind. Meanwhile, the higher the concentration of PM2.5, the smoke and dust particles are in the form of smog, which in humid conditions will become thicker, reducing visibility [26], and the more dangerous it is for health. [28], the difference in the gas content produced from burning peat occurs due to differences in the maturity level of peat, fuel and the environment, the amount of smoke from fires is determined by the potential of the fuel, the area burned and the fuel consumed. In addition, it also produces volatile chemical products, forming a partial or complete chain of complete oxidation of organic matter.

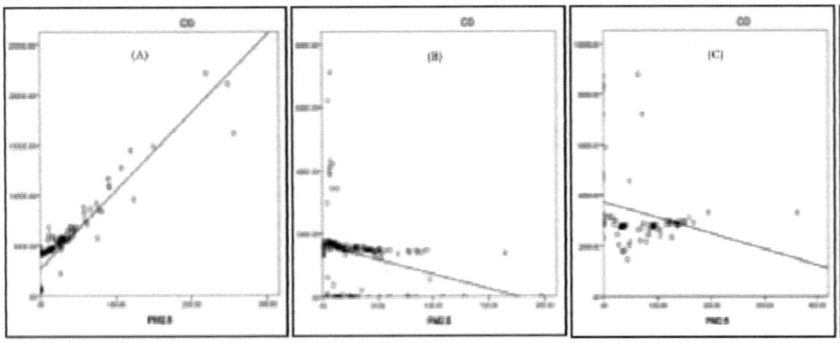

Figure 8. Coefficient correlation of CO and PM2,5; (A) Tumbang Nusa Village; (B) Buntoi Village; and (C) Perum Pemda Km 10 Palangka Raya

4. Conclusion

Central Kalimantan peatland forest fires in 2019 caused of concentrations CO and PM2.5 to be in the ***medium – very dangerous***, so a serious mitigation program is needed to overcome this, while in 2021, nothing peatland forest fires occur, the CO and PM2,5 concentration was low in ***good*** category. When the peatland forest fire occurs in 2019, wind direction and vegetation make the CO and PM2,5 concentration become lower at Buntoi village.

References

[1] Balai Besar Penelitian dan Pengembangan Sumber Daya Lahan Pertanian 2019 *Peta Lahan Gambut Indonesia Skala 1: 250.000.* (Jakarta: Kementerian Pertanian)

[2] Broich M, Hansen M, Stolle F, Potapov P, Margono B A and Adusei B 2011 Remotely sensed forest cover loss shows high spatial and temporal variation across Sumatera and Kalimantan, Indonesia 2000 – 2008

[3] Kementrian Lingkungan Hidup dan Kehutanan 2015 *Deforestasi Indonesia Tahun 2013-2019.* (Jakarta: Kementerian Lingkungan Hidup dan Kehutanan)

[4] Harrison M and Page S 2015 The global impact of Indonesian forest fires

[5] Randerson J, Leeuwen T and Chen Y 2017 Global fire emissions estimates during 1997 – 2016

[6] Kementerian Lingkungan Hidup 2020 *Peraturan Menteri Lingkungan Hidup dan Kehutanan No.: P.14/MENLHK/SETJEN/KUM.1/7/2020 tentang Indek Standar Pencemaran Udara (ISPU).* (Jakarta: Kementerian Lingkungan Hidup)

[7] Mcclellan R O 2002 Setting ambient air quality standards for particulate matter **182** 329–47

[8] Kunii O 2010 Archives of Environmental Health: An International The 1997 Haze Disaster in Indonesia: Its Air Quality and Health Effects The 1997 Haze Disaster in Indonesia: Its Air Quality and Health Effects 37–41

[9] Fujii Y, Iriana W, Oda M, Puriwigati A, Tohno S, Lestari P, Mizohata A and Setiyo H 2014 Characteristics of carbonaceous aerosols emitted from peatland fire in Riau, Sumatra, Indonesia *Atmos. Environ.* **87** 164–9

[10] Crippa P, Castruccio S, Lebron G B, Kuwata M and Thota A 2016 Population exposure to hazardous air quality due to the 2015 fires in Equatorial Asia *Nat. Publ. Gr.* 1–9

[11] Kim S, Koplitz S N, Mickley L J, Marlier M E, Buonocore J J and Kim P S 2016 Public health impacts of the severe haze in Equatorial Asia in September – October 2015: demonstration of a new framework for informing fire management strategies to reduce downwind smoke exposure public health impacts of the severe haze in Equatorial Asia in September – October 2015: demonstration of a new framework for informing fi re management strategies to reduce downwind smoke exposure.

IOP Conf. Series: Earth and Environmental Science 1111 (2022) 012006 doi:10.1088/1755-1315/1111/1/012006

[12] Federal Register 2006 *National Ambient Air Quality Standards for Particulate Matter: Final Rule. In: 40 CFR Parts 50, 53, and 58, vol. 62(138). U.S. EPA, Office of Air and Radiation, Office of Air Quality Planning and Standards* (Triangle Park, North Carolina: Federal Register)

[13] Chay K Y and Greenstone M 2003 The Impact Of Air Pollution On Infant Mortality: Evidence From Geographic Variation In Pollution Shocks Induced by a Recession *J. Econ.* **118** 1121–67

[14] Chretien J and Nebut M *Environmental Injuries of the Airways* (New York: Marcel Dekker.)

[15] Kumar S R, Arumugam T, Anandakumar C R, Balakrishnan S and Rajavel D S 2013 Properties Use of Plant Species in Controlling Environmental Pollution- A Review *Bull. Environ. Pharmacol. Life Sci.* **2** 52–63

[16] Shannigrahi A S, Sharma R and Fukushima T 2015 Air Pollution Control By Optimal Green Belt Development Around The Victoria Memorial Monument **7233** 20–9

[17] Grey G and Deneke F 1978 *Urban forestry.* (New York: John Wiley and Sons, Inc.)

[18] Harris, RW J, Clark R and Matheny N 1999 *Arboriculture* (New Jersey: Prentice Hall, Inc.)

[19] De Kok L and Stulen 1998 *Responses of plant metabolism to air pollutions and global change.* (Backhuys: Leiden)

[20] Omasa K, Tobe K and Kondo T 2022 *Absorption of Organic and Inorganic Air Pollutants by Plants.* (Berlin/Heidelberg, Germany: Springer)

[21] Xin B, Lee Y and Hadibarata T 2020 Phytoremediation Mechanisms in Air Pollution Control : a Review

[22] Popek R, Wrochna M and Gawronski S 2015 International Journal of Phytoremediation Particulate Matter on Foliage of 13 Woody Species: Deposition on Surfaces and Phytostabilisation in Waxes – a 3- Year Study 2013–6

[23] Herdina H U, Titiek S and Sadjono 1990 *Fisiologi Tumbuhan,* (Jakarta: Rajawali Press)

[24] Jones H 1992 *Plants microclimate, 2nd edition* (Cambridge: Cambridge University Press)

[25] Mansfield T and Pearson M 1996 *Can stomata play a role in protecting plants against air pollutants?* (Berlin)

[26] Begum B A, Biswas S K, Pandit G G, Saradhi I V and Waheed S 2011 Atmospheric Pollution Research Long – range transport of soil dust and smoke pollution in the South Asian region

[27] Kusin K, Sulistyanto Y, Usup A, Ardianor, Yulianti N, Fengky, F A, Adji F., Gawei A B . and Susetyo K E 2022 Smoke Free Room Model Of The Palangka Raya University For Mitigation Peatland Forest Fires In Central Kalimantan *Seybold Rep.* 312–22

[28] Nurhayati A D, Aryanti E and Saharjo B H 2010 Kandungan Emisi Gas Rumah Kaca Pada Kebakaran Hutan Rawa Gambut Di Pelalawan Riau *J. Ilmu Pertan. Indones.* **15** 78–82

Decoding Spatial Configuration: Resiliency in the Periphery (Case Study: Sedati Sidoarjo)

Fairuz Mutia[1]*, Muchlisiniyati Safeyah[2] and Wiwik Dwi Susanti[3]
[1] Department of Architecture, Faculty of Architecture and Design, Universitas Pembangunan Nasional "Veteran" Jawa Timur, Surabaya, 60296, Indonesia, 0000-0002-3976-7925
[2] Department of Architecture, Faculty of Architecture and Design, Universitas Pembangunan Nasional "Veteran" Jawa Timur, Surabaya, 60296, Indonesia, 0000-0001-5242-7186
[3] Department of Architecture, Faculty of Architecture and Design, Universitas Pembangunan Nasional "Veteran" Jawa Timur, Surabaya, 60296, Indonesia, 0000-0002-9967-4632
[1]*fairuzmutia.ar@upnjatim.ac.id, [2]muchlisiniyati.ar@upnjatim.ac.id, [3]wiwikds.ar@upnjatim.ac.id

Abstract. Rural and urban communities merge. Sedati's districts are vulnerable due to a lack of infrastructure, dispersed populations, disaster management capacities, and subsistence. A city is resilient if its citizens can survive and recover from a disaster. These relationships allow for implementing programs that promote urban and rural resilience. How do urban spatial patterns affect resilience determinants? This research examines the typo-morphological urban layout of Sedati District to discover the periphery's resilience and current situation. By using typo – morphological analysis and space syntax analysis researcher seeks the configuration of typology and morphology of the area. The results show that global connectivity and integration exist only in the center of Sedati District (HH3). Disconnecting two nodes won't stop the network if there are multiple pathways. Sedati's axial lines cease in space. Decreases road strength. Controlling expansion and using land efficiently, fostering urban and peri-urban agriculture, and making urban hubs inaccessible to nature are local solutions. Sedati lacks mixed-use clusters and pedestrian mobility. Urban livelihood, activity, and total life resilience reflect a 'center-periphery' spatial structure with expanding and declining gradients. Looping spatial syntactic configurations are needed for a robust urban fabric. Mixed-use land helps urban clusters' peripheries.

1. Introduction

Interoperability between urban and rural areas is crucial for socio-economic performance, as their advantages are frequently distinct and complimentary. Aspects such as population, labor, public services, and the environment may serve as potential links [1]. The rural and urban communities are not distinct from one another. Even under normal conditions, the link between the two is close. They are interdependent because they require one another. The city relies on the village to provide its residents with food and other raw materials. These relationships cannot be managed by local governments alone to grow urban and rural populations. Urban communities in developing nations cannot amass sufficient resources to sustain a shock, and pressures gradually degrade resilience and

increase the population's vulnerability over time. Similar to the growth of the industrial sector in the Sidoarjo Regency. Geographically, Sidoarjo Regency is part of the Kertosusila gate area. As a gateway, the entrance gate to the city of Surabaya from the surrounding regencies/cities, namely Mojokerto, Malang, and Pasuruan, must travel through Sidoarjo Regency. Sidoarjo Regency will have a fantastic potential to progress because it will be able to profit from establishing relationships through enhanced accessibility supported by infrastructure, transportation, and communication. One of the adjacent districts is Sedati, which lies between Surabaya and Sidoarjo. Simultaneously, Sedati Districts consist of 16 villages, growing more vulnerable due to their lack of infrastructure, dispersed populations, disaster management capacities, and restricted means of subsistence [2].
Moreover, a city is only resilient if most citizens can endure and recover from the effects of a crisis [3]. Similarly, a region is resilient if its population (whether they reside in urban, peri-urban, or rural areas) can withstand the detrimental effects of catastrophes and recover to their pre-disaster condition within a short timeframe. These linkages offer the potential to prepare for increased urban resilience through policies that can contribute to urban and rural resilience. However, resilience is a term that emerged from ecology in the 1970s to describe the capacity of a system to maintain or restore functionality in the event of a disturbance or disturbance [4,5].

Urban resilience is a burgeoning topic that is attracting strong interest. It is defined by the capacity of an urban system to withstand all stressors, preserve continuity, and recover while adapting and reforming toward sustainability [6]. Although urban resilience has developed dramatically in terms of the number of publications, the recent concentration of papers in a few journals indicates that the field is still in its infancy. Resilient cities must anticipate natural disturbances and interruptions and respond to constant change with a dynamic architecture [7,8]. The many themes highlighted in this study by Su [9] demonstrate a shift in recent years from more theoretical to more applied research (e.g. by giving resilience strategies). The data also suggests that urban resilience experts embrace socio-ecological resilience more than engineering and ecological interpretations. The issue, therefore, becomes, what is the urban configuration of Sedati's periphery? Thus, we extend the studies to analyze the impact of cities' different spatial configurations on resilience determinants. The primary contribution of this article is an analysis of the typo-morphological urban structure of Sedati District to identify some characteristics of resiliency in the periphery area and the current state of that area.

2. Methods

This research falls under the category of qualitative research. Researchers employ subjectivity to uncover the "essence" of place-based experience [8]. Phenomenology can be viewed as a method for identifying sense-perceivable symptoms as phenomena [10]. The site comprises Sidoarjo's Sedati District. The research was conducted in most of the Sedati District, but mostly in the disputed area between Surabaya and Sidoarjo. The Resilience Alliance has characterized the resilience of an ecosystem, and the qualities listed can easily be applied to quantify the resilience of a built environment. Planning solutions to provide urban development can be summed up in eight guiding principles [11]: compact densification, scaling, functional flexibility, acceptable grain diversity, highly connected networks, synergy, greening and water recycling, and urban-rural integration. By using typo – morphological analysis [12] and space syntax analysis [13], we seek the configuration of typology and morphology of the area [14,15]. DepthMapX assesses the already-drawn axial map to identify existing spatial configuration patterns, which will then be evaluated based on the sustainability principle. The urban area is then analyzed and characterized based on its typology and morphology to make conclusions and obtain research results. The urban area was evaluated in person and qualitatively with Depthmap X.

3. Results and Discussion

There are numerous ways in which cities might influence the development dynamics of the regions in which they are located [16]. These channels result from how a city influences the spatial layout of economic and social activity and the effects that this spatial distribution has on the growth and development of a region [1]. This section will begin by introducing the key concepts from the economic geography literature, which is part of the urban economics research about cities. It will then examine how these concepts relate to our empirical approach. The primary premise of the concept of rural-urban links is the complementary and reciprocal roles of cities and villages within the regional context [3]. The concept of rural-urban connectivity must be grounded in both spatial (spatial) and a-spatial (economic and policy) thought. It implies rural economic development that focuses on the agricultural sector must be interconnected and interdependent and influence the development of the urban economy that focuses on the industrial sector, commercial, and services. This will be accomplished by constructing or developing service centers at each level of the regional hierarchical structure following their distinct comparative advantages (the economic base).

3.1. Morphology: Spatial Configuration

The morphology of a city is shaped by the fabric of human interactions that create meeting spaces and spatial patterns. Highly connected networks are a significant factor in determining urban resilience. A set of links whose strength follows a scale-free distribution promotes system resilience by avoiding abrupt and catastrophic variations from rapidly disorganizing the system. Few long-distance connections should exist, and these should be weak enough to avoid spreading disruptive fluctuations. In contrast, many robust short-range connections maintain the system's resilience through local clustering and various alternate paths. The phrase spatial area is often addressed when we want to map and plan the path of urban growth. One of the concerns that need consideration is the phenomena of suburban areas and the spatial and socio-economic transformation process in this area [17]. To establish a pattern of interpretation of spatial patterns in space, it is necessary to watch movement from one place to another throughout the entire system of spatial structures.

According to research on space syntax, spatial configuration accounts for a significant variation in the rates of movement of human aggregates in different sites. By examining the configuration form, the behavior pattern of a specific area can be determined. Using an integrated method, it is possible to evaluate the configuration type of a set-up. Integration value is a unit for calculating the strength of the relationship between each element in a space. Knowing the strengths and limitations of a form's order will make evaluating a system's accessibility easier. The value of Sedati's space integration study is as shown in figure 1:

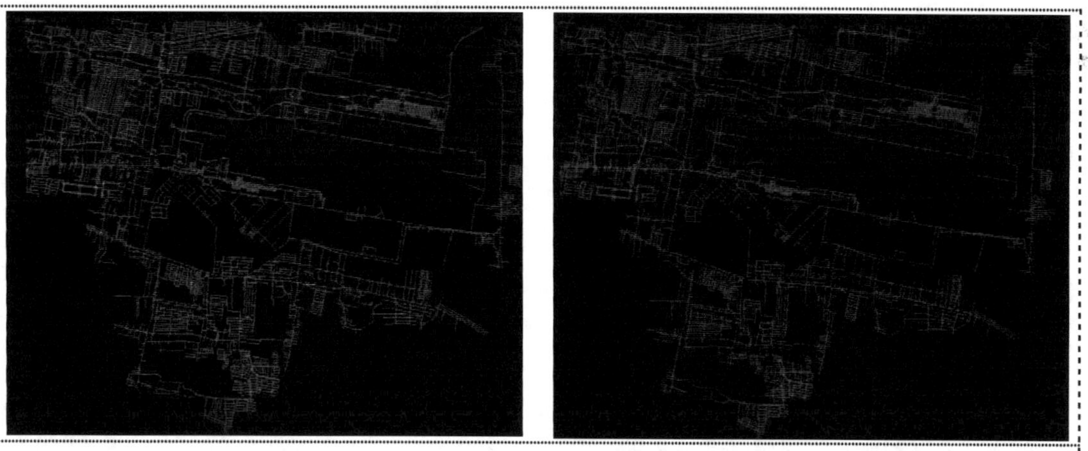

Figure 1. Global Integration Value HH (right) and Pedestrian Integration Value HH3 (left) of Sedati

Figure 1 shows the line analysis of axial line integrity [18]. The value of integration is the value of a person's ease in reaching a particular space. In the Sedati District area, the DepthmapX analysis results show that the global integration value is quite evenly distributed in the middle of the region.

This is indicated by data in red, orange, and yellow colors, which indicate that the connectivity between urban spaces in the Jabon area has a probability of being more crowded in that area. The red-yellow space area is the way to Juanda Airport, the commercial area comprised of lodging facilities (ranging from no-star to five-star hotels) and many commercial uses. In addition, because this area is the primary connection route between Surabaya and Sidoarjo, this movement is not only driven by land use but also by the mobilization of employees who reside in Sidoarjo and work in Surabaya, particularly in the eastern portion of the city. In resilient urban systems, social and street networks exhibit traits called "small world" features. In particular, they exhibit a high amount of clustering, suggesting that a significant fraction of surrounding nodes are similarly connected. Complex developed street patterns exhibit the small world property when a street is seen as a node and its intersection with another street as a link.

Figure 2. Mean Depth of Sedati District

On the other hand, a climate-resilient city [19] must be able to change and adapt to new circumstances, which necessitates the modification of its original concept. In a continuous interplay between urban design and adaptive self-organization processes that make the initial organization's typical characteristics more complicated, resilient cities develop.

3.2. Typology : Compact Densification, Scaling, Fine Grain Diversity

High density and mixed-use can be viewed as significant strategic advantages of urban areas that aid in the more efficient use of energy through synergistic techniques based on the integration of energy systems and compact, energy-efficient housing. Infill redevelopment solutions make it possible to deploy innovative strategies to reduce urban communities' energy and resource intensity by bolstering urban density and mixed-use. The locally diverse building uses, and building types generate locally diverse demand loads. According to solid-void area mapping (Figure 2), the distribution of buildings appears to be more concentrated in large pockets with commercial and services. Some residential in small solid clusters tend to be functionally segregated in these commercial and service zones because the provided products are not daily needs. Appropriately pairing various energy-demand categories with energy-conversion technologies is crucial to enhancing the effectiveness of urban energy systems. Residential, commercial, retail and public buildings have extremely distinct load profiles. This variety of load profiles facilitates the deployment of synergy strategies. In the Sedati region, the spatial arrangement of the periphery appears to be divided and more densely populated by small lots dominated by residential areas. Since consumption peaks in different building types do not occur at the same time of day, high levels of local mixed-use help to reduce consumption peaks.

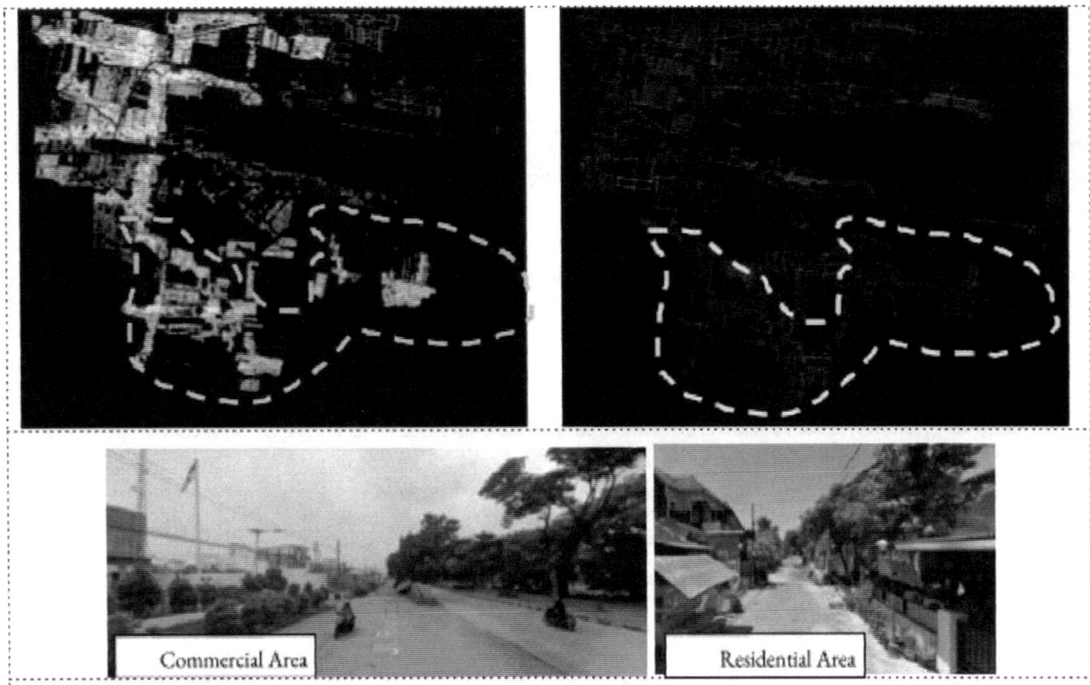

Figure 3. Solid – Void and Segregation of Axial Map of Sedati District

3.3. The resilience of Urban Periphery

Sedati District tends to be less resilient because connectivity and integration are only focused in the core area (globally) and are not connected when seen from the pedestrian radius (HH3). Resilient cities have networks with a high degree of redundancy, similar to the veins of leaves [12,20]. If multiple routes cross through different nodes from one part of the city to another, then disconnecting two nodes will not prevent the network from functioning. In the spatial structure of the Sedati region, connections between axial lines result in dead ends at the edge of space. This reduces the resilience of the road network. In the variable of Urban-Rural Integration, local responses include preserving available land by limiting sprawl and by efficient land use, support for urban and peri-urban agriculture, and immersion of compact urban nodes (less than 10 minutes walk) within accessible, continuous green spaces. The calculation of the space syntax analysis in Figure 1 demonstrates that pedestrian movement is inadequate. Within the boundaries of the Sedati District alone, there was not a single clustering region with mixed-use structures. This would make them vulnerable since if there had been an area that was sealed down, they wouldn't be able to replenish themselves. As demonstrated by the red-er line in Figure 3, the resilience area may be farther out and lower than the main axial map.

4. Conclusion

The Urban layouts determine the resilience of urban structures. The higher the integration value, the more durable the urban structure can be. The more integrated the configuration, the greater the integration value of the urban fabric. The spatial divergence between urban livelihood, activity, and natural life resilience is essential and exhibits a 'center-periphery' spatial structure with gradients exhibiting both growing and decreasing trends. Low to high degrees of negative spatial autocorrelation are predominant in urban livelihood and activity resilience. For resilient urban fabric, the spatial syntactic configuration must be looping and contain no dead ends. Mixed-use land is helpful in restoring every urban cluster's periphery. The plan for strengthening the resilience of urban life should emphasize priority, pertinence, cross-regional dynamic guidance, collaborative governance of interconnected urban challenges, and the impact of new science and technology and urban development concepts.

Acknowledgments
This research was supported by the Ministry of Education, Culture, Research, and Technology on contract number 08/UN63.8/LT-Kontrak/III/2022. We thank our colleagues from UPN "Veteran" Jawa Timur, who provided insight and expertise that greatly assisted the research. However, they may not agree with all of the interpretations/conclusions of this paper. We would also like to show our gratitude to the LPPM UPN "Veteran" Jawa Timur for sharing their pearls of wisdom with us during this research, and we thank reviewers for their so-called insights.

References
[1] Carriazo F, Berdegue J A, Jara B, Modrego F and Soloaga I 2015 Cities , Territories , and Inclusive Growth : Unraveling Urban – Rural **73** 56–71
[2] Mutia F, Yunisya A N and Elviana E 2020 Spatial Planning for Future Risk Reduction in Coastal Area (Case Study : Kalanganyar Village , Sidoarjo) **MATEC Web** 1–10
[3] Holis Y M 2017 Local Economic Linkages pada Kota Kecil / Menengah Sebagai Instrumen Konsep Keterkaitan Desa-Kota (Rural-Urban Linkage) *Pros. Semin. Nas. Perenc. Pembang. Inklusif Desa-Kota* 817–26
[4] Wilk J, Jonsson A C, Rydhagen B, Callejo I, Cerruto N, Chila G, Encinas S, Kumar A and Rani A 2018 Assessing Vulnerability in Cochabamba , Bolivia and Kota , India : How do Stakeholder Processes Affect Suggested Climate Adaptation Interventions ? *Int. J. Urban Sustain. Dev.* **10** 32–48
[5] Silvia J Da 2015 *City Resilience Framework* (The Rockefeller Foundation ARUP)
[6] Statistik B P 2019 *Lingkungan hidup indonesia 2019* (Jakarta)
[7] Mitoulis A S and Argyroudis S A 2021 Restoration Models of Flood Resilient Bridges : Survey Data *Data Br.* **36** 107088
[8] Groat L N and Wang D 2013 *Architectural research methods* (John Wiley & Sons)
[9] Bautista-Puig N, Benayas J, Manana-Rodriguez J, Suarez M and Sanz-Casado E 2022 The Role of Urban Resilience in Research and its Contribution to Sustainability *Cities* 1–13
[10] Creswell J W, Hanson W E, Clark V L P and Morales A 2007 Qualitative Research Designs : Selection and Implementation **35** 236–64
[11] Salat S and Bourdic L 2014 Spatial Planning Principles & Assessment Framework For Climate Adaptive & Resilient Cities in China *Int. Work. Urban Adapt. to Clim. Chang.* 1–19
[12] Ramadanta A and Darjosanjoto E T S B 2012 Application of Space Syntax as Presentation and Analysis Technique in the Study of Spatial Integration in Contoured Landform *J. Basic Appl. Sci. Res.* **2** 6850–6
[13] Yamu C and van Nes A 2021 *Introduction to Space Syntax in Urban Studies* (Cham, Switzerland: Springer)
[14] Ye Y and van Nes A 2014 Quantitative tools in urban morphology : combining space syntax , spacematrix and mixed-use index in a GIS framework *Int. Semin. Urban Form* **18** 97–118
[15] Basee D H and Abdulla Z R 2022 Transformation of urban morphology , vulnerability and resilience : Haifa Street Area , as a case study *Ain Shams Eng. J.* **13** 101718
[16] Nengroo Z A, Bhat S M and Kuchay N A 2017 Measuring urban sprawl of Srinagar city , Jammu and *J. Urban Manag.* **6** 45–55
[17] Mutia F, Avenzoar A and Zahroh F 2021 The Effect of Urban Spatial Configuration on Sustainable Tourism The Effect of Urban Spatial Configuration on Sustainable Tourism Development in Jabon District , Sidoarjo *EMARA Indones. J. Archit.* **7** 68–74
[18] Sherlia, Jordan N A and Syafitri E D 2021 Space Syntax Analyses in Drfining the Connection of Development Centers in Balikpapan *J. Archit. Built Environ.* **48** 1–8
[19] Zari M P 2018 *Regenerative Urban Design and Ecosystem Biomimicry* (Routledge)

[20] Navastara A . and Mandasari V 2019 Sustainable pedestrian ways in Central Business District of Tunjungan Surabaya : Can principles of new urbanism be applied ? Sustainable pedestrian ways in Central Business District of Tunjungan Surabaya : Can principles of new urbanism be applied ? *IOP Conf. Ser. Earth Environ. Sci. 340*

Spatial Study of Oil Palm Plant Health Using Sentinel 2 Imagery at PTPN VIII Cikasungka Plantation, Bogor, West Java

Zakiy Akmal[1]*, Adi Wibowo[1], Revi Hernina[1], Tjiong Giok Pin[1]
[1]Department of Geography, Faculty of Mathematics and Natural Science, University of Indonesia, Depok, 16424, Indonesia
[1]*zakiy.akmal@ui.ac.id, [2]adi.w@sci.ui.ac.id, [3]revie12@yahoo.com, [4]tjiong.giok@gmail.com

Abstract. Oil palm (*Elaeis guineensis Jacq.*) is a plantation commodity that is the highest producer and exporter in Indonesia and has high economic value. Increasing oil palm productivity requires several factors, one of which is plant health. Therefore, detecting the health condition of oil palm plants requires research with spatial analysis and a remote sensing method that can monitor plantation data collection strategies efficiently and effectively. One of them is Sentinel-2 using the NDVI algorithm. The algorithm can estimate the relationship between vegetation and health conditions. This study resulted in very healthy plants having NDVI values ranging from 0.80-0.85, healthy plants having NDVI values ranging from 0.76-0.80, critical plants having NDVI values ranging from 0.72-0.76, and unhealthy plants having NDVI values ranging from 0.68-0.72. The distribution of the NDVI value at the PTPN VIII Cikasungka Plantation in January 2022 has a high NDVI value and is spread over every area of the district.

1. Introduction

Indonesia is one of the largest producers and exporters of Crude Palm Oil (CPO) in Asia and the world [1], [2]. Oil palm plants (*Elaeis guineensis Jacq.*) are tropical plants cultivated for the oil industry's production and have become one of the most economical crops in several countries, especially in Indonesia [3]. The most significant palm oil production is in Riau Province at 9.87 tons, but the highest oil palm productivity is in Central Kalimantan Province at 4908 kg/ha [4]. For palm oil exports in Indonesia in 2019 amounted to 28279350 tons and exports of palm kernel amounted to 1937238 tons [4]. The high production and export of palm oil in Indonesia make the Indonesian economy driven by the industrial sector, one of which is oil palm plants owned by the private sector, the people, and the government (BUMN) [5]. The increase in oil palm productivity is also in line with the increase in the palm oil industry, which can help improve the regional economy, provide jobs, and improve the welfare of oil palm smallholders [6]. The existence of oil palm plants can help the Indonesian economy, which causes the palm oil industry every year, and new land is opened to plant oil palm to increase its productivity [7].

Remote sensing is a method that could be used for extensive plantation monitoring with an efficient and effective strategy for collecting plantation data [8]. One of the remote sensing images that could be used for monitoring plantation areas is the Sentinel-2 image with a resolution of 10 meters, 20 meters, and 60 meters [8], [9]. Normalized Difference Vegetation Index (NDVI) is a vegetation index often used in research to estimate the relationship of vegetation to the age, health condition, productivity, and production of a plant [10]. When a plant has a low NDVI value, the plant is unhealthy, whereas when the NDVI value is high, the plant is healthy [11]. Remote sensing, such as Citra Sentinel-2 can analyze plant health using the NDVI algorithm [9]. Sentinel-2 and NDVI image data help to see the distribution of plant health and analyze the health of oil palm plants.

PT Perkebunan Nusantara (PTPN) VIII is a plantation company with various commodity businesses such as tea, rubber, palm oil, and coffee [12]. PTPN VIII was located in West Java and Banten. PTPN VIII has a plantation area of 21331.08 Ha, and there are 9 plantation units, one of which is Cikasungka Plantation. Cikasungka Plantation is one of the oil palm plants located in Bogor, West Java, and there is a palm oil processing factory on the plantation. This research was conducted at PTPN VIII Cikasungka Plantation, which aims to analyze the distribution of health conditions of oil palm plants.

2. Methods

This research was conducted in Afdeling Cikasungka and Toge, located at PTPN VIII Cikasungka Plantation, Bogor, West Java. Afdeling Cikasungka is located in Cigudeg District and Sukajaya District. Afdeling Cikasungka in Cigudeg District is located at 106°31'31"-106°31'34" East Longitude and 6°32'2"-6°35'2" South Latitude. Meanwhile, Afdeling Cikasungka in Sukajaya District is located at 106°29'46"-106°29'47" East Longitude and 6°36'38"-6°37'27" South Latitude. Meanwhile, Afdeling Toge is located in Cigudeg District and Jasinga District. Afdeling Toge in Cigudeg District is located at 106°29'44"-106°31'15" East Longitude and 6°28'25"-6°31'33" South Latitude. Meanwhile, Afdeling Toge in Jasinga District is located at 106°29'29"-106°29'30" East Longitude and 6°30'57"-6°31'36" South Latitude.

Data collection in this study consisted of primary and secondary data collection. The primary data collected in this study were photos of the physical condition of oil palm plants. Meanwhile, secondary data collected were Sentinel-2, administrative boundaries, and afdeling boundaries.

2.1. Sentinel-2 Imagery

Sentinel-2 imagery consists of two constellation satellites, Sentinel-2A and Sentinel-2B [13].The satellites have medium spatial resolutions of 10, 20, and 60 meters with a temporal resolution of 10 days for one satellite or 5 days for two satellites [14]. Sentinel-2 imagery has 13 spectral bands consisting of visible bands, Visible and Near Infrared (VNIR), and Short-Wave Infrared (SWIR) [14].

2.2. NDVI Algorithm

Sentinel-2 imagery consists of two constellation satellites, Sentinel-2A and Sentinel-2B [13].The satellites have medium spatial resolutions of 10, 20, and 60 meters with a temporal resolution of 10 days for one satellite or 5 days for two satellites [14]. Sentinel-2 imagery has 13 spectral bands consisting of visible bands, Visible and Near Infrared (VNIR), and Short-Wave Infrared (SWIR) [14]. The NDVI formula can be seen in equation 1.

$$NDVI = \frac{NIR - Red}{NIR + Red} \quad (1)$$

Where:
NDVI = Normalized Difference Vegetation Index
NIR = Band 8
Red = Band 4

2.3. Relationship between NDVI with Plant Health

Data processing using ArcGIS 10.8, ENVI, QGIS, and Microsoft Excel software. Sentinel-2 image data processing using ArcGIS 10.8 and QGIS software. In QGIS software, atmospheric correction on sentinel-2 image eliminates errors caused by atmospheric effects on the image and processes the image using the NDVI algorithm (Equation 3). Next, the image is clipped based on the research area using ArcGIS 10.8 software. The NDVI value is used to analyze the health of oil palm plants. The classification of NDVI values with plant health saw in Table 1.

Table 1. NDVI Value Classification with Oil Palm Plant Health [8,11]

Health Classification	NDVI Values
Very Healthy	0.80-0.86
Healthy	0.76-0.80
Critical	0.72-0.76
Unhealthy	0.68-0.72

3. Results and Discussion

3.1 Distribution of Oil Palm Plant Health in Afdeling Cikasungka and Toge

Based on the results of processing Sentinel-2B image data on January 4th, 2022, using the NDVI algorithm, the NDVI distribution of oil palm plants at PTPN VIII Cikasungka Plantation is dominated by high NDVI. Still, some areas have low NDVI (Figure 1). Afdeling Cikasungka has the highest NDVI of 0.85, and the lowest is 0.13. Meanwhile, Afdeling Toge has the highest NDVI of 0.86 and the lowest of 0.03.

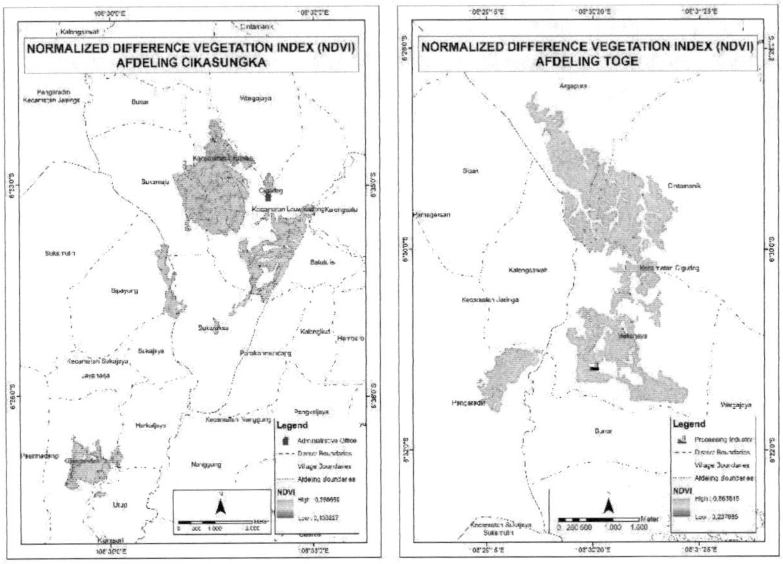

Figure 1. Normalized Difference Vegetation Index Map of Oil Palm Plants in Afdeling Cikasungka and Toge

(Source: Authors. 2022)

The health distribution of oil palm plants in Afdeling Cikasungka is dominated by very healthy plants with an area of 307.87 Hectares (Table 2). Health areas for oil palm plants that are very healthy, healthy, critical, and unhealthy are scattered in every area of Afdeling Cikasungka in Cigudeg District and Sukajaya District (Figure 1). Still, the health area for very healthy and healthy oil palm plants has a large coverage area. Meanwhile, critical and unhealthy plant areas have a small coverage area. Areas of critical and unhealthy oil palm plants are close to each other. This oil palm sick may indicate that unhealthy oil palm plants can spread to several other oil palm plants that are near unhealthy plants. Based on the field survey results, a total of 100 sample points in Afdeling Cikasungka contained 38 very healthy plants from 44 samples, 39 healthy plants from 43 samples, 6 critical plants from 8 samples, and 4 unhealthy plants from 5 samples. However, of the 4 unhealthy plants, one of them was infected with Ganoderma disease on Grid GK400. It was validated from the field survey that there was fungus on the stems. For plants infected with Ganoderma disease, the NDVI value is 0.66. Samples of plants that are not healthy but have no fungus on the stems could mean that the plant lacks water or nutrients, and the plant will have yellow or brown leaves causing the plant to have a low NDVI value.

The health distribution of oil palm plants in Afdeling Toge is dominated by very healthy plants with an area of 363.56 Hectares (Table 2). Very healthy plants are scattered in every area of Afdeling Toge in Cigudeg District and Jasinga District (Figure 2). The area of healthy plants in Afdeling Toge is more significant than in Afdeling Cikasungka, but Afdeling Cikasungka has a larger area of healthy plants than in Afdeling Toge. Meanwhile, the critical and unhealthy plant area in Afdeling Toge has a smaller area than Afdeling Cikasungka. Based on the field survey results, there were 60 very healthy plant samples from 67 samples, 22 healthy plants from 26 samples, 3 critical plants from 4 samples, and 4 unhealthy plants from 4 samples. The samples of unhealthy plants in Afdeling Toge were all infected with Ganoderma disease.

Table 2. Oil Palm Plant Health in Afdeling Cikasungka and Toge

Health Classification	NDVI Values	Afdeling Cikasungka		Afdeling Toge	
		Area (Ha)	Percentage	Area (Ha)	Percentage
Very Healthy	0.80 - 0.86	307.87	44.81%	363.56	67.57%
Healthy	0.76-0.80	294.64	42.88%	140.17	26.05%
Critical	0.72-0.76	54.82	7.98%	19.39	3.60%
Unhealthy	0.68-0.72	29.76	4.33%	14.90	2.78%
Total		687.09	100%	538.02	100%

(Source: Authors. 2022)

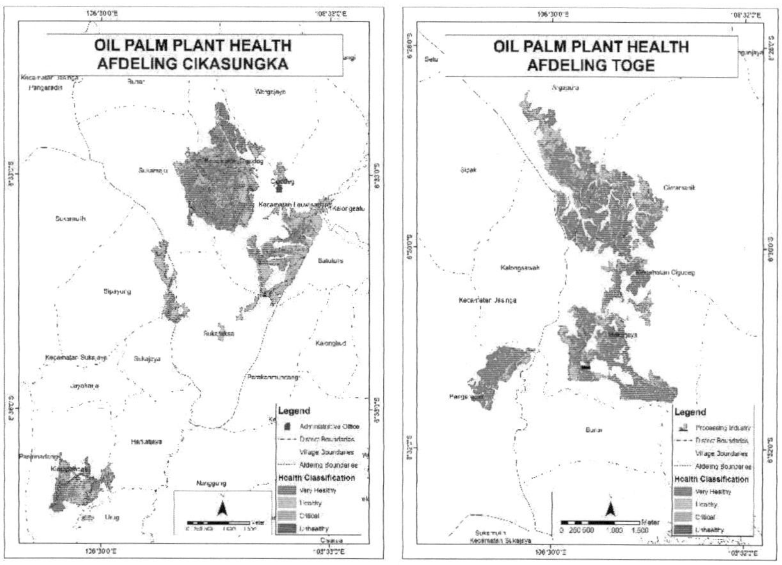

Figure 2. Map of Health Distribution of Oil Palm Plants at Afdeling Cikasungka and Toge in 2022
(Source: Authors. 2022)

Based on Table 3, the health of oil palm plants in each health classification has different plant characteristics, but there are also some similarities. Oil palm plants that are very healthy and healthy have several similarities in plant characteristics that are somewhat difficult to distinguish, such as having many between 18-25, a dense crown density, and an upright and straight stem. However, the difference can be seen clearly from the broken leaves and midrib. Characteristics of a healthy plant are the midrib that is broken or hanging down from the tree and has a few yellow spots or brown leaves. Meanwhile, the characteristics of critical and unhealthy oil palm plants can be clearly distinguished on the part of the stem with or without fungi.

Table 3. Pictures of Oil Palm Plant Characteristics in Afdeling Cikasungka and Toge

Health Classification	Pictures of Plant Characteristics			
Very Healthy				
Healthy				
Critical				
Unhealthy				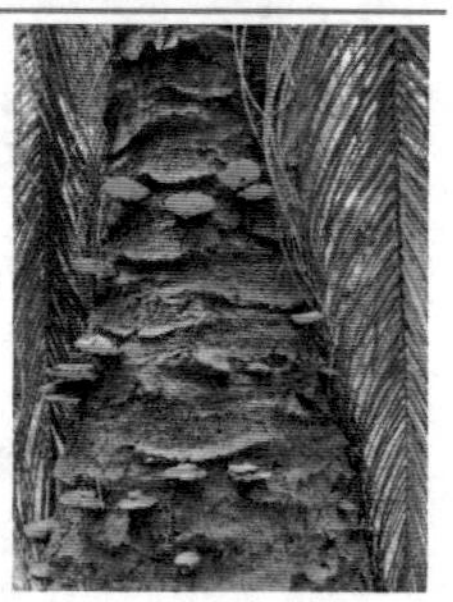

(Source: Documentation of Field Survey. 2022)

3.2. *Accuracy Test of Oil Palm Plant Health*

The accuracy test was carried out on Afdeling Cikasungka and Toge with 100 sample points. Sample points from field survey results that correspond to each health classification in Afdeling Cikasungka are 87 points, while in Afdeling Toge, there are 88 points. Then, the overall accuracy value for Afdeling Cikasungka is 87%, and Afdeling Toge is 88%. The results of the Kappa Accuracy value at

Afdeling Cikasungka have a good level of accuracy with a value of 79%. Meanwhile, Afdeling Toge has a good level of accuracy with a kappa accuracy value of 76%. The results of the accuracy test from both afdeling have good accuracy in providing accuracy so that the health classification of Sentinel-2B image processing using the NDVI algorithm can be used in analyzing the health of oil palm plants.

4. Conclusion

The distribution of NDVI values using Sentinel-2 imagery at PTPN VIII Cikasungka Plantation can use to determine the health of oil palm plants by 79% in Afdeling Cikasungka and 76% in Afdeling Toge. The distribution of NDVI values at PTPN VIII Cikasungka Plantation had a high NDVI value. It spread in every district area because healthy plants in Cikasungka and Toge dominated it. Very healthy plants have good tree characteristics based on the number of midribs >18, dense crowns, and dark green leaves. Meanwhile, unhealthy plants have precise characteristics on the stems of Ganoderma fungi, but unhealthy and critical plants are not necessarily diseased. It could be that the plant lacks water or nutrients that make the leaves yellow, and the NDVI value is low.

Acknowledgments

Research Grant supported this paper publication from Penelitian Dasar Unggulan Perguruan Tinggi Kementrian Pendidikan dan Kebudayaan, Riset dan Teknologi Tahun Anggaran 2022 based on contract number: NKB-958/UN2.RST/HKP.05.00/2022.

References

[1] M. A. S. Untoro *et al.*, "Estimating Oil Palm Production Using Sentinel 2: By-age Separation Method," *IOP Conf. Ser. Earth Environ. Sci.*, vol. 165, no. 1, 2018, doi: 10.1088/1755-1315/165/1/012026.

[2] V. V. Taufik, A. Sukmono, and H. S. Firdaus, "Estimasi Produktivitas Kelapa Sawit Menggunakan Metode NDVI (Normalized Differnce Vegetation Index) dan ARVI (Atmospherically Resistant Vegetation Index) dengan Citra Sentinel-2A (Studi Kasus : Beberapa Wilayah di Provinsi Riau)," *J. Geod. Undip*, vol. 10, no. 1, pp. 153–162, 2021, [Online]. Available: https://ejournal3.undip.ac.id/index.php/geodesi/article/view/29636.

[3] H. Kano and R. Saleh, "Nationalism , Globalization , and Transnational Movement : A Case of Oil Palm Plantation Business in Indonesia," *J. Sosiol.*, vol. vol 24, no. 2, p. 211, 2019, doi: 10.7454/M.

[4] BPS, "Indonesian Oil Palm Statistics 2019," *Badan Pus. Stat.*, pp. 1–155, 2020.

[5] B. Saragih, "Produktivitas sumber pertumbuhan minyak sawit yang berkelanjutan," *WPLACE-World Plant. Conf. Exhib.*, pp. 1–7, 2017.

[6] W. S. Saragih, E. Purba, and K. Tampubolon, "Analisis Hara Cu dan Zn pada Vegetasi Gulma sebagai Penanda Keberadaan Jamur Ganoderma dari Kebun Kelapa Sawit," *J. Agrotek Trop.*, vol. 7, no. 3, p. 519, 2019, doi: 10.23960/jat.v7i3.3237.

[7] P. H. Khiabani and W. Takeuchi, "Assessment of Oil Palm Yield and Biophysical Suitability in Indonesia and Malaysia," *Int. J. Remote Sens.*, vol. 41, no. 22, pp. 8520–8546, 2020, doi: 10.1080/01431161.2020.1782503.

[8] R. D. Handrian, B. H. Trisasongko, and D. R. Panuju, "Discriminating the Severity of Basal Stem Rot Disease in Oil Palm (Elaeis guineensis Jacq.) Plantation Using Sentinel-2," *IOP Conf. Ser. Earth Environ. Sci.*, vol. 950, no. 1, p. 012043, 2022, doi: 10.1088/1755-1315/950/1/012043.

[9] T. R. Putri and S. Supriatna, "A Spatial Study of Rubber Plant Health Using Sentinel 2-A Imagery in Cibungur Plantation, Sukabumi, West Java," *IOP Conf. Ser. Earth Environ. Sci.*, vol. 623, no. 1, 2021, doi: 10.1088/1755-1315/623/1/012078.

[10] O. Danylo *et al.*, "A Map of The extent and Year of Detection of Oil Palm Plantations in Indonesia, Malaysia and Thailand," *Sci. Data*, vol. 8, no. 1, pp. 1–9, 2021, doi: 10.1038/s41597-021-00867-1.

[11] R. Malinee, D. Stratoulias, and N. Nuthammachot, "Detection of Oil Palm Disease in

Plantations in Krabi Province, Thailand with High Spatial Resolution Satellite Imagery," *Agric.*, vol. 11, no. 3, 2021, doi: 10.3390/agriculture11030251.
[12] PTPN8, "Bisnis Kelapa Sawit PT Perkebunan Nusantara VIII," 2021. https://www.ptpn8.co.id/kelapa-sawit/ (accessed Jun. 30, 2021).
[13] ESA, "Sentinel-2 - Satellite Description - Sentinel Online - Sentinel," 2021. https://sentinel.esa.int/web/sentinel/missions/sentinel-2/satellite-description (accessed Apr. 28, 2021).
[14] ESA, "Sentinel-2 - Missions - Resolution and Swath - Sentinel Handbook - Sentinel," 2021. https://sentinel.esa.int/web/sentinel/missions/sentinel-2/instrument-payload/resolution-and-swath (accessed Apr. 28, 2021).

Spatial and Prevalence Analysis of Behavioral Non-Communicable Diseases on Soppeng Regency: Results of Health Survey 2016-2018

Basri[1]*, Tasrifin Tahara[2] and Andi Tilka Muftiah Ridjal[3]
[1] Department of Public Health, Sekolah Tinggi Ilmu Kesehatan Makassar, Makassar, 90232, Indonesia, 0000-0002-2398-524X
[2] Department of Anthropology, Hasanuddin University, Makassar, Indonesia, 0000-0002-2026-5474
[3] Department of Public Health, Sekolah Tinggi Ilmu Kesehatan Makassar, Makassar, 90232, Indonesia
[1]*basrikesmas@gmail.com, [2]tasrifin@unhas.ac.id, [3]tilka.muftiah@gmail.com

Abstract. Behavioral non-communicable disease as a chronic disease is a health challenge in the 21st century because it is a major contributor to global mortality. Soppeng Regency, one of the regencies in the South Sulawesi region, faces the problem of the high prevalence of behavioral non-communicable diseases (NCDs) in the future. The study analyzes the spatial distribution and prevalence of behavioral NCDs in the Soppeng Regency. Primary data was obtained from the compilation of monthly reports from 17 public health centers in the administrative area of the Soppeng district. Meanwhile, spatial data was obtained from Soppeng district spatial plans (RTRW) maps, BPS of Soppeng district, RBI map, and the National DEM Image. Hypertension and diabetes data sets were applied to the prevalence rate formula and analyzed using Microsoft Office and SPSS ver. 21.0. The results showed the fifth highest prevalence rate of hypertension 2016-2018 occurred in Barang village, Jampu village, Marioriaja village, Maccile village, and Kessing village by an average of 19.19%, 3.55%, 2.73%, 6.24%, and 3.78% respectively. Meanwhile, distribution of the prevalence rate of diabetes in 2016-2018 was high in the five villages of Maccile (6.24%), Barang (5.27%), Botto (4.93%), Lemba (4.35%), and Soga (1.08%), respectively. Analysis showed variations in levels between sub-districts in the two types of behavioral non-communicable diseases.

1. Introduction

Non-communicable diseases (NCDs), also known as chronic diseases, tend to be of long duration and are the result of a combination of genetic, physiological, environmental[1], and behavioral factors[2]. The main types of NCD are cardiovascular diseases (such as heart attacks and stroke), cancers, chronic respiratory diseases (such as chronic obstructive pulmonary disease and asthma), and diabetes[3]. NCDs are the main health challenges of the 21st century as the major contributor to global mortality[2]. NCDs are responsible for 41 million of the 57 million deaths in the world (71%), of which 15 million of these deaths are premature (30 to 70 years)[2]. Low-and middle-income countries contribute approximately 80% of deaths from non-communicable diseases (NCDs)[4], including cancer, respiratory disease, and diabetes[5]. The risk factors for death are mostly caused by tobacco use, low sports activity, excessive alcohol use, and an unhealthy diet. Changes in the environment, technology, smoking prevalence, and lifestyle have changed the disease pattern in Indonesia, which is dominated by NCDs [6]. Other studies have shown that behavioral risk factors acquired during adolescence and then adopted in adulthood also contribute to the incidence of NCDs [7].

NCDs are estimated to contribute to 73% of all mortality in Indonesia in 2018. The most common diseases are cardiovascular disease (35%), cancer (12%), chronic respiratory disease (6%), diabetes (6%), and other NCDs (15%)[8]. Soppeng is one of the regencies in South Sulawesi with a total area of 1557 km^2 with various demographics and disease prevalence. Health Profile of South Sulawesi by 2017 indicated that Soppeng Regency is the second highest hypertension case by 24.92% in South Sulawesi. Moreover, the number of DM sufferers was 3,987 by 2019[9].

Reviews of Pengpid and Peltzer (2019) indicated some factors associated with individual and/or multiple behavioral NCD risk factors among adolescents, including sex of male, older age, lower socioeconomic status, insufficient physical activity, not attending physical education classes, overweight/obesity, psychological distress, school truancy, and lack of peer and parental support[10]. Currently, spatial analysis is one of the methods for describing and analyzing geographically indexed health data based on demographic, environmental, behavioral, socioeconomic, genetic, and infectious or non-infectious risk factors[11]. Spatial analysis using an epidemiological approach extends the rich tradition of ecological studies that use explanations of the distribution of disease in different places better to understand the etiology of disease[11]. This article focuses on a small area analysis of PTM behavior, especially for hypertension and diabetes, for which there is considerable interest in spatial analysis and epidemiology[11]. The results of data analysis are used as solutions to some challenges facing local or regional decision makers and practitioners of spatial epidemiology, including issues of data availability and quality, confidentiality, disease prevalence, and study design[11]. Based on the challenge of overcoming the increasing health problems triggered by NCDs, this study was designed to analyze and describe the spatial distribution and prevalence of PTM behavior in the Soppeng Regency.

2. Methods

2.1. Data collection and procedure

The data used in this study are primary and secondary data obtained from various sources. Primary data was obtained through the collection of monthly disease data forms and monthly data from 2018-2016 at the Community Health Centers within the Soppeng district. NCDs data entered include data on hypertension and diabetes. The selected health centers included Baringeng, Batu-batu, Cabenge, Cakkuridi, Cangadi, Citta, Ganra, Goarie, Leworeng, Malaka, Pacongkang, Panincong, Salotungo, Sewo, Tajuncu, Takalala and Tanjonge. NCDs data is compiled based on the distribution of eight sub-districts, including Marioriwawo, Lilirilau, Lalabata, Liliriaja, Ganra, Citta, Donri-donri, and Marioriawa.

The Secondary data in the form of Soppeng district administrative maps and Soppeng district spatial plans (RTRW) maps were obtained from the Soppeng district BAPPEDA. Meanwhile, the Soppeng Regency population and socioeconomic data were accessed from the Central Bureau of Statistics (BPS), Soppeng Regency. The distribution map of hypertension and diabetes mellitus is displayed on the RBI map, and the National DEM Image was obtained from the Indonesian Geospatial Information Agency (BIG). The tool used in this research is a computer equipped with mapping software and Microsoft Office. The tool used in this research is a computer equipped with mapping software, SPSS ver. 21.0 and Microsoft Office[12].

2.2. Data analysis

Non-communicable disease data is displayed based on the prevalence rate value[6], which is analyzed based on the following formula:

$$Prevalence\ rate = \frac{All\ new\ and\ pre-existing\ cases\ of\ a\ specific\ disease\ during\ a\ given\ period}{.The\ total\ population\ during\ the\ same\ period}\ 1000$$

Each category's range of prevalence rate values is determined based on the annual average village prevalence rate (from 2016 to 2018). After getting the level of disease distribution in each village, the data is presented in the form of a Disease Distribution Map in Soppeng Regency using mapping software. The color gradation represents the prevalence rate: red is very high; orange is high; yellow is medium; light green is low, and dark green is very low.

3. Results and Discussion

3.1. Distribution of cases and prevalence rate of NCDs

Figure 1 shows the distribution trend of hypertension cases in 70 villages in Soppeng Regency. There are five villages that show the highest accumulated number of hypertension cases in the 2016-2018 range, including Barang, Jennae, Tettikenrae, Botto, and Panincong villages, with 440, 405, 325, 251, and 232, respectively. The highest prevalence rate of hypertension sequentially occurred in five villages, including Barang village, Jampu village, Marioriaja village, Barang, Jennae, Panincong, Maccile, and Tetewatu villages, with 19.2%, 7.6%, 6.4%, 6.3%, and 6.2%, respectively (Fig. 1).

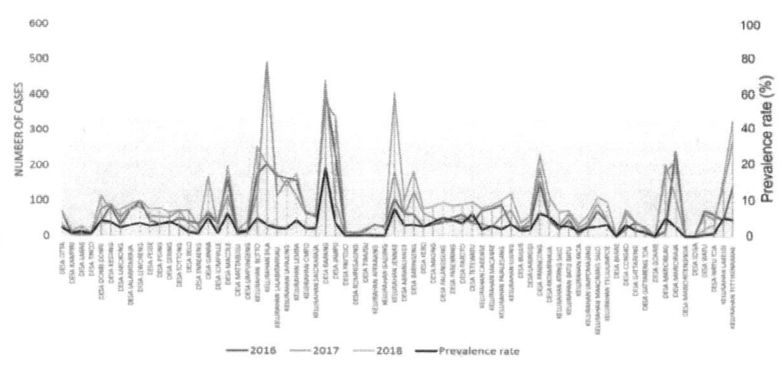

Figure 1. Cases trend and prevalence rate of hypertension based on village distribution in Soppeng district in the year 2016, 2017 and 2018

Meanwhile, figure 2 indicates five villages with the highest number of diabetes cases in the range of 2016-2018, including Maccile, Pattojo, Barang, Jampu, and Botto villages (251, 195, 177, 177, and 149, respectively). Similarly, the prevalence rate of diabetes was sequentially found in five villages, including Maccile, Pattojo, Barang, Jampu, and Botto Villages (62.5%, 56.6%, 52.8%, 50.7%, and 49.4%, respectively) (Fig. 2).

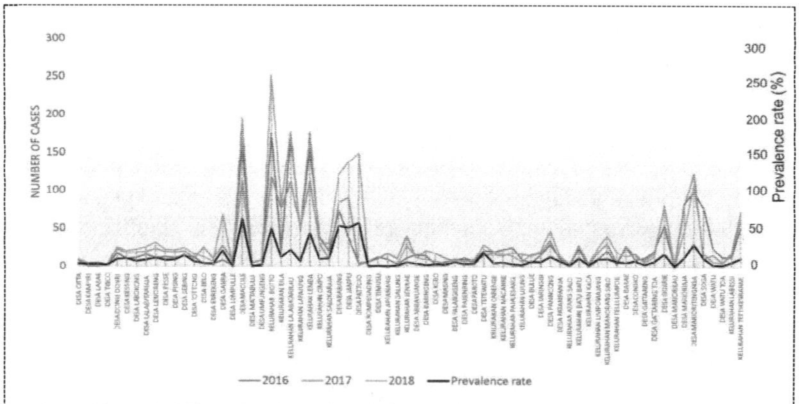

Figure 2. Cases trend and prevalence rate of diabetes based on village distribution in Soppeng district in the year 2016, 2017 and 2018

3.2. Spatial distribution and analysis of NCDs

The spatial distribution and analysis of the incidence of hypertension used three categories of prevalence values: low, medium, and high. If the prevalence rate is applied per 10,000 population, the spatial distribution will be divided into categories 0-384 (low category), 385-769 (medium category), and >1540 (high category) (Fig. 3A). The incidence of hypertension with a high category is found in Barang Village. Furthermore, the incidence of hypertension in the moderate category was found in Jampu Village. Meanwhile, other villages and sub-districts are included in the incidence of hypertension in the low category. Trends in prevalence rates vary between villages based on 2016-2018.

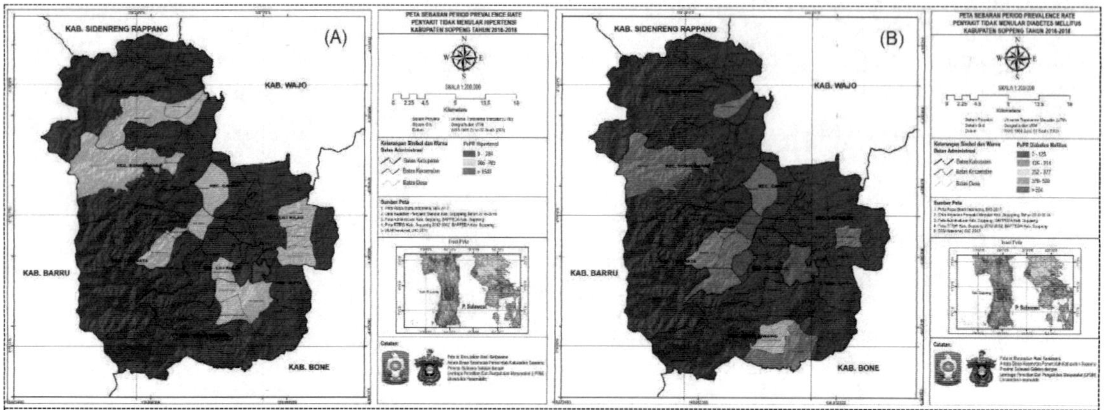

Figure 3. Spatial distribution of behavioural NCDs in Soppeng district for the year of 2016, 2017 and 2018. (A) spatial analysis based on the distribution of prevalence rate value of hypertension; (B) spatial analysis based on the distribution of prevalence rate value of diabetes

Meanwhile, Figure 3B shows the grouping of the prevalence rate of Diabetes Mellitus divided into three categories, namely very high, high, medium, low, and very low. As with hypertension, the spatial distribution of prevalence rate values is applied per 10,000 population so that it shows the following grouping of prevalence rate values: above 504 (very high category), 378 - 503 (high category), 252 - 377 (low category), 126 - 251 (high category), low) and 0 -125 (very low category). The incidence of Diabetes Mellitus with a high category was found in Maccile Village, Barang, Botto Village, and Lemba. Furthermore, the incidence of Diabetes Mellitus with a moderate category was found in Jampu Village, Pattojo, Marioriaja, Marioritengnga, Soga, and Lalabatarilau Village. Meanwhile, other villages and sub-districts are included in the incidence of Diabetes Mellitus with a low category.

Disease mapping is one of the first steps to understanding a disease's spatial structure and dynamics [13]. Spatial disease maps depict disease distribution and intensity, facilitate identification of heterogeneous distributions of disease in a given region, and describe high-burden areas where the disease is concentrated in space non-communicable diseases (NCDs), also known as chronic diseases, tend to be of long duration and are the result of a combination of genetic, physiological, environmental and behavioral factors[14]. Figure 3 and 4 presents a spatial analysis of NCDs load by sub-district within the Soppeng district. The color pattern shows a spatial pattern, where darker colors indicate a higher prevalence rate and lighter colors indicate a lower prevalence rate. The results show strong geographic disparities between regions among almost all variables, along with gradual spatial development from high to low levels of behavioral non-communicable diseases.

The spatial map for hypertension cases shows the prevalence level, which varies by sub-district. The highest prevalence occurred in Liliriaja sub-district, with the number of cases from 2016 to 2018 respectively 834, 988, and 952, with the average point prevalence reaching 31%, 36%, and 35%, respectively. Donri-Donri sub-district is in the second highest prevalence rate period, with the number of cases in 2016-2018 being 438, 612, and 736, respectively, with an average point prevalence of 21%, 26%, and 32%, respectively. Meanwhile, the Citta sub-district had the lowest average prevalence period of 14%, with the distribution of cases in the same three years, namely 82%, 40%, and 114%, respectively. Five other sub-districts, including Lili Rilau, Mario Riawa, Lalabata, Mario Riwawo, and Ganra, had prevalence periods under Donri-Donri, namely 3.1%, 3.1%, 2.9%, 2.2%, and 2.0%, respectively.

As with hypertension, diabetes mellitus in the Soppeng district increased steadily from 1835 cases in 2016 to 1909 cases in 2017, then increased almost twice in 2018, as many as 1497 cases. The prevalence rate of diabetes mellitus was highest in the Lili Riaja sub-district with a significant increase in cases from 2016 to 2018 (155, 259, and 438, respectively), which was followed by an increase in the point prevalence rate of 0.6%, 1.0%, and 13.0%, respectively. The second highest prevalence rate occurred in the Lalabata sub-district, with the number of cases falling from 2016 to 2017 but increasing in 2018 (787, 632, and 990 cases). The lowest prevalence rate occurred in the Citta sub-district, only 2%, with a very low number of diabetes cases in the same three years (10, 7, and 20 cases). Meanwhile, five other sub-districts, including Donri-donri, Mario Riwawo, Ganra, Mario Riawa, and Lili Rilau had almost the same prevalence rate of diabetes mellitus (0.9%, 0.9%, 0.8%, 0.8%, and 0.4%, respectively).

It is widely recognized that the risk factors for behavioral non-communicable diseases, including hypertension and diabetes, can include an unhealthy diet, lack of exercise, excessive alcohol consumption and smoking, and being overweight or obese [15,16]. The sub-districts of Liliriaja, Donri-donri, and Lalabata are areas with a high prevalence of hypertension and diabetes, which support the increasing prevalence of smokers. In addition, the number of elderly people who increases yearly will increase the number of vulnerable groups who have unhealthy lifestyles, such as less fiber consumption and sedentary lifestyles.

The prevalence of hypertension in the sub-districts of Liliriaja and Donri-donri (3.5% and 3.2%) was still higher than the prevalence of diabetes in South Sulawesi (1.8%) and Indonesia (2%) and lower than DKI Jakarta (3.1%)[17]. In general, the prevalence of hypertension is much higher in developing countries including Egypt, China, Mozambique, Turkey, Nigeria, Mexico, Paraguay, Venezuela, Thailand, Algeria, Chile and Vietnam (26.3%, 24.0%. 33.1%, 31.7%, 16.0%, 32.0%, 32.5%, 36.9%, 20.5%, 35.5%, 33.7%, and 25.1%)[18]. The prevalence of diabetes is 75% in low- and middle-income countries, indicating that the diabetes burden has considerable regional variation. Meanwhile, in developing countries, the impact of the global diabetes epidemic is disproportionate. The population's income level determines the general distribution pattern of the prevalence of DM in a place. Ass a developed unitary country, the West European region shows a much higher prevalence of DM, which continues to increase even though several public health efforts have been implemented. The comparison of the prevalence of DM in 2019 by age sequentially is the Middle East and North Africa, the West Pacific and Southeast Asia, North America and the Caribbean, South and Central America, Europe and Africa with percentages (12.2, 11.4, 11.1, 8.5, 6, 3 and 4.7 respectively)[19].

4. Conclusion

This study was conducted to analyze and describe the spatial distribution and prevalence of behavior NCDs in Soppeng Regency, including hypertension and diabetes, in the 2016-2018 period. Spatial and prevalence analysis showed variations in levels between village and sub-districts in the two types of behavioral non-communicable diseases. Sub-districts with high population density and the number of older people affect the consumption of cigarettes and alcohol and sedentary lifestyle behavior. The tendency to increase cases of hypertension and diabetes should be a major concern in controlling health problems, especially behavioral non-communicable diseases.

Acknowledgment
This research was supported by the Regional Planning and Development Agency (BAPPEDA) and the Soppeng Health Office, Indonesia. We would like to thank to the Soppeng District Government for allowing us to access the Puskesmas for data collection, compilation, and analysis.

References

[1] Basri; Sakakibara, M.; Sera, K. Current Mercury Exposure from Artisanal and Small-Scale Gold Mining in Bombana, Southeast Sulawesi, Indonesia—Future Significant Health Risks. Toxics 2017, Vol. 5, Page 7 2017, 5, 7, doi:10.3390/TOXICS5010007.
[2] WHO Noncommunicable Diseases Available online: https://www.who.int/news-room/fact-sheets/detail/noncommunicable-diseases (accessed on 2 August 2022).
[3] World Health Organization WHO Package of Essential Noncommunicable (PEN) Disease Interventions for Primary Health Care. Geneva, World Health Organization 2020.
[4] Basri; Sakakibara, M.; Ratnawati Economic Features of the Artisanal and Small-Scale Gold Mining Industry in Bombana, Southeast Sulawesi, Indonesia. IOP Conference Series: Earth and Environmental Science 2017, 71, 012016, doi:10.1088/1755-1315/71/1/012016.
[5] World Health Organization Noncommunicable Diseases: Mortality Available online: https://www.who.int/data/gho/data/themes/topics/topic-details/GHO/ncd-mortality (accessed on 2 August 2022).
[6] World Health Organization Noncommunicable Diseases: Country Profiles 2018 Available online: https://www.who.int/health-topics/noncommunicable-diseases#tab=tab_1 (accessed on 2 August 2022).
[7] Das, J.K.; Salam, R.A.; Thornburg, K.L.; Prentice, A.M.; Campisi, S.; Lassi, ZS; Koletzko, B.; Bhutta, Z.A. Nutrition in Adolescents: Physiology, Metabolism, and Nutritional Needs. Annals of the New York Academy of Sciences 2017, 1393, 21–33, doi:10.1111/NYAS.13330.
[8] Kristina, S.A.; Santosa, K.A. An Estimated Mortality and Disability Adjusted Life Years (DALYs) of Non-Communicable Diseases in Indonesia.| International Journal of Pharmaceutical Research 2020, 12, doi:10.31838/ijpr/2020.SP2.370.
[9] Dinas Kesehatan Sopppeng Profil Kesehatan Kabupaten Soppeng 2018 Available online: https://soppeng.go.id/dinas/dinas-kesehatan/ (accessed on 2 August 2022).
[10] Pengpid, S.; Peltzer, K. Behavioral Risk Factors Of Non-Communicable Diseases Among A Nationally Representative Sample Of School-Going Adolescents In Indonesia. International Journal of General Medicine 2019, 12, 387, doi:10.2147/IJGM.S226633.
[11] Werneck, G.L. Georeferenced Data in Epidemiologic Research. Ciência & Saúde Coletiva 2008, 13, 1753–1766, doi:10.1590/S1413-81232008000600010.
[12] Basri; Sakakibara, M.; Sera, K. Mercury in Soil and Forage Plants from Artisanal and Small-Scale Gold Mining in the Bombana Area, Indonesia. Toxics 2020, Vol. 8, Page 15 2020, 8, 15, doi:10.3390/TOXICS8010015.
[13] Basri; Tahara, T.; Malim, DDLO; Munafi, L.O.A. Spatial Distribution of Infectious Disease Epidemic in The Soppeng Regency, 2016-2018. IOP Conference Series: Earth and Environmental Science 2021, 940, 012017, doi:10.1088/1755-1315/940/1/012017.
[14] Real, LA; Biek, R. Spatial Dynamics and Genetics of Infectious Diseases on Heterogeneous Landscapes. Journal of the Royal Society Interface 2007, 4, 935, doi:10.1098/RSIF.2007.1041.
[15] Ng, R.; Sutradhar, R.; Yao, Z.; Wodchis, W.P.; Rosella, L.C. Smoking, Drinking, Diet and Physical Activity—Modifiable Lifestyle Risk Factors and Their Associations with Age to First Chronic Disease. International Journal of Epidemiology 2020, 49, 113, doi:10.1093/IJE/DYZ078.
[16] Ridjal, A.T.M.; Kasma, A.Y.; Aminullah; Basri Study of Rhodamine B Dyes Content in Snacks of Karuwisi Traditional Market Makassar, South Sulawesi, Indonesia. IOP Conference Series: Earth and Environmental Science 2022, 1027, 012012, doi:10.1088/1755-1315/1027/1/012012.
[17] Ridwanah, A.A.; Megatsari, H. Hypertension in Indonesia in 2018: An Ecological Analysis. Indian Journal of Forensic Medicine & Toxicology 2021, doi:10.37506/ijfmt.v15i2.14669.

[18] Mohsen Ibrahim, M.; Damasceno, A. Hypertension 3 Hypertension in Developing Countries. The Lancet 2012, 380, 611–619, doi:10.1016/S0140-6736(12)60861-7.
[19] Diabetes by Region - IDF DIABETES ATLAS - NCBI Bookshelf Available online: https://www.ncbi.nlm.nih.gov/books/NBK581937/ (accessed on 3 August 2022).

The Role of Food Waste Management in Green City Development

Salma Mar'atus Sholihah[1*], Tia Rizka Nuzula Rachma[2], Uswatun Khasanah[3]
[1] School of Environmental Science, Universitas Indonesia, Jakarta, 10430, Indonesia, 0000-0001-8751-8512
[2] Geospatial Information Agency, Bogor, 16911, Indonesia, 0000-0001-6523-2003
[3] School of Environmental Science, Universitas Indonesia, Jakarta, 10430, Indonesia
[1*]salma.maratus11@ui.ac.id, [2]tia.rizka@big.go.id, [3]uswatun.khasanah02@ui.ac.id

Abstract. Food waste generation is a severe problem in Indonesia today. Providing food waste processing facilities that are evenly distributed and reach service areas is essential. This step can lead to developing green cities that harmonize the circular economy and ecological civil aspects. This study aims to identify the location of food waste processing and the range of services to the service area (community settlements). The method used is spatial analysis. The research location is in Ciamis Regency. The results showed that currently, there are 58 locations spread over 24 sub-districts in Ciamis Regency. Food waste processing facilities have not been spread evenly because three districts still do not have waste processing facilities. The number of existing waste processing facilities is not evenly distributed. If viewed from the coverage of the service area (settlement), the existing food waste processing facilities can only meet 23.58% of the existing residential area. Collaboration between the government, the community, and the private sector is needed to build food waste processing facilities that are evenly distributed and reach all service areas. The concept of a circular economy in the food waste management system must be considered to create sustainable food waste management.

1. Introduction

The development of urban areas causes a change from natural ecosystems to artificial ecosystems. Contemporary urban conditions led to overconsumption of natural resources and increased waste generation [1]. The increase in waste generation is currently a severe problem in Indonesia. The amount of national waste generation reaches 15.8 million tons [2]. The source with the most significant amount is produced by households with a percentage of 42.8%, while the composition with the most significant amount is food waste with a percentage of 26.12% [2]. According to a study by the Ministry of National Development Planning/Bappenas, food waste wasted in Indonesia from 2000-2019 reached 23-48 million tons per year or the equivalent of 115-184 kilograms per capita per year [3]. The household is the leading actor who is the source of food waste [4].

Ciamis Regency is one area that has a household waste problem. The Department of Public Housing, Settlement Areas and the Environment of Ciamis Regency in 2017 recorded that household production waste in Ciamis Regency reached 100 tons per day. Still, only 20% of that amount was managed by the community. The rest ended up in the TPA (Final Disposal Site) [5]. Based on its composition, food waste becomes waste, with the most significant total waste generation in Ciamis Regency, reaching 34.4% [2]. This problem is exacerbated by the condition of the two TPAs in

Ciamis, which have not been able to accommodate the total waste generation in Ciamis and have created social conflicts due to their land ownership status. The concept of a green city that includes waste management is an alternative solution expected to overcome these problems.

The implementation of waste management policies in building a green city is the provision of waste banks and 3R TPS facilities [5,6,7]. 3R TPS may include recycling or composting food waste. This policy is proven to be able to reduce the waste generation in the landfill and increase the profile of the waste management system itself [8], support the realization of the zero waste concept in the context of a green city [7], and increase the level of community satisfaction with environmental management and services [9]. It can increase community participation [7] and income [10]. In the concept of sustainable development, implementing this policy can fulfill three aspects: environmentally sustainable, socially acceptable, and economically profitable.

The construction of the processing facility must be linked to the city's construction and enhanced by its existing information mechanism [11]. In addition, the availability of public information balanced with science and diplomacy can lead to a better change in the situation in waste management [12]. The design of a waste management facility needs to consider the connectivity of the source. Connectivity considerations for household/settlement actors are critical in managing food waste. This is for distribution cost efficiency in transportation activities. The design intended to minimize the total costs of the management system, such as costs of establishment, operation, transportation, uncollected waste, and pollution [13]. Dividing the existing ecological units in the city and recording the relevant spatial data and data attributes is a further implementation of the primary goal of green city development [14]. This study aims to identify the location of food waste processing and the range of services to community settlements.

2. Methods

The research location is Ciamis district, West Java province. This research approach is quantitative with a spatial analysis method. The data used in this study are Indonesian topographic maps, village boundary data from BIG, and composting location data from the Ministry of Environment and Forestry in 2022. There are several data analysis processes to produce a map description related to the location and service range of food waste processing facilities:

2.1 Identify the location of the waste treatment facility

The location data for food waste processing facilities are sourced from the National Waste Management Information System of the Ministry of Environment and Forestry in 2022. The facilities in question are organic waste processing facilities and composting facilities. These two facilities were chosen because food waste is classified as organic waste and can be processed in these two types of facilities. Table 1 shows the distribution of composting facilities in Ciamis Regency.

Table 1. Distribution of Composting Facilities [15]

No	Subdistrict	Number of Facilities
1	Banjaranyar	2
2	Banjarsari	2
3	Ciamis	13
4	Cidolog	1
5	Cihaurbeuti	2
6	Cijeungjing	10
7	Cikoneng	1
8	Cimaragas	1
9	Cipaku	1
10	Cisaga	3
11	Imbanagara	2
12	Jatinagara	1
14	Kawali	1
15	Lumbung	2
16	Pamarican	2
17	Panawangan	1
18	Panjalu	1

19	Panumbangan	1
20	Rajadesa	2
21	Rancah	2
22	Sadana	2
23	Sindang Kasih	1
24	Sukadana	1
25	Sukamantri	1
26	Tambaksari	1

2.2 Estimated area of waste processing service area for settlements

After the location distribution is known spatially, a search for settlements is carried out, which is located within a radius of 1 km from the composting location. The distance of one kilometer is assumed to be the maximum reach of the population served by the composting facility based on [ss]. The settlement was chosen as the object of service because The leading actor who is the source of food waste is the household [4]. Furthermore, the calculation and analysis of the settlements not covered by the composting service are carried out.

3. Results and Discussion

Ciamis Regency has several waste management facilities, such as a waste bank, 3R TPS, and maggot cultivation, both managed by the government and the private sector. From the data processing results, 58 locations for processing organic waste in Ciamis Regency are spread across various sub-districts. The sub-district with the most waste processing is the Ciamis sub-district with 13 composting locations, followed by the Cijeungjing sub-district with ten composting locations. Meanwhile, out of 27 sub-districts in all districts, three sub-districts still do not have composting facilities at all, namely Baregbeg, Lakbok, and Purwadadi sub-districts. These data indicate that the availability of food waste processing facilities is not evenly distributed. Figure 1 shows a map of the location of food waste processing facilities:

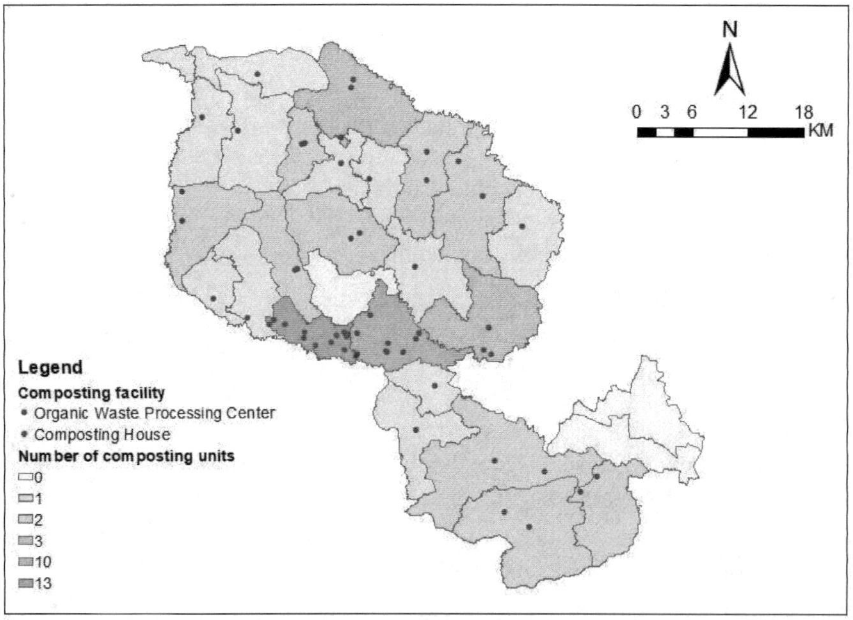

Figure 1. Map of the location of food waste processing facilities

Actually, in Ciamis Regency, there is a practice of maggot cultivation which can be a potential food waste processing facility because the output of the cultivation is in the form of dry larvae for animal feed, and Ciamis Regency is a base area for breeding chickens and beef cattle. Suppose food waste processing facilities, especially maggot cultivation, are available evenly in each sub-district by considering their affordability to service areas (settlements). In that case, the implementation of a circular economy as one of the critical concepts in the development of Green City has a great opportunity and possibility to be realized.

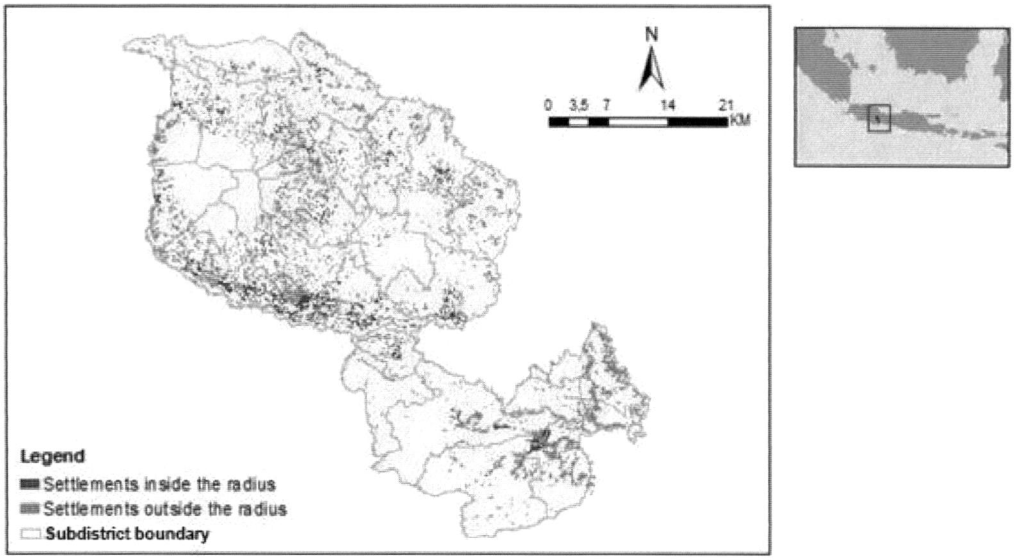

Figure 2. Service coverage of food waste processing facilities to settlements

Figure 2 shows the range of services provided by food waste processing facilities to residential areas. Blue color marks are settlements within a 1 km radius of the food waste processing facility. In comparison, the brown color mark is settlements located outside a 1 km radius of the food waste processing facility. The area of settlements within a service radius of 1 km is 45.7 km^{2}, with 23.58% of the total residential area in Ciamis Regency 193.9 km^{2}. This means that there are still 74.42% of settlements that do not have access to a maximum distance of 1 km. Distance and connectivity are critical in determining the location of waste processing facilities to service areas because the principle of efficient *origin-destination* can lead to efficiency in transportation costs and increased community participation in food waste management.

4. Conclusion

Spatial data related to waste management is essential in developing a green city. This research describes the location of food waste processing facilities in Ciamis Regency, 58 locations spread over 24 districts. Food waste processing facilities have not been spread evenly because three districts still do not have waste processing facilities. The number of existing waste processing facilities is not evenly distributed. If viewed from the coverage of the service area (settlement), the existing food waste processing facilities can only meet 23.58% of the existing residential area. Collaboration between the government, the community, and the private sector is needed to build food waste processing facilities that are evenly distributed and reach all service areas. The concept of a circular economy in the food waste management system must be considered to create sustainable food waste management.

Acknowledgement

The author would like to thank Dr. Hayati Sari Hasibuan, Himpasiling, and all parties who have provided input to this research.

References

[1] Marín-Beltrán, I., Demaria, F., Ofelio, C., Serra, LM, Turiel, A., Ripple, WJ, Mukul, SA, & Costa, MC (2022). Scientists' warning against the society of waste. In Science of the Total Environment (Vol. 811). Elsevier BV https://doi.org/10.1016/j.scitotenv.2021.151359

[2] Directorate General of Waste, Waste and B3 Management. (2021). National Waste Management Information System. Jakarta: Ministry of Environment and Forestry.
[3] Katadata. (2021). Food Waste in Indonesia. Available at https://katadata.co.id/ariayudhistira/infografik/61cd12d75181b/indonesia-kegunungan-sampah-food, 29 March 2022 pk. 15.00.
[4] Chaerul, M., & Zatadini, SU (2020). Food Waste Disposal Behavior and Food Waste Management in Various Countries: A Review. Journal of Environmental Science, 18(3), 455–466. https://doi.org/10.14710/jil.18.3.455-466
[5] Rukhayati, R., Haryani, S., & Pariyati, P. (2021, October 19). Development of a Green City Concept Through Waste Management to Increase Community Income. https://doi.org/10.4108/eai.19-12-2020.2309246
[6] Regards, R. (2021). Waste Management Policy Based on Community Empowerment in Supporting South Tangerang City as a Green City.
[7] Suyanto, E., Soetarto, E., Hardjomidjojo, H., Natural Resources and Environmental Management IPB, P., Rural Sociology FEMA IPB, P., Communication and Development FEMA IPB, P., & Agricultural Industrial Technology FATETA IPB, P (2015). "Green Community" Participation-Based Waste Management Policy Model Supports Green Cities. In MIMBAR (Vol. 31, Issue 1).
[8] Thushari, I., Vicheanteab, J., & Janjaroen, D. (2020). Material flow analysis and life cycle assessment of solid waste management in urban green areas, Thailand. Sustainable Environment Research, 30(1). https://doi.org/10.1186/s42834-020-00057-5
[9] Shahsavar, MM, Akrami, M., Kian, Z., Gheibi, M., Fathollahi-Fard, AM, Hajiaghaei-Keshteli, M., & Behzadian, K. (2022). Bio-recovery of municipal plastic waste management based on an integrated decision-making framework. Journal of Industrial and Engineering Chemistry, 108, 215–234. https://doi.org/10.1016/j.jiec.2022.01.002
[10] Pamuji, K., Nasihuddin, AA, Sukirman, K., Kartika, W.-N., Muflichah, S., Asyik, N., Sukolegowo, P., & Darmawan, AR (2021). A Juridical Study on the Role of Waste Bank in Domestic Waste Management in Banyumas Regency. Journal of Legal Dynamics, 21. https://doi.org/10.20884/1.jdh.2021.21.3.3126
[11] Wang, Q., & Li, S. (2021). Experience of Urban Solid Waste Management in Russia under the Concept of Smart City and Its Enlightenment to Shenyang. IOP Conference Series: Earth and Environmental Science, 719(4). https://doi.org/10.1088/1755-1315/719/4/042021
[12] Tulebayeva, N., Yergobek, D., Pestunova, G., Mottaeva, A., & Sapakova, Z. (2020). Green economy: Waste management and recycling methods. E3S Web of Conferences, 159. https://doi.org/10.1051/e3sconf/202015901012
[13] Tirkolaee, EB, Mahdavi, I., Esfahani, MMS, & Weber, GW (2020). A robust green location-allocation-inventory problem to design an urban waste management system under uncertainty. Waste Management, 102, 340–350. https://doi.org/10.1016/j.wasman.2019.10.038
[14] Sun, X. (2021). Green city and regional environmental economic evaluation based on entropy method and GIS. Environmental Technology and Innovation, 23. https://doi.org/10.1016/j.eti.2021.101667
[15] Directorate General of Waste, Waste and B3 Management. (2022). National Waste Management Information System. Jakarta: Ministry of Environment and Forestry.

IOP Conf. Series: Earth and Environmental Science 1111 (2022) 012011 doi:10.1088/1755-1315/1111/1/012011

Utilization of Denim Waste into Debric Board

Winda Oktaviani[1]*, Yunita Ismail Masjud[2]
[1] Environmental Engineering Department, Faculty of Engineering, President University, Cikarang, 17500, Indonesia, 000-0001-299-7916
[2] Environmental Engineering Department, Faculty of Engineering, President University, Cikarang, 17500, Indonesia, 0000-0002-3297-8850
[1]*winda.oktaviani@student.president.ac.id, [2]yunitaismail@student.president.ac.id

Abstract. More than one billion pieces of denim are sold every year. The consumptions from North America are 39%, West Europe 20%, East Asia 20%, and the rest of it, including Indonesia, is 13%. This study aims to know the comparation between denim waste become particle boards toward SNI as a national standard through several variables. The variables are water content, thickness increment, density, acidity, and alkalinity resistance. Making the sample refers to the particle board-making process, such as preparation, mixing, gluing, pressing,and drying. On the other hand, this study's method for the test is based on SNI 03-2105-2006 about Particle Board and SNI 01-7201-2006 Plywood and Beautiful Paper-faced Blockboard. The sample test divided by four ratios, 100%:100% (a), 75%:25% (b), 50%:50% (c), and 25%:75% (d). The result of this study showed that the water content test is the only variable influenced by the sample's ratio. At the same time, sample A is the perfect sample with a ratio of 100% because sample A complied with the water content, thickness test, density test, and minor damage on acidity and alkalinity resistance test. It can conclude that denim waste has the potential to be improved as interior materials such as particle board.

1. Introduction

The fashion industry has been named the second largest industry in the world, line with this high ranking causes the textile industry to contribute a large amount of solid waste [1]. In 2021 nationally in Indonesia, the percentage of fabric waste generated is 6.81% of the amount of other waste. These problems are increased compared to 2020, about 2.62% [2]. Denim is a fabric from naturalresources from cotton crude fibber [3]. About more than one billion types of denim are sold every year. The consumption from North America is 39%, West Europe 20%, East Asia 20%, and the rest of it, including Indonesia, are 13% [4]. Nearly 5% of textile waste, including denim, is filling landfill space [5]. The recycling of denim as non-biodegradable textiles must be recognized as a solution to thisproblem [6]. Unfortunately, the textile waste treatments are buried with other household waste or incinerations [7]. If this action continues without any preventive measures, it willlead to severe environmental damage.

This study's motivation is further to discuss the more appropriate use of fashion waste, focusing on denim waste for other products. The study to be carried out is to explore and analyze the utilization of denim waste from unused clothes as the interior materials, thus the same asparticle board. The resulting product will be a particle board named "DEBRIC-BOARD."

2. Methods

The method of the study is collected on the preliminary experimental result. The experimenttest also refers to SNI 03-2105-2006 regarding Particle Board and SNI 01-7201-2006 regarding Plywood and Beautiful Paper-faced Blockboard [8, 9].

2.1. DEBRIC-Board Method

(a) preparation of sawdust and denim particles, this process is separating the particles that are not proper in terms of size, (b) mixing, the process of mixing done with scale the weight of each composition towards the ratio, (c) gluing is the process after mixing by mix the denim, sawdust, and the glue becomes the dough, (d) molding, the molding process are by put the dough on the mold, (e) pressing, are the process of pressing the dough into the mold. In this research, the pressing process is the cold pressing process. So, the glue can flow around the particle, and the particlecan stick to each other, (f) drying. This process is needed because the pressing process is not enough to remove the water content on the mat.

2.2. Experimental Methods and Materials

The variable test used in the study are:
Water content is the variable to measure the amount of water content on the sample.

$$\text{Water Content } (\%) = \frac{Ba - Bk}{Bk}\, x\ 100$$

Where *Ba* is the initial weight of the sample and *Bk* is the sample's weight after drying in the oven in periodical time. Thickness increase is water absorption thickness expansion rate.

$$\text{Thickness Increasment} = \frac{T_2 - T_1}{T_1}\, x\ 100$$

Where T_2 and T_1 are sample conditions before and after immersion in the water.
Density is the variable to measure the density or solidity of the sample.

$$\text{Density } (gram/cm^3) = \frac{\mathrm{B}}{I}$$

Where B is the weight of the sample, and I is the calculation that comes from the sample's volume, thickness, and width. Then, acidity and alkalinity are the variables to measure the resistance of the sample toward the acid andalkali conditions. The analysis of this test is by visual observation, by looking at the sample changessuch as blistering, discolorations, and softening.

The variable test has a different standard. The standard of each variable is stated in table 1:

Table 1. Characteristic Standard

Characteristic	Test	Standard
Water Content	Water Content Test	Not allowed to exceed 14%
Thickness	Thick Swelling TestAfter Soaking In Water	Maximum 12%.
Density	Density Test	0,40 gram/cm^3 – 0,90gram/cm^3
Acidity & Alkalinity	Acidity & Alkalinity ResistanceTest	Record all the changes (blister, discoloration, softening)

The primary materials in this study are shown in figure 1. This research has different ratios with different compositions on each sample. The aim is to know the perfect composition of the experiment. Table 2 shows the detail of the material composition of each sample.

Figure 1. Denim Fibers

Table 2. Materials Composition

Sample	Ratio	Denim Fiber *(gr)*	Material Sawdust *(gr)*	Adhesive *(gr)*
A	100%	50	0	70 grams
B	75% : 25%	37,5	12,5	70 grams
C	50% : 50%	25	25	70 grams
D	25% : 75%	12,5	37,5	70 grams

2.3. Analysis Method

ANOVA analysis used to analyze the influence of the independent variable, in this case, is the ratio towardsthe dependent variable. The method to find the result of the hypothesis is by comparing the P-Value with α. TheT-test function compares the variable measurement results toward thestandard of SNI by covering all of the population.

3. Results and Discussion

The result of an experimental result is explained in table 3 and table 4.

Table 3. Experimental Result

Water Content			Thickness Increasement			Density		
Sample	Result	Standard	Sample	Result	Standard	Sample	Result	Standard
A1	4.35	<14%	A1	9.52	<12%	A1	0.43	0.4 gr/cm^3 – 0,90gr/cm^3
A2	4.17		A2	10.0		A2	0.41	
B1	4.00		B1	**25.0**		B1	0.59	
B2	4.35		B2	5.56		B2	0.48	
C1	5.26		C1	10.0		C1	0.49	
C2	5.00		C2	**12.5**		C2	0.50	
D1	4.76		D1	**15.8**		D1	0.40	
D2	4.55		D2	**13.9**		D2	0.43	

Table 3 shows the result of the experimental test. The result comes up with variation value, on water content are about 4% until 5,26%. The thickness increment are about 5,56% until 25% and density are about 0,4 gram/cm3 until 0,59 gram/cm3. Table 4, as a result of visual observation of acidity and alkalinity resistance, the sample shows the reaction.

Table 4. Acidity Resistance Test

Sample	Before	After	Visual Observation Result
A			Softening and discoloration
B			Softening
C			Nothing changes
D			Softening and discoloration

Table 5. Alkalinity Resistance Test

Sample	Before	After	Visual Observation Result
A		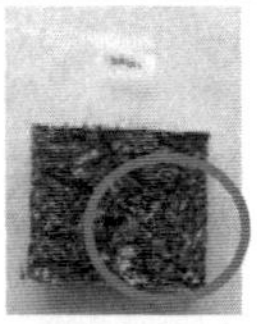	Softening and discoloration
B			Softening
C			Nothing changes
D			Softening and blister

On the acidity test, samples A, B, and D run into softening and discoloration, while sample C does notshow the reaction. While on the alkalinity test, samples A, B, and D also show the reaction and softening, discoloration, and blistering. At the same time, sample C does not react to anything.

3.1. Statistical Analysis Result

ANOVA tests compare the p-value to alpha. The result in table 6 concludes that thickness increment and density test are more than alpha, which means that the ratio does not influence thetwo parameters. While water content tests, the p-value is less than alpha, which means the ratio influences the test.

Table 6. ANOVA Single Factor Test Result

Water Content (*p-value*)	α	Thickness Increasement (*p-value*)	α	Density (*p-value*)	α
0.019	0.05	**0.8**	0.05	**0.10**	0.05

Table 7. T-test One-Tailed Result

	Water Content (*p-value*)	α	Thickness Increasment (*p-value*)	α	Density (*p-value*)	α
A	**4.2**	0.05	0.005	0.05	**0.09**	0.05
B	0.0001	0.05	**0.3**	0.05	**0.06**	0.05
C	0.0001	0.05	**0.3**	0.05	0.001	0.05
D	**6.3**	0.05	0.04	0.05	**0.2**	0.05

On the test to know the comparison between dependent variable toward the standard by using T-test analysis. Table 7 on water content samples A and D shows no significant differences or means to acceptthe standard. While on samples B and C, there are significant differences between the sample to the standard, which means they do not pass the standard. On the thickness increased test, samples B and C, there is no significant difference between the standard, which means accepting the standard. In contrast, thereare differences in samples A and D that mean not accepting the standard. On the density test, samples A, B, and D accept the standard, while sample C are not accepting the standard. The previous study about denim waste utilization also found that denim waste shows the potential to be used as materials for wood cabinets, desks, and other furniture [10].

4. Conclusion

This study concludes that the water content test is the only variable influenced by the sample ratio. At the same time, sample A is the perfect sample with a ratio of 100% because sample A complied with the water content, thickness test, density test, and minor damage on acidity and alkalinity resistance test. This result showed that DEBRIC BOARD is an excellent alternative to changing the particle board in general with denim materials. So, the denim waste is managed, and the populations are controlled.

Acknowledgments

This research can not be done without advice and guidance from Environmental Engineering President University lecturer, especially Dr. Ir. Yunita Ismail Masjid, M.Si, and others giving much meaningful feedback so the author could finish this study. Hope this study will assist the following research as innovations to bring the next step to a sustainable environment.

IOP Conf. Series: Earth and Environmental Science 1111 (2022) 012011 doi:10.1088/1755-1315/1111/1/012011

References

[1] Deepti Sethi, Amita Walia D S, A W 2017 Product Development using Textile Waste *Int. J. Text. Fash. Technol.* **7** 33–40

[2] Indonesia K L H dan K R 2022 Garbage Composition, SIPSN (Sistem Informasi Pengelolaan Sampah Nasional) 2022. 2022

[3] Annapoorani S G 2017 *Introduction to denim* (Elsevier Ltd)

[4] Santoso, Felicia., Wijaya, Florentina Tamariska. Ibrahim S 2017 Management of Waste and Used Denim Fabrics into Complementary Products for Fashion and Interior Elements *Simp. Nas. RAPID XVI – 2017 FT UMS* 214 – 215

[5] Uncu Akı S, Candan C, Nergis B, and Sebla Önder N 2020 Understanding Denim Recycling: A Quantitative Study with Lifecycle Assessment Methodology *Waste Text. Leather Sect.*

[6] Mazibuko M, Ndumo J, Low M, Ming D and Harding K 2019 Investigating the natural degradation of textiles under controllable and uncontrollable environmental conditions *Procedia Manuf.* **35** 719–24

[7] M. Juciene, V. Dobilaitė, D. Albrektas and R B 2022 Investigation and evaluation of the performance of interior finishing panels made from denim textile waste *Textile Res. J.*

[8] Anon 2006 Beautiful Plywood and Paper Blackwood

[9] BSN 2006 SNI 03-2105-2006: papan partikel *Standar Nas. Indonesia. Badan Stand. Nas. Republik Indonesia.* 1–23

[10] Meng X, Fan W, Wan Mahari W A, Ge S, Xia C, Wu F, Han L, Wang S, Zhang M, Hu Z, Ma N L, Van Le Q and Lam S S 2021 Production of three-dimensional fiber needle-punching composites from denim waste for utilization as furniture materials *J. Clean. Prod.* **281** 125321

IOP Conf. Series: Earth and Environmental Science 1111 (2022) 012012 doi:10.1088/1755-1315/1111/1/012012

The Strategy of Organic Farming Development for Environment, Social, and Economic Improvement from Farmers Perspective

M. Octorayanti[1], K. Mizuno[2]* and S. Setiawati[3]
[1] School of Environmental Science, Universitas Indonesia, Jakarta, 10430, Indonesia, [1]0000-0002-2714-1706
[2] Environmental Economic, School of Environmental Science, Universitas Indonesia, Jakarta, 10430, Indonesia, 0000-0002-7411-8074
[3] School of Environmental Science, Universitas Indonesia, Jakarta, 10430, Indonesia, 0000-0002-8944-3750
[1]maria.octorayanti@ui.co.id, [2]*kosuke.mizuno@ui.co.id, [3]sri.setiawati11@ui.ac.id

Abstract. The increase in the sustainability of environmental and health awareness is a huge driving factor for the increasing demand of organic products in various countries. Thus, many countries are competing to expand organic agricultural land to accommodate market needs. However, in Indonesia, the practice of organic farming is relatively low, as indicated by the land conversion for organic production is relatively small compared to several countries in Southeast Asia. The key factors and challenges for organic farming development are explored, and an effective strategy is formulated as the main objective of this study. This study deals with a case study in Bandung Regency as one of the main agricultural producer areas in Java. A qualitative approach is employed to ensure in-depth information and analysis. This study utilized SWOT (Strength Weakness Opportunity Threat) to formulate the strategy for the development of organic farming to improve the environment, economic and social aspects. The interview and survey were conducted with a random sampling with a total of 30 respondents from a total of 87 population organic farmers. The result of the SWOT analysis shows that the S-W value is -0.29 and the O-P value is 0.04. The analysis result shown in the cartesian coordinate confirms that the most relevant strategy for Organic Agriculture Development is to enhance guidance and assistance from the government and private parties related to funding and certification, as well as the establishment of a reliable quality assurance system with the enforcement of regulations or policies related to organic farming.

1. Introduction

The increase in the sustainability of environmental and health awareness is a huge driving factor for the increasing demand for organic products in many countries [1]. In 2018 the market for organic products was estimated to reach USD 161.5 billion with a growth rate of 15% per year [2]. The retail value in the world reached 92,074 million euros in 2017 and 96,683 million euros in 2018[3]. Although there is no comprehensive statistical data regarding the exact amount of organic agricultural production, the organic market in Indonesia tends to continue to increase, both domestic and export [5]. Thus, the implementation of organic farming is increasing along with the demand for organic products worldwide.

However, the conversion of land from conventional to organic areas is considered stagnant, compared to other countries, especially in Southeast Asia, for instance, the Philippines and Vietnam. Thus, it can be overviewed by comparison of the shared land between conventional and organic farming. From 2016 to 2018, the organic farming area remained at the same level of 0.4% [4]. Vietnam has shown an impressive development in conversion area, indicated by the expansion of the share land area of organic farming from 0.5% in 2017 to 2.5% in 2018. The shared organic farming land in the Philippines is also increasing from 1.6% in 2017 to 1.8% in 2018 [4].

Indonesia has the largest agriculture area among ASEAN countries. It shows that Indonesia has a high potential to develop organic farming and expand its market. Considering the market potential as well as intention toward environment preservation, the Indonesian government has initiated an organic farming movement through the "Go-Organic" Program. The case of developing organic farming in Bandung Regency as one of the major agricultural producers in Java can be considered an example. Several farmer groups have converted their land into an organic agricultural area. One of them has received certification from an international certification agency, namely Control Union (CU), the USDA type of certification (United States Department of Agriculture). With the certificate, these farmer groups can export their products to the United States [6] and Singapore [7]. However, the fulfillment of the quantity, quality, and continuity of organic products in Bandung Regency still faces some obstacles, including inconsistency among farmers in organic cultivation because conventional farming behavior is still fervent [7].

Most agricultural practices still use the intervention of synthetic chemicals that have an impact on environmental pollution. The excessive use of pesticides in Indonesia during the 1970s and 1980s caused serious environmental problems, such as acute and chronic human pesticide poisoning, animal poisoning, the contamination of agricultural products, the destruction of both beneficial natural parasites and pest predators, and pesticide resistance in pests [8] The decline in the quality of groundwater has decreased in Bandung Regency with an increase in nitrate and ammonium levels above the quality standard as a result of the use of chemicals in agricultural activities [9,10]. As well as air pollution by organochlorines at several points near agricultural areas due to the use of pesticides [11]. Several pesticide residues were found in vegetables with levels above the minimum residual threshold and impacted the farmers' health [12]. The reason many farmers are reluctant to convert to organic farming is crucial to study. Therefore, this study aims to discuss the key factors of organic conversion and its challenge to find the reason behind it as a basis to formulate an effective strategy for developing the organic farming system.

2. Methods

This study employed a qualitative approach with a mixed- method. The qualitative approach is purposed to obtain in-depth analysis from respondents to gather the information that can portray the root of the problem so that the factors behind the low implementation rate of the organic farming system will emerge. To obtain a comprehensive data analysis, the research method is a mixed-method combination method between quantitative and qualitative. The interview and survey were conducted with random sampling with a total of 30 respondents from a total of 87 population organic farmers. This study employed SWOT analysis to reveal the key factors for organic conversion and its challenges. SWOT analysis has several benefits, including; (1) Helps analyze the problem comprehensively in terms of 4 sides of the analysis; (2) With a 4-sided analysis compartment, a sharp analysis can be obtained to provide directions or recommendations to maintain strength while increasing profits based on the existing opportunities and minimizing shortcomings as well as avoiding threats; (3) Help re-identify things that could be overlooked in the previous analysis; (4) Can be a quite powerful instrument in carrying out a strategic analysis that can help to find the right and best steps according to the current situation.

2.1 Farmers Interview
To reveal the key factors for organic conversion and its challenge, an interview approach with the horticulture farmers was adopted to gather information. The farmer's interview is fundamental as it is considered data collection from the issue's root. The participants had a vast experience in organic agricultural practices. The interview result was utilized for encoding the key point for SWOT analysis.

2.2 SWOT analysis
SWOT analysis is a classic strategic planning instrument that provides a simple way to estimate the best way to determine a strategy [13]. The strategy formulation in SWOT analysis uses a SWOT matrix consisting of IFAS (Internal Strategic Factor Analysis Summary) and EFAS (External Strategic Analysis Summary). From IFAS and EFAS, a strategy is developed or formulated with four analysis compartments: SO-WO (Strength Opportunity - Weakness Opportunity) and ST-WT (Strength Threat-Weakness Threat). The SWOT matrix framework appears as follows:

	Strength **Positives characteristic and advantages of the issue, situation, or technique**	**WEAKNESS** **Negative characteristics and disadvantages of the issue, situation, or technique**
OPPORTUNITIES **Factors, situations that can benefit, enhance or improve the issue, situation, or techniques**	**S-O Strategy/Analysis** **Using strength to take advantage of opportunities**	**W-O Strategy/Analysis** **Overcoming weaknesses by taking advantage of opportunities**
THREATS **Factors, situations that can hinder the issue, situation, or technique**	**S-T Strategy/ Analysis** **Using strengths to avoid threats**	**W-T Strategy/Analysis** **Minimize weaknesses and avoid threats**

Figure 1. A conceptual matrix framework of SWOT Analysis [14]

2.3 Data Analysis
Participants were given a set of questionnaires, where they were requested to answer the question by rating the highest to the least point to each category in the above list. Hence, the participants also requested to put the weightage for each point to measure the priority list of each category. These rating and weightage results are displayed in SWOT Diagram.

3. Results and Discussion

The interview result revealed 9 Strengths, 6 Weaknesses, 10 Opportunities, and 5 Threats. The elaboration of each factor is discussed in the following parts.

3.1 Strength
The strength of Organic Agriculture is shown in table 1. Farmer Strategic Internal Factor (Strength). From the table, it was identified that Farmer considers that the most valuable factor for organic farming development is idealism and awareness of the environment of the farmers, with the weightage for this factor being 0.13. Meanwhile, the most urgent factor that needs to be promoted is establishing an agriculture cooperative or company to support managerial production as well as marketing. However, considering both the weightage and rating for each factor, the availability of seeds, fertilizers, and another organic production support is considered the most fundamental to be fulfilled.

Table 1. Farmer Strategic Internal Factor for Strength

NO	Farmer Strategic Internal Factor (**Strength**)			
	Item	Weightage	Rating	W*X
1	Idealism and awareness of the environment of the farmers	**0.13**	2.27	0.30
2	Freshwater is sufficiently available for organic farming practices and not contaminated by industrial or domestic waste and pollution	0.12	2.27	0.27
3	Drainage system to ensure the sustainable supply of water	0.12	2.73	0.33
4	A forum for coordination of both technical production issues as well as marketing	0.12	2.20	0.26
5	Willingness for continuous learning of the farmer	0.11	2.07	0.23
6	Establishing agriculture cooperative or company to support managers in production and marketing	0.10	**2.93**	0.30
7	Availability of seeds, fertilizers, and other organic production support	0.12	2.67	**0.32**
8	Direct communication between farmer and consumer	0.09	2.53	0.23
9	Proactive participation of the farmer in attending organic product exhibitions and events to open marketing channel	0.09	2.73	0.24
Total Strength		1.00	2.53	0.23

3.2 Weakness

The most urgent issue considered as a Weakness factor is limited land ownership with a total value of 0.57, and the 2nd urgent issue is lack of facility and equipment with a total value is 0.56. The current condition is that most farmer use rented land, with the average rental cost equivalence to 50% of the total product sales. Hence, the limited access to financial assistance also hinders the improvement of production, especially for the irrigation facility where farmers could not secure sufficient water during the dry season. The data for Weakness is shown in table 2.

Table 2. Farmer Strategic Internal Factor (Weakness)

NO	Farmer Strategic Internal Factor (Weakness)			
	Item	Weightage	Rating	W*X
1	Limited land ownership	0.22	2.60	**0.57**
2	Limited land area	0.16	2.53	0.41
3	Lack of Human Resources	0.15	2.80	0.43
4	Farmer's knowledge of business	0.13	2.93	0.38
5	Production yield was low during the transition and adaptation period to convert to the organic farming method	0.14	2.93	0.42
6	Lack of facility and equipment	0.20	2.87	**0.56**
Total Weakness		1.00	16.67	2.73

3.3 Opportunity

The government has implemented assistance for guidance and certification processes for some organic farmers. However, the number is considered limited. Therefore, enhancement in these areas and issues need to be considered as it is identified as prioritized with a total value of 0.34. Establishing cooperation and collaboration with various stakeholders to open marketing channels is also crucial, with a total value of 0.31. The 3rd priority issue is building a trusted organic quality assurance as it guards the trust-ability against the producer to consumer. In addition, the involvement of private financial institutions also plays a vital role as financial support is considered still limited. Both quality assurance systems and private financial institution involvement scored 0.27.

Table 3. Farmer Strategic External Factor (Opportunity)

NO	Farmer Strategic External Factor (Opportunity) Item	Weightage	Rating	W*X
1	Organic Certification	0.13	2.43	**0.30**
2	Capital support and assistance from the government	0.09	3.67	**0.34**
3	Access to capital from a private financial institution such as a bank	0.11	2.40	**0.27**
4	Platform to accommodate farmer participation in exhibitions and events to open marketing channel	0.09	2.67	0.24
5	Technical assistance from government and non-government organizations such as training and extension as well as equipment	0.08	3.20	0.26
6	Cooperation and collaboration channel for marketing channel	0.11	2.80	**0.31**
7	The contract between farmer and buyer to ensure price stability	0.10	2.47	0.25
8	Periodical term of payment to enable measured cash flow (weekly or 3 times in a month)	0.10	2.60	0.25
9	Trusted organic quality assurance system from the government and or organic certification body	0.11	2.60	**0.27**
10	Support from society	0.08	2.60	0.22
Total Strength		1.00	27.40	2.71

3.4 Threats

As a result, weightage and certification cost are still considered big issues for farmers. It is indicated by the highest threat score given in the questionnaire, with a total score of 0.58. Another eminent issue is organic product counterfeiting as the 2nd highest threat with a score of 0.54. Therefore, enhancing the trusted organic quality assurance system is indispensable as it also arose in the Opportunity factor as an effort to guard the trust-ability against the producer to consumer.

Table 4. Farmer Strategic Internal Factor (Threats)

NO	Farmer Strategic Internal Factor (Threats) Item	Weightage	Rating	W*X
1	Cost for certification	0.21	2.73	**0.58**
2	The non-organic farming area around the organic area	0.14	3.20	0.46
3	Consumer awareness of the urgency of organic product consumption for the environment and health	0.22	2.60	**0.56**
4	Organic Product Counterfeit	0.21	2.60	**0.54**
5	Security (thief etc.)	0.22	2.40	0.52
Total Threats		1.00	13.53	2.67

3.5 SWOT Diagram

Taking into account the analysis of Strengths, Weaknesses, Opportunities, and Threats that arise from farmers' voices through the interview, as well as the calculation result of each factor from the questionnaire taken from the participant in this research, it can be formulated several strategies to be considered as a priority to be taken to further developing organic agriculture practice in each element as shown in the below figure 3.

	STRENGTH (S)	WEAKNESS (W)
	STRATEGY S-O	STRATEGY W-O
OPPORTUNITY (O)	Supporting the idealism and awareness of farmers in the form of technical and non-technical assistance such as funding, organic certification, infrastructure (irrigation system), basic production needs, and providing guidance for the formation of cooperatives or companies that can assist farmers in terms of production management and marketing supported by a reliable quality assurance system	Enhance guidance and assistance from the government and private parties related to funding and certification, as well as the establishment of a reliable quality assurance system with the enforcement of regulations or policies related to organic farming. This can be projected to address deficiencies in land ownership and limited facilities and infrastructure
	STRATEGY S-T	STRATEGY W-T
THREATS (T)	Increasing consumer awareness of the importance of consuming organic products and strengthening organic quality assurance to prevent counterfeiting of organic products is significant support for organic farmers even though farmers have to increase production costs with organic certification and conventional agriculture is still common	Support for facilities and infrastructure, land ownership, certification as well as increasing consumer awareness of the importance of consuming organic products for both the environment and health

Figure 2. Matrix SWOT for Development Strategy of Organic Agriculture

To identify the most relevant strategies based on the farmer's perspective, the questionnaire result is calculated between weightage and rating. The total score for the internal factor element is 2.47 for strength, followed by Weakness, with a total score is 2.76. Thus, the final result between the Strength and Weakness score is -0.29. The negative score indicated that the spot between Strength and Weakness lies in the left bottom quadrant. The calculation result for the internal factor element is shown in figure 4. Moreover, the total score for external factors is 2.71 for Opportunity, followed by Threats, with a total score is 2.67. The final result between Opportunity and Threats is 0.04. The positive value indicated that the spot lies in the right upper quadrant, as shown in figure 5.

	Farmer Stratetic Internal Factor	Weightage	Rating	W X R
Strength	Item			
	Total Strength (S)	1.00	22.40	2.47
	Farmer Strategic Internal Factor	Weightage	Rating	W X R
Weakness	Item			
	Total Weakness (W)	1.00	16.67	2.76
	S-W		-0.29	

	Farmer Strategic External Factor	Weightage	Rating	W X R
Opportunity	Item			
	Total Opportunity (O)	1.00	27.40	2.71
	Farmer Strategis External Factor	Weightage	Rating	W X R
Threats	Item			
	Total Threats (T)	1.00	13.53	2.67
	O - T		0.04	

Figure 3. Matrix Farmer Strategic Internal Factor

According to the calculation mentioned above, the finding shows that the grid line that shows the position of organic agriculture from the farmer's perspective lies between Weakness and Opportunity or 3rd quadrant, as shown in figure 6.

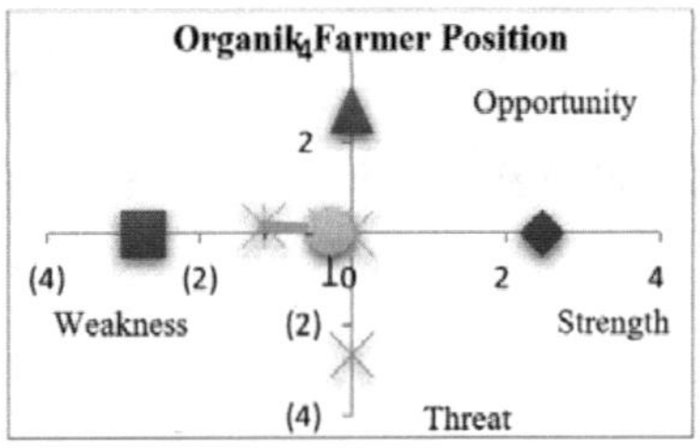

Figure 4. Diagram SWOT Result of Organic Farmer Position

The analysis result shown in the cartesian coordinate confirms that the most relevant strategy for Organic Agriculture Development is to enhance guidance and assistance from the government and private parties related to funding and certification, as well as the establishment of a reliable quality assurance system with the enforcement of regulations or policies related to organic farming. This can be projected to address deficiencies in land ownership and limited facilities and infrastructure. This strategy was formulated based on the current situation considering the farmer voices taken by interview. Some program has been implemented so far related to guidance and assistance related funding and certification. However, the number is still considered limited. Likewise, the regulation related to quality assurance of organic agriculture system also has been executed through regulation from Ministry of Agriculture No.64 of 2013. However, the enforcement of this regulation is considered weak. Thus, this confirms that counterfeit organic agriculture products are still found in the market, which is reported by the farmer.

The assistance for the funding, along with enhancement in the regulation enforcement for the organic agriculture system, will strengthen the stability of the organic agriculture business [15-17]. The support related to land ownership or ensuring that the farmer has secure land for their organic farming practice is also become imperative to support the development of organic farming [18-19].

4. Conclusion

Based on the calculation of the SWOT analysis conducted in this study, it can be concluded that the most relevant strategy for organic farming development from farmers' perspective is enhancing guidance and assistance from the government and private parties related to funding and certification, as well as the establishment of a reliable quality assurance system with the enforcement of regulations or policies related to organic farming. This can be projected to address deficiencies in land ownership and limited facilities and infrastructure that still become a hindrance.

Farmer voices are the root of organic agriculture practices. However, developing a tactical strategy must be completed with other key elements from various stakeholders along the systems, including distributor, government, certification body, and consumer. Therefore, further study with various stakeholders must be undertaken to formulate a more comprehensive strategy. In addition, the measurement of Organic Farming's impact on the improvement of Social, Environment, and Economic also need to be elaborated further with the exact number and calculation. Thus, it need to be studied further.

Acknowledgments

We acknowledge the support received from Organic Farmer Group in Bandung Regency especially to the head of the Farmer Group, Mr. Waryudin and Mr. Thohir for given the utmost support in providing accuracy of the data and rising the voice of the organic agriculture farmer.

References

[1] K. Ditlevsen, P. Sandøe, and J. Lassen, "Healthy food is nutritious, but organic food is healthy because it is pure: The negotiation of healthy food choices by Danish consumers of organic food," *Food Qual. Prefer.*, vol. 71, no. November 2017, pp. 46–53, 2019, doi: 10.1016/j.foodqual.2018.06.001.

[2] O. of C. Ecconomis, "Daily Economic and Market Review," *Dly. Econ. an Maret Rev. Bank Mandiri*, pp. 2017–2018, 2018.

[3] A. M. Ha, N. America, M. Ha, A. M. Ha, A. At, and W. W. W. Ifoam, *GLOBAL POLICY TOOLKIT ON PUBLIC SUPPORT TO ORGANIC AGRICULTURE Looking for professional information?* 2018.

[4] FiBL, *The World of Organic Agriculture*. 2019.

[5] W. David and Ardiansyah, "Organic agriculture in Indonesia: challenges and opportunities," *Org. Agric.*, vol. 7, no. 3, pp. 329–338, Sep. 2017, doi: 10.1007/s13165-016-0160-8.

[6] R. Andriani, B. Kusumo, A. Charina, Y. Sukayat, and G. W. Mukti, "Kapasitas Petani Dalam

Agribisnis Sayuran Organik Di Kabupaten Bandung Barat Capacity of Farmers in Organik Vegetable Farming in Bandung Barat Regency," vol. 6, pp. 123–130, 2018.

[7] D. Supyandi, Y. Sukayat, and M. A. Heryanto, "Pertanian (Studi Kasus Di Kabupaten Bandung Propinsi Jawa Barat)," pp. 190–201.

[8] M. J. Mohd. Fuad *et al.*, "The impact of pesticides on paddy farmers and ecosystem," *Adv. Nat. Appl. Sci.*, vol. 6, no. 1, pp. 65–70, 2012.

[9] R. Maria, D. Mulyadi, and H. Lestiana, "Pengaruh kondisi lingkungan di pagelangan kabupaten bandung," *Geologi*, pp. 577–589, 2014.

[10] A. F. Rusydi, W. Naily, and H. Lestiana, "Pencemaran Limbah Domestik Dan Pertanian Terhadap Airtanah Bebas Di Kabupaten Bandung," *J. Ris. Geol. dan Pertamb.*, vol. 25, no. 2, p. 87, 2015, doi: 10.14203/risetgeotam2015.v25.201.

[11] R. Prananditya and K. Oginawati, "Identifikasi Dan Distribusi Pencemar Pestisida Organoklorin Pada Udara Ambien Di Daerah Pertanian Hulu Sungai Citarum," *J. Teh. Lingkung.*, vol. 22, no. 1, pp. 73–82, 2016, doi: 10.5614/j.tl.2016.22.1.8.

[12] E. Amilia, B. Joy, and S. Sunardi, "Residu Pestisida pada Tanaman Hortikultura (Studi Kasus di Desa Cihanjuang Rahayu Kecamatan Parongpong Kabupaten Bandung Barat)," *Agrikultura*, vol. 27, no. 1, pp. 23–29, 2016, doi: 10.24198/agrikultura.v27i1.8473.

[13] J. Yuan, H. Xie, D. Yang, X. Xiahou, M. J. Skibniewski, and W. Huang, "Strategy formulation for the sustainable development of smart cities: A case study of Nanjing, China," *Int. J. Strateg. Prop. Manag.*, vol. 26, no. 4, pp. 379–399, 2020, doi: 10.3846/ijspm.2020.13345.

[14] K. Skinner, R. M. Hanning, C. Sutherland, R. Edwards-Wheesk, and L. J. S. Tsuji, "Using a SWOT analysis to inform healthy eating and physical activity strategies for a remote first nations community in Canada," *Am. J. Heal. Promot.*, vol. 26, no. 6, 2012, doi: 10.4278/ajhp.061019136.

[15] P. Thiers, "From grassroots movement to state-coordinated market strategy: The transformation of organic agriculture in China," *Environ. Plan. C Gov. Policy*, vol. 20, no. 3, pp. 357–373, 2002, doi: 10.1068/c2v.

[16] Y. Xie, H. Zhao, K. Pawlak, and Y. Gao, "the Development of Organic Agriculture in China and the Factors Affecting Organic Farming," *J. Agribus. Rural Dev.*, vol. 9, no. 2, 2016, doi: 10.17306/jard.2015.38.

[17] Y. Qiao, F. Martin, X. He, H. Zhen, and X. Pan, "The changing role of local government in organic agriculture development in Wanzai County, China," *Can. J. Dev. Stud.*, vol. 40, no. 1, pp. 64–77, 2019, doi: 10.1080/02255189.2019.1520693.

[18] M. W. Akram *et al.*, "Impact of land use rights on the investment and efficiency of organic farming," *Sustain.*, vol. 11, no. 24, pp. 1–18, 2019, doi: 10.3390/su11247148.

[19] S. Ferreira *et al.*, "Assessment of Factors Constraining Organic Farming Expansion in Lis Valley, Portugal," *AgriEngineering*, vol. 2, no. 1, pp. 111–127, 2020, doi: 10.3390/agriengineering2010008.

A Study the Relationship of Soil Quality with Land Suitability for Coconut (*Cocos nucifera L.*) Development in Kebonagung Sub-district, Pacitan District

Suko Irawan[1], Ernoiz Antriyandarti[2*], Dea Hagania Laia[3]
[1] Department of Soil Science, Faculty of Agriculture, University of Sebelas Maret, Surakarta, 57126, Indonesia
[2] Department of Agribusiness, Faculty of Agriculture, University of Sebelas Maret, Surakarta, 57126, Indonesia
[3] Department of Agribusiness, Faculty of Agriculture, University of Sebelas Maret, Surakarta, 57126, Indonesia
[1]sukoirawan9@student.uns.ac.id, [2*]ernoiz_a@staff.uns.ac.id, [3]deahagania7@gmail.com

Abstract. This study aims to assess the soil quality index (SQI) under the shade of coconut (*Cocos nucifera L.*) and its relationship to land suitability to provide recommendations for improvement to increase coconut productivity. This research conducted in Kebonagung District, Pacitan District using descriptive survey method. Land Map Units (LMU) are determined by overlaying soil type, land use, slope, and rainfall maps. Determination of survey using purposive sampling with 10 sample and three replications. The samples were analyzed in the laboratory with analysis method Bulk Density (Pycnometer), Porosity (Sample ring), Textures (Pipet), pH (Electrometric), Available P (Olsen method), Available K (NH4Oac 1 N extraction), Kjeldahl (Total N), Base saturation (Ammonium acetate extraction), SOC (Walkey and Black), CEC (Ammonium acetate method). Soil quality assessment was carried out using Principal Component Analysis (PCA) to obtain Minimum Data Sets (MDS). Land suitability was evaluated through the matching method with the conditions for growing coconut plants. The analysis showed that the soil quality index had a significant positive correlation with land suitability (r=0.768, p=0.004). The limiting indicators based on the correlation test were BV, pH, available P, organic C, and CEC. Suggested improvement recommendations include applying land cover crops, dolomite, manure, and balanced P fertilization.

1. Introduction

Agricultural development is a multidimensional process that includes fundamental changes in social structures, people's attitudes, and state institutions to accelerate economic growth, address income inequality, and alleviate poverty. Indonesia's agricultural development paradigm is inseparable from the success of the green revolution, the implementation of the green revolution since the mid-1970s has changed farmers' cultivation systems from conventional to more modern, and the application of agricultural technology includes essential fertilizers (NPK), compound fertilizers, pesticides, superior seeds, has gradually succeeded in increasing Indonesia's agricultural productivity. The success of the green revolution, on the other hand, harms environmental health, the massive, intensive, and long-term use of chemical pesticides and inorganic fertilizers can cause a decrease in soil quality and changes in physical, chemical, and biological properties and reduces soil quality [1]. Indonesia is a tropical country that has high potential as a coconut producer. Coconut plants are widespread in yards and plantations in almost all parts of Indonesia. The coconut area in 2018 reached 3,417,951 hectares, with a percentage of 99% or an area of 3,385,085 hectares cultivated by farmers using the monoculture or intercropping method [2]. Coconut (*Cocos nucifera L.*) has the potential to become a leading commodity with promising prospects in terms of international market demand. Based on data from the [2] coconut derivative products, including coconut fiber, copra, coconut oil, dry grated coconut, charcoal, coconut shell, coconut sugar, and Virgin Coconut Oil (VCO) is famous exports to America, Netherlands, Malaysia, Thailand, China, and Vietnam. Judging from price developments, during 2014-2017, coconut prices at the producer level continued to increase. The cost of coconut grains in the country tends to increase every year, the highest coconut price reaches Rp. 283.492/100 grains.

Kebonagung District is the second-highest coconut production center in Pacitan Regency, area of coconut plantations reaches 2,850 ha with a production of 2,880.58 tons, only behind Tulakan District, which has 3,043 ha with a production of 2,910.51 tons. The development of coconut as a leading commodity is faced with various challenges, including low productivity, development policies that have not optimally supported the coconut plantation sector, downstream industries that have not developed so that most of the products are sold in primary form, lack of awareness of farmers about the prospects of coconut, and inappropriate coconut cultivation practices. The development of coconut plantations must refer to aspects of land carrying capacity and soil quality index. The land's carrying capacity is the level of land suitability for developing certain commodities. At the same time, the soil quality index is the ability of the soil to provide an ideal ecosystem that ensures plant productivity, environmental quality, and human, animal, and plant health in a sustainable manner [3]. Assessment of land support is carried out through land suitability. This study examines the relationship between soil quality and land suitability for coconut development in Kebonagung Sub-district to provide recommendations for appropriate land management to increase coconut productivity.

2. Methods

The research was conducted in Kebonagung Sub-district, Pcitan Regency, East Java. Soil analysis was carried out at the Physics and Soil Conservation Laboratory and the Chemistry and Soil Fertility Laboratory, Faculty of Agriculture, Sebelas Maret University, Surakarta. The study took place from September 2020 to March 2021. This study used an exploratory, descriptive method with a survey approach. The survey was carried out by direct observation in the field, and then soil samples were taken for analysis in the laboratory. Furthermore, analysis method used in this study is presented in table 1 below.

Table 1. Soil Physical, Chemical and Biological analysis

No	Parameter	Analytical method
1	Bulk Density	Pycnometer
2	Porosity	Sample ring
3	Textures	Pipet
4	pH	Electrometric
5	Available P	Olsen method
6	Total N	Kjeldahl
7	Available K	NH4Oac 1 N extraction
8	Soil organic Carbon (SOC)	Walkey and Black
9	Base saturation	Ammonium acetate extraction
10	CEC	Ammonium acetate method
11	C-biomass	Fumigation

The data obtained from the laboratory and field analysis results are then matched with the requirements for growing coconut plants. The determination of the sampling point was carried out by purposive sampling based on the Land Map Unit in the Kebonagung Sub-district. Data analysis used

Minitab 18. Minitab was used to select Minimum Data Sets (MDS), Principal Component Analysis (PCA), and correlation test. Selected data is an indicator that represents the closest soil condition and has the highest influence value on the soil quality index, the selected data then look for the weighting index (Wi) for each indicator of soil, the weighting index is obtained by multiplying the proportion by the cumulative. The selected data are then assessed based on the soil quality index [4–6]. Data analysis continued with the calculation of the soil quality index according to the formula below:

$$SQI = \sum_{n=1}^{n} Wi\, x\, Si \qquad (1)$$

Description :
SQI : Soil Quality Index
Si : Score for the selected indicator
Wi : Weighting index
n : Sum of indicator

3. Results and Discussion

The average Bulk Density (BD) of soil at the study site is about 1.46 g/cm3, soil with high BD causes lower porosity. Porosity in the study area ranged from 31.17% to 41.18%, including in bad-to-bad conditions. Clay textured soil has a higher water holding capacity because it has a more significant number of micropores; micropores function to store water in the soil. The number of pores plays a vital role in water retention and determines the movement of water in the soil. The higher the number of pores, the higher the ability of the soil to provide water availability.

Table 2. Soil physical, chemical, and biological properties for each land map unit

LMU	pH	SOC (%)	Tot-N (%)	Av-P (ppm)	Av-K (cmol(+)/kg)	CEC (cmol(+)/kg)	BS (%)	BD	Porosity	C-bio
1	5,23	1,98	0,21	6,93	0,14	23,23	51,66	1,37	36,70	0,05
2	5,97	1,47	0,31	15,95	0,54	20,73	59,22	1,59	31,68	0,02
3	6,99	2,15	0,34	16,67	0,74	43,10	26,18	1,43	36,69	0,18
4	6,37	2,92	0,38	16,18	0,79	28,11	50,64	1,44	35,99	0,18
5	6,41	2,49	0,16	16,86	0,63	20,51	54,96	1,52	32,96	0,17
6	6,17	1,64	0,15	16,42	0,51	25,85	53,27	1,38	39,14	0,04
7	5,38	3,67	0,25	8,08	0,47	22,14	54,93	1,45	37,31	0,30
8	6,62	1,48	0,35	18,02	0,69	23,59	52,59	1,54	31,17	0,09
9	6,43	1,15	0,27	16,81	0,64	42,91	25,83	1,36	41,18	0,07
10	5,99	1,06	0,28	17,01	0,55	41,11	20,59	1,52	33,12	0,08

CEC (Cation Exchange Capacity) ranges from 20.73 to 43.10 cmol (+)/kg, CEC value is also influenced by clay and organic matter content. Soils with a high clay content that dominates colloids will have a higher binding capacity and cation exchange value than sandy or silty soils. N ranged from 0.16 – 0.21%, classified as moderate to high. P-available went from 6.93 – 18.02 ppm, [7] explained that P in the soil could be sourced from P-inorganic and organic P, the availability of P is closely related to soil use, fertilization, and organic. Affairs. Potassium (K) is an essential macronutrient for plants, and potassium plays a role in metabolic processes and enzyme activation. Potassium at the study site ranged from 0.14 to 0.79. The low content of available K was caused by very high K fixation.

Organic matter is vital in improving soil's physical, chemical, and biological characteristics, research results [1,8,9] prove that applying organic matter can increase soil pH depending on the level of maturity and the type of soil. Adding immature organic fertilizers will slow down the process of increasing soil pH because the organic matter is still not appropriately decomposed and releases organic acids. C-organic at the study site ranged from 1.15 to 3.67, which low to moderate. Microbial biomass is a very sensitive indicator of soil changes, [10] stated that C-microbial biomass is very sensitive to changes in the natural cycle of the soil caused by land use, tillage, and fertilization. C-Microbial biomass is also closely related to the chemical properties of the soil (availability of N and P nutrients) and the physical properties of the soil (aeration). The analysis showed that the C-biomass of microbes at the research site ranged from 0.02 to 0.18, which was very low.

The first stage of soil quality assessment is the Pearson Correlation Test, which aims to obtain correlations between physical, chemical, and biological soil indicators. The correlation test is carried out at the 5% level ($\alpha = 0.05$), correlated indicators have a p-value of less than 0, 05 and the Pearson correlation value (r) is close to 1 or -1 [11]. The correlation of pH with available-P, Total-N, and available -K was explained as the availability of N, P, and K in the soil at a specific pH range. Low available P is a common problem. [12] explains that the maximum availability of P is found in the pH range between 5.5 - 7. The availability of P decreases at a soil pH lower than 5.5 or higher than 7 due to P fixation by acidic or basic cations, at low pH, the availability of P is lower this is because P in the soil is in a fixed form in the form of Al-P, Fe-P, Ca-P, and P-bound [7,13,14]

The availability of K is affected by the saturation of Ca^+ and Al^{3+}, in the acid soil reaction, the soil colloid will be saturated with acidic cations (Al^{3+}), while in the alkaline soil reaction, the colloidal soil will be saturated with cations (Ca^+). Thus, K can be adsorbed by Ca^+ and Al^{3+} so that the available K becomes limited. The negative correlation of Bulk density (BD) with porosity is explained as an inversion, the higher the BD, lower the porosity, porosity is the proportion between the number of soil pores and the weight of the soil volume. The positive correlation of soil organic carbon with C-Biomass is explained the role of carbon as source of nutrition to support the activity of microorganisms, the higher the levels of C-organic, the higher the C-microbial. The abundance of organic matter that has undergone a decomposition process and produces organic C will increase microbial C.

The stage of soil quality assessment is determining the indicator with the highest sensitivity, carried out through Principal Component Analysis (PCA), and weighting index for each selected data, results of the soil quality assesment are presented in table 3 below

Table 3. Soil quality index

MDS	Wi	Si									
		1	2	3	4	5	6	7	8	9	10
pH	0,194	2	4	5	5	5	5	3	5	5	4
Av-P	0,194	1	3	3	3	3	3	1	3	3	3
Av-K	0,194	2	3	4	4	4	3	3	4	4	3
BD	0,267	4	2	4	4	2	4	3	2	4	2
BS	0,267	4	4	2	4	4	4	4	4	2	2
C-bio	0,111	1	1	1	1	1	1	1	1	1	1
SOC	0,111	3	3	3	3	3	3	4	3	3	3
CEC	0,124	3	3	1	4	3	3	3	3	1	1
	SCI	3,92	4,36	4,50	5,40	4,74	5,08	4,15	4,74	4,50	3,58
	Class	2	2	2	2	2	2	2	2	2	2

Soil quality index in Kebonagung Sub-district ranges from 0.39 to 0.58, which is medium class. Medium class can be defined as moderate soil quality where the land can be used for the development of certain commodities supported by the best production inputs, as well as land management in accordance with the limiting factors of soil quality. In the strategic plan for coconut development, information about the level of land suitability in each Land Map Unit is needed.

The land suitability class for coconut plants is determined using land characteristics such as temperature, water availability, oxygen availability, rooting media, nutrient retention, available nutrients, erosion hazard, flood hazard, and land preparation. The results of the analysis of land characteristics were then matched with the growing requirements of coconut plants based on the land evaluation guidelines of the ministry of agriculture. The land suitability class for coconut development in Kebonagung Sub-district is in the S2 (fairly appropriate) and S3 (marginally appropriate) classes. The level of land suitability of S2 and S3, the focus of coconut development is based on the limiting factors, at the level of land suitability of S2 the limiting factors include tc (temperature), wa (Dry season and Rainfall), and na (Available P). Temperature, Rainfall and Dry season are permanent limiting factors that are difficult to repair, generally improvements can be made by modifying the microclimate. Available P is not a permanent limiting factor, so it can be improved through balanced inorganic fertilization manure, or compost. Efforts to improve the availability of P will increase the land suitability status from S3 (marginally appropriate) to S2 (fairly appropriate).

Soil quality is an important environmental factor to be managed and must be considered about land use. Intensive land use and ignoring conservation rules are the main causes of land degradation, therefore it is necessary to study the correlation between land suitability and soil quality index as a reference in determining limiting factors and potentially producing large productivity gaps. The following table presents the results of the correlation analysis between SQI and land suitability:

Table 4. Correlation between land suitability and SQI

		SQI
Land suitability	Pearson Correlation	0,768
	P-Value	0,004

Based on the table 9, SQI has a significant relationship with land suitability (r=0.768, p=0.004). The correlation results show that the higher the SQI value, would better the land suitability. This finding shows that the proper management of land suitability limiting factors in each LMU will also improve the soil quality index, an in-depth study of the limiting factors of soil quality index and land suitability is presented in table 5 below

Table 5. Relationship between SQI and land suitability limiting factor

LMU	Low MDS Score	Limiting factor
1	Availability P	S3, Availability P
2	Bulk Density	S2, Availability P
3	Base saturation	S2, Textures, Availability P, Erosion
4	SOC and Availability P	S3, Availability P
5	Bulk Density	S2, Textures, Availability P
6	Availability P, Availability K, SOC, CEC	S3, Textures, Availability P
7	Availability P	S3, Availability P
8	Bulk Density	S2, Erosion
9	CEC	-
10	CEC	-

The low MDS score for each LMU reflects the indicator with the lowest score so that it must be managed appropriately to increase the soil quality index and the carrying capacity of the land for the development of coconut commodities. In some LMU Bulk Density and texture are limiting factors, soils with a sandy texture generally have low CEC and BS, causing low nutrient availability. Bulk density that is too high causes the soil to become compact and difficult to cultivate, this is due to improper land management such as maximum tillage and no addition of organic matter. Rainfall factor also affects the increase in bulk density, ground cover plants are helpful for protecting soil aggregates from raindrops, suppressing erosion, accelerating water infiltration rate, and increasing biological activity and soil fertility. Ground cover plants that are often used are Calopogonium caeruleum and Arachis pintoi, a consideration in choosing ground cover plants for coconut is the nature of plant propagation so that it can interfere with the growth and development of the main crop and competition for nutrients.

For specific nutrient improvement such as P, Ca, and K, dolomite or other agricultural lime can be the solution. Dolomite is a soil enhancer often used by farmers to increase soil pH. Dolomite can increase the availability of basic cations, especially Ca^{2+} and Mg^{2+}, so it is expected to reduce the concentration of H^{+} and the concentration of Al^{3+} and Fe^{2+} so that P can be available in the soil solution.

4. Conclusion

The soil quality index in Kebonagung Sub-district is included in the medium class, the results of the correlation test between the soil quality index and land suitability showed a significant relationship ($r = 0.768$, $p = 0.004$), these results indicate that the higher the SQI value, the better the land suitability. From the results of a comprehensive study, it can be concluded that for the development of coconut commodities, efforts to achieve optimal productivity levels are through recommendations for appropriate improvements according to limiting factors, limiting indicators based on correlation tests are Bulk Density, pH, available-P, Soil Organic Carbon, and CEC. Suggested improvement recommendations include cover crops, dolomite, manure, and balanced P fertilization. Suggested improvement recommendations include cover crops, dolomite, manure, and balanced P fertilization.

References

[1] Irawan S and Antriyandarti E 2021 Physical deterioration of soil and rice productivity in rural Java *J. Phys. Conf. Ser.* **1825**

[2] Ministry of Agriculture 2021 *Statistical of National Leading Estate Crops Commodity* (Jakarata: Ministry of Agriculture)

[3] Supriyadi S, Vera I L P and Purwanto P 2021 Soil Quality at Rice Fields with Organic, Semi-organic and Inorganic Management in Wonogiri Regency, Indonesia *Caraka Tani J. Sustain. Agric.* **36** 259

[4] Chandel S, Hadda M S and Mahal A K 2018 Soil Quality Assessment Through Minimum Data Set Under Different Land Uses of Submontane Punjab *Commun. Soil Sci. Plant Anal.* **49** 658–74

[5] Andrews S S and Carroll C R 2001 Designing a soil quality assessment tool for sustainable agroecosystem management *Ecol. Appl.* **11** 1573–85

[6] Andrews S S, Karlen D L and Mitchell J P 2002 A comparison of soil quality indexing methods for vegetable production systems in Northern California **90** 25–45

[7] Hartati S, Minardi S, Hartatik W and Haniati I L 2018 The Effects of Inorganic Fertilizer and Mineral Leucite Residues on K Uptake and Maize Yields (Zea mays L.) in Oxisols *SAINS TANAH - J. Soil Sci. Agroclimatol.* **15** 115

[8] Irawan S and Antriyandarti E 2021 BIOTRICO: A Breakthrough Fertilizer for Sustainable Agriculture *IOP Conf. Ser. Earth Environ. Sci.* **940**

[9] Irawan S and Antriyandarti E 2021 Fertilizer application, climate change and rice production in rural Java *IOP Conf. Ser. Earth Environ. Sci.* **755**

[10] Kusumawati I A and Prayogo C 2020 Impact of Land Use Change at UB Forest on Microbial Biomass Carbon and Total Bacterial Population *J. Sumberd. lahan* **6**

[11] Supriyadi S, Septiana R, Aktavia H and Purwanto P 2018 Soil Quality Assesment of The Rained Lowland Ricefields Under Organic and Conventional Farming System in Kaliwungu (Central Java) *Polish J. Soils Sci.* **1**

[12] Firnia D 2018 Dinamika Unsur Fosfor Pada Tiap Horison Profil tanah Masam *Jur. Agroekotek* **10** 118–38

[13] Minardi S, Hartati S, Widijanto H and Wulandari D 2016 The Use of Extract Banana Corm and Phosphate Rock to Increase Available-P in Alfisols *Sains Tanah - J. Soil Sci. Agroclimatol.* **12** 76

[14] Minardi S, Harieni S, Anasrullah A and Purwanto H 2017 Soil Fertility Status, Nutrient Uptake, and Maize (Zea mays L.) Yield Following Organic Matters and P Fertilizer Application on Andisol *IOP Conf. Ser. Mater. Sci. Eng.* **193**

Ensuring Environmental Sustainability of Infrastructure Projects: A Perspective Review in A Case Of Indonesia

Alti Kusmayadi[1], Raldi Hendro Koestoer[1]*
[1]School of Environmental Science, Universitas Indonesia, Jakarta, 10430, Indonesia.
[1]nurmuhariaty@gmail.com, [1]*raldy.hk@ui.ac.id

Abstract. Infrastructure is crucial to the development process and is intimately related to productivity, economic growth, and well-being. Infrastructure also plays a vital role in environmental contamination and degradation which leads to high financial cost to recover. This article aims to highlight how a green bond is a potential tool to be used as an alternative source of financing for infrastructure development which can also ensure environmental sustainability in Indonesia. Using a literature review from various available sources, green bond positively impacts environmental sustainability. Indonesia is on the right track in directing future development to maintain the balance between economic growth and environmental sustainability. The green bond ecosystem also has been established to support the growing market. The green bond is a promising financing tool in Indonesia due to its attention to the environmental aspect, the established framework and regulation, and its opportunity to finance potential green projects available in Indonesia.

1. Introduction

Infrastructure is crucial to the development process and is intimately related to productivity, economic growth, and well-being [1]. However, environmental effects of infrastructure development, such as those on biodiversity, water management, and land quality, can significantly impact people's quality of life [2]. Infrastructure has a vital role in environmental contamination and degradation, which has high financial costs to recover the environmental damages; neglecting environmental concerns results in decline in environmental quality while promoting rapid growth [3].

In Indonesia, infrastructure development is one of the main directions of the President, as stated in the National Medium-Term Development Plan (*Rencana Pembangunan Jangka Menengah Nasional*/RPJMN) 2020-2024. It aims to connect production areas with distribution areas, facilitate access to tourist areas, increase new job opportunities, and accelerate the increase in added value to the people's economy [4]. The RPJMN 2020 – 2024 also states that the direction for future development is to maintain the balance between economic growth, emission reduction targets, and intensity, as well as the carrying capacity of natural resources and the carrying capacity of the environment today and in the future. Thus, Indonesia is on the right track to implementing green economy.

Given how urgent it is for governments to develop infrastructure, many may have trouble finding funding to keep up with demand [5], including Indonesia. The amount of infrastructure spending required to meet the objective of GDP growth as targeted in the RPJMN 2020–2024 is IDR 6,445 trillion, while the government can only afford IDR 2,385 trillion (only 37 percent of the total need) [4]. To encourage public and corporate entities to invest in infrastructure development, innovative initiatives are required. To direct funding towards initiatives that assist the shift to a low-carbon economy, increase resource efficiency, and lessen environmental impacts, a variety of cutting-edge financial instruments and vehicles have been created [1]. For instance, corporations, national and subnational governments, and development banks have issued more and more green bonds to attract private funding for green initiatives. In 2014, Indonesia developed Sustainable Finance Roadmap, which includes the need to develop a framework and regulations on green bonds [6]. The Indonesia green bond market is growing but is still in the early stage.

Even though green bonds have been increasingly issued, according to ADB research in 2021 [7], there are only 2% or fewer green bonds in the whole bond market. Indonesia's green bond market is overgrowing but is still in its early stage. This condition could become an obstacle for Indonesia in achieving sustainable infrastructure development. Green bond has the potential to be utilized in Indonesia to balance the need for infrastructure development and maintain environmental sustainability, as stated in the RPJMN 2020 – 2024. This article aims to highlight how green bond is a potential tool to be used as an alternative source of financing for infrastructure development which can also ensure environmental sustainability in Indonesia.

2. Methods

This article is developed using a qualitative method through a systematic literature review. The first step is gathering all the relevant literature and data on the impact of green bonds and environmental sustainability from journals and international reports. The second step is to review the gathered literature and data, including the Indonesian infrastructure development issues and the green bonds ecosystem in Indonesia. The third step analyzes how to ensure the environmental sustainability of infrastructure projects in Indonesia by utilizing green bonds.

3. Results and Discussion

As previously explained, infrastructure development requires a large amount of capital and financing. On the contrary, infrastructure development has a negative impact on the environment. How can the financial aspect of infrastructure development ensure environmental sustainability? The research found that the financial sector can provide financial support for businesses and projects that promote environmental protection. In parallel, financial development can encourage the modernization of industrial structures, which is crucial in lowering energy consumption and carbon emissions [8,9,10]. More fundamental research about green finance's role in promoting environmental protection has been conducted, including how green finance delivers to the environment [11], how to increase private green capital's participation in environmental protection projects [12,13], and government involvement in the growth of green financing [14]. Green finance does not directly affect the environment but could support environmentally friendly enterprises and projects, thereby improving environmental quality [15].

One of the strategies for raising funds for sustainable infrastructure development, for example, to meet the Paris Agreement's carbon reduction goals, is using green bonds [16]. By 2030, 4.7 Gt of carbon dioxide emissions can be avoided if green bonds are implemented as planned, and due to the increase in green financing, the share of non-fossil energy power will rise from 42 to 46 percent [17].

Indonesia still needs alternative financing outside the government budget for its infrastructure development. To achieve the GDP growth target in the 2020-2024 RPJMN, the need for infrastructure spending reaches IDR 6.445 trillion, while the government's ability to fund is only IDR 2.385 trillion (only 37 percent of the total needs). As stated previously, the future development direction should consider the balance between economic growth and environmental sustainability. For this reason, innovative efforts are needed to encourage innovative financing schemes that could support economic growth and environmental sustainability.

Regarding green bonds, Indonesia is considered one of the top three green bond issuing countries in ASEAN, alongside Singapore and Malaysia [18]. The Indonesian government's concrete steps to support the green bond concept are shown by the issuance of the Sustainable Finance Roadmap

2015-2019 issued by the Financial Services Authority (*Otoritas Jasa Keuangan*/OJK) on December 5, 2014. OJK also issued regulations regarding green bonds through OJK Regulation (POJK) Number 60/POJK.04 /2017 concerning Issuance and Requirements for Green Bonds. According to POJK, green bonds define as debt security whose issuance proceeds are used to finance or refinance all or part of Environmentally Friendly Business Activity *(Kegiatan Usaha Berwawasan Lingkungan*/KUBL*)*. KUBL is defined as business activities and/or other activities aimed at protecting, repairing, and/or improving the quality or function of the environment. The eligible sectors that can be financed through green bond issuance according to POJK 60/2017 are:

- Renewable energy
- Energy efficiency
- Pollution prevention and control
- Management of natural resources and sustainable land use
- Conservation of terrestrial and aquatic biodiversity
- Environmentally friendly transportation
- Sustainable water and wastewater management
- Climate change adaptation
- Products that can reduce resource use and cause less pollution (eco-efficient)
- Environmentally friendly buildings
- Business activities and/or other environmentally friendly activities.

Indonesia's Ministry of Finance also issued the framework of green bonds to support the green bond ecosystem in Indonesia. Indonesia's green bond framework consists of four main principles, which follow the international Green Bond Principles developed by the International Capital Market Association (ICMA). The main principles are compiled in Table 1.

Table 1. Indonesia's primary principles of the green bond [19]

Green Bond Principles	Explanation
Use of proceeds	At least 70% of the green bond proceeds must be used exclusively to finance and/or refinance eligible green projects.
Project evaluation and selection	The National Development Planning Agency and the Ministry of Finance will review and approve projects/budget allocation/subsidies to be included in the State Budget.
Management of proceeds	Each green bond's proceeds will be managed in the government's general account in line with a solid and responsible treasury management strategy.
Reporting	A green bond report shall be created and published by the Ministry of Finance every year, initially on the day that is no more than a year after the green bond was issued.

Indonesia's first corporate green bond was issued by PT. Sarana Muti Infrastruktur (PT. SMI) in 2018. It was part of the Sustainable Public Offering *(Penawaran Umum Berkelanjutan*/PUB*)* worth IDR 3 trillion with an emission value of IDR 500 billion in Phase I. PT. SMI is a State-Owned Enterprise (SOE) whose entire capital is owned solely by the Republic of Indonesia under the Ministry of Finance according to Minister of Finance Regulation No. 100/PMK.010/2009 concerning Infrastructure Financing Companies. In carrying out its role, PT. SMI is committed to sustainable development by providing financing facilities for various environmentally friendly infrastructure projects in Indonesia. The proceeds from the green bond issuance were used to finance two renewable energy projects and one clean transportation project with a potential estimated environmental impact of 193,928-ton CO_2e greenhouse gas emissions avoided during 2021 [20].

Several companies have issued green bonds in Indonesia, including Star Energy Geothermal (Wayang Windu) Ltd green bonds with a principal amount of USD 580 million, Star Energy Geothermal (Darajat II) Ltd green bonds with a principal amount of USD 1.1 billion, and the Tropical Landscape Finance Facility Pte (TLFF) green bonds with a principal amount of USD 96 million [21]. While Star Energy uses the green bond proceeds to finance renewable energy projects, TLFF uses it to finance a natural rubber plantation in two Indonesian provinces concentrating on the reforestation of severely degraded land. The targeted environmental impact of the TLFF green bond is 27,000 hectares of land for the conservation and protection of wildlife [22]. Not only does the use of green bonds positively impact the environment, but the social aspects of the project being funded are also positively impacted, for example, in PT. SMI's Green Bond Report 2022 states that one of their mini-hydro renewable energy projects has other positive impacts that benefit society. It has the potential to contribute to SDG 7's goal of providing access to clean and affordable energy for 41,676 families by reducing oil imports by USD 1,804,813 in 2021, training 18 employees, distributing IDR 143 million in Corporate Social Responsibility (CSR) funding, and employing 40 people [20].

From the perspective of potential green projects, Indonesia has many potential green projects available that green bonds can fund. The potential projects are available in various sectors, such as low carbon and/or sustainable transport, renewable energy, sustainable water management, and sustainable waste management. For example, the Indonesian Electricity Company (*Perusahaan Listrik Negara*/PLN) has developed a pipeline of renewable energy projects to meet Indonesia's target of net zero emission. The pipeline stated in the Electricity Supply Business Plan (*Rencana Usaha Penyediaan Tenaga Listrik*/RUPTL) 2021-2030. The potential of renewable energy in Indonesia is shown in Table 2.

Table 2. Indonesia's renewable energy potential [23]

Energy Source	Potential (MW)	Installed Capacity (MW)	Utilization (%)
Geothermal	29,544	1,438.5	4.9
Hydro	75,091	4,826.7	6.4
Mini-micro-Hydro	19,385	197.4	1.0
Bioenergy	32,654	1,670	5.1
Solar	207,898 (4.80 kWh/m2/day)	78.5	0.04
Wind	60,647 (> 4 m/s)	3.1	0.01
Sea Wave	17,989	0.3	0.002

Another potential green project is sustainable water management, as stated in RPJMN 2020-2024. For example, eighteen multipurpose dams are planned to be built to ensure water supply availability, electricity supply for the society, and water supply for irrigation to support food security.

There are many options to finance potential green projects, and green bonds can be one of them. It can impact economic growth as the result of infrastructure development and ensure environmental sustainability as the result of green bond use of proceeds. The potential green projects, as mentioned above, are stated in the medium-long term government documents such as RPJMN 2020-2024 and RUPTL 2021-2030. Therefore, one could encourage using creative financing tools such as green bonds to attract local and international potential investors.

4. Conclusion

Green bonds can be used as an alternative for infrastructure financing that balances economic and environmental aspects. From the literature review, several studies state that there is a relationship between green bonds and environmental sustainability, although they do not contribute directly. How can green bonds help to ensure the environmental sustainability of infrastructure projects in Indonesia? As discussed above, several points are worth noting. First, the concept of green bond itself is to finance eligible green infrastructure projects. The concept indicates that environmental sustainability is a priority and the project's main feature. Second, the established framework and regulation enable the stakeholders to focus on the project's environmental aspects. Failing to fulfill the requirements and principles may lead to various sanctions, from administrative sanctions and penalties to cancellation of the approval and registration. Third, the potential green infrastructure available in Indonesia may bring opportunities for green bonds and, therefore more sustainable environment both regionally and nationally. Suggestion for further research is to conduct a projection of green bond growth and how its impact on the environment in Indonesia.

Acknowledgments
The authors wish to acknowledge assistance or encouragement from colleagues and direction from the co-author.

References

[1] OECD 2019 *OECD Reference Note on Environmental and Social Considerations in Quality Infrastructure* p 2
[2] IEG World Bank 2007 *The Nexus Between Infrastructure and Environment* p 1
[3] Phimphanthavong H 2013 The Impacts of Economic Growth on Environmental Conditions in Laos *Int.J.Buss.Mgt.Eco.Res.* Vol **4(5)** 766-774
[4] National Development Planning Agency (Bappenas) 2020 Presidential Decree No. 18 Year 2020. Natl Mid-Term Dev Plan 2020-2024
[5] UNOPS 2019 *The critical role of infrastructure for the Sustainable Development Goals* p 12
[6] Otoritas Jasa Keuangan 2014 *Roadmap Keuangan Berkelanjutan di Indonesia 2015-2019.*
[7] Asian Development Bank 2021 *Policy Support in Promoting Green Bonds in Asia* (1275):75–75.
[8] Chang SC 2014 Effects of financial developments and income on energy consumption *Int Rev Econ Financ* **35**:28–44
[9] Mahdi Ziaei S 2015 Effects of financial development indicators on energy consumption and CO2 emission of European, East Asian and Oceania countries *Renew Sustain Energy Rev* **42**:752–9
[10] Nasreen S, Anwar S, Ozturk I 2017 Financial stability, energy consumption and environmental quality: Evidence from South Asian economies *Renew Sustain Energy Rev* **67**:1105–22.
[11] Wang Y, Zhi Q 2016 The Role of Green Finance in Environmental Protection: Two Aspects of Market Mechanism and Policies *Energy Procedia* **104**:311–6
[12] Ruiz JDG, Arboleda CA, Botero S 2016 A Proposal for Green Financing as a Mechanism to Increase Private Participation in Sustainable Water Infrastructure Systems: The Colombian Case *Procedia Eng.* **145**(May):180–7.
[13] Taghizadeh-Hesary F, Yoshino N 2019 The way to induce private participation in green finance and investment *Financ Res Lett.* **1**;31:98–103.
[14] Owen R, Brennan G, Lyon F 2018 Enabling investment for the transition to a low carbon economy: government policy to finance early stage green innovation *Curr Opin Environ Sustain* **31**:137–45
[15] Zhou X, Tang X, Zhang R 2020 Impact of green finance on economic development and environmental quality: a study based on provincial panel data from China *Environ Sci Pollut Res.* **27**(16):19915–32.
[16] Gianfrate G, Peri M 2019 The green advantage: Exploring the convenience of issuing green bonds. *J Clean Prod.* **219**:127–35.
[17] Glomsrød S, Wei T 2018 Business as unusual: The implications of fossil divestment and green bonds for financial flows, economic growth and energy market *Energy Sustain Dev* **44**:1–10.
[18] Azhgaliyeva D, Kapoor A, Liu Y 2020 Green bonds for financing renewable energy and

energy efficiency in South-East Asia: a review of policies *J Sustain Financ Invest* **10**(2):113–40
[19] Kementerian Keuangan Republik Indonesia 2018 *THE REPUBLIC OF INDONESIA GREEN BOND AND GREEN SUKUK FRAMEWORK:* djppr.kemenkeu.go.id
[20] PT Sarana Multi Infrastruktur 2022 *Laporan Obligasi Berwawasan Lingkungan 2022*
[21] CBI 2018 *Green Infrastructure Investment Opportunities Indonesia* 32
[22] Convergence 2019 *Tlff Inaugural Transaction : Corporate Sustainability Bond for Natural Rubber Production* 1–5.
[23] PT. Perusahaan Listrik Negara 2019 *Rencana usaha penyediaan tenaga listrik 2021-2030*

What Should We Do with the Oil Palm Solid Waste?

Illyin Abdi Budianta[1], and Misri Gozan[1]*
[1]Biorefinery Lab., Department of Chemical Engineering, Faculty of Engineering, Universitas Indonesia, Kampus Baru UI, Depok, 16424, Indonesia
[1]illyin.a.b@gmail.com, [1*]mgozan@ui.ac.id

Abstract. The valorisation of oil palm solid wastes is always a fascinating subject. The palm oil industries produce millions of tons of these high lignocellulose materials in empty fruit bunch, palm kernel shells, fronds, and others. The technological considerations are always based on selecting the best pretreatment methods, fermentation, and chemical processing technologies. Nevertheless, the selected method faces multiple concerns, such as practical, environmental, and economic ones. For example, can we get highly value-added products from these resistant materials without compromising the environment? This mini-review paper discusses various alternatives for solid waste treatment of solid waste to produce high-added-value materials by collecting the previous study from other publications. The development of oil palm biomass biorefinery is expected to increase the sustainability and profitability of the palm oil industry. Further research related to combined technology, economic assessment, and life cycle analysis is required to show the feasibility of the available technology.

1. Introduction
Oil palm biomass is a natural material that contains lignocellulose which is very widely available and can be a low-cost raw material for conversion into higher-value products. The journey from palm fresh fruit bunches to various products and wastes is depicted in Figure 1.

Oil Palm Empty Fruit Bunches (OPEFB) are a by-product of sterilisation and milling fresh fruit bunches. Mesocarp fibre (MF) is produced from mesocarp and pericarp oil extraction. Palm Kernel Shell (PKS) is a by-product of crushing palm kernels. In 2030, Indonesia estimated that production of OPEFB, MF, and PKS would reach 54, 31, and 15 million tons per year, respectively [1]. Oil palm fronds (PS) and oil palm trunks are the leaves of oil palm trees cut down from oil palm plantations to keep the plants productive. It is estimated that oil palm plantations yielded 115 million tons per year frond and 59.7 million tons per year trunk [1, 2]. Palm Oil Mill Liquid Waste (POME) is a liquid by-product of sterilisation and oil extraction, reaching 141.9 million tons annually in Indonesia [1].

This paper aims to elaborate on the various uses of oil palm biomass. This is an important step to understanding the valorization of oil palm biomass, as different purposes or products may require a very different approach to pretreatment, operation and process conditions.

2. Method
This mini-review paper is performed by collecting the previous study from other publications. In the beginning, different references were collected to discuss the products of oil palm empty fruit bunches (OPEFB) conversion. The same discussion will follow to products of palm kernel shell (PKS), palm trunk, mesocarp and frond conversion. The content discuss also the utilisation and products of those solid waste.

3. Results and Discussion

Around 10% of the solid waste of the palm oil industry is used as boiler fuel and fertiliser, while the rest is not utilised [3-5]. In addition, PKS has been used commercially as boiler fuel for other industries such as cement processing, clay brick making, rubber glove production, and kernel crusher plant [6]. Mixing oil palm biomass and fossil fuels allows for lowering the cost of the combustion process).

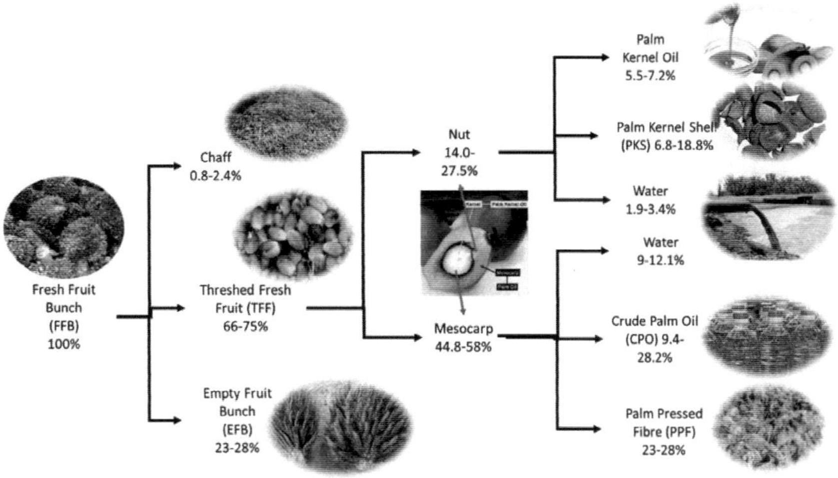

Figure 1. Products and wastes from the palm oil production process (modified from [7])

The main content of palm oil waste is presented in Table 1

Table 1. Composition of Cellulose, Hemicellulose, Lignin, and Ash from Oil Palm Biomass.

Part of waste	Cellulose	Hemicellulose	Lignin	Ash	Ref.
OPEFB	34.0–40.4	17.2–22.4	23.1–29.6	5–6.5	[8]
MF	23.0–28.8	25.3–30.5	25.5–28.97	2.6–5.8	[9]
PKS	20.8–27.2	21.6–22.7	44–50.7	8.6–16.3	[9]
OPF	31.0–42.8	12.5–22.5	15.2–25	5–5.8	[10]
Trunk	40.3 –50.78	18.7 –30.36	17.9 –26.8	2.4 –2.9	[11]

3.1. Products of Oil Palm Empty Fruit Bunches (OPEFB) Conversion

OPEFB can be processed into activated carbon, recycled paper additives, biochar, paper pulp, biocomposites, polymer graphs, graphene nanosheets, lignin graft copolymers, and composite boards. The processes for synthesising some precious products (e.g., furfural and levulinic acid) are described in Table 2.

Table 2. Products of OPEFB Conversion.

Products	Conversion methods	Ref.
Aromatics	Alkali pretreatment is followed by lignin recovery from black liquor and nitrobenzene oxidation from lignin.	[12]
Acetone–Ethanol–Butanol (ABE)	Simultaneous saccharification and fermentation of pre-alkaline biomass by *Clostridium acetobutylicum* ATCC 824.	[13]
Polylactic Acid	*Candida rugosa* Lipase as a Biocatalyst for L-lactide Ring-Opening Polymerization	[14]
Biochar	Torrefaction at 230 - 270 °C for 40-60 min. Microwave-assisted pyrolysis	[15]

Polyhydroxybutirate	Hydrolysates polymerisation by Bacillus cereus Quaeda B-001	[16]
Paper Pulp	Pulping by diethanolamine, followed by bleaching	[17]
Lactide Acid	Hydrothermal treatment with Pb(II) ion catalyst. Acid hydrolysis and fermentation by *Bacillus coagulans* JI12	[18, 19]
Poli-of Lignin	Liquefaction with polyethylene glycol and glycerol, and sulfuric acid catalyst	[20]
Bio-oil	Microwave-assisted pyrolysis. Hydrothermal Supercritical Liquefaction at 390 °C and 25 MPa for 1 h.	[15, 21]
Cellulase	Alkaline Pretreatment. Fermentation by *Thermobifida fusca* and *E. coli* EgRK2 recombinant	[22-24]
Lignin Graft Copolymer	Acidification of highly alkaline black liquid for extraction of Kraft lignin (KL), KL grafted with acrylic acid (AA) using p-toluene sulphonic acid,	[25]
Graphene Nanosheet	Hydrothermal pretreatment of biomass at 250 °C for 20 minutes, followed by pyrolysis of hydrocarbon-	[26]
p-Hydroxy-Benzoylated lignin	Successive extraction with water, ethanol, acetone, and chloroform, followed by grinding	[27]
Polybutylene Adipate-co-terephthalate (PBAT)	Melted biomass mixing and PBAT with succinic anhydride and di cumyl peroxide	[28]
Microcrystalline Cellulose (MCC)	Total chlorine-free pulping and bleaching (TCF) of OPEFB-pulp, followed by acid hydrolysis, MCC hydrolysed by 64% H_2SO_4	[29]
Cellulose Phosphate	Biomass water hydrolysis, followed by soda-anthraquinone pulping, Oxygen-Ozone-Peroxide bleaching process and acid hydrolysis, Cellulose phosphorylated by H_3PO_4 / P_2O_5 / Et_3PO_4 / hexanol method	[30]
Biomethane	Cofermentation with activated sludge from POME under mesophilic and fungal conditions	[31]
Biodiesel	Pretreatment with dilute sulfuric acid, followed by enzymatic hydrolysis of the solid residue. Cultivation with Mucor plumbeus on residual enzymatic hydrolyzate (EFBEH) and liquid hydrolyzate (EFBLH) pretreatment.	[8]
Bioethanol	Alkaline-pretreated and simultaneous saccharification and fermentation (SSF) by adding *Saccharomyces cerevisiae* and cellulose enzyme or encapsulated Rhizopus oryzae.	[32, 33]
Biohydrogen	Pretreatment by ammonia fibre expansion (AFEX). Fermentation by *Enterobacter* sp. KBH6958.	[34]
Formic Acid, Furfural & Levulinic Ccid	Alkali Pretreatment followed by acid hydrolysis using H_2SO_4.	[35, 36]
Rhamnolipid	Produced by *Pseudomonas aeruginosa* and OPEFB as substrate	[37]

3.2. Products of Palm Kernel Shell (PKS) Conversion

PKS can be processed into activated carbon, biochar, and biocomposite products. The conversion processes of POM into these products are described in Table 3.

Table 3. Products of Palm Kernel Shell Conversion

Products	Conversion methods	Ref.
Acetone–Ethanol–Butanol (ABE)	Fermentation by *Clostridium saccharoperbutylacetonicum* N1-4.	[38]
Activated Carbon	Chemical activation using KOH impregnation, Impregnation of biomass by $ZnCl_2$ followed by heating at 500 °C, and activation by carbonisation at 900 °C by CO_2.	[39]
Biochar-Bio oil	Hydrothermal carbonisation at 180 °C for 30 minutes. Microwave-assisted pyrolysis	[40, 41]
Polyester Biocomposite	Melting MCC powder blended with unsaturated polyester isophthalate and methyl ethyl ketone peroxide.	[42]
Phenol, Acetone, and Butanone	Catalytic cracking with FeOOH catalyst supported on ZrO_2	[43]

3.3. Products of Palm Trunk Conversion

Palm trunks can be processed into products such as activated carbon, biochar, and composite boards. The processes for converting oil palm trunks into some high-value-added products are described in Table 4.

Table 4. Products of Palm Trunk Conversion

Products	Conversion methods	Ref.
Syngas	Steam-assisted gasification at 800 °C.	[44]
L-lactide acid	Pretreatment by hot water, followed by a sprinkling of enzymes on the entire sludge undergoing the pretreatment. Fermentation by engineered Lactobacillus paracasei.	[45]
Biohydrogen	Microwave-assisted H_2SO_4 pretreatment. Fermentation by a microbial community consisting of Clostridium sp., Thermoaerobacterium sp., and Lactobacillus sp.	[46]
Cellulase and Xylanase	The biomass is ground to a fine powder. Solid state fermentation by Aspergillus fumigatus SK1.	[47]

3.4. Products of Palm Mesocarps Fibre (MF) Conversion

MF can be processed into various products, such as biodiesel and polymer grafts (poly butyl acrylate). The processes for converting MF into these products are described in Table 5.

Table 5. Products of Palm Mesocarps Fibre Conversion

Products	Conversion methods	Ref.
Graft polymer (poly (butyl acrylate))	Graft polymerisation using butyl acrylate as a monomer by H_2O_2 and ferrous ammonium sulfate,	[48]
Acetone–Ethanol–Butanol (ABE) and Polyhydroxy-butyrate	Acid hydrolysis and fermentation by *Clostridium beijerinckii* TISTR 1461. Clostridium acetobutylicum DSM 1731. Further fermentation is fermented by *Cupriavidus necator* for the production of polyhydroxybutyrate.	[49, 50]
Biomethane	Biological pretreatment by oyster mushroom followed by co-fermentation with cow dung and POME as inoculum	[51]

3.5. Products of Oil Palm Frond (OPF) Conversion

OPF can be processed into various products such as paper pulp, adsorbent, and composite board. The processes for converting OPF into these products are described in Table 6.

Table 6. Products of Oil Palm Frond (OPF) Conversion

Products	Conversion methods	Ref.
Paper Pulp	Pulping using NaOH and ethanol followed by hand sheet making,	[52]
2,3-Butanediol	Acid hydrolysis and fermentation by *Enterobacter cloacae* SG1.	[53]
Bioethanol	Pretreatment by the organosolv method catalysed by NaOH with ultrasonic assistance using ethanol. Hydrolysis of residue from pretreatment and fermentation by Saccharomyces cerevisiae	[54]
Biochar- Bio-oil	Pyrolysis at 500 °C for 60 min.	[55]
Gallic Acid (GA), Vanillic Acid (VA) and Quercetin (QU)	Pretreatment with the organosolv method was catalysed by NaOH using ethanol recovered from the pre-treated liquid with ultrasonic assistance.	[56]

4. Conclusion

The high cellulose and hemicellulose content in oil palm solid waste can be converted into various chemical, biochemical, and fine chemical products, such as biomaterials, fuels, animal feed, catalysts, and other high-quality chemicals. A series of pretreatment, physical, chemical, or biological processes are required to produce higher-value products. The strategies of the technologies determined the cost and quality of the products. Nevertheless, using oil palm biomass for various products has opened alternative solutions to environmental problems. The development of oil palm biomass biorefinery is expected to increase the sustainability and profitability of the palm oil industry. Further research related to combined technology, economic assessment, and life cycle analysis is required to show the feasibility of these available technologies.

Acknowledgement
The authors acknowledged Badan Pengelola Dana Perkebunan Kelapa Sawit (BPDPKS) for the funding of this study. This research was supported by The Ministry of Research and Higher Education of Indonesia's financial support through the research grant "Hibah Penelitian Disertasi Doktor" (PDD) 2022 (NKB-1012/UN2.RST/HKP.05.00/2022).

References

[1] Hambali E and Rivai M 2017 *IOP Conf. Ser.: Earth Environ. Sci.* IOP Publishing.
[2] Awalludin M F, Sulaiman O, Hashim R, and Nadhari W N A W 2015 *Renew. Sustain. Energy Rev.*. **50**: 1469-84.
[3] Dewanti D P 2018 *Jurnal Teknologi Lingkungan* **Vol. 19 No. 1**.
[4] Fitrianto A C, Darmawan A, Tokimatsu K, and Yoshikawa K 2019 *Energy Procedia*. **158**: 3565-71.
[5] Nasution M A, Herawan T, and Rivani M 2014 *Energy Procedia*. **47**: 166-72.
[6] Hosseini S E and Wahid M A 2014 *Renew. Sustain. Energy Rev*. **40**: 621-32.
[7] Ohimain E I, Izah S C, and Obieze F A 2013 *Adv. J. Food Sci. Technol*. **5** (3): 289-94.
[8] Ahmad F B, Zhang Z, Doherty W O, and O'Hara I M 2016 *Biofuel. Bioprod. Biorefin*. **10** (4): 378-92.
[9] Abnisa F, Arami-Niya A, Daud W M A W, and Sahu J N 2013 *Bioenergy Res*. **6** (2): 830-40.
[10] Tan J, Jahim J, Wu T, Harun S, and Mumtaz T 2016 *IOP Conf. Ser.: Earth Environ. Sci.* IOP Publishing.
[11] Lai L W and Idris A 2016 *Renew. Energy*. **99**: 738-46.
[12] Tang P-L, Hassan O, Maskat M Y, and Badri K 2015 *Biomed Res. Int*. **2015**.
[13] Ibrahim M F, Abd-Aziz S, Yusoff M E M, Phang L Y, and Hassan M A 2015 *Renew. Energy*. **77**: 447-55.
[14] Rahmayetty, Whulanza Y, Sukirno, Rahman S, Suyono E, Yohda M, and Gozan M 2018 *Biocatal. Agric. Biotechnol*. **16**.
[15] Salema A A and Ani F N 2012 *Bioresour. Technol*. **125**: 102-7.
[16] Yustinah, Hidayat N, Alamsyah R, Roslan A M, Hermansyah H, and Gozan M 2019 *Biocatal. Agric. Biotechnol*. **18**: 101019.
[17] Jiménez L, Serrano L, Rodríguez A, and Ferrer A 2009 *Bioresour. Technol*. **100** (3): 1478-81.
[18] Ye L, Hudari M S B, Zhou X, Zhang D, Li Z, and Wu J C 2013 *Appl. Microbiol. Biotechnol*. **97** (11): 4831-8.
[19] Chin S X, Tasirin S M, Chan C H, Chia C H, Chook S W, Zakaria S, and Sajab M S 2016 *BioResources*. **11** (1): 2186-201.
[20] Faris A H, Ibrahim M N M, Rahim A, Hussin M H, and Brosse N 2015 *BioResources*. **10** (4): 7339-52.
[21] Chan Y H, Yusup S, Quitain A T, Uemura Y, and Sasaki M 2014 *J. Supercrit. Fluids*. **95**: 407-12.
[22] Harun N A F, Baharuddin A S, Zainudin M H M, Bahrin E K, Naim M N, and Zakaria R 2013 *BioResources*. **8** (1): 676-87.

[23] Limoes S, Rahman S, Setyahadi S, and Gozan M 2018 *IOP Conf. Ser.: Earth Environ. Sci.* IOP Publishing.
[24] Amraini S Z, Ariyani L P, Hermansyah H, Setyahadi S, Rahman S F, Park D-H, and Gozan M 2017 *Biotechnol. Bioprocess. Eng.* **22** (3): 287-95.
[25] Ibrahim M N M, Ahmed-Haras M R, Sipaut C S, Aboul-Enein H Y, and Mohamed A A 2010 *Carbohydr. Polym.* **80** (4): 1102-10.
[26] Parshetti G K, Chowdhury S, and Balasubramanian R 2014 *RSC Advances.* **4** (84): 44634-43.
[27] Lu F, et al. 2015 *BioEnergy Res.* **8** (3): 934-52.
[28] Siyamak S, Ibrahim N A, Abdolmohammadi S, Yunus W M Z B W, and Rahman M Z A 2012 *Molecules.* **17** (2): 1969-91.
[29] Haafiz M M, Hassan A, Zakaria Z, and Inuwa I 2014 *Carbohydr. Polym.* **103**: 119-25.
[30] Wanrosli W, Zainuddin Z, Ong P, and Rohaizu R 2013 *Ind. Crops Prod.* **50**: 611-7.
[31] Purnomo A, Romli M, and Hasanudin U 2018 *IOP Conf. Ser.: Earth Environ. Sci.* IOP Publishing.
[32] Harahap A F P, Panjaitan J R H, Curie C A, Ramadhan M Y A, Srinophakun P, and Gozan M 2020 *Appl. Sci.* **10** (17): 5987.
[33] Sahlan M, Hermansyah H, Wijarnako A, Gozan M, Lischer K, Ahmudi A, and Pujianto P 2020.
[34] Abdul P M, Jahim J M, Harun S, Markom M, Lutpi N A, Hassan O, Balan V, Dale B E, and Nor M T M 2016 *Biores. Tech.* **211**: 200-8.
[35] Panjaitan J R H and Gozan M 2017 *Int. J. Appl. Eng.* **12** (14): 4382-90.
[36] Gozan M, Panjaitan J R H, Tristantini D, Alamsyah R, and Yoo Y J 2018 *Int. J. Chem. Eng.* **2018**: 1920180.
[37] Fauzi A G, Sari C N, Harahap A F P, Ramadhan M Y A, and Gozan M 2021 *Adv. Sci. Technol.* Trans Tech Publ.
[38] Shukor H, Al-Shorgani N K N, Abdeshahian P, Hamid A A, Anuar N, Rahman N A, Isa M H, and Kalil M S 2014 *Open Access.*
[39] Guo J, Luo Y, Lua A C, Chi R-a, Chen Y-l, Bao X-t, and Xiang S-x 2007 *Carbon.* **45** (2): 330-6.
[40] Nizamuddin S, Mubarak N, Tiripathi M, Jayakumar N, Sahu J, and Ganesan P 2016 *Fuel.* **163**: 88-97.
[41] Mushtaq F, Abdullah T A T, Mat R, and Ani F N 2015 *Biores. Tech.* **190**: 442-50.
[42] Nabinejad O, Sujan D, Rahman M, and Davies I J 2015 *J. Therm. Anal. Calorim.* **122** (1): 227-33.
[43] Masuda T, Kondo Y, Miwa M, Shimotori T, Mukai S, Hashimoto K, Takano M, Kawasaki S, and Yoshida S 2001 *Chem. Eng. Sci.* **56** (3): 897-904.
[44] Nipattummakul N, Ahmed I I, Kerdsuwan S, and Gupta A K 2012 *Appl. Energy.* **92**: 778-82.
[45] Eom I-Y, Oh Y-H, Park S J, Lee S-H, and Yu J-H 2015 *Biores. Tech.* **185**: 143-9.
[46] Khamtib S, Plangklang P, and Reungsang A 2011 *Int. J. Hydrog. Energy.* **36** (21): 14204-16.
[47] Ang S, Shaza E, Adibah Y, Suraini A, and Madihah M 2013 *Process Biochem.* **48** (9): 1293-302.
[48] Teh C C, Ibrahim N A, and Yunus W M Z W 2013 *BioResources.* **8** (4): 5244-60.
[49] Komonkiat I and Cheirsilp B 2013 *Biores. Tech.* **146**: 200-7.
[50] Sangkharak K, Yunu T, Paichid N, and Prasertsan P 2016 *J. Oil Palm Res.* **28** (1): 52-63.
[51] Saidu M, Yuzir A, Salim M R, Azman S, and Abdullah N 2014 *Int. Biodeterior. Biodegradation.* **95**: 189-94.
[52] Zainuddin Z, Wan Daud W R, Pauline O, and Shafie A 2011 *Biores. Tech.* **102** (23): 10978-86.
[53] Hazeena S H, Pandey A, and Binod P 2016 *Renew. Energy.* **98**: 216-20.
[54] Ofori-Boateng C, Lee K T, and Saad B 2014 *Energy Convers. Manag.* **81**: 192-200.

[55] Abnisa F, Arami-Niya A, Daud W W, Sahu J, and Noor I 2013 *Energy Convers. Manag*. **76**: 1073-82.
[56] Ofori-Boateng C and Lee K T 2014 *Fuel*. **119**: 285-91.

Determining A Model for Municipal Solid Waste Treatment using Mixture Design

Christia Meidiana[1*], Ratan Kumar Gosh[2], Kartika Eka Sari[1]
[1]Department of Regional and Urban Planning, Faculty of Engineering, Universitas Brawijaya, Malang, 65147, Indonesia, 0000-0001-8416-2266
[2]Bangladesh Energy and Power Research Council, Dhaka, 1000, Bangladesh, 0000-0002-9153-407X
[1]*c_meideiana@ub.ac.id, [2]member.inno@eprc.gov.bd, [3]kartikaeka@ub.ac.id

Abstract. The purpose of the study is to propose an optimum combination of waste management to reduce GHG emissions while ensuring feasible economic benefits through the calculation of BCR value using the MD technique of A Simplex Centroid Design consisting of three possible waste treatments, which are composting, reuse, and recycling. The recovery factor of each waste treatment component was analyzed to calculate the GHG emission. The result showed that the MD approach to formulating the possible mixture components of MSW treatment is feasible. Referring to the objective of each response variable, which is to produce the least GHG emission and achieve the highest BCR value, the best value of each mixture component is 95.36 m^3/day for composting, 322.29 m^3/day for recycling, and 2.35 m^3/day for reuse. Therefore, this proposed combination could produce the most negligible GHG emission by 0.029 Gg CO2eq /day while achieving the feasible BCR value of 1.36.

1. Introduction

The recent development of municipal solid waste (MSW) treatment has made noticeable progress. It is revealed that the progression of MSW treatment involves more efficient waste sorting systems by utilizing the fuzzy multi-criteria model [1], improvement of waste treatment as part of the solid waste management system (SWMS) through mixed-integer linear programming (MIL), which resulted in more optimized number, locations, waste flow and net daily cost of SWMS [2], MSW treatment scenarios using ReCiPe 2016 Midpoint (H) method in regards to reducing the greenhouse gas (GHG) emission which ultimately could reduce the GHG emission by 55% to 58% [3], as well as the discovery of the most effective strategies of plastic waste treatment through tar and concrete production, and the road construction which provided significant benefits to reduce the GHG emission and ease of localization [4]. Nevertheless, these progressions in MSW treatment can be regarded as part of the commitment to achieve sustainability for MSW management which considers minimizing social and environmental impacts [5].

On the other hand, the utilization of the mixture-design (MD) approach, as a tool of Design of Experiment, regarding modeling the MSW treatment still has noticeable gaps as it is not widely recognized for the application. Some previous studies have used the MD in various contexts, but limited numbers were focused on solid waste management [6]. For example, the study of [7] determined the formulation for SWM in the regional context, while [8] defined the best model for SWM on an urban scale. Those studies used MD to produce the optimum compositions from the multi-mixture components. Initially, the method was applied to generate the model having the optimum solid feedstock [9]. In this study, typical MSW treatments such as composting, recycling, and reuse are regarded as multivariable systems which could be mixed to produce the most optimum combination to promote the sustainability concept from the environmental and economic perspective.

2. Methods

2.1. The area of study

The boundary system for the study covers all domestic waste flow in Malang City which is transported to the landfill. The mass flow of domestic waste was calculated after collecting the data on waste transportation from all transfer stations in Malang City, which is a total of 64 units comprising 61 temporary waste disposal (TPS) and three 3R-TPS. A seven-day survey for data collection on waste transportation was conducted in February 2022. During data collection, questionnaires were distributed to the cleaning officer to obtain more detailed information required for the study.

2.2. Calculation of waste volume and reduction

The study focused on the domestic waste transported daily from transfer stations to the Supit Urang Landfill, Malang City. The load-count analysis [10] was performed to calculate the total waste volume (V_{tot}) entering the landfill by multiplying the number of trucks (n) with its daily capacity (V_t), as described in Equation 1. The measurement was conducted seven days long, with the operational hours consisting of 7 am – 11 am and 3 pm – 4 pm.

$$V_{tot} = V_t \times n \tag{1}$$

Using the existing waste composition of Malang City, the total volume of each waste type ($V_{tot\text{-}i}$) is further calculated, including organic waste, paper, fabric, rubber, plastic, metal, glass, and others. Referring to the data of specific recovery factor (RF) of solid waste from the Malang municipality, the potential waste reduction through composting of organic waste, recycling of plastic and paper waste, and reusing of fabric, rubber, metal, glass, and other wastes are calculated. The RF from each waste was previously surveyed in 2021 in all transfer stations in Malang City and used to describe the potential percentage of treated MSW. The potential waste reduction was calculated using the assumption that the waste transported to the landfill is proportional to the population growth. Furthermore, the RF for each waste type was determined by dividing the reduced waste volume of each type ($V_{t\text{-}i}$) with the total waste volume of each type ($V_{r\text{-}i}$), as described in Equation 2. The RF reflects the potential percentage of waste that can be treated into composting, recycling, and reusing.

$$Rf_i = V_{t-i}/V_{r-i} \tag{2}$$

2.3. Greenhouse gas emission

The GHG emission calculation comprises composting, recycling, and waste transporting emissions. Tier 2 of IPCC 2006 Guideline for National Greenhouse Gas Inventories [11] is used due to no field measurement data and the shortage of local data.

2.3.1 Composting emission

Methane (CH_4) and Nitrous Oxide (N_2O) are produced during the composting process. Therefore Equations 3 and 4 (IPCC, 2006) are used to calculate the methane and nitrous oxide, respectively, which are equivalent to the Carbon Dioxide (CO_2) in the Giga gram (Gg) unit daily.

$$E_m = \Sigma_a((V_{organic} \times M \times 10^{-6} \times EF_m) \times 10^{-3}) \times 21 \tag{3}$$

$$E_n = \Sigma_a((V_{organic} \times M \times 10^{-6} \times EF_n) \times 10^{-3}) \times 310 \tag{4}$$

Where E_m and E_n are methane and nitrous oxide emissions (Gg CO2 eq), $V_{organic}$ is the organic waste volume accounted for the composting process (m^3), M is the organic waste density which is equal to 288,6 kg/m^3, EF_m is the emission factor of methane which is equal to 4 g CH_4/kg waste, and EF_n is the emission factor of nitrous oxide which equals to 3 g N_2O/kg waste.

2.3.2 *Recycle emission*

The recycling process uses the shredder machine, which incorporates fuel combustion, thuscontributing to GHG emissions. Equation 5 is used to calculate the fuel consumption for the shredder machine to operate relative to the daily potential recycled waste and is further used in Equations 6 to 8 for calculating the emission from each GHG gas which is composed of Carbon Dioxide (CO_2), Methane (CH_4) and Nitrous Oxide (N_2O) respectively.

$$F = \frac{M_{plastic\,\&\,paper}}{C} \times F_C \quad \textbf{(5)}$$

$$E_C = F \times 36.10^{-6}\,\frac{TJ}{L} \times EF_C \quad \textbf{(6)}$$

$$E_m = F \times 36.10^{-6}\,\frac{TJ}{L} \times EF_m \quad \textbf{(7)}$$

$$E_n = F \times 36.10^{-6}\,\frac{TJ}{L} \times EF_n \quad \textbf{(8)}$$

Where F is the fuel consumption of the shredder machine (TJ), $M_{plastic\&paper}$ is the mass of plastic and paper waste used for the recycling process, C is the shredder machine capacity which is equivalent to 400 kg per hour, and F_c is the fuel consumption factor of shredder machine per hour which is 3 liters per hour. Additionally, E_c is the emission of carbon dioxide, EF_c is the CO_2 emission factor which is 69.3 kg CO_2/TJ, E_m is the emission of methane, EF_m is CH_4 emission factor which is 33 kg CH_4/TJ, and lastly, E_n is the emission of nitrous oxide, EF_n is N_2O emission factor which is 3,9 kg N_2O/TJ.

2.3.3 *Waste transporting emission*

Transporting the collected waste from the waste transfer station to the landfill consumes diesel. Equation 9 calculates the diesel consumption, while Equations 10 and 11 calculate the emission from GHG gas.

$$F_{DT} = S \times F_{DTC} \quad \textbf{(9)}$$

$$E_{Consumption} = F_{DT} \times 36.10^{-6}\,TJ/L \quad \textbf{(10)}$$

$$E_C = E_{Consumption} \times 74.1\,Kg/Tj \quad \textbf{(11)}$$

Where F_{DT} is the solar fuel consumption from the diesel engine of the dump truck, S is the total distance required to transport organic waste from the waste transfer station to the Supit Urang Landfill, F_{DTC} is the solar fuel consumption of the dump truck, which is 0,56 liters per km [12], while $E_{Consumption}$ stands for the energy consumption. The coefficient of 36.10^{-6} stands for the calorific value, while 74.1 is the emission factor for solar fuel. In addition, the S value from the proposed MD approach has a lower value than the actual data due to the increasing volume of MSW treatment. It is assumed that MSW is treated directly using incinerator that there is no emission. Other fractions for recycling and reuse is assumed not to be transported to the landfill as it is possible to treat the potential waste at the transport station, thus reducing the overall GHG emission from waste transporting activity.

2.4. *Benefit Cost Ratio for municipal solid waste treatment*

The modeled MSW treatment through the mixture design approach consists of three components: composting, recycling, and reuse, which all have economic value. The economic value is further assessed by calculating the benefit-cost ratio (BCR) value by dividing the benefits by the cost value, as explained in Equation 12. For identifying the cost and benefit value for each kilogram mixture component processed, previous feasibility research on the MSW treatment business is utilized [13,14]. Therefore, for both composting and recycling or reuse cost is Rp. 1,503/kg and Rp. 1,864/kg respectively, whereas the benefit value for composting and recycling or reuse is Rp. 2.000/kg and Rp. 2,064/kg.

$$BCR = \frac{Benefit}{Cost} \quad \textbf{(12)}$$

2.5. *Mixture design for modeling municipal solid waste treatment*

The mixture-design (MD) approach was utilized to formulate the optimum mixture of MSW treatment in Malang City, which could further contribute to minimizing GHG emissions. The statistical principles of MD [15] are composed of the Design Expert software, which consists of the necessary steps to operate. In this research, the GHG emission (Gg CO_2/day) is the response variable, while the mixture components of MSW treatment consist of composting, recycling, and reuse.

The first step is to consider the maximum combined value of three components, which is 420 m^3/day, equivalent to the 24% MSW reduction target by 2021 based on Regulation No. 34/2018. In addition, the minimum value of each component should be specified according to the actual daily volume of MSW treatment in Malang City by 2021, which composes of 95.35 m^3, 63,1 m^3, and 2.35 m^3 for composting, recycling, and reuse, respectively. At this point, the proposed MD, a simplex centroid design, may be warned that it cannot fit the maximum combined value of three components. Therefore some adjustment ranges are proposed, which later define the maximum value of each mixture component to meet the maximum combined value of 420 m^3/day. The adjusted ranges of a maximum value of each mixture component are composed of 354.6 m^3, 322.3 m^3, and 261.5 m^3 for composting, recycling, and reuse. Subsequently, the model is later constructed using a simplex centroid design with quadratic order, which produces the response surface contour plots and the regression model equation fitted to a quadratic expression. The model is evaluated using the analysis of variance (ANOVA).

3. Results and Discussion

Totally there are 14 tries run for the calculation with three mixture components and each try is calculated for its GHG Emission and BCR values as the responses. The overall runs fit the maximum combined value of three components, 420 m^3/day, which matches the Malang City MSW reduction target by 2021. Due to this, the reuse component has the highest increase compared to its actual value of 2.35m^3/day up to 260.83 m^3/day. The composting and recycling components have a rational increasecompared to their initial range, up to 3 to 5 times the actual value. Regarding the GHG emission, the composting activity produces the highest value compared to the other two activities, while in the BCR perspective, all the potential combination is feasible (>1).

Test of normality is later conducted to evaluate if the modelled MD could fit the normal distribution and to calculate how a random variable underlying the dataset to normally distributed, and later projected using the standard plot of residuals as described in Figure 1 for GHG Emission and BCR value. The plot shows that most of the model fit the normal distribution. The independent and response variables were all tested to fit into potential models such as linear, quadratic, special cubic, and cubic models. Furthermore, the coefficient of determination (R2) and ANOVA were utilized to evaluate the constructed models. The ANOVA results showed that the models are significant. Thus the models could describe the systems. Moreover, the value of both coefficient of determination (R2) and the adjusted R2 are relatively high, above 0.94, while the sequential p-value is below 0.005 only for linear model. It indicates that the experimental data used in linear program fits with the response functions. Therefore, the liner model was used for further step in determining the best combination for waste treatment.

The result of MD analysis composed of three MSW treatment components with 14 iterations produced regression equations on each response variable (GHG Emission and BCR) as explained in equations 13 and 14. Where C is composting, Rc is recycled, and Ru is reused.

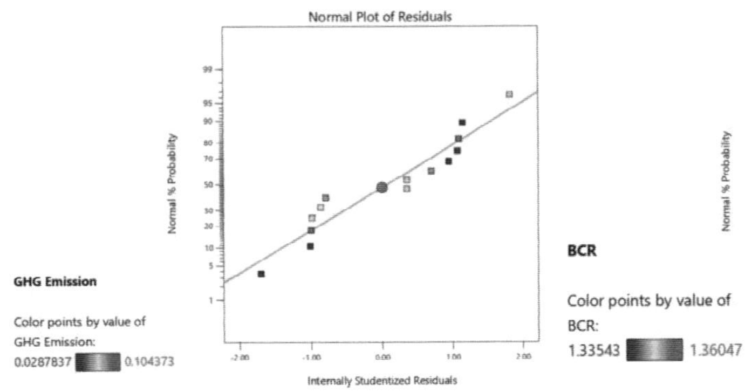

Figure 1. Standard Probability Plots of the Residuals for GHG emission and BCR value

Furthermore, Figure 2 explains the mixture surface plots for both GHG emission and BCR value, which shows the optimal combination of each mixture component to align with the objective value of each response variable. Forinstance, GHG Emission aims to produce the minimum value of GHG emission. Therefore,the mixture design favoured the dark blue area, mainly composed of recycled and reuse components. This result is aligned with Table 1, as composting produces the most GHG emission. In addition, it explains the desirability of surface plots and the optimal combination for producing the minor GHG emission while ensuring the highest BCR value achieved. Therefore, the bestvalue of each mixture component to achieve this target is 95.36 m^3/day for composting, 322.29 m^3/dayfor recycling, and 2.35 m^3/day for reuse. These combinations of mixture components will produce a minor GHG emission of 0.029 Gg CO_2 Eq/day while ensuring the treated MSW could still be feasible toconduct with a BCR value of 1.36.

$$GHG_{Emis} = 0.1043\,C + 0.0291\,Rc + 0.0288\,Ru + 0.0001\,C * Rc - 0.0003\,C * Ru + 0.0001\,Ru * Rc \quad \textbf{(13)}$$

$$BCR = 1.331\,C + 1.3726\,Rc + 1.3645\,Ru - 0.015\,C * Rc - 0.031\,C * Ru + 0.006\,Ru * Rc \quad \textbf{(14)}$$

From the social perspective, the result of the mixture components to achieve the minor GHG emission while ensuring the economic benefit could be proposed as the strategy of the local authority to promote municipal waste reduction through composting, recycling, and reuse. The possibility of achieving the economic benefit presented in this study could become the main attractive force to encourage the involvement of the local community. Therefore, the development of the local economy could be achieved by ensuring environmental sustainability from the GHG emission perspective and changing the local perspective on handling the MSW.

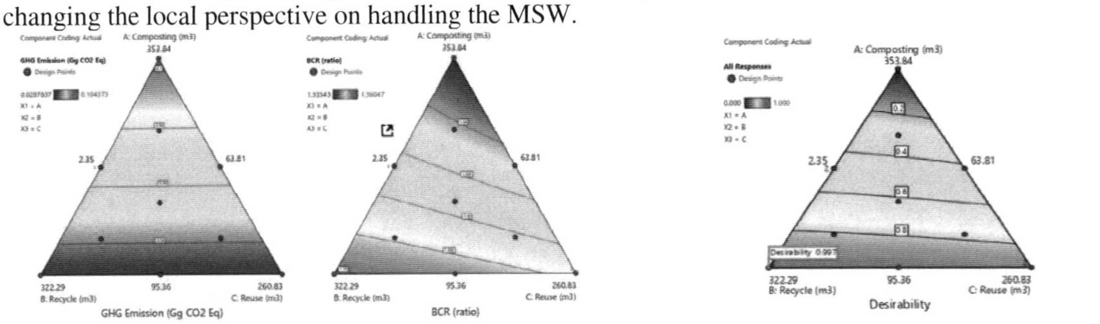

Figure 2. Mixture Surface Plots for the GHG Emission and BCR Value and Desirability Surface Plots and Optimal Combination

4. Conclusion

In this study, the simplex centroid design was used to formulate the desirable mixture components consisting of composting, reuse, and recycling with the design objectiveto produce the least GHG emission and highest BCR value. The result showed that the proposed model could achieve a high coefficient of determination (R2) by 0.9996. Furthermore, the best value of each mixture component is 95.36 m^3/day for composting, 322.29 m^3/day for recycling, and 2.35 m^3/day for reuse which could produce low GHG emission by 0.029 Gg CO_2 Eq/day while achieving the feasible BCR value of 1.36.

Acknowledgment

The authors want to overall remark the clear contribution of the Institute of Research and Community Services Brawijaya University (LPPM) which has supported this research

References

[1] Giel R and Kierzkowski A 2022 A Fuzzy Multi-Criteria Model for Municipal Waste Treatment Systems Evaluation including Energy Recovery *Energies* **15** 1–16

[2] Shaban A, Zaki F, Afefy I H, Gravio G Di, Falegnami A and Patriarca R 2022 An Optimization Model for the Design of a Sustainable Municipal Solid Waste Management System *Sustainability* **14** 1–19

[3] Aldhafeeri Z M and Alhazmi H 2022 Sustainability Assessment of Municipal Solid Waste in Riyadh , Saudi Arabia , in the Framework of Circular Economy Transition *Sustainability* **14**

[4] Huang S, Wang H, Ahmad W, Ahmad A, Vatin N I, Mohamed A M, Deifalla A F and Mehmood I 2022 Plastic Waste Management Strategies and Their Environmental Aspects : A Scientometric Analysis and Comprehensive Review *Int. J. Environ. Res. Public Health* **19** 1–31

[5] Dat T, Ming F, Tseng M and Helmi M 2020 Identifying Sustainable Solid Waste Management Barriers in Practice Using the Fuzzy Delphi Method *Resour. Conserv. Recycl.* **154** 104625

[6] Lee B C Y, Mahtab M S, Neo T H, Farooqi I H and Khursheed A 2022 A Comprehensive Review of Design of Experiment (DOE) for Water and Wastewater Treatment Application - Key Concepts, Methodology and Contextualized Application *J. Water Process Eng.* **47** 102673

[7] Najm M A, El-Fadel M and Ayoub G 2002 An Optimisation Model for Regional Integrated Solid Waste Management II . Model Application and Sensitivity Analyses *Waste Manag. Res. J. Int. Solid Wastes Public Clean. Assoc.* **20** 46–54

[8] Meidiana C, Wahyu I and Ari I D 2020 Simplex-Centroid Mixture for Municipal Solid Waste Treatment Optimisation *IOP Conf. Ser. Earth Environ. Sci.* **471**

[9] Sathish T, Lakshmi G S, Rao C S, Brahmaiah P and Prakasham R S 2008 Mixture Design as First Step for Improved Glutaminase Production in Solid-State Fermentation by Isolated Bacillus *Lett. Appl. Microbiol.* **47** 256–62

[10] Tchobanoglous G 1993 *Integrated solid waste management : engineering principles and management issues / George Tchobanoglous, Hilary Theisen, Samuel Vigil (S. A. Vigil & H. Theisen (eds.))* (McGraw-Hill)

[11] Japan I for G E S (IGES) 2008 *2006 IPCC Guidelines for National Greenhouse Gas Inventories* (Hayama: Institute for Global Environmental Strategies (IGES)

[12] Yunita A T and Ali M 2017 Analisis Sistem Transportasi Sampah Kota Tuban Menggunakan Dynamic Programming *J. Ilm. Tek. Lingkung.* **6** 45–52

[13] Wahana S and Savitri M I 2020 Harga Pokok Produksi Kompos Potensial Limbah Media Jamur Merang Kampung Padamaran Kecamatan Susukan Kabupaten Cirebon *Paradig. Agribisnis* **2** 34–42

[14] Kustanti R, Rezagama A, Ramadan B S, Sumiyati S, Prasetyo B and Hadiwidodo M 2020 Tinjauan Nilai Manfaat pada Pengelolaan Sampah Plastik Oleh Sektor Informal (Studi Kasus : Kecamatan Purwodadi , Kabupaten Grobogan) *J. Ilmu Lingkung.* **18** 495–502

[15] Cornell J A 2011 *Experiments with Mixtures: Designs, Models, and the Analysis of Mixture Data. Wiley*

IOP Conf. Series: Earth and Environmental Science 1111 (2022) 012017 doi:10.1088/1755-1315/1111/1/012017

Study of The Efficiency of A Drip Irrigation System Model Based on the Internet of Things (IoT)

Eko Noerhayati[1*], Difadila Alif Tasarlik[1], Anita Rahmawati[1], Soraya Norma[2]
[1] Department of Civil Engineering, Faculty of Engineering, Islamic University of Malang
[2] Department of Electrical Engineering, Faculty of Engineering, State University of Malang
[1]*eko.noerhayati@unisma.ac.id,[2]difadila.alif@gmail.com, [3]anita.rahmawati@unisma.ac.id, [4]soraya.norma.ft@um.ac.id

Abstract. Farmers have used drip irrigation technology extensively, particularly for fruit and vegetable crops. However, investigations on the effectiveness of this irrigation system have not been widely conducted. IoT (Internet of Things)-based drip irrigation is a drip irrigation network design that can be employed automatically with remote control. The effectiveness of the IoT-based drip irrigation system model will therefore be investigated in this study using two distinct types of pipes with various widths. The study's goal was to evaluate the effectiveness of an IoT-based drip irrigation system model with various diameters. Direct observation was employed as the study method in the lab, and the data collected were examined using the Minitab 18 program to get the ANOVA regression value. The findings indicated that the average headloss on ½-inch pipes was 0.7657 m, and the average headloss on 1-inch pipes was 0.6126787 m, indicating that the headloss decreased with increasing pipe diameter. On a 1-inch pipe, the correlation between discharge and headloss is 0.998. It is envisaged that a productive drip irrigation system can assist farmers in resolving infrastructural issues in water-scarce regions.

1. Introduction

Better facilities and infrastructure are required as the industrial era 4.0 transitions to the industrial era 5.0, including agriculture. Farmers extensively utilize drip irrigation for fruit and vegetable crops [1–5]. IoT-based drip irrigation is a network design that can be operated automatically using a remote control outfitted with sensors to assess agricultural land factors, including soil moisture, temperature, and water flow, making it easier for farmers to run [6]. Technical considerations and appropriate drip irrigation management patterns (emitter flow rate, lateral depth, irrigation amount, and fertilizer amount) must be considered to minimize environmental risks [7]. Drip irrigation can easily transport nutrients and water to the root zone and deeper into the soil. In order to prevent plant moisture stress during the period of plant growth, water management is essential [8]. As a result, both the adverse environmental effects of irrigation and the high demand for water supplies are minimized [9]. There have been numerous studies of drip irrigation system models. For instance, irrigation hoses transfer water from storage tanks situated above agricultural land to the root zones of each plant [10]. Another effective drip irrigation model has been created to forecast the wetting pattern's dimensions utilizing elements like emitter output, water volume, and soil hydraulic parameters. His research yielded MAE values ranging from 0.87 to 10.43 cm, which evaluated the precision of the wetting zone dimensions by comparing predictions with field and laboratory data [11].

However, the effectiveness of this irrigation technique has not been the subject of much research. In order to determine the most effective diameter, this study will analyze two distinct irrigation system diameters and assess the IoT-based drip irrigation system model's effectiveness on two types of pipes with varying diameters. The study's goal was to evaluate the effectiveness of an IoT-based drip irrigation system model with various diameters. In order to help farmers in water-scarce areas overcome irrigation infrastructure issues, this research aims to create an effective drip irrigation model design. What is the link between discharge and headloss in an IoT-based test irrigation system pipe, and how does it affect the overall energy loss generated in an IoT-based drip irrigation system due to variations in irrigation pipe diameter?

2. Methods

2.1 Material Preparation

The materials created for this study are as follows, in agreement with [12–14] the solenoid valve, relay, and mcb, the submersible pump Outsell plc, web design, and app design mobile phone, tendon, pipe with a dim diameter, a dim pipe, a knee pipe, a PE hose, a valve, Turn off the faucet.

2.2 Research Stage

The first step in the research is to install soil moisture sensors, submersible pump discharge sensors, solenoid valves, relays, and MCBs. Next, automatic water drip pipelines are installed by design, and pipe lengths and emitter distances are arranged by the length of the bunds using the design in Figure 1. The following stage is to simulate an automated drip irrigation system using IoT before operating drip irrigation equipment in the lab. The last step involves measuring the water volume in the reservoir, the irrigation water volume for ½-inch and 1-inch pipes, the volume of water on each drip, the readings of the ½-inch and 1-inch pipe discharge sensors, and the temperature sensor readings on ½-inch and 1-inch pipe diameters. Each pipe diameter was measured ten times.

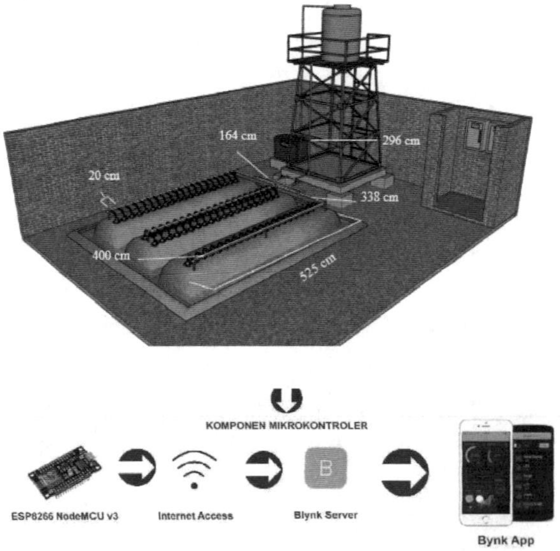

Figure 1. IoT-Based Drip Irrigation System Design

2.3 Data Analysis

The dependent variable in this study is the water discharge that travels through the pipe, while the pipe's head loss is the independent variable. Simple linear regression analysis will be used to calculate these variables' effects [15,16]. We made ten observations of each variable. Additionally, this study employed ANOVA to ascertain the association between water loss in the pipe utilizing two different diameters and water discharge flowing in the pipe. The hypothesis is carried out by simple linear regression using the following equation or function: $Y = b1\ X1 + b_2\ X2 + e$: error.

3. Results and Discussion

3.1 Headloss Pipe ½ Inch

Depending on the experiments conducted, further energy losses, both big and slight, are experienced in drip irrigation circuits. by using the Darcy-Weisbach equation, the headloss results are displayed in Table 1 as follows:

3.1.1 Loss of energy out of the reservoir

$$H = \frac{v^2}{2g}$$

Where :
H = Loss of energy in the reservoir (m)
V = flow rate (m/s)
g = Gravity (9.81 m/s^2)

3.1.2 Energy loss due to turns, connections, valves (Minor)

$$Hp = \frac{k-v^2}{2g}$$

Where :
Hp = power loss due to pipe fitting (m)
K = pipe type coefficient and average velocity in the pipe
V = flow rate (m/s)
g = gravity (9.81 m/s^2)

3.1.3 Energy loss due to roughness in the pipe (major)

$$Hf = \frac{4f.l.v^2}{2.g.d}$$

Where :
Hf = Loss due to pipe roughness (m)
F = friction factor
V = flow rate (m/s)
L = pipe length (m)
D = pipe diameter (m)
g = gravity (9.81 m/s^2)

An accurate assessment of the total head losses, represented by friction losses throughout the pipe and emitter, is necessary for accurate hydraulic design results in lateral drops. Using ten observations, Table 1. displays the headloss value for pipes with 0.5 and 1-inch diameters placed in drip irrigation systems. Table 2 shows the results of the normalization test performed on this data using the Minitab 18 program. According to [17], the total head loss values measured on 15 commercially available coextruded laterals were compared with those discovered using the now-proposed methodology. For the scenarios analyzed, the relative inaccuracy in the pressure head estimation is consistently 2.4% [17].

Table 1. Results of ½ Inch and 1 Inch Headloss Pipes

No	Trial Name	Volume Water (Liter)	Water discharge Pipe ½ Inch (m^3/s)	Water discharge Pipe 1 Inch (m^3/s)	Total Headloss Pipe 1/2 Inch (M)	Total Headloss Pipe 1 Inch (M)
1	Trial Name 1	10	0.00000454	0.0000079	0.765230	0.611995
2	Trial Name 2	20	0.00000363	0.0000100	0.765008	0.612205
3	Trial Name 3	30	0.00000387	0.0000120	0.765072	0.612460
4	Trial Name 4	40	0.00000411	0.0000132	0.765137	0.612631
5	Trial Name 5	50	0.00000460	0.0000137	0.765284	0.612717
6	Trial Name 6	60	0.00000662	0.0000135	0.766048	0.612684
7	Trial Name 7	70	0.00000698	0.0000136	0.766211	0.612704
8	Trial Name 8	80	0.00000701	0.0000138	0.766229	0.612736
9	Trial Name 9	90	0.00000732	0.0000139	0.766373	0.612737
10	Trial Name 10	100	0.00000741	0.0000140	0.766419	0.612762
		Total Headloss			7.657010	6.125630
		Average Headloss			0.7657	0.61256

The 1-inch pipe yielded eight average data, as shown in table 2, following the headloss data normalization test carried out using the Minitab 18 program.

Table 2. Normalization Results of 1 Inch Pipe Headloss Data

No	Trial Name	Volume Water (Liter)	Water discharge Pipe 1 Inch (m^3/s)	Total Headloss Pipe 1 Inch (M)
3	Trial Name 3	30	0.0000120	0.612460
4	Trial Name 4	40	0.0000132	0.612631
5	Trial Name 5	50	0.0000137	0.612717
6	Trial Name 6	60	0.0000135	0.612684
7	Trial Name 7	70	0.0000136	0.612704
8	Trial Name 8	80	0.0000138	0.612736
9	Trial Name 9	90	0.0000139	0.612737
10	Trial Name 10	100	0.0000140	0.612762
		Total Headloss		4.901
		Average Headloss		0.6126787

Figures 2 illustrate the link between headloss and flow rate in a drip irrigation system using a diameter of 0.5 inches, while Figure 3 illustrates the relationship using a diameter of 1 inch. The graphic shows that the value of water loss or head loss is lower in this drip irrigation system because it employs a bigger diameter. It has been demonstrated that the research findings are consistent with the current formula [18–20].

3.2 Regression and Correlation

With a pipe value of ½ inch 0.998 and 1-inch pipe 0.988 and a p-value of 0.05, the Minitab 18 software was used to test the relationship between discharge and headloss. This result indicates that the relationship between discharge and headloss is highly significant and that discharge affects headloss [21,22]. Figures 2 and 3 display the outcomes of the Minitab software's linear regression test.

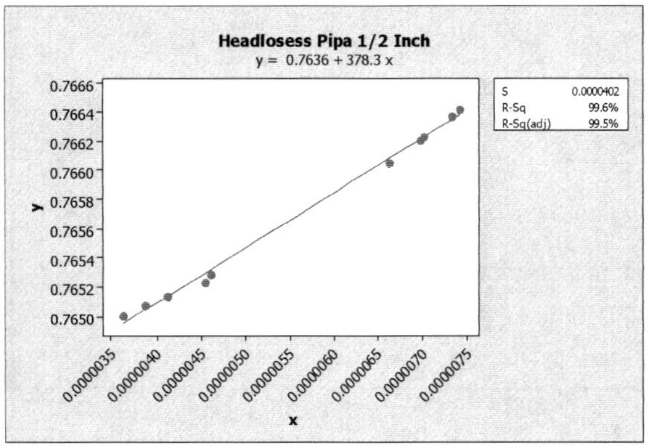

Figure 2. Results of ½ Inch Pipe Linear Regression

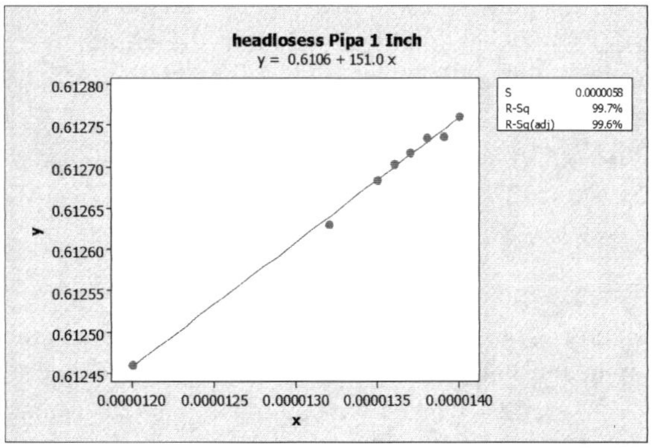

Figure 3. Results of 1 Inch Pipe Linear Regression

3.3 Emitter discharge varying

The coefficient of manufacturing variation for the emitter expresses the rate of variation of the emitter discharge (coefficient of manufacturing for the emitter).
The following equation is used to calculate the value of the emitter discharge variation:

$$V = \frac{\left(\frac{\sqrt{(q1^2+q2^2+q3^2....+qn^2-qa^2)}}{n-1}\right)}{qa}$$

Where:
q1 + q2 + q3... + qn = discharge per emitter (M^3/s)
N = quantity of emitters
qa = Average Emitter Discharge (M^3/s)

The results are displayed in Table 3 after determining the emitter discharge variation (v) for ½ inch and 1 inch pipes.

Table 3. Variation of Emitter Discharge Pipe Inch And 1 Inch

No	Volume	Pipe Variation A 1 Inch	Pipe Variation A 1 Inch	Pipe Variation B ½ Inch	Pipe Variation B 1 Inch	Pipe Variation C ½ Inch	Pipe Variation C 1 Inch
1	10	0.0453	0.093	0.0728	0.024	0.0823	0.052
2	20	0.0768	0.042	0.0856	0.059	0.0903	0.066
3	30	0.0607	0.042	0.0524	0.052	0.0408	0.038
4	40	0.2117	0.044	0.0409	0.056	0.0423	0.035
5	50	0.1399	0.063	0.1064	0.061	0.1968	0.068
6	60	0.0891	0.150	0.0611	0.061	0.0467	0.057
7	70	0.0656	0.056	0.0492	0.053	0.0552	0.063
8	80	0.0592	0.042	0.0560	0.032	0.0476	0.045
9	90	0.0725	0.158	0.0712	0.072	0.0454	0.061
10	100	0.0765	0.186	0.0455	0.061	0.0508	0.036

3.4 ANOVA with Regression

ANOVA (Analysis of Variance) is a statistical theory technique for examining how one variable behaves compared to another [23–25]. ANOVA regression analysis with Minitab 18. Results for head loss indicate that p-value (0.000) 0.05, which indicates that the volume of water impacts the headloss value, and p-value (0.04) 0.05, which indicates that pipe differences impact the headloss. According to the test results on the value of (v) ½ inch pipe, neither the amount of water used for the pipe's variances had an impact on the value of the emitter discharge variation produced. The 1-inch pipe demonstrates that pipe variances impact the emitter discharge variation while the amount of water used has no bearing on the value of the emitter discharge variation created.

4. Conclusion

According to the research study, larger pipe diameters have lower headloss values than smaller ones. While the total energy loss in inch pipes is typically 0.7657 m, it is often 0.4901 m in 1-inch pipes. With the linear regression equation for inch pipe y = 0.7636 + 378.6 x and 1 inch pipe y = 0.6106 + 151.0 x, the correlation coefficient (r) for inch pipe is 0.998 and for 1 inch pipe is 0.998. The linear regression formula represents the link between discharge and the amount of energy lost in a drip irrigation system.

References

[1] Alcon F, Miguel M D de and Burton M 2011 Duration analysis of adoption of drip irrigation technology in southeastern Spain *Technol. Forecast. Soc. Chang.* **78** 991–1001

[2] Dahnil D P, Hood Z, Adam A, Razak M Z A and Ismail A G 2021 Drip Irrigation Detection for Power Outage-Prone Areas with Internet-of-Things Smart Fertigation Managemant System *Int. J. Adv. Comput. Sci. Appl.* **12** 745–55

[3] Kaur A B, Inamdar D, Raut V, Patil A and Patil N 2016 a Survey on Smart Drip Irrigation System *Int. Res. J. Eng. Technol.* **3** 100–2

[4] Rao F, Abudikeranmu A, X. Shi N H and Ma X 2021 Impact of participatory irrigation management on mulched drip irrigation technology adoption in rural Xinjiang, China *Water Resour. Econ.* **33** 100170

[5] Shrestha R B and Gopalakrishnan C 1993 Adoption and diffusion of drip irrigation technology: an econometric analysis *Econ. Dev. Cult. Change* **41** 407–418

[6] Ouni R and Saleem K 2022 Framework for Sustainable Wireless Sensor Network Based Environmental Monitoring *sustainability* **14** 1–26

[7] Touill S, Richa A, Fizir M, García J E A and Gómez A F S 2022 A review on smart irrigation management strategies and their effect on water savings and crop yield *Irrig. Drain.*
[8] Qiu Z and Sun M 2022 Mechanisms of impact on the soil environment of drip irrigation applying treated sewage effluent: a review in China *Water Supply* **22** 5785–99
[9] Salokhe V M, Babel M S and Tantau H J 2005 Water requirement of drip irrigated tomatoes grown in greenhouse in tropical environment *Agric. water Manag.* **71** 225–242
[10] Chaer M S I, Abdullah S H, Priyati A and others 2016 Aplikasi Mikrokontroler Arduino Pada Sistem Irigasi Tetes Untuk Tanaman Sawi (Brassica Juncea)(Application of Arduino Microcontroller on Drip Irrigation for Mustard Plant (Brassica juncea) *J. Ilm. Rekayasa Pertan. dan Biosist.* **4** 228–38
[11] Kandelous M M and Šimůnek J 2010 Comparison of numerical, analytical, and empirical models to estimate wetting patterns for surface and subsurface drip irrigation *Irrig. Sci.* **28** 435–44
[12] Abioye E A, Abidin M S Z, Mahmud M S A, Buyamin S, AbdRahman M K I, Otuoze A O, Ramli M S A and Ijike O D 2021 IoT-based monitoring and data-driven modeling of drip irrigation system for mustard leaf cultivation experiment *Inf. Process. Agric.* **8** 270–83
[13] Ani A and Gopalakrishnan P 2020 Automated Hydroponic Drip Irrigation Using Big Data *Proc. 2nd Int. Conf. Inven. Res. Comput. Appl. ICIRCA 2020* 370–5
[14] Jain R K, Gupta B, Ansari M, and Ray P P 2020 IoT Enabled Smart Drip Irrigation System Using Web/Android Applications *2020 11th Int. Conf. Comput. Commun. Netw. Technol. ICCCNT 2020*
[15] Q.Ren 2022 Seismic acoustic full waveform inversion based on the steepest descent method and simple linear regression analysis *J. Appl. Geophys.* 104686
[16] Sarathy J and Vadhan S 2022 *Analyzing the Differentially Private Theil-Sen Estimator for Simple Linear Regression* vol 1 (Association for Computing Machinery)
[17] Provenzano G, Pumo D and Dio P Di 2005 Simplified procedure to evaluate head losses in drip irrigation laterals *J. Irrig. Drain. Eng.* **131** 525–532
[18] Hashemi S, Filion Y, Speight V and Long A 2020 Effect of Pipe Size and Location on Water-Main Head Loss in Water Distribution Systems *J. Water Resour. Plan. Manag.* **146** 1–11
[19] Jamil R 2019 Frictional head loss relation between Hazen-Williams and Darcy-Weisbach equations for various water supply pipe materials *Int. J. Water* **13** 333–47
[20] Jianhua W, Wanzheng A and Qi Z 2010 Head loss coefficient of orifice plate energy dissipator *J. Hydraul. Res.* **48** 526–30
[21] Malejka S, Vadillo M A, Dienes Z and Shanks D R 2021 Correlation analysis to investigate unconscious mental processes: A critical appraisal and mini-tutorial *Cognition* **212** 1–20
[22] Wan F, Liu W and Bretz F 2022 Confidence Sets for a level set in linear regression 1–20
[23] Bual C L C, Cunanan R D, Bedruz R A R, Kerwin Billones R and Vicerra R R P 2019 Microcontroller-based Irrigation System for Experimental Design of Mung Bean Sprouts Cultivation *2019 IEEE 11th Int. Conf. Humanoid, Nanotechnology, Inf. Technol. Commun. Control. Environ. Manag. HNICEM 2019* 0–5
[24] de Aguiar R A, da Cunha M G and Lobo Junior M 2014 Management of white mold in processing tomatoes by Trichoderma spp. and chemical fungicides applied by drip irrigation *Biol. Control* **74** 1–5
[25] Judd C M, McClelland G H and Ryan C S 2017 *Data analysis: A model comparison approach to regression, ANOVA, and beyond* (New York: Routledge)

Review of Environmental Management System Implementation in The Oil and Gas Company in Indonesia

Adonis Muzanni[1], Tri Edhi Budhi Soesilo[1*], Dwi Nowo Martono[1], Udi Syahnoedi Hamzah[1], Wezia Berkademi[1], David Febraldo[1]
[1]School of Environmental Science, Universitas Indonesia, Jakarta, Indonesia, 10430
[1]muzannilestari@gmail.com, [2*]tri.edhi@ui.ac.id, [3]dwinowo11@ui.ac.id, [4]syahnoedi@yahoo.com, [5]wezia.berkademi91@ui.ac.id, [6]davidfebraldo@gmail.com

Abstract. Managing the environment is vital in guaranteeing beneficial effects to environmental action for oil and gas companies. This study aimed to analyze the programs and efforts to prevent environmental pollution by applying the Environmental Management System (EMS) in the oil and gas industry. The most popular, frequently applied EMS is the ISO 14001 and Amdal. This study used a literature review method from 26 literature and three databases: Google Scholar, ScienceDirect, and Scopus. Inclusion criteria in this study are articles related to the environmental management system in Indonesia's oil and gas company. This study looked forward to giving insights to the actors in this industry in using the opportunity to invest while implementing efforts to prevent environmental damages and their impacts. Six underlying parts of the choice cycle have been created and utilized in using EMS: strategy/arranging, project appraisal, execution, technical thoroughness, interview, and institutional trustworthiness.

1. Introduction

The oil and gas business activity are one of the vital industries as it produces energy to supply the world's increasing needs. Energy consumption keeps growing. The oil and gas industry has another characteristic from other industries, such as the manufacturing industry or mining industry, in the sense that this industry has a high economic risk during the exploration stage with a high potential to fail in achieving the production stage and a high potential for losing a considerable amount of the capital [1]. During the production phase, the time needed to reach the break-even point is very long, but the profit is very high [1,2]. Expanding consciousness of ecological issues and enhancements in realistic training programs has impacted the interest for and advancement of environment amicable strategic policies [3]. Although a significant part of the previous exploration has zeroed in on the effect of enormous organizations on the climate, it has been proposed that the assessed aggregate effect of little and moderate ventures on the climate is generous [4] and exceeds the consolidated ecological effect of huge organizations. Consequently, it could be contended that more critical consideration should be compensated to this area in environmental management research [5].

Even though the corporate profit growth is increasing, and the local state economic growth is also rising (SE), the escalation of social and environmental crises (ESEC) is also intensifying at the same time. Social and economic issues become more complex and dangerous. This phenomenon is often called the "economic growth paradox" [6,7]. Several examples of environmental pollution incidents in the oil and gas industries have presented a substantial financial loss to the company in Indonesia. The oil spill in Balikpapan city on Mei 29, 2006, is one example [8,9]. Another example is the oil spill in the Balikpapan gulf that causes seawater pollution with a potential ecological loss of around IDR 111.7 trillion, equal to USD 7,9 billion. Globally, environmental pollution incidents due to the oil and gas production process are also seen, such as the explosion of BP's soil drilling rig in Mexico Gulf, the United States, which led to an enormous crude oil spill that polluted the Mexico Gulf. BP lost USD 40.9 million, or around IDR 384 trillion, in this incident [10].

Implementing an environmental management program can reduce the production cost for a company as it is considered to give a tangible profit in the form of reduced environmental damage and risk recovery [11,12]. Currently, numerous environmental management tools are being used, for example, natural effect evaluation, ecological assessments, substance flow examination, environmental management systems, and life-cycle assessments (LCA) [13,14]. These apparatuses give ecological data and help natural administration make associations all the more suitable for eco-friendly action [14]. By these apparatuses, an Environmental Management System (EMS) is perhaps the likeliest usually utilized in associations worldwide [15]. In addition, intangible benefits are gained from such programs, including increased productivity, future potential efficiency, access to bank loans and insurance premiums, better company image in the eye of the public, absence of court process costs related to environmental pollution cases, and higher environmental quality. A living system cannot be sustained when the needs for materials and energy exceed the environmental system's capacity to provide inputs for the necessary resources [16]. Natural disasters can also cause environmental pollution. Analysis of the probability of extreme storms in various countries will be valuable for identifying costs when applied to determine the investment priorities in Indonesia [17].

1.1. Sustainable Development Theory
Economic development to achieve a prosperous society often has to face a challenging dilemma between the demands to be met and the efforts to maintain sustainability. Natural resource exploitation can negatively impact the environment, which has a limited carrying capacity if it is not well-conserved. The sustainable development concept was initiated by Malthus at 1798, who was concerned about the availability of land from the perspective of population explosion. In 1972, Meadow et al. published a book entitled The Limit to Growth that described the notion that the availability of natural resources limits economic growth. This becomes the basis for the Bruntland Report–Our Common Future at 1987. The United Nations established the Bruntland commission in 1983 to prevent social and economic development regress and to reflect on ways to save human natural resources and the environment. Economic and social development aims to create an equitable relationship. It is expected that economic and environmental development should be viable. Development can be considered sustainable if performed in an overlap area between the economic, ecological, and social dimensions as the preferable condition.

1.2. Environmental management system
Management is the way of utilizing, making plans, taking the lead, and taking control of the member's efforts of the organization with the process of using company resources to achieve the predetermined organizational goals. Meanwhile, management is a procedure that includes action to plan, organize, and move human resources and other kinds of resources to achieve predetermined goals. EMS includes aspects of implementing the environmental policies of the overall management function (including making plans) that lead and determine (BBS 7750, in ISO 14001). Generally speaking, an EMS is a comprehensive and efficient administration framework with methodical cycles and strategies that permit an association to design its natural exercises evaluate and deal with its ecological viewpoints and effects, produce and execute its ecological arrangements and accomplish its ecological targets.

What is referred to as the environment in this study is the environment as defined in the ISO 14001 EMS, which relates to the internal and external environment. The essential elements of environmental management based on the above definition consist of the environmental aspects and impacts. In ISO

14001, company organizations must conduct audits to implement, inspect, and correct the planned environmental management system when irregularities are found. The audit program is scheduled on the importance of the environmental impacts documented in the assessment. Planning extended for the natural administration framework can be evaluated together with related exercises and the aftereffects of past reviews. The review methodology includes Scope of Audit, Methodology, Person-in-Charge, Reporting Implementation Requirement, and Documentation. The ISO 14001 management system includes ten integrated clauses. In ISO 14001, the critical elements used are presented in table 2.1 below [18].

Current studies have put effort into assessing an EMS's adequacy and ecological execution compared with the ISO14001 standard instead of researching its appropriation and advancement in an association [15,19,20]. The EMS is a continuous cycle of planning, implementation, evaluation, and process improvement that is organized so that the business objectives of the company/government can be achieved and the environmental objectives are well-integrated and synergized [18]. Evaluation of an Environmental Management program implementation through an economic approach has evolved to determine a significant socioeconomic burden in environmental pollution. For the economic evaluation results to be informative, cost data must be relevant and comparable to the consequences. Natural disasters usually occur as complex events of occasions that can affect the nearby economy badly and, in extreme cases, the economy of a nation [9]. Calculating damages caused by such events can become challenging because the costs related to natural disasters are eventually linked to various factors and, most importantly, vary according to the type of disaster [9].

The Economic Evaluation, developed from the company's perspective, aims to find answers to whether the limited resources have been used optimally and to see whether the results have been worth the number of costs already spent [21–23]. A quantitative assessment of what is edesired by the community is a definition of economic evaluation in investing in a project or program, where outcomes are valued in terms of expenses and consequences where the consequences are positive results or benefits gained from a program [24]. Environmental Management System implementation evaluation can be performed using economic analysis technics. The principle of the analysis is based on the financing and consequences of a service program or other activities. It is related to the community's life is generally identified through calculating the program's monetary value. However, the limitation of the economic assessment is that it does not consider the value of the environmental damage impacts or the suffering experienced, which is expressed in monetary terms.

The BP oil and gas disaster analysis in the Mexico Gulf stated that the process of producing oil and exploration is very challenging. However, they offer significant results, which are also balanced by major risks. The total damage to BP, the climate, and the US beachfront economy is estimated at USD 36.9 billion. The harm to the network, organizations, and people groups influenced by the oil slick can, in any event, be halfway estimated with a claim of the harm out cases that BP has paid. Economic analysis of public health programs is generally done by calculating the program's monetary value. One limitation of the economic assessment is that it does not consider the value of environmental damage or the suffering experienced in monetary terms. In the decision-making process, these values are considered, but in economic analysis that focuses on cost accounting, these are not considered [25].

2. Methods

This study is a methodical literature review with electronic writing academic journals with specific standards. This study used a literature review method from 3 databases: Google Scholar, ScienceDirect, and Scopus. Inclusion criteria in this study are articles linked to EMS in Indonesia's oil and Gas Company. Writing on EMS is viewed as necessary on the off chance that it meets the consideration standards, which are 1) articles identified with EMS implementation in oil and gas companies in Indonesia; 2) articles as results acquired from the catchphrases utilized; 4) articles have been distributed since 2000-2019; 5) the article is a literature review; 6) English language articles. The number of articles discovered was 533. All references that have been found are overseen utilizing Mendeley programming. Study determination measure incorporates 4 phases: ID, screening, qualification, and incorporate. Catchphrases in this study were EMS implementation in oil and gas companies in Indonesia used to look through articles on the information base. The articles acquired were then screened to get articles qualified to be integrated. Appraisal of this research quality utilized standard measures that test for misclassification, choice, and revealing by assessing components of examining technique, the sufficiency of tests, expectation of predisposition, focal point of intercession and correlation gatherings, investigation, appropriateness of objective tests, the portrayal of mediation strategies, assurance of consideration rules and prohibition, constraints of examination and announcing of result information.

3. Results and Discussion

3.1. Result

The public authority of Indonesia has given and authorized different laws and guidelines, organizations, and strategies that had the option to help the execution of feasible advancement projects. Indonesia instituted Act. Number 4 in 1982 concerning the necessary arrangement of the natural administration. In order to get these commitments, in 1986, the public authority had guidelines in line with EMS rehearses by the "Amdal framework." Essentially, the Amdalframework may be separated into an arrangement: EMS and Monitoring Measures (for the introductory advance of venture improvement), Environmental Impact Assessment (for program development and improvement plan), and EMS and Monitoring Plan (for the undertaking activities). Moreover, to further use the framework, the public authority has set ecological quality norms and specialized rules, announcing and archiving frameworks, management methodology, and a natural review framework. The Amdal framework was intended to give direction to any estimate organization in the company on the best way to oversee, estimate, upgrade, and impart the natural parts of their activities exercises.

ISO 14001 is an example of a progression of worldwide natural administration guidelines pointed toward advancing persistent improvement in organization ecological execution by the selection and usage of EMS. ISO 14001 characterizes a general EMS and includes the components: strategy, arranging, usage and activity, checking and remedial, and occasional administration survey. The EMS laid out in ISO 14001 gives an organized cycle to accomplishing continuous improvement, the rate and degree dictated by the association considering monetary and different conditions. Essentially, the Amdal framework has firmly amassed the model of ISO 14001, even though the utilization of different components in the Amdal framework is still earliest stages and may contrast. Moreover, the Amdal framework is restricted to just managing the related activities or elective relief projects of natural effect that can be executed. Given the said general components from assessing the Amdal framework has included up to 70 percent of the ISO 14001 framework necessities. Most of the components included in the Amdal framework are centered on activities steps, as it were. Item dispersion and administration steps were not canvassed in the Amdal framework. PERTAMINA and other oil and gas companies in Indonesia are yet to set up the ISO 14001 rules. By utilizing the Amdal framework as a flight highlight to arrive at the ISO 14001, the Author looks forward to the ISO 14001 can be executed for natural oil and gas companies from Indonesia.

3.2. Discussion

The fundamental idea of using an EMS for oil and gas tasks on the field is the takeoff from the Amdal framework in the blend of administrative responsibility. Center parts of EMS rehearses are the thought of the biological, innovative, and bureaucratic, with economic evaluation and social values on the chosen cycle. This bearing is an unsafe ruse if the arranging cycle is limited and data is fragmented. The system is also prone to deliver extreme unevenness between the operational from one viewpoint, and EMS rehearses on the other. Accordingly, by endeavoring to adjust the methodology. Six underlying parts of the choice cycle have been created and utilized in using EMS: strategy/arranging, project appraisal, execution, technical thoroughness, interview, and institutional trustworthiness. The six segments are emphatically connected, interrelated, and commonly steady.

The usage of EMS for petrol activities in the field territory depended on the endorsed Amdal archive and the organization's specialized rules for Health, Environmental, and Safety. Three significant advances engaged using the EMS in the field: natural administration plan, ecological checking and control, and honest reviews. The ecological plan has been set up to guarantee consistency with indicated working conditions. Some control systems have been associated with consistent observing projects, while others included nearby observation and investigation. Natural checking is performed to figure out which impacts of tasks occur. Checking gives two significant wellsprings of data: the natural effect of the movement and the viability of moderation measures. Both sorts of data give a significant contribution to use in other comparative ventures inside the field block. Honest reviews have been created to formalize the assortment and assessment of the ecological moderation data. Review likewise has been utilized to assess the viability of the organization's norms and systems in natural insurance and safeguarding, the association, innovation implementation, wagering and preparing, recording, and other problems.

4. Conclusion

Every one of the primary segments of EMS: strategy, project evaluation, and execution should be established on specialized sound premises, including data, assessment, and understanding, yet should likewise oblige esteem evaluation and public inclusion. To oblige their technical meticulousness, EMS usage in field territories already enhanced to a few specialized components such as improved ecological mindfulness and information for every concerned individual or colleague by EMS, training, gathering, preparing, counsel, and group discussion, it is significant that the workplace laborers and their bosses comprehend what is needed on EMS program.

Recommendations for further research that can be done in the field such as coordination and communication to the government official and other same type offices to carry out formal workplace investigation to guarantee the tasks already agreed to the regulations and guidelines like expressed on the affirmed Amdal record, recognition of possible effect or risks be for changing the nature or activities offices, the fixed project including the assessment and examination of the possible issue was created to limit activity's effects on the climate, arrangement of a crisis reaction plan to decrease the likely for crises with the seriousness, endeavors company and activities to perceive and react to any worries or attentiveness of authority, network, or the social platform, implementation of normal ecological checking, examination, documenting, and revealing which is needed by Amdal report, official guidelines, and corporate standard.

Acknowledgments

We would like to express our gratitude to the people who helped in this research. This study is funded through the Doctoral Publication Grant Program of the Ministry of Education, Culture, Research and Technology Financial Year 2022, under the Contract Number of NKB-1044/UN2.RST/HKP.05.00/2022

References

[1] Suda K A, Nazatul S A R, Hamzah A R and Chen W 2015 A Review on Risks and Project Risks Management: Oil and Gas Industry Machine learning and Big Data analytics-swarm intelligence approaches View project A Review on Risks and Project Risks Management: Oil and Gas Industry *Int. J. Sci. Eng. Res.* **6** 938–43

[2] Obro-Adibo D, Agbodohu G and Kumi W 2013 Analysis of Risk Management Practices in the Oil and Gas Industry in Ghana. Case Study of Tema Oil Refinery (Tor) *Eur. J. Bus. Manag.* **5** 2222–839

[3] Newton P, Baum W S, Bhatia K, Brown S, Cameron A S, Foran B, Grant T, Mak S, Memmott P and Mitchel S 2001 *Human Settlements, Australia State of the Environment Report 2001* (Canberra: Department of the Environment and Heritage)

[4] Hillary R 2000 *Small and Medium-sized Enterprises and the Environment: Business*

Imperatives (Shefield: Greenleaf Publishing)
[5] Hamzah A, Saleh A A, Subagio M and Santoso B 1998 Environmental management system integration on natural gas operational practices: Experience in the Kangean block area, Indonesia *Soc. Pet. Eng. - SPE Int. Conf. Heal. Safety, Environ. Oil Gas Explor. Prod. 1998, HSE 1998*
[6] Lako A 2014 *Green Economy, Menghijaukan Ekonomi, Bisnis & Akutansi* (Jakarata: Erlangga)
[7] Nurmi V and Ahtiainen H 2018 Distributional Weights in Environmental Valuation and Cost-benefit Analysis: Theory and Practice *Ecol. Econ.* **150** 217–28
[8] Drake P 2016 Multiple visions of Indonesia's mud volcano: Understanding representations of disaster across discursive settings *Disasters* **40** 346–64
[9] Ratnatunga J and Sopanah A 2015 Disaster Financing: A Contingent Valuation Approach. *J. Appl. Manag. Account. Res.* **13** 1–27
[10] Smith L C, Smith M and Ashcroft P 2012 Analysis of Environmental and Economic Damages from British Petroleum's Deepwater Horizon Oil Spill *SSRN Electron. J.*
[11] Brown J P, Fitzgerald T and Weber J G 2019 Does resource ownership matter? Oil and gas royalties and the income effect of extraction *J. Assoc. Environ. Resour. Econ.* **6** 853–78
[12] Barbier E B and Burgess J C 2017 The sustainable development goals and the systems approach to sustainability *Economics* **11**
[13] Ramos T B and Joanaz De Melo J 2005 Environmental management practices in the defence sector: Assessment of the Portuguese military's environmental profile *J. Clean. Prod.* **13** 1117–30
[14] Emilsson S and Hjelm O 2002 Implementation of standardised environmental management systems in Swedish local authorities: Reasons, expectations and some outcomes *Environ. Sci. Policy* **5** 443–8
[15] Nawrocka D and Parker T 2009 Finding the connection: environmental management systems and environmental performance *J. Clean. Prod.* **17** 601–7
[16] Miller G T and Spoolman S E 2014 *Environmental Science (Fifteenth Edition)* (USA)
[17] Haryanto B, Lestari F and Nurlambang T 2020 *Extreme Weather Events and Human Health. Extreme Weather Events and Human Health* (Springer International Publishing)
[18] BSI 2015 BSI Standards Publication Environmental management systems — Requirements with guidance for use *BSI*
[19] Dawson M 2004 Environmental Management Systems (EMS) in the Military Sector *Def. Environ. Eff. Sci. Commun.* 75–83
[20] Hertin J, Berkhout F, Wagner M and Tyteca D 2008 Are EMS environmentally effective? The link between environmental management systems and environmental performance in European companies *J. Environ. Plan. Manag.* **51** 259–83
[21] Riaño-Casallas M I and Tompa E 2018 Cost-benefit analysis of investment in occupational health and safety in Colombian companies *Am. J. Ind. Med.* **61** 893–900
[22] Tompa E, Dolinschi R and De Oliveira C 2006 Practice and potential of economic evaluation of workplace-based interventions for occupational health and safety *J. Occup. Rehabil.* **16** 375–400
[23] Tompa E, Dolinschi R, De Oliveira C and Irvin E 2009 A systematic review of occupational health and safety interventions with economic analyses *J. Occup. Environ. Med.* **51** 1004–23
[24] Uegaki K, De Bruijne M C, Van Der Beek A J, Van Mechelen W and Van Tulder M W 2011 Economic evaluations of occupational health interventions from a company's perspective: A systematic review of methods to estimate the cost of health-related productivity loss *J. Occup. Rehabil.* **21** 90–9
[25] Adams A 2015 Summary of Information concerning the Ecological and Economic Impacts of the BP Deepwater Horizon Oil Spill Disaster *Natl. Resour. Def. Counc.* 1–9

IOP Conf. Series: Earth and Environmental Science 1111 (2022) 012019 doi:10.1088/1755-1315/1111/1/012019

Diagenesis of Subang Formation in Cibeet River, Bogor, West Java

R S Nasution[1], A Luthfiyyah [1], Rezky Aditiyo[1]*, and M R Septyandy[1]
[1]Geology Study Program, Faculty of Mathematics and Natural Sciences, Universitas Indonesia, Depok, 16424, Indonesia
[1]rifa.saifanah@ui.ac.id, [2]afifah.luthfiyyah@ui.ac.id, [3]*rezkyaditiyo@sci.ui.ac.id, [4]m.rizqy@sci.ui.ac.id

Abstract. Subang Formation is one of many turbidite flow-induced sediment that built Bogor Trough, Java. Despite its marine origin, the exposure to the meteoric environment and burial process caused this formation to go under diagenesis, thus altering its primary characteristics. Intensive cementation by equant calcite, feldspar alteration, and grain dissolution indicates diagenesis in a meteoric environment. The cementation process filled primary and secondary pores, decreasing the primary porosity. On the other hand, the dissolution of the grain formed new secondary porosities. The presence of stylolites and fractured fragments indicated the formation had experienced burial. Although compaction often decreases rock's porosity, the fracturing of the rocks due to compaction leads to fracture porosity formation, increasing rock's porosity. Analyzing the compaction signs that destroyed the cementation and dissolution features in the rock samples, it is assumed that the diagenesis had occurred in the meteoric environment before it occurred in the burial environment. This study showed that the heterogeneity of the Subang Formation in the Cibeet River resulted from all these diagenesis processes mentioned above. Furthermore, the diagenesis study of the Subang Formation can be implemented as a preliminary stage to determine the quality of the petroleum reservoir. Diagenesis of Subang Formation in Cibeet River, Bogor, West Java.

1. Introduction

Subang Formation is one of many formations that built Bogor Trough, West Java [1]. During the Miocene epoch, Subang Formation was deposited as clastic sediment by turbiditic flow in the marine environment, overlaid the Jatiluhur Formation and Klapanunggal limestone [3]. Today, Subang Formation is exposed to the surface on the eastern side of Bogor Trough [2], including the Cibeet River, Bogor, and West Java. The Subang Formation in the study area is located in the northwestern part of the Cianjur Quadrangle, represented by Late Miocene claystone consisting of nodules and marl, marly limestone beds, and beds of dark limestone. In some places, intercalations of glauconitic sandstone were found [3].

Diagenesis refers to processes that occur to rocks after deposition [5,6] that can alter the rock's primary characteristics [7]. Some diagenesis processes are cementation, compaction, dissolution, recrystallization, and replacement [5,8]. These diagenesis processes are recorded in the rocks and can be analyzed using several methods, including petrography, x-ray diffraction, and scanning electron microscopy (SEM).

The previous study of the Subang Formation was mainly focused on the claystone member, focusing on the rock's durability [4]. The environment evolution of the Subang Formation, from its deposition until today, makes it an interesting object for diagenesis study. This paper aims to analyze diagenesis processes that occurred in Subang Formation and their effects on the rock's primary characteristics. The result obtained from this research can be implemented as a preliminary stage to determine the quality of petroleum reservoirs. Hence, it became the novelty of this study.

2. Materials and method

Field data was collected with sixteen rock samples taken from four outcrops: carbonaceous sandstones, sandstones, carbonaceous siltstones, and siltstones. The research's scope of the problem is limited to sandstone and carbonaceous samples as the dominant lithology of the Subang Formation in the Cibeet River. Three laboratory tests were conducted for about five months before the Community Activities Restriction Enforcement by the government to prevent COVID-19.

2.1. Petrographic analysis

Petrography analysis was carried out to determine diagenesis features, namely cementation, dissolution, mineral alteration, compaction, and porosity. Thin sections were prepared using blue-dye resin to mark the porosity of the rocks. Petrography analysis was conducted in Optical Mineralogy Laboratories of Universitas Indonesia using a polarized light microscope, with 2-, 4-, 10-, and 40-times magnification under plane-polarized light (PPL) and cross-polarized light (XPL) microscopic views.

2.2. X-ray diffraction (XRD) analysis

Three samples were prepared and underwent the XRD test further to quantify the rocks' mineralogy composition using x-rays [9, 10]. The samples were chosen based on the low visibility of their mineralogy from the petrography analysis that can be determined by reading the results of the XRD diagram, such as the mineral group of quartz, feldspar, and clays. The quantification then being compared with petrography and SEM data.

2.3. Scanning electron microscopy (SEM) analysis

SEM analysis was carried out to observe the morphology of minerals in three dimensions on a minor scale [11], analyze the cementation process, and define rock porosity on a smaller scale. This analysis was done on three rock samples, namely samples 1.2, 2.1, and 2.2. 26 SEM pictures were taken using 1200 – 8500 times magnification. The pictures were then studied and analyzed to complement the results from previous methods.

3. Results and discussion

Previous studies divided Subang Formation into three members: breccia member, sandstone member, and clay member [3]. Its units' distributions differ from place to place, hence the lack of a certain lithotype [1]. This research is the first sequence to identify the stages of the diagenesis process and history, starting from determining lithofacies assemblages at the Cibeet River. The previous diagenesis study was conducted by Abdurrokhim & Ito [12] based on the distribution of rocks in the Jatiluhur Formation, Cipamingkis River, overlaid by the Subang Formation in the southern part of the research area. Therefore, this study can add information on the Subang Formation's diagenesis process located at Cibeet River.

3.1. Lithofacies assemblages

At Cibeet River, Subang Formation was exposed as an intercalation of limestone, sandstone, and siltstone, as seen in Figure 1 and Figure 2. Although all samples taken are reactive to hydrochloric acid (HCl), the limestones have more than 50% of calcite, whereas the sandstones have less than 50%, as shown in Table 1. The sandstones mainly consisted of quartz, albite, muscovite, and calcite.

The lowest part of the outcrop is a gray, thick, massive bed of splendid sandstone containing glauconite and molluscs shells. It is overlain by several intercalations of limestone and sandstone (Figure 2A&E) before turning into a thick, massive bed of limestone with calcite veins. The thick bed of limestone contained glauconite, lithic fragments, quartz, and plagioclase feldspar. The limestone itself is beige colored grainstone with an excellent sand-sized particle (Figure 2F).

Thickness	Lithology	Description
4 m		Massive, gray siltstone with fractures and quartz veins. Contains lithic fragment, quartz, and carbonate minerals.
5 m		Massive, brown, very fine sandstone with fractures and quartz vein. Contains foraminifera shells
5 m		Massive, gray siltstone with fractures and quartz veins. Contains lithic fragment, quartz, and carbonate minerals.
0,8 m		Massive, brown, very fine sandstone with fractures and quartz vein. Contains foraminifera shells.
10 m		Massive, beige limestone with fractures and calcite veins. Contains shell fragments, quartz, plagioclase feldspar, and lithic fragments.
1 m		Massive, gray very fine sandstone. Contains glauconite and shell fragments.
1 m		Massive, beige limestone. Contains shell fragments, quartz, plagioclase feldspar, and lithic fragments.
2 m		Massive, gray very fine sandstone. Contains glauconite and shell fragments.

Figure 1. Simplified illustration of outcrop's stratigraphy in Cibeet River, Bogor.

The limestone bed is overlain by a massive bed of splendid sandstone containing foraminifera shells. The massive sandstone bed is overlain by parallel bedded, intercalated sandstone and siltstones (Figure 2B-D). The sandstone beds are massive, with some fractures and quartz veins. The siltstone beds are gray, with several fractures and quartz veins (Figure 2H), containing lithic and quartz fragment, with small amount of carbonate minerals.

3.2. Cementation

Calcite and silica cement can be found in the samples. The calcite cement is mostly even, equant to blocky shaped, characterizing meteoric diagenesis [5]. The evenly distributed cement growth indicated that it is likely that the cementation took place in the phreatic zone. Massive, uneven poikilitic calcite cement also grew unevenly in the sandstones (Figure 3F), indicating burial diagenesis process [6]. Silica cement can only be found in the sandstone samples (Figure 4C&D), which appear as a thin, dark margin surrounding the mineral grains.

3.3. Dissolution

Most of the carbonate grain in the rocks has been completely dissolved, forming secondary porosities, or filled with calcite, as shown in Figure 3F and Figure 4A. Carbonate shells usually underwent replacement from aragonite into calcite because calcite is more stable than aragonite. These prove that the rocks underwent meteoric diagenesis [5].

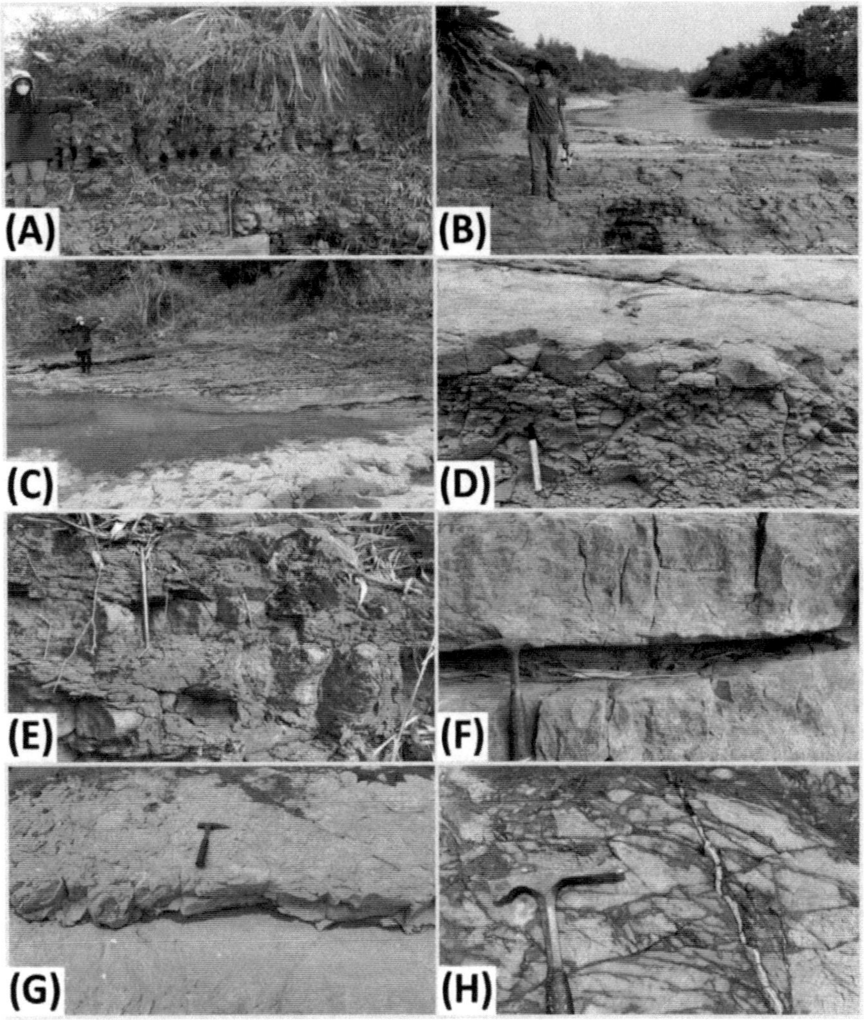

Figure 2. Subang Formation outcrops in Cibeet River, Bogor. (A) Intercalation of sandstone and siltstone; (B) Parallel bedding of sandstone intercalating with limestone and siltstone; (C) Intercalation of sandstone and siltstone; (D) A closeup of facies change from siltstone to sandstone; (E) A closeup of intercalation of grey siltstone and beige sandstone; (F) Parallel bedding of limestone; (G) Erosional surface between two beds of sandstone; (H) A bed of siltstone with fractures and quartz vein.

Table 1. X-ray diffraction (XRD) data of the samples

	%Calcite	%Albite	%Muscovite	%Quartz
Limestone	52.26	24	17	7.7
Sandstone 1	8.7	5.9	79.8	5.41
Sandstone 2	24	17	12	47

Figure 3. Thin section pictures of Subang Formation samples. Red bar scale = 0.5 mm. (A) PPL-Grainstone with moderately compacted grain packing, cemented by equant calcite, showing point contacts and concavo-convex contacts of the grains. Some dissolutions in form of intraparticle and vuggy porosity also exist; (B) XPL-Stylolite separates pre-existing equant to blocky calcite cements and grains, creating discontinuity. Vuggy porosity formed around the stylolite; (C) PPL-Moderately-compacted grainstone with several dark insoluble mineral in the center of the photograph. The grains are surrounded by cloudy calcite micrite. Some grains have been dissolved, while some are filled with micrite; (D) XPL-Sandstone with fracture porosity due to dissolution. Leftover insoluble mineral filled the porosity. Note the altered feldspar surrounding the grains; (E) PPL- Sandstone with floaty texture and large interparticle porosity. Small amount of silica cement surrounding the grains; (F) XPL-Sandstone with fractured feldspar grains, some are filled with calcite. Calcite cement in grew unevenly between the grains (green arrow).

3.4. Feldspar alteration

Most of the feldspar minerals contained in the samples are dissolved at the margin or, sometimes, the center of the grain. The outer part of the feldspar is then surrounded by layers of clay minerals that appear brown in thin sections. Through SEM pictures, it can be inferred that the feldspar has undergone alteration to form pseudohexagonal kaolinite (Figure 4B). The alteration from feldspar to this type of kaolinite is sometimes associated with heavy weathering (Figure 2G) [13].

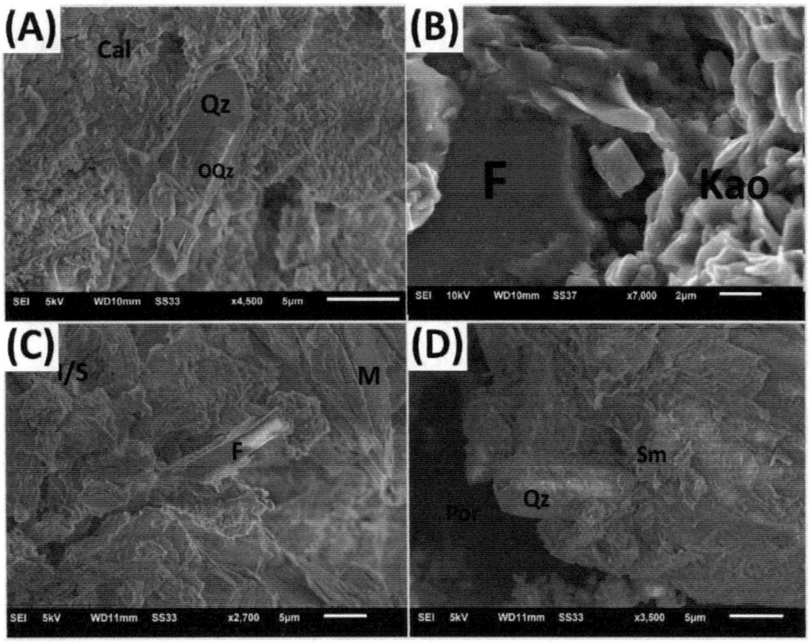

Figure 4. SEM pictures of samples of Subang Formation, showing close-up morphology of cements. (A) Quartz grain (Qz) with overgrowth (OQz), shown as cleavage on the quartz. The quartz is surrounded by equant, subhedral calcite cement (Cal); (B) Feldspar grain (F) surrounded by kaolinite (Kao) which appear pseudohexagonal shaped. This might be a sign of feldspar alteration; (C) Feldspar grain (F) surrounded by illite/smectite (I/S) clay, connecting the grain with presumably euhedral muscovite clay (M); (D) Quartz grain (Q) surrounded by flame-like shaped smectite clay. It also shows interparticle porosity (Por) that has not been filled with cement nor matrix.

3.5. *Compaction*
All rock samples show a fair amount of mechanical and chemical compaction signs. Chemical compaction in Subang Formation leads to the formation of stylolites and grain contacts. Mechanical compaction event is shown by the packing of grains, fracturing, and deformation of grains.

Chemical compaction occurs in the form of a pressure solution, a process where the solubility of the grains is increasing due to an increase in pressure mainly at the grains' contacts [14], thus allowing the grains to overlap (Figure 4B) [5]. Sandstone units' grain contacts are primarily point contact and concavo-convex (Figure 3A, 3C, 4B). Floaty contact and point contact indicates the early stage of pressure solution. As the pressure solution process proceed, the point contacts developed into concavo-convex contacts. The leftover insoluble minerals [5] then formed a dark, curvy line called stylolites. These two features indicate a late-stage burial diagenesis.

Fractures in Subang Formation sandstones, appear outside and inside the grains (Figure 3D). These fractures then act as secondary porosities or filled with calcite minerals. Some of the samples, although having floaty fabrics (Figure 3E), also developed several fractures (Figure 3F).

3.6. *Porosity*
Sandstone porosity can be seen as the blue parts in thin sections and void spaces shown in the SEM analysis. At least three dominant pore types existed in the samples: intraparticle, interparticle, and fracture. Intraparticle pores formed due to grain dissolution by meteoric waters during diagenesis,

creating a new void inside the grain [6]. Intraparticle porosity can be seen in all samples. Interparticle porosity formed around the grains (Figure 3E), while fracture porosity occurs inside and around the mineral grains (Figure 3F, 3D).

4. Conclusions

After its deposition in the marine realm, Subang Formation underwent several diagenesis processes in meteoric and burial environments. The meteoric diagenesis caused cementation by blocky to equant calcite cement, dissolution and replacement of skeletal grains, and feldspar alteration. Subsequently, burial diagenesis caused Subang Formation to compact mechanically and chemically, resulted in the formation of fractures, grain contacts, and stylolites. The dominant diagenesis processes of Subang Formation are compaction which is dominated by mechanical and chemical compaction, cementation by calcite and clays, feldspar alteration, and grain dissolution.

Acknowledgments

The authors would like to express gratitude to Ori Zahra Sativa as fieldwork and analysis partner for this research. Universitas Indonesia financially supports this research through the International Indexed Publication (PUTI) Saintekes Fiscal Year 2020, contract number NKB-4891/UN2.RST/HKP.05.00/2020.

References

[1] Martodjojo S. Evolusi Cekungan Bogor Jawa Barat. Penerbit ITB Bandung. Bandung: Penerbit ITB; 2003. 239 p.
[2] Alam S. No Title. *J Geol dan Sumberd Miner*. 2020;**21**(199–206).
[3] Sudjatmiko. Peta Geologi Lembar Cianjur, Jawa, Skala 1:100.000. Bandung: Pusat Penelitian dan Pengembangan Geologi; 1972.
[4] Misbahudin M, Sadisun IA. Analisis Ketahanan (Durability) Batulempung Formasi Subang Di Daerah Ujungjaya Dan Sekitarnya, Kabupaten Sumedang, Jawa Barat. *Bull Geol*. 2018;**2**(1):163–74.
[5] Adams AE, MacKenzie WS. A Color Atlas of Carbonate Sediments and Rocks Under the Microscope. London: Manson Publishing Ltd; 1998. 101–156 p.
[6] Scholle PA, Ulmer-Scholle DS. A Color Guide to the Petrography of Carbonate Rocks. Tulsa: The American Association of Petroleum Geologists; 2003.
[7] Al-Juboury A, Howard JP, Vincent SJ, Nichols G. Marine and Petroleum Geology. 2021;**123**:1–18.
[8] Li J, Zhang X, Tian J, Liang Q, Cao T. Effects of deposition and diagenesis on sandstone reservoir quality: A case study of Permian sandstones formed in a braided river sedimentary system, northern Ordos Basin, Northern China. *J Asian Earth Sci*. 2021;**213**:2–16.
[9] Bunaciu AA, Udriştioiu EG, Aboul-Enein HY. X-Ray Diffraction: Instrumentation and Applications. *Crit Rev Anal Chem*. 2015;**45**:289–99.
[10] Wei W, Zhu X, Azmy K, Zhu S, He M, Sun S. Depositional and compositional controls on diagenesis of the mixed siliciclastic-volcaniclastic sandstones: A case study of the Lower Cretaceous in Erennaoer Sag, Erlian Basin, NE China. *J Pet Sci Eng*. 2020;**188**:1–12.
[11] Chen L, Xu J, Chen J. Applications of scanning electron microscopy in earth sciences. *Sci China Earth Sci*. 2015;**58**:1768–1778.
[12] Abdurrokhim, Ito M. The role of slump scars in slope channel initiation: A case study from the Miocene Jatiluhur Formation in the Bogor Trough, West Java. *J Asian Earth Sci* [Internet]. 2013;**73**:68–86. Available from: http://dx.doi.org/10.1016/j.jseaes.2013.04.005
[13] Santos AE de A, Rossetti D de F. Origin of the Rio Capim Kaolin based on optical (petrographic and SEM) data. *J South Am Earth Sci*. 2008;**26**(3):329–41.
[14] Bruna PO, Lavenu APC, Matonti C, Bertotti G. Are stylolites fluid-flow efficient features? *J Struct Geol*. 2019;**125**:270–7.

Facies, Facies Association, and Depositional Environment of The Klapanunggal Formation, Klapanunggal District, West Java

Desy Rinikasari[1], Rezky Aditiyo[1*], Khalisha Shafa Yumnanisa[1] and Natasya Alia Zahran[1]
[1] Geology Study Program, Faculty of Mathematics and Natural Sciences (FMIPA), Universitas Indonesia, Depok 16424, Indonesia
[1]desyrinikas9@gmail, [2]*rezkyaditiyo@sci.ui.ac.id, [3]khalisha.shafa91@ui.ac.id, [4]natasya.alia@ui.ac.id

Abstract. The research focuses on the Klapanunggal Formation, composed of limestone in the West Java Basin, specifically in the Klapanunggal area, Klapanunggal District, Bogor Regency, West Java Province. A massive limestone outcrop was found in this location. The method used is field data collection, stratigraphic cross-section measurement, and petrographic analysis. It aims to determine the facies characteristics, depositional environment, and association of the limestone facies of the Klapanunggal Formation. The results obtained from macroscopic and microscopic descriptions resulted in six (6) facies: Skeletal Wackestone, Skeletal Packstone, Benthic Foram Grainstone, Skeletal Rudstone, Coral Rudstone, and Bindstone. Then, the facies are classified into facies associations, resulting in three (3) facies associations: Platform Interior (Open Marine), Platform Margin Reef, and Slope.

1. Introduction

The North West Java Basin is a sedimentary basin consisting of limestone formations, one of them is Klapanunggal Formation, which was deposited in the northern part of West Java during the Middle Miocene - Upper Miocene Age [1]. Currently, this formation is formed in the back-arc basin as a response to the subduction between the Eurasian Plate and Indo-Australian plate [2] shown in figure 1A. The Miocene Carbonate in the Klapanunggal Formation is one of the indications of carbonate development in Java-Bali, which is part of the Southeast Asian equatorial carbonate formed in shallow marine. The formation was formed in the Oligocene - Miocene when the sea level decreased and spread to the back-arc basin of West Java [3]. This resulted in carbonate growth in the back-arc basin consisting of the productive hydrocarbon reservoirs during the Neogene [1]. Miocene carbonate development occurred north and south of the volcanic arc, bounded by structural highs [4]. The Klapanunggal Formation with a thickness of up to 500 m onshore and offshore from a few feet to 400 feet and it is part of the Northwest Basin and North Bogor Trough [5].

In addition, fossils had a dominant change from large foraminifera to coral. Stratigraphically, the Klapanunggal Formation was deposited above the Jatiluhur Formation, which is composed of clastic sediments, followed by the deposition of the Subang Formation and the Cantayan Formation [5] that is shown in figure 1B.

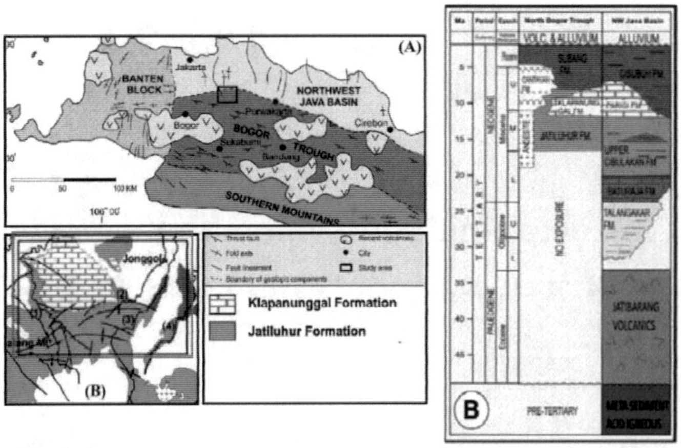

A B

Figure 1. (A) Tectonic elements of the West Java Basin & geological map of the north side of Bogor Through. The research area is shown by red mark; (B) Regional stratigraphy of the Northwest Java Basin and North Bogor Trough [6].

2. Methods

The method used is field data collection, stratigraphic cross-section measurement, and petrographic analysis. The data was collected using the field observation method on well-exposed outcrop and measuring sections. The total vertical thickness of each section is about 20 m – 25 m. Rock samples taken were 37 samples based on each representative facies in the measuring section. Furthermore, as many as 36 rock samples were made into thin sections based on the differences in the presence of fossils in rocks to see the overall fossils in these rocks. The thin sections were given a solution of blue dye and alizarine red with 4X magnifications to determine the carbonate component and facies characteristics based on classification Embry & Klovan (1971) [7]. Information from the field and laboratory were combined to determine facies characteristics, depositional environment, and diagenesis history.

3. Results and Discussion

3.1. Field data

Based on the results of field data collection, rock samples were identified using the classification approach by Embry & Klovan (1971) [7]. Figure 2 shows one of the outcrops in the field. Furthermore, the stratigraphic cross-sectional measurements resulted in four log sections, namely the KLP 1.1-1.3 log section located in the northwest, the KLP 1.4 log section in the southwest, the KLP 2.1, 2.3-2.9 log section in the east, and the KLP 2.2 log section in the northeast. Those data collections are shown in figure 3.

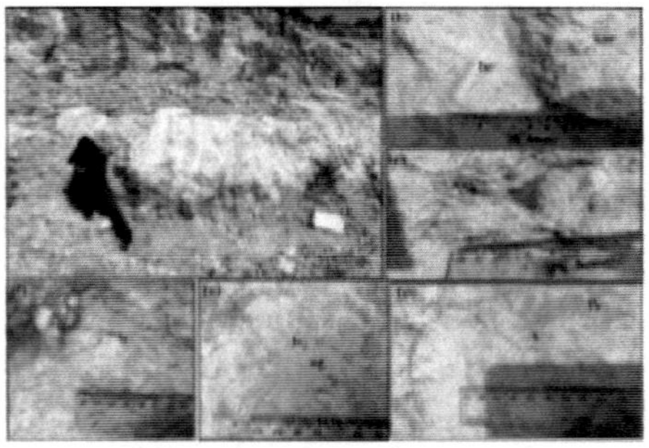

Figure 2. Shows (A) outcrop Appearance at KLP 2.1 B; (B) Boring (br) and Coral (kor); (C) High Porosity Cemented; (D) Bivalve Fossils (b); (E) Bivalve Fragments (b) and Algae (ag); (F) Benthic Foram (fb) and Bivalve Fraction (b)

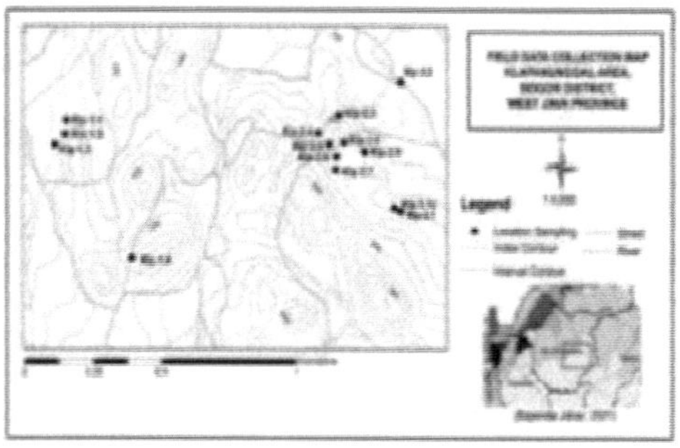

Figure 3. Field Data Collection Map.

3.2. Facies description

Rock facies can be determined by analyzing macroscopic descriptions in the field and petrographic analysis. Differences in lithological characteristics, fossil composition, matrix, and cement produce various facies. Based on the classification of Embry & Klovan (1971) [7], the results of macroscopic and microscopic descriptions produced six (6) facies which is stated as F.

3.2.1. F1 - Skeletal Wackestone, F2 - Skeletal Packstone, & F3 - Benthic Foram Grainstone. The Skeletal Wackestone Facies has a gray color with a mud composition of 40 %, which shows dominance compared to the fossil composition. Fossils in these facies include red algae, corals, and large foraminifera of about 0.2mm – 2mm, less than 10 %. There is no dominant fossil composition found. Quartz and Opaque minerals are also found with amounts of 4 % and 0.67 %. The Skeletal Packstone Facies has a white to gray color with a dominance of 31.5 – 35 % of fossil fragments compared to 31.67 – 32.5 % of the mud matrix. The fossils found included red algae, bivalves, and corals measuring 1– 2 mm with a percentage of less than 10 %. There is no dominant fossil material found. These facies also found 2 – 7 % Quartz and 5 – 7.5 % Opaque. The Benthic Foram Grainstone Facies is white and creams with a sand and mud matrix. The composition of the fossils found were red algae, large foraminifera, and undetermined fossils. Fossils of large foraminifera with a size of 0.4 mm to 2.2 mm dominate compared to other fossils. There are also 1.67 % Quartz minerals.

3.2.2. F4 - Skeletal Rudstone, F5 - Coral Rudstone, & F6 - Bindstone. The Skeletal Rudstone Facies has chalky white, cream, and gray colors with a 41 – 55.9% of fossil composition is more dominant than 16.67 – 26.67 % of the matrix. The fossils were small foraminifera, large foraminifera, corals (platy and massive), red algae, bivalves, bivalves, gastropod fragments, bryozoans, radiolaria, and echinoids with a size of about 0.2 – 3 mm. Also, there is bioturbation in the form of boring (Skolithos) and burrowing (Thalassinoides). In addition, there is 0.33 % to 3.33 % Quartz and 3.33 % to 5 % Opaque. The Coral Rudstone Facies has chalky white, white, gray, and cream colors. This facies is allochthonous and dominated by 35.01 – 62 % of fossil composition rather than bivalves, red algae, large foraminifera, and bryozoans with a size of about 0.2 – 6 mm. In these facies, coral fossils are found dominantly. Furthermore, bioturbation was found in the form of a boring (Skolithos). In this facies also found 0.67 % to 3.33 % Quartz and 3.33 % to 5 % Opaque. The Bindstone Facies has a beige and gray color. This facies is autochthonous and dominated by 40 – 45 % of coral fossils rather than 14.99 % of other fossils, including large foraminifera, gastropods, bivalves, and red algae. There are also found that 1 % to 3.33 % of Quartz minerals.

3.3. Facies associations and depositional environment

Furthermore, the facies are classified into facies associations [8] in [9], resulting in three (3) *ASF – Association of Facies*: the Platform Interior (Open Marine) facies association, the Platform Margin Reef facies association, the Slope facies association.

3.3.1. ASF 1 – Platform Interior (Open Marine). This facies association consists of F1 – Skeletal Wackestone and F2 – Skeletal Packstone, which has a dominant mud matrix composition. F1 dominated by a mud matrix compared to the identified fossil composition, which comprises coral and red algae in small quantities. The F1 layer thickness is up to 3 meters. Based on the petrographic analysis result, the bioclasts are of algae, corals, and large foraminifera (Milliolids). Meanwhile, F2 is dominated by fossil composition, but some mud matrix still exists. Some fossils were identified, including bivalves, algae, and corals. Based on the results of the macroscopic and petrographic description of F1 and F2 that formed a mud matrix dominant facies association with (1) fossil composition, includes bivalves, algae, corals, and red algae; and (2) minerals in KLP 1.1 A sample, include quartz 3 %, calcite 60 %, hornblende 1 %, and chloritoid 36 %. A mud matrix and siliciclastic minerals indicate a low-energy depositional environment with low abrasion energy levels, low water currents, and average salinity [10]. Based on Wilson's model (1975) [8] in Flugel (2004) [9], these facies are associated with Platform Interior (Open Marine).

3.3.2. ASF 2 – Platform Margin Reef. Based on the results of the macroscopic and petrographic description of F3 – Benthic Foram Grainstone, F4 – Skeletal Rudstone, F5 – Coral Rudstone, and F6 – Bindstone compiling this facies association indicates the presence of bivalves fossils, red algae (Crustose Coralline Algae, Branching Coralline Algae, and Rodholith), large foraminifera (Nummulites sp, Tricites sp, Miliolids, and others), bryozoan, echinoid, gastropods, and corals are common indicators of the Reef Margin Platform environment [9]. Some minerals are also found in KLP 1.3 B, KLP 1.4 D, KLP 2.2 C, KLP 2.8 B, and KLP 2.9, which comprise quartz 1 % and calcite 99 %. In addition, there is Skolithos bioturbation indicating high energy and shallow marine environment [11]. This facies association is composed of mud matrix and sand with a dominant component of coral and in situ coral, indicating an environment in reef areas and a small supply of siliciclastic minerals. Based on Wilson's model (1975) [8] in Flugel (2004) [9], these facies are associated with Platform Margin Reef. This association indicates medium to high abrasion energy and water currents [12].

3.3.3. ASF 3 – Slope. This facies association consists of F4 – Skeletal Rudstone, found in KLP 1.4 A, KLP 1.4 B, KLP 2.6 B, and KLP 2.6 C. This facies association is composed of fractions fragment produces very poor sorting with mud matrix. The presence of small foraminifera and radiolarian bioclasts in small quantities with a mud matrix shows the presence of an internal environment mixture. Some minerals are also found in these facies, including large quantities of calcite and small quantities of quartz and opaque. In addition, Thalassinoides bioturbation indicates a shallow marine environment until deep-sea [13], and identification of Skolithos shows high energy and shallow marine environment [11]. High energy on Skolithos bioturbation is caused by the vertical shape of the burrow, which is affected by the gravity force [14], and there are small amounts of siliciclastic minerals. Based on Wilson's model (1975) [8] in Flugel (2004) [9], these facies are associated with Slope.

The facies associations can be seen in more detail from each log section in figures 4, 5, & 6.

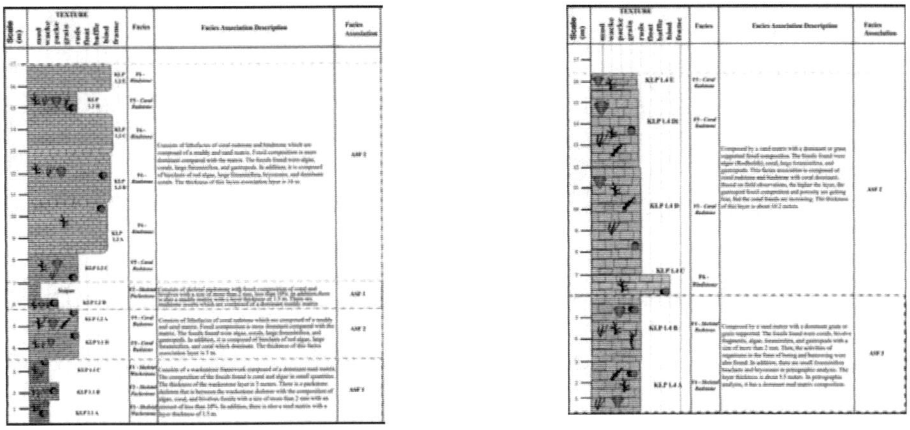

A B

Figure 4. (A) Facies association of KLP 1.1-1.3 log section; (B) Facies association of KLP 1.4 log section.

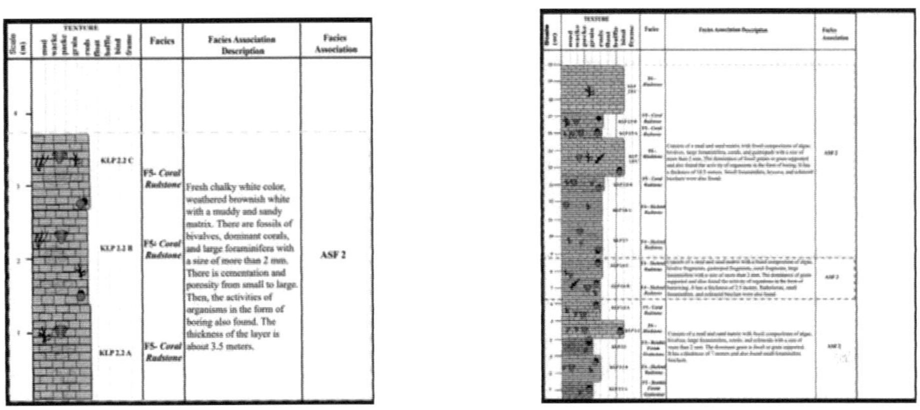

A B

Figure 5. (A) Facies association of KLP 2.2 log section; (B) Facies association of KLP 2.1, 2.3-2.9.

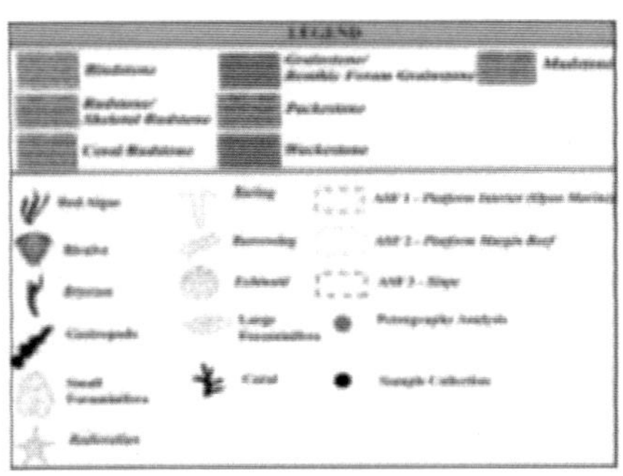

Figure 6. The log section's legend.

4. Conclusion

Based on the results of petrographic analysis and macroscopic descriptions, six facies (6) resulted: Skeletal Wackestone, Skeletal Packstone, Benthic Foram Grainstone, Skeletal Rudstone, Coral Rudstone, and Bindstone. These facies produced three facies associations (3): Platform Interior (Open Marine), Platform Margin Reef, and Slope. The writer would like to suggest further studies about The Klapanunggal Formation since it has interesting aspects to discuss and comprises some fossils. The expansion study area is needed to acquaint the lateral variation of the facies.

Acknowledgments

We want to thank Universitas Indonesia for financially supporting this research through the International Indexed Publication (PUTI) Saintekes the Fiscal Year 2020 No. NKB-4892/UN2.RST/THKP.05.00/2020.

References

[1] Satyana A H and Purwaningsih M E M 2003 Oligo-Miocene carbonates of Java: Tectonic setting and effects of volcanism Proceedings of Joint Convention Jakarta **2003** The 32nd IAGI and 28th HAGI Annual Convention and Exhibition pp 1–27

[2] Hall R 2012 Late Jurassic-Cenozoic Reconstructions of the Indonesian Region and the Indian Ocean Tectonophysics **570–571** pp 1–41

[3] Wilson M E 2002 Cenozoic Carbonates in Southeast Asia: Implications for Equatorial Carbonate Development pp 295–428

[4] Bishop M G 2000 Petroleum Systems of The Northwest Java Province, Java and Offshore Southeast Sumatra, Indonesia North

[5] Abdurrokhim and Ito M 2013 The role of slump scars in slope channel initiation: A case study from the Miocene Jatiluhur Formation in the Bogor Trough, West Java Journal of Asian Earth Sciences **73** 68–86

[6] Wilson M E 2008 Global and Regional Influences on Equatorial Shallow-Marine Carbonate during the Cenozoic pp 262–274

[7] Embry A F & Klovan J E 1971 A Late Devonian reef tract on Northeastern Banks Island Canadian Petroleum Geology Bulletin **19** 730–781

[8] Wilson J L 1975 Carbonate Facies in Geologic History (Springer-Verlag)

[9] Flugel E 2004 Microfacies of Carbonate Rocks: Analysis, Interpretation, and Application. (Springer Verlag)

[10] Oluwajana O A and Ehinola O A and Ofiwe C U and Akhayere E and Egunjobi K 2020 Depositional Environment and Diagenesis of Late Cretaceous-Early Paleogene Carbonates on the Benin flank, Southwestern Nigeria Journal of African Earth Sciences pp 163 https://doi.org/10.1016/j.jafrearsci.2020.103762

[11] Knaust D 2017 Atlas of Trace Fossils in Well Core (Switzerland: Springer International Publishing AG)

[12] Adrian F and Putra A and Muhni A 2020 Depositional Environment Analysis Carbonate Rocks in Montasik District **9(3)** 72–77 https://doi.org/10.24815/jacps.v9i3.17028

[13] Monaco P and Caracuel J E and Giannetti A 2009 Thalassinoides and Ophiomorpha as Cross-Facies Trace Fossils of Crustaceans from Shallow-to deep- Water Environments: Mesozoic and Tertiary examples from Italy and Spain 3rd Symposium on Mesozoic and Cenozoic Decapod Crustaceans pp 79–82

[14] Bromley R G & Asgaard U 1991 Ichnofacies: A Mixture of Taphofacies and Biofacies Lethaia **24** 153–163

Projection of The Effect of Climate Change on Crop Water Requirements for Rice Plants in Majalengka Regency

Adam Hastara Aji[1], Adi Wibowo[1*], Mangapul Parlindungan Tambunan[1], Dennish Ari Putro
[1]Department of Geography, Faculty of Mathematics dan Natural Sciences, University of Indonesia, Depok 16424, Indonesia
[1]adam.hastara@ui.ac.id, [2*]adi.w@sci.ui.ac.id,
[3]mangapul.parlindungan@ui.ac.id, [4]aridennish@gmail.com

Abstract. Climate change significantly impacted climate conditions and crop water requirements in the Majalengka Regency, a super-priority location related to climate resilience in the agricultural sector. In dealing with the impacts of climate change, the relationship between crop water requirements to climate change could be projected using the General Circulations Model with scenarios of RCP 4.5 (medium-light change) combined with the CROPWAT 8.0 model. This study aimed to determine the effect of climate change and the crop water requirements for rice plants in the three districts with the broadest rice fields in the Majalengka Regency from 2021-2025. This research found that climatic conditions in Majalengka Regency between the year 2021-2025 experienced changes in the average temperature, where the average temperature decreased in 2022 but continued to increase in 2023-2025 with an increase of 0.1°C- 0.24°C. Then from 2023-2025, the average maximum temperature increases by 0.39°C-0.47°C. On the contrary, the total rainfall each year decreased compared to 2021. Based on the calculations, the crop water requirement was always higher than adequate rainfall, so irrigation water was needed. The highest irrigation water requirement had a discharge value of 10.45 l/sec/ha.

1. Introduction

Rice is a food crop that requires 600–1200 mm of water for 90–120 days from planting to harvesting [1]. Calculating the amount of crop water requirements for plants is crucial because if the amount of crop water requirements for rice plants is not fulfilled, it will cause the rice to wither and die. In contrast, if the need for the amount of water received by rice plants is excessive, it will be a waste and can cause the leaching of nutrients in the land of rice fields so that the soil becomes less fertile [2]. The amount of crop water requirement for rice plants could be determined through the calculation of plant evapotranspiration which describes the amount of water loss in plants due to two processes, namely evaporation which is a process of water loss on the soil surface, and transpiration, which is a process of water loss in plants which was strongly influenced by various meteorological elements [3].

The condition of meteorological elements in an area will be closely related to the hydrological cycle, which will be strongly influenced by changes in energy due to global warming, whose impacts vary locally in influencing changes in patterns in climate parameters such as temperature, rainfall, wind, humidity, and evaporation [4].

Global warming causes a tendency to increase global temperature and is closely related to climate change [5]. Climate change has impacted various sectors, including the food and beverage sector. Climate change has a vulnerable impact on the harvested area and rice production, where the decline can reach 1.37% per year, as seen in South Sumatra and Malang Raya [6]. Rice productivity is sensitive to rainfall and temperature, especially maximum temperature, where rainfall and maximum temperature have an optimum point to increase rice productivity, namely for rainfall of 10,177 inches/year and maximum temperature of 31,35 °C. Rice productivity will decrease if it passes the optimum point [7].

Since the industrial revolution, world temperature has increased from 0.8°C to 1.2°C, and it will continue to increase from 0.1°C – 0.3°C per decade, where the possibility of a temperature increase will reach 1.5°C between 2030 and 2052 [8]. This climate change phenomenon was further strengthened by official data from the Meteorology, Climatology and Geophysics Agency (BMKG), where the temperature trend in Indonesia showed an increase of 0.03°C every year, which meant around 0.9°C within 30 years, in addition to the frequency of extreme rainfall as well increase. The climate change phenomenon could not be avoided, and its impact was increasingly felt, including in Indonesia, one of which was in the Majalengka Regency.

Majalengka Regency is one of the regencies that have become a rice barn in West Java, with an area of rice fields in 2020 reaching 50.281 Ha with a rice production level in 2019 reaching 572.058,83 tons [9] designated as Bappenas in 2021 as a related super-priority location climate resilience in the water supply and agriculture sectors [10]. The climate change phenomenon in the Majalengka Regency has changed the conditions of the existing meteorological elements and will undoubtedly affect the crop water requirements for rice plants. The absence of information about the projected effects of climate change will potentially have a detrimental impact on the agricultural sector in Majalengka. The projected effect of climate change on the crop water requirements for rice plants in the Majalengka Regency can be used as an initial basis for stakeholders to determine plans for managing the agricultural sector.

In projecting the relationship between crop water requirements for rice plants and climate change, previous studies used the General Circulations Model (GCM) [1,11]. GCM is the most effective tool for simulating the current climate and projecting future climates. The output of the GCM model with various spatial resolutions can be used as initial data to conduct hydrological model simulations to assess the impact of climate change on the hydrology of water resources [12].

Based on the background described, this study aimed to determine the effect of climate change on climate conditions and the amount of crop water requirements for rice plants in the Majalengka Regency from 2021 to 2025, where the research location focused on the northern part of Majalengka Regency including Kertajati District, Ligung District, and Jati Tujuh District, which are the three districts with the broadest rice fields. The climate change model projection used in this study was the HadGEM2-ES model with the RCA4 regional climate model, which had been processed through the coordinated regional climate downscaling experiment (CORDEX), which had been reviewed and analyzed using Representative Concentration Pathway (RCP) scenario, namely RCP 4.5 (medium-mild change).

2. Methods

Based on climate change projection data, this study used quantitative and spatial analysis conducted in Majalengka Regency, especially Kertajati, Ligung, and Jati Tujuh Districts. Analysis of the projected crop water requirements related to climate change became the general description of this research. The water requirements of rice plants in the Majalengka Regency depend on climatic elements such as temperature, rainfall, wind speed, humidity, and solar radiation, so when climate elements change, they are used as a basis for calculating the water requirements of rice plants. This phenomenon was analyzed quantitatively by looking at the trendline of the projected changes.

The spatial analysis obtained from the output of the GCM model was limited to the area of the Majalengka Regency. Based on this spatial analysis, whether the value of crop water requirements would be the same for all rice fields in Majalengka Regency could be seen. The spatial analysis took a sample of rice fields in the three districts with the most extensive rice fields to analyze the projected changes in crop water requirements in that area.

2.1. *Research location*

This research was conducted in Majalengka Regency, located at coordinates to the west of 108° 03' – 108° 19' east longitude, east of 108° 12' – 108° 25' east longitude, north of 6° 36' – 5° 58' south latitude and south of 6° 43' – 7° 44' South Latitude. In general, Majalengka had an area of 50.323 hectares of rice fields in 2019, spread across every district, of which Kertajati District was the district that has the most extensive rice field area, reaching 5.258 hectares, followed by Ligung and Jati Tujuh Districts with 4.891 hectares and 3.541 hectares of rice fields [9]. The district focused on research related to the projected crop water requirements.

Based on the GCM model output, Majalengka Regency was divided into seven grid points with the resolution of each grid of 25 km x 25 km, as shown in Figure 1. Kertajati, Ligung, and Jati Tujuh Districts, special districts focused on the research, were in grid points 2 and 5.

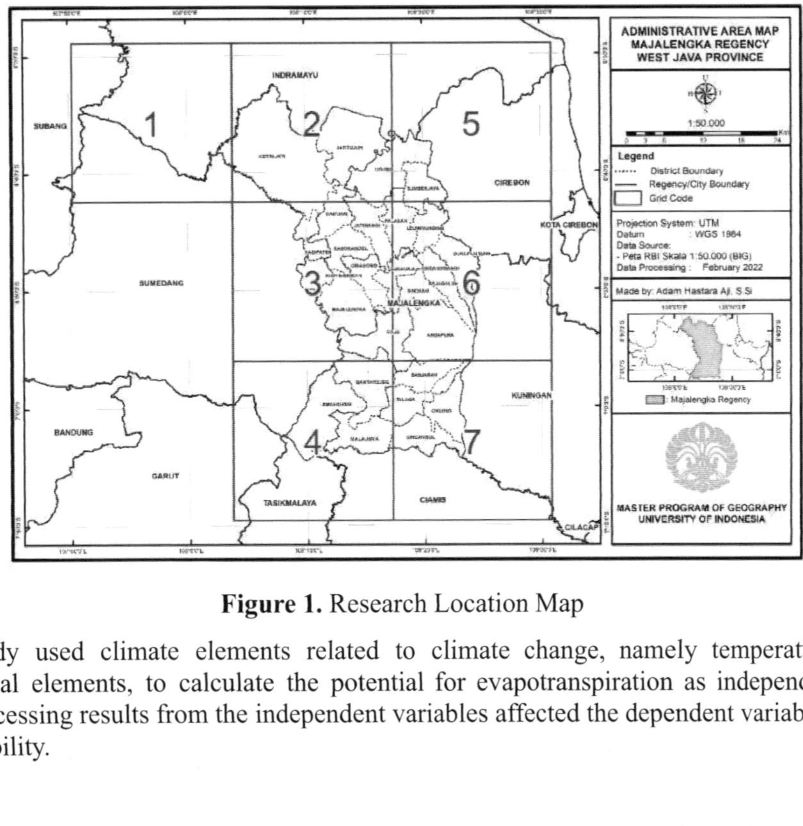

Figure 1. Research Location Map

The study used climate elements related to climate change, namely temperature and other meteorological elements, to calculate the potential for evapotranspiration as independent variables. The data processing results from the independent variables affected the dependent variable, namely the water availability.

2.2. *Data*

The data used in this study is climate data from climate change projections in the form of temperature projections and other meteorological elements in 2021–2025 obtained from the HadGEM2-ES model output with the RCA4 regional climate model that has been developed. Through the CORDEX downscaling process, the projection data became the primary processing data to get the projected water needs of the plants from the processing results through the CROPWAT 8.0 model.

2.3. *Data processing*

2.3.1. *Climate change projection*

Projection of climate data at the study site in 2021–2025 was determined from the output of the HadGEM2-ES model with the regional climate model RCA4 by using RCP 4.5 scenario analysis, which had been processed through the CORDEX downscaling. This data processing aimed to see climate change projections in the Majalengka Regency area. Furthermore, the projection data were combined to obtain the value of crop water requirements. The data processing results were analyzed using quantitative analysis by looking at the trendline of the projected changes. Projected temperature changes are displayed in the form of monthly average temperatures in 2021– 2025.

2.3.2. The projected crop water requirements

The value of crop water requirements was known by looking at the value of plant evapotranspiration. The method for calculating the value of plant evapotranspiration was calculated using the CROPWAT 8.0 model in 2021–2025. This method used projection data of meteorological elements such as temperature, rainfall, wind speed, humidity, solar radiation, information on the beginning of the planting season, and the location of the rice fields to analyze the results of this calculation. Crop water requirements were analyzed using the CROPWAT 8.0 model calculation results. Based on this analysis, it was seen that the crop water requirements of rice plants in Majalengka, especially in Kertajati, Ligung, and Jati Tujuh Districts.

3. Results and discussion

3.1. Temperature change

In the range of 2021–2025, the average change in temperature in Majalengka Regency saw in Figure 2. The average annual temperature in Majalengka decreased in 2022 but continued to increase from 2023– 2025 to 2021. However, the average temperature change in Majalengka occurred during 2021–2025 did not reach 1°C, only around 0.1°C – 0.24°C.

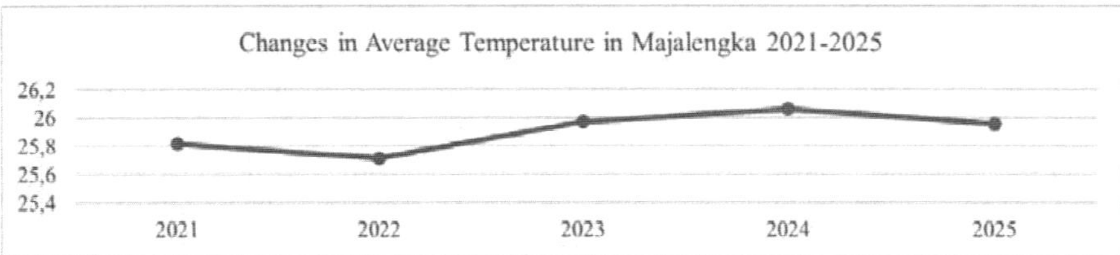

Figure 2. Changes in Average Temperature in Majalengka 2021–2025

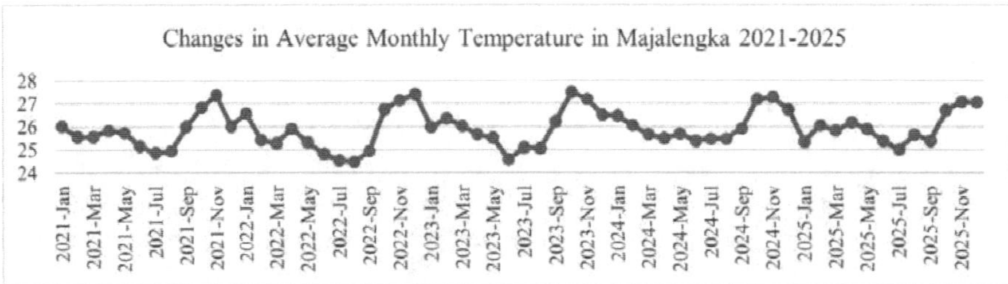

Figure 3. Changes in Average Monthly Temperature in Majalengka 2021–2025

Figure 3 showed that the highest temperature was 27°C – 27.5°C and had a repeating pattern, where the highest average temperature occurred between October–December. This temperature was partly influenced by the sun's apparent motion, which began to move from the equator to the southern hemisphere at the end of September.

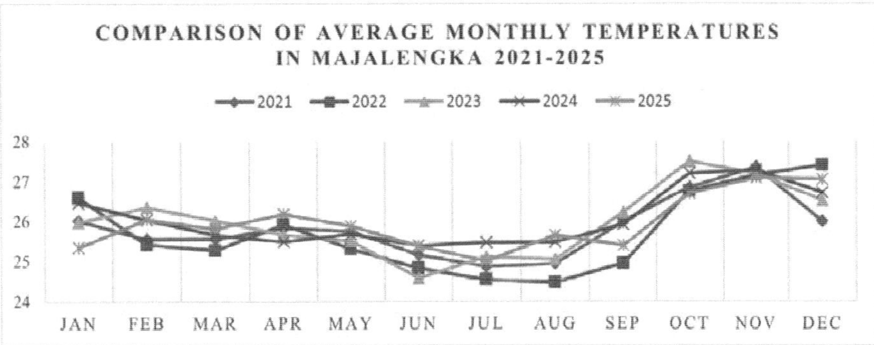

Figure 4. Comparison of Average Monthly Temperatures in Majalengka 2021–2025

On the other hand, viewed the average temperature in Majalengka every month, only in 2022 the average temperature was much below the 2021 average, namely in February, March, May, June, July, August, September, October, November, while for other years the majority of the monthly average was above the monthly average in 2021, as shown in Figure 4. Climate change also affected the average maximum temperature that would occur in Majalengka in 2021–2025.

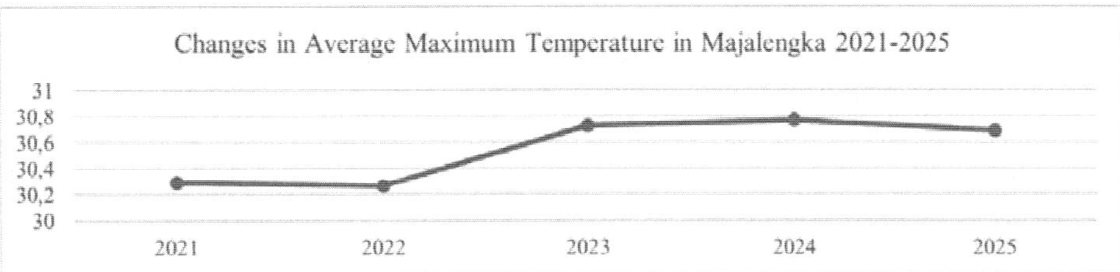

Figure 5. Changes in Average Maximum Temperature in Majalengka 2021–2025

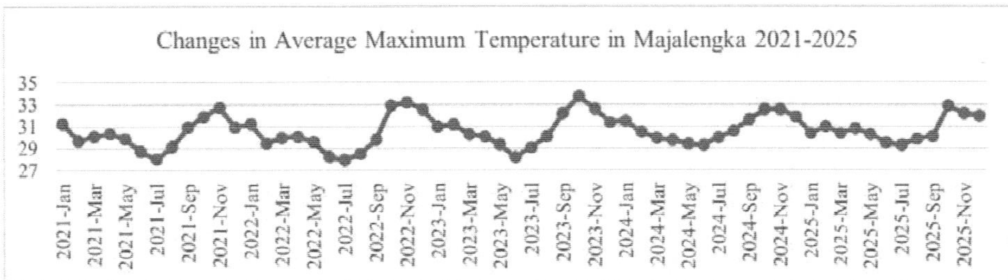

Figure 6. Changes in Average Maximum Temperature in Majalengka 2021–2025

Figure 5 and Figure 6 showed that the average maximum temperature in Majalengka from 2023–2025 continued to increase compared to 2021 and 2022, with a temperature increase range of 0.39°C – 0.47°C. The peak average maximum temperature per month had consistently increased yearly since 2021 and 2024, when the peak average maximum temperature per month was slightly below the peak average maximum temperature per month in 2021. The highest average maximum temperature repeats itself in October and November every month. The magnitude of the peak average maximum temperature is shown in Table 1.

Table 1. Peak Maximum Average Temperature in Majalengka 2021–2025

Highest Average Temperature Per Year				
Nov-2021	Nov-2022	Oct-2023	Nov-2024	Oct-2025
32.72°C	33.17°C	33.71°C	32.53°C	32.82°C

3.2. Rainfall change

Figures 7 and 8 show a downward trend, but, based on monthly data, the peak of monthly average rainfall in 2022 and 2023, which usually occurs from January to March, was higher than in 2021. This significant issue needed to be watched with strategic management of water irrigating rice fields.

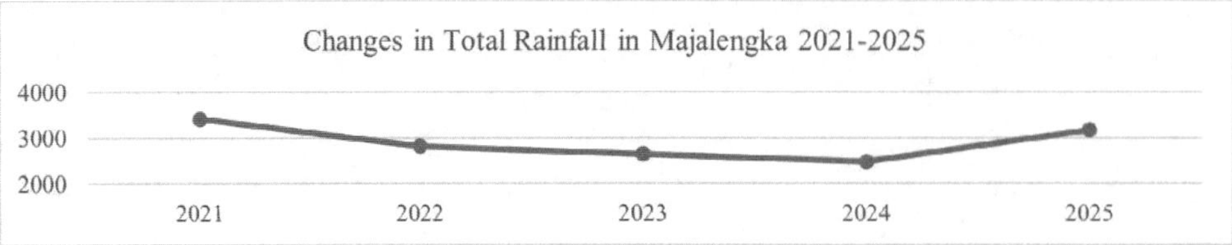

Figure 7. Changes in Total Bulk in Majalengka 2021–2025

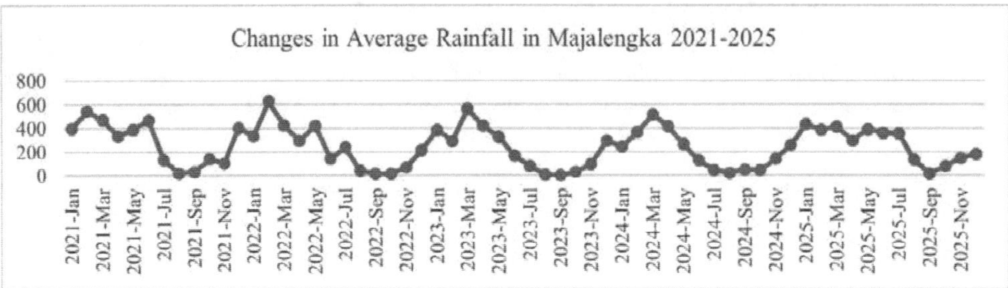

Figure 8. Changes in Average Rainfall in Majalengka 2021–2025

Based on a review of the results of climate change scenarios in the Majalengka Regency, generally, there was a change in temperature in the Majalengka area. However, the change did not reach 1°C, only in the range of 0.1°C-0.24 °C. The temperature change that occurred dominantly indicated an increase from 2021 to 2025, only in 2022 when the temperature decreased by 0.1°C. This temperature change had a maximum temperature, with an increase in the maximum average temperature from 2023-2025, which continued to increase compared to 2021 and 2022. Meanwhile, regardings changes in rainfall, each year has decreased compared to 2021. When viewed monthly in 2022 and 2023, the peak of the average monthly rainfall, which usually occurs in January-March, was higher than the peak of the average monthly rainfall in 2021, which meant that in that year, there was an uneven rain distribution. This rainfall showed that Majalengka Regency was experiencing the impact of climate change in the form of an increase in temperature, a decrease in rainfall, and an uneven distribution of rainfall every month.

3.3. Crop water requirement

Based on the planting calendar application from the Ministry of Agriculture, it was known that the rice-growing season (GS) in Majalengka begins on the third ten days of November. Therefore, based on FAO's Cropwat 8.0 application, it was estimated that the harvest period would occur in the second ten days of March. The rice-growing season in Majalengka was known as routine and held every November–March. In analyzing crop water requirements, a comparison was made between crop water requirements and adequate rainfall during the five growing seasonings from November 2021–March 2026, as shown in Figure 9 and Figure 10. The sample area taken was Kertajati District, Ligung District, and Jati Tujuh District, which was located on grid points 2 and 5 from the output of the projection model.

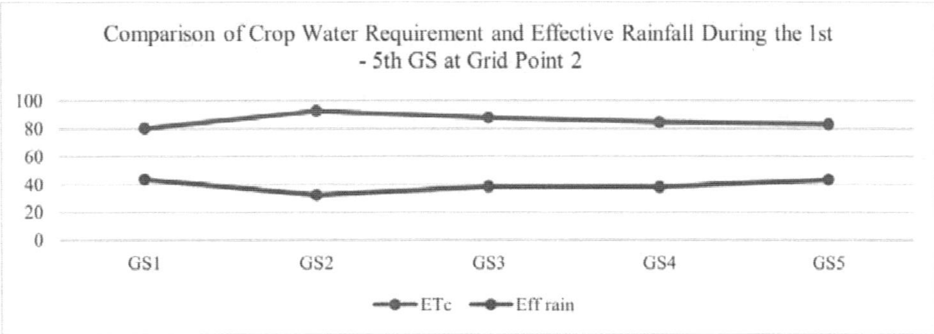

Figure 9. Comparison of Crop Water Requirement and Effective Rainfall During the 1st – 5th GS at Grid Point 2

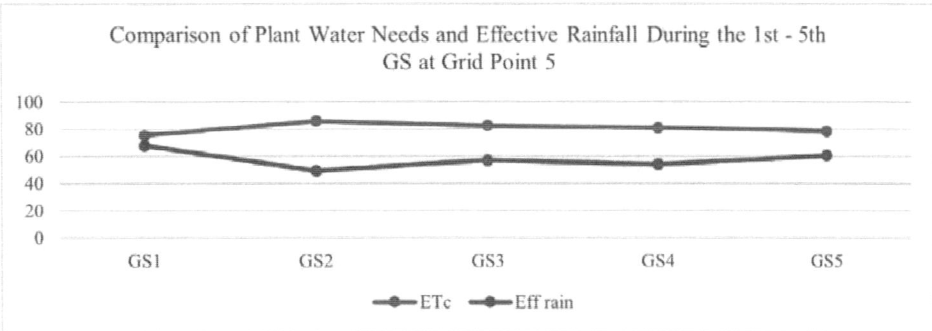

Figure 10. Comparison of Crop Water Requirement and Effective Rainfall During the 1st – 5th GS at Grid Point 5

Based on the calculations using the CROPWAT 8.0 model, it was known that the crop water requirements during the 2nd GS to 5th GS on grid points 2 and 5 indicated higher crop water requirements than the 1st GS. The crop water requirements were inversely proportional to the adequate rainfall at grid points 2 and 5. At the same time, the adequate rainfall during the 1st GS to 5th GS has always been lower than the crop water requirements in each growing season and tends to be lower than the 1st GS. This phenomenon needed to be anticipated by providing irrigation to the rice fields in grid points 2 and 5. The total irrigation needs in these areas are as follows Figure 11, Figure 12, Table 2, and Table 3.

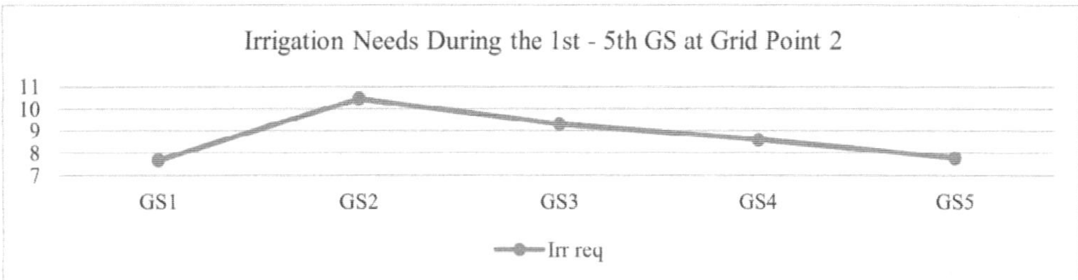

Figure 11. Irrigation Needs During the 1st – 5th GS at Grid Point 2

Table 2. Irrigation Needs During the 1st – 5th GS at Grid Point 2 (in l/sec/ha)

GS1	GS2	GS3	GS4	GS5
7.67	10.45	9.3	8.59	7.78

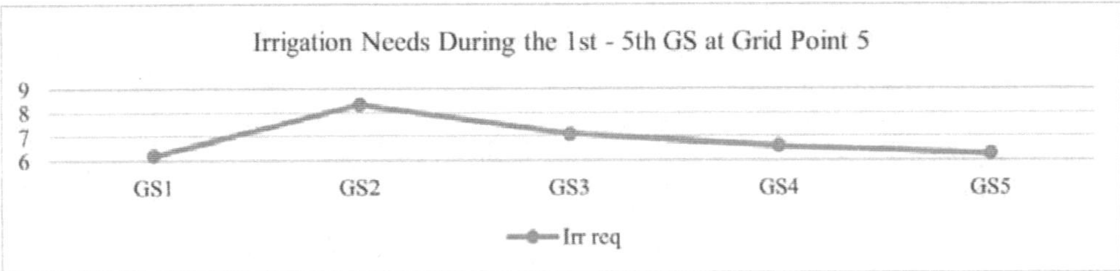

Figure 12. Irrigation Needs During the 1st – 5th GS at Grid Point 5

Table 3. Irrigation Needs During the 1st – 5th GS at Grid Point 5 (in l/sec/ha)

GS1	GS2	GS3	GS4	GS5
6.2	8.32	7.08	6.58	6.28

The calculation results showed the irrigation needs at grid points 2 and 5 in GS 1 and GS 5. In general, irrigation needs at grid point 2 were higher than at point 5, and irrigation needs at each grid point have always been higher than in GS 1. The highest irrigation needs were found at grid point 2, namely Kertajati District and part of Jati Tujuh District in the 2nd GS, with the required irrigation flow rate of 10.45 l/sec/ha. Plant water needs and irrigation needs had to be considered because they would affect the productivity of rice plants.

Changes in temperature and rainfall would be closely related to fulfilling crop water requirements. The increased temperature would increase the value of plant evaporate, impacting the high crop water requirements. This crop water could be seen at the 2nd GS, where the increase in temperature affected increasing crop water requirements. Crop water requirements could naturally be met with sufficient rainfall in the rice fields. However, this study found that during the 2nd GS in the rice fields of Kertajati, Ligung, and Jati Tujuh districts, crop water requirements increased, and rainfall during the growing season decreased, so additional water was needed through irrigation to meet the crop water requirements.

The research showed that Kertajati, Ligung, and Jati Tujuh Districts required more irrigation than in 2021. The most significant increase in irrigation needs occurred in the 2nd GS, with the required irrigation flow rate of 10.45 l/sec/ha. This rate was due to the increase in crop water requirements and a projected decrease in the effectiveness of rainfall. The existence of a projection of a decrease in rainfall using the output of the GCM model was also shown by previous research in the north of Majalengka Regency, namely the Indramayu Regency. The analysis results using the RegCM3 model with SRES A1B emission change scenarios illustrated that in the projection period of 2021–2040, rainfall would experience a decrease ranging from 9–12.6% from the rainfall in the 1980–1999 period. [13]. This rainfall needed to be mitigated by ensuring sufficient irrigation in the 2nd GS, for example, by conducting irrigation management by utilizing existing dams and irrigation canals.

4. Conclusion

Based on the GCM HadGEM2 data processing results using the RCP 4.5 scenario analysis, it was known that global climate change affects climate conditions in the Majalengka Regency area, especially the Kertajati, Ligung, and Jati Tujuh Districts. Climatic conditions in Majalengka Regency year 2021– 2025 were projected to experience changes in the average temperature, whereas the average temperature decreased in 2022 but continued to increase in 2023–2025, with an increase of around 0.1°C – 0.24°C. When viewed from the monthly average temperature in Majalengka, only 2022 is the average. The temperature was much below the 2021 average, namely in February, March, May, June, July, August, September, October, and November, while for other years, the majority of the monthly average was above the monthly average in 2021. Furthermore, the maximum average temperature in Majalengka from 2023 to 2025 continued to increase compared to 2021 and 2022, with a temperature increase of 0.39°C – 0.47°C. On the other hand, the total rainfall each year decreased compared to 2021.

Projections of climate change in the Majalengka Regency, especially the Kertajati, Ligung, and Jati Tujuh Districts, showed an increase in the average temperature and a decrease in total rainfall and affected the amount of crop water requirements. Based on the calculations using the CROPWAT 8.0 model, it was known that the increase in crop water requirements was directly proportional to the increase in the average temperature. The average temperature from the 2nd GS to the 5th GS at grid points 2 and 5, the average temperature has always been higher than the 1st GS, and the crop water requirements at that grid point also showed an increase. The highest average temperature was projected to occur in the 2nd GS, impacting GS's high increase in crop water requirements. Based on the research results above, the projected increase in crop water requirement has always been higher than adequate rainfall, so it was necessary to calculate irrigation water needs. Based on the CROPWAT.8.0 calculation, it was known that the highest irrigation water needs were at grid point -2 in the 2nd GS with a discharge value of 10.45 l/sec/ha.

Further research related to climate change needs to be developed with various methods that can increase the accuracy of projections spatially and temporally. In calculating the water needs of rice plants, further research is still needed to determine what meteorological parameters are influencing them. In addition, the development of the CROPWAT 8.0 model can continue to be carried out by using more specific rice crop coefficient values for certain varieties, using the actual conditions of the planting schedule, or entering specific parameters for the type of soil used.

Research related to various impacts of climate change in the agricultural sector, such as an increase in extreme weather events, the emergence of plant pest diseases, and extended droughts, needs to be developed so that it can assist the community in responding to the phenomenon of climate change, both in the form of adaptation and mitigation responses according to environmental conditions and needs.

Acknowledgments

The author would like to thank the lecturer and supervisor who have provided inspiration and guidance in making this research journal. Furthermore, the author would like to thank the Head of the Geography Masters Study Program, who has provided input and direction throughout the research process.

References

[1] Estiningtyas W and Syakir M 2018 Pengaruh Perubahan Iklim Terhadap Produksi Padi di Lahan Tadah Hujan *J. Meteorol. dan Geofis.* **18** 83–93

[2] Harjadi S S 2019 *Dasar-Dasar Agronomi* (Jakarta: PT Gramedia Pustaka Utama)

[3] Allen R, Pereira L, Raes D and Smith M 1998 *Crop Evapotranspiration Guidelines for Computing Crop Water Requirements* (Roma)

[4] Aldrian E, Karmini M and Budiman B 2011 Adaptasi dan Mitigasi Perubahan Iklim di Indonesia (Jakarta: Badan Meteorologi, Klimatologi, dan Geofisika (BMKG))

[5] Sumastuti E and Pradono N S 2016 Dampak Perubahan Iklim Pada Tanaman Padi di Jawa Tengah *J. Econ. Educ.* **5** 31–8

[6] Ruminta R, Handoko H and Nurmala T 2018 Indikasi Perubahan Iklim dan Dampaknya Terhadap Produksi Padi di Indonesia (Studi kasus : Sumatera Selatan dan Malang Raya) *J. Agro* **5** 48–60

[7] Nurhayanti Y and Nugroho M 2016 Sensitivitas Produksi Padi Terhadap Perubahan Iklim Di

Indonesia Tahun 1974-2015 *Agro Ekon.* **27** 183

[8] Delmotte P ., Zhai H O, Pörtner D, Roberts J, Skea P ., Shukla A, Pirani W, Moufouma-Okia C, Péan R, Pidcock S, Connors J B R, Matthews Y, Chen X, Zhou M I, Gomis E, Lonnoy T, Maycock M, Tignor T and Waterfield 2018 *Global Warming Of 1.5°C. An IPCC Special Report On The Impacts Of Global Warming Of 1.5°C Above Pre-Industrial Levels And Related Global Greenhouse Gas Emission Pathways, In The Context Of Strengthening The Global Response To The Threat Of Climate Change,* vol 3

[9] Badan Pusat Statistik 2021 *Kabupaten Majalengka Dalam Angka Tahun 2021* (Jakarta)

[10] Badan Perencanaan Pembangunan Daerah 2021 *Daftar Lokasi dan Aksi Ketahanan Iklim* (Jakarta)

[11] Nugroho S, Febriamansyah R, Ekaputra E G and Gunawan D 2019 Simulasi Kebutuhan Air Untuk Tanaman Padi Pada Skenario Perubahan Iklim di Daerah Aliran Sungai Lembang-Sumani *J. Sumber Daya Air* **15** 15–26

[12] Duong D T, Tachikawa Y and Yo K 2014 Changes in River Discharge in the Indochina Peninsula Region Projected Using Mri-Agcm and Miroc5 Datasets *J. Japan Soc. Civ. Eng. Ser. B1 (Hydraulic Eng.* **70** 115–20

[13] Jadmiko S D and Faqih A 2014 Dynamical Downscaling Luaran Global Climate Model (Gcm) Menggunakan Model Regcm3 Untuk Proyeksi Curah Hujan di Kabupaten Indramayu **28** 9–16

IOP Conf. Series: Earth and Environmental Science 1111 (2022) 012022 doi:10.1088/1755-1315/1111/1/012022

Environmental Impact Assessment on Plastic Waste Recycling Business in Bengkulu City, Indonesia

Sisilia Dwi Puspita[1]*, Ahyahudin Sodri[2] and Suyud Warno Utomo[3]
[1]School of Environmental Science, Universitas Indonesia, Jakarta, 10430, Indonesia
[1]0000-0002-0030-5790
[2]School of Environmental Science, Universitas Indonesia, Jakarta, 10430, Indonesia
0000-0001-5708-1356
[3]School of Environmental Science, Universitas Indonesia, Jakarta, 10430, Indonesia
0000-0003-4169-761X
[1]*sisilia.dwi01@ui.ac.id, [2]ahyahudin.sodri@ui.ac.id, [3]suyudwarno@gmail.com

Abstract. In Indonesia, small-scale plastic recycling businesses dominate the plastic waste recycling business. The practice of plastic waste recycling in developing countries also does not pay attention to the Environment and is not energy and material efficient during the process. Andusti Simpari company in Bengkulu City is one of the small-scale plastic waste recycling companies that face common problems such as the use of human labor, inefficient energy, and lack of liquid waste management, which has the potential to cause eutrophication and produce greenhouse gases in the form of CO_2 which can contribute to global warming. This study aims to assess the environmental impact of eutrophication and the global warming potential (GWP) of businesses using the Life Cycle Assessment (LCA) method. Cradle to grave LCA is applied to small-scale plastic waste recycling, with 1000kg of shredded plastic/day production capacity. From the LCA analysis, it is known that to produce 1 kg of shredded plastic can produce as much as 0.97363 $KgCO_2eq$ of pollutants that contribute to the Global Warming Potential (GWP) and eutrophication of 0.01492 kg PO_4.

1. Introduction

After China, Indonesia is the world's second-largest producer of plastic waste [1]. Statistical data from the Indonesia Solid Waste Association in 2014 showed that plastic waste was the second most significant type of waste from the total waste, which was 5.4 million tons per year or 14% of the total waste production [2]. The increase in the volume of waste is in line with the increase in the industrial sector. The increase in the amount of plastic waste also occurred in Bengkulu City, Bengkulu Province. Data from the Environment and sanitation service in Bengkulu City show that every day Bengkulu City produces 260-280 tons of waste to the Air Final Processing Site, Bengkulu City. Of the total waste, 11% comprises plastic waste [3].

Plastic waste, depending on the quality of the recovered waste portion, has the potential to be recycled numerous times while preserving its value and functionality, which would considerably aid the circular economy [4]. In Indonesia, the small-scale industry dominates the plastic recycling industry [5]. A small medium enterprise (SME) is a sector of the economy that uses low levels of technology, human resources, inefficient energy consumption, and low levels of energy efficiency. It has management, financial, and personnel issues [6]. The small-scale industry often produces waste and pollutants. [7]. One company that recycles plastic waste in Bengkulu City, Indonesia, is Andusti Simpari company. This company was chosen because the plastic recycling industry in Bengkulu is proliferating, and the Andusti Simpari Company is classified as a medium-scale plastic recycling company. Another reason is that this company is the only plastic recycling company in Bengkulu registered in the Indonesian Plastic Recycling Association and has been operating for more than ten years. Furthermore, based on observations, this company does not have waste management, and the location of this company is in the middle of a residential area and is close to a swamp where wastewater from the production potentially causes adverse environmental effects.

The Andusti Simpari company uses 1000 kg of plastic waste as raw material to produce about 900 kg of shredded plastic daily. In this company, wastewater containing detergent is produced from shredded plastic washing, and the wastewater is discharged into the Environment (soil) without prior treatment before being disposed of. Characterization of detergent wastewater has high levels of COD (chemical oxygen demand) and phosphate. Excess phosphate in water will negatively impact water by increasing algae growth, eutrophication, and reducing water quality [8]. In addition, solid waste recycling can also generate greenhouse gases in the form of $CO_{2,}$ which can contribute to global warming due to the use of enormous energy [9].

Life Cycle Assessment (LCA) is a method for assessing the inputs, outputs, and potential environmental effects of a product system over its life cycle. LCA can determine whether a process or product has a positive or negative impact on the Environment, identify areas for process and product development, compare and contrast various processes based on the environmental effects they cause, and quantitatively calculate the environmental impact of a product. [10]. This study aims to evaluate the environmental impact of eutrophication and Global Warming Potential (GWP) on Andusti Simpari company as a small-scale plastic waste recycling. Various studies have been performed. However, to the best knowledge of the authors, research that explicitly addresses and analyzes a small-scale plastic waste business's environmental impact and aims to analyze eutrophication and GWP from its business has never been carried out.

2. Methods

2.1. Sample

The study's sample is a small-scale plastic waste recycling business in Bengkulu City, Bengkulu Province. It uses 1000 kg of plastic waste daily as raw material and has five employees that do all tasks, from preparation to packaging. It operates eight hours, from 8:00 am to 5:00 pm, and travels 840 miles to the manufacturing company.

2.2. Methods

This study uses a quantitative approach, and the method used to analyze the environmental impact of a plastic waste recycling company in Bengkulu City uses the Life Cycle Assessment (LCA) method. LCA is a tool with international standard 14040 to assess environmental impacts that arise throughout the life cycle of a product from raw materials, manufacturing, transportation, use, and final disposal [11]. This study uses software that allows users to model and analyze complex life cycles systematically and transparently. Based on ISO 14040, the LCA stage consists of the following four stages:

2.2.1. Goal Definition and Space

This phase aims to formulate and describe the objectives, system boundary, functional unit, and assumptions related to impacts throughout the product life cycle. A functional unit measures how well the product system's functional outputs perform and is often based on the mass of the product being studied [12]. A functional unit's primary goal is to serve as a reference point for the inputs and outputs [13]. The functional unit in this study is 1 kg of shredded plastic produced from a plastic waste recycling system. This study aims to analyze the environmental impacts (Global Warming Potential/GWP and Eutrophication) of shredded plastic production by Andusti Simpari company in a year. This study's scope of LCA or system limits covers the production process from raw materials to the distribution process (cradle to grave).

2.2.2. Inventory Analysis

The Life Cycle Inventory phase is inventorying the required data, namely materials and energy, during each scenario's processing. The data were inventoried by collecting secondary and primary data. This study collected all data regarding energy, raw materials consumption, and chemical substance usage consumption. Therefore, the number of basic needs, energy, and emissions produced can be obtained by producing one kilogram of shredded plastic. In this study, the data collected is the data for shredded plastic production throughout 2021 (one year).

2.2.3. Impact Assessment Analysis

Environmental impact analysis in the LCA study will be done by grouping and assessing the environmental effects of the plastic or production process based on the data obtained at the Life Cycle Inventory (LCI) stage. Based on the LCA method's eleven impact categories in the environmental impact analysis, only two categories will be analyzed in the shredded plastic production process: global warming potential/GWP (kg CO2 eq) and eutrophication (kg PO_4eq) categories. These two categories were chosen because the production process of shredded plastic uses energy and has the potential to cause global warming (GWP). Waste has the potential to cause eutrophication. In this study, the impact assessment method using the CML IA Baseline was developed by the Institute of Environmental Science (CML) at Leiden University.

2.2.4. Interpretation

The interpretation stage of the results based on the ISO 14044:2006 guidelines states that the results of the LCI or Life Cycle Impact Assessment phase must be interpreted following the goals and parameters of the study. Interpretation must consider the sensitivity of the inputs, outputs, and methodological choices to comprehend the degree of uncertainty in the results. At the interpretation stage, the main element is evaluating the results and formulating conclusions and recommendations from the entire study.

3. Results and Discussion

3.1. Shredded Plastic Production Process

The shredded plastic production process consists of six central processing in separated working stations as follows:

a. Sorting
   At this stage, plastic waste is separated from dirt, chemical substances, and other organic waste, then grouped with similar plastics. This separation can be done manually by workers.
b. Shredding
   At this stage, the plastic waste separated by type is put into a chopping machine with a capacity of 350-500 kg per hour. This machine is driven by a large electric motor or by using a diesel engine. After leaving the chopping machine, the plastic waste is crushed into pieces size of 1 cm^2 and goes straight into the washing tub. In one day, the company chops 1000 kg and produces 900 kg.
c. Washing
   At this stage, the shredded plastic goes directly into the washing tub. The washing tub contains water and detergent used to remove adhering dirt. This washing process uses water from a local water supply utility and a water pump.
d. Drying
   After the washing process, the plastic pieces are aerated or dried in an open place directly exposed to the sun.
e. Packaging
   After the chopped plastic is dry, the chopped plastic will be collected in the sack of packaging and ready to be sent to the plastic factory.
f. Distribution

The pre-packaged shredded plastic will be distributed to manufacturing companies, and the distance is about 840 km using a truck with a 20-tonne load.

3.2. Input and Output

This study's life cycle inventories (LCI) analysis involves assessing and compiling inputs and outputs of shredded plastic production throughout its life cycle. Data used in this research is the shredded plastic production process data by Andusti Simpari company from January to December 2021. The data collected is primary data obtained from the Andusti Simpary company using interview techniques and secondary data obtained from supporting information sources in the plastic recycling industry. The overview of shredded plastic production and its mass balance is depicted in figure 1.

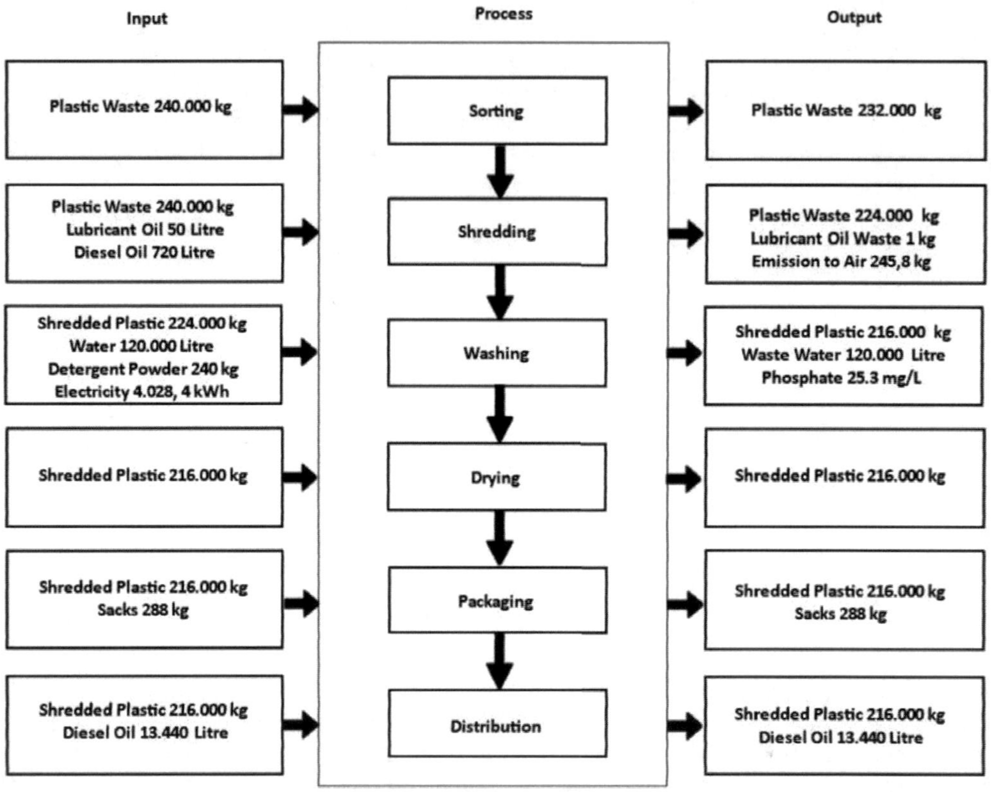

Figure 1. The system boundary of shredded plastic production.

This industry uses plastic crushing machines that use diesel and lubricant oil to produce shredded plastic. Another source was electricity for running the water pump in the washing process for producing shredded plastic, and it often ran for eight hours each day. A truck that used diesel fuel was used to transfer shredded plastic from the Andusti Simpari Company to the plastic factory, and the distance was around 840 kilometers. The mass balance showed that process of distribution, shredding, and washing contributed the highest energy usage. All energy sources contribute to pollutants and emissions, and the result of the emission is shown in table 1.

Table 1. Environmental impact of producing 1 kg shredded plastic.

Category of Environmental Impact	Process	Value	Total	Unit
Global Warming Potential (GWP)	Shredding	0.26992	0.97363	kg CO_{2}-eq
	Distribution	0.70371		
Eutrophication	Shredding	0.00028	0.01492	kg PO_{4}-eq
	Washing	0.01408		
	Distribution	0.00056		

Taking into table 1, the method for manufacturing 1 kg shredded plastic produces 0.97363 kg CO2-eq, released to the Environment, and led to global warming potential (GWP) and eutrophication of 0.01492 kg PO_4-eq. Previous research states that the environmental impact of the plastic recycling process is less than the production of raw plastic [14] [15]. However, the research conducted by Vita in 2019 stated that the Global Warming Potential from the plastic recycling industry is more than 2 kg CO2-eq. The research conducted by Vita has a higher global warming impact than the case in this study due to different business scales. Vitas's research uses an industrial scale as the research sample [16]. Based on table 1, it was shown that the distribution process had the highest impact on Global Warming Potential (GWP) and the washing process had the highest impact on eutrophication. The research by Hendrawan also stated that the shredding process contributes as much as 11% to producing carbon dioxide [17].

4. Conclusion

The conclusion of this research was as follows:

a. Based on Life Cycle Assessment studies on the small-scale plastic industry, Andusti Simpari company, the value of Global Warming Potential (GWP) and eutrophication were still low compared to environmental impact results on an industry scale. However, if the plastic waste recycling business produces shredded plastic in massive amounts, it can contribute significantly to global warming and eutrophication.
b. This research recommends the industry's waste treatment scenarios to protect the Environment's quality from the impact of wastewater.
c. Alternative improvements that can be made to reduce greenhouse gases include using a plastic chopper based on an electric motor. Meanwhile, in the transportation system, the use of gas-fuelled transportation modes such as CNG (compressed natural gas) and LNG (Liquid Natural Gas) can reduce greenhouse gas emissions.

Acknowledgments

We acknowledge the support received from Andusti Simpari company, the small-scale plastic waste recycling company in Bengkulu Province, for given the utmost support in providing accuracy of the data and Ailesh Green Consulting company for providing a database set for LCA software.

References

[1] Maramis R 2020 Tanggung jawab negara dalam menanggulangi pencemaran lingkungan akibat sampah plastik di era revolusi industri *J. Lex Privat.* 4 219-228
[2] Widiyatmoko H, Purwaningrum P, and Putri A P F 2016 Analisis karakteristik sampah plastik di permukiman Kecamatan Tebet dan alternatif pengolahannya *J. urb & environ tech.* 24
[3] Sari L P 2016 Analisis kebijakan pengelolaan sampah pada dinas lingkungan hidup *J. Pub. & Pol. Adm* 2 1-23
[4] Leonidas M, Lena H C, David M, Camilla C, Marie K R, and Mikael H E 2018 Plastic recycling in the Nordics: A Value chain market analysis *J Wast. Manag.* 76 180-189

[5] Kemenperin 2021 Menperin Resmikan Pabrik Daur Ulang Plastik Terbesar di Indonesia. *kemenperin.go.id*
[6] Bayu S, Isti S, Muryantoc, Arief A R, and Edi I W 2019 Global warming impacts study on tofu products in Mampang Prapatan small and medium enterprises with life cycle assessment methods *J. IJoLCAS.* 9-24
[7] Kurniawati, Supartono, Suyantohadi 2019 Life cycle assessment on a small-scale tofu industry in Baturetno village, Bantu district Yogyakarta (IOP Publishing) p 365
[8] Fitria I 2017 Identifikasi karakteristik limbah cair industri daur ulang plastik pada CV Majestic Buana Group dan pengolahannya menggunakan sequencing batch reactor (SBR) (thesis)
[9] Baigo H, Rosye H R T, Suwito, Hendra K, Maury, and Alianto 2018 Kajian kualitas air laut dan indeks pencemaran berdasarkan parameter fisika-kimia di perairan distrik depapre, Jayapura *J. Ilm. Ling*. 35-43
[10] Chaerul M, Allia V 2019 Tinjauan kritis studi Life Cycle Assessment (LCA) di Indonesia *J. Seram. Eng*. 5 816-823
[11] International Standard Organisation 1997 *Environmental assessment - Principles and framework* (Switzerland: International Organization for Standardization) p 3
[12] Cederberg C, Mattson 2000 Life cycle assessment of milk production – a comparison of conventional and organic farming *J. Clean. Prod.* 8 49-60
[13] International Standard Organisation 2006 *Environmental management- Life Cycle Assessment-Requirements and Guidelines* (Switzerland: International Organization for Standardization) p 8
[14] Li S, Ernst W, Martin K P, 2010 Open-loop recycling: a LCA case study of PET bottle-to-fibre recycling *J. Res. Cons. Recyc*le 34-52
[15] Rizqi P M, 2014 Daur ulang dan konservasi energi dalam alternatif pengelolaan sampah dengan metode LCA (Banjarmasin: Universitas Lambung Mangkurat)
[16] Rosmiati V 2019 Kajian Dampak Lingkungan Pada Proses Daur Ulang Sampah Plastik Dengan Pendekatan Life Cycle Assesment (thesis)
[17] Hendrawan 2010 An analysis of the CO2 emission abatement in plastic recycling system using life cycle assessment (LCA) methodology: a case study of Bandung city, Indonesia *J. Man. Tekn*.

Rice Plant Health Spatial Study with Sentinel and SPOT Imagery in Parakansalak District

Muthia Tri Octavia[1], S. Supriatna[1]*
[1]Department of Geography, Faculty Mathematics and Natural Sciences, Universitas Indonesia
[1]muthia.tri@ui.ac.id, [2]*ysupri@sci.ui.ac.id

Abstract. The Rice plant is a food plant cultivated through lowland, one of the world community's basic needs and source of livelihood for most Indonesian people. However, the amount of rice production in Indonesia is still relatively low compared to the consumption rate of the people, so it needs to get more attention in supporting the realization of food self-sufficiency. One of the efforts to achieve this is made by monitoring, such as health conditions. The purpose of this research is to identify the health condition of rice plants in Parakansalak District, Sukabumi Regency, West Java, by applying high-resolution satellite imagery, Sentinel-2 and SPOT-6 processed with NDVI (Normalized Difference Vegetation Index) algorithm to the monitoring plants health. The result concluded that rice plants have good health with an area of 197 hectares or 52 % of the existing paddy field area by Sentinel-2, and the distribution of NDVI values in higher resolution images SPOT-6 has a similar spatial pattern. This percentage shows that the distribution of NDVI values is relatively high and indicates that the plants have a high density.

1. Introduction

Rice Plant (Oryza sativa, sp) is a type of food plant that is one of the basic needs of the world community, most of which is cultivated through lowland [1]. Based on the Center for Agriculture data, rice also serves as a staple food for more than 90 % of Indonesians, with a high carbohydrate content to meet energy needs. Indonesia also relies on the agricultural sector as a source of livelihood for its residents. However, the amount of rice production in Indonesia is still relatively low compared to the consumption rate of the people, where the total rice production in 2020 is 54 million tons [2]. In addition, there are other problems, namely changes in land function, resulting in reduced agricultural yields. In this regard, the agricultural sector needs to get more attention from the government and the community in supporting the realization of food self-sufficiency. One of the efforts to achieve this can be made by monitoring the condition of rice plants, such as the health conditions of plants in the production center area [3], [4].

Along with advances in science and technology, monitoring rice plants does not need to conduct a direct survey but can be done through remote sensing imagery [5], [6]. One of the remote sensing images in observing plant health areas is the Sentinel-2 [7], [8], and SPOT images. In knowing the health condition of rice plants through satellite imagery, the NDVI algorithm or the Normalised Difference Vegetation Index can be used [3], [9], [10]. The principle of the NDVI algorithm is based on green plants that reflect near-infrared radiation and absorb radiation in visible light, so we get a combination of bands from the NDVI algorithm that can show the greenness of plants as a measure of relative plant health [11], [12]. The density of a plant can affect the high or low value of NDVI, where a low value indicates that plants are rare or not dense and vice versa. Therefore, the purpose of this study is to identify the health condition of rice plants according to the NDVI algorithm using satellite imagery.

Parakansalak Sub-district is among the sub-districts in Sukabumi Regency, with the most extensive total agricultural land among other sub-districts in the district [2]. According to data from the Central Statistics Agency for 2020, Parakansalak District has an area of 694 ha of rice fields and 1,047 ha of non-rice fields, which makes the sub-district included in the total land area in the Sukabumi Regency. According to Tucker [12], one of the essential aspects of advancing rice production is effective plant health monitoring. By doing this, farmers can reduce the potential harm to plants, such as pests or plant diseases, to optimize rice production. In this study, the remote sensing imagery used is Sentinel 2 and SPOT-6, which have spatial resolution and real-time data. These two types of satellite imagery were used to identify plant indices, such as leaf area, chlorophyll, and water content, by looking at the village's uniqueness in using extensive land use for agriculture. Therefore, the research on "Spatial Study of Rice Plant Health Using Sentinel Imagery and SPOT in Parakansalak Village, Sukabumi Regency" was carried out with other hopes that it could help develop studies on the use of remote sensing in agriculture in Indonesia.

2. Methods

This study area was located in Parakansalak District, Sukabumi Regency, at 6°47' S latitude and 106°40' L latitude. Parakansalak District has an area of 5,626 km^2 with the use of the area as agricultural land covering an area of 6.5 km^2, and most of the village areas are on plains and slopes or ridges. This study extracts NDVI values to be carried out to monitor the health of rice plants in Parakansalak District, Sukabumi Regency, by using Sentinel-2 and SPOT imagery.

2.1. Sentinel-2 imagery

Sentinel-2 is a satellite launched by the *European Space Agency* (ESA) as an optical imaging vehicle to monitor and provide an explanation of the condition of the earth's surface [13]. Sentinel-2 consists of two constellations, namely Sentinel-2A and Sentinel-2B, with various spectral resolutions with 13 spectral channels and three different resolutions, namely 10 meters, 20 meters to 60 meters [14]. The spatial resolution of the Sentinel-2 image consists of 4 bands with a resolution of 10 meters, namely Red, Green, Blue, and Near Infra-Red, then used to analyze the health of rice plants.

2.2. SPOT-6 imagery

SPOT is a satellite launched to provide high-resolution optical images and a wide area to increase knowledge or manage earth resources and other phenomena on the earth's surface with high spatial resolution, such as SPOT-6. In the SPOT-6 image, there are two types of bands, one panchromatic band, visible light and near-infrared, of which the panchromatic band has a resolution of 1.5 meters. In contrast, the three visible and near-infrared bands have a spatial resolution of 6 meters.

2.3. NDVI algorithm

The vegetation index is a mathematical transformation that simultaneously involves several channels, which can produce a new image that is more representative in presenting the vegetation phenomenon [15]. In general, the identification of vegetation in digital image data uses red and near-infrared bands, which have long been used as indicators of the existence and condition of vegetation [16]. The vegetation index algorithm used to calculate the vegetation index of rice plants is the Normalised Difference Vegetation Index (NDVI), with NDVI values ranging from -1.0 to 1.0 [17]. A value of 0 is assumed to be the pixel boundary between vegetation and non-vegetation, and the higher the index value is defined as, the healthier vegetation cover [18,19]. Mathematically algorithm for NDVI is defined by applying the formulation below:

$$NDVI = (NIR - Red) / (NIR + Red)$$

The NDVI calculation uses the principle that green plants absorb radiation in the Photosynthetically Active Radiation spectrum, while green plants reflect near-infrared radiation [20], [21]. Plants will emit and absorb unique waves. Therefore this condition can be associated with wave emission with other objects that can distinguish between vegetation and non-vegetated objects [22].

2.4. Detecting rice plant health by NDVI values

The distribution of NDVI values is applied to observed the health condition of plants with a corrected image based on the vegetation index algorithm. Processing the NDVI algorithm aims to determine the distribution of NDVI values in Parakansalak District, so the results are then used again to compare the NDVI value with the health of rice plants that can be saw in Table 1.

Table 1. Classification of Plant Health Levels by NDVI

Plant Health & Plant Density	NDVI Values
Very Good	0.72 – 0.92
Good	0.42 – 0.72
Normal	0.22 – 0.42
Bad	0.12 – 0.22
Very Bad	-0.10 – 0.12

Source: http://endeleo.vgt.vito.be/dataproducts.html.

The principle of the NDVI algorithm is based on green plants reflecting near-infrared radiation and absorbing radiation in visible light, so a combination of bands from the NDVI algorithm can show the greenness of plants as a measure of relative plant health [11,12]. Thus, a high NDVI value indicates that an area has healthy and dense plants [23].

3. Results and discussion

Processing was carried out using images taken on September 16 year 2021, with a spatial resolution of 10 meters. There are two bands used to see the value of the NDVI distribution, namely the NIR (Near-Infrared) and RED bands, which on Sentinel-2 are band 8 and band 4.

Table 2. Relationship between NDVI value and rice plant health

Plant Health & Plant Density	NDVI Values	Area (ha)
Very Good	0.72 – 0.92	74
Good	0.42 – 0.72	197
Normal	0.22 – 0.42	96
Bad	0.12 – 0.22	11
Total		**378**

The results of the health classification of rice plants have a different area for each classification and amount to 378 hectares of the total rice fields in Parakansalak District. The classification of good rice plant health is the most significant area compared to other health classifications, with an area of 197 hectares or equivalent to 52 % of the percentage of the existing paddy field area. The distribution of this good classification was spread over the paddy fields of the Parakansalak District. This percentage shows that the distribution of NDVI values is relatively high, indicating that the plants have a high density and are getting healthier with active photosynthesis. So with a reasonably high NDVI value, it shows that the distribution of rice plant health in Parakansalak District in 2021 is considered good. The

IOP Conf. Series: Earth and Environmental Science 1111 (2022) 012023 doi:10.1088/1755-1315/1111/1/012023

health classification of typical rice plants also counts a lot, with 96 hectares spread across the north to south of the sub-district.

Meanwhile, the area is not much different for the classification of excellent rice plant health, namely 74 hectares. The poor or unhealthy rice plant health classification has an area of at least 11 hectares. In addition, after testing the correlation coefficient, it is known that the correlation coefficient (R) is 0.929, which indicates a solid relationship between the health of the rice plant and the NDVI value from the field survey.

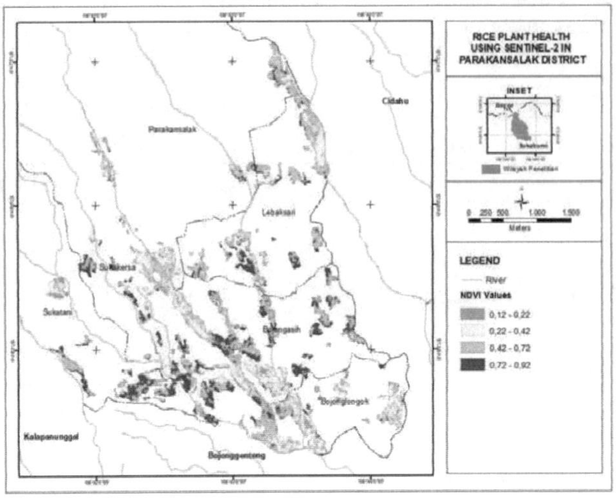

Figure 1. Rice Plant Health Using Sentinel-2 in Parakansalak District

The classification of rice plant health was carried out on another image, SPOT-6, to see variations in rice plant health that were captured in a higher resolution. The SPOT-6 recording time used was April 22, 2020, with a spatial resolution of 6 meters, and the bands used were also visible light and near-infrared.

Almost the same as the sentinel-2 image, the health classification of rice plants in Parakansalak District was still dominated by good health, with an area of 240 hectares or 63 % of the total available rice fields. Meanwhile, for the second place, there is the health of normal rice plants with an area of 82 hectares and an area of very good with 39 hectares. The classification of bad health covers an area of 17 hectares.

Table 3. Relationship between NDVI value in SPOT-6 and rice plant health

Plant Health & Plant Density	NDVI Values	Area (ha)
Very Good	0.72 – 0.92	39
Good	0.42 – 0.72	240
Normal	0.22 – 0.42	82
Bad	0.12 – 0.22	17
Total		**378**

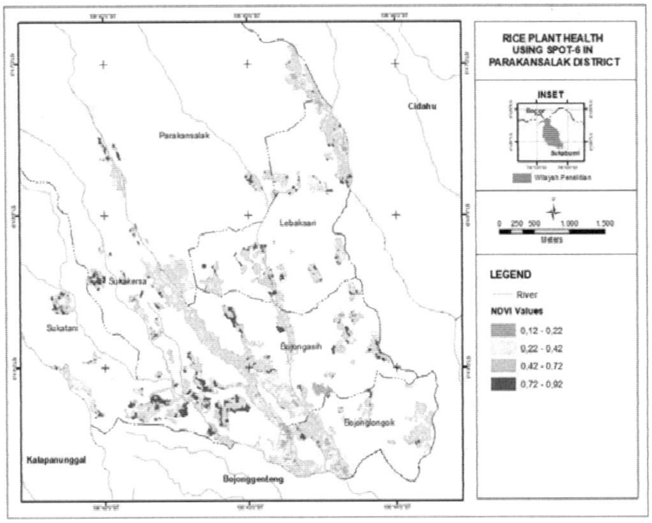

Figure 2. Rice Plant Health Using SPOT-6 in Parakansalak District.

4. Conclusions

The distribution of NDVI values could be applied to define the health condition of rice plants in the paddy fields of Parakansalak District. From image processing, rice plants in 2021 were have good health, with 197 hectares or 52 % of the existing paddy field area. So with this percentage then shows that the distribution of NDVI values is relatively high, indicating that the plants have a high density and the distribution of rice plant health in Parakansalak District is considered good. In addition, the distribution of NDVI values in higher resolution images or SPOT-6 has a similar spatial pattern so that it indicates the relation between the NDVI value and plant health.

Acknowledgments

The authors thanks to Department of Geography, Faculty of Mathematics and Natural Sciences, Universitas Indonesia.

References

[1] E. L. Subandi, Widiatmaka, and M. Ardiansyah, "Spatial modelling of paddy fields in Gorontalo Regency, Indonesia," *IOP Conf. Ser. Earth Environ. Sci.*, vol. 399, no. 1, 2019, doi: 10.1088/1755-1315/399/1/012037.

[2] BPS, "Luas Panen dan Produksi Padi di Indonesia 2020," *Ber. Resmi Stat.*, no. 22, pp. 1–20, 2021.

[3] J. Wijitdechakul, S. Sasaki, Y. Kiyoki, and C. Koopipat, "UAV-based multispectral image analysis system with semantic computing for agricultural health conditions monitoring and real-time management," in *2016 International Electronics Symposium (IES)*, 2016, pp. 459–464.

[4] Rokhmatuloh, Supriatna, Tjiong Giok Pin, Revi Hernina, Ronni Ardhianto, Iqbal Putut Ash Shidiq, Adi Wibowo. "Paddy field mapping using UAV multispectral imagery," *Int. J. GEOMATE*, vol. 17, no. 61, pp. 242–247, 2019, doi: 10.21660/2019.61.icee408.

[5] S. Rapinel, C. Mony, L. Lecoq, B. Clément, A. Thomas, and L. Hubert-Moy, "Evaluation of Sentinel-2 time-series for mapping floodplain grassland plant communities," *Remote Sens. Environ.*, vol. 223, no. September 2018, pp. 115–129, 2019, doi: 10.1016/j.rse.2019.01.018.

[6] Rokhmatuloh, Supriatna, A. Wibowo, I. P. A. Shidiq, T. G. Pin, and R. Hernina, "Spatial

analysis of rice phenology using Sentinel-2 and UAV in Parakan Salak, Sukabumi Regency," *IOP Conf. Ser. Earth Environ. Sci.*, vol. 500, no. 1, pp. 205–210, 2020, doi: 10.1088/1755-1315/500/1/012072.

[7] Supriatna, Rokhmatuloh, A. Wibowo, and I. P. Ash Shidiq, "Rice productivity estimation by Sentinel-2A imagery in Karawang Regency, West Java, Indonesia," *Int. J. GEOMATE*, vol. 19, no. 72, pp. 49–53, 2020, doi: 10.21660/2020.72.5622.

[8] Supriatna *et al.*, "Spatial distribution of phenology stage on a paddy field in Karawang Regency," *IOP Conf. Ser. Earth Environ. Sci.*, vol. 561, no. 1, 2020, doi: 10.1088/1755-1315/561/1/012051.

[9] R. Rosle, N. N. Che'ya, N. A. Roslin, R. M. Halip, and M. R. Ismail, "Monitoring Early Stage of Rice Crops Growth using Normalized Difference Vegetation Index generated from UAV," *IOP Conf. Ser. Earth Environ. Sci.*, vol. 355, no. 1, 2019, doi: 10.1088/1755-1315/355/1/012066.

[10] F. Khoirunnisa, Supriatna, and A. Wibowo, "Using NDVI algorithm in Sentinel-2A imagery for rice productivity estimation (Case study: Compreng sub-district, Subang Regency, West Java)," *IOP Conf. Ser. Earth Environ. Sci.*, vol. 481, no. 1, 2020, doi: 10.1088/1755-1315/481/1/012064.

[11] L. Ryan, "Creating a normalised difference vegetation index (NDVI) image using multispec," *Univ. New Hampsh.*, 1997.

[12] R. Tucker, J. A. Callaham, C. Zeidler, A. L. Paul, and R. J. Ferl, "NDVI imaging within space exploration plant growth modules – A case study from EDEN ISS Antarctica," *Life Sci. Sp. Res.*, vol. 26, no. January, pp. 1–9, 2020, doi: 10.1016/j.lssr.2020.03.006.

[13] N. Oktaviani and H. A. Kusuma, "Pengenalan Citra Satelit Sentinel-2 Untuk Pemetaan Kelautan," *Oseana*, vol. 42, no. 3, pp. 40–55, 2017, doi: 10.14203/oseana.2017.vol.42no.3.84.

[14] C. Corbane *et al.*, "A global cloud-free pixel-based image composite from Sentinel-2 data," *Data Br.*, vol. 31, p. 105737, 2020, doi: 10.1016/j.dib.2020.105737.

[15] N. W. Sudarsono, B. Sudarsono, and A. P. Wijaya, "Analisis Fase Tumbuh Padi Menggunakan Algoritma NDVI, EVI, SAVI, dan LSWI pada Citra Landsat 8." Diponegoro University, 2016.

[16] K. S. Fassnacht, S. T. Gower, M. D. MacKenzie, E. V. Nordheim, and T. M. Lillesand, "Estimating the leaf area index of North Central Wisconsin forests using the landsat thematic mapper," *Remote Sens. Environ.*, vol. 61, no. 2, pp. 229–245, 1997, doi: 10.1016/S0034-4257(97)00005-9.

[17] R. E. Arhatin, "Pengkajian Algoritma Indeks Vegetasi Dan Metode Klasifikasi Mangrove Dari Data Satelit Landsat-5 TM Dan Landsat-7 ETM+ (Studi Kasus Di Kabupaten Berau, Kalimantan Timur)," 2007.

[18] W. J. Carper, T. M. Lillesand, and R. W. Kiefer, "The use of intensity-hue-saturation transformations for merging SPOT panchromatic and multispectral image data," *Photogramm. Eng. Remote Sens.*, vol. 56, no. 4, pp. 459–467, 1990.

[19] P. Rahaldi, H. H. Handayani, and A. Wibowo, "Analisa Kesehatan Tanaman Padi Berdasarkan Nilai Normalized Difference Vegetation Index (Ndvi) Menggunakan Citra Aster (Studi Kasus : Kabupaten Indramayu - Jawa Barat)," *Geoid*, vol. 8, no. 2, p. 107, 2013, doi: 10.12962/j24423998.v8i2.720.

[20] X. Jin, S. Liu, F. Baret, M. Hemerlé, and A. Comar, "Estimates of plant density of wheat crops at emergence from very low altitude UAV imagery," *Remote Sens. Environ.*, vol. 198, pp. 105–114, 2017, doi: 10.1016/j.rse.2017.06.007.

[21] Y. Xie, Z. Sha, and M. Yu, "Remote sensing imagery in vegetation mapping: a review," *J. Plant Ecol.*, vol. 1, no. 1, pp. 9–23, 2008, doi: 10.1093/jpe/rtm005.

[22] N. Horning, P. T. Mcphearson, and O. C. Wallace, "An Introduction to Remote Sensing (Synthesis)," no. 98210, 2009.

[23] A. Y. Nafi, "Estimasi Produktivitas Padi Menggunakan Teknik Penginderaan Jauh Dalam Mendukung Program Swasembada Pangan," *Geo-Image*, vol. 5, no. 2, pp. 1–8, 2016, doi: 10.15294/geoimage.v5i2.13552.

Spatial Study of Land Cover Changes and Land Surface Temperature in Bekasi Regency, West Java, Indonesia

Ariella Novita Hot Asi[1], S. Supriatna[1]*, and Faris Zulkarnain[1]
[1]Department of Geography, Faculty Mathematics and Natural Sciences, Universitas Indonesia
[1]ariella.novita@gmail.com, [2]*ysupri@sci.ui.ac.id, [3]faris.zulkarnaen@sci.ui.ac.id

Abstract. The increase in the built-up area in the Jakarta and its surrounding area influences the surrounding area so that there is the propagation of the built-up area and an increase in ground surface temperature. This study aims to analyze the pattern of land cover and land surface temperature in 2004, 2012, and 2020, identify the relationship between NDBI, NDVI, and land surface temperature, and predict land cover and land surface temperature in 2028 in Bekasi Regency. The method used is statistical analysis and prediction analysis with MLP-MC. The results are that the ground surface temperature increase occurred in Bekasi Regency from 2004 – 2020. The increase occurred in the types of built-up land cover such as settlements, industry, and roads. The biggest land cover change occurred in rice fields into settlements. From the statistical results, the mean land surface temperature, settlement land cover, industry, and roads also experienced an increase in temperature. NDBI can be positively correlated with land surface temperature, while NDVI is negatively correlated. In 2028, there will be an increase in the settlement area, which also results in a decrease in the areas of rice fields. This also has an impact on increasing the temperature of the land surface in the area that is experiencing changes.

1. Introduction

Built-up land is widely developed in urban areas due to urbanization [1-3]. Generally, changes in land use in urban areas that develop into built-up areas are accompanied by an increase in land surface temperature (LST) due to reduced vegetation cover and the increasing number of buildings with artificial materials such as asphalt and concrete that reduce evapotranspiration and increase temperatures [1]. Jakarta and its surrounding area is a metropolitan area in Indonesia that has experienced urban development in a short time [4,5]. In urban development, Jakarta becomes the center and affects the surrounding areas, namely Bogor, Depok, Tangerang, and Bekasi [4]. One area that is part of Jakarta and its surrounding area is Bekasi Regency. Bekasi Regency is designed for industrial and agricultural development [6]. Based on research by [4] during 2000–2015, the development of urban areas in Bekasi Regency can be identified from the increase in the built-up area followed by a decrease in the areas of vegetation. According to [7], the factors that underlie changes in land include political and institutional factors, economic, technological, cultural, and natural physical factors.

Land surface temperature and land cover patterns have been widely used in studies related to climate in urban areas such as Urban Heat Island (UHI). To assess land cover changes, it is necessary to analyze the model. The land cover model of the Multi-Layer Perceptron-Markov Chain (MLP-MC) is used to simulate the prediction of urban development [8]. Problems are related to land surface temperature and changes in land cover in urban areas that cause UHI to need to be studied in areas that are experiencing development, such as Bekasi Regency. The effects of UHI can cause health problems in urban communities, such as respiratory and heart problems, especially in children and the elderly [1]. Studies on changes in land cover and land surface temperature in urban areas are needed to order predict future urban development. The purpose of this study was to analyze the pattern of land cover and land surface temperature in 2004, 2012, and 2020, identify the relationship between NDBI, NDVI, and land surface temperature, and predict land cover and land surface temperature in 2028 in Bekasi Regency.

2. Materials and Methods

2.1 Land cover analysis and Analysis of land surface temperature

Landsat images in 2004, 2012, and 2020 are classified using the digitized method. The land use accuracy test is carried out with data from the field to be retrieved with Google Earth and by using the confusion matrix, as well as the Kappa Index of Agreement (KIA). Proportional random sampling is used as testing or accuracy test samples. Analysis of land cover change from 2004–2020 using identified changes in area statistics. Thermal channels in Landsat images in 2004, 2012, and 2020 were extracted using the single-channel method by [9] to obtain land surface temperature values. Identification of changes in land surface temperature is carried out by statistical analysis of the mean surface temperature of each land cover class [9] [10].

2.2 Analysis of the relationship between NDBI, NDVI, and land surface temperature

Thermal channels in Landsat images in 2004, 2012, and 2020 were extracted using the single-channel method by [11] to obtain land surface temperature values. Identification of changes in land surface temperature is carried out by statistical analysis of the mean surface temperature of each land cover class. The NDBI and NDVI indices were extracted from 2004, 2012, and 2020 Landsat images with the following equation:

$$NDBI = \frac{MIR-NIR}{MIR+NIR} \quad (1)$$

Where MIR is the middle infrared band (Band 5 on Landsat 5 and 7, Band 6 on Landsat 8) and NIR is the near-infrared band (Band 4 on Landsat 5 and 7, Band 5 on Landsat 8).

$$NDVI = \frac{NIR-RED}{NIR+RED} \quad (2)$$

Where NIR is the near-infrared band (Band 4 on Landsat 5 and 7, Band 5 on Landsat 8) and RED is the red band (Band 3 on Landsat 5 and 7, Band 4 on Landsat 8). Pearson correlation was carried out on each land cover class to determine the effect of NDBI and NDVI on land surface temperature in each land cover class.

2.3 Land cover prediction in 2028

Prediction of land cover using the MLP-MC method using Cramer's V analysis to determine the relationship between land cover change and the driving factors obtained from DEMNAS, Regional Planning Office of Bekasi Regency, in the form of secondary data and plotting. Land cover predictions for 2020 were validated with a 2020 land cover map using a confusion matrix and calculating the Kappa Index of Agreement. Furthermore, land cover predictions are made in 2028 [12] [13].

Kappa Index of Agreement. Kappa index calculation is done with the following equation [14]:

$$k = \frac{N\sum_{i=1}^{r} xii - \sum_{i=1}^{r} (xi+ * xi+i)}{N^2 - \sum_{i=1}^{r} (xi+ * xi+i)} \quad (3)$$

where

$$k = \frac{(Total\ sum\ of\ correct) - sum\ of\ the\ all\ the (row\ total\ column\ total)}{Total\ squared - sum\ of\ the\ all\ the (row\ total\ column\ total)}$$

The driving factors for land cover change that represent economic factors are the distance from administrative offices, distance from CBD, distance from health facilities, distance from toll gates, distance from markets and shops, distance from the bus stops, distance from the train station, distance from the river, distance from the road, and NDBI to show the effect of the density of the area on the surroundings. Natural factors are represented by elevation and slope.

2.4 Prediction of land surface temperature in 2028

Predictive analysis of land surface temperature using the MLP-MC method using Cramer's V analysis to determine the relationship between land surface temperature classes and the driving factors, namely NDBI and NDVI. The prediction of land surface temperature in 2020 is validated with a map of land surface temperature in 2020 using a confusion matrix and the calculation of the Kappa Index of Agreement. Furthermore, the land surface temperature prediction in 2028 is carried out.

3. Results and Discussion

3.1 Bekasi Regency land cover

In the land cover classification of Bekasi Regency in 2004, 2012, and 2020, each Kappa index was obtained at 0.83. From the classification results, it was found that in those three years, the type of paddy field cover was the land cover that dominated the area, and the smallest land cover was mangrove

(figure 1). However, rice fields continued to experience a decline in the area from 865.646 km^2 in 2004, 778.740 km^2 in 2012, and 746.993 km^2 in 2020.

It was found that there was an increase in the areas of settlement land cover by 77.48 km^2, industry by 10.166 km^2, road network by 8.97 km^2, grass and shrubs by 16,315 km^2, open land by 6.98 km^2, and mangroves by 0.68 km^2 in the period 2004–2012. However, there was a decrease in the areas of rice fields by 86.89 km^2, mixed gardens by 30.79 km^2, and aquacultures by 3.14 km^2. Many areas of rice fields turned into settlements. Between 2004–2012, the area of paddy fields turned into settlements was 90.83 km^2. There is an increase in the areas of settlement land cover by 21.21 km^2, industry by 17.31 km^2, road network by 17.72 km^2, and mangroves by 4.81 km^2 in 2012–2020. There was a decrease in the area of rice fields by 31.72 km^2, mixed gardens by 12.62 km^2, grass and shrubs by 6.09 km^2, open land by 2.61 km^2, aquacultures by 7.74 km^2, and water bodies by 0.25 km^2. The biggest change in area between 2012–2020 occurred in rice fields into settlements, which was 49.52 km^2.

3.2 Land surface temperature in Bekasi Regency

The land surface temperature in 2004, which was analyzed from Landsat 7 ETM+ imagery, has a range of values from 20.17-33.27°C (figure 2). Based on the land cover type, the highest mean temperature was found in the type of industrial land cover and roads, namely 27.33°C and 26.43°C.
The lowest mean temperatures for aquacultures and mixed garden land cover types were 23.76°C and 24.04°C, respectively. The land surface temperature in 2012, which was analyzed from Landsat 5 TM images, had a range of values between 14.9–35°C and an average temperature of 23.77°C. Based on the type of land cover, the highest mean land surface temperature was found in industrial land cover and roads, namely 27.22°C and 26.47°C (figure 4). The lowest mean temperature was found in aquaculture and mangrove land cover types, namely 22.08°C and 22.08°C. The land surface temperature in 2020, which was analyzed from Landsat 8 OLI/TIRS images, has a range of values from 18.44–34.96°C, and the average temperature is 25.33°C. Based on the land cover type, the highest land surface temperature was found in industrial land cover and roads, namely 28.39°C and 27.60°C (figure 4). The lowest mean land surface temperature was found in aquaculture and mangrove land cover, 23.59°C and 23.61°C.

In built-up areas, settlement, industry, and roads have increased in the mean value from 2004–2020. The mean settlement surface temperature increased from 25.77 in 2004, 25.79 in 2012, and 27.11 in 2020. The mean industrial surface temperature increased from 27.337 in 2004, decreased slightly to 27.22 in 2012, and increased to 28,397 in 2020. The mean road surface temperature increased from 26.126 in 2004, 26.47 in 2012, and 26.60 in 2020.

3.3 Relationship between NDBI, NDVI, and land surface temperature

The correlation test using the Pearson's Correlation method found that NDBI and NDVI from Landsat 8 image processing in 2020 significantly affected land surface temperature in most land cover classes, except for grass and shrub cover and mangroves. The Pearson NDBI correlation value with land surface temperature ranges from 0.118 to 0.675, which is from a weak to strong relationship. The lowest correlation value was found in the grass and shrub land cover class and the highest in the industry. In addition, NDBI has a strong correlation on settlement cover (0.630) and moderate correlation on road cover (0.529), rice fields (0.555), mixed gardens (0.551), open land (0.589), mangroves (0.508). Pearson NDVI correlation value to land surface temperature has a range from -0.219 to -0.641, which is weak to a strong relationship. The lowest correlation value is in the mangrove land cover class, and the highest is in the industry. In addition, NDVI has a moderate correlation with settlement land cover (-0.400), roads (-0.508), rice fields (-0.472), mixed gardens (-0.517), open land (-0.465), aquacultures (0.573), water bodies (-0.553).

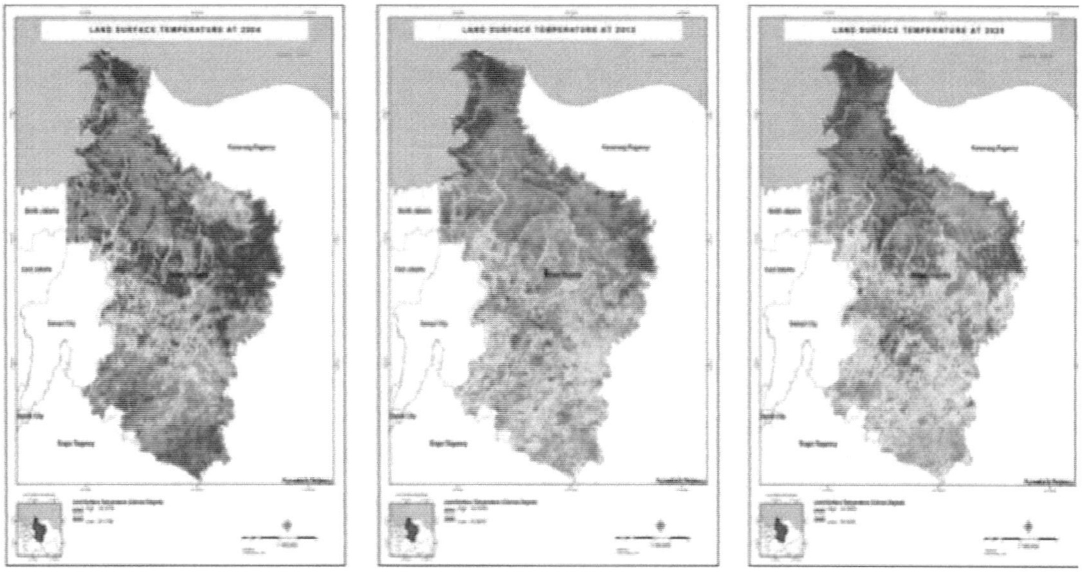

Figure 1. Land surface temperature of Bekasi Regency a) Year 2004; b) Year 2012; c) Year 2020

The results of the F test with a significance value of <0.05 were found in the settlement, industrial, rice fields, and aquaculture land cover classes. NDBI and NDVI of the land cover class together affect the surface temperature of the land. The coefficient of determination (R^2) NDBI and NDVI of each land cover class is 0.423; 0.523; 0.397; and 0.382. This means that NDBI and NDVI affect 42.3% of the land surface temperature in settlements, 52.3% of the land surface temperature in industry, 39.7% in rice fields, and 38.2% in surface temperatures in aquacultures.

Table 1. Correlation of NDBI, NDVI, and land surface temperature in 2020

Land Cover	N	Mean			Correlation coeficient		Sig.		F	Sig. (F)	R^2	Regression
		SPT	NDBI	NDVI	NDBI	NDVI	NDBI	NDVI				
Settlement	70	27,21	-0,02	0,34	0,63	-0,40	0,00	0,00	24,67	0,00	0,42	y = 26.4 + 12.657 (NDBI) + 2.972 (NDVI)
Industry	14	28,61	0,05	0,11	0,68	-0,64	0,00	0,01	6,02	0,02	0,52	y = 28.959 + 5.648 (NDBI) + (-5.601) (NDVI)
Road	14	27,33	-0,21	0,51	0,53	-0,51	0,03	0,03	2,16	0,16	0,28	y = 29.396 + 6.629 (NDBI) + (-1.29) (NDVI)
Paddy field	235	24,75	-0,32	0,59	0,56	-0,47	0,00	0,00	76,39	0,00	0,40	y = 26.953 + 3.655 (NDBI) + (-1.762) (NDVI)

Mixed agriculture	10	24,79	-0,28	0,81	0,55	-0,52	0,05	0,06	1,60	0,27	0,31	y = 27.123 + 3.144 (NDBI) + (-1.775) (NDVI)
Grass and bush	10	26,02	-0,27	0,69	0,12	-0,35	0,37	0,16	1,31	0,33	0,27	y = 30.456 + (-10.639) (NDBI) + (-10.596) (NDVI)
Open land	10	26,25	-0,10	0,38	0,59	-0,47	0,04	0,09	1,91	0,22	0,35	y = 26.684 + 3.012 (NDBI) + (-0.375) (NDVI)
Mangrove	10	23,76	-0,49	0,74	0,51	-0,22	0,07	0,27	1,24	0,35	0,26	y = 24.806 + 2.449 (NDBI) + 0.214 (NDVI)
Aquaculture	36	23,43	-0,46	0,04	0,33	0,57	0,02	0,00	10,21	0,00	0,38	y = 23.818 + 0.927 (NDBI) + 0.875 (NDVI)
Water body	10	24,26	-0,45	-0,25	-0,27	-0,55	0,22	0,05	1,96	0,21	0,36	y = 23.767 + (-0.521) (NDBI) + (-0.998) (NDVI)

3.4 Land cover and land temperature prediction in 2028

The magnitude of the influence of the driving factors on land cover changes can be seen in the value of Cramer's V. The results of the land cover prediction accuracy test with the 2020 classification map show the Kappa index of 0.86, which means it is almost perfect with the 2020 classification map. In the prediction of land cover in 2028, settlement areas will be increasingly dense, and expansion will occur from built-up areas to non-built rice fields. In predicting land surface temperature, NDBI and NDVI are used as driving factors. From the results of the Cramers' V test, it was found that NDBI has a large influence on changes in land surface temperature, but the Cramer's V value of the two factors has shown a good effect for modeling (table 2).

Table 2. Cramer's V value for land surface temperature change is driving factors

Driving Factor	Cramer's V
NDBI	0.331
NDVI	0.253

The results of the 2020 land surface temperature prediction accuracy test with the 2020 land surface temperature classification map show the Kappa index of 0.86, which means it is almost perfect. The land cover prediction map in 2028 shows that areas that were previously non-built, such as rice fields, have changed into settlements. From the predicted land surface temperature in 2028, the area also experienced an increase in temperature (figure 2). This means that the increase in temperature mostly occurs in areas that were previously non-built, such as rice fields.

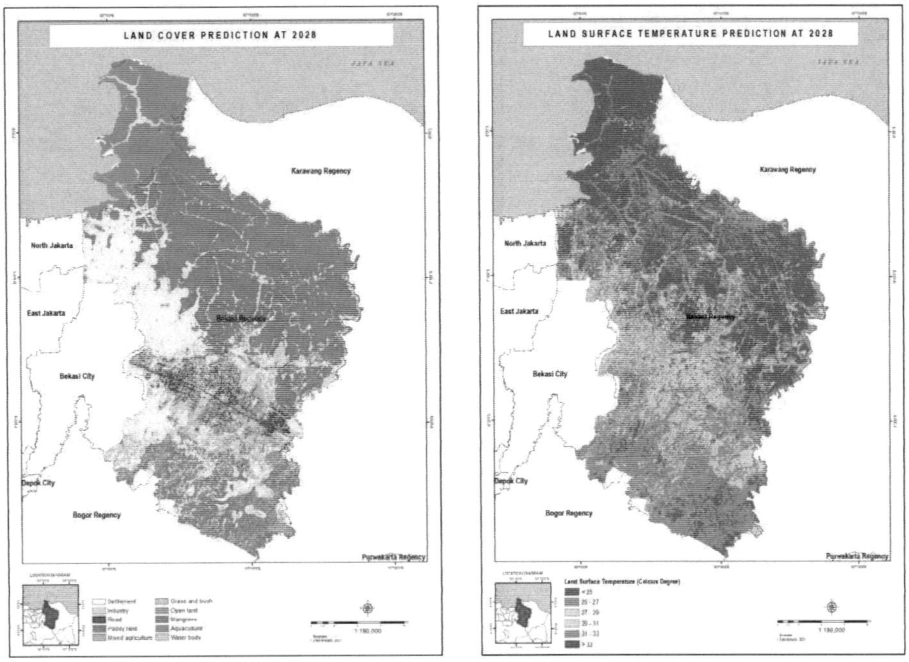

Figure 2. Land cover prediction and land temperature prediction in 2028

4. Conclusions

The increase in ground surface temperature occurred in Bekasi Regency from 2004-2020. The increase occurred in the types of built-up land cover such as settlements, industry, and roads. The biggest change in land cover occurred in rice fields into settlements between 2004-2012 and 2012-2020. From the statistical results of the mean land surface temperature, the land cover for settlement, industrial, and roads also experienced an increase in temperature. The correlation between NDBI and NDVI to land surface temperature shows that most NDBI positively correlates with land surface temperature. Meanwhile, NDVI is negatively correlated. Some land cover types, such as open land and roads, do not correlate well with the land surface temperature. In 2028, there will be an increase in the settlement area, which will also decrease the area of rice fields. This also has an impact on increasing the temperature of the land surface in the area that is experiencing changes.

Acknowledgments

The authors thanks to Department of Geography, Faculty of Mathematics and Natural Sciences, Universitas Indonesia.

References

[1] M. Bokaie, M. K. Zarkesh, P. D. Arasteh, dan A. Hosseini, "Assessment of Urban Heat Island based on the relationship between land surface temperature and Land Use/Land Cover in Tehran," *Sustain. Cities Soc.*, vol. 23, p.94–104, 2016.

[2] Y. Feng, H. Li, X. Tong, L. Chen, dan Y. Liu, "Projection of land surface temperature considering the effects of future land change in the Taihu Lake Basin of China," *Glob. Planet. Change*, vol. 167, no. May, p. 24–34, 2018.

[3] A. Nurwanda dan T. Honjo, "Analysis of land use change and expansion of surface urban heat island in Bogor city by remote sensing," *ISPRS Int. J. Geo-Information*, vol. 7, no. 5, 2018..

[4] S. Arifin, N. Mukhoriyah, dan D. Yudhatama, "Analysis of Land Use Spatial Pattern Change of Town Development Using Remote Sensing," *Int. J. Remote Sens. Earth Sci.*, vol. 15, no. 1, p. 93, 2018.

[5] T. Firman dan F. Z. Fahmi, "The Privatization of Metropolitan Jakarta's (Jabodetabek) Urban Fringes:

The Early Stages of 'Post-Suburbanization' in Indonesia," *J. Am. Plan. Assoc.*, vol. 83, no. 1, p. 68–79, 2017.

[6] A. A. Kurnia, E. Rustiadi, dan A. E. Pravitasari, "Characterizing industrial-dominated suburban formation using quantitative zoning method: The case of bekasi regency, Indonesia," *Sustain.*, vol. 12, no. 19, p. 1–19, 2020.

[7] T. Plieninger *et al.*, "The driving forces of landscape change in Europe: A systematic review of the evidence," *Land use policy*, vol. 57, hal. 204–214, 2016.

[8] D. Ozturk, "Urban growth simulation of Atakum (Samsun, Turkey) using cellular automata-Markov chain and Multi-layer Perceptron-Markov chain models," *Remote Sens.*, vol. 7, no. 5, p. 5918–5950, 2015.

[9] P. A. Shidiq, M. H. Ismail, S. Supriatna, and A. Wibowo. Spatial-temporal analysis of changes in land-cover and land surface temperature (LST) within Universiti Putra Malaysia campus area. IOP Conf. Series: Earth and Environmental Science, 561, (2020) 012031.

[10] Diki Nurul Huda, Supriatna, Iqbal Putut Ash Shidiq, and Taqyuddin. Spatiotemporal analysis land surface temperature in relation to earthquake occurrence around the cimandiri fault. IOP Conf. Series: Earth and Environmental Science, 540, (2020) 012069.

[11] J. C. Jiménez-Munoz dan J. A. Sobrino, "A generalized single-channel method for retrieving land surface temperature from remote sensing data," *J. Geophys. Res. Atmos.*, vol. 108, no. 22, 2003.

[12] S. Supriatna, Sri Fauza Pratiwi, Kuswantoro Marko, Masita Dwi Mandini Manessa, and Yoanna Ristya. Spatial Dynamics of Tsunami Prone Areas in Pariaman City, West Sumatera. Journal of Comput. Theor. Nanosci. 17, 1474–1491 (2020). ISSN 1546-1955 (Print); ISSN 1546-1963 (Online).

[13] Della Ayu Lestari, Dewi Susiloningtyas, Supriatna. Spatial Dynamics Model of Land Availability and Population Growth Prediction in Bengkulu City. International Journal Geography Vol. 52, No. 3 (2020).

[14] S. Pal dan S. Ziaul, "Detection of land use and land cover change and land surface temperature in English Bazar urban centre," *Egypt. J. Remote Sens. Sp. Sci.*, vol. 20, no. 1, p. 125–145, 2017.

Identification of Potential Groundwater Recharge Zone Using Remote Sensing and GIS in Upstream Cibeet Sub-watershed, Bogor, West Java

Jessica Ivana[1*], L. Supriatna[2], and Tri Rani Puji Astuti[3]

[1,3]Geology Study Program, Faculty of Mathematics and Natural Sciences (FMIPA), Universitas Indonesia, Depok 16424, Indonesia
[2*]Study Program of Geology, Faculty Mathematics and Natural Sciences, Universitas Indonesia.
[1*]jessica.ivana@ui.ac.id, [2]ysupri@sci.ui.ac.id, [3]trirani@sci.ui.ac.id

Abstract. The Upstream Cibeet sub-watershed is part of the Citarum watershed and Cibeet sub-watershed with an area of 69.35 km^2. Hydrologically, the upstream part of the watershed usually functions as a recharge area, and its development should be geared towards land conservation and erosion mitigation. Parameters of rainfall, soil type and texture, geology, lineament density, drainage density, geomorphology, and land cover are used in predicting the most potential catchment area. Thematic maps were made for each parameter. Weighting is carried out using the Analytical Hierarchy Process (AHP) with the following results: rainfall 0 %, soil type and texture 34.62 %, geology 24.98 %, lineament density 17.66 %, drainage density 8.57 %, geomorphology 9.34 %, and land cover 4.83 %. The integration of each thematic map resulted in a map with four zonations, namely zone of poor (4.93 %), fair (32.67 %), good (60.58 %), and very good (1.81 %) recharge potential. Based on the groundwater table contour, groundwater mimics the surface topography and flows from the south to the north and northeast part of the study area. Geological structures in the study area possess a NE-SW and NW-SE trend with a normal oblique regime. Structural damage zones are predicted to function as conduits for water movement, especially in volcanic and intrusion lithology.

1. Introduction

Groundwater is a very important resource in human life. The need for groundwater will continue to increase along with population growth and urbanization, but sometimes groundwater can be a non-renewable resource [1]. This is caused by an imbalance in the hydrological cycle, where groundwater exploitation exceeds an aquifer's recharge rate. Therefore, the spatial and temporal variation of groundwater recharge is essential in the balance of the hydrological cycle and groundwater system [2].

The upstream Cibeet sub-watershed is part of the Citarum watershed, and the Cibeet sub-watershed is located in the Bogor physiographic zone. Hydrologically, a watershed's upstream part is suitable for catchment areas. Therefore, development priorities should be in the form of land conservation and erosion control [3]. Currently, the Bogor area is a region with limited published information regarding water resources [4] and a high level of urbanization. Therefore, it is necessary to map the productive catchment area in the upstream Cibeet sub-watershed as accurately as possible as a reference for future development.

Previous studies in groundwater recharge potential [1, 5, 6, 7] have shown that the following parameters effectively predict recharge potential: geology, geomorphology (slope and topography), lineament density, drainage density, land cover, rainfall, and soil type. Each parameter should be weighted according to local conditions that can potentially affect the hydrogeology of the study area, e.g., climate, physiography, and degree of land degradation in the area.

Based on the Regional Geological Map of Indonesia, Cianjur Sheet [8], and Bogor Sheet [9], the stratigraphy of the study area consists of the Jatiluhur Formation marl and quartz sandstones, the Cantayan Formation breccia members, the Cantayan Formation claystone members, Tuffacious sandstones and conglomerates, as well as Mount Kencana and Mount Limo breccias and lava. The structure in Java is the result of tectonic activity from the collision of the Eurasian Plate with the Indian-Australian Plate, which resulted in four structural patterns, including Meratus Pattern (NE – SW), Sunda Pattern (N – S), Java Pattern (W – E), and the Sumatra pattern (NW – SE). The study area is controlled by a northeast-southwest trending fault structure (Meratus Pattern) and a folding pattern that produces sedimentary hills with W – E orientation. In addition, it is also known that there are minor strike-slip faults in the area around Jonggol [5, 10].

The hydrogeology of the research area has been mapped on the Hydrogeological Map of Indonesia, Cianjur Sheet, and Bogor Sheet with a scale of 1:100,000. In general, this area is characterized by young indivisible volcanic rocks as well as claystone, marl with sandstone, tuff, and limestone inserts. Young volcanic rock acts as a non-aquifer or sometimes produces a well discharge of less than 5 l/second. The lithology of claystone, marl with inserts of sandstone, tuff, and limestone forms areas of scarce groundwater or non-aquifer with a pass between 10-4 and 1 m/day [11, 12].

2. Materials and method/Experimental

In general, three methods are carried out in collecting and processing research data:

2.1. Remote sensing

Remote sensing is a surface method that aims to identify the presence, difference, and distribution of lineaments, slopes, topography, lithology, drainage patterns (DEMNAS), and land cover (Landsat-8). The results of each analysis are depicted in thematic maps.

2.2. Geographic Information System (GIS)

GIS is used in the process of rasterization, reclassification, and integration of thematic maps in the weighted overlay feature. Field data is also integrated into the form of isophreatic, spring, and structural maps that are used to identify local hydrogeological flows and system controls.

2.3. Geologic and hydrogeologic field mapping

Field sampling was carried out as validation for the remote sensing analysis. The samples obtained were lithological, groundwater table measurements from local wells, geological structure measurements, and geomorphological data. Lithological samples were obtained through observation and sampling from outcrops using random cluster sampling for each lithological unit in the study area. Measurement of geological structures is carried out using a geological compass in the form of fault and joint orientations.

2.4. Ranking, weighting, and integration

The evaluation of recharge potential is obtained through classification and assignment of ranks for each parameter class with a scale of 1-3. The largest value describes the highest potential. Thematic maps are then integrated into a GIS environment according to the AHP weighting technique described in figure 1. The AHP weighting is made with reference to various literature and author modifications adapted to the research area's geological setting. AHP weighting is carried out using the AHP-OS software [13]. Based on the integration of reclassified thematic maps and parameter weightage, a map of recharge potential can be obtained using Equation 1 [5]. The recharge potential map will then be assessed along with a map of the geological structure, groundwater table, and springs to provide a complete picture of the hydrogeological conditions of the research area.

Intensity	Definition	Explanation
1	Equal importance	Two elements contribute equally to the objective
3	Moderate importance	Experience and judgment slightly favor one element over another
5	Strong Importance	Experience and judgment strongly favor one element over another
7	Very strong importance	One element is favored very strongly over another, it dominance is demonstrated in practice
9	Extreme importance	The evidence favoring one element over another is of the highest possible order of affirmation
2,4,6,8 can be used to express intermediate values		

(a)

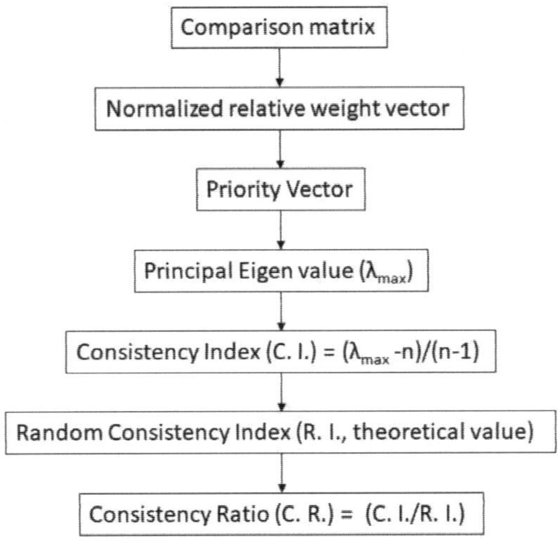

(b)

Figure 1. (a) Scale of pair-wise comparison based on importance for AHP; (b) Basic workflow of AHP weighting method [14]

$$Recharge\ Zone\ (R) = \sum_{i=1}^{n} (Pi * Wi) \quad (1)$$

n = total class for each parameter

Pi = weight of each parameter

Wi = Rank of each class

3. Results and discussion

3.1. *Thematic Map Analysis*

Based on the seven research parameters, processing, and reclassification of each thematic map yield the following results:

3.1.1. *Rainfall.*

Rainfall in the upstream area of the Cibeet sub-watershed was measured through Cibongas Station, Gunung Mas Station, Kranji Station, and Panjang Station [3], which were interpolated with Thiessen polygons to cover the entirety of the study area (table 1). Rainfall was calculated over 8 – 10 years and showed a uniform result of high – very high precipitation rate (table 2).

Table 1. Annual rainfall and ranking for each rain station

No	Rain Station	Area (km^2)	Percentage (%)	CHT (mm/year)	Rank
1	St. Cibongas	2.48	3.58	3828.25	3
2	St. Gunung Mas	51.52	74.27	3098.27	3
3	St. Kranji	14.98	21.60	3097.79	3
4	St. Panjang	0.38	0.55	2792.13	3

Table 2. Rainfall classification for Cibeet sub-watershed (modified from [3])

No	Rainfall (mm/year)	Category
1	< 1500	Very low
2	1500 - < 2000	Low
3	2000 - < 2500	Moderate
4	2500 - < 3000	High
5	>= 3000	Very high

3.1.2. Soil Type and Texture

Soil type and texture data are based on [15] divided soil types of the area according to the [16] classification (table 3). The ranking is based on the descriptive ability of soil to transmit and store water.

Table 3. Soil type, texture, and rank

No	Soil Type and Texture	Area (km^2)	Percentage (%)	Rank
1	Gleisol District – Fine	5.04	7.26	1
2	Mediterran Haplic – Fine	8.95	12.90	2
3	Litosol – Medium Coarse	11.16	16.09	3
4	Andosol District – Fine	14.92	21.50	3
5	Cambisol Oxic - Fine	29.31	42.25	3

3.1.3. Geology.

Geological data were mainly obtained through field observations and sampling at 37 stations. Eight lithologies were identified, then divided into five unofficial geological units based on lithological characteristics and distribution (table 4).

Table 4. Geological units and rank

No	Geological Unit	Area (km^2)	Percentage (%)	Rank
1	Andesit intrusion	0.10	0.15	1
2	Tuf breccia	24.15	34.68	2
3	Lava and lava breccia	17.37	24.95	1
4	Sandstone and limestone	1.83	2.63	3
5	Claystone and siltstone	26.18	37.60	2

3.1.4. Lineament density.

The results of auto-lineament extraction upstream of the Cibeet Sub-watershed have produced some lineaments with the main orientations of NE – SW and NW – SE (table 5). This is in accordance with the regional tectonic trends operating in the Bogor Basin, namely the Meratus Pattern and the Sumatra Pattern, respectively. Thus, it can be assumed that most of the lineaments obtained have associations with tectonism. A high lineament density value is considered to have a positive impact on the porosity and secondary permeability of an area. Therefore, the higher the FFD value, the better the recharge potential.

Table 5. Lineament density and rank

No	Lineament Density	Area (km²)	Percentage (%)	Rank
1	0 - 2.68	22.97	33.12	1
2	2.68 - 4.60	23.12	33.34	2
3	4.60 - 10.67	23.26	33.54	3

3.1.5. Geomorphology.
Based on the classification of [17], the Upstream Cibeet sub-watershed can be divided into three geomorphological units (table 6). Steeper slopes and topography, especially in mountainous areas, reduce the quantity of water that can be absorbed since the water flow tends to be faster and turns into run-off rather than infiltrating the soil. Meanwhile, gently sloping areas tend to give water sufficient time to absorb when precipitation occurs.

Table 6. Geomorphological units and rank

No	Geomorphological Unit	Area (km²)	Percentage (%)	Rank
1	Gently sloping – moderately steep hills	22.01	31.74	3
2	Moderately steep – steep hills	43.78	63.14	2
3	Steep – very steep mountains	3.54	5.11	1

3.1.6. Drainage Density.
Drainage density is a parameter inversely proportional to the area's productivity as a catchment area. High drainage density is an indicator of high run-off or low infiltration. Thus, the higher the drainage density, the worse the infiltration potential in an area is (table 7).

Table 7. Drainage density and rank

No	Drainage density (km/km²)	Area (km²)	Percentage (%)	Rank
1	0 - 0.6	24.61	35.48	3
2	0.6 - 1.5	37.78	54.46	2
3	1.5 - 2.9	6.98	10.06	1

3.1.7. Land Cover. A land cover map was produced through NDVI analysis (table 8). Water bodies, settlements, and barren land are areas that are not productive for water infiltration. The increase in vegetation density increases the ability of water infiltration and percolation. Areas of sparse vegetation, rice fields, and shrubs land cover are assumed to have moderate infiltration potential, while dense and very dense vegetation forests will have the best infiltration (table 9).

Table 8. Classification of NDVI [18, 19]

No	NDVI Value	Land Cover
1	-1 – 0.3	Water bodies, settlements, and barren land
2	0.3 – 0.6	Sparse vegetation, rice fields, and shrubs
3	0.6 – 0.8	Dense rainforest
4	0.8 – 1	Very dense rainforest

Table 9. Land cover and rank

No	Land Cover	Area (km^2)	Percentage (%)	Rank
1	Sparse vegetation, rice fields, and shrubs	15.05	21.70	2
2	Dense rain forest	50.59	72.94	3
3	Very dense rainforest	3.11	4.49	3
4	Water bodies, settlements, and barren land	0.60	0.01	1

3.2. AHP Weight Assignment

Several factors become the main considerations in the weighting of the seven research parameters:

- In tropical climates, the main infiltration mode is direct recharge. The large influence of exogenous factors in the form of weathering in a humid climate makes soil type the main parameter in a catchment area [20].
- Furthermore, the aquifer beneath it needs to have the ability to facilitate water storage and flow [21]. This is a function of the primary permeability component in geology, followed by the presence of secondary permeability from faults and fractures observed through lineament density.
- The area is located at the base of mountains, with 5.11 % of the area composed of Steep – very steep mountains, so geomorphology becomes the fourth factor to be considered.
- Drainage density is a very good indicator of surface permeability and infiltration ability [22]. This parameter reflects local climate, slope, geology, soil properties, and landscape evolution [23] and are used in this study as a supporting parameter for these factors.
- Land cover has the smallest weight due to the limited human influence on the land cover based on the NDVI analysis (< 21.71 % of the total study area). Primary forest ecosystems that are the best recharge areas remain fairly undisturbed in this area.

These considerations are translated into a pair-wise comparison matrix to generate normalized weights presented in figure 2 and table 11.

Matrix		Soil type and texture	Geology	Lineament Density	Geomorphology	Drainage Density	Land Cover	Normalized Principle Eigenvector
		1	2	3	4	5	6	
Soil type and texture	1	1	2	3	3	3	5	34.62 %
Geology	2	1/2	1	2	3	3	5	24.98 %
Lineament Density	3	1/3	1/2	1	3	3	3	17.66 %
Geomorphology	4	1/3	1/3	1/3	1	1	3	9.34 %
Drainage Density	5	1/3	1/3	1/3	1	1	2	8.57 %
Land Cover	6	1/5	1/5	1/3	1/3	1/2	1	4.83 %
Consistency Ratio = 0.37								

Figure 2. Pair-wise comparison matrix for groundwater recharge parameters

3.3. Zonation of Potential Recharge Area

The integration of thematic maps results produces four areas with different recharge potential values (figure 3 and table 10). In general, poor recharge potential is distributed around the main Cipamingkis river, with an area of 4.93 % of the upstream Cibeet sub-watershed. The eastern area is characterized by fair recharge potential, with an area of 32.67 %. Meanwhile, good and very good recharge potentials can be found on the south and west sides with an area of 60.58 % and 1.81 %, respectively. Good and very good catchment potential areas tend to dominate in Sukamulya Village, Sirnajaya Village, and Wargajaya Village. Good infiltration potential is characterized by lithological units of Tuff breccia, Sandstone – limestone, as well as Lava and lava breccia, geomorphology of Moderately steep – steep hills, and land cover of Very dense to dense rainforests. A very good recharge potential zone shows the same characteristics but has a higher lineament density value.

Table 10. Recharge potential zonation

No	Recharge Potential	Area (km^2)	Percentage (%)
1	Poor	3.40	4.93
2	Fair	22.53	32.67
3	Good	41.78	60.58
4	Very good	1.24	1.81

3.4. Hydrogeology and Geological structure

The contour of the groundwater table in the Upstream Cibeet sub-watershed is a reflection of the surface contour of this area. It does not show an anomaly of depression that could arise due to over-exploitation of the aquifer (figure 4). In general, groundwater flows from high to low hydraulic heads, so water that has seeped into the aquifer system will tend to flow from south to north and northeast directions. The springs on the south side are springs that are identified as depression springs, or springs caused by the meeting of the groundwater level with the surface contour so that it can be used as groundwater level data on the south side, where no housing or wells are found. Depression springs are also found on the east side of this sub-watershed (blue arrow). Meanwhile, the other three springs can be associated with the damage zone of the tentative structure due to the close proximity to local structures. Fault and joint analysis showed a normal oblique mechanism. Therefore, faults in this area are believed to be conduits for recharge or discharge of groundwater. However, calcite veins can be found filling faults and fractures consistently in sedimentary lithologies, resulting in a significant decrease in permeability. These veins do not appear to form and affect the secondary permeability in volcanic and intrusion lithologies.

Table 11. The weight assigned for each parameter through AHP

No	Parameter	Class	Rank	Weight
1	Soil type and texture	Gleisol District – Fine	1	34.62 %
		Mediterran Haplic – Fine	2	
		Litosol – Medium coarse	3	
		Andosol District – Fine	3	
		Cambisol Oxic - Fine	3	
2	Drainage density	0 - 0.6	3	8.57 %
		0.6 - 1.5	2	
		1.5 - 2.9	1	
3	Lineament density	0 - 2.68	1	17.66 %
		2.68 - 4.60	2	
		4.60 - 10.67	3	
4	Geology	Andesite intrusion	1	24.98 %
		Tuff breccia	2	
		Lava and lava breccia	1	
		Sandstone – limestone	3	
		Claystone – siltstone	2	
5	Geomorphology	Gently sloping – moderately steep hills	3	9.34 %
		Moderately steep – steep hills	2	
		Steep – very steep mountains	1	
6	Land cover	Sparse vegetation, rice fields, and shrubs	2	4.83 %
		Dense rainforest	3	
		Very dense rainforest	3	
		Water bodies, settlements, and barren lands	1	
7	Rainfall	2792	3	0 %
		3098	3	
		3829	3	

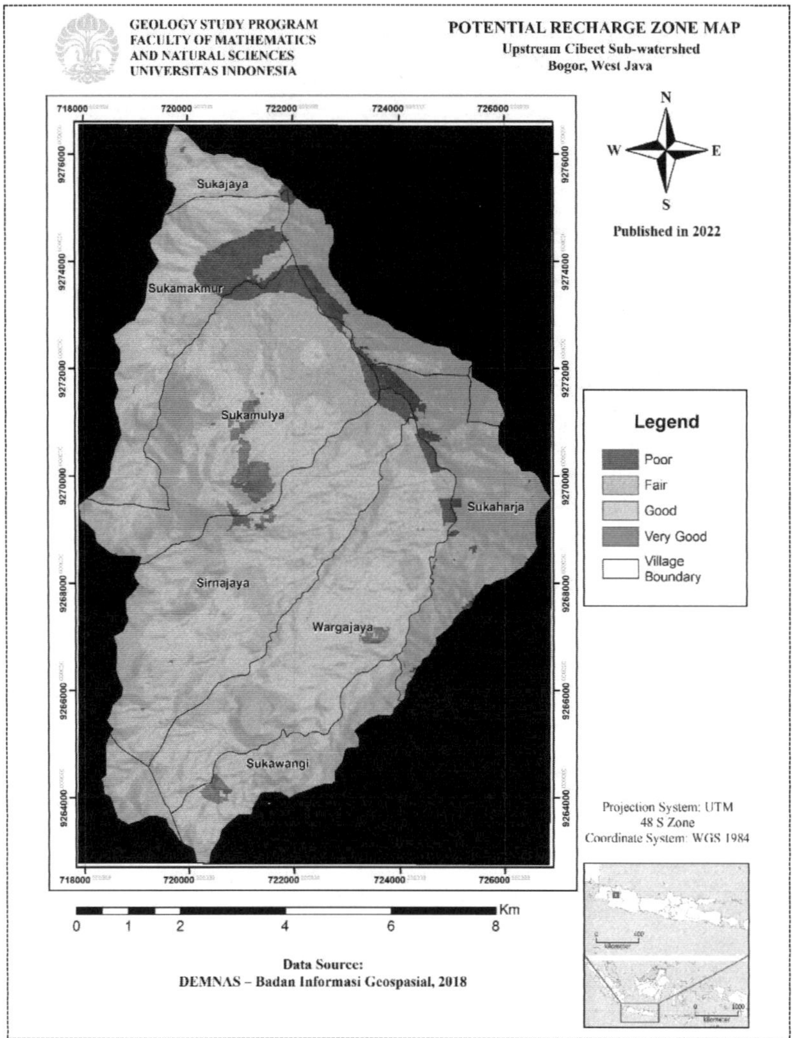

Figure 3. Recharge potential map

4. Conclusion

Based on the AHP method, the weights for each parameter are as follows: soil type and texture 34.62 %, Geology 24.98 %, Line Density 17.66 %, Geomorphology 9.34 %, Drainage density 8.57 %, and Land Cover 4.83 %. Potential catchment areas can be divided into four zonations. Poor recharge potential is distributed around the Cipamingkis River with an area of 3.40 km^2. Fair recharge potential is distributed over an area of 22.53 km^2. Good and very good recharge potential can be found in the south and west parts of the study area and spans 41.78 km^2 and 1.24 km^2, respectively. Zonation of good and very good recharge potentials (mainly located in Sukamulya, Sirnajaya, and Wargajaya Villages) should be developed for conservation and erosion control. Groundwater in the Upstream Cibeet sub-watershed tends to flow from the south to the north and northeast direction. Geological structures in the study area possess a NE-SW and NW-SE trend with a normal oblique regime. The structurally damaged zones are predicted to function as conduits for water movement, especially in volcanic and intrusion lithology.

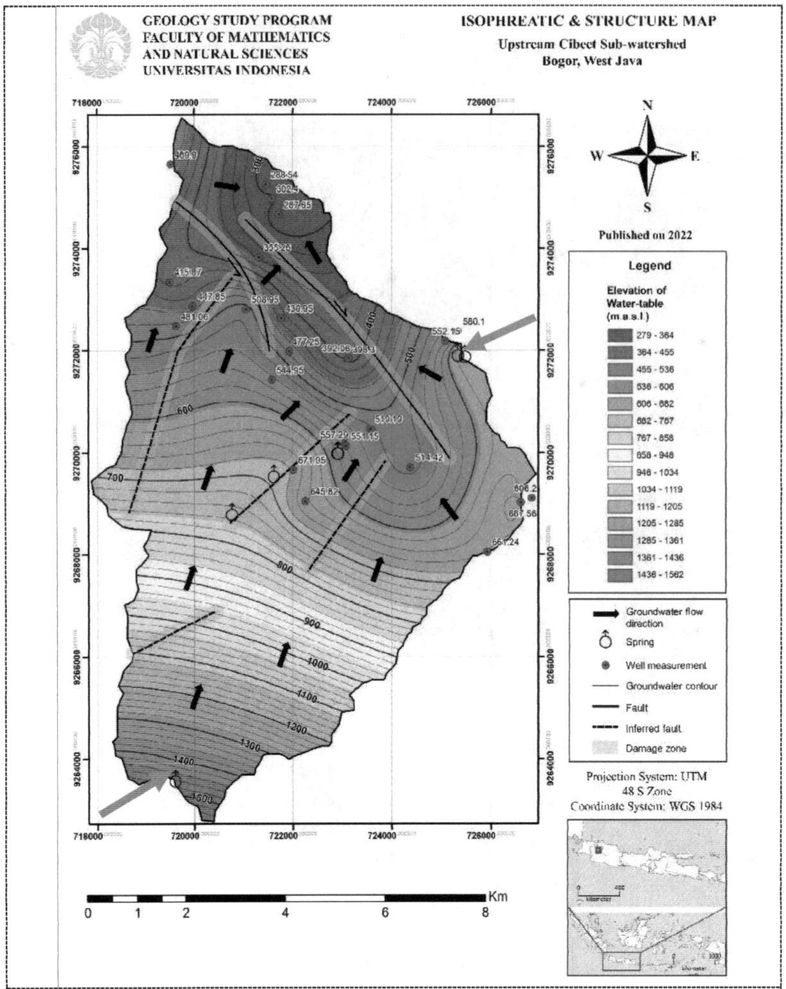

Figure 4. Isophreatic and structure map

Acknowledgments

Authors wish to acknowledge the assistance and encouragement provided by colleagues of the Geoscience Department, insights provided by the lecturers of Geology Universitas Indonesia, as well as data provided by Balai Besar Litbang Sumberdaya Lahan Pertanian (BBSDLP) Kementerian Pertanian Republik Indonesia and Direktur Jenderal Bina Pengelolaan DAS.

References

[1] Dar, T., Rai, N., & Bhat, A. (2021). Delineation of potential groundwater recharge zones using analytical hierarchy process (AHP). Geology, Ecology, and Landscapes, 5(4), 292-307.

[2] Delin, G. N., & Risser, D. W. (2007). Ground-water recharge in humid areas of the United States—A summary of ground-water resources program studies, 2003–06. USGS fact sheet fs–2007, 3007, 4p.

[3] Direktur Jenderal Bina Pengelolaan DAS. (2013). Peraturan Direktur Jenderal Bina Pengelolaan Daerah Aliran Sungai dan Perhutanan Sosial. Nomor: P. 4/V-SET/2013 Tentang Petunjuk Teknis Penyusunan Data Spasial Lahan Kritis. Jakarta [ID]: Dirjen BPDAS.

[4] Sumartoyo, S. (2010). ESTIMASI POTENSI AIRTANAH MELALUI PENDEKATAN TIPOLOGI BENTUKLAHAN WILAYAH BOGOR PROVINSI JAWA BARAT. MAJALAH

ILMIAH GLOBE, 12(1).
[5] Gopi, K., Rajesh, & M., Shirisha, K. (2021). Identification of Groundwater Recharge Potential Zones of Hyderabad Region using Analytical Hierarchy Process in Arcgis. National Conference on Recent Advances in Civil Engineering Infrastructure (rACEi-2021)
[6] Pathak, D. and Shrestha, S.R., (2016). Delineation of groundwater potential zones in rocky aquifers of central Nepal. Jour. Nepal Geol. Soc., v. 50, pp. 161–169.
[7] Patil, S. G., & Mohite, N. M. (2014). Identification of groundwater recharge potential zones for a watershed using remote sensing and GIS. International journal of geomatics and geosciences, 4(3), 485-498.
[8] Sudjatmiko. (1972). Peta Geologi Lembar Cianjur, skala 1: 100.000.
[9] Effendi, A. C. (1974). Peta Geologi Lembar Bogor. Jawa, Pusat Penelitian dan Pengembangan Geologi, Bandung.
[10] Pulunggono, A., & Martodjojo, S. (1994). Perubahan tektonik paleogen-neogen merupakan peristiwa tektonik terpenting di Jawa. Proc. Geologi dan Geoteknik Pulau Jawa, Yogyakarta, h, 37-49.
[11] Murtianto, E. (1994). Peta Hidrogeologi Indonesia Lembar Bogor 1:100.000. Direktorat Geologi Tata Lingkungan; Bandung.
[12] Pasaribu, M., Mudiana, W., dan Sunarya, Y. (1994). Peta Hidrogeologi Indonesia Lembar Cianjur 1:100.000. Direktorat Geologi Tata Lingkungan; Bandung.
[13] Goepel, K.D. (2018). Implementation of an Online Software Tool for the Analytic Hierarchy Process (AHP-OS). International Journal of the Analytic Hierarchy Process, Vol. 10 Issue 3 2018, pp 469-487, https://doi.org/10.13033/ijahp.v10i3.590
[14] Saaty, T.L. (1980). The Analytic Hierarchy Process; McGraw-Hill: New York, NY, USA.
[15] Balai Besar Litbang Sumberdaya Lahan Pertanian (BBSDLP) Kementerian Pertanian Republik Indonesia. (2016). Jenis Tanah Kabupaten Bogor Skala 1:50.000.
[16] FAO. (2012). Harmonized World Soil Database (version 1.2). FAO, Rome, Italy and IIASA, Laxenburg, Austria.
[17] Van Zuidam, R. A. (1985). Aereal photo-interpretation in terrain analysis and geomorphologic mapping. Netherlands.
[18] Januar, D., Suprayogi, A., & Prasetyo, Y. (2016). Analisis Penggunaan Ndvi Dan Bsi Untuk Identifikasi Tutupan Lahan Pada Citra Landsat 8 (Studi Kasus: Wilayah Kota Semarang, Jawa Tengah). Jurnal Geodesi Undip, 5(1), 135-144.
[19] Weier, J. dan Herring, D. (2000). Measuing Vegetation (NDVI & EVI). NASA accessed through https://earthobservatory.nasa.gov/features/MeasuringVegetationCaplar R and Kulisic P 1973 *Proc. Int. Conf. on Nuclear Physics (Munich)* vol 1 (Amsterdam: North-Holland/American Elsevier) p 517 ISBN/DOI number
[20] Sophocleous, M., & Perry, C. A. (1985). Experimental studies in natural groundwater-recharge dynamics: the analysis of observed recharge events. Journal of Hydrology, 81(3-4), 297-332.
[21] Knutsson G. (1988) Humid and Arid Zone Groundwater Recharge — A Comparative Analysis. In: Simmers I. (eds) Estimation of Natural Groundwater Recharge. NATO ASI Series (Series C: Mathematical and Physical Sciences), vol 222. Springer, Dordrecht. https://doi.org/10.1007/978-94-015-7780-9_32
[22] Horton, R. E. (1932): Drainage basin characteristics. Transactions A. Geophysical Union, 13: 350-361
[23] Hilmi, F., & Haryanto, I. (2008). Pola Struktur Regional Jawa Barat. Bulletin of Scientific Contribution, 6(1), 57-66.

JESSD-2022 IOP Publishing
IOP Conf. Series: Earth and Environmental Science 1111 (2022) 012026 doi:10.1088/1755-1315/1111/1/012026

Effect of Control Temperature on The Performance of PV Modules Using Active Water-Spray Cooler

Le Vinh Thang[1], Ngo Xuan Cuong[2], Nguyen Duc Minh[3], Do Nhu Y[4*]
[1]Department of the automatic control system and control, National Research University of Electronic Technology, Moscow, 124498, Russia
[2]School of Engineering and Technology, Hue University, Thua Thien Hue, 530000, Viet Nam, 0000-0002-0571-2168
[3]Institute of Energy Science - Vietnam Academy of Science and Technology, Vietnam
[4]Department of Electrification, Hanoi University of Mining and Geology, Hanoi, 100000, Viet Nam, 0000-0001-6395-2875
[1]levinhthangevn@gmail.com, [2]ngoxuancuong@hueuni.edu.vn, [3]minhnguyenduc.ies@gmail.com, [4*]donhuy@humg.edu.vn.

Abstract. As a result of the greenhouse effect and climate change, renewable energy is gradually replacing traditional electricity sources today. Photovoltaic (PV) technology, in particular, is gaining popularity worldwide, but low efficiency is a challenge that must be overcome. This paper describes an experimental method for determining the effect of control temperature on the performance of PV modules equipped with an active water-spray cooler. The study chose two polycrystalline silicon PV modules with identical specifications, and the experimental evaluation results show that the capacity difference is less than 1%. Through experimental results, this study has shown that on sunny days the daily energy output of PV modules using an active water-spray cooler improves over reference PV modules by 3% to 5%. With increasing temperature control, the performance improvement becomes less. When the control temperature is lower, the working time of the pump motor during the day increases, increasing its power consumption and affecting the proposed system's performance improvement. In addition, with the control temperature at 45°C, the performance improvement was highest in the experimental results. The findings of the study can be used to build future experimental models.

1. Introduction

Currently, there are warnings about power shortage [1]. However, some renewable energy sources, such as solar power, have not been effective. Compared to other solar technologies, the electricity generated by a solar photovoltaic (PV) system not only causes no pollution during operation but also reduces the problems of global warming and operation and maintenance costs [2]. PV technology, in general, and grid-tied solar power, in particular, have some common problems that need to be addressed, such as low efficiency and being negatively affected by hail, dust, and surface temperature [3]. The operating temperature of PV modules has a significant influence on their performance. Therefore, developing a research model to improve the performance of PV systems is essential. In addition, the increase in PV power system efficiency contributes to increasing grid-connected electricity output, reduces the investment return time of solar power projects, saves system installation area, and uses energy economically and efficiently.

PV cell technology is typically divided into three generations. First-generation PV panels are the most commonly used in conventional settings and are made of monocrystalline silicon or polysilicon. Second-generation thin film PV cells are used primarily for photovoltaic power stations, integrated into buildings, or smaller solar systems. Third-generation technologies include thin film technologies and concentrated PV cells. Some use organic materials to generate electricity, while others use inorganic substances. A typical PV module comprises around 36 or 72 PV cells, and module efficiency can range from 5% to 24%, depending on the type of PV cell technology used [4,5]. However, exposure to solar radiation combined with high ambient temperatures causes a significant increase in PV module operating temperature. The working temperature of PV modules is valued between 27 °C and 52 °C when the ambient temperature is 24.5 °C [6]. This working temperature can reach temperatures above 90 °C. It is well known that reducing module temperature can improve electrical efficiency in PV systems. This issue has been thoroughly studied in recent years through the development of different cooling techniques. For PV technologies currently on the market, the electrical performance degradation is due to increased module temperature, ranging from 0.25 %/°C to 0.5 %/°C, so improvements in electrical efficiency can be achieved with a proper cooling technique, keeping in mind that each cooling technique must be proven viable.

In an open circuit PV module cooling system, where the cooling liquid is directed at the PV module, it runs over its surface without the support of ducts. The liquid injection is mainly done by spraying or dripping onto the top surface of PV modules, cooling water runs on it, and cleaning and removing heat from PV modules. Research on water flow over the front of photovoltaic panels was conducted during a clear day at the PV-Labs in Rio de Janeiro [7]. The results have shown that the water flow has a cooling effect and also helps to reduce reflection by 2-3.6%, reduce PV cell temperature by up to 22 degrees Celsius, and power yield can return a surplus of 10.3%; net profit can be obtained from 8-9% even taking into account the power required to run the pump.

A system consisting of tubes allows water to flow on the surface of the PV module with a capacity of 60W [8]. The tests were carried out in Australian cities with different latitudes. The results show that the system output power increase ranges from 4% to 10% when the cooling system is running. In addition, the PV module surface temperature on a typical summer day is around 58 °C, which can be reduced to 33 °C [8]. Two 130W polycrystalline PV modules were used, one with a direct current (DC) water pump cooling system and the other as a reference [9]. The experiment was conducted in March at the center of Excellence Renewable Energy. The controller works in on/off mode with a temperature of 35 °C. Experimental results have shown that, when the water-cooling system is working, the average temperature of a PV module with a cooling system is 37.65 °C. In comparison, the average temperature of PV without a cooling system was 45.74 °C, average power increased to 38.98%. At the same time, research has also shown that a cooling water system is better than a forced ventilation system [9].

In the study [10], the authors present a set of experimental measurements to select a PV-cooled arrangement to ensure the best plausibility in water film uniformity, reduce module temperature, minimize water consumption and maximize module power output. Experiments were conducted with a 230W photovoltaic module in the Laboratory of Fluid Machines and Energy Systems in the Department of Industrial Engineering at the University of Padova. The results show that the cooling system is equipped with three characteristic nozzles with an injection angle of 90°, operating with an inlet pressure of 1.5 bar, remains active for 30 seconds, and shuts down for 120 seconds, which can reduce module temperature down to 28 °C and improve module efficiency by about 14% [10].

The above analysis shows the improved effect of coolers on PV module performance. Although the results discuss the effect of temperature on the performance of PV modules, very few studies evaluate the effect of temperature control on the performance of PV modules using an active water-spray cooler. This study contributes to the following main issues. First, an active water spray cooler for PV modules is proposed to improve their performance. Second, the article analyzes the influence of control temperature on the output power of PV modules. The research results are presented in the next section.

IOP Conf. Series: Earth and Environmental Science 1111 (2022) 012026 doi:10.1088/1755-1315/1111/1/012026

2. Methods

2.1 System description

Two polycrystalline PV modules are used as outdoor test systems with PV modules installed at 17 °N in Hue City, Vietnam. One PV module is used with an active water-spray cooler, and the other is used for reference. Both modules are connected to a grid-tied inverter with independent MPPT inputs. The research PV modules have a capacity of 330 Wp, with dimensions of 1.95x0.99m.

The cooler in this study is designed with six nozzles placed at the high part of the PV module width so that they do not always cast shadows. Each nozzle is spaced 16cm apart, as shown in Figure 1. Cooling water is contained in a 100-liter tank and is supplied by a pump motor with a power of 24W, a maximum water flow of 3.5 liter/min, a pressure of 0.48 bar, and the cooling effect of the spray stream directly over the entire area of the PV modules. The L298n driver with 3.3V Voltage controls the pump motor of the cooler through the NodeMCU ESP32S board. The water-spray cooler works in on/off mode with preset control temperature.

The more excellent controller uses four temperature sensors placed on the back surface of the PV module to measure the average temperature of the PV module. The cooling status and temperature of the PV module are recorded with a sampling time of the 20s. The DC power of each PV module is collected from the solar data logger built into the microinverter with a sampling interval of 5 min.

Figure 1. PV module model using an active water-spray cooler.

2.2 Research method

The following methods were used to conduct the research: (a) Experiment: Set up a PV module model using an active water-spray cooler with the variable setting temperature at 35°C, 40°C, and 45°C. Measurement parameters include PV module temperature and DC power output of cooled and uncooled PV modules; (b) Data analysis: The study was conducted from April 2022 to May 2022. Selecting typical days to analyze the performance of cooled versus uncooled PV modules. Evaluation of the effect of control temperature on the performance of PV modules using active water-spray coolers.

2.3 Performance evaluation

To evaluate the difference between the two studied PV modules, under the same conditions, the daily electrical yield of the PV2 module (E_{PV2}) and PV1 module (E_{PV1}) is measured, and the average deviation is calculated according to the following formula:

$$\varepsilon = 100.\frac{E_{PV1}-E_{PV2}}{E_{PV2}} \quad (1)$$

For comparing the improved performance of the active water-spray cooler, the daily electrical yield by the PV1 module using the proposed system (E_{PV1}); The electrical yield of the PV module not using the proposed system (E_{PV2}), and the total energy consumed by the control circuit and by the water pump is also needed (E3). The average daily improvement of the active cooling system can be calculated as a percentage using the following formula [11]:

$$\eta = 100.\frac{E_{PV1}-E_{PV2}-E_3}{E_{PV2}} \quad (2)$$

The formula calculates the energy consumption of the water pump motor:

$$E_3 = t.\ P_{pumb} \quad (3)$$

Where t - working time of active water-spray cooler; P_{pumb} – the power of the water pump.

3. Results and Discussion

3.1 Difference evaluation

The study selected five days in April to compare the homogeneity of 2 PV modules and the average output deviation of the PV1 module with the reference module. Because the average deviation is less than 1%, the average electric yield of two PV modules is very similar. Two PV modules can be used to assess the proposed system's performance.

Table 1. Electric yield and average deviation of PV modules.

Date	EPV1, Wh	EPV2, Wh	ε, %
18/4/2022	1192	1181	0.95
19/4/2022	392	395	-0.69
20/4/2022	538	543	-0.92
21/4/2022	886	886	0.03
22/4/2022	1499	1495	0.28
Total	4507.1	4499.2	0.17

Figure 2 shows the PV modules' power output and electrical yield over a typical sunny day, with PV modules peaking at 267 W at 12.7 h. Electrical yield for the day reaches 1.5 kWh for each PV module.

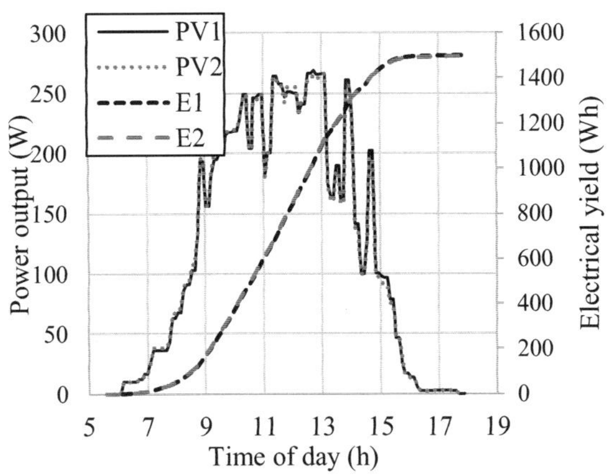

Figure 2. The PV module's power output (PV1, PV2) and electrical yield (E1, E2) were on April 22, 2022.

3.2 *Performance improvement*

The study conducts the study of the cooler under the actual operating conditions of the active water-spray cooler with a preset control temperature. The selected control temperature values are 45, 40, and 35°C. Corresponding to a controlled temperature, the study selected by three typical sunny days for analysis. The obtained results are summarized in Table 2. The results have shown that the electrical yield of the PV1 module using active water-spray coolers is improved by 3-5% compared to the PV2 module. When the control temperature is lower, the motor's working time during the day increases, leading to its power consumption also increasing, which affects the improved performance of the proposed system. The above results are close to 4% to 10% in the study [8]. However, it is lower than the result of 14% in the study [10].

Table 2. Aggregate electrical yield, water pump energy consumption, and performance improvement.

Control temperature, °C	Date	E_{PV1}, Wh	E_{PV2}, Wh	E_3, Wh	ϵ, %	η, %
45	25-Apr	1715.7	1653.0	12.96	3.79	3.01
45	26-Apr	1833.3	1758.2	13.44	4.27	3.50
45	27-Apr	1724.6	1664.9	13.20	3.58	2.79
40	6-May	1310.7	1259.5	38.40	4.07	1.02
40	7-May	1426.7	1366.5	39.60	4.40	1.50
40	8-May	1445.4	1385.9	41.04	4.29	1.33
35	9-May	1500.7	1427.7	69.60	5.11	0.24
35	10-May	1342.5	1279.0	64.80	4.96	-0.10
35	23-Apr	1876.3	1775.1	74.40	5.70	1.51

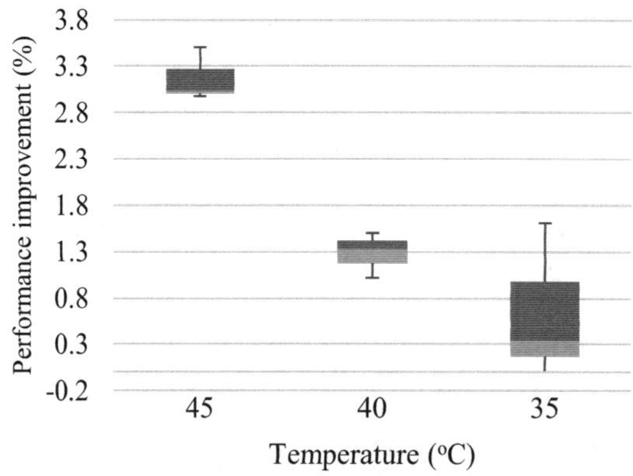

Figure 3. Dependence of performance improvement on control temperature.

In Figure 3, showing the dependence of performance improvement on control temperature, it is clear that the performance is lower as the control temperature decreases. There are times when it is still negative. Thus, with the control temperature at 45 °C, the performance improvement is the highest in the experimental results. To further analyze the cooling mode of the proposed system, the study selects a typical day (April 25, 2022) for analysis. The controlled temperature is 45 °C, and when the temperature drops to 44 °C, the controller will turn off the water pump motor. Figure 4 shows the generating power and electrical yield of 2 PV modules. It is clear that the PV1 module has a higher power output than the PV2 module, and the electrical yield is also higher.

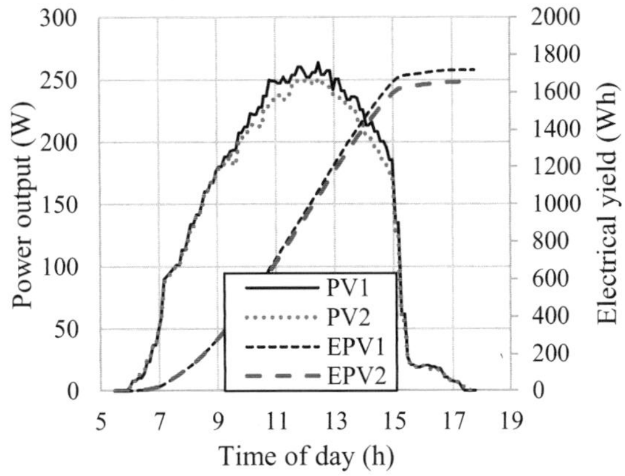

Figure 4. Power output and electrical yield of PV modules on April 25, 2022.

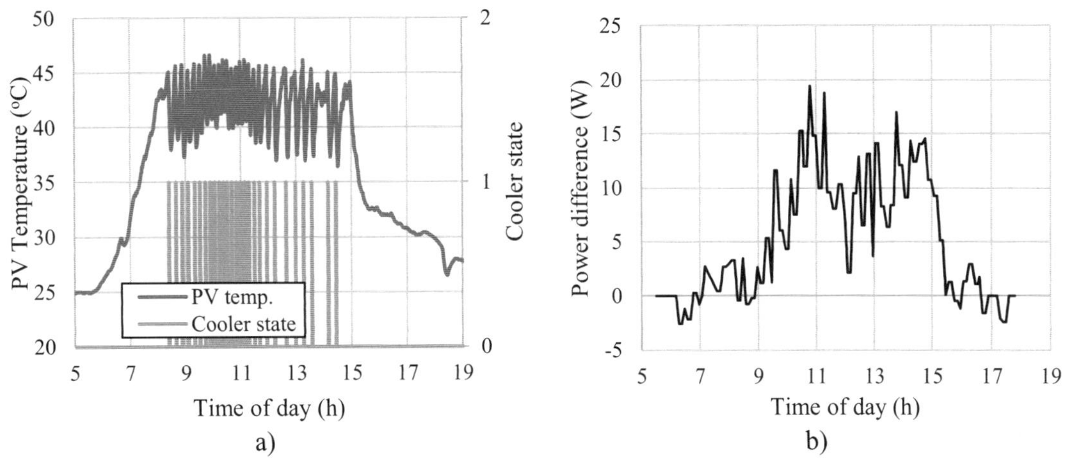

Figure 5. a) PV module temperature and cooling state of active water-spray cooler; b) Power difference between a cooling module and standard module.

Figure 5 shows the active water-spray cooler's PV module temperature and cooling state. The results show that the temperature of the PV module is maintained at 45 °C, and under the effect of cooling, that temperature drops below 40 °C. The proposed system's working time is from 8:20 am to 2:27 pm. During the working time, the generating power output of the PV1 module using the cooler is significantly improved than PV2.

4. Conclusion

This study designed and built an experimental model to evaluate the influence of the active water-spray cooler is controlled temperature on the PV modules' performance. Research to select two PV modules with the exact specifications. The experimental evaluation results show that their power difference is less than 1%. Experimental results have shown that the electrical yield of PV modules using active water-spray coolers is improved by 3% to 5% over reference PV modules. The higher the control temperature is, the less improvement there will be. When the control temperature is lower, the motor's working time during the day increases, leading to its power consumption also increasing, which affects the improved performance of the proposed system.

Furthermore, the performance improvement was most significant in the experimental results when the control temperature was set to 45 °C. These experimental results provide essential data for future experimental models. In addition, it is necessary to study accurate models in many regions and other conditions to better understand the enhanced performance of PV modules using the proposed active water-spray cooler.

References

[1] Wang B, Wang L, Zhong S, Xiang N and Qu Q 2022 Low-Carbon Transformation of Electric System against Power Shortage in China: Policy Optimization *Energies* **15** 1–18
[2] Ngo X C and Do N Y 2022 The Impact of Electrical Energy Consumption on the Payback Period of a Rooftop Grid-Connected Photovoltaic System: A case Study from Vietnam *Int. J. Renew. Energy Dev.* **11** 581–9
[3] Cuong N X, Hong N T and Nhu D Y 2022 A Comprehensive Assessment of a Rooftop Grid-Connected Photovoltaic System: A Case Study for Central Vietnam *Int. Energy J.* **22** 13–24
[4] da Silva V O, Martinez-Bolanos J R, Heideier R B, Gimenes A L V, Udaeta M E M and Saidel M A 2021 Theoretical and experimental research to development of water-film cooling system for commercial photovoltaic modules *IET Renew. Power Gener.* 206–24
[5] Aste N, Del Pero C and Leonforte F 2017 Water PVT Collectors Performance Comparison *Energy Procedia* **105** 961–6
[6] Kumar A, Baredar P and Qureshi U 2015 Historical and recent development of photovoltaic thermal (PVT) technologies *Renew. Sustain. Energy Rev.* **42** 1428–36
[7] Krauter S 2004 Increased electrical yield via water flow over the front of photovoltaic panels *Sol. Energy Mater. Sol. Cells* **82** 131–7
[8] Odeh S and Behnia M 2009 Improving photovoltaic module efficiency using water cooling *Heat Transf. Eng.* **30** 499–505
[9] Irwan Y M, Leow W Z, Irwanto M, Fareq M, Hassan S I S, Safwati I and Amelia A R 2015 Comparison of solar panel cooling system by using dc brushless fan and dc water *J. Phys. Conf. Ser.* **622**
[10] Benato A and Stoppato A 2019 An experimental investigation of a novel low-cost photovoltaic panel active cooling system *Energies* **12** 1–24
[11] Ngo X C, Nguyen T H, Do N Y, Nguyen D M, Vo D V N, Lam S S, Heo D, Shokouhimehr M, Nguyen V H, Varma R S, Kim S Y and van Le Q 2020 Grid-connected photovoltaic systems with single-axis sun tracker: Case study for central Vietnam *Energies* **13**

Processing Efforts and Resource Efficiency in Achieving Green Campus at the State University of Surabaya, Indonesia

Tjitik Rahaju[1], Suci Megawati[2*], Meirinawati[1], Indah Prabawati[1], Eva Hany Fanida[1], Kurniawan Fery Arianto[1] and Faza Saila Salsabila[1]
[1]Department of Public Administration, Faculty of Social Sciences and Law, Universitas Negeri Surabaya, Ketintang Campus, Surabaya, Indonesia 60231
[2]Department of Public Administration, Faculty of Social Sciences and Law, Universitas Negeri Surabaya, Ketintang Campus, Surabaya, Indonesia 60231, 0000-0002-1798-6879
[1] tjitjikrahaju@unesa.ac.id, [2]*sucimegawati@unesa.ac.id, [3]meirinawati@unesa.ac.id, [4]indahprabawati@unesa.ac.id, [5]evahanyfanida@unesa.c.id, [6]kurniawanfery.20050@mhs.unesa.ac.id ,[7]fazasaila.20055@mhs.unesa.ac.id

Abstract
Efforts to reduce global warming based on green campuses are still not widely realized. A green campus is one of the concepts to build a sustainable campus that is environmentally friendly in higher education institutions. A long-term strategy is needed to create a green campus. This study aims to describe efforts to manage and utilize resources to create a green campus area and promote it as an alternative to sustainable campus management at the State University of Surabaya (Unesa). This research is descriptive. The data collection technique was carried out using a literature study approach, collecting, classifying, analyzing, visualizing, validating data, and drawing conclusions. This study found that in realizing a green campus in universities, management actions and resource utilization were needed to initiate the presence of Unesa's green campus, namely an eco-campus with efficient governance. The management and efficiency of resources at the State University of Surabaya have not only succeeded in bringing about efforts to manage a green campus at the institutional level but also proved that a green campus can have implications for the concept of sustainable development. This study has limitations on the sources of data obtained by the author. However, this limitation opens up opportunities for other researchers to fill it. Realizing a green campus can be an alternative solution amid global efforts to achieve SDGs goals.

1. Introduction

Global warming is an increase in temperature on the earth's surface caused by the trapping of greenhouse gases more than normal conditions, such as carbon dioxide (CO2), methane (CH4), water vapor (H2O), ozone (O3), and CFCs in the atmosphere [1]. Increasing Earth's temperature is evidenced by warming oceans, melting glaciers, reduced snow cover in the Northern Hemisphere, and rising sea level [2]. Rising sea levels will cause small islands to sink. In addition, an increase in sea temperature will result in decreased fishery yields, an increase in air temperature will increase the development of disease, and an increase in rainfall will increase flooding, landslides, and changes in the growing season. Lack of awareness of environmental management is a major factor causing environmental damage and global warming [3].

One of the efforts to reduce the impact of global warming is to invite the academic community to play an active role in minimizing global warming by creating a green campus. A green campus is a concept for building environmentally friendly, sustainable living in higher education institutions. A green campus is often the first step taken by universities towards sustainability [4]. Sustainability is the deployment, conservation, and reuse of resources in responsible ways [5]. The contribution of universities to the implementation of sustainable development goals (SDGs) is carried out in at least three ways, namely through ecologically clean society (eco-community), increasing understanding and concern (eco-literacy), and environmentally friendly design of the built environment (eco-design) [6]. Given the importance of green areas for sustainability, universities can contribute while increasing the use of green spaces on campus. Green campus is an initiative from academics that aims to contribute to efforts to reduce global warming [7] by increasing the use of green and clean energy sources to minimize gases formed from the greenhouse effect and other emissions from conventional energy use. Greening also aims to stop climate change and environmental pollution [8].

Universities compete to apply green concepts to buildings and their environment, including the State University of Surabaya. This study aims to describe efforts to manage and utilize resources to create a green campus area and promote it as an alternative to sustainable campus management at the State University of Surabaya. Thus, universities also actively participate in sustainable development goals (SDGs), in particular SDGs number 4, which aims to ensure comprehensive education by promoting a culture of environmental awareness in a student environment. Based on previous research, the novelty of this research is to prove that the management and efficiency of resources at the State University of Surabaya have not only succeeded in bringing about efforts to manage a green campus at the institutional level but also proved that a green campus can have implications for the concept of sustainable development.

2. Methods

This research used a literature study approach and was implemented in August 2022. In the first stage, researchers assign search keywords, then collect relevant data about green campus based on integrated data from academic books, research reports, scientific journals, encyclopedias, Unesa websites, and other written and electronic sources. The keywords set are "Green Campus", "Unesa Green Campus", and "Environmental Sustainability Campus." In the second stage, documents are classified based on the level of relevance and need. Documents used to refer to the year 2011-2022. In the third stage, the researcher analyzes the data, which examines the essence of the research objectives through documents classified according to the level of relevance and need. In the fourth stage, the researcher reanalyzed the collected data as raw data to be developed. The fifth stage is data validation, and the sixth stage concludes [9].

3. Results and Discussion

Resources that are not appropriately managed will cause new problems ranging from environmental pollution and unequal distribution of resources to the depletion of resources that are no longer available. The campus, with its innovations, creates stability and sustainability of resources and contributes to creating a sustainable environment in the campus area. The concept of sustainability in a campus environment or eco-campus must be supported by resource management, including resource efficiency, to maintain the existence of resources available in the future. Unesa, one of the universities in Surabaya, began to develop the concept of a sustainable campus in campus life. Eco-campus is a program to increase public awareness and concern as a collection of scientific communities that participate and are responsible for reducing environmental damage and global warming. The existence of the eco-campus is motivated by, among others, a campus environment that is clean, beautiful, green, healthy, and comfortable for learning; a campus environment that has a role and contribution to increasing or reducing global warming; actions of the campus community that can implement science and technology in the environmental field in a concrete way [10]. Efforts that can be made include cultivating the use of tumblers to reduce plastic waste, disposing of waste in its place and sorting according to its type, recommending walking and cycling in the campus environment, realizing a green campus, saving water and electricity use, reducing paper use, and increasing innovation renewable energy sector in the campus.

3.1. Civilize the use of tumblers to reduce plastic waste

To reduce plastic waste, awareness is needed to start switching from using plastic once to non-reusable equipment. The equipment in question is using a tumbler or drinking bottle to reduce the plastic waste of drinks. The use of drinking bottles can reduce the use of plastic in the campus environment [11]. However, on the other hand, using these bottles makes it difficult for students when they want to buy drinks on campus because many beverage sellers still use closed bottles. Hence, students continue to produce waste by buying these packaged drinks [12]. Despite the existing weaknesses, drinking bottles can still reduce the use of single-use plastics in the campus environment. In addition, it makes users more efficient and healthier because drinks brought from home are more guaranteed in quality. This is where the campus, as the owner of the policy for campus residents, must implement a policy or recommendation to use tumbler bottles to support the eco-campus. With recommendations from the campus, it is hoped that later it can be used as a culture for residents in the campus environment. In line with the goal of reducing plastic waste, UPN Veterans Jakarta is also conducting a "Go Green, No Plastic" campaign, one of the main points of which contains the plastic bag diet and the recommendation to bring your place to eat and drink [13]. The updating and strengthening of the latest regulations regarding the use of drinking bottles will make the entire Unesa academic community change their habit of using single-use bottles and switch to drinking bottles.

3.2. Dispose of garbage in its place and sort it according to its type

Unesa has provided integrated trash, which can be separated into two types, namely organic and inorganic waste bins. This simplification aims to make the existing waste can be processed easily. Organic waste will be processed into compost, and inorganic waste will be recycled. Also, there are compost houses [14] where waste is used as compost, hydroponic areas, and plastic [10]. In addition, Unesa strives for the 3R concept, namely reduce, reuse, and recycle waste in its management that has economic value through the creation and provision of waste banks for waste exchange at the faculty level, such as the FMIPA waste bank, FBS waste bank and FISH waste bank [10]. On the other hand, Unesa found limitations related to waste disposal efforts, namely the minimal number of trash bins [15] and low awareness of sorting waste so that trash cans are still found that are filled but not properly. As a result, janitors have to re-select the waste and make the waste sorting process inefficient. Based on this, Unesa as a facility provider, is expected to add facilities to facilitate waste sorting. In addition, students who care about waste management in the campus environment are expected to participate in waste management by socializing on campus internally. When it comes to waste management, the University of Nottingham produces about 3,000 tonnes of solid waste annually. However, the University of Nottingham provides various recycling facilities to recycle glass, clothing, paper, cards, plastics and cans and also provides an interactive map to find the nearest external recycling facility from campus [16]. These several examples of green campuses mean that Unesa is far behind in terms of waste management compared to other campuses. Unesa can improve waste management by providing various recycling facilities to recycle clothes, glasses, cans, and paper so that they can be used better later.

3.3. Encouraging walking and cycling on the campus

The mobility of motorized vehicles on and off campus will impact the creation of pollution, which is contrary to the eco-campus concept with the aim of reducing pollution. In supporting decarbonization in the campus area, things can be done: walking and cycling for mobility. Besides being cheap, cycling and walking will be healthy for the body. So it can be interpreted that cycling and walking on campus have a positive impact on the campus environment and themselves [17]. In facilitating the mobilization of various regions in Unesa, transportation is used to overcome pollution and save energy in the campus environment, namely campus buses, school buses, and bicycles. Unesa also provides sidewalks on the shoulder of the road as pedestrian paths and special roads for people with disabilities to walk more safely[14]. The limitation, in this case, is that there is no special path for bicycle users, in addition to the lack of facilities for providing bicycle units in the campus environment. In line with the eco-campus concept, namely reducing pollution from motor vehicle emissions, Sultan Ageng Tirtayasa University (Untirta) also provides free bicycle services to create a pollution-free campus environment. Untirta encourages the daily use of bicycles as a means of transportation to improve the safety of road users, improve environmental quality and help reduce the use of motorized vehicles [18]. The culture of walking and cycling on campus will help Unesa reduce the carbon footprint generated by motorized vehicles so that it will improve the quality of the environment at Unesa.

3.4. Creating a green campus

Campus with all its facilities needs to pay attention to the factors that support the pattern of implementing a green campus. Unesa, with its resources, tries to do various things, especially in the eco-campus program, by developing conservation areas such as forests and lakes. It is not only a medium for reforestation or supporting the eco campus, but the conservation area is also a medium for student learning and recreation for the wider community. The conservation area of concern is the campus forest, where the mayor appointed Unesa as a place for preserving urban forests in Surabaya. The Unesa campus forest is known as the "Laboratorium Merdeka Belajar" [19]. In addition, there are also various types of animals ranging from deer, parrots, green peacocks, and rabbits to pigeons [19]. Unesa maintains biodiversity in soil ecosystems by installing biopore infiltration holes to increase the soil's ability to absorb water and preserve the microbes in it [16]. A limitation in terms of realizing a green campus at Unesa is the lack of greenery because the average building in Unesa still uses glass buildings compared to environmentally friendly buildings. In addition, as reported by the Unesa eco-campus site, the plant collection is also minimal, with only rambutan, mango, star fruit, sapodilla, and ambarella [14]. In line with realizing a green campus, Mansoura University in Egypt is also making efforts to create a green campus, including several faculty buildings with courtyard components that function like thermostats to control temperature. Most of the building windows are oriented to the north to improve air circulation and lighting. In addition, Mansoura University also took the initiative to plant it on the roof of the university building [16]. In order to support a green campus, Unesa creates a modern environment that is environmentally friendly while still prioritizing the balance of natural ecosystems.

3.5. Save on water and electricity usage

Energy and natural resources will have a negative impact on humans if their use is not wise. Energy use, especially electricity and water, must be considered to support the Unesa eco-campus program. In addition, it can be in the form of replacing tools that consume large amounts of electrical energy. Unesa has transitioned from non-LED lamps to LED lamps that are more environmentally friendly and use less electricity. Students, staff, and lecturers participate in these savings, including using electricity as needed, turning off water faucets, lights, and air conditioners when not in use, and other activities. Meanwhile, water savings are realized through maximum water utilization, such as the management of drinking water produced by Unesa based on reverse osmosis technology [14]. These activities can sustainably save electricity and water use in order to realize a sustainable campus life. To save water and electricity, the University of Nottingham in the UK also uses street lighting in University Park using LEDs, solar electricity in its buildings, and solar water heaters installed in its buildings [16]. The development of technology such as LED lights will make Unesa more environmentally friendly in order to realize a sustainable campus life.

3.6. Reduce paper usage

Paper will boomerang when its use is too wasteful. Nowadays, the use of technology in campus life must be increased to reduce the use of paper. Deforestation for printing paper must be reduced because deforestation is very far from the concept of a sustainable campus. It took a long time for trees to grow and become forests again. So in campus life, it is necessary to switch to using technology [20]. The limitations, in this case, are that Unesa conducts no paper test cycle, and there are still requests for the use of paper in assigning assignments. In line with reducing paper use, Mansoura University has been recycling exam papers. Some students have also taken the initiative to distribute trash bins to collect paper waste, which can be a good first step towards realizing a green campus [16]. The use of technology is the key to reducing the use of paper. Unesa has done this by shifting the collection of tasks from paper to digital as a form of seriousness for nature conservation. Unesa still needs to provide a place to recycle used paper in the campus environment so that the paper can still be used after being recycled.

3.7. Increase innovation in the field of renewable energy

Innovation is a step that universities can take to support the eco-green pattern in the campus environment. Unesa, with its various innovations, has succeeded in recycling existing waste into handicrafts, including turning banner waste into multifunctional tote bags. Another form of innovation, namely Unesa, succeeded in creating energy-efficient electric motors for people with disabilities, which was initiated by Unesa lecturers [21]. In addition, Unesa has also used solar cells in street lights [14]. This is evidence of Unesa's move to focus on developing renewable innovations, especially in energy. The University of Nottingham is also making innovations in the field of renewable energy, such as using solar electricity installed in 10 buildings, solar water heaters installed in 3 buildings, a ground source heat pump and a lake source heat pump [16]. Adoption of innovation tools from other parties can be done to deal with innovation development at Unesa.

4. Conclusion

Managing and using resources efficiently is important, especially in the green campus program environment. In order to use environmentally friendly energy, renewable innovations need attention to continue to be developed. The State University of Surabaya as the education provider for the green campus program has carried out various forms of innovation as a pilot medium for the broader community regarding the use of resources. Resource management focuses on using land as effectively as possible, especially regarding land conservation in urban forests and tree planting aimed at decarbonizing the campus environment. However, there is a need for elaboration in managing other resources that need to be considered and improved, such as trash can facilities, bicycle facilities, and other facilities that support the green campus program. The contribution of this research to the program developer, the State University of Surabaya, suggests developing clearer regulations in the life of the campus environment. On the other hand, Unesa needs to bring in other campuses that have succeeded in building a green campus-based university life in order to motivate, improve, and as an alternative to promote sustainable campus management in Unesa. Regarding environmental issues, especially the eco-campus program can become a model medium for other educational institutions. The sustainability of this research still needs to be carried out, especially studies on Unesa with linear topics such as green campus strategy and SDGs.

References

[1] R. Pratama and L. Parinduri, "Penaggulangan Pemanasan Global," *Bul. Utama Tek.*, vol. 15, no. 1, pp. 91–95, 2019.

[2] G. D'Amato and C. A. Akdis, "Global warming, climate change, air pollution and allergies," *Allergy Eur. J. Allergy Clin. Immunol.*, vol. 75, no. 9, pp. 2158–2160, 2020, doi: 10.1111/all.14527.

[3] B. Leu, "Dampak Pemanasan Global Dan Upaya Pengen-Daliannya Melalui Pendidikan Lingkungan Hidup Dan Pendidikan Islam," *At-Tadbir*, vol. 5, no. 2, pp. 1–15, 2021, doi: 10.51700/attadbir.v1i2.207.

[4] G. Sonetti, P. Lombardi, and L. Chelleri, "True green and sustainable university campuses? Toward a clusters approach," *Sustainability*, vol. 8, no. 1, pp. 1–23, 2016, doi: 10.3390/su8010083.

[5] A. Malhotra, N. P. Melville, and R. A. Watson, "Spurring Impactful Research on Information Systems for Environmental Sustainability," *MIS Q.*, vol. 37, no. 4, pp. 1265–1274, 2013.

[6] M. Chaerul and A. Susangka, "Selection of City Waste Composting Technology Using Analytic

Hierarchy Process Approach," *J. Purifikasi*, vol. 11, no. 2, pp. 71–78, 2011.
[7] I. Karim and M. Makmur, "Program Green Campus melalui Penanaman Pohon Ketapang Kencana (Termenelia mantily) dan Ki Hujan (Samanea saman) dalam Upaya Mengurangi Global Warming," *CARADDE J. Pengabdi. Kpd. Masy.*, vol. 2, no. 1, pp. 1–7, 2019, doi: 10.31960/caradde.v2i1.103.
[8] G. Kayakutlu and E. Mercier-Laurent, *Greening and Innovating with Renewable Energies*. London: ISTE Press, 2017. doi: 10.1016/b978-1-78548-039-3.50004-3.
[9] S. Megawati, T. Rahaju, M. A. Mahdiannur, and B. Kurniawan, "Integrated Data-based Poverty Alleviation to Achieve SDGs," *IOP Conf. Ser. Earth Environ. Sci.*, vol. 940, no. 1, pp. 8–13, 2021, doi: 10.1088/1755-1315/940/1/012069.
[10] Ecocampusfmipa, "Eco Campus," 2021.
[11] R. pranandita Putra, "5 Keuntungan Bawa Botol Minum Sendiri Saat ke Luar Rumah," 2020. https://www.idntimes.com/life/inspiration/rivandi-pranandita-putra/bawa-botol-minum-c1c2
[12] T. Uehara and A. Ynacay-Nye, "How water bottle refill stations contribute to campus sustainability: A case study in Japan," *Sustainability*, vol. 10, no. 9, pp. 1–19, 2018, doi: 10.3390/su10093074.
[13] Q. Anandra, U. Uljanatunnisa, and I. P. Cahyani, "Analisis Elaboration Likelihood Theory Pada Kampanye 'Go Green, No Plastic' Universitas Pembangunan Nasional Veteran Jakarta," *J. Komunika J. Komunikasi, Media dan Inform.*, vol. 9, no. 2, pp. 96–104, 2020, doi: 10.31504/komunika.v9i2.3421.
[14] HumasUnesa, "Eco Campus Unesa," 2022. https://ecocampus.unesa.ac.id/
[15] Gema, "Taman Cantik, Taman Pembuangan Sampah," 2018. https://gema.unesa.ac.id/2018/04/06/taman-cantik-taman-pembuangan-sampah/
[16] A. E. Abdallah, "Toward Sustainable Campuses in Egypt Case Study Mansoura University," *Int. J. Sci. Eng. Res.*, vol. 9, no. 6, pp. 1050–1058, 2018.
[17] S. N. Aeni, "15 Manfaat Bersepeda untuk Kesehatan Tubuh yang Perlu Diketahui," 2021. https://katadata.co.id/sitinuraeni/berita/6194d5c48eee9/15-manfaat-bersepeda-untuk-kesehatan-tubuh-yang-perlu-diketahui
[18] R. T. Bethary, A. Budiman, D. E. Intari, and D. Ramdhan, "Evaluation of the Green Transportation Concept in the Sultan Ageng Tirtayasa University Campus," *IOP Conf. Ser. Earth Environ. Sci.*, vol. 1000, no. 1, pp. 1–6, 2022, doi: 10.1088/1755-1315/1000/1/012015.
[19] HumasUnesa, "Unesa Paparkan Kampus Hijau Berkelanjutan di Hadapan Asesor UI Greenmetric," 2022. https://www.unesa.ac.id/unesa-paparkan-kampus-hijau-berkelanjutan-di-hadapan-asesor-ui-greenmetric
[20] HumasUnesa, "Unesa Bersama Enam Universitas Terapkan E-Learning," 2018. https://www.unesa.ac.id/unesa-bersama-enam-universitas-terapkan-e-learning
[21] HumasUnesa, "Dosen Teknik Rancang Motor Listrik Sistem Recycle Energy untuk Disabilitas," 2021. https://www.unesa.ac.id/dosen-teknik-rancang-motor-listrik-sistem-recycle-energy-untuk-disabilitas

The Management Concept of Mountain-Based Destination through Carrying Capacity Analysis

Fitri Rahmafitria[1], and Ghoitsa Rohmah Nurazizah[2*]
[1] Management of Resort & Leisure, Faculty of Social Science Education, Universitas Pendidikan Indonesia, Bandung, 40154, Indonesia, ORCID ID: 0000-0001-7865-4184
[2] Doctoral Student of Natural Resources and Environmental Management Science, Postgraduate School, IPB University, Bogor, 16029, Indonesia, ORCID ID: 0000- 0001-5715-6055
[1]rahmafitria@upi.edu, [2*]ghoitsanurazizah@apps.ipb.ac.id

Abstract. The increasing number of tourists and the intensive development of infrastructure can lead to the decline of the carrying capacity, especially if ignoring the concept of sustainable management. The preservation aspect embellishes the focal concern in mountainous areas, while the increasing number of tourists has become the main reason to gain financial benefits. The research aims to assess the carrying capacity level in Situ Patengang Nature Park, Bandung. The study was carried out using a descriptive-explanatory approach with three different analytic methods: 1) GIS analysis to map the carrying capacity zones, 2) Important-Performance comparison analysis to assess the management performance, and 3) Triangulation analysis to arrange the tourism development concept. The results show that most Situ Patengang Nature Park areas have low carrying capacity, medium quality of the environment, high tolerance of natural resources, less dispersion of tourist activities, fewer benefits of natural resources, and low environmental management. The development concept of Situ Patengang Nature Park should apply the five principles of ecotourism which lead to ecologically, financially, and socially sustainable activities. For the zonation, the management should bind the core zone in high carrying capacity zones, a service zone in the medium carrying capacity zone, and a buffer zone in the low carrying capacity zone.

1. Introduction

Bandung is one of the popular tourist destinations and the capital city of West Java, Indonesia [1]. It has a variety of tourist attractions such as natural, cultural, and unique interest attractions. The uniqueness of Bandung is in the typology aspects, in which the area is primarily mountainous and hilly, with heights between 500m - 1800m above sea level [2]. In addition, the city also has a lot of forests, plantations, as well as wide-long rivers. While viewed from the climate, Bandung has a refreshing temperature range from 120C to 240C [3]. Those features are favorable for tourism development, especially natural tourism, due to the physical condition of Bandung, which presents mountains, craters, lakes, and waterfalls. Bandung Regency offers various types of natural tourist attractions. Mostly, the tourist destination is in the form of natural attractions. The prime natural attractions in Bandung Regency consist of diverse types, such as Natural Park Situ Patengang, which provides the beauty of a lake at the foot of mountains.

The tourists' arrival numbers in Bandung increase each year. In 2022, the tourist arrivals reached 1.5 million persons and triplet at the end of 2010, reaching 4.6 million [4]. The average growth of tourist arrivals in Bandung grew by 17% per year [4]. Primarily tourists are likely to visit nature destinations [3,5]. This upsurge of tourist arrival can increase water consumption, litter and waste amounts, air pollution, and disturbance of flora and fauna [6]. With the combination of high nature potential and the enormous number of tourists attracted, the carrying capacity assessment is essential to keep the sustainability of ecology, economy, and social aspects.

As mentioned earlier, the growth of tourist numbers will impact the economic sector and the local community. The increase of tourism business activities upsurges the community welfare and regional income toward tax and retributions. Nevertheless, the uncontrolled tourists' arrival will influence the quality of the environment, especially the destination that provides natural potential, such as Natural Park Situ Patengang. The environmental quality will determine the level of tourist satisfaction in such attractions. Therefore, well-planned and credible management in a destination is needed to preserve natural biodiversity and visitors' satisfaction to maintain their loyalty [7]. Based on those backgrounds, this research aims to analyze the carrying capacity level and tourists' satisfaction, as well as develop a suitable management concept for Natural Park Situ Patengang.

2. Methods

The research location was in the Natural Park Situ Patengang, Bandung. This study was based on Kusumoarto, Gunawan, and Nurazizah's notion [8] that "the excessive use of natural tourism will reduce tourism quality, generally damaging the surrounding landscape." The study used a descriptive-explanatory approach with three different analytic methods.

2.1. *GIS Analysis (Carrying capacity assessment by map overlaying technique)*

The researchers mapped and described the condition of tourism carrying capacity through field observations in Natural Park Situ Patengang. Observations were carried out on aspects of environmental quality, resource tolerance, tourists, benefits of resources, and environment management [8] to measure the quality of carrying capacity. Carrying capacity was analyzed through a scoring technique based on weight standards, as seen in Table 1.

Table 1. Weight and score standard of quality of carrying capacity

No	Variable	Parameter	Weight
1	Environment quality	Type of land covering	15
2	Resource tolerance	% slope	30
3	Tourists' number	The standard number of tourists per hectare	17.5
4	Benefits of resources	Benefits level of resource as a destination	15
5	Environment Management	Amount of environmental management programs	22.5

The weight values were obtained by comparing the importance of five indicators that affect the carrying capacity [9]. The weighting matrix is as follows (Table 2).

Table 2. Weight matrix of influence variable of carrying capacity value

Variable	A	B	C	D	E	Rating	Weight (%)
A: Environmental quality (timber)	0	1	2	2	1	6	15.0
B: Tolerance of resources	3	0	3	3	3	12	30.0
C: Number of tourists	2	1	0	2	2	7	17.5
D: Benefit of resources	2	1	2	0	1	6	15.0
E: Environmental management	3	1	2	3	0	9	22.5

The scoring results determined the quality of the carrying capacity and were processed to the overlaying step. Overlaying step was done by mapping the indicator score and the considered thematic map. This thematic map was overlaid using Arc GIS 8.0 program to obtain the overall quality of carrying capacity.

2.2. *Importance-Performance Analysis*

The researchers surveyed the tourists' satisfaction. The close-ended questionnaires were distributed to 250 respondents by incidental-cum random sampling techniques. The respondent was asked about their perception of the environment, management services, and amenities. The collected data were then analyzed by comparing the Important Performance score resulting from four quadrants of the scatter plot, then counting the Customer Satisfaction Index.

2.3. *Triangulation Analysis*

The researchers developed a suitable management concept based on the carrying capacity and tourist satisfaction. Triangulation analysis was used by combining the results of two previous analyses and the principles of ecotourism perspectives. With this method, the reliability and validity of the data are more assured because various approaches produce complete and comprehensive findings.

3. Results and Discussion

3.1. *Quality of Carrying Capacity Factor on Natural Park Situ Patengang*

Here are the results of the overlaid analysis (Figure 1) on the five parameters of the function of carrying capacity assessment Natural Park Situ Patengang.

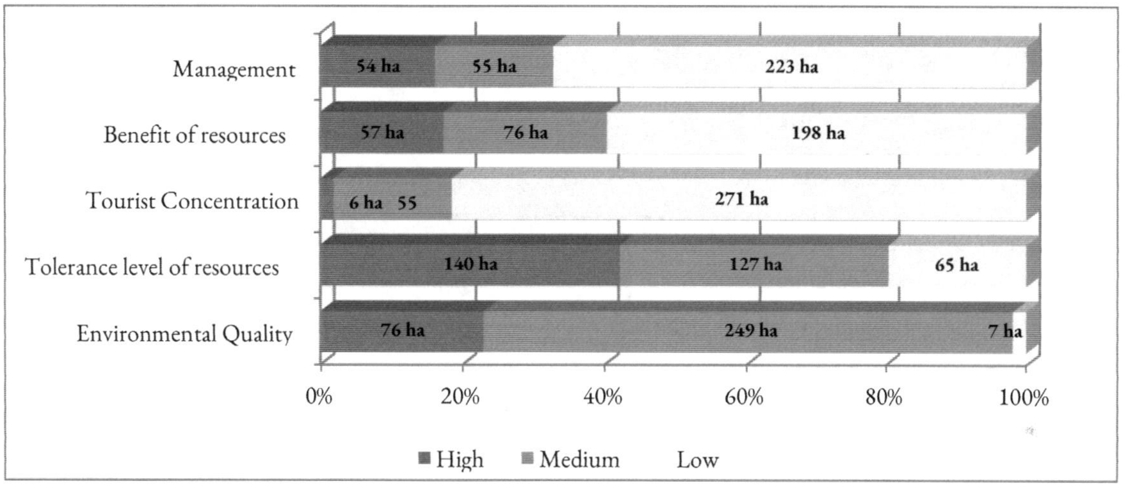

Figure 1. The score of a function of carrying capacity parameter in Natural Park Situ Patengang

The value obtained from this study results from a calculation of attributes that affect the carrying capacity, such as environmental quality, natural resources tolerance, concentrations of tourists, the benefits of natural resources, and existing environment management in Natural Park Situ Patengang. As for the value of its carrying capacity formula is:

Carrying capacity value = F {(score of environmental quality x 15) + (tolerance level of resources 30) + (tourist concentrations x 17.5) + (benefit of resources x 15) + (tourism management x 22.5)}. **(1)**

The formulation will obtain three categories of carrying capacity, i.e. low, medium, and high carrying capacity. Low carrying capacity is ranged from 100-170, while the middle is 170-240, and the high is more than 240. Based on the result, the carrying capacity map produced (see Figure 2):

a. Areas with high carrying capacity value: This area is marked with a dark green color. This area is described as having a high tolerance of natural resources with some concentrations of tourists and has the highest advantages for generating satisfaction and protecting the environment. The high carrying capacity zone lies along the water-body area of the lake, which coincidently is the primary attraction of the destination.
b. Areas with medium values: This area is marked with a yellow color. This area includes an intensive

development zone on its north side. Compared to the existing condition, this area has been built with pavilions, souvenir stalls, docks, and a raft harbor. Thus, coincidentally the current zonation already fits the carrying capacity analysis. The area has a moderate environmental quality, high natural resource tolerance, high tourist density with the natural resources that have medium function for tourism activities, and low environment management.

c. Areas with low carrying capacity value: This area is in light green color. This area includes nature reserve forest areas, plantations, and community residents. The area is dominated by a medium level of environmental quality, medium and low tolerance of natural resources, low tourist concentration, fewer benefits of natural resources, and low management.

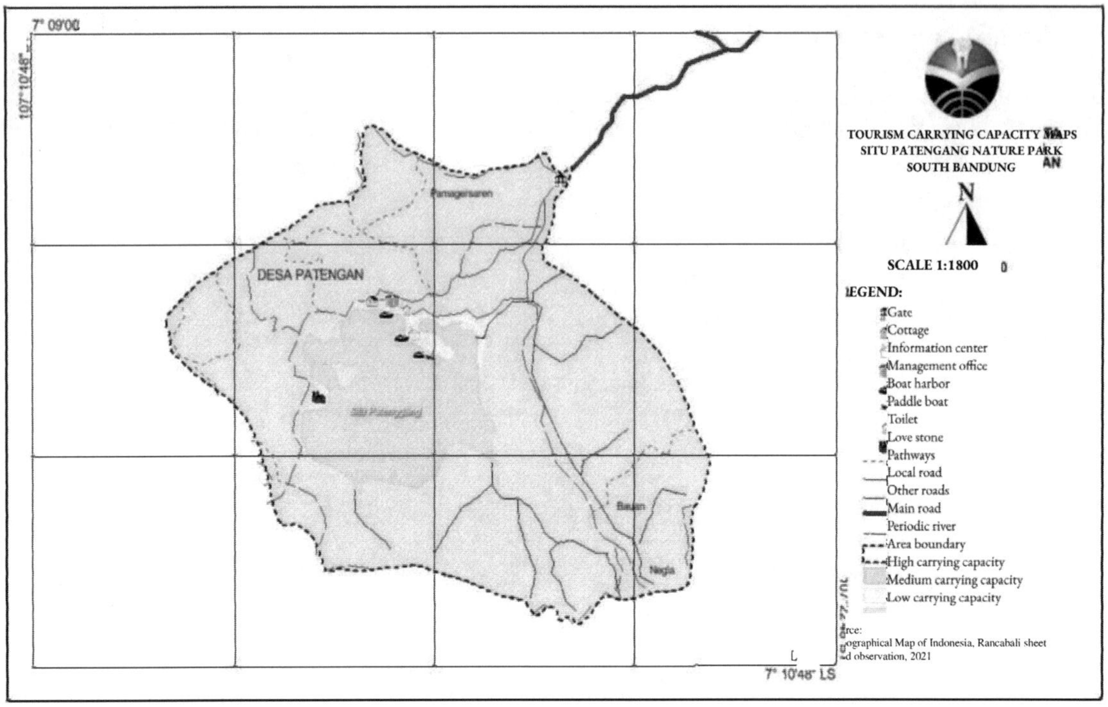

Figure 2. Map of tourism carrying capacity in Natural Park Situ Patengang

Overlay maps divide three levels of the carrying capacity area in Nature Park Situ Patengang. The total area is 331.8 hectares wide and has 82.75% low carrying capacity area, 1.91% medium carrying capacity area, and 15.34% high carrying capacity area. Tourism development in Nature Park Situ Patengang is limited to zones with a high tourism carrying capacity.

3.2. *Tourist Satisfaction of Natural Park Situ Patengang*

The Importance-Performance Analysis can be seen in Figure 3. From the aspect of environmental quality, indicators that have been considered essential and have met the expectations of tourists are related to natural beauty and air quality. In contrast, the expectations that have not been met are the physical safety aspects of tourist objects. From the aspect of management services, what has met visitors' expectations is the management of nature for tourism activities and improving the community's economy, with less performance in adding educational elements in traveling. Moreover, from the aspect of infrastructure, nothing meets the expectations of tourists. These tourism facilities are considered lacking and need to be improved, such as the condition of the road infrastructure, information service center, signing boards, toilets, and water must be repaired. From the third assessment of the quality attributes, the CSI shows that the average tourist satisfaction is only on the value of 3.26 (somewhat satisfied) while the index is 65%. The index means that managers meet 65% of the tourists' needs and expectations, and it is interpreted that tourists are already satisfied with the manager's performance.

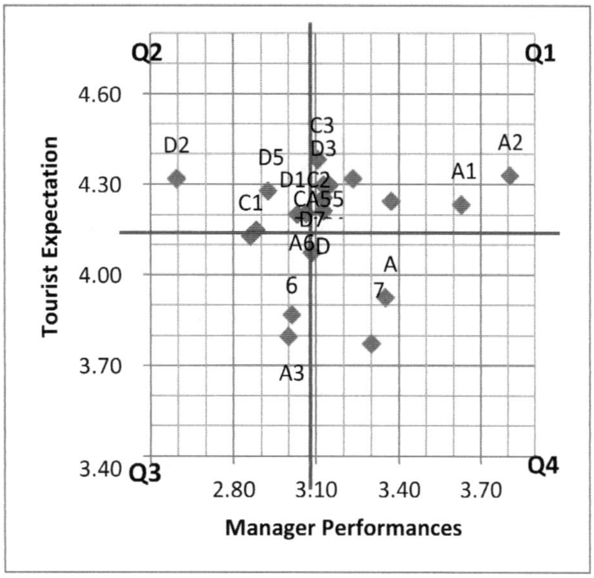

Figures 3. Quadrant analysis of the factor of carrying capacity in Natural Park Situ Patenggang

3.3. *Development Concept of Tourist*

The development concept of Nature Park Situ Patengang is preferably directed by the sustainability principle of ecotourism, such as providing water-based recreational activities that are ecologically, financially, and socially sustainable, namely:

a. The nature-based: The core natural resource that can be utilized for tourism activities is the lake area which has the highest value of carrying capacity. Tourism development can be carried out by adding various tourist activities that provoke tourists' participation, such as water-based outbound activity, traditional fishing, bird-watching, and lake canoeing to see and learn about water birds' behavior.
b. The ecologically sustainable: Tourist attractions' continuity in natural area course relies heavily on the management skills in conserving resources. Therefore, the manager should pursue ecology sustainability, such as monitoring and controlling natural resources regularly, restricting access to sites with low carrying capacity, conducting an interpretation program, and these dispersing facilities and tourist activities.
c. The environmentally educative: improving satisfaction and experience traveled by giving value-added services such as improving treasures, knowledge, and concern for the environment by developing an interpretation and interpretation program.
d. The locally beneficial: involving local communities in tourism management and entrepreneurship opportunities in the provision of the object.
e. It generates tourist satisfaction: by improving quality tours by improving the quality of human resources and facilities.

The zoning concept of Nature Park Situ Patengang better adapts the level of tourism carrying capacity in each area. The Patengang zoning should be divided into three zones, namely the core zone, buffer zone, and service zone. The core zone is a recreation zone located. The service zone is a zone for tourism amenities and facilities, while the buffer zone serves as the water catchment area for the life of flora and fauna. The buffer zone should also be kept in a pristine condition that is not disturbed. The criteria for the division of the zone are:

a. Core zone ~ The location of the main attraction. Tourists motivate to travel and enjoy the main activity in this area. Thus, this zone must be equipped with essential facilities. A high-capacity zone can be used as the core zone of intensive tourism development. This zone is located right on the lake. Thus, development should adopt education-based activities.
b. Service zone ~ An area where all the activity and support facilities are placed. The medium carrying

capacity functions as a service zone for building supporting facilities. From the analysis, that zone is located on the lake's north bank. It is constructed with parking lots, ticket booths, an information center, security posts, souvenir stalls, food stalls, shelters, toilets, and the pier for the boat. The development in this zone must focus on the area utilization and the maintenance of existing facilities to cope with tourist dissatisfaction in facilities maintenance, arrangement of stalls, and parking space availability.

c. Buffer zone ~ A buffer area is between two zones, i.e., the main activity and support facilities. This area's primary function is to maintain the image of the destination and keep the comfortability of the area. The low carrying capacity zone is better used as a buffer zone to ensure the stability of the ecological cycle in Situ Patengang Nature Park. The analysis shows the low carrying capacity zone in the nature reserve forest areas and plantations.

3.4. *Development of Activities and Facilities*

The result of the analysis indicates that the water area has a high value of carrying capacity. One of the causes of this high capacity is the excellent tolerance of natural resources, fewer tourist concentrations, and the potential benefits of natural resources for intensive development. This result is similar to Riniwati *et al.* notion [10] that even though the sea always attracts tourists more than other terrestrial objects, the limitation of tourists' skills in contact with water activities has caused less concentration. The Importance-Performance analysis results show that the density and queue variable is included in Quadrant IV. That means that Situ Patengang Nature Park has the potential to accommodate more tourists' arrival because today's assessment results in a high value of positive perception regarding crowding and comfort perception. Fewer recreational activities in the area also contribute to the positive crowding perception of tourists. The only available activity is paddle boating.

Tourist activities may have implications for the impact on quality and environmental sustainability [8]. The tourist activity development standard divides the area into three activity zones. Adding more activities can accommodate the needs of both group and individual tourists. Furthermore, the manager should monitor and control tourists' behavior by providing guided tours and interpretation boards.

4. Conclusion

The results obtained in this study are: Analysis of tourism carrying capacity and the analysis of the overlaying maps determine that areas in the Situ Patengang Nature Park area dominated with the low carrying capacity for tourist use. Importance-Performance analysis shows that the manager should improve the security and comfortable conditions for tourism activities, add educational value to the activities, and repair some facilities like roads, information centers, board of directions, toilets, and clean water. The CSI shows that the management can meet 65% of tourists' expectations. The development concept of Natural Park Situ Patengang must implement the sustainable principle of ecotourism: nature-based, ecologically sustainable, environmentally educative, locally beneficial, and generates satisfaction. Hence it can create water-based nature tourism activities that are ecologically, financially, and socially sustainable. The applied concept of zoning development must adapt to the carrying capacity overlay results. Further research needs to be done, primarily related to carrying capacity, resulting in recommendations for the maximum number of tourists the destination can accommodate.

References

[1] Komaladewi R 2017 The Representation of Culinary Experience as the Future of Indonesian Tourism Cases in Bandung City, West Java *Int. J. Bus. Econ. Aff.* **2**

[2] Anranur Uwaisy M, Baizal Z K A and Yusza Reditya M 2019 Recommendation of scheduling tourism routes using tabu search method (case study Bandung) *Procedia Comput. Sci.* **157** 150–9

[3] Marwa S and Rahmafitria F 2018 A Factor Analysis of Visitors' Motivation in Visiting the Geology Museum of Bandung *IOP Conf. Ser. Earth Environ. Sci.* **145**

[4] Department of culture and tourism 2022 *Perkembangan Wisatawan ke Kabupaten Bandung Tahun 2003-2022* (Bandung: Department of culture and tourism)

[5] Rahmafitria F and Misran 2018 Disaster risk and travel decision of Middle Eastern tourists to natural destination in Indonesia *IOP Conf. Ser. Earth Environ. Sci.* **179**

[6] Wall G 2019 *Part I : Overtourism : theoretical positioning and implications* (Leiden: Universiteit Leiden)

[7] Munandar A, Gunawan A, Nurazizah G R and Kusumuarto A 2020 Analysis on Willingness to Pay at the Ecotourism Destination, Cigamea Waterfall, Halimun-Salak National Reserve *J. Environ. Manag. Tour.* **9** 42

[8] Kusumoarto A, Gunawan A and Nurazizah G R 2017 Landscape Potential Analysis for Ecotourism Destination in the Resort II Salak Mountain, Halimun-Salak National Park *IOP Conf. Ser. Earth Environ. Sci.* **91**

[9] Kaptan Ayhan Ç, Cengiz Taşlı T, Özkök F and Tatlı H 2020 Land use suitability analysis of rural tourism activities: Yenice, Turkey *Tour. Manag.* **76**

[10] Riniwati H, Harahab N and Abidin Z 2019 A vulnerability analysis of coral reefs in coastal ecotourism areas for conservation management *Diversity* **11**

Study the Relationship of Soil Fertility with Land Suitability For Arabica Coffee (*Coffea arabica L.*) Development in Bandar Sub-district, Pacitan District

Suko Irawan[1], Ernoiz Antriyandarti[2*], Dewi Nawang Suprihatin[2], Anggityas Werdining Pangesti[2]

[1]Department of Soil Science, Faculty of Agriculture, University of Sebelas Maret, Surakarta, 57126, Indonesia,

[2]Department of Agribusiness, Faculty of Agriculture, University of Sebelas Maret, Surakarta, 57126, Indonesia

[1]sukoirawan9@student.uns.ac.id, [2*]ernoiz_a@staff.uns.ac.id,
[3]dewinawang11@student.uns.ac.id, [4]anggit02@student.uns.ac.id

Abstract. This study aims to assess the soil fertility index (SFI) under the shade of coffee and its relationship with land suitability to provide recommendations for improving arabica coffee productivity. This research was conducted in Bandar Sub-district, Pacitan District, East Java, using an exploratory, descriptive survey method. Land Map Units (LMU) are determined by overlaying soil type, land use, slope, and rainfall maps. Determination of survey points using purposive sampling with 10 sample points and three replications. The samples were analyzed in the laboratory to determine the soil's physical, chemical, and biological characteristics. Soil quality assessment was carried out using Principal Component Analysis (PCA) to obtain Minimum Data Sets (MDS). Land suitability was evaluated through the matching method with the conditions for growing arabica coffee. The analysis showed that the soil fetility index had a significant positive correlation with land suitability (r=0.806, p=0.002). The limiting indicators based on the correlation test were SOC, pH, available P, and Base Saturation. Suggested improvement recommendations include applying land cover crops and balanced P fertilization.

1. Introduction

Agriculture has a strategic role in developing the national economy, including providing food, industrial materials, feed, bioenergy, and the employment sector. Furthermore, agriculture is the sector with the second highest contribution to Gross Domestic Product (GDP), reaching 13.7%, in more detail, the plantation sector is the highest contributor reaching 3.63 % or more than 25% of the total agricultural GDP [1]. The agricultural development roadmap is directed toward sustainable agriculture as a milestone in sustainable development, the paradigm of sustainable agriculture must be a practical solution in facing the challenges of food security, nutrition, and the environment, on the other hand, conventional agricultural practices with industrial approaches, capital intensive and technological innovation, planting varieties uniform and mass yields, as well as dependence on production inputs (fertilizers and pesticides) are still the leading choices for farmers, but conventional farming practices in the long term are reported to have adverse side effects on environmental sustainability [2].

Excessive use of agricultural ecosystems and exceeding the ecological carrying capacity can trigger land degradation, [3] states that degradation is a severe problem for ecology, the environment, and agricultural ecosystems, land degradation is defined as a temporary or permanent decrease in land quality so that productivity decreases significantly. Gradually, the decline in land quality is marked by a reduction in the carrying capacity of the soil's physical, chemical, and biological properties. The study's results [4] show that excessive production inputs (inorganic fertilizers and pesticides) in the long term negatively impact crop yields in the following growing season. On the other hand, the level of adaptation of farmers to enriched organic fertilizers is still very minimal, even though organic fertilizers enriched with biological agents can be a profitable solution for farmers [5]

Indonesia is the fourth largest coffee production centre after Brazil, Vietnam, and Colombia, with output in 2017 reaching 654,000 tons or controlling 6.54% globally. Indonesian coffee has vast international market potential and is very popular with European countries, followed by Asia and the United States, estimates that the volume of coffee exports will increase along with the level of world coffee consumption. It is estimated that global coffee consumption will reach 10 million tons by 2023. Bandar Sub-district is one of the best coffee producers in East Java, based on BPS data, the coffee plantation area is 542 ha, with production reaching 102 tons, but the productivity of coffee plants in Bandar Sub-district is decreasing every year [6].

The land is the essential production factor for plants, so that growth can reach the optimum point of Arabica coffee requires supportive land conditions, Arabica coffee can grow optimally at 1000 – 1200 masl, temperature 16 – 20° C, the slope of less than 30%, SOC more than 3.5%, CEC 15 cmol(+)/kg, C/N ratio 5-12, pH 5.5 – 7, and Base Saturation more than 35%. Soil fertility is defined as the ability of the soil to provide essential macronutrients to support plant growth. Furthermore, land management is currently a serious challenge for farmers, from various studies show that soil fertility is decreasing due to land degradation and fertility depletion resulting from poor management practices. Arabica coffee, as a leading commodity in Bandar Sub-district, is faced with various challenges such as decreasing productivity every year, policy-making that has not been fully profitable for smallholders, massive attacks of pests and diseases and improper handling, downstream industries that have not developed so that most of the products are sold in primary form, and inappropriate coffee cultivation practices.

The development of arabica coffee plantations must refer to aspects of land carrying capacity, soil quality and fertility. The land's carrying capacity is the level of land suitability for developing certain commodities. At the same time, the soil fertility index is the ability of the soil to provide an ideal nutrient that ensures plant productivity. This study examines soil fertility and land suitability for coffee development in Bandar Sub-district to recommend appropriate land management.

2. Methods

The research was conducted in Bandar Sub-district, Pacitan District, East Java. Soil analysis was carried out at the Physics and Soil Conservation Laboratory and the Chemistry and Soil Fertility Laboratory, Faculty of Agriculture, Sebelas Maret University, Surakarta. The study took place from February 2022 to July 2022. The survey was carried out by direct observation in the field, in addition, determination of the sampling point was carried out by purposive sampling based on the Land Map Unit, for each LMU, soil samples were taken 3 times, called a composite. Furthermore, the analysis method used in this study is presented in table 1 below.

Table 1. Soil Physical, Chemical and Biological analysis

No	Parameter	Analytical method
1	Textures	Pipet
2	pH	Electrometric
3	Available P	Olsen method
4	Total N	Kjeldahl
5	Available K	NH4Oac 1 N extraction
6	Soil organic Carbon (SOC)	Walkey and Black
7	Base saturation	Ammonium acetate extraction

8	CEC	Ammonium acetate method
9	C/N ratio	C/N ratio

The selection of soil fertility indicators is based on ease of analysis and replication and the laboratory's availability of tools and materials. Furthermore, the indicators used should affect the capacity of the soil to perform crop production or environmental functions. Soil attributes are useful in determining soil fertility criteria and serve as indicators of changes in fertility. Soil fertility index assessment is carried out through several stages: Principal Component Analysis (PCA), Minimum Data Set (MDS) selection, and correlation test. From the PCA, selected data will be obtained containing the indicators that have the most influence on soil fertility, referred to as Minimum Data Sets (MDS), the MDS requirement is to have an eigenvalue of more than 1 or contribute a minimum proportion of 75% of the data. The Selected indicator then looks for each indicator's weighting index (Wi). The weighting index is obtained by multiplying the proportion by the cumulative. The selected indicators are then assessed using the soil fertility index [7–11]. Data analysis continued with the calculation of the soil fertility index according to the formula below

$$SFI = \left(\frac{Sci}{N}\right) \times 10 \quad (1)$$

$$Pc = \frac{1}{Nc} \quad (2)$$

$$Cj = Wi \times Si \quad (3)$$

$$Sci = Cj \times Pc \quad (4)$$

SFI : Soil Fertility Index
Sci : Score of each indicator
N : Sum of indicator
Wi : weighting index
Cj : Sum of Weighting index
Pc : Score of class
Nc : Sum of class
Si : Scoring index

3. Results and Discussion

The indicators used for soil fertility index are pH, Soil Organic Carbon (SOC), total N, C/N ratio, available P, available K, CEC (Cation Exchange Capacity), Base saturation (BS), and soil texture, the results of the analysis of each LMU are presented in table 2, below.

Table 2. Soil properties for each Land Map Unit

LMU	pH	C-org (%)	Tot-N (%)	C/N ratio	Av-P (ppm)	Av-K (cmol(+)/kg)	CEC (cmol(+)/kg)	BS (%)	Textures
1	5.16	1.09	0.19	5.73	2.06	1.33	21.97	36.42	clay
2	5.25	0.79	0.18	4.38	2.04	0.29	19.53	52.04	clay
3	5.07	0.83	0.21	3.95	2.09	0.22	21.16	47.43	clay loam
4	5.22	1.20	0.18	6.66	2.29	0.23	29.68	49.09	silty clay
5	5.56	1.68	0.19	8.84	0.72	0.98	21.25	39.36	clay loam
6	6.14	2.26	0.30	7.53	1.08	0.66	20.38	30.87	clay
7	5.42	1.20	0.20	6	9.56	1.23	18.90	43.56	clay loam
8	5.83	2.28	0.27	8.44	0.75	2.32	21.56	51.95	clay
9	5.59	0.81	0.16	5.06	1.85	0.27	16.93	31.77	clay
10	5.02	1.04	0.16	6.5	2.12	0.48	15.24	32.67	clay

Source: Primer data

Table 2 shows that the pH ranged from 5.02 to 5.83, the highest at LMU 6 and the lowest at LMU 10. pH was closely related to nutrient solubility and nutrient availability in the soil. The pH at the research site is in the slightly acidic category. Nutrients are not easily absorbed by plant roots at neutral soil pH because under these conditions, nutrients are not easily soluble in the air. Furthermore, the solubility of Al and Fe is highest in acidic to slightly acidic conditions, causing the solubility of some nutrients such as P to decrease, in some plant varieties, it can cause Al and Fe poisoning. The decrease in pH at certain soil depths is caused by the dominance of secondary minerals such as acidic Fe, this is exacerbated by the reduced amount of organic matter [12]. pH is a significant indicator of nutrient availability, for arabica coffee, the recommended pH ranges from 5.6 to 6.6. CEC in Bandar Sub-district is classified as low to high, ranging from 15.24 cmol(+)/kg – 29.68 cmol(+)/kg.

The CEC value in SPL 10 is in a low category, SPL 1, 2, 3, 5, 6, 7, 8, and 9 are in the medium category, while SPL 4 is in the high category. CEC is strongly influenced by parent material, soil texture and soil pH. The addition of materials can increase the cation exchange capacity (CEC) because the decomposition of organic matter will release organic acids containing reactive functional groups, functional groups act as a negative source of soil [13]. Base saturation (BS) is closely related to soil pH, at an acidic pH, the colloid adsorption complex is filled with acidic cations (Al^{3+} and H^{+}) at a specific concentration, AL^{3+} can cause toxicity to plants, at an alkaline pH, the colloid adsorption complex is filled with basic cations (Ca^{2+}, Mg^{2+}, K^{+}, and Na^{+}), showed an analysis which showed that the base saturation ranged from 30.87 to 52.04. In

The available P values ranged from 1.08 ppm – 9.56 ppm, with the highest at LMU 3 and the lowest at LMU 4. The available P values in all LMU were in the very low to low category. The low content of P available in the soil can be caused by parent material from the ground that is poor in P elements and the low range of P in organic matter. [14] explained that P's availability is in the pH range between 5.5 – 7. The availability of P will decrease at a pH of more than 5.5 or higher than 7. The available K have a value range of 0.22 cmol(+)/kg - 2.32 cmol(+)/kg which low – very high, availability of K was influenced by the saturation of Ca^{+} and Al^{3+} in the soil).

Organic matter dramatically affects the activity of microorganisms, organic matter serves as an energy source for heterotrophic microorganisms. The analysis showed that C-organic ranged from 0.79% - 2.28%, which very low to moderate. Organic matter is very closely related to total N, the C/N ratio explains this, the C/N ratio is the proportion of Nitrogen used to convert organic matter, the higher the C/N ratio the organic matter is not completely decomposed. The C/N ratio indicates bacterial activity at the study site, generally, the C/N ratio ranges from 10 – 20, but in all study sites the C/N ratio is below 10, a lower C/N ratio increases nitrogen loss through leaching (nitrate mobilization) and ammonia evaporation; while a high C/N percentage will require a longer composting time because Nitrogen is a limiting nutrient for microbes.

The next stage to determine the soil fertility index is the selection of indicators with the highest sensitivity in the assessment of soil fertility index, the choice of these indicators is made through Principal Component Analysis (PCA) and Weighting index. The Principal Component is a linear combination of various variables as the maximum representation with high sensitivity derived from the variance data set [7]. Result of Soil Fertility Index are presented in table 3.

Table 3. Soil Fertility Index (SFI) for each Land Map Unit (LMU)

No	MDS	Pro	Cum	Wi	LMU									
					1	2	3	4	5	6	7	8	9	10
1	pH	0.54	0.74	0.37	0.74	0.74	0.74	0.74	1.11	1.85	1.11	1.48	1.11	0.74
2	SOC	0.54	0.74	0.37	0.74	0.37	0.37	1.11	1.11	1.11	1.11	1.11	0.74	1.11
3	Tot-N	0.54	0.74	0.37	0.74	0.74	1.11	0.74	0.74	1.11	1.11	1.11	0.74	0.74
4	CEC	0.19	0.74	0.26	0.26	0.26	0.26	0.26	0.52	0.26	0.26	0.52	0.26	0.26
5	C/N	0.19	0.74	0.26	0.78	0.78	0.78	1.04	0.78	0.78	0.78	0.78	0.52	0.52
		SFI			0.41	0.36	0.41	0.49	0.53	0.64	0.55	0.63	0.42	0.42
		Class			L	L	L	L	M	M	M	M	L	L

Source: Primer data
Description: L (Low), M (Moderate)

The analysis results show that the soil fertility index in Bandar ranges from 0.36 – 0.64 belonging to the low to moderate class. LMU 1,2,3,4,9 and 10 have a low soil fertility index, while the land map units 5,6,7 and 8 have a moderate soil fertility index. Related to the development of arabica coffee as a leading commodity in the Bandar Sub-district, need information on the carrying capacity of the land, the carrying capacity of the land can be assessed through land evaluation in each LMU, this evaluation is based on technical instructions from the Ministry of Agriculture of the Republic of Indonesia.

Based land suitability evaluation, class actual for arabica coffee development in all land map units is S3 (marginally inappropriate), In the class row, there is a codification of nr, which reflects nutrient retention (nr) and nutrient availability (na) as limiting factors, then nr and na are not single factors but includes several soil indicators, nr dan na is a non-permanent barrier so it is easy to repair, corrected properly can increase the potential for land stability to S2 (quite suitable) even S1 (appropriate). Indicators influencing coffee commodities' development in Bandar Sub-district, the toughest limiting being available P. Available P is not a permanent barrier so that it can be increased through balanced inorganic fertilization, manure, or compost. The addition of compost from coffee grounds can be an appropriate alternative considering the number of coffee grounds that reaches 35% of coffee production, compost has a pH of 5.4 – 6.6, a C/N ratio of more than 6, a total N of 0.2%, available P was 8 ppm, and CEC ranged from 15 cmol(+)/kg.

Soil fertility is an important environmental factor to be managed holistically. Then in the framework of sustainable development, agricultural activities must maintain ecosystems, biodiversity, and productivity to remain socially, economically, and environmentally beneficial. Land suitability evaluation studies with soil fertility index for developing arabica coffee can be a point of view and validation in policy making by stakeholders. Soil fertility is a crucial environmental limiting factor to be considered and managed according to the purpose of land use. The results of the correlation between land suitability and SFI are presented in the following Table 4

Table 4. Correlation between Land Suitability and Soil Fertility Index

		Soil Fertility Index
Land Suitability	Pearson Correlation	0,806
	P-Value	0,002

Based on table 4, it is known that land suitability is very closely correlated with soil fertility index, this shows that the higher the fertility index, the more land suitability for arabica coffee development will also increase. This correlation study is critical as a reference in providing recommendations for improving soil fertility, it is hoped that suggestions for improvement are not only oriented to the number of nutrients but include increasing the status of land suitability for coffee plants because fertile soil cannot always be manifested as suitable soil for development arabica coffee. Table 5 presents recommendations for improving soil fertility.

Table 5. Recommendation for improving soil fertility

No	Indicator	Recommendation
1	SOC, pH, Base saturation	Cover crop
2	Available P	Balanced fertilizer,

Cover crops protect soil aggregates from raindrops, suppressing erosion, accelerating water infiltration rates and increasing biological activity and soil fertility levels. Ground cover plants that are often used are Calopogonium caeruleum and Arachis pintoi. Considerations in choosing ground cover plants for cocoa are the propagation properties of these plants so that they can interfere with the growth and development of the main crop and competition for nutrients. Cover crops can minimize the rate of leaching and reduce nutrient loss due to the erosivity of rainwater, soil physical compaction, and low pH. Improvements to the status of essential macronutrients P are also needed at all research sites, these improvements can be pursued through fertilizer recommendations. Fertilization aims to encourage vegetative growth and increase arabica coffee productivity. Determination of the type and dose of fertilizer is generally based on soil or leaf analysis results. Still, if the analysis is not carried out, a tentative recommendation can be an alternative to fertilization. Fertilization must be based on four principles: the correct dose, the right time, the type of fertilizer, and the right stocking target. Arabica coffee fertilization can be divided into the immature plant phase (IPP) and the mature plant phase (MPP).

4. Conclusion

Soil Fertility Index (SFI) in Bandar Sub-district is classified as low to moderate, and land suitability for arabica coffee is marginally inappropriate, improper management of coffee plantation in Bandar Sub-district will cause economic and environmental losses, the results of the correlation test between soil fertility index and land suitability showed a significant relationship ($r = 0.806$, $p = 0.002$), these results indicate that the higher the SFI value, the better the land suitability. The limiting indicators based on the correlation test are pH, available P, Soil Organic Carbon (SOC), and Base Saturation. Suggested improvement recommendations include cover crops and balanced P fertilization of immature plant phase (IPP) and mature plant phase (MPP).

References

[1] Ministry of Agriculture 2021 *Statistical of National Leading Estate Crops Commodity* (Jakarata: Ministry of Agriculture)

[2] Riwthong S, Schreinemachers P, Grovermann C and Berger T 2017 Agricultural commercialization: Risk perceptions, risk management and the role of pesticides in Thailand *Kasetsart J. Soc. Sci.* **38** 264–72

[3] Irawan S and Antriyandarti E 2021 Physical deterioration of soil and rice productivity in rural Java *J. Phys. Conf. Ser.* **1825**

[4] Irawan S and Antriyandarti E 2021 Fertilizer Application , Climate Change and Rice Production in Rural Java

[5] Irawan S and Antriyandarti E 2021 BIOTRICO: A Breakthrough Fertilizer for Sustainable Agriculture *IOP Conf. Ser. Earth Environ. Sci.* **940**

[6] BPS 2019 *Kecamatan Bandar Dalam Angka* (BPS Kabupaten Pacitan)

[7] Chandel S, Hadda M S and Mahal A K 2018 Soil Quality Assessment Through Minimum Data Set Under Different Land Uses of Submontane Punjab *Commun. Soil Sci. Plant Anal.* **49** 658–74

[8] Andrews S S, Karlen D L and Cambardella C A 2004 The Soil Management Assessment Framework *Soil Sci. Soc. Am. J.* **68** 1945–62

[9] Mulat Y, Kibret K, Bedadi B and Mohammed M 2021 Soil quality evaluation under different land use types in Kersa sub-watershed, eastern Ethiopia *Environ. Syst. Res.* **10**

[10] Andrews S S and Carroll C R 2001 Designing a soil quality assessment tool for sustainable agroecosystem management *Ecol. Appl.* **11** 1573–85

[11] Armenise E, Redmile-Gordon M A, Stellacci A M, Ciccarese A and Rubino P 2013 Developing a soil quality index to compare soil fitness for agricultural use under different managements in the mediterranean environment *Soil Tillage Res.* **130** 91–8

[12] Kátai J, Oláh Á, Tállai M and Alshaal T 2020 Ecotoxicology and Environmental Safety Would fertilization history render the soil microbial communities and their activities more resistant to rainfall fluctuations ? *Ecotoxicol. Environ. Saf.* **201** 110803

[13] Hartati S, Minardi S, Hartatik W and Haniati I L 2018 The Effects of Inorganic Fertilizer and Mineral Leucite Residues on K Uptake and Maize Yields (Zea mays L.) in Oxisols *SAINS TANAH - J. Soil Sci. Agroclimatol.* **15** 115

[14] Firnia D 2018 Dinamika Unsur Fosfor Pada Tiap Horison Profil tanah Masam *Jur. Agroekotek* **10** 118–38

IOP Conf. Series: Earth and Environmental Science 1111 (2022) 012030 doi:10.1088/1755-1315/1111/1/012030

Environmental Awareness Based on Community Knowledge, Attitude, and Behavior of The Environmental Impact of Plastic Packaging Use in Urban Areas

Herdis Herdiansyah[1*], Salma Mar'atus Sholihah[2], Evi Frimawaty[3]
[1]School of Environmental Science, Universitas Indonesia, Jakarta, 10430, Indonesia, ORCID ID 0000-0003-2684-3551
[2]School of Environmental Science, Universitas Indonesia, Jakarta, 10430, Indonesia ORCID ID 0000-0001-8751-8512
[3]School of Environmental Science, Universitas Indonesia, Jakarta, 10430, Indonesia ORCID ID 0000-0002-9016-4062

[1*] herdis@ui.ac.id, [2]salma.maratus11@ui.ac.id, [3]evi.frimawaty11@ui.ac.id

Abstract. The problem of plastic packaging is a threat to human life. This problem is illustrated by high plastic consumption and poor plastic waste management. The purpose of this paper is to discuss community environmental awareness regarding the use of plastic packaging. This paper uses a qualitative approach with an analysis of the literature study. From the results of the study, it was found that community environmental awareness is still low. It is necessary to increase awareness to encourage environmentally friendly plastic usage behavior. Behavior toward the use of plastic starts from minimizing its consumption and the willingness to sort and recycle. Many factors influence the willingness to participate in plastic management. By manipulating these factors, a sustainable plastics management strategy can be achieved. Raising awareness through increasing knowledge, increasing guilt, optimizing facilities, and supporting regulations. Strategy cannot work without the cooperation of various stakeholders. The common goal of creating a healthy and clean environment and an empowered community is the impetus for improving plastic management.

1. Introduction

Plastic packaging is a popular material in human life, but its use is excessive [1]. Plastic is used as packaging for food, beverages, clothing, or others. Plastic is a popular material because it is affordable and easy to use. The use of plastic continues to increase even though it causes problems that threaten human life and the environment. Garbage is a by-product of the consumption of plastic packaging. Plastic waste impacts global warming and threatens marine ecosystems [2]. In the future, the challenge humans face in environmental problems is waste, especially inorganic waste. According to some literature, plastic is a non-biodegradable material that causes environmental and health problems. Plastics contain heavy metals and chlorine [3], cause a decrease in the quality of agricultural soil and produce emissions [1,3]. It can be concluded that the notion of plastic as an environmental problem is unquestionable [4].

The annual plastic consumption in Indonesia reaches 15.96 kg/person, and as much as 83% of plastic waste is unmanaged [5]. Plastic management in Indonesia is not collected properly, even though the collection is the most crucial part of the waste management chain [6]. The behavior of littering and the culture of burning garbage have been embedded in the community. Disposal of waste in its place also faces the challenge of mixing all organic and inorganic waste; there is no separation between the two [7]. The lack of waste facilities is one of the causes, especially since the existing waste bins do not allow for sorting. Even in high-income counties, waste is a problem that has not been managed optimally. These countries export plastic waste to reduce the presence of waste in their country. Exporting is a form of diverting the waste problem to developing countries. Plastic consumption in developing countries is low, but poor management makes the environment unhealthy. This phenomenon is illustrated in table 1, which shows a significant difference between the percentage of unmanaged waste in developed and developing countries.

Table 1. Total plastic consumption and percentage of unmanaged waste in various countries [5]

Country	Total consumption (kg/year)	Percentage of Unmanaged Waste (%)
Belgium	169.72	2
Korea	126.92	2
USA	90.64	2
Indonesia	15.96	83
India	8.64	87

Increasing awareness is essential to create a healthy and sustainable environment about the negative impacts of plastic so that people can limit the use of plastic in their daily lives. The community can also be willing to manage the plastic waste produced. Many types of research on increasing awareness of plastics have been carried out, such as community attitudes about plastics [4,8-10], the effectiveness of plastic waste management [11-13], and the improvement of environmental education [14].

This paper discusses the awareness of urban communities about the use of plastic. Previous research has stated that urban communities use plastic more than rural communities [1,7]. The consumerist lifestyle and the more durable plastic quality make plastic popularly used. The overuse illustrates the urban community's low awareness of plastic use [8]. Excessive plastic consumption with minimal management causes further problems. This paper aims to discuss community awareness of plastic packaging use. Awareness is one of the critical factors for optimizing plastic waste management strategy so that environmental problems can be minimized. Hopefully, this paper can provide recommendations for plastic waste management strategies in urban environments.

2. Methods

The approach in this study is a qualitative approach using the literature review analysis method. The researcher conducted an extensive review by limiting the themes to community awareness, plastic use, and plastic waste management. The review was carried out by looking for journal reference sources in preparing papers with the keyword "attitude toward plastic". The author ensures the process of data adoption, and a thorough review is obtained from journals and databases relevant to the scope of the required research theme. The literature was also classified to determine the analytical structure used to process and analyze the data. The data generated from various articles are then synthesized and explained descriptively. The results provide an overview of community awareness and awareness raising for plastic use and waste management.

3. Results and Discussion

Plastic waste management must be done with a systemic approach because it involves various parties. Plastic waste management cannot be done without support from the community as plastic consumer agents and household sorting agents [15]. Individuals and households play a significant role in increasing recycling. The community is one of the stakeholders that must be empowered toward sustainable management. Environmental management aims to reduce the environmental footprint by reducing the demand for raw materials [8]. Embracing the community to have pro-environmental attitudes and behaviors is difficult. Based on the theory of planned behavior, people's attitudes and behaviors related to environmental protection are influenced by concerns and behavioral control factors

[16]. The strategy for intervention in environmental management is carried out by modeling social and environmental changes and changing behavior based on environmental changes [15].

3.1 Understanding awareness based on attitude and behavior

The willingness to participate in sorting waste by the community is influenced by individual perceptions [17]. The community perception sees how much management can provide benefits to human life. If management efforts result in better environmental conditions, the incentive to participate in plastic management will increase [7]. According to the theory of planned behavior, attitudes are represented in memory, so they strongly influence behavior. Moreover, there are individual experiences that increase support for pro-environmental behavior.

Most studies agree that behaviors and attitudes positively affect waste management [2,7,11]. Attitude is a reflection of perception and is a subjective assessment of citizens. The more positive the community's perception of management, the higher the willingness of the community to participate in management. The community's view of plastics as an environmental problem can improve management. The implications of what will happen in the future can also make people want to protect the environment. Because plastic consumption is more, it has implications for the higher amount of waste produced, so people are willing to minimize its use. Pro-environmental behavior is associated with environmental concern, which is also associated with management motivation [18]. Motivating pro-environmental behavior can be done through social pressure, stimulation, or pro-environmental intervention. [10]. Individual awareness increases social pressure; individuals tend to reduce plastic bags use.

The results of the study [19] are contrary to previous research. Linh argues that attitude does not have a significant impact on plastic consumption. The influencing factors are environmental knowledge, subjective norms, behavioral control, and intentions. In research [4], the community desires to reduce the use of plastic bags, but this desire is not reflected in real action. Many factors may influence this condition. Factors that affect community awareness include knowledge, guilt, and the availability of facilities.

3.1.1 Knowledge

Knowledge is an essential factor in increasing awareness and pro-environmental behavior. However, community knowledge regarding waste management is still not sufficient. There are still many people who do not know. Education carried out from an early age can increase environmental knowledge. Transfer of knowledge related to waste management at the primary level has been transferred. However, many obstacles still stand in the way [20]. Many studies on the importance of knowledge have been carried out [1,10,11,21]. In some studies, increased knowledge does not affect pro-environment behavior due to the lack of inculcation of environmental awareness [7,13]. Knowledge transfer is good enough, but the education sector may only instill theoretical formalities without supporting awareness and practice. The knowledge transfer process has a weakness. In the transfer process, knowledge loss is caused by transferring knowledge and the transfer instrument [20]. So in providing knowledge, it is necessary to pay attention to the methods and instruments used to minimize knowledge transfer loss.

3.1.2 Guilty feeling

Instilling a sense of guilt into the community has quite an impact on the willingness to participate. This guilt fosters community awareness to behave better towards the environment [2]. Internalizing guilt remains effective even when individuals are in conditions where societal behavior is contradictory [12]. Individuals remain consistent in pro-environmental behavior even though there is no one. [22] supports the study's results, stating that guilt acts as a mediator, connecting environmental concerns with individual intentions to recycle plastic. Guilt also plays a role in people's willingness to pay more for the use of plastic bags. People are willing to use environmentally friendly plastics to protect the environment [2].

3.1.3 Facility

The existence of waste management facilities affects pro-environmental behavior. Behavior is associated with the facilities that support it and the accessibility to the facilities [5,18]. The distance of the trash can makes it easier for people to dispose of their garbage in its place. They can throw garbage at any time. Management practices will be hampered when the facilities are insufficient [11]. The closer the distance to the trash can encourages the disposal of plastic waste in its place. Facilities are also related to the existence of plastic substitute goods. The existence of packaging materials that are more environmentally friendly and affordable allows people to switch from using conventional plastics.

3.2 Social movement to build community awareness
Increased community awareness regarding the negative impact of waste has been carried out. Increasing community awareness has even started early [14,20,23]. Raising awareness is done by providing knowledge regarding the types of recycled materials, environmental sustainability, attitudes, and sustainable waste management practices. School children are also given an understanding of the impact of their behavior on the environment. The promotion of self-awareness is to increase their role in the environment related to waste management [23].

In developing countries, management activities have shifted to informal management [7]. Plastic management starts with a community-based participatory approach from the local community level [6]. The community has succeeded in influencing community opinion regarding waste [24]. Creating a clean and healthy environment. The form of participatory efforts in waste management is the formation of the waste bank movement. A waste bank is a form of effort to increase knowledge, skills, and community awareness. The existence of waste bank research has inspired other regions to participate in building waste banks. Garbage banks aim to minimize the amount of waste generated, sorted, and recycled. The concept of the existing waste bank is to make the community customers deposit their waste into the waste bank. The resulting waste products are then sold for profit.

Frugal consumption, reuse, and use of environmentally friendly products [12]. Awareness of the importance of environmental management began to be imprinted on individuals. This is proven by research from [11]. People are more willing to sort waste in every household. Awareness Sorting waste is very helpful in managing and improving the quality of waste, thus facilitating recycling. This is due to the satisfaction level of management in the communal waste collection system [11]. Community participation is a form of positive self-image and moral obligation to the environment [21]. The separation of plastic waste by the community affects the amount of waste and costs required, as well as the efficiency of waste management services [17].

There are still pros against the community regarding waste sorting. Nevertheless, it is undeniable that some people are willing to pay more than they have to participate; the cause is economic factors [21]. Opinions about management are that it only wastes time when the community must continue to work to meet needs. Even some people still do not know the benefits and importance of recycling.

3.3 Problem related to plastic management
Problems often occur because people do not feel they are responsible for the environment. The community places the responsibility on the government [4], so they are reluctant to participate. The challenge in management is how to encourage people's willingness to participate in minimizing the use of plastic, sorting waste, and recycling. The community's reluctance to participate can be attributed to the absence of incentives, lack of access to waste management facilities, and perceptions of less than optimal management [11]. The absence of perceived benefits is often an obstacle in managing plastic waste [8]. Plastics that have less economic value make people reluctant to recycle. Because no reciprocity is obtained, especially the economy is one of the factors of willingness to participate. This shows that the balance between environmental management and the incentives obtained is essential for the sustainability of community participation-based plastic waste management.

The economic aspect of community benefits has been proven to affect people's willingness to manage plastic waste. However, various other research sources prove that the main inhibiting factor is the lack of community awareness and knowledge [18], so there are limited skills in sorting and recycling. Waste sorting from the household requires more effort and needs more storage space at home. This lack of support hinders the willingness to sort plastic waste [13]. Management also takes a long time to sort plastic waste to recycle it [16].

3.4 Community awareness raising

The low level of stakeholder involvement in waste management, especially in the community, is still a highlighted issue [16]. Other challenges that await include inadequate resources and poor management. The community cannot do the management on its own. Support from other stakeholders is needed to encourage community behavior. The government, the private sector, and educational institutions can help the community. The government and the private sector play a role in coordinating management [11]. These agents play a role in increasing participation, increasing knowledge, providing facilities as well as supervising program implementation.

Management cannot only expect recycling efforts from the community. The government and industry must continue to support minimizing the use of plastic bags because, after all, the problem of plastic waste must begin to be addressed from the source. Excessive use will burden management efforts with more amounts. Meanwhile, reducing usage will facilitate recycling efforts. Every country needs the policy to minimize household plastic waste [21].

The government is an influential policy-setting agent. Regulations can include avoiding plastic packaging, adopting environmentally friendly packaging materials, and increasing plastic packaging recycling. The government can support the minimization of consumption by the community by providing plastic subsidies or the use of environmental taxes [19]. Provide alternative material substitution for another packaging. In some cases, using paid plastic regulations has reduced plastic consumption [9]. People are turning to the use of cloth bags. These efforts can occur with the cooperation between the government, industry, and society [8].

We can conclude that solid institutions, binding regulations, and awareness raising are strengths in waste management. This is supported by [13], who argue that institutions are agents in transitioning willingness to become actual actions. Awareness-raising and sustainability practices must be introduced to the community through awareness campaigns. The campaigns toward sustainable management can be conducted by stakeholders from the community, companies, and educational institutions.

Packaging management strategy is seen from several aspects, namely social, environmental, and legislative aspects. The first strategy is to minimize the amount of plastic waste produced. Sending waste to other places without being processed has increased plastic consumption [5]. This is because the pollution in the environment of origin has shifted to other areas so that the area of origin of the waste does not feel the impact of the waste. So consumption continues and increases. The strategy on the social aspect is carried out by increasing awareness through education. The environmental aspect strategy is providing optimal facilities and ease of accessibility in waste management. Meanwhile, from the legislative aspect, it is the effectiveness of policies.

A psychological strategy that had become a mainstay was manipulating guilt [22]. Manipulating is carried out with situational factors by providing complete facilities for easy management and access. Completeness of facilities can create a sense of guilt when they do not throw garbage in its place and do not sort their trash, even though the trash can has been classified. This guilt manipulation contributes more to sustainable management.

Several studies have stated that age affects management willingness [16]. Older people are more willing to do plastic management than younger people. This implies the need for a strategy to encourage the younger generation to improve pro-environmental behavior by minimizing and managing plastic waste. Educational institutions can play a role in increasing knowledge. Educational curricula can teach awareness and practice of environmental sustainability, so the younger generation has increased awareness. In addition, gamification-based digital approaches can be used efficiently to support changes in the behavior of young consumers. Previous research has proven that a pilot project based on digitalization boosted recycling through a 3R game application can increase the percentage of plastic waste recycling activities [25]. After the pilot, the recycling rate of plastic also increased significantly: 84% of the inhabitants currently recycle plastic, while only 25% did so before the pilot [25].

Informal management, especially scavengers, participate in the management of plastic waste. Scavengers sort the waste thrown away by households and then sell it to collectors of used goods. Garbage is one of the community's potential income and employment opportunities [18,24]. Some communities have even built waste banks, which are operated by local communities to increase plastic waste recycling. Weak community attitudes in management can be grown by providing incentives [18]. Incentives as a form of appreciation for individuals. However, individuals with high environmental sensitivity will continue participating even though no reward is obtained. Waste management requires various stakeholders who work together. Each agent needs good communication to increase understanding in management [18]. Good communication is also a medium for introducing government policies to the community.

4. Conclusion

Environmental awareness in urban communities regarding the use of plastic is still low. Urban communities still use plastic as the primary packaging material because it is easy to obtain at a very affordable price. The increase in the use of plastic produces a by-product, namely waste. The amount of plastic consumption not proportional to the amount of recycling threatens human life. Awareness as an effective strategy in increasing pro-environmental behavior is manipulated with guilt and supported by affordable policies and facilities. Awareness of plastic use starts from minimizing plastic use, looking for plastic substitute products, and managing plastic waste. The collaboration of various stakeholders is one of the strengths of sustainability.

Acknowledgments

This study was funded by the Ministry of Education, Culture, Research And Technology (KEMDIKBUDRISTEK) grant number [021/E5/PG.02.00.PT/2022] and grant number [NKB-1059/UN2.RST/HKP.05.00/2022], Research and Development (Risbang), Universitas Indonesia.

References

[1] MA Zambrano-Monserrate and M. Alejandra Ruano, "Do you need a bag? Analyzing the consumption behavior of plastic bags of households in Ecuador," *Resour. conserve. Recycle.* , vol. 152, no. June 2019, p. 104489, 2020, doi:10.1016/j.resconrec.2019.104489.

[2] MV Zwicker, HU Nohlen, J. Dalege, GJM Gruter, and F. van Harreveld, "Applying an attitude network approach to consumer behavior towards plastic," *J. Environ. Psychol.* , vol. 69, no. December 2019, p. 101433, 2020, doi:10.1016/j.jenvp.2020.101433.

[3] O. Alam, M. Billah, and D. Yajie, "Characteristics of plastic bags and their potential environmental hazards," *Resour. conserve. Recycle.* , vol. 132, no. January, pp. 121–129, 2018, doi:10.1016/j.resconrec.2018.01.037.

[4] LS Dilkes-Hoffman, S. Pratt, B. Laycock, P. Ashworth, and PA Lant, "Community attitudes towards plastics," *Resour. conserve. Recycle.* , vol. 147, no. January, pp. 227–235, 2019, doi:10.1016/j.resconrec.2019.05.005.

[5] SJ Barnes, "Out of sight, out of mind: Plastic waste exports, psychological distance and consumer plastic purchasing," *Glob. environment. Chang.* , vol. 58, no. October 2018, p. 101943, 2019, doi:10.1016/j.gloenvcha.2019.101943.

[6] G. MacRae and L. Rodic, "The weak link in waste management in tropical Asia? Solid waste collection in Bali," *Habitat Int.* , vol. 50, pp. 310–316, 2015, doi: 10.1016/j.habitatint.2015.09.002.

[7] JX Xiao, MJ Luo, and W. Li, "Evaluation of models for household recycling behavior in high-rise buildings: A Chinese case study in urban Guangzhou," *Waste Manag.* , vol. 131, no. January, pp. 126–135, 2021, doi: 10.1016/j.wasman.2021.06.002.

[8] K. Cai, Y. Xie, Q. Song, N. Sheng, and Z. Wen, "Identifying the status and differences between urban and rural residents' behaviors and attitudes toward express packaging waste management in Guangdong Province, China, " *Sci. Total Environ.* , vol. 797, p. 148996, 2021, doi:10.1016/j.scitotenv.2021.148996.

[9] L. Xu, Y. Zhong, X. He, X. Shi, and Q. Song, "Perception and Behavioral Changes of Residents and Enterprises under the Plastic Bag Restricting Law," *Sustainability* , vol. 14, no. 13, p. 7792, 2022, doi:10.3390/su14137792.

[10] E. Arı and V. Yılmaz, "Consumer attitudes on the use of plastic and cloth bags," *Environ. Dev. sustain.* , vol. 19, no. 4, pp. 1219–1234, 2017, doi:10.1007/s10668-016-9791-x.

[11] A. Almasi *et al.* , "Assessing the knowledge, attitude and practice of the kermanshahi women towards reducing, recycling and reusing of municipal solid waste," *Resour. conserve. Recycle.* , vol. 141, no. May 2018, pp. 329–338, 2019, doi:10.1016/j.resconrec.2018.10.017.

[12] JJ Escario, C. Rodriguez-Sanchez, and LV Casaló, "The influence of environmental attitudes and perceived effectiveness on recycling, reducing, and reusing packaging materials in Spain,"

Waste Manag. , vol. 113, pp. 251–260, 2020, doi:10.1016/j.wasman.2020.05.043.
[13] I. Muise, M. Adams, R. Côté, and GW Price, "Attitudes to the recovery and recycling of agricultural plastics waste: A case study of Nova Scotia, Canada," *Resour. conserve. Recycle.* , vol. 109, pp. 137–145, 2016, doi:10.1016/j.resconrec.2016.02.011.
[14] H. Mahat, M. Hashim, Y. Saleh, N. Nayan, and SB Norkhaidi, "Environmental Sustainability Knowledge, Attitude and Practices among Pre-school Students," *IOP Conf. Ser. Earth Environment. science.* , vol. 286, no. 1, 2019, doi:10.1088/1755-1315/286/1/012003.
[15] A. Varotto and A. Spagnolli, "Psychological strategies to promote household recycling. A systematic review with meta-analysis of validated field interventions," *J. Environ. Psychol.* , vol. 51, pp. 168–188, 2017, doi:10.1016/j.jenvp.2017.03.011.
[16] H. Reijonen, S. Bellman, J. Murphy, and H. Kokkonen, "Factors related to recycling plastic packaging in Finland's new waste management scheme," *Waste Manag.* , vol. 131, pp. 88–97, 2021, doi:10.1016/j.wasman.2021.05.034.
[17] P. Sukholthaman and A. Sharp, "A system dynamics model to evaluate the effects of source separation of municipal solid waste management: A case of Bangkok, Thailand," *Waste Manag.* , vol. 52, pp. 50–61, 2016, doi:10.1016/j.wasman.2016.03.026.
[18] I. Roche Cerasi *et al.* , "Household plastic waste habits and attitudes: A pilot study in the city of Valencia," *Waste Manag. res.* , vol. 39, no. 5, pp. 679–689, 2021, doi: 10.1177/0734242X21996415.
[19] DH Linh, DTT Cam, DTH Chi, LTB Ngoc, HP Nhi, and HP Nguyen, "Factors Influencing Consumers' Behavioral Intentions to Reduce Plastic Waste: Empirical Research with The Case of Vietnam," *South East Asia J. Contempary Business , Econ. Law* , vol. 18, no. February, pp. 1–9, 2019.
[20] T. Ferdous and T. Das, "A Study about the Attitude of Grade Eight Students for the Use of Plastic in Gwarko, Balkumari, Lalitpur District," *Procedia - Soc. Behav. science.* , vol. 116, pp. 3754–3759, 2014, doi:10.1016/j.sbspro.2014.01.836.
[21] M. Czajkowski, T. Kadziela, and N. Hanley, "We want to sort! Assessing households' preferences for sorting waste," *Resour. Energy Eco.* , vol. 36, no. 1, pp. 290–306, 2014, doi:10.1016/j.reseneeco.2013.05.006.
[22] L. Elgaaied, "Exploring the role of anticipated guilt on pro-environmental behavior - a suggested typology of residents in France based on their recycling patterns," *J. Consum. Mark.* , vol. 29, no. 5, pp. 369–377, 2012, doi: 10.1108/07363761211247488.
[23] OM Owojori, R. Mulaudzi, and JN Edokpayi, "Student's Knowledge, Attitude, and Perception (KAP) to Solid Waste Management: A Survey towards a More Circular Economy from a Rural-Based Tertiary Institution in South Africa," *Sustain* . , vol. 14, no. 3, 2022, doi:10.3390/su14031310.
[24] MA Budihardjo, SY Ardiansyah, and BS Ramadan, "Community-driven material recovery facility (CdMRF) for sustainable economic incentives of waste management: Evidence from Semarang City, Indonesia," *Habitat Int.* , vol. 119, no. November 2021, p. 102488, 2022, doi:10.1016/j.habitatint.2021.102488.
[25] U. Santti, A. Happonen and H. Auvinen, "Digitalization boosted recycling: Gamification as an inspiration for young adults to do enhanced waste sorting," *AIP Conf. Proc* ., vol. 2233, 050014, 2020, doi: 10.1063/5.0001547

Coastal Settlement Planning based on the Network Density in Small Island: A Study on Sulabesi Island, Indonesia

L M Lautetu[1], H S Hasibuan[1] and R P Tambunan[1]
[1] School of Environmental Science, Universitas Indonesia, Central Jakarta, 10430, Indonesia
[1] lautetulisa@gmail.com,[2*] hayati.hasibuan@ui.ac.id, [3]rudyptamb@gmail.com

Abstract. Sulabesi Island, as a small island, is faced with challenges in developing road infrastructure to facilitate community mobility, especially to meet their settlements' needs. With most settlements located on the coast, this study aims to analyze the road network's density level based on the building distribution and plan settlements based on connectivity between regions. The method used is a road network density (ND) analysis using ArcGIS 10.5 software. The analysis results show that the density level of the Sulabesi Island road network is still low and only concentrated in one area, Sanana District. There are still unfinished road and bridge network constructions in other areas, so some areas are not yet connected. This condition causes connectivity between regions to be low, and community mobility becomes difficult, especially to meet their basic needs, including distributing their harvests. Even with a road network that cannot connect between regions, it can increase the isolation of a settlement. From this research, it can be a description of the characteristics of settlements on small islands from the aspect of land transportation.

1. Introduction

Small islands have unique characteristics regarding regional conditions and the lives of the people in them. Several studies have stated that small islands generally have a life that depends on coastal resources with a high threat of climate disasters [1–4]. Even small islands are the most disadvantaged areas from climate change-related disasters [5]. Moreover, people on small islands generally choose the coast as a place to live because it is a sloping plain, and it is easy to reach their needs. A study conducted in the Salomon Islands also stated that the coast became a place for people to live because of their significant dependence on coastal resources [6].

The main challenge faced by small island coastal settlements is realizing coastal settlements to support life in them [7,8]. As a place to live, a settlement must be a place that is more than just a house but also contains an integrated interest and harmony with the environment [9]. In the theory of ekistic elements, there are five main elements in settlements: nature, humans, society, buildings, and infrastructure [10], so settlements are places where humans live and live [11]. In realizing small island coastal settlements that can meet the lives of the people in them, it must be seen comprehensively, especially in the community's ability to be independent and recover when faced with various threats and limitations in meeting their basic needs.

Small island communities have specific characteristics because they are influenced by the area in which they live. Several previous studies have explained that small island communities highly depend on coastal ecosystems, climate, and markets as the center for meeting their needs [12,13]. Therefore, it is explained that the small island community is identical to the local wisdom; they have to be able to survive and fulfill their needs from the resources they have [14–18]. Economically, it has an economic characteristic related to imports from outside because it has been unable to produce its goods or services [4]. In addition, in meeting their needs, the community is faced with limitations in transportation [19], which is not only between islands but also between regions on the island plain. Connectivity is a crucial aspect of meeting the needs of small island communities. With the condition of separate islands and limited natural resources, it is difficult for small islands to improve the quality of settlements and the people living there. Some studies say small islands are usually synonymous with isolated areas [20]. With adequate infrastructure, it can improve connectivity between regions on small islands, especially for developing economic activities for the people [21].

In looking at the connectivity between regions on small islands, many studies have focused on inter-island connectivity, especially sea transportation [19,22–24]. However, few studies discuss the connectivity between regions on small island plains, even though the mobility of people living on small islands depends on road infrastructure to reach activity centers, such as markets, banks, government offices, and educational facilities. And health, especially in rural areas. Therefore, this research will assess connectivity between regions in small island plains, with the road network as the leading indicator. This research focuses on meeting the needs and selling natural products owned by small island communities, so they need a good road network to facilitate community mobility.

This study tries to provide an overview of connectivity in small island plains, especially on isolated small islands, and its relationship with community independence to meet their needs. This research was conducted on Sulabesi Island, Sula Islands Regency, the southernmost part of North Maluku Province. Sulabesi Island has a land area of <2,000km2 consisting of 6 sub-districts and 42 villages, with a population of 65,421 [25]. Climatically, Sulabesi Island is influenced by two seasons, the west and east monsoons [26]. As a small island, the morphological condition of Sulabesi Island has an altitude ranging from 0-1500 meters above sea level, starting from the flat plains on the coast to very steep in the island's interior. So that many community settlements are located in coastal areas. Settlements are scattered in coastal areas due to socio-ecological interactions [6]. Several previous studies have explained that the condition of coastal settlements on Sulabesi Island is faced with slum conditions and has limited infrastructure [27], especially the road network. The road network is the primary means of transportation for the people on Sulabesi Island to reach the center of activity and fulfill their needs.

Thus, this study aims to analyze the level of connectivity between regions on the plains of Sulabesi Island, with the road network as the primary indicator for settlement planning and increasing the independence of the Sulabesi Island community. This research is critical to increasing the community's mobility to meet their needs. By looking at the level of connectivity between regions on small islands, especially on island plains, it is hoped that it can help provide new insights regarding the condition of small islands with isolated regional characters in the realization of the sustainability of the lives of the people who live in them.

2. Methods

2.1. Study area

The research was carried out for three months, from May to July 2022, with the research location on Sulabesi Island as one of the inhabited islands in the Sula Islands Regency. There are 6 sub-districts: Sanana, Central Sulawesi, East Sulabesi, South Sulabesi, West Sulabesi, and North Sanana. It is called a small island because it has an area of 1,196 km^2, which according to the Indonesian government's policy in Law no. 27/2007 states that small islands are islands that have an area of <2,000 km^2 and there are also previous studies that explain small islands are areas with an area of <3,000 km^2 [28]. Sulabesi Island, in the Regional Spatial Plan (RTRW) of the Sula Islands Regency in 2011-203, was designated as the center of district activities, namely Sanana District and North Sanana, which became the regional activity center (PKW). Sulabesi Island has a flat coastal area with an altitude of 0-200 meters above sea level. It is used as a residential area for people who spread randomly in almost the entire coastal area. This distribution then causes people to travel long distances to meet their needs. With this condition, it is essential to see how the connectivity between settlements in the coastal area of Sulabesi Island is so that an overview is obtained to support settlement planning in realizing community independence.

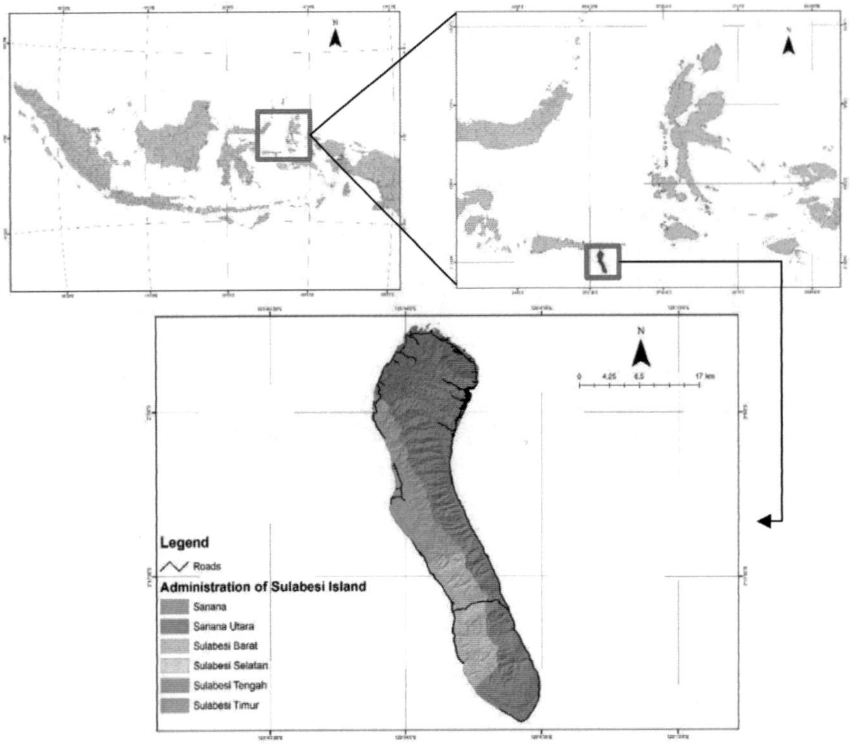

Figures 1. Location of Sulabesi Island

2.2. Data collection and analysis

The data used in this study were collected secondary and primary. Secondary data from the existing road network was collected from local governments in 2017, and the Open Street Map (OSM) data through the BBBike website in 2022. Cross-checking from this data was conducted in the field to see if they matched the existing conditions, including plotting each road node. In this study, the road network has not been divided in more detail based on the function or type of road because the complexity of the road network is still low, so it is only focused on the available network in connecting between regions on Sulabesi Island.

The analysis used the ND to answer the research objectives using ArcGIS 10.5. This analysis analyzes road network connectivity based on road density, namely the calculation of the ratio between the total length of the road divided by the land area in square kilometers [29]. By determining the density level, namely, the higher the score, the higher the level of connectivity [30]. Furthermore, the results of road density are input into the raster data to be able to classify the level of road density.

3. Results and Discussion

3.1. Assessment of the density of the Sulabesi Island Road network

The ND assessment in Sulabesi Island aims to see the current state of networking available. Based on the analysis results, the density of the road network on Sulabesi Island is still low, with the majority of roads on Sulabesi Island being concentrated in the northern part of the island. The results of the ND analysis on Sulabesi Island are presented in Figure 2. Regionally, most areas on Sulabesi Island have a low level of road density, including the Districts of East Sulabesi, South Sulabesi, and West Sulabesi.

Meanwhile, sub-districts with high ND values are only found in the Sanana and North Sanana sub-districts.

Table 1. Results of ND Analysis by Region on Sulabesi Island

District area	Road Length (km)	Areas (km^2)	ND		Class
			Score	%	
Sanana	65,45	116,70	0,56	20	High Density
Sulabesi Tengah	38,69	75,29	0,51	18	Medium Density
Sulabesi Timur	23,52	81,84	0,29	10	Low Density
Sulabesi Selatan	31,72	90,06	0,35	13	Low Density
Sulabesi Barat	35,81	91,47	0,39	14	Low Density
Sanana Utara	51,91	77,06	0,67	24	High Density

Based on the analysis results in Table 1, it can be seen that the Districts of Sanana and North Sanana have a high level of road density compared to the other four regions. It is because the two areas are the center of activities for the Sula Islands Regency; Sanana District is the district capital that provides education, health, and trade services, while North Sanana District is the administrative center of the Sula Islands Regency. Therefore, the road network in these two areas is denser than in the other areas. In addition to regional function factors, there are uneven road infrastructure development factors. Based on the analysis in Figure 2, the areas located in the east, south, and west have intermittent road conditions; besides that, the types of roads found in these areas are only primary roads and environmental or local roads. These conditions eventually caused the ND level in the districts of East Sulabesi, South Sulabesi, and West Sulabesi to be the areas with the lowest ND.

The low density of the road network in the Districts of East Sulabesi, South Sulabesi, and West Sulabesi compared to the Districts of Sanana and North Sanana. The activity centers of the Sula Islands Regency are due to the region's function and the road network infrastructure development that has not been completed. Other research says the road network's low density is mainly located outside the city because the road construction has not yet been realized [31]. Actually, in the eastern area of Sulabesi Island, a ring road has been built to link the South Sulabesi District and some villages in West Sulabesi District. However, based on the results of field visits, it was found that this ring road has not been fully able to support the fulfillment of community needs because an adequate road network has not fully served some villages. The condition of road density is still low, causing the people on Sulabesi Island to face challenges in meeting their needs. In addition to the road network that has not been able to connect between regions, the existing road conditions also make it difficult for the community to meet their needs and distribute their natural products. Based on the results of field visits, it is known that people depend on shopping centers in Sanana District for their needs. It then causes many villages on Sulabesi Island to be isolated, and the ability of the community to meet their needs becomes difficult.

3.2. Sulabesi Island settlement planning

The remote condition of Sulabesi Island as a small island increases the challenges of the sustainability of settlements in it because it has limitations in terms of development, both in terms of natural and social resources. In line with this, small islands have characteristics, one of which is accessibility and low quality of human resources [4]. Based on the analysis of connectivity between regions on Sulabesi Island mentioned in Figure 2, it also shows a low level of connectivity due to the low density of the road network. Some areas on Sulabesi Island have road conditions that are cut off due to topographical conditions and from the aspect of development that has not been fully realized (Figure 3).

With the road network condition, the accessibility of the community in meeting the needs of activity centers such as markets, banks, and health services is relatively low. Even the Districts of South Sulabesi and West Sulabesi have poor connectivity because their road network is not yet connected. Whereas meeting the needs and reaching the center of community activities is very dependent on the condition of the road network. Therefore, building a ring road network is necessary to connect the area behind Sulabesi Island, namely West Sulabesi and South Sulabesi Districts, to facilitate community accessibility further. In addition, road repairs and bridge construction also need to be improved and implemented, considering the level of community dependence on the district capital in Sanana District as the primary source of community needs.

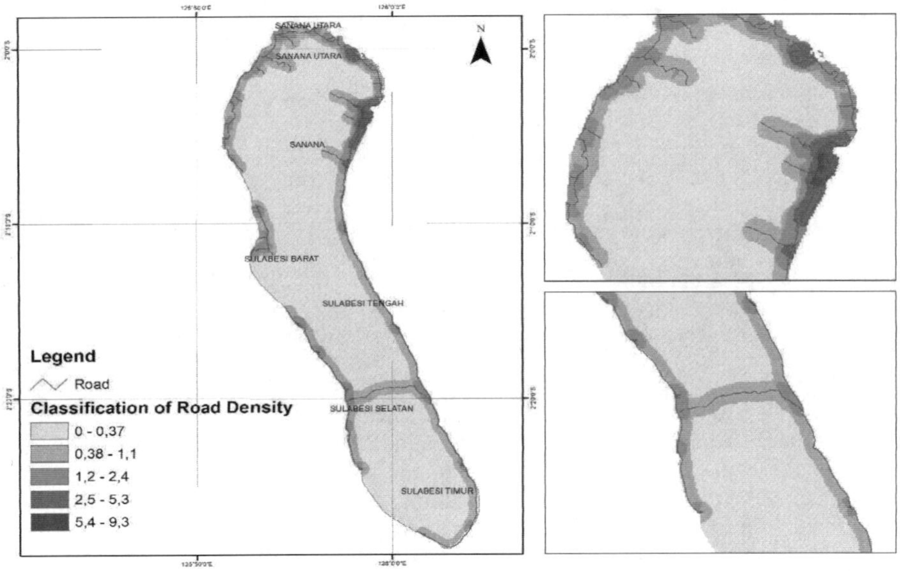

Figure 2. Distribution ND on Sulabesi Island

Figure 3. The road network condition on Sulabesi Island in the west and south.

In improving small island settlement planning, it is vital to increase connectivity between regions because connectivity between regions can increase people's access to basic settlement facilities, especially for health, education, trade, and services. With strong interaction between regions, the community will also find it easier to meet their needs. The misjudgment of community resilience was from adequate infrastructure [12]. With poor infrastructure, especially on the road aspect, it will be difficult for the community to meet their needs, let alone recover when a disaster occurs with activity centers that are difficult to reach.

Improving the road network's quality and significantly accelerating road and bridge infrastructure construction can create settlements that can support life in them. Regional connectivity on small islands can develop the community's economic activities [21]. A low connectivity level will impact the people of Sulabesi Island, especially in reaching activity centers such as markets, banks, and health facilities.

Therefore, connectivity is one of the main drivers to realizing the sustainability of coastal settlements on Sulabesi Island.

4. Conclusion

Road density on Sulabesi Island is still very low, causing the community to face challenges in meeting their residential needs. High road density is only found in the Sanana District, the center of Sulabesi Island activities. The condition of the road network is low due to the existing bridges and roads that have not been completed. As a small island, the center of activity is only focused on one area, so in the future, it is essential to develop adequate road infrastructure to facilitate community mobility to meet their needs and distribute natural products owned for trade. This research focuses on the road network in general, which has not been divided further based on the type of road, primary, secondary and local, with community mobility still being seen spatially. Therefore, in further research, a more in-depth analysis can be carried out according to the type of road network and see the accessibility of the community based on distance and travel time from settlement to the center of activity to meet their needs.

Acknowledgments

This research is funded by the Directorate of Directorate General of Higher Education, Research and Technology, Ministry of Education, Culture, Research and Technology with Grant number: NKB-1040/UN2.RST/HKP.05.00/2022.

References

[1] Chien L K, Huang W P, Hsu C H, Hsieh C C and Fang C Y 2017 Coastal disaster risk assessment and designation of protection zoning in Taiwan using GIS *WIT Trans. Built Environ.* **170** 173–84

[2] Narayan S, Esteban M, Albert S, Jamero M L, Crichton R, Heck N, Goby G and Jupiter S 2020 Local adaptation responses to coastal hazards in small island communities: insights from 4 Pacific nations *Environ. Sci. Policy* **104** 199–207

[3] Wunas S 2019 The disaster resolving on tourism settlements in small islands *IOP Conf. Ser. Earth Environ. Sci.* **235**

[4] Pamungkas A 2014 Indikator Kemandirian Pulau Kecil: Studi Kasus Poteran Sumenep *J. Penataan Ruang* **9** 1–10

[5] Connell J 2018 Effects of Climate Change on Settlement and Infrastucture reevant to the Pacifiv Islands *pacific Mar. Clim. Chang. Rep. Card* 159–76

[6] Basel B, Goby G and Johnson J 2020 Community-based adaptation to climate change in villages of Western Province, Solomon Islands *Mar. Pollut. Bull.* **156** 111266

[7] Cao W, Zhou Y, Li R, Li X and Zhang H 2021 Monitoring long-term annual urban expansion (1986–2017) in the largest archipelago of China *Sci. Total Environ.* **776** 146015

[8] Chen Q and Dong H 2019 Sustainable Development Indicator Systems for Ssland Cities: The Case of Zhoushan Maritime Garden City *Isl. Stud. J.* **14** 137–56

[9] Sadana A 2014 *Perencanaan Kawasan Pemukiman* (Yogyakarta: Graha Ilmu)

[10] Mandal P and Sen J 2018 The Inclusive Role of Ekistics Elements in Earmarking Innovation Zones through a Balanced Distribution of Smart Development and Local Expression: Case of Kolkata Metropolitan Area (KMA) *Expand. Cities - Diminishing Sp. Are Smart Cities Solut. or Part Probl. Contin. Urban. Around Globe? (Real Corp 2018)* 461–70

[11] Winarso H 2013 *Teori Ekistics dan Penataan Ruang di Indonesia* (Sejarah Penataan Ruang)

[12] Nurzaman A, Shaw R and Roychansyah M S 2020 Measuring community resilience against coastal hazards: Case study in Baron Beach, Gunungkidul Regency *Prog. Disaster Sci.* **5** 100067

[13] Wahyudin Y 2003 Sistem Sosial Ekonomi Dan Budaya Masyarakat Pesisir *Makal. Pelatih. Pengelolaan Kaw. Konserv. Perairan, Pus. Diklat Kehutan.* 25

[14] Ali N and Sinilele A 2019 Kearifan Lokal Dalam Melestarikan Pulau Dan Laut Di Pulau

Bonetambu Sulawesi Selatan *Hasanuddin J. Sociol.* **1** 101–15
[15] Ekawati E 2018 Tulude: Antara Modernitas Dan Tradisi Masyarakat Pulau Marore *J. Masy. dan Budaya* **19** 385
[16] Firdaus A M, Pelupessy J M and Tampubolon J R 2016 Strategi Penyelesaian Masalah Sosial Ekonomi Masyarakat Pesisir Di Kepulauan Banda Neira, Kabupaten Maluku Tengah *J. Sos. Ekon. Kelaut. dan Perikan.* **11** 55
[17] Ohleky M P, Pattiselanno A E and Kaplale R 2018 Namlai Kerne: Kearifan Lokal Dan Ketahanan Pangan Masyarakat Desa Werwaru Kecamatan Moa Kabupaten Maluku Barat Daya *Agrilan J. Agribisnis Kepul.* **5** 114
[18] Fathi Royyani M, Efendy O, Farida Hasanah I, Rambu Ana R and Bintoro R 2020 Tanaman Pekarangan dan Ketahanan Pangan pada Masyarakat Sumba Timur *J. Biol. Indones.* **16** 205–15
[19] Watloly A, Henderika L and Christina S 2018 Potret Kemiskinan Masyarakat Pulau Masela Di Ladang Migas Terkaya Blok Masela Di Kabupaten Maluku Barat Daya *Sosio Konsepsia* **7** 15–30
[20] Danson M and Burnett K A 2017 Enterprise and Entrepreneurship on Islands *J. Entrep. Innov.* **18** 25–35
[21] Sappe A R, Manaf M and Syafri S 2020 Peran dan Strategi Transportasi Laut Terhadap Konektivitas Antarwilayah di Kabupaten Banggai Laut Provinsi Sulawesi Tengah *Urban Reg. Stud. J.* **2** 1–7
[22] Amin M and Jufrin 2020 Peranan Pengangkutan Laut Sebagai Sarana Transportasi Masyarakat Indonesia *Fundam. J. Ilm. Huk.* **9** 191–207
[23] Mattiro S, Nasrullah and P R 2021 Pengembangan Ekowisata Berbasis Budaya Bahari Untuk Mendukung Ketahanan Ekonomi Masyarakat Pesisir Di Pulau Kerayaan Kabupaten Kotabaru Provinsi Kalimantan Selatan *Pros. Semin. Nas. Lingkung. Lahan Basah* **6** 1–8
[24] Jusna and Nempung T 2016 Peranan Transportasi Laut Dalam Menunjang Arus Barang Dan Orang Di Kecamatan Maligano Kabupaten Muna *J. Ekon. Uho* **1** 189–200
[25] Statistik Badan Pusat 2022 *Regional Statistic of Kepulauan Sula Regency in 2022* (Jakarta: Badan Pusat Statistik)
[26] Badan Pengembangan Daerah 2018 *Laporan Draft Akhir Rencana Pembangunan dan Pengembangan Perumahan dan Kawasan Permukiman (RP3KP) Kabupaten Kepulauan Sula* (Kepulauan Sula: Badan Pengembangan Daerah)
[27] Than F, Suparjo S and Takumansang E D 2018 Studi Pengembangan Kebutuhan Air Minum Di Permukiman Desa Bajo Kecamatan Sanana Utara Kabupaten Kepulauan Sula *Spasial* **5** 1–9
[28] Rodrigues M 2015 A spatial typology for settlement pattern analysis in small islands *GeoFocus. Rev. Int. Cienc. y Tecnol. la Inf. Geográfica* **0** 3–26
[29] Maity B, Mallick S K and Rudra S 2021 Integration of urban expansion with hybrid road transport network development within Haldia Municipality, West Bengal *Egypt. J. Remote Sens. Sp. Sci.* **24** 471–83
[30] Tresidder M 2005 Using GIS to measure connectivity: An exploration of issues *Sch. Urban Stud. Planning, Portl. State Univ.*
[31] Tini N H 2019 Appraisal of Road Network Spatial Connectivity in Abuja City, Nigeria *Adamawa State Univ. J. Scentific Reseacrh* **7** 1–13

The Usage of Public Open Spaces in Candi Baru, Semarang

Maria Damiana Nestri Kiswari[1]*, Nany Yuliastuti[1] and Budi Sudarwanto[1]
[1]Doctoral Program of Architecture and Urban Engineering, Faculty of Engineering, Diponegoro University, Semarang, 50275, Indonesia
[1]*nestri_kiswari@unika.ac.id, [2] n.yuliastuti@pwk.undip.ac.id,
[3]budisudarwanto@lecturer.undip.ac.id

Abstract. Public open space has a role in making settlements more comfortable and liveable. However, they were generally found deserted, unmaintained, and dull. Candi Baru (New Candi) was a residential area in Semarang located in the southern area, which is a highland. The residential area was designed by Thomas Karsten, a Dutch architect and urban planner during the colonial administration. The concept design of Candi Baru implemented by Thomas Karsten was similar to Garden City's concept. From the previous discussions about the Candi Baru area, there was a gap in the use of public open spaces based on the spatial justice aspect. So, this paper aims to identify the use of public open spaces based on spatial justice for three parks in Candi Baru: the Sudirman, Merapi, and Diponegoro parks. For the identification, a qualitative method was implemented to describe spatial justice in the current usage of the three public open spaces. Therefore, the result of the paper will equip the understanding of the existence of public open spaces in the colonial urban area, which was inclusive, safe, resilient, and sustainable.

1. Introduction

Public open space affects the quality of an area's environment and life. In the 11th point of the Sustainable Development Goals document, it is stated to make cities and human settlements inclusive, safe, resilient, and sustainable [1], that public open space is an environmental facility that can be used safely and comfortably by the general public, namely men and women including groups of children, adults and the elderly as well as persons with disabilities. Public open spaces would be good places if they had social inclusion that was integrated, connected, accessible, and provided a sense of security from outside disturbances, including crime, as well as being a comfortable place to sit, rest and meet to support the sustainability of the urban environment [2]. These environmental facilities made the residential environment more humane to inhabit. On the other hand, there was a problem that public open spaces looked dirty with an impression of a slum [3], and public open space facilities were unmaintained or even damaged [4][5][6]. Low safety and less comfort for users were due to disturbance from motorized vehicle traffic passing around public open spaces [7]. Public open spaces had limited accessibility and hardly any spaces for people with disabilities and the elderly [8]. The other problem was that the open space area was used for selling. The problem often found that the public open space was dominated as a place to vend street vendors [9]. The areas in the public open spaces used for selling lower the public quality of the space [10].

The Candi Baru area was one of the settlements in the city of Semarang, which was developed during the colonial administration. The settlement was the work of Herman Thomas Karsten, a Dutch architect in Indonesia. Thomas Karsten planned to build the Candi Baru settlement in the southern part of Semarang, a hilly area [11]. The landscape condition of the settlement was steeply contoured with a spiral environmental road network pattern, and the open spaces responded to the area's contours. [12].

Karsten considered three elements in planning and designing the Candi Baru: the road system, city parks, attractions, and vegetation. [13]. Candi Baru was still known as an elite residential area. The physical conditions of the settlements were changed. Some residential buildings with the colonial architectural style have already transformed into new modern buildings, and their functions have also changed to become hotels, cafes or restaurants, and other commercial functions. Meanwhile, the structure of the area, which was the pattern of roads and open spaces, still existed. Some public open spaces were developed to accommodate public activities and beautify the area.

This paper aims to identify the usage of public open spaces in the Candi Baru area based on spatial justice. Three public open spaces as a case study in the area were discussed. How does the public open space at Candi Baru use to consider spatial justice from a placemaking approach? An elaboration of the interrelation of those topics discussing the Candi Baru was to show the novelty of this paper. Thomas Karsten applied a discussion of the concept of Garden City in planning the Candi Baru in Semarang city. The implementation of the Garden City concept in the area was seen in the city's shape, the building masses' layout, the road pattern, and the green layout [14]. The Garden City concept, which Thomas Karsten applied, was different from the Garden City concept by Howard because there was a response from the topographical aspects. Thomas Karsten paid attention to local aspects: economy, environment, and Javanese concept (culture) in his works, including the Candi Baru. [15]. Locality insight related to the significance of the economic base, environmental context, and aspects of Javanese thought, namely spatial zoning, was visible in the Candi Baru[16]. Diponegoro Park (*Raadsplein*), which Thomas Karsten designed, is considered one of Semarang's urban expansion centers. The topic was discussed in The Sustainability Concept of Alun-Alun as a Model of Urban Design in the Future [12]. The urban open spaces in Semarang, which one was Taman Diponegoro, were proof that the concept of an open space system was a tool to integrate old and new areas into the urban environment. Candi Baru was already known as one of the residential areas that still had green spaces. The article on the area was a case study discussing the evolution of the concept of public green space in the city of Semarang in the early 20th century [17]. The evolution of the green space concept in the area resulted in a reduction in the quantity and quality of public green space, which impacted air pollution, the city's air temperature, floods, and landslides. The Sudirman Park was one of the public open spaces in Candi Baru, crowded with people. The research was on people's preferences for Sudirman Park as a playground for children [18]. As an area designed with a concept inspired by the Garden City concept, the public open spaces had a significant role. From the existing discussions about Candi Baru, there was a gap that has not been discussed: the use of public open spaces in the area based on aspects of spatial justice. Besides Sudirman Park being discussed most, the other two parks were not discussed from the usage, and the spatial justice aspects were Diponegoro Park and Merapi Park. Therefore, the novelty of this paper is spatial justice in the use of public open space in the Candi Baru, Semarang, which is comprehended from the Placemaking approach. Placemaking reinforced spatial justice within an urban area [19].

2. Methods

The method used was based on the research objective, which was to identify the use of public open space based on spatial justice with a Placemaking approach in Candi Baru. The research was conducted in three public open spaces: Sudirman Park, Merapi Park, and Diponegoro Park. Those parks were developed and utilized by the public. There were two studies which were in Candi Baru. First, the Evolution of the Public Green Space Concept in the City of Semarang in the Early 20th Century Until Now - Public Green Space in the Candi Baru Area, had method used descriptive qualitative to obtain the evolution of green public spaces and the impact on its ecological function [17]. Second, the Locality in the Area of Candi Baru Semarang had a case study method that was tried out empirically on Karsten's works in Semarang. That research aimed to reveal local aspects of Karsten's concept of thought in the case study. The collected data was interpreted qualitatively [16].

In the research on the usage of public open space in Candi Baru, the method was qualitative by implementing Bailey's steps [20] [21], as follows: 1) research questions 2) research design, which supported the theories of public open space, spatial justice, and placemaking; 3) data collection included primary data from three public open spaces and secondary data; 4) data analysis; 5) interpretation of the results. Bailey's steps were a cycle, so the results of interpreting the results would be research questions for further research. The method described the use of public open spaces from the aspect of spatial justice with the placemaking approach. The data collection consisted of primary and secondary data. The primary data collection was through observation of conditions in the field and strengthened by the results of interviews with residents or activity actors, which are then recorded or made a note. Secondary data were obtained from literature studies on the history and development of the area, as well as government policies related to the Candi Baru area. Observations were made several times on different days and times for the three public open spaces, which were on weekdays and weekends, in the mornings and afternoons.

The data analysis utilized two related theories, placemaking and spatial justice. The Placemaking approach in public open spaces helped shape patterns of creative use of space, paying attention to physical, cultural, and social identities in determining places and following changes in their environment. [22]. The principle of Placemaking was to create a good place. St, the elements of a good place were used as the principle of Placemaking: Activity, Form, and Image. Activity in a place results from a combination of vitality and diversity. Activities that make a place live and distinguish it from other urban areas. Activities carried out by individuals or several people (groups) in one place, such as walking stiffly, sitting, certain celebrations, and other activities whose intensity can be routine every day, at certain hours, or occasionally done. The form was explored to achieve a quality urban design so that the created form could produce a sense of place in the area [23]. The components of form included scale, intensity, permeability, landmarks, building ratio, and other components related to shape. Image related to the identity of a place. Every place had an identity, but it was not the same. The image of a place was a combination of the identity of the place with the perception of the place. Thus, what was related to the image was cognition, perception, and information. For example, a person's knowledge and understanding of a place will form perceptions and impressions of the place, or through the information collected, someone would get a picture of the image of a place.

Spatial justice began with the need for environmental justice for social groups, such as workers, youth, and community groups who receive social discrimination because of race, class, gender, sexual preferences, and other inequalities. This issue formed the basis for the concept of spatial justice [24]. Spatial justice theory focuses on logical spatial aspects of social justice and fair distribution of public open space [25]. Spatial justice in public open spaces had critical influencing factors: Access and Management, Sociability and Diversity, Social Stratum and Information, Social Inclusion, Demand, and Provision [26]. Access and Management considered the physical elements in and around the public open spaces and their locations. The walkability, affordability, and safety components are considered part of this Access and Management factor. Sociability and diversity allowed public open space to become a space for community interaction. This could be seen from the community interactions that occurred and the activities that supported these interactions. Social stratum and information related to equal opportunity to participate in the process. The two factors, Social Inclusion, and Demand and Provision, were interrelated because they related to the provision of needs and access for each group that differed from their ability and socioeconomic background. [24] [26].

There were three main concerns in the analysis, which were usage, placemaking, and spatial justice. From the explanation of each of these components, the identification analysis carried out through observation was as follows: Physical elements were the availability of public open space facilities. Affordable, safe, and accessible facilities, especially for people with disabilities, the elderly, and children.

1. Types of activities and their intensities were carried out in public open spaces, including the surrounding space. Whether they were inclusive activities or not.
2. The doers of these activities, individuals or groups. These activities were carried out by children, adolescents, adults, the elderly, people with disabilities, or all general people so that all groups become actors in these activities.

IOP Conf. Series: Earth and Environmental Science 1111 (2022) 012032 doi:10.1088/1755-1315/1111/1/012032

The narrated analysis was complemented by maps, photos, and tables that explained and summarized the discussion. Thus, an overview of the use of public open space and its spatial justice is obtained from the placemaking approach in Candi Baru, along with potential problems.

3. Result and Discussions

The discussion on the use of public open space in Candi Baru consists of three sub-topics, as follows:

3.1 The history of Candi Baru as a residential area in Semarang city

During the colonial administration, the center of Semarang was in the old city, the lowlands. Then, the city area was extended by developing residential areas outside the city center. The development of the city areas was planned by Thomas Karsten, an architect and urban planner. Thomas Karsten planned to develop some residential areas in Semarang, including the upper area, which was Candi Baru[27]. In 1916, Thomas Karsten developed a residential area of Candi Baru in the south of the city which was a hilly area [28] [29] [12] [15]. Candi Baru was planned by considering the socioeconomic aspect, not the racial aspect of the occupants that was common at that time. This concept supports relations between ethnic groups in a residential area [29]. The unique spatial arrangement of Candi Baru was that there was a space gradation where the high classes, which were the majority of Europeans, occupied areas along the main road, and then on the back there were the middle classes until the backmost was the low-classes as urban kampongs [30]. Figure 1 shows those zones and the streets. Areas inhabited by Europeans at that time were located, shown in Fig. 1, at S. Parman street, Gajahmungkur Selatan (South Gajahmungkur) street, Slamet street, Guntur street, Merapi street, Dieng street, and Raung street, etc. At the same time, the residential area for the lower community was inhabited by local people, located in Lempongsari [31], Kintelan [32], and dan Kalilangse [15].

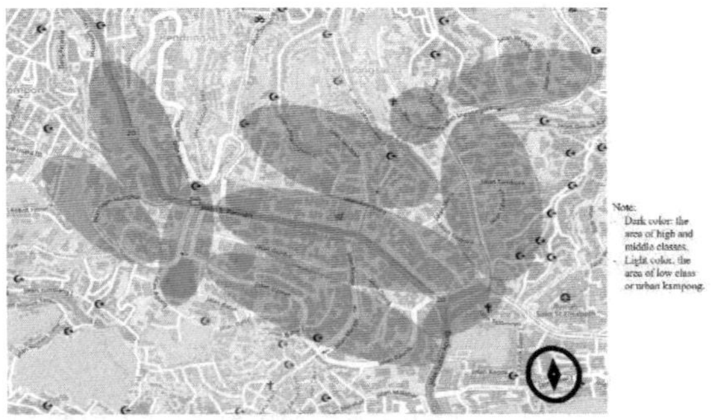

Figure 1. Distribution of high, middle, and low classes in the Candi Baruarea
Source: Dinas Penataan Ruang Kota Semarang [33] developed by authors, 2022

Candi Baru was planned in detail, considering the natural conditions of the steep contours [34] [15]. The planning of the area responded to the condition of the landscape, with the pattern of the road network, which bends and the shape of the public open space followed the contours. [12].

The existence of public open spaces as parks by adapting the Garden City concept was significant for its characteristics [17]. At present, Candi Baru still functions as a residential area. However, more residential buildings have become commercial, such as hotels, cafes, restaurants, or other commercial. The situation gave influenced the usage of the public open space nearby.

The location of Sudirman, Merapi, and Diponegoro parks are shown in figure 2. The parks had similar shapes, which tended to elongate because of the contoured landscape. The main busy streets surrounded Diponegoro park. Meanwhile, Sudirman park lay at the main street, S. Parman street. However, it elongated inward, and the Merapi park was a narrow and long park, surrounded by neighborhood streets, less busy than the streets at Diponegoro park. However, the contour in Merapi park was the steepest of the three parks. The Semarang Municipal government managed these three parks based on the local regulation of Semarang Municipality number 8 of 2016, concerning the

Management of trees in public green open spaces, including parks that had social and aesthetic functions as a means of recreational, educational, or other activities [35].

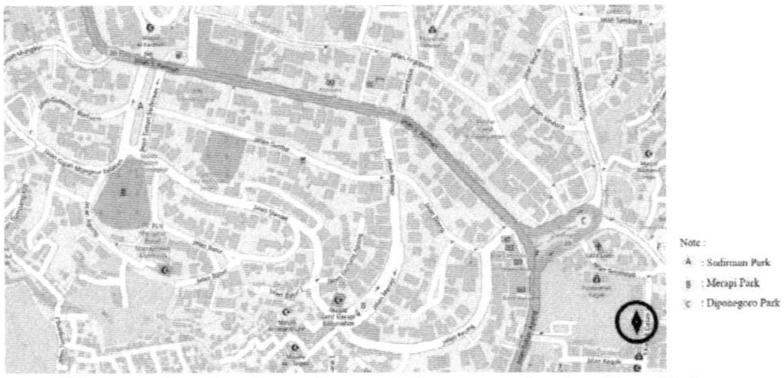

Figure 2. The location of three parks in Candi Baru
Source: Dinas Penataan Ruang Kota Semarang [33] developed by authors, 2022

3.2 Placemaking of public open spaces in Candi Baru at present

Public open space was a place to accommodate the activities of the general public in an open environment. Public open space consisted of complex elements: pedestrian paths, roads, and other pavement surfaces, and soft elements, namely plants and water, as softening elements in the landscape [36]. The areas in public open spaces, the pavement areas, were used for community activities. The placemaking approach used three public open spaces in Candi Baru elaborated on three elements, activity, form, and image [37].

A. Sudirman Park

Activities were carried out by one person or in groups in the park, and the surrounding area, such as sitting – eating in the food vendor area across the street near one side of the park. Relaxing, chatting, and enjoying the garden atmosphere. Even though there were no garden chairs, people sat on the edges of the park, on the stairs, and even on the pedestrian area under the leafy tree branches. Playing was carried out by children because there are swing facilities in both park areas, the North and the South side parks. While playing on the swing, children were accompanied by their parents or older people who usually sat around the playground. In the pavement area (small plaza), the older children generally played football in groups, and on weekends, there was a battery car rental for children about 3 - 5 years old to ride. The exercise was often done in the morning or evening. People did walk or jog around or through the park. Parking for motorbikes and cars. Because people from outside the neighborhood visited this park, they rode vehicles to go to the park. Besides that, there were traders selling food and drinks in parked cars or motorbikes. The vehicles of online transportation also occupied the parking area while waiting for orders to come in. Janitors carried out Park maintenance. They swept the gardens and picked up the trash. Temporary activities, for example, were public services organized by the Gajahmungkur District Office, Semarang Municipality. There were stands temporarily erected for public services on weekends and evenings. In addition, films by local filmmakers were also shown so that the public was interested in coming and taking advantage of the public service [38]. The forms of Sudirman Park were elongated and divided into two parts: the North and the South parks, which were separated by a street, as seen from the layout Figure. 3. The park's north side was by S. Parman street, and the south side was behind it. The areas inside the park had different heights. The south side park is higher than the north side, and the kid's playground was the highest area. The entrance to the north park is shown in Figure. 4a. The elevation was slightly higher than the street level. The entrance to the south park is shown in Figure 4b, marked by a pair of gates and the stairs like natural stones. Around the park were overgrown trees and shrubs that served as shade and aesthetic elements. On the south park were pedestrian paths placed around it. The pavement area was a small plaza, and a pair of stairs connected the playground to the plaza. Between those stairs laid a grandstand for seating covered by natural stone tiles. The seating area looked like a temple from a distance, as shown in Figure. 5a. On the north park, the pedestrian path inside the site was a linear pavement with fewer steps, 2-3 steps, shown in Figure

IOP Conf. Series: Earth and Environmental Science 1111 (2022) 012032 doi:10.1088/1755-1315/1111/1/012032

5b, side by side with the playground. The shape of Sudirman Park in the south was more varied than in the North, near S. Parman street. Image of Sudirman park as a leafy area because there were many trees and a place to enjoy local street foods. From online searching, it was more common to find discussions of Sudirman park as a culinary place with a leafy atmosphere because of the many trees and the other park facilities.

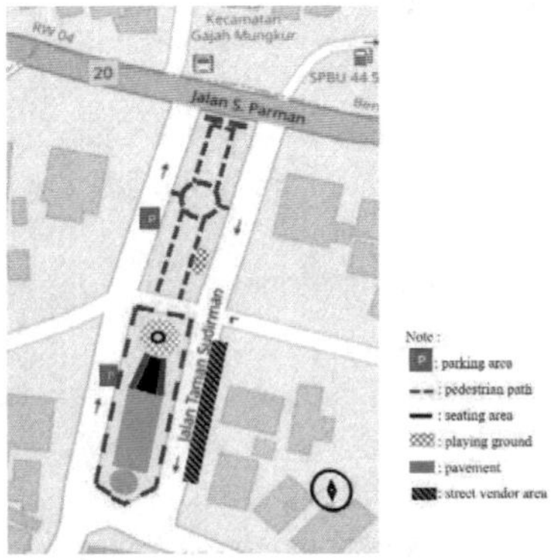

Figure 3. Layout of Sudirman Park
Source: Dinas Penataan Ruang Kota Semarang [33] developed by authors 2022

a b

Figure 4. Sudirman Park (a) The north park, (b) The south park
Source: authors, 2022

a b

Figure 5. (a) small plaza, stairs, and seating area; (b) pedestrian path and playground inside the park
Source: authors, 2022

B. Merapi Park

Activities carried out at Merapi Park were sitting and using the accessible WIFI facility or chatting, taking a rest, including waiting for the children playing in the playground. Sometimes a man was sleeping in this park. Playing for children: swings, slides, see-saws, and climbing playgrounds. Parking for cars and motorbikes. Park visitors were from outside of the neighborhood, so they rode vehicles and parked them. Parking activities were also carried out by online transportation drivers who rested and waited for orders to come in. Individual exercise, walking or jogging around the park or through the park. Volleyball was a sport carried out by a group of people from the settlement (urban kampong) behind Merapi street. Park maintenance by the janitor, who swept and collected garbage and dry leaves on site. When Friday prayers were held, and the number of people of a mosque, Merapi *Baiturrohim* Mosque, could not be accommodated at the mosque site, people used the pavement area in Merapi park.

The form of Merapi park was elongated and followed the contours of the landscape shown in the layout in figure 6. The park's shape decreased towards the south, and the area was more expansive than the front. The park was overgrown with large and leafy trees, as shown in figure 7a. Trees' roots that appeared on the ground and some were hanging add to the natural impression of this park. The seating area is shown in figure 7b. It was in the middle of the park, and the playground showed in figure 8b. At the southern end of the park, a paved open area is shown in figure 8a. There was a height difference between the park (green area) and the open area, so the stairs connected it, and there was a gate for the space identity. Paths for pedestrians around the park, mixed with the street on both sides of the park. The pedestrian path was a footpath in the playground and the paved open area. The image of Merapi Park was less well known than Sudirman Park. Its location was further from the main street, so this park was less visible. The park's appearance had a calmer and more relaxed vibe with large and leafy trees. However, the park gave a spooky impression at night because the neighborhood was quiet and dark.

a b

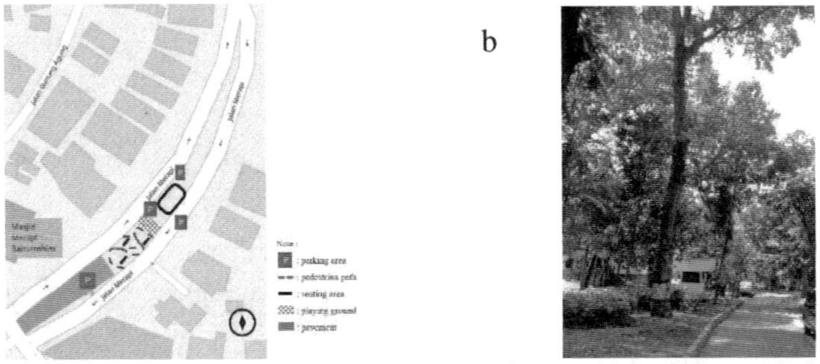

Figure 6. (a) Layout of Merapi Park (b) narrow, elongated, and descended following the landscape contour
Source: Dinas Penataan Ruang Kota Semarang [33] developed by authors 2022

a b

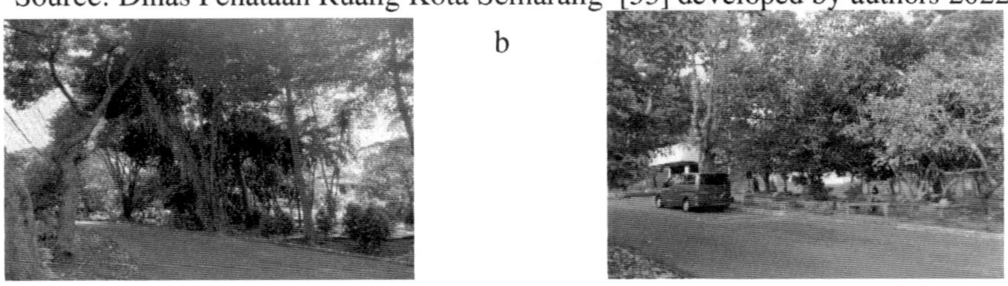

Figure 7. The Merapi Park (a) the big old trees (b) the seating area
Source: authors, 2022

a b

Figure 8. The Merapi Park (a) pavement area (b) the playground
Source: authors, 2022

C. Diponegoro Park

Activities at Diponegoro Park were sports, walking and jogging, sitting and relaxing, and park maintenance. Few activities were held in this park because the facilities and accessibility were limited. The park was in the middle of a busy main street. The location and layout of the park are shown in Figure 9. Most activities were sports, which were more commonly done in the morning before the traffic

IOP Conf. Series: Earth and Environmental Science 1111 (2022) 012032 doi:10.1088/1755-1315/1111/1/012032

volume increased. Meanwhile, sitting and taking a rest day and night were rarely carried out, as shown in Figure 10 a. The form of Diponegoro park extended into a street island. The elevation of the park is higher than the surrounding streets. The park was accessed from two sides. On each side, there were three spots of stairs to enter the park from the pedestrian path that surrounds the park. A row of *glodokan tiang* trees (*Polyalthia Longifolia*) rose high outside the park, as shown in figure 11a. There were two seating areas placed at both ends of it. In another seating area, as shown in Figure 11b, there were red poles as vocal points and aesthetic elements. The image of the park was as a protected public open space amid the crowds of traffic around it. The position of the park, which was higher than the street, and the giant trees on all sides resembling a closed urban forest are shown in figure 10 b.

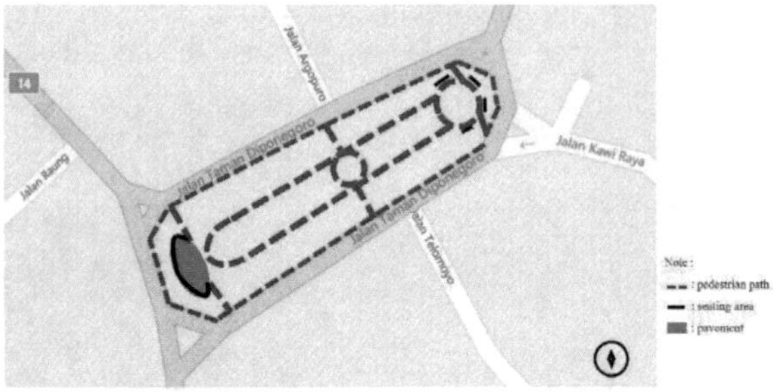

Figure 9. Layout of the Diponegoro Park
Source: Dinas Penataan Ruang Kota Semarang [33] developed by authors 2022

a

b

Figure 10. (a) chatting activity (b) outside view
Source: authors 2022

a

b

Figure 11. (a) the other outside view (b) another seating area inside the park
Source: authors 2022

3.3 Spatial justice in the usage of public open space

The basic understanding was that public space was a place for social relations between communities, individuals, or groups. The existing space was able to facilitate or hinder the relationships. Space was also the result of social interaction [39]. Spatial justice is discussed from the aspect of the availability of physical elements of public open space that are affordable, accessible, safe, and comfortable for all community groups. This meets the accessible and inclusive public open space [40].

The result of the observation from three parks was analyzed. The analysis assessed the use of physical elements of public open space with spatial justice. It is shown in Table 1.

Table 1. Physical Elements and Inclusivity of Three Parks

No	Parks	Social interaction	Physical elements	Affordable	Accessible	Safe	Inclusive
1	Sudirman	Yes	Pedestrian path	Yes	Yes	Yes	No
			Plaza	Yes	Yes	Yes	No
			Stairs	Yes	Yes	Yes	No
			Seating area	Yes	Yes	No	No
			Playground	Yes	Yes	No	No
			Vegetation	Yes	Yes	Yes	No
2	Merapi	Yes	Pedestrian path	Yes	No	No	No
			Plaza	Yes	Yes	Yes	No
			Stairs	-	-	-	-
			Seating area	Yes	Yes	No	No
			Playground	Yes	Yes	No	No
			Vegetation	Yes	Yes	No	No
3	Diponegoro	No	Pedestrian path	Yes	No	No	No
			Plaza	-	-	-	-
			Stairs	Yes	Yes	Yes	No
			Seating area	Yes	Yes	No	No
			Playground	-	-	-	-
			Vegetation	Yes	Yes	No	No

Source: authors, 2022

Sudirman and Merapi parks fulfilled the social interaction aspect, which meant that users of public open spaces could interact with each other in various activities without any barriers. Meanwhile, Diponegoro Park was considered not to meet the social interaction aspect because of the limited activities carried out in the space.

The physical elements assessed four characteristics: affordable, accessible, safe, and inclusive. The affordable nature existed for the three parks, which were no fees charged when visiting the park because the government managed the park. While the accessible and safe characters related to whether the physical elements were available and accessible to the public or not. At the same time, the safe character meant no interference or hazard in accessing and using these physical elements. Disturbances and hazards at Merapi Park were from the traffic of cars and motorbikes passing around the park, while on the outside of the park, there were no pedestrian paths. The number of pedestrian paths in the parking area was limited, the shape was uneven and not maintained, and the presence of tree roots spread on the ground, endangering visitors when walking. Disturbance or other hazards was from plants that were not maintained. The leaves were thick and became a place to live for ants, wasps, and other insects that disturbed and endangered visitors. The Diponegoro park was in the middle of the confluence of busy main roads, endangering visitors to access the park. The existence of a narrow pedestrian path, with a width of less than one meter which surrounded the outside of the park, gave a risk for pedestrians to be hit by vehicles that passed at high speed. One of the seating areas was full of branches and leaves from trees which could reduce the comfort and safety in the area.

Meanwhile, the inclusive character of the three public open spaces was not fulfilled yet. Although there were playground facilities for children in both parks, there were no safety aspects for children, such as hard, even rocky floors mixed with plant roots, and there were no barriers to protect children from vehicular traffic around the place. Meanwhile, facilities for the elderly and people with disability were not available, such as the stairs area without handrails, and there was no ramp to access higher or lower areas. The elevation of the pedestrian path surrounding the park from the highway had unequal heights, making it difficult to access the park. The distance between the park and the vehicle street was relatively close, so it was dangerous for children, the elderly, and people with disabilities.

4. Conclusion

Answering the discussion's aim, the public open space in Candi Baru was for recreational and social activities. Taman Sudirman has the most significant recreational and social elements from the placemaking approach. The Diponegoro Park, a big street island located among the busy city transportation routes, had low accessibility and high accident risk. So that Diponegoro park was the least fulfilled in the recreational and social elements. Meanwhile, Merapi Park required an arrangement for its physical elements so that the park could be used safely and optimally. Candi Baru was a residential area for the elite class; fortunately, its public open spaces were enjoyed by all economic classes of society. Spatial justice is concerned with walkable routes in an urban area [41]. However, the public open spaces in the Candi Baru area were hardly walkable, and the inclusive character was not fulfilled. There were no complete supporting facilities for children, the elderly, and people with disabilities. They needed help from others who are stronger physically to access and enjoy public open spaces. The existence of local food vendors selling in the Taman Sudirman area added to the attractiveness of the area but also affected the spatial justice in the area. The crowds of visitors and vehicles parked around the park hindered pedestrian accessibility. The image of Candi Baru as a Garden City area developed by Thomas Karsten was essential to improve its quality. Although the buildings in it were already turned into commercial functions, the area's morphology that responds to the environmental landscape should be maintained. It will equip the understanding of public open spaces in the colonial urban area, inclusive, safe, resilient, and sustainable. Thus, the next step would be to develop the physical elements of public open spaces that support spatial justice in the area.

Acknowledgment

We are grateful for the support from friends, colleagues, and lecturers on S3 Architectural and Urban Engineering (PDIAP), Engineering Faculty, Diponegoro University, and on Architecture Program, Faculty of Architecture and Design, Soegijapranata Catholic University. Natalia Suwarno supported the maps. Bayu Widiantoro gave the information. Hanung Habsoro and Hendri Teramulsetya assisted with the photographs. Eventually, all the persons in the locations were able to be interviewed.

References

[1] UN, "THE 17 GOALS | Sustainable Development," *United Nations*. https://sdgs.un.org/goals (accessed Oct. 16, 2021).

[2] P. Garau, *Global Public Space Toolkit: From Global Principles to Local Policies and Practice*. Nairobi: UN Habitat, 2015.

[3] Y. Liem and R. C. Lake, "The Meaning of Public Space of Kupang City Nostalgia Park," *ARTEKS J. Tek. Arsit.*, vol. 2, no. 2, pp. 149–158, Jun. 2018, doi: 10.30822/arteks.v2i1.48.

[4] M. D. N. Kiswari and R. Tarigan, "Identifikasi Fungsi Pada Ruang Publik di Perumahan Mlaten," in *Prosiding Temu Ilmiah IPLBI 2016*, 2016, p. G 033-040, doi: ISBN 978-602-17090-1-6.

[5] E. F. Porajouw, R. J. Poluan, and F. Mastutie, "Efektivitas Ruang Terbuka Publik di Kota Tomohon," *Spasial*, 2017, [Online]. Available: https://ejournal.unsrat.ac.id/index.php/spasial/article/view/15546.

[6] D. Arlistasari and H. Weni Rosdiana, "EVALUASI PENATAAN RUANG TERBUKA HIJAU DI KAWASAN KOTA KECAMATAN KEBOMAS KABUPATEN GRESIK," *Publika*, vol. 7, no. 2, Mar. 2019, Accessed: Jul. 15, 2020. [Online]. Available: https://jurnalmahasiswa.unesa.ac.id/index.php/publika/article/view/27116.

[7] J. Hasim and A. O. Samili, "EFEKTIFITAS PEMANFAATAN RUANG PUBLIK MELALUI RUANG TERBUKA HIJAU (RTH) DI KOTA TERNATE," *J. Geocivic*, vol. 4, no. 1, 2021, doi: http://dx.doi.org/10.33387/geocivic.v4i1.3061.

[8] Q. Aini, "Evaluation of Accessibility for People with Disability in Public Open Space," *IOP Conference Series: Materials Science and Engineering*, vol. 506, no. 1. 2019, doi: 10.1088/1757-899X/506/1/012018.

[9] I. Sudarisman, "Kajian Pedagang Kaki Lima Di Taman Tegalega, Bandung, Jawa Barat," *J. Arsir*, vol. 1, no. 2, pp. 161–174, 2017, doi: 10.32502/arsir.v0i0.3647.
[10] D. Hantono and D. Pramitasari, "Aspek perilaku manusia sebagai makhluk individu dan sosial pada ruang terbuka publik," *Nat. Natl. Acad. ...*, vol. 5, no. 2, pp. 85–93, 2018, doi: https://doi.org/10.24252/nature.v5i2a1.
[11] J. Tio, *Kota Semarang Dalam Kenangan*. Semarang, 2002.
[12] R. S. Rukayah and T. Roesmanto, "The Sustainability Concept of Alun-alun as a Model of Urban Design in the Future," *Procedia-Social Behav. Sci.*, 2013, doi: https://doi.org/10.1016/j.sbspro.2013.08.390.
[13] Y. Wisnu, "Candi Baru, Jejak Orang Kaya di Semarang Tempo Dulu," *Solopos.com*, 2022. https://www.solopos.com/candi-baru-jejak-orang-kaya-di-semarang-tempo-dulu-1251986 (accessed Aug. 09, 2022).
[14] W. Kurniawati, M. Apriana, N. Silvia, R. Dewi, and A. Febrianto, "Penerapan Konsep Garden City di Kawasan Candi Baru Semarang," *J. Arsit. Zo.*, vol. 4, no. 3, pp. 320–326, 2021, doi: doi.org/10.17509/jaz.v4i3.34827.
[15] A. S. Muljadianta, Antariksa, and P. Salura, "The Application of Local Concepts on Herman Thomas Karsten's Town Planning," *Psychol. Educ. J.*, vol. 58, no. 2, pp. 930–939, 2021, doi: 10.17762/pae.v58i2.2032.
[16] A. S. Muljadinata and A. Ardianto, "Lokalitas Pada Tatanan Kawasan Candi Baru Semarang," *Tesa Arsit.*, vol. 19, no. 2, pp. 75–84, 2022, doi: 10.24167/tesa.v19i2.3699.
[17] K. Wulansari, "Evolusi Konsep Ruang Hijau Publik di Kota Semarang pada Awal Abad ke 20 Hingga Sekarang (Ruang Hijau Publik di Kawasan Candi Baru)," *J. Pembang. Wil. Kota*, vol. 11, no. 1, p. 1, 2015, doi: 10.14710/pwk.v11i1.8653.
[18] R. F. Jamila, "Preferensi masyarakat terhadap Taman Sudirman Semarang sebagai playground," *Vitruvian*, vol. 5, no. 1, pp. 50–57, 2015.
[19] S. Kneeshaw, "Gender sensitive public space? Placemaking and spatial justice through the perspective of gender," *URBACT*, 2019. https://urbact.eu/gender-sensitive-public-space-placemaking-and-spatial-justice-through-perspective-gender (accessed Jul. 28, 2020).
[20] M. A. Yusuf, *Metode Penelitian Kuantitatif, Kualitatif dan Penelitian Gabungan*, Ke-4. San Fransisco: KENCANA, 1998.
[21] P. Kristjanson, W. Bank, and C. Jost, "CCAFS Gender Training and Strategizing Workshop Report. CGIAR Research Program on Climate Change, Agriculture and Food Security," 2013. [Online]. Available: https://www.researchgate.net/publication/263587815.
[22] Project for Public Spaces, "What is Placemaking?," *Project for Public Spaces*, 2020. https://www.pps.org/article/what-is-placemaking (accessed Aug. 12, 2022).
[23] K. Lynch, *A Theory of Good City Form*, vol. 15, no. 2. Cambridge: The Massachusetts Institute of Technology Press, 1981.
[24] E. Soja, *Seeking Spacial Justice*, vol. 37 nro.111. 2011.
[25] I. Y. Jian, E. H. W. Chan, Y. Xu, and E. K. Owusu, "Inclusive public open space for all: Spatial justice with health considerations," *Habitat Int.*, vol. 118, no. March, p. 102457, 2021, doi: 10.1016/j.habitatint.2021.102457.
[26] I. Y. Jian, "Spatial justice in public open space planning: Accessibility and inclusivity," *Habitat Int.*, vol. 97, 2020, doi: 10.1016/j.habitatint.2020.102122.
[27] A. S. Muljadinata, "CITY'S ARCHITECTURAL CONSERVATION IN SEMARANG," in *International Conference Urban Heritage and Sustainable Infrastructure Development 5*, 2016, vol. 5, p. 271, [Online]. Available: https://www.researchgate.net/profile/Eko-Nursanty/publication/318721495_Book_of_Proceedings_UHSID5_2016_International_Conference_International_Conference_Urban_Heritage_and_Sustainable_Infrastructure_Development/links/5989c2c0a6fdcc7562637f8a/Book-of-Proc.
[28] H. J. Essup, "Dutch architectural visions of the Indonesian tradition," *Muqarnas*, vol. 3, pp. 138–161, 1985, doi: 10.1163/22118993-90000202.

[29] J. L. Cobban, "Exporting planning: The work of Thomas Karsten in colonial Indonesia," *Plan. Perspect.*, vol. 7, no. 3, pp. 329–344, 1992, doi: 10.1080/02665439208725753.
[30] B. Setyohadi K, "Tipologi Pola Spasial dan Segregasi Sosial Permukiman Candi Baru," *J. Tek. SIPIL Perenc.*, vol. 9, no. 2, pp. 97–106, 2007, doi: https://doi.org/10.15294/jtsp.v9i2.1612.
[31] V. Dyah P and N. Yuliastuti, "Penilaian Keberlanjutan Permukiman Kampung Lama di Kelurahan Lempongsari," *Tek. PWK (Perencanaan Wil. Kota)*, vol. 3, no. 4, pp. 766–775, 2014, [Online]. Available: https://ejournal3.undip.ac.id/index.php/pwk/article/view/6734.
[32] R. S. Wijono, "Public housing in Semarang and the modernization of kampongs, 1930–1960," in *Cars, Conduits, and Kampongs*, C. Freek and J. Coté, Eds. Brill, 2015, pp. 172–192.
[33] Dinas Penataan Ruang Kota Semarang, "Dinas Penataan Ruang Kota Semarang," 2020. https://distaru.semarangkota.go.id/semarang/ (accessed Aug. 14, 2022).
[34] A. Sidharta Muljadinata, A. Antariksa, and P. Salura, "The role of localities in karsten's works in architecture and city of Semarang," *IOP Conf. Ser. Earth Environ. Sci.*, vol. 126, no. 1, 2018, doi: 10.1088/1755-1315/126/1/012007.
[35] DISPERKIM Kota Semarang, "Taman Kota Semarang," *DISPERKIM Kota Semarang*, 2020. http://disperkim.semarangkota.go.id/index.php/post/read/142/Profil-Taman-Kota-Semarang (accessed Feb. 12, 2022).
[36] E. Budihardjo and D. Sujarto, *Kota Berkelanjutan (Sustainable City)*, Cetakan ke., vol. 1999, no. December. Bandung: PT Alumni, 2013.
[37] J. Montgomery, "Making a city: urbanity, vitality and urban design," *J. Urban Des.*, vol. 3, no. 1, pp. 93–116, 1998, doi: 10.1080/13574809808724418.
[38] A. Firhannusa, "Buka di Taman Sudirman, Pelayan Publik Kecamatan Gajahmungkur Alami Peningkatan," *AYO SEMARAANG.COM*, 2022. https://www.ayosemarang.com/semarang-raya/pr-774089899/buka-di-taman-sudirman-pelayanan-publik-kecamatan-gajahmungkur-alami-peningkatan.
[39] H. Molotch, "The Space of Lefebvre," *Theory Soc.*, vol. 22, no. 6, pp. 887–895, 1993.
[40] I. Y. Jian, J. Luo, and E. H. W. Chan, "Spatial justice in public open space planning: Accessibility and inclusivity," *Habitat Int.*, vol. 97, no. January, p. 102122, 2020, doi: 10.1016/j.habitatint.2020.102122.
[41] I. Y. Jian, E. H. W. Chan, M. J. Luo, T. Y. Yao, and ..., "A spatial justice study of pedestrian-friendly routes to privately owned public open space in Hong Kong," *ISUF 2020 Virtual ...*. 128.110.184.59, 2021, [Online]. Available: https://128.110.184.59/index.php/ISUF2020/article/download/4188/3570.

IOP Conf. Series: Earth and Environmental Science 1111 (2022) 012033 doi:10.1088/1755-1315/1111/1/012033

The Drivers of Food Loss at the Farm Level: A Study of Farmers in Salatiga, Central Java, Indonesia

Liska Simamora[1], Damara Dinda Nirmalasari Zebua[2], Widhi Handayani[3*]
[1] Faculty of Agriculture and Business, Universitas Kristen Satya Wacana, Salatiga, 50711, Indonesia, ORCID ID: 0000-0002-3830-040X
[2] Faculty of Agriculture and Business, Universitas Kristen Satya Wacana, Salatiga, 50711, Indonesia, ORCID ID: 0000-0003-1038-2332
[3] Faculty of Interdisciplinary Studies,Universitas Kristen Satya Wacana, Salatiga, 50711, Indonesia, ORCID ID: 0000-0002-3242-5535
[1]liska.simamora@uksw.edu, [2]damaradinda.zebua@uksw.edu,
[3*]widhi.handayani@uksw.edu

Abstract. Food loss and waste (FLW) contribute to adverse economic and environmental impacts. In Indonesia, the agriculture sector is the primary source of income for 38,78 million people, and food loss is a significant issue. At the farm level, food loss falls into two categories: food that is never harvested and food that is lost between the point of harvest and sale. The term "food loss" is frequently used to refer to agricultural production that is lost unintentionally. This research aims to identify the drivers of food loss at the farm level in Indonesia using a case study of farmers in Salatiga City, Central Java Province of Indonesia. The farmers' survey was distributed in Salatiga's most significant urban village (Sidorejo Lor), with n=22 farmers. Using descriptive quantitative analysis, the study found that farmers are not well informed about the issue of food loss. The food loss drivers at the farm level include the presence of pests and disease, human error during manual harvesting, and gaps in appropriate harvesting technology, which results in substantial crop loss.

1. Introduction

Food loss and waste (FLW) have many negative economic and environmental impacts. Economically, FLW represents a wasted investment that reduces farmers' incomes and increases consumers' expenses. Environmentally, food loss and waste inflict a host of impacts, including unnecessary greenhouse gas emissions and inefficient water and land use, leading to diminished natural ecosystems and disruption in the services they provide[1]. Food loss is the decrease in the quantity or quality of food resulting from decisions and actions by food suppliers in the chain, excluding retail, food service providers, and consumers. Food waste is the decrease in the quantity or quality of food resulting from decisions and actions by retailers, food services, and consumers [2].

At the farm level, food loss falls into two categories: food that is never harvested and food that is lost between harvest and sale[3–7]. The term "food loss" is frequently used to refer to agricultural production lost unintentionally because of various factors, including market conditions, poor infrastructure, poor agricultural practices, pests, disease, natural disasters, and weather events. The drivers of food loss at the farm level could be divided into two groups, namely direct and indirect drivers [7]. Direct drivers come from (1) biological and environmental factors: pest and disease, weather, climate, soil, and water

(2) technology and infrastructure factors: harvest and post-harvest technologies, storage, container, packaging, and infrastructure connection to the market. Indirect drivers come from (1) market structure, governance, investment, and trade: regulation and standard, access to finance, contractual arrangements, and (2) human factor: culture of land ownership, training and outreach, and labor (availability, quality, cost).

Wasted food data is provided, such as the estimation that around 931 million tonnes of food waste were generated in 2019, 61% of which came from households, 26% from food service, and 13% from retail, but the food loss data is not found [8]. Consumer food waste is expensive and environmentally damaging as it travels the length of the supply chain and ends up in landfills. Most research and campaigns emphasize the consumer level, which has resulted in the omission of data collection and the development of solutions for producers of fruit and vegetable crops [6]. On average, approximately 65% of the unharvested crop that remained in the field was of wholesome, edible quality, although the appearance may not meet buyers' specifications for specific markets [9]. Another reason for farm-level losses is overproduction due to power imbalances, gaps in infrastructure, rejected produce due to stringent aesthetic values, precarious labor, and economic and environmental [10].

Sustainable Development Goal Indicator 12.3 states, "By 2030, to halve per capita global food waste at the retail and consumer levels and reduce food losses along production and supply chains, including post-harvest losses". This can be achieved if the food loss volume is well-quantified. Reducing FLW is key to making food systems significantly more sustainable [11]. It is key to improving farmers' economic well-being and the sustainable management of water, energy, labor, and other resources [10].

Indonesia relies on agriculture as a sector, with the Central Bureau for Statistics Indonesia reporting that as many as 38.78 million people, or around 29.59% of the population, work in the agriculture, forestry, and fishery sectors [12]. The agricultural sector in Indonesia contributes 13.28% to the national GDP[13]. Despite being a major sector, Agriculture in Indonesia have the challenge of coping with the smallholder farmers. The number of farm households by farm size in 2018 is 27.682. 117 farms consisted of 16.257.430 (58,72%) farms categorized as smallholder farmers with the land size < 0,5 ha [14]. Indonesia is the second largest FLW-producing country globally, reaching 115-184 kg per capita per year [15].

The Indonesian Ministry of National Development Planning report categorized the FLW drivers into two groups, namely direct cause (D) and indirect driver (I) [15]. A study on food waste has been conducted in Indonesia, which is related to the management of food waste [16], recommendations on food waste control [17], and policies and strategies to reduce food loss and waste in Indonesia [18]. However, the study is minimal in terms of food loss and waste, particularly in Salatiga. The study on waste is mainly related to the management of municipal solid waste [19] instead of food waste specifically, while the study on food loss has not yet been found. More studies are required to provide baseline data on food loss and waste in Salatiga. In addition, research in Indonesia place end consumers (households) as the main contributors to the accumulation of food waste [20], culinary business [21], [22], whereas according to [23], the number of food loss and waste generated from vegetables (31%) followed by rice (20%), meat (11%), milk product (10%) and fish (10%).

The type of food wasted addresses this research to identify the reasons for vegetables wasted. It is assumed that this is caused not only by consumer behavior but also by agricultural practice. Therefore, this paper will identify the reasons behind food loss at the farm level and explore the barriers and opportunities to reduce food loss. As previously explained, this study will provide data on food loss in Salatiga at the farm level, which the government could use to conduct further research on the topics. In the long term, this research would bring opportunity to the stakeholders in formulating policy to reduce food loss at the farm level and reduce food loss and waste in Salatiga.

2. Methods

This study was conducted at Salatiga City, Central Java Province of Indonesia, using a farmer survey questionnaire. The survey was conducted in Sidorejo Lor, the largest urban village in Salatiga. According to the Government of Sidorejo Lor Urban Village, Sidorejo Lor is located in the Sidorejo sub-district, with an area of 271.600 hectare, of which 32.768 hectare is used for paddy cultivation. 14,390 residents of this urban village spread over 16 *Rukun Warga* or hamlets. Among the residents, there are 3,170 unemployed; 2,555 people employed in the private sector; 2,497 people work as students in schools or universities; 1,896 people are housekeepers; 1,266 people who work as entrepreneurs, and 1,001 people work as daily workers.

Figure 1. The geographical map of the Sidorejo sub-district and its urban villages, including the Sidorejo Lor (Source: https://sidorejo.salatiga.go.id/kondisi-geografis/)

There are three farmer groups in this urban village, i.e., Sri Sadani, Tunas Mekar, and Subur Makmur, with varying membership for each group, with 25 members in total based on the data gathered from the Food Crop and Agriculture Office of Salatiga. The survey was conducted from May to July 2022 by distributing questionnaires to the farmers of all groups, i.e., 22 respondents of 25 farmers. The questionnaires were completed through one-on-one interviews for about 30 minutes to 1 hour. The questions covered the following main themes: (1) the agricultural commodities cultivated and the production scale; (2) farmers' understanding of the issue of food loss; (3) and the drivers of food loss at the farmer level. Following data collection is tabulation and data analysis using descriptive quantitative analysis. This study works with a small number of farmers, which were only 25 farmers according to the data gathered from the Food Crop and Agriculture Office of Salatiga, while based on the data from the Government of Sidorejo Lor Urban Village, 24 farmers live in the area. Based on those two sources, there is a bit of difference in the number of farmers. However, this indicates that less than 30 farmers live in this urban village, which is the limitation of the study.

3. Results and Discussion

3.1. General description of respondents and cultivated commodities

Most of the respondents are elderly (nonproductive age), which is around 58 years. The eldest is 79 years old, while the youngest is 27. As of January 2022, the retirement age of Indonesians is announced to be 58 years old by *Badan Penyelenggara Jaminan Sosial Ketenagakerjaan* (National Social Security Agency for Employment). Only nine farmers are of productive age, while the rest are of nonproductive age. Furthermore, all respondents were male paddy field farmers whose most (12 respondents) experienced primary school as their formal education, five graduated from senior high school, three graduated from junior high school, and one did not experience formal education. Those farmers are workers instead of the owner of their land. Only one respondent graduated from university and is also the paddy field owner.

Out of 22 respondents, 21 farmers plant paddy as their main cultivated crop, although some also plant other commodities such as maize, peanuts, banana, sweet potato, and taro. However, these commodities are usually grown to fill some areas that could not be possibly planted by paddy. Therefore, the farmers do not put much effort and concern into how much the harvest of those commodities. Their main concern and efforts are put into their paddy production. Among all respondents, only one cultivates decorative plants instead of a common commodity, i.e., paddy. This is mainly related to the fact that the respondent has only an area of 150 m^2, which is challenging to be planted by paddy. Therefore, instead of cultivating paddy, he tends to cultivate decorative plants to earn income to fulfill his household's daily needs.

3.2. The understanding of farmers on food loss and waste

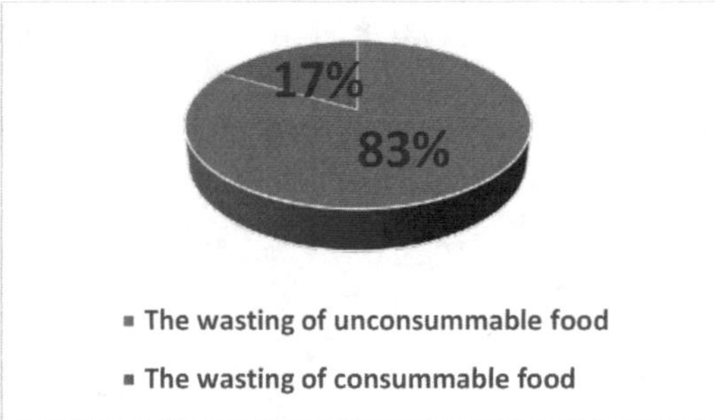

Figure 2. Definition of food loss and waste according to respondents

Figure 2 shows that among all respondents, 83% define food waste as food that is not fit for consumption, while the rest also consider edible food to be food waste if it is not consumed. This understanding might relate to the social-economic situation of farmers' households who will not dispose of edible food, except if the food is not fit to consume. Hence, they will throw it away. This finding is different from previous research found that education significantly impacts food waste production, in terms of the higher the education level, the lower the possibility of people disposing of foods [24]. Furthermore, based on our collected data, not all respondents dispose of non-edible food, as some process these foods into compost that can be used for fertilizer. This finding is similar to that reported farmer households in Klaten process non-edible food as animal feed or organic fertilizer [24].

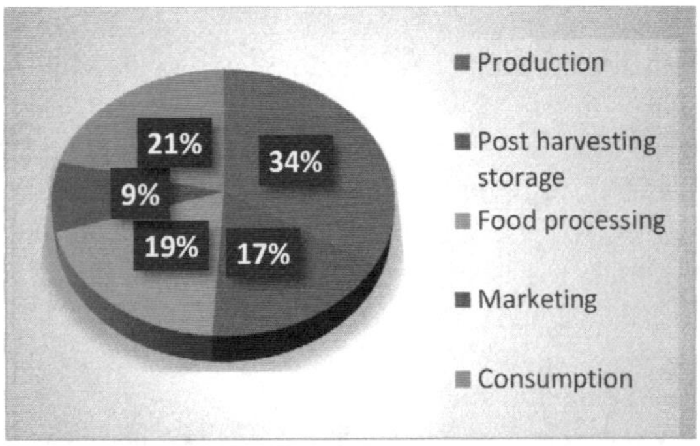

Figure 3. Sources of food loss and waste

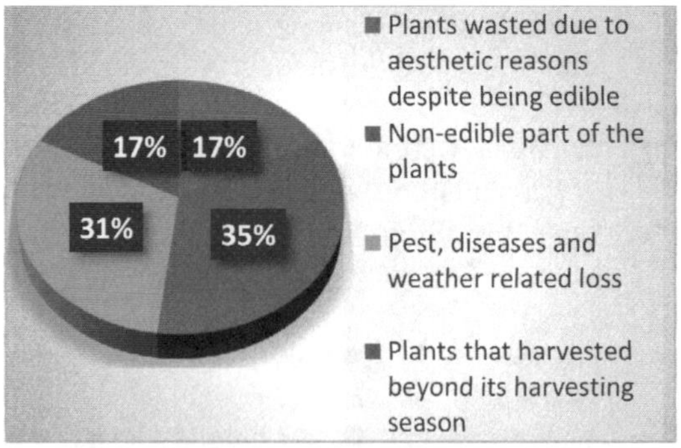

Figure 4. Drivers of food loss at the farm level

During the study, respondents were expected to show their opinion regarding food loss in some stages, i.e., from production to consumption. This means that respondents will give answers based on their understanding of each stage and experience. Based on Figure 3, according to respondent opinion, the highest source of food loss and waste was in the production stage, accounting for 34% above other stages. This might relate to their experience, particularly how they experience food loss during harvest time.

During post-harvesting storage, the food loss might relate to the decrease in the quality of the grains during storage, for example, because of mold, etcetera. Since the farmers have experience in food processing, such as cooking and consuming the foods, it is understandable that they also respond to these stages. It was found that only 9% of the respondents, or 4 out of 22 respondents, chose that marketing is also a stage that contributes to food loss. This might relate to not all farmers selling their rice directly to consumers. Some farmers sell their harvest to intermediaries, while there is also a farmer who sells their grains to the rice mill. Other farmers prefer to save their rice from fulfilling their daily needs instead of selling their grains to other parties.

In addition, 35% of the respondents agree that food loss in their field is caused by the plants' non-edible part that becomes waste (Figure 4). Moreover, 31% of respondents also identified that food loss is caused due to pests and diseases. However, this study has not yet calculated the volume of rice loss on the farm. A previous study indicates the volume of rice loss during harvesting time in Karawang, i.e., 298.26 kg/hectare of 5,962.56 kg/hectare of harvested dry grain (*Gabah Kering Panen*) or equal to 5% of total dry grain harvested [24]. Another study in Jombang found that rice grain loss during harvesting time was 7.2% of total harvested dry grain, while in collecting, threshing, and drying were 1.27%, 1.74%, and 2.48%, respectively, and in total, the loss was 12.7% [20].

Based on those studies, total rice grain loss during harvesting time is 5.0% – 7.2%, or 6.1% on average. Meanwhile, only 17% of the respondents agree that plants harvested beyond their harvesting season could lead to food loss (Figure 4). It is possible that the farmers are not accustomed to harvesting the plants beyond the harvesting time, or they do not yet experience this. Therefore, only 18% of respondents chose this option. Proper harvesting, in terms of how it is conducted and at the right season, will influence food loss [24]. However, another study found that harvesting time does not significantly affect food loss; instead, some factors that significantly affect food loss were field area, farmers' behavior, variety of paddy, and threshing tools [25]. Technology's importance in reducing food loss, particularly for chili and rice, has been reported [18]. Furthermore, in the case of consumer FLW, some efforts will be required to change the behavior of consumers, including intensive and massive education, socialization, and communication among stakeholders [18].

Table 1. The yield of rice grain of respondents

Respondent number	Field area (m^2)	Yield per harvesting season (kg)
1	1,800	700
2	10,000	200
3	6,800	2500
4	5,000	900
5	8,000	1500
6	3,000	1300
7	2,500	800
8	2,500	2000
9	10,000	6000
10	1,800	600
11	5,000	800
12	7,000	900
13	1,700	n.d
14	5,000	1700
15	2,000	800
16	2,500	1000
17	1,600	700
18	5,000	1200
19	900	500
20	2,500	1400
21	3,000	1,000

Source: Primary Data (2022)

Based on Table 1, it can be calculated that the average field area is 4,557 m^2 or 0,456 hectare, while the average yield is 1,325 kg per harvesting season or equal to 2,905.70 kg/hectare. By applying the 6.1% loss of rice grain, it can be calculated that the rice grain loss will equal 177.25 kg/hectare. However, further research will be required to confirm the calculated result.

All respondents agree that the main challenge to paddy cultivation is pests and diseases, which are connected to the influence of climate. Particularly in this era of climate change, farmers' traditional knowledge, which has been used to detect seasonal variations accurately, is now becoming irrelevant. The farmers in this study face difficulties in dealing with pests and diseases.

The situation and the experience of the farmers, who predominantly suffer from food loss due to pests and disease attacks, have formed a mindset that food loss is unavoidable. Therefore they tried to accept this reality as part of their life. Finally, they consider that the loss of grains is an ordinary situation usually faced by the farmers.

Their yield loss might correspond to the technology they used to reap the grains. All the farmers use the sickle to thresh the grains instead of using a threshing machine because the land contour in the area is unsuitable for a conventional threshing machine. However, using the sickle might influence yield loss [26]. The use of sickle causes the grains to be scattered on the ground and, therefore, could not be picked by farmers. It is also indicated that if power threshers can be changed to utilize combined harvesters, food loss could be reduced. This method could reduce or minimize the loss of grains by 3.593 points [25]. However, based on the interview, the farmers stated that it is possible to reduce loss by preparing a broader space to collect the grains traditionally.

4. Conclusion

The study found that farmers in Salatiga were not well informed about the definition of food loss and waste.-The findings indicate that the driving factors for the occurrence of food loss at the farm level are mainly caused by inedible plant parts; followed by pest and disease attacks causing plants not to produce edible fruit; the agricultural products pass the harvesting period, and quality display of agricultural products do not meet consumer demand. Minimizing food loss at the farm level is quite difficult for farmers who have the mindset that the presence of pests and disease lead to the creation of non-edible plants. This study leads the farmers to a new comprehension that food loss at the farm level creates a loss in income. Thus, farmers will improve farming techniques so that the crops are produced to meet the quality consumers expect. This study has a weakness, namely only collecting information from farmers who cultivate rice plants, so it is not representative of all types of farmers who cultivate crops that produce food. Therefore, further research is expected to examine food loss at the level of farmers who cultivate horticultural crops.

Acknowledgments

The authors are grateful to acknowledge The Ministry of Education, Culture, Research and Technology of the Republic of Indonesia (Kementerian Pendidikan, Kebudayaan, Riset, dan Teknologi Republik Indonesia) who support this research by Hibah Penelitian Dasar Kompetitif Nasional with contract no 001/LL6/PB/AK.04/2022 and 173/SPK-PDKN/PR V/5/2022. The authors would also like to acknowledge Dr. Tammara Soma from the School of Resource and Environmental Management, Simon Fraser University, who brings precious insights to the writing of this paper.

References

[1] Lipinski B, Hanson C, Lomax J, Kitinoja L and Waite R 2013 *Reducing Food Loss and Waste* (Washington, DC)

[2] Food and Agriculture 2019 *The State of Food and Agriculture. Moving Forward on Food Loss and Waste Reduction*

[3] Gunders D 2012 Wasted: How America is Losing Up to 40 Percent of Its Food from Farm to Fork to Landfill *NRDC Issue Pap.* 1–26

[4] Gustavsson J, Cederberg C, Sonesson U, Otterdijk R van and Meybeck A 2019 *Food Loss and Food Waste: Extent, Causes and Prevention* (Rome: FAO)
[5] Food and Agriculture Organization 2013 *Food Wastage Footprint*
[6] Johnson L K, Dunning R D, Gunter C C, Dara Bloom J, Boyette M D and Creamer N G 2018 Field measurement in vegetable crops indicates need for reevaluation of on-farm food loss estimates in North America *Agric. Syst.* **167** 136–42
[7] Julian P, Anna B and Tim C 2021 Driven to Waste: The Global Impact of Food Loss and Waste on Farms *WWF-UK*
[8] Forbes H, Quested T and O'Connor C 2021 Food Waste Index Report 2021 *The UNEP Food Waste Index Report* pp 3–99
[9] Johnson L K, Dunning R D, Bloom J D, Gunter C C, Boyette M D and Creamer N G 2018 Estimating On-Farm Food Loss at The Field Level: A Methodology and Applied Case Study on a North Carolina Farm *Resour. Conserv. Recycle.* **137** 243–50
[10] Soma T, Kozhikode R and Krishnan R 2021 Tilling food under: Barriers and opportunities to address the loss of edible food at the farm-level in British Columbia, Canada *Resour. Conserv. Recycle.* **170** 105571
[11] Sethi G, Bedregal L P A, Cassou E, Constantino L, Hou X, Jain S, Messent F, Morales X Z, Mostafa I and Pascual J C G 2020 *Addressing Food Loss and Waste : A Global Problem with Local Solutions* vol 34 (Washington D.C)
[12] Badan Pusat Statistik 2020 *Statistik Indonesia 2020* (Jakarta)
[13] Bandan Pusat Statistik 2021 *Pertumbuhan Ekonomi Indonesia 2021* (Jakarta)
[14] Kurniawan E, Brunori G and Prosperi P 2020 Small Family Farms; A Perspective from Indonesia, Challenges and Investment
[15] Ministry of National Development Planning of the Republic of Indonesia 2021 Executive Summary for Policy Makers: Food Loss & Waste in Indonesia 20
[16] Hadiningrat G 2020 Women's Role in Food Waste Management in Indonesia (Study Case in Bandung) **31** 31–5
[17] Rachman I and Septiana A I 2020 Food Waste Control Recommendations in Indonesia Based on Public Opinion Related To the Target Sdgs *J. Community Based Environ. Eng. Manag.* **4** 25–30
[18] Saliem H P, Mardianto S, Samedi, Suryani E and Widayanti S M 2021 Policies and Strategies for Reducing Food Loss and Waste in Indonesia *IOP Conf. Ser. Earth Environ. Sci.* **892**
[19] Budiana M N and Maryono 2017 Inisiatif Standarisasi Waktu Pengumpulan Sampah Di Kota Salatiga *J. Pembang. Wil. Kota* **13** 353–67
[20] Chaerul M and Zatadini S U 2020 Perilaku Membuang Sampah Makanan dan Pengelolaan Sampah Makanan di Berbagai Negara: Review *J. Ilmu Lingkung.* **18** 455–66
[21] Wulansari D, Ekayani M and Karlinasari L 2014 Kajian Timbulan Sampah Makanan Warung Makan *Ecotrophic* 125–34
[22] Brigita G and Rahardyan B 2013 Analisa Pengelolaan Sampah Makanan Di Kota Bandung *J. Teh. Linking.* **19** 34–45
[23] Rizaty M A 2021 Indonesia Kebanjiran Sampah Makanan
[24] Saputro W A, Purnomo S and Salamah U 2021 Study of Food Waste of Farmers' Households in Klaten to Support Food Security *Anjoro Int. J. Agric. Bus.* **2** 58–64
[25] Hidayat S I, Parsudi S and Putri G L A M 2021 Komoditas Padi: Telaah Kehilangan Hasil Saat Panen Di Kabupaten Jombang *Mimb. Agribisnis J. Pemikir. Masy. Ilm. Berwawasan Agribisnis* **7** 577
[26] Hastuti, Syaukat Y, Hardjanto A, Raswatie F D, Amanda D, Nasrullah N and Falatehan A F 2021 Analisis Volume dan Nilai Kerugian dari Food Loss Komoditas Beras di Kabupaten Karawang *J. Manaj. Agribisnis* **9** 315–29

Monitoring 20 Years of Land Cover Change Dynamics in The Satellite Cities of Jakarta, Indonesia

Tia Rizka Nuzula Rachma[1,2]*, Florence Elfriede Sinthauli Silalahi[3] and Nadya Oktaviani[4]
[1]Geospatial Information Agency, Bogor, 16911, Indonesia, 0000-0001-6523-2003
[2]School of Environmental Science, University of Indonesia, Jakarta, 10430, Indonesia
[3]Geospatial Research Center, Research Organization of Earth and Maritime, National Research and Innovation Agency, Bogor, 16912, Indonesia, 0000-0001-9817-2190
[4]Geospatial Information Agency, Bogor, 16911, Indonesia. 0000-0003-4563-2071
[1]*tia.rizka@big.go.id, [3]flor006@brin.go.id, [4]nadya.oktaviani@big.go.id

Abstract. As the capital city of Indonesia, Jakarta has eight satellite cities as the supporting areas, namely Bogor Regency, Bogor City, Depok City, Tangerang City, Tangerang Regency, South Tangerang City, Bekasi City, and Bekasi Regency. The rapid urbanization in Jakarta is driving the growth of these areas. One of the impacts of urban growth is the land cover change that can be observed through multitemporal satellite imagery. This study analyses the land cover change within 20 years in urban and non-urban areas. The data used are Landsat satellite imagery and Indonesian base maps validated by ESRI World Imagery. In this study, satellite imagery is processed into predetermined classes. The analysis process by comparing land cover classes between 2001 and 2021 and calculating the area of each class in each regency/city. The results show that dense and evenly distributed urban growth occurs throughout the "City" area. Attention must be given to the government of Bekasi City, Tangerang City, South Tangerang City, and Depok City because the built area already exceeds 70% of the city area.

1. Introduction

Around the world, urbanization is placing significant pressure on ecosystems. In general, urbanization threatens the loss of biodiversity, such as the loss of most bird species [1]. Urbanization is also one of the factors causing the loss of wetland habitat in Bons Sinais Estuary, Mozambique [2]. Land clearing for plantation areas has been the cause of the extinction of endemic species in Costa Rica [3]. Although some species (such as birds) have resilience in the face of habitat loss, land changes affect the extinction of local species [4]. Outside urban areas, some species have higher and different functional diversity [5].

Several previous studies have been conducted to analyze the phenomenon of urbanization through the analysis of land cover changes [6–10]. One method that can be used is comparing multi-temporal satellite imagery [11,12].

In Indonesia, there is rapid urbanization in the capital and surrounding areas. Jakarta, as the center of government and the center of the Indonesian economy, is getting support from Bogor, Depok, Tangerang, and Bekasi (Bodetabek) [13]. The Greater Jakarta area (Jakarta, Bogor, Depok, Tangerang, Bekasi) is now the most significant urban center in Indonesia [10]. Urban development is concentrated in suburban metropolitan areas [14]. Cities in Bodetabek have grown to become part of the ten largest cities in Indonesia, even though they are not provincial capitals [15].

Concerning the urbanization of the Jabodetabek area, the Indonesian government issued Presidential Regulation Numbers 6 of 2020 concerning Spatial Planning for the Urban Areas of Jakarta, Bogor, Depok, Tangerang, Bekasi, Puncak, and Cianjur. The regulation states that each regency/city must have green open space for 30% of its area. Green Open Space is an elongated or clustered area as a place to grow plants, both naturally and intentionally planted. Based on this rule, the number of other areas in a city, such as built-up land, water, and others, may not exceed 70% of the city area. However, research shows Jakarta reached 69.9% of its built-up area in 2012, a very slim figure with a limit of 70%, and increased by 11.1% in just 12 years [16]. Other research shows that all cities in Jakarta exceeded the 70% limit in 2014 [17].

On the west side of the capital, Tangerang City shows an increase in the built-up area by 22.6% in almost the same time, so in 2012 the built-up land reached 60.6% of the total city [18]. This fact raises the question of whether the built-up land in the cities supporting the capital has reached its maximum area. This study aims to evaluate land cover changes and the area of urban and non-urban areas in Jakarta's supporting areas, namely Bogor, Depok, Tangerang, and Bekasi.

2. Methods

The satellite cities of Jakarta cover eight administrative areas. On the south side of Jakarta, there are Bogor Regency, Bogor City, and Depok City. Tangerang Regency, Tangerang City, and South Tangerang City are on the west side, and Bekasi Regency and Bekasi City are on the east side. The map of the study location can be seen in Figure 1.

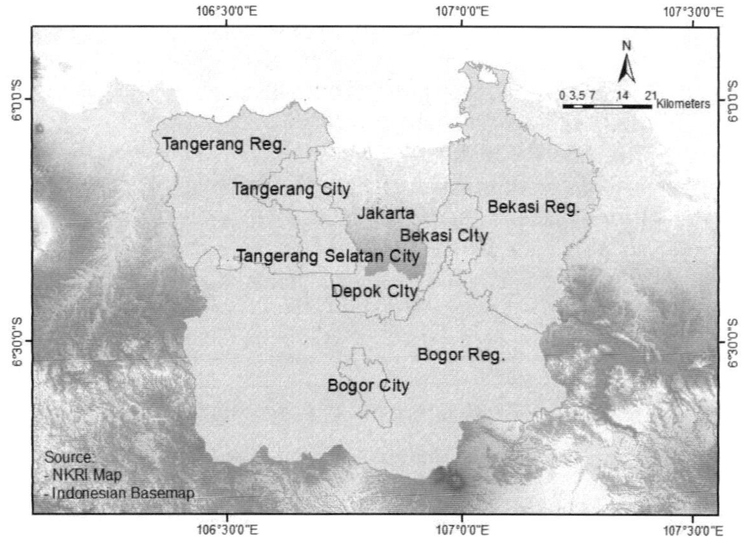

Figure 1. The bodetabek (Bogor, Depok, Tangerang, and Bekasi) area.

The primary data in this study are Landsat satellite imagery in 2001 and 2021 (Table 1). Administrative boundaries are taken from the Indonesian base map of the Geospatial Information Agency. The classification validation is carried out using ESRI World Imagery which has a higher resolution.

IOP Conf. Series: Earth and Environmental Science 1111 (2022) 012034 doi:10.1088/1755-1315/1111/1/012034

Table 1. The list of satellite imagery

	Imagery	Path/Row	Date
2001	Landsat 7 ETM+	122/064 & 122/065	17 September 2001
2021	Landsat 8 OLI/TIRS	122/064 & 122/065	11 Mei 2021

Land cover is classified using the supervised classification method into eight classes: built-up, vegetation, agricultural, ground, water, pond, cloud, and shadow. Since the analysis in this study only requires the land cover, i.e., vegetation, agriculture, and built-up area, the other classes are combined into "others". The built-up class in red refers to urban areas characterized by buildings (residential, industrial) and roads. Agriculture class refers to the land used for agriculture, colored light green. The vegetation class includes forests, trees, shrubs, meadows, and colored dark green. The last class, others, in grey, is a collection of other land covers such as waters, ponds, open land, and others. After being classified, the area of each class is extracted for each regency/city. Furthermore, data from 2001 and 2021 are compared to obtain land cover changes in the Bodetabek area.

3. Results and Discussion

The accuracy test was conducted using the stratified random sampling method with 200 test points. The calculation of accuracy uses error matrix (confusion matrix) and kappa index value. From the test results, the overall accuracy is 79.6%, and the kappa index is 0.69.

Since 20 years ago, the “city” administrative areas (Tangerang City, South Tangerang City, Depok City, Bogor City, and Bekasi City) have a denser proportion of built-up areas than “regencies”. The built-up area can be seen in the red area in Figure 2. Even though it is far from Jakarta, Bogor City has had many built-up areas over the last twenty years. In 2021, the building area will be expanded, almost evenly distributed throughout the region.

The difference that looks most prominent occurs in Tangerang Regency. The land cover in Tangerang Regency was initially dominated by agriculture (52.8%), but in 2021 the agricultural area decreased to 34%. The decline in the area is relatively high and also occurred in the Tangerang Regency. Built-up land has significantly increased and now dominates the area of Tangerang Regency.

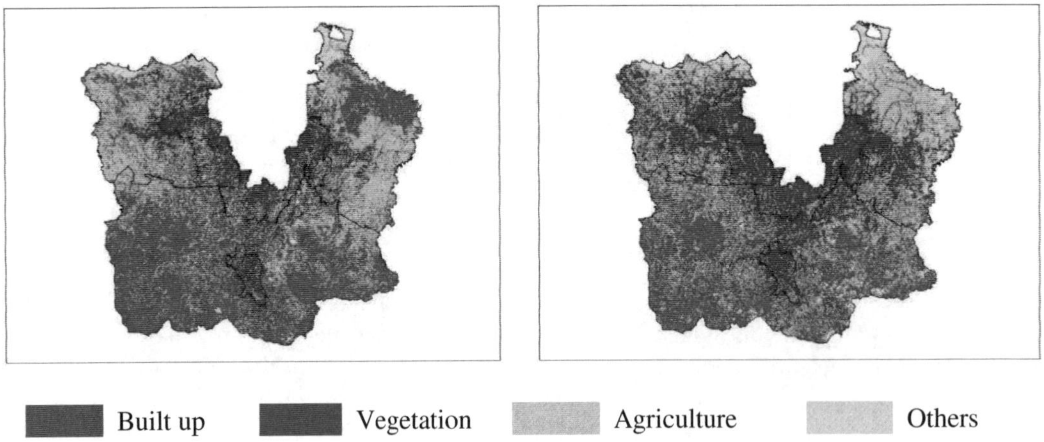

Figure 2. Land cover change in Bodetabek.

In contrast to “City”, the “Regency” area (Bogor Regency, Bekasi Regency, and Tangerang Regency), which has a larger area, still has a variety of land cover functions. The land cover change that stands out is the loss of vegetation areas on the north side of the Bekasi Regency in 2001 to become agricultural areas in 2021. Similar conditions occur in Tangerang Regency, especially on the north side, which was initially a vegetation area, some have become agricultural areas, and some have become built-up areas. In the Bogor Regency, agriculture and built-up areas have fragmented vegetation. Utami & Hendarto (2020) stated that development in suburban areas is faster than in the core area [9].

Based on Figure 2, it is known that the distribution of the built-up area has exceeded the administrative limit. The distribution of the city center with the surrounding cities forms a network. The urban area between Jakarta, Depok, and Bogor, previously a network around the main road, has developed to form a denser network extending to the surrounding area. This result is in line with research conducted by Murtadho et al. (2018), which states that the development of Jakarta and Bandung encourages conurbation, or the merging of two urban areas impacts the growth of other cities.

Research by Rusdiati et al. (2015) stated that land use changes in Greater Jakarta are dominantly caused by land conversion into urban activities (residential, industrial, commercial services, and others) [10,19]. One of the impacts is the change in the culture of an agrarian society to a non-agrarian culture which also causes changes in social, economic, and cultural structures [7]. Changes in land use are shown in a comparison chart of the built-up area, vegetation cover, and agriculture (Figure 3).

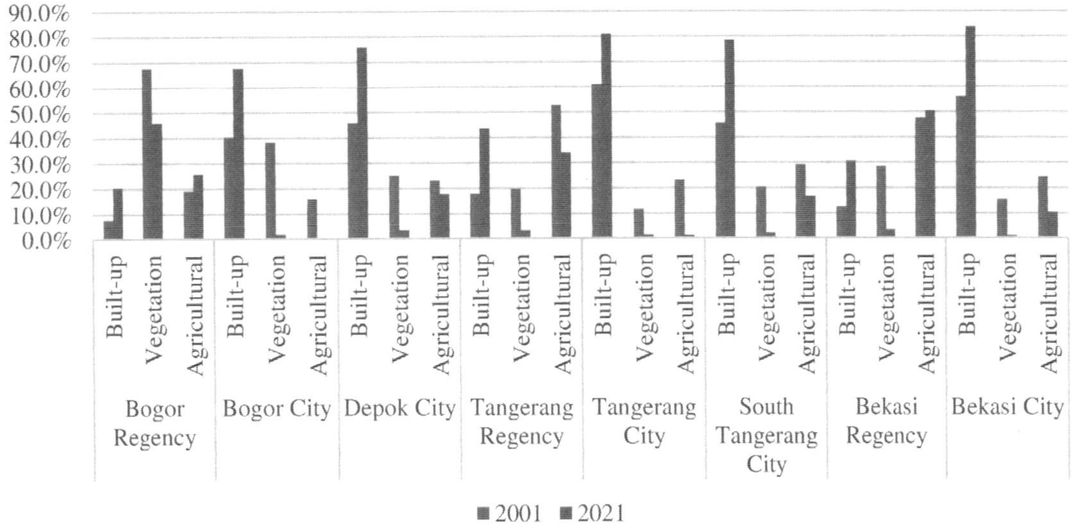

Figure 3. The comparison chart of the built-up area, vegetation cover, and agricultural area.

Among all areas, Bogor Regency had the highest percentage of vegetation area in 2001, and it will still survive even though the area decreased by 20% in 2021. The area of vegetation that experienced the most decline was Bogor City, which was initially 38.2% and became 1.6%, followed by Bekasi Regency, which was initially 28.5% and became 3.2%. Bogor's most visible green open space is the Bogor Botanical Gardens complex, a significant source of ecosystem services [20].

The percentage of built-up land in all regions has increased. The highest increase was in South Tangerang City by 33% in twenty years. In 2021, the area built in South Tangerang reached 78.5% of the city area. This result contradicts the predictions of Lamonda et al. (2019), which stated that the growth of built-up areas in South Tangerang is predicted to reach a safe limit in 2026, where the ratio of open space: to the built-up area is 3:7 [21].

The land increase in South Tangerang's built-up area can be attributed to regional economic factors. South Tangerang City is one of the areas in Banten province whose economy is growing fast. One of the factors driving the South Tangerang economy is the rapid industrial growth [22]. South Tangerang City is now one of the centers of industrial estates in the province of Banten.

Land cover of each city by 2021:

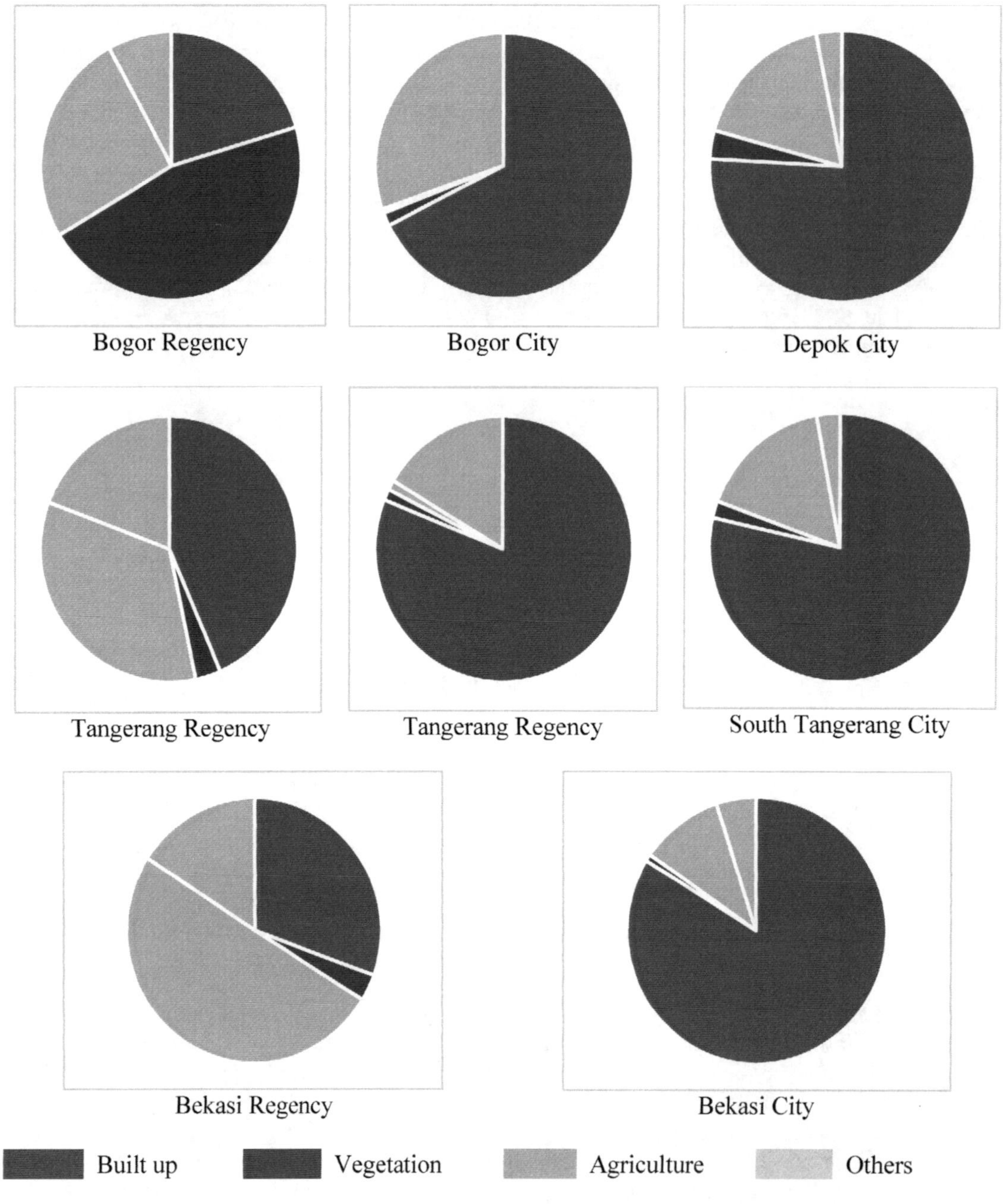

The increase in built-up land causes a decrease in vegetation and the agricultural area. The increase in the agricultural area only occurred in Bogor Regency and Bekasi Regency. Agriculture is the dominant land cover in the Bekasi Regency (50.5% of the total area). While Bogor City almost does not have a good agricultural area because there is only 0.5% agriculture in the entire area. The condition is the same as Bekasi City, which only has 0.9% of the city area functioned by agriculture. The decrease in the agricultural area impacts food security [23].

4. Conclusion

In the past 20 years, land use has significantly changed in all supporting areas of the capital. The dense and evenly distributed urban area growth occurred in Bogor City, Depok City, Tangerang City, South Tangerang City, and Bekasi City. The Indonesian government has issued a rule for regencies and cities supporting the capital city to have 30% green open space in the city area. For that, the government needs to pay attention to cities that have a built area of more than 70% of the city, namely Bekasi City (83.8%), Tangerang City (81%), South Tangerang City (78.5%), and Depok City. (75.8%). The loss of vegetation density and various land-use changes needs to be a concern of the local government in spatial planning.

Acknowledgments
The author would like to thank Dr Hayati Sari Hasibuan, Himpasiling, and all parties who have provided input to this research.

References

[1] Bilal A 2021 Urbanization Causing Habitat Destruction and Loss of Birds Diversity in District Sargodha *Open Access J. Biog. Sci. Res.* **10**
[2] Furaca N B, Hoguane A M, MacKay F, Willemse M and Langa A A 2021 Exploring urbanization and critical habitat loss through land cover change around the Bons Sinais Estuary, Mozambique *WIO J. Mar. Sci.* **Special Is** 43–58
[3] Muñoz P, García-Rodríguez A and Sandoval L 2021 Urbanization, habitat extension and spatial pattern, threaten a Costa Rican endemic bird *Rev. Biol. Trop.* **69**
[4] Dri G F, Fontana C S and Cristian de Sales Dambros 2021 Estimating the impacts of habitat loss induced by urbanization on bird local extinctions *Biol. Conserv.* **256**
[5] Banaszak-Cibickaa W and Dylewskib Ł 2021 Species and functional diversity — A better understanding of the impact of urbanization on bee communities *Sci. Total Environ.* **774** 145729
[6] Hermanto S S A, Makalew A D N and Sulistyantara B 2018 Hubungan Antara Perubahan Tutupan Lahan Terhadap Total Penduduk Yang Dipengaruhi Oleh Fenomena Urbanisasi Di Bogor, Jawa Barat *J. Lanskap Indones.* **10** 7–11
[7] Prihatin R B 2015 Alih Fungsi Lahan di Perkotaan (Studi Kasus di Kota Bandung dan Yogyakarta) *Aspir. J. Masal. Sos.* **6** 105–18
[8] Rukmana S N and Shofwan M 2020 Fenomena Urbanisasi: Transformasi spasial di Kawasan Pinggiran Kabupaten Sidoarjo *Tataloka* **22** 393–9
[9] Utami F V and Hendarto M 2020 The Speed of Land Use Changes in The Region of Depok City *Econ. Dev. Anal. J.* **9** 318–27
[10] Rustiadi E, Pribadi D O, Pravitasari A E, Indraprahasta G S and Iman L S 2015 Jabodetabek Megacity: From City Development Toward Urban Complex Management System 421–45
[11] Pokhariya H S, Singh D P and Prakash R 2021 Investigating the impacts of urbanization on different land cover classes and land surface temperature using GIS and RS techniques *Int. J. Syst. Assur. Eng. Manag.*
[12] Deng Y, Jiang W, Wu Z, Ling Z, Peng K and Deng Y 2022 Assessing Surface Water Losses and Gains under Rapid Urbanization for SDG 6.6.1 Using Long-Term Landsat Imagery in the Guangdong-Hong Kong-Macao Greater Bay Area, China *Remote Sens.* **14** 881
[13] Rustiadi E, Zain A M, Trisasongko B H and Carolita I 2002 Land Cover Change In Jabotabek Region
[14] Octifany Y 2020 The History of Urbanization in Java Island: Path to Contemporary Urbanization *TATA LOKA* **22** 474–85
[15] Mardiansjah; F H and Rahayu P 2019 Urbanisasi dan Pertumbuhan Kota-kota di Indonesia: Suatu Perbandingan Antar-kawasan Makro Indonesia *J. Pengemb. Kota* **7** 91–110
[16] Rushayati S B, Prasetyo L B, Puspaningsih N and Rachmawati E 2016 Adaptation Strategy Toward Urban Heat Island at Tropical Urban Area *Procedia Environ. Sci.* **33** 221–9

[17] Rushayati S B, Shamila A D and Prasetyo L B 2018 The Role of Vegetation in Controlling Air Temperature Resulting from Urban Heat Island *Forum Geogr.* **32** 1–11
[18] Wibowo A 2013 Spatial - Temporal Analysis of Urban Heat Island in Tangerang City *Indones. J. Geogr.* **45** 101–15
[19] RUSTIADI E and KITAMURA T 1998 Analysis of Land Use Changes in City Suburbs. A Case Study on Some Subdistricts of the Bekasi Area of West Java, Indonesia. *J. Rural Plan. Assoc.* **17** 20–31
[20] Estoque R C and Murayama Y 2016 Quantifying landscape pattern and ecosystem service value changes in four rapidly urbanizing hill stations of Southeast Asia *Landsc. Ecol.* **31** 1481–507
[21] Lamonda R, Supriatna, Hernina R, Manessa M D M and Ristya Y 2019 Spatial Dynamics Model for Land Carrying Capacity Prediction in Tangerang Selatan City *E3S W eb of Conferences 125* (Semarang, Indonesia) p 01006
[22] Irkham M 2019 Analisis ketimpangan wilayah *Akuntabel* **16** 98–110
[23] Rasool R, Fayaz A, Shafiq M ul, Singh H and Ahmed P 2021 Land use land cover change in Kashmir Himalaya: Linking remote sensing with an indicator based DPSIR approach *Ecol. Indic.* **125** 107447

Synergizing Food Security With Forest Conservation: Preliminary Insights From Social Forestry Implementation In Riau, Indonesia

Rizky Januar[1]*, Dean Affandi[2] and Smita Tanaya[3]
[1] Research, Data, and Innovation, World Resources Institute, Jakarta, 12170, Indonesia, ORCID ID: 0000-0002-9591-1936
[2] Research, Data, and Innovation, World Resources Institute, Jakarta, 12170, Indonesia, ORCID ID: 0000-0002-1946-322X
[3] Agriculture, Forests, Land Use, World Resources Institute, Jakarta, 12170, Indonesia, ORCID ID: 0000-0001-5847-0230
[1]*rizky.haryanto@wri.org, [2]dean.affandi@wri.org, [3]smita.sitanggang@wri.org

Abstract. Amidst food security challenges faced by Indonesia, policies to address them may come at the expense of forest ecosystems. Social Forestry (SF) program provides a potential solution to synergize forest conservation and food security objectives in Indonesia by allowing local communities to manage and protect the local forests. However, there is limited evidence on how the synergy is realized within the SF implementation on the ground. The study aims to fill this knowledge gap by investigating community-based practices, efforts, and challenges in line with the pathways connecting forest conservation and food security in three SF sites in Riau, Indonesia. Relevant data are collected through stakeholder interviews, field observation, and spatial analysis; and further triangulated through literature review. While SF ecosystems at the study sites provide food and income sources and help build climate resilience for the local and indigenous communities, challenges remain in improving the value added and market access of local forest products; incentivizing forest-based food subsistence potential in line with the site-specific practices; and, in some areas, reducing the rate of forest loss. The policy implications and recommendations for improvement are provided.

1. Introduction

Amidst food security challenges faced by Indonesia, existing policies to address them may come at the expense of forest ecosystems. Indonesia ranked 65th out of 113 countries in the Global Food Security Index 2020 [1]. Meanwhile, Global Hunger Index 2020 put Indonesia 70th out of 132 countries [2]. To strengthen food security amidst Covid-19, Indonesia began implementing the Food Estate policy: a large-scale food production program that includes agriculture and livestock. However, the Food Estate allocation plan covers approximately 1.5 million hectares of natural forests [3], raising questions about deforestation's risk and its threat to Indonesia's climate mitigation commitment. Aligning forest conservation and food security in Indonesia remains a challenge to be addressed.

Various studies have offered alternative approaches to synergize forest conservation measures and food security. Forests may contribute to strengthening food systems mainly through three pathways: providing direct food sources for consumption (*i.e.*, direct pathway), income from the sale of forest products leading to purchase of food items (*i.e.*, income pathway), and ecological services, which sustain the forest environment including the food and income sources (*i.e.*, ecological pathway) [4-7]. In

Indonesia, among other countries – the agroforestry practice, where forest trees are retained along with the cultivation of crops and animal farming, may fulfill these pathways [7]. Nonetheless, it might also produce a potential threat to forests if profitable cropping proves more economically beneficial [8].

Among existing policies in Indonesia, Social Forestry (SF) provides a potential alley to connect forest conservation with communities' food security. SF is designed to accelerate the provision of forest management permits for local and indigenous communities and provide opportunities to manage forests, thereby reducing tenurial conflicts, supporting community-based forest protection, and improving community welfare. As of August 2022, SF policy has issued forest management permits of up to five million hectares [9], a fraction of the 12.7-million-hectare target [10]. SF implementation is also relevant to Indonesia's Nationally Determined Contributions [11], in which the engagement of local and indigenous communities are expected to support environmental services' protection.

Bolstering the connection between forest conservation and local food security provides the highest benefit to the local and indigenous communities, considering that almost 50 million people live within and nearby forest areas, which includes 35 percent of the poor population in the country [12,13]. A recent study denotes the potential of conserved forests as a vital component toward food security and nutrition and calls for increasing their relevance within global and national discourses [7]. Enhancing local food security through sustainable forest management also aligns with the Sustainable Development Goals (SDGs) #15 (Life on Land) and #2 (Zero Hunger).

However, there is currently limited knowledge on how the implementation of Social Forestry can support the synergy of food security and forest conservation objectives at the community level in Indonesia. Thus, the study aims to investigate community-based practices, efforts, and challenges in line with the forest conservation and food security pathways within SF areas and to shed light on the relevant policy implications in line with the status and trends of current SF governance. This is key to giving much-needed insights for post-permit facilitation and policy improvements around SF implementation, which may drive welfare improvement for forest communities and ensure sustainable use of forest ecosystems.

2. Methods

In response to the research objective, a case study approach is implemented in several SF sites to collect the body of evidence related to direct, income, and ecological pathways (see section Introduction). Implementation of this approach builds upon previous studies linking forest conservation and food security [4–7]. The case study is applied in Riau province, where the SF policy targets 1.3 million hectares of permit provision (10.2 percent of the total national target). The case study is conducted in three sites within the province (Table 1). The choice of locations is informed by prior consultations with local experts and forest agencies, considering the research objective.

Table 1. Study sites in Riau

Location	**Characteristics**
Village Forest of Sapat	A predominantly mangrove forest in the Indragiri Hilir district, Riau, with an area of 4.249 ha (3.538 ha for conservation, 315 ha for production forest, and 396 ha for conversion). The village forest was legalized on 11th December 2017.
Village Forest of Sungai Piyai	A predominantly mangrove forest in the Indagiri Hilir district, Riau, with a total area of 299 ha. The village forest was legalized on 31st December 2017. The forest location takes about 90 minutes by boat from the main island
Customary forest of Rantau Baru	A lowland tropical rainforest in the Pelalawan district, Riau, has an area of 871.99 ha. Currently, working closely with local NGOs, the

indigenous people of Rantau Baru are seeking recognition of their customary forest covering an area of 200 ha. Indigenous people in Rantau Baru adopt a matrilineal system, where women own and manage Rantau Baru's natural resources.

To collect data on the existing practices, challenges, and community efforts in line with the three pathways, field observation, and interviews are conducted with a total of 31 individuals consisting of local community members, government employees, and civil society organization (CSO) representatives. The fieldwork data on community-based forest conservation efforts are triangulated with forest cover loss (FCL) assessment [10]. Accordingly, spatial analysis of annual FCL data is applied using Global Forest Watch (GFW) datasets [14]; the data are then overlaid with the forest extent data from the Ministry of Environment and Forestry of Indonesia.

Finally, interviews on the key trends and challenges in SF governance in Riau are conducted to investigate the policy implications of the fieldwork data. We ensure cross-sectoral representation at national and provincial levels in Riau to minimize results bias and enrich perspectives of the data. Overall, we interviewed 21 individuals from the government institutions, research institutes, and CSOs. Data completion and triangulation are further conducted through a literature review. Based on the findings, rooms for policy improvement and further research are highlighted.

3. Results and Discussion

3.1 Social Forestry contribution to household diet and income

3.1.1 Rantau Baru.

In Rantau Baru, we found that a significant proportion of local customary members rely on forest resources for their daily food consumption and contribution to household income. Rattan shoots taken from the forest are part of the regular local diets—mainly consumed as vegetables. Communities also leverage capture fisheries from the watershed river within the forest zone for daily consumption and income. The fish is smoked or salted before being sold to the closest market (mostly in Langgam or Pangkalan Kerinci). Depending on the species, the price range of salted fish is 8,000-30,000 rupiah per kilogram (kg), whereas smoked fish are priced at 60,000-200,000 rupiah per kg.

Forest-sourced honey is also a key income source; to a small extent, it is consumed by local households. However, honey farming is exclusive to community members owning a fraction of the forest containing Sialang trees, where the honeycombs are produced. Honey can be sold to intermediaries (usually coming from the provincial capital city Pekanbaru) or directly to the market at a price range of 70,000-120,000 rupiah per kg. The Covid-19 pandemic reduced sales prices in 2020, despite the rebound in 2021. Some efforts to improve sales are made through digital platforms, with the support of the local government – albeit with limited success. Meanwhile, there is yet limited agroforestry practice within the Rantau Baru forest. The forest-based livelihood sources, e.g., rattan and honey, are mainly based on the existing forest trees, with no additional cultivation of crops.

3.1.2 Sungai Piyai and Sapat.

Sapat and Sungai Piyai are two SF sites with similar ecological characteristics—mangrove forests dominate both. According to MoEF data, Sapat and Sungai Piyai have around 93 percent and 86 percent of mangrove cover compared to their total SF areas. We also found significant similarities in the utilization model of forest resources in both sites.

The primary commodities utilized from Sapat and Sungai Piyai forests are fishes and crabs that are endemic to the mangrove habitat. Captured fish are consumed mainly by communities as part of the local diet, and the rest are sold to the local traditional markets. Meanwhile, crabs are a vital income source due to their high price: up to 170,000 rupiah per kg, with a daily catch of up to 2-3 kg. Some are also exported to Singapore at a much higher price. Nonetheless, the crab market chain was impacted by the supply chain disruption due to the Covid-19 pandemic.

Sungai Piyai and Sapat forests are dominated by Nypafruticans (Nypa) and *Rhizophora sp* (locally named *bakau*). In both villages, nypa is usually processed into dry leaves and brooms. The price of dry Nypa leaves reaches 7,000 rupiahs per kg, whereas nypa-based brooms are priced around 5,000-5,500 rupiah per kg. On average, despite the occasional fluctuation, monthly household income from selling nypa may reach 6 million rupiahs. Meanwhile, utilization of *bakau* remains limited. Community's plan to develop *bakau*-based processed foods, such as syrup, are still restricted by limited capacity and training support. Accordingly, in 2022, the Peat and Mangrove Restoration Agency (BRGM), the national-level institution mandated to restore mangroves in Indonesia, conducted training on mangrove-based commodity development for the local communities. The training includes crab farming and syrup processing, among others. Meanwhile, the mangrove crab and fish farming practices in both villages remain limited.

To support business development and income generation from SF commodities, communities in Sapat and Sungai Piyai established their commodity-based business units called *Kelompok Usaha Perhutanan Sosial* (KUPS), facilitated by the government. Based on the commodity differentiation, each village has two types of KUPS: one KUPS focuses on nypa, and the other mainly on fish, crabs, and shrimps. In Sapat, women are further encouraged to develop a specific KUPS in which all members are female. The purpose is to improve the role of women in local SF management. In the unit, the local women mainly develop plates made of nypa to be sold to the local markets.

3.1.3 Conserving Social Forestry areas and their ecological services: practices and challenges.
In Rantau Baru, some of the forest areas are based in peatlands. Peatland canals for oil palm plantations constructed by a plantation company adjacent to the SF area had led to drainage, making the area more vulnerable to fires, especially during the dry season. Based on GFW data analysis, in 2015, the FCL in Rantau Baru was 4.3 times higher than in 2014 (3.14 ha FCL in 2014 and 13.49 ha in 2015)—presumably due to the large-scale peat fires in 2015. Respondents confirmed this harmed bee pollination and production of forest-sourced honey, a key income source for Rantau Baru communities.

Since then, with support from the state agency, the locals established the Fire-Care Community (*Masyarakat Peduli Api* / MPA), a community-based institution leading patrol activities for preventing and mitigating forest fires. This effort, among others, contributed to controlling forest loss – the average FCL from 2016 to 2020 was 2.9 ha, whereas the dry season in 2019 did not lead to a significant FCL increase (FCL in 2019 was 1.76 ha). Furthermore, in response to forest fires, a customary regulation was established to forbid cutting trees where bee pollination takes place—violations could lead to a fine of up to tens of millions of rupiah.

Meanwhile, in Sapat and Sungai Piyai, communities have also relied on coconut plantations for their staple income. Plantation activities were disrupted during the coastal abrasions of 2017, mainly due to the destruction of mangrove ecosystems, which could otherwise have buffered the sea waves. The deterioration of mangroves can be linked to the communities' use of the woods as building materials.

Since the abrasion, the communities have aimed to prioritize sustainable use of mangroves, especially within the SF areas. Both Sapat and Sungai Piyai applied for SF management permits in 2017. In both villages, the local communities conducted regular patrols to prevent illegal mangrove logging. Communities in Sapat and Sungai Piyai have also been contributing to the mangrove rehabilitation program initiated by BRGM. In Sungai Piyai, BRGM developed a pilot for mangrove nurseries—the resulting seeds were transferred to neighboring coastal villages to support mangrove rehabilitation. Further, the local consensus has also been developed in which transaction of juvenile carbs is prohibited.

Nonetheless, spatial analysis shows that SF conservation remains challenging in Sapat and Sungai Piya. Since the SF permits were handed in 2017 for both sites, forest loss has continued to increase. The annual FCL in Sungai Piyai in 2018, 2019, and 2020 are, respectively, 1.2, 24, and 30 times higher than the FCL in 2017, i.e., 0.43 ha. Meanwhile, in Sapat, FCL in 2018, 2019, and 2020 are 1.5, 2, and 1.85 times higher than the FCL in 2017, i.e., 1.21 ha. Our data indicate that the pertinent deforestation threats in both sites include housing and local infrastructure development using the mangrove woods, making mangrove charcoal for sale, and forest conversion into coconut plantations. Both villages are based in the Indragiri Hilir district, the largest coconut-producing district in Indonesia. Given the forest authorities' insufficient resources, local forest utilization practice supervision has been limited.

3.2 Policy discussion and implications

SF ecosystems allow local and indigenous communities to manage their food and income sources through sustainable forest use. This study shows that SF ecosystems in Rantau Baru, Sapat, and Sungai Piyai contribute to sustaining local diets and providing alternative livelihoods. However, challenges remain in improving local forest products' value added and market access to increase household income contribution. In addition, conservation attempts in these sites are still a work in progress. In Rantau Baru, based on the annual forest cover loss trend, forest loss has generally remained under control with the help of local community efforts. On the other hand, in Sapat and especially Sungai Piyai, despite ongoing restoration and conservation efforts, continuous guidance and supervision are still needed to minimize the threat of post-permit deforestation.

Opportunities and challenges from SF on food security and forest conservation in Riau suggest the need to augment SF's resource allocation and improve SF's mainstreaming across local policies. In Riau, there is limited prioritization for SF within local policy planning—as observed in, among others, Riau's 2019-2024 Regional Medium Term Development Plan and the Environment and Forestry Agency Workplan 2019-2024. Limited policy mainstreaming contributes to the limited budget available for SF. In Riau, the SF-related budget within the 2017 fiscal year covers only 2.6 percent of the total direct expenses of the Riau Environment and Forestry Agency. Filling the resource gaps will be relevant to upgrade the markets and value chain of the non-timber forest products (NTFPs); support capacity building on forest protection and sustainable use; and bolster the potential of agroforestry and commodity farming in the study sites and beyond – in relevance to the biophysical and social site diagnostics [8,10], and also with caution as to carefully manage the potential deforestation threat [8].

Further, the need for improving synergy between conservation, food security, and economic livelihoods calls for strengthened multi-stakeholder collaboration. An improved, dedicated task force consisting of government agencies, communities, community developers, and the private sector can provide a breakthrough in the post-permit SF facilitation process. The proposed task force might build upon lessons learned from the existing Working Group on the Acceleration of Social Forestry (*Kelompok Kerja Percepatan Perhutanan Sosial*) in Riau, which, according to our interviews, is limited in effectiveness and coordination. Also, enhancing collaborative mechanisms with small-medium enterprises and corporations with solid sustainability commitment could help create a sustainable supply chain and knowledge transfer for SF-based communities, such as on non-timber forest commodities' management and marketing.

Finally, our findings suggest that the importance of forest resources toward local subsistence (*i.e.*, direct pathway) should be acknowledged. In line with this study, various studies suggest that forest-based subsistence can provide a safety net in the face of external shocks [4,7,15]. However, the SF policy scheme has not yet fully incentivized the direct pathway. For instance, the Minister of Environment and Forestry Regulation 9/2021 suggests that the evaluation of forest communities' social welfare is solely determined by evidence of income increase, as measured by the amount of income and assets. Policies may thus be refined to recognize and incentivize site-specific practices which promote forest-based food subsistence in an environmentally sustainable manner. Further research can quantitatively explore the extent to which SF areas contribute to the local food security pathways, including subsistence.

4. Conclusion

This study investigates community-based practices, efforts, and challenges within Social Forestry (SF) areas, in line with the direct, income, and ecological pathways connecting forest conservation and local food security. The case study, committed in Riau, confirms that SF provides a potential alley in realizing the three pathways. Community participation in local conservation and restoration efforts also helps sustain local livelihoods while bringing benefits against pertinent climate threats, such as abrasion and forest fires. However, challenges remain in improving the value added and market access of local forest products, incentivizing forest-based food subsistence potential in line with site-specific practices, and reducing the forest loss rate in some locations. Beyond accelerating the issuance of SF licenses, policy and institutional breakthroughs are needed to further align the conservation and food security objectives within the SF ecosystems, ensuring a meaningful contribution of the SF policy toward the local community welfare, SDGs, and Indonesia's climate ambition.

Acknowledgments
This study is supported by the International Development Research Centre (IDRC), Ottawa, Canada, under the Think Climate Indonesia project. The views expressed herein do not necessarily represent those of IDRC or its Board of Governors.

References

[1] Economist Intelligence Unit 2021 Global Food Security Index 2020
[2] Welthungerhilfe 2021 Global Hunger Index 2020
[3] Yayasan Madani Berkelanjutan 2021 *Menakar Ancaman terhadap Hutan Alam dan Ekosistem Gambut di Balik Rencana Pengembangan Food Estate di Papua, Kalimantan Tengah, Sumatera Utara, dan Sumatera Selatan*
[4] HLPE 2017 *Sustainable Forestry for Food Security and Nutrition* (Rome)
[5] Baudron F, Tomscha S A, Powell B, Groot J C J, Gergel S E and Sunderland T 2019 Testing the Various Pathways Linking Forest Cover to Dietary Diversity in Tropical Landscapes *Front. Sustain. Food Syst.* **3**
[6] Wezel A, Herren B G, Kerr R B, Barrios E, Gonçalves A L R and Sinclair F 2020 Agroecological principles and elements and their implications for transitioning to sustainable food systems. A review *Agron. Sustain. Dev.* **40**
[7] Ickowitz A, Rowland D, Powell B, Djoudi H, Sunderland T, Nurhasan M, Gitz V, Meybeck A M, Nasi R, Agroforestry W, McMullin K S, Rosenstock T, Dawson I, March K, Novak A, Jamnadass R, Ickowitz A, McMullin S, Rosenstock T, Dawson I, Rowland D, Powell B, March K, Djoudi H, Sunderland T, Nurhasan M, Novak A, Gitz V, Meybeck A, Jamnadass R, Guariguata M R, Termote C and Nasi R 2022 Transforming food systems with trees and forests *Lancet Planet. Heal.* **6** e632–9
[8] Reith E, Gosling E, Knoke T and Paul C 2020 How much agroforestry is needed to achieve multifunctional landscapes at the forest frontier?-Coupling expert opinion with robust goal programming *Sustain.* **12**
[9] Ministry of Environment and Forestry 2022 Hingga 1 Agustus 2022, Capaian Perhutanan Sosial Tembus 5 Juta Hektar
[10] Januar R, Daeli W, Tobing S and Firmansyah R 2021 *Village Forest and Deforestation in Indonesia: Analysis of a Case Study on the Implementation of Social Forestry Management in Jambi* (Jakarta)
[11] Government of Indonesia 2021 Updated Nationally Determined Contribution Republic of Indonesia 2021 32
[12] Bisnis 2016 KEMISKINAN: 48,8 Juta Orang Indonesia Tinggal di Kawasan Hutan
[13] Firdausi R Z 2021 Hari Hutan Sedunia 21 Maret: Hutan untuk Kesejahteraan Masyarakat
[14] Hansen M C, Potapov P V., Moore R, Hancher M, Turubanova S A, Tyukavina A, Thau D, Stehman S V., Goetz S J, Loveland T R, Kommareddy A, Egorov A, Chini L, Justice C O and Townshend J R G 2013 High-resolution global maps of 21st-century forest cover change *Science (80-.).* **342** 850–3
[15] McCarthy J F, Winarto Y T, Sitorus H, Kutanegara P M and Budianto V 2020 COVID-19 and food systems in Indonesia *COVID-19 and food system in the Indo-Pacific: An assessment of vulnerabilities, impacts and opportunities for action* pp 41–92

Positive Water Impact Validation: Study Case at Kubang Water Sources AQUA Operation Sukabumi, West Java, Indonesia

Rachmat Fajar Lubis[1]*, Asep Mulyono[2], Arif Fadillah[3], Hendra Bakti[2], Gumilar Utama Nugraha[2], Azwar Satya Muhammad[3], Raditya Pramudiantoro[3], and Arman Abdurrohman[3]

[1] Research Center for Limnology and Water Resources, National Research and Innovation Agency (BRIN), Cibinong, 16911, Indonesia, 0000-0002-5430- 7169
[2] Research Center for Environmental and Clean Technology, National Research and Innovation Agency (BRIN), Bandung, 40135, Indonesia
[3] Danone Aqua Group, Jakarta, Indonesia
[1]*rach007@brin.go.id, [2]asep033@brin.go.id, [3]Arif.fadillah@danone.com, [4]hend009@brin.go.id, [5]g.utamas.n@gmail.com, [6]Azwar.muhammad@danone.com, [7]Raditya.Pramudiantoro@danone.com, [8] Arman.ABDURROHMAN@danone.com

Abstract. W*ater stewardship* is an effort to maintain the availability of water for humankind. AQUA water stewardship program "*good water stewardship*" that Danone has carried out is called *"positive water impact.* " PWI has various efforts to get a good impact on its water sources and environment. This program focuses on water stewardship to achieve quantity and quality as well as good water hygiene/sanitation and become the first initiative in Indonesia. Validation of these activities and their impact on Aqua's Kubang groundwater resource recharge areas by PT AGM Mekarsari and PT AGM Babakanpari. Kubang water source is located on the southeast side of Mount Salak, Sukabumi, West Jawa. This activity was intended for the period 2011-now. The validation process would be carried out by referring to the division of 3 significant groups of water stewardship activities carried out :(1) The restoration of land conversion recharge area, (2) Restoration of recharge area-artificial recharge, and (3) Water access sanitation and hygiene. Validation was carried out from 2011 – 2020. The validation results of the positive water impact program as a water stewardship activity at Kubang water sources indicated that the calculation method and activity planning could be well validated during the period.

1. Introduction

Positive water impact was an AQUA Danone water stewardship program to promote the principles of Integrated Water Resource Management (IWRM) through collective community action. This program was the first initiative in Indonesia, and one of the main focuses on water stewardship to achieve quantity, quality, and good water hygiene/sanitation. Water stewardship is defined as using water in a socially, environmentally sustainable, and economically beneficial way. This is achieved through an inclusive stakeholder process involving the site and catchment-based actions. Water stewardship is one-way companies can manage the complexity of balancing water use with the community and ecosystem needs. It stated that water stewardship means responsible planning and management of water resources [1].

Water stewardship activities have already been conducted at the Kubang water source since 2011. Kubang water source for PT AGM Mekarsari and PT AGM Babakanpari is located on the southeast side of Mount Salak West Jawa, Indonesia. The artesian aquifer water source operated by AQUA in Kubang geologically consists of volcanic product formations [2] and alluvial clay covering aquifer formations with volcanic sandstone type with slope values ranging between 1.12 and 44.68° (Figure 1).

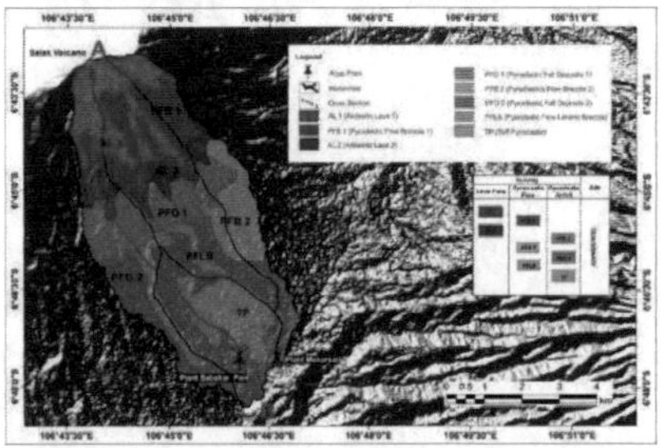

Figure 1. Geology and topography of Kubang water resources area [2]

The program focused on Kubang groundwater resource recharge areas. The activities can be divided into three main programs: recharge area restoration, water quality, water accessibility, sanitation, and hygiene (WASH). These efforts are carried out to compensate for the water extraction permit (SIPA) granted with a total permit (Figure 2).

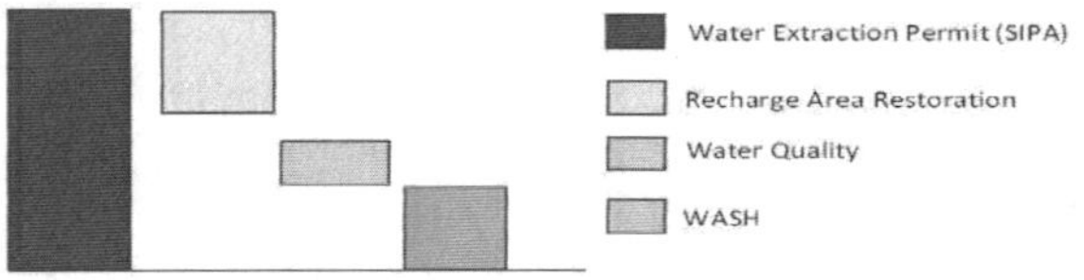

Figure 2. Positive water impact focus program.

AQUA water stewardship activities, carried out in PT AGM Mekarsari and PT AGM Babakanpari, needed to be validated to evaluate the achievement of the expected goals. This study aims to validate the Positive Water Impact of the Aqua Water stewardship program at Kubang Water sources with VWBA methodology.

2. Methods

The planning and calculation of the water stewardship target referred to the Volumetric Water Benefit Accounting (VWBA) guide issued by the water research institute [3]. Based on the activity targets, the activities were divided into several main activities, namely: (a) Restoration of Land Cover in Recharge Areas, (b) Restoration of Artificial Recharge-Catchment Areas, (c) Water Accessibility, Hygiene, and Water Sanitation (*WASH*). The detailed potential calculation will be validated using validation calculation based on field observation after the program runs from 2011 until 2020.

2.1 Restoration of Land Cover in Recharge Areas

Restoration of Land Cover in Recharge Areas, also known as the tree planting program, already starts in 2011-now. The formulation of the recharge calculation applied an equation that was also used in modeling using the SWAT model [4], which was volume per unit time, as shown in the equation below:

$$\frac{Number\ of\ trees}{400} \times VWB \times 31.536 \tag{1}$$

Where $VWB = Runoff\ saved$ ($CN\ Number$ method), with:

$$VWB = Q_{"baseline"} - Q_{"with-project"} \quad (2)$$

$$Q = \frac{(P-0.25)^2}{(P+0.85)} \quad (3)$$

$$S = \frac{25400}{CN} - 254 \quad (4)$$

The baseline CN value is 73, while the CN value after planting is 58. Validation activities were carried out on two main parameters: the number of trees and the amount of recharge from 2011-2020. The method used to validate the number of trees was based on the tree database owned by AQUA Mekarsari and Babakanpari, the Jejakin CSR database, and field observations. Validation of the amount of recharge due to the number of tree plantings was carried out through an analytical model equation.

The first step of the presumption was to calculate the ratio of the amount of runoff water, water that leached into the soil, and water that had evaporated back due to interception. The first stage of presumption was carried out using the Curve Number method [5,6], based on the relationship between the infiltration capacity of each soil type and the amount of rainfall that falls each time it rains. The total annual rainfall (P) was divided into three components; Runoff water (Q), actual groundwater retention (F), and initial abstraction (Initial Abstraction: Ia). Conceptually, the relationship between P, Qsurf, and Ia is assumed as follows:

$$F/S = Qsurf/(P - Ia) \quad (5)$$

Where S is the maximum potential of the soil to retain water. Actual soil retention (F) can also be formulated as follows:

$$F = (P - Ia) - Qsurf \quad (6)$$

Substitutions (5) and (6) will obtain:

$$((P - Ia) - Qsurf)/S = Qsurf / (P - Ia) \quad (7)$$

or

$$Qsurf = (P - Ia)2/ (P - Ia) + S \quad (8)$$

From the above equation, Q and P could be obtained from the measurement results. At the same time, S and Ia were unknown parameters that should be primarily determined by soil hydrological factors such as recharge, interception, depression storage, and soil antecedent moisture. According to the results of practical experience in many places (in the US), it is obtained:

$$Ia = 0{,}2\,S \quad (9)$$

Furthermore, from the relationship above, the value of Q can be obtained from the input P based on the equation:

$$Qsurf = (P - 0{,}2S)2/ (P + 0{,}8S) \quad (10)$$

To solve equation (10), it was necessary to estimate the value of S. SCS developed a runoff curve number (CN) as an index that reflected the combination of soil hydrological factors. Empirical studies showed that CN was a function of three factors: soil type (texture), land cover, and initial soil moisture. SCS classifies soil into four classes (A, B, C, and D), classifies land cover into 21 classes, and classifies antecedent moisture classes (AMC) into three classes. CN values were in the range from 0 to 100, and the relationship between S and CN values was formulated empirically:

$$S = (25400/CN) - 254 \text{ ;S in millimeter unit} \quad (11)$$

Equations (9) and (10) could be used to estimate Qsurf if the values of P and CN were known. If $P \leq 0{,}2\,S$, it is assumed that Qsurf = 0. CN was an index that showed the ratio of rainfall and runoff for each type of soil and certain land cover determined based on the hydrological group. If P and Qsurf were known, it could be assumed that S was also known so that he and F could be calculated by equations (9) and (5). The amount of recharge was calculated based on the water balance equation below:

$$Recharge = Rainfall\,Volume - Evapotranspiration - Runoff \quad \textbf{(12)}$$

Each rainfall volume, evapotranspiration, and runoff were calculated by the equation below:

$$Rainfall\,Volume = Rainfall\ x\ Block\ area \quad \textbf{(13)}$$

$$Evapotranspiration = Total\ Tree\ x\ Canopi\ x\ Rainfall\ Average\ x\ Tree\ Coefficient\ (Kc) \quad \textbf{(14)}$$

$$Runoff = \Delta\, Qsurf\ x\ Block\ Area \quad \textbf{(15)}$$

Where:

$$\Delta\, Qsurf = Qsurf_{\,(baseline)} - Qsurf_{\,(tree\ planting)} \quad \textbf{(16)}$$

$\Delta Qsurf$ was the difference in the runoff before conservation actions (baseline) and after tree planting activities. CN baseline used a value of 82 and CN tree planting of 70 (Figure 3) in group C hydrological conditions based on AQUA data collected for the 2012-2015 SWAT program [7].

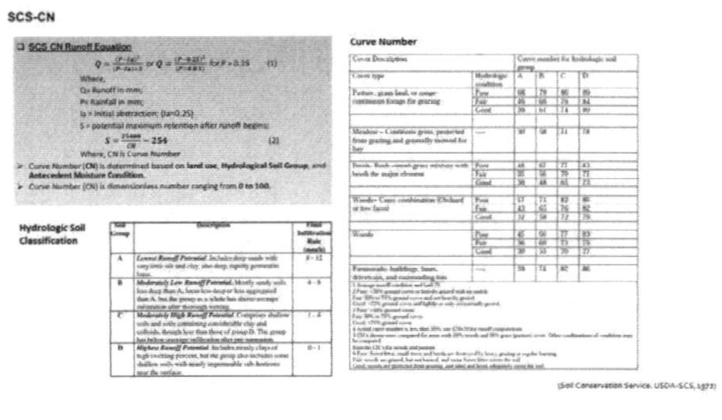

Figure 3. SCS curve number validation applied [5]

2.2 Restoration of Artificial Recharge-Catchment Areas

This activity included five activities with potential calculations referring to the VWBA concept [3]. The types of activities carried out were: (a) infiltration well, (b) infiltration well from rainwater harvesting on the roof (Infiltration Well RWH), (c) infiltration pond (Water pond), (d) bio pores, (e) trench. Calculations were carried out from 2014 – 2020. The formula for calculating volume per unit time is as shown in the equation below [6]:

$$Recharge\ Volume = Total\ each\ activity\ x\ Volume\ x\ (f\ inf)\ x\ Total\ rain/year \quad \textbf{(17)}$$

Where: f inf was the number of capacity times filled, has a value of 3, and n rain/year varies for each year following the rainfall pattern in that location. Validation activities were carried out by comparing the targets made with the realization of each technical conservation effort. The method used in evaluating infiltration wells was based on the water level fluctuation (WTF) installed in resident wells in the study area.

The WTF method could be implemented to estimate recharge stages using a measured groundwater table hydrograph with a specific time resolution [8,9]. Groundwater recharge resulted from type transfer (specific yield=Sy) and groundwater fluctuations at particular time intervals [9,10]. The groundwater recharge volume could be calculated based on the following formula:

$$Recharge\ volume = s_y\ \Delta h/\Delta t \quad \textbf{(18)}$$

Where R was the volume of groundwater recharge in the observation time interval (cm), Sy was the specific yield (dimensionless), Δh was the difference in water level as an attribute of recharge in a certain period (cm), and t is time. The quantity of groundwater recharge in volume could be obtained by multiplying the recharge amount by the recharge area. In this case, the estimated area was based on the shape of the infiltration well (Figure 4).

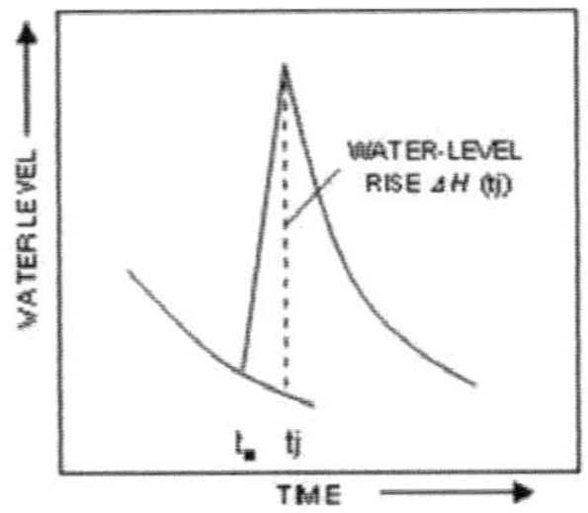

Figure 4. Water level fluctuation (WTF) method to estimate groundwater recharge [11]

2.3 Water Accessibility, Sanitation, and Hygiene (WASH)

The arrangement activity water accessibility, sanitation, and hygiene (WASH) was not included in the activities with the target of restoration of catchment areas but was part of water stewardship activities. These WASH activities would provide suitable activities for the output targets of decent quality and water status.

Validation activities were performed by comparing the targets made with their realization in the field. WASH activities in the area were divided into two major activities, validated annually from 2011- 2020. This activity included treating factory wastewater treatment plants and providing clean water for residents around the location (water access). The validation method was to test the water meter installed in the field and ensure it had been calibrated.

3. Results and Discussion

Based on the target of planting trees for ten years, the tree planting target was 1,163 million liters, with the immense recharge amount in 2012 amounting to 693 million liters and the lowest in 2011 amounting to 5 million liters. The validation target shows the number of trees planted over ten years (2011-2020), 460,000 trees, with the highest planting in 2012 and the lowest in 2011. During 2018 – 2020 no tree planting activity was carried out. The results of database validation show the total number of trees in 50 conservation blocks (Figure 5) that had been planted as many as 112,359 trees with 59 tree species.

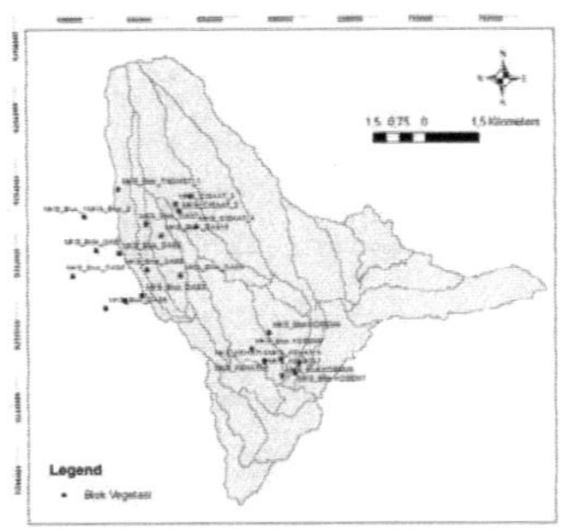

Figure 5. 50 Tree planting block conservation activities for Kubang Water Source

The validation results showed a difference/gap in the recharge amount over ten years (Table 1). The difference comes from tree species type and age, assuming that the previous year's tree planting activity recharge value will undoubtedly contribute to the year when new conservation activities occur. There is an additional contribution from the previous year in other blocks (non-new planting blocks) so that the contribution of each tree planting in each year will be calculated.

The artificial recharge activity validation target showed a different value between the target data and the database in the field (Table 1). Field validation revealed that several infiltration wells were damaged or had changed functions. The entire Biopores had been buried. The groundwater recharge volume was calculated based on these data and equation (13). In calculating the cross-sectional area of the well, it was assumed that there was a linear reduction in the depth of the well. This depth reduction was due to adding an average of 50 cm thick sediment in the wells inspected in the field.

Table 1. Validation result compared with potential PWI calculation

Positive Water Impact (PWI)	**2011**	**2012**	**2013**	**2014**	**2015**	**2016**	**2017**	**2018**	**2019**	**2020**	**M Liter/ Year**	**Liter/Year**
Tree Planting (Ha)	1,011	1,011	1,011	1,011	1,011	1,011	1,011				1,011	1.011.175
Infiltration Well (Unit)				1,947	1,521	2,322	2,079	1,998	1,653	1,764	1,898	1.897.629
Infiltration well direct from rainwater harvesting (Unit)							0,215	0,207			0,211	211.050
Water pond (Unit)							31,536				31,536	31.536.000
Biopores (Unit)					0,005		0,006				0,005	5.325
Trench (Unit)								0,148	0,148		0,148	148.200
Validation	**2011**	**2012**	**2013**	**2014**	**2015**	**2016**	**2017**	**2018**	**2019**	**2020**	**M Liter/ Year**	**Liter/Year**
Tree Planting (Ha)	1,065				1,074	1,074	1,074	1,076			1,073	1.072.576
Infiltration Well (Unit)							2,194	2,108	1,744	1,862	1,977	1.976.825
Infiltration well direct from rainwater harvesting (Unit)							0,215	0,207	0,171		0,198	197.750
Water pond (Unit)							31,536				31,536	31.536.000
Biopores (Unit)					0,000033		0,000024				0,000	29
Trench (Unit)								0,678	0,561		0,620	619.822

Sedimentation carried by surface runoff was a serious problem encountered in the field. This was due to the loss of pyroclastic sedimentary rock and soil conditions and agricultural activities, which were the people's livelihoods in this area. This condition occurred not only in infiltration wells and infiltration wells from rainwater harvesting but also in bio pores and trench installations. The infiltration pond did not experience silting because it was continuously treated as an alternative to meet the residents' water needs.

Biopores were carried out in 2015 in as many as 3500 units, and in 2017, as many as 500 units. In 2015 assuming the Biopores were full of sediment for 30 days, the efficiency was only 30 days. If the rainy day in 2015 was 150 days, the efficiency factor was 30/150. The year 2017: Assuming bio pores was full of sediment for 30 days, the efficiency was only 30 days. If the rainy day in 2017 was 205 days, then the efficiency factor was 30/205.

The results of the Water Accessibility, Hygiene, and Water Sanitation (WASH) validation focus on the waste from the Waste Water Treatment Plant (WWTP) installation and water access facilities built by AQUA PT AGM Mekasrsari and PT AGM Babakanpari. Validation was carried out on the quality and quantity of these facilities using a water quality checker and water meters installed in each facility. There were 2 WWTP facilities, 1 in PT AGM Mekarsari and 1 in PT AGM Babakanpari. From the potential value base on the permit granted for 500 m3/day, field validation showed that the WWTP outputs in both locations with an average of 200 and 290 m3/day. Discharge is well recorded, and momentary physical-chemical quality measurements show that the pH value of the wastewater is still below the required quality standard. 47 units of clean water supply facilities for the community and installed water meters with a total of 1,189 million liters/year

4. Conclusion

The results of validating the positive water impact program at Kubang water sources indicated that the potential calculation method and activity planning could be well validated. The difference in the total calculation of potential program plans with the validation in 2011-2020 occurred due to differences in the number of activities validated in the field. As the first initiative in Indonesia, the positive water impact activities validation showed a good result and needed to be continuously developed. This activity must be implemented in other water source locations to maintain continuous discharge production and sustainable water environmental services.

Acknowledgments
The authors would like to thank the Head of Research Center for Limnology and Water Resources, BRIN, the Head of Research Center for Environment and Clean Energy, BRIN, and Mr. Nico Anatoly for the advice and discussion while writing the paper. Also, all the reviewers for giving many inputs to enhance the paper. The authors gratefully acknowledge the financial support from PT. Aqua Golden Mississippi Research Organization for Earth and Maritime, Indonesia National Research and Innovation Agency (ORKM BRIN).

References
[1] United Nations Industrial Development Organization 2021 Water stewardship
[2] UNPAD-Sorbonne University 2021 *KUBANG Conceptual Model Summary*
[3] Reig P, Larson W, Vionnet S and Bayart J B 2019 *Volumetric Water Benefit Accounting (VWBA): a Method for Implementing and Valuing Water Stewardship Activities*
[4] Neitsch S ., Arnold J ., Kiniry J . and Williams J . 2011 *Soil & Water Assessment Tool Theoretical Documentation Version 2009*
[5] (SCS) S C S 1972 *National Engineering Handbook, Section 4: Hydrology* (Department of Agriculture Year)
[6] McCuen R H 1982 *A Guide to Hydrologic Analysis Using SCS Methods* (Universitas Michigan: Prentice-Hall)
[7] Arifjaya, N.M., Prastowo, Soedomo, S., Risdianto, I., Ardiyanto, Y., Feligi, F.G., Hidayat, T., Hartono, S., Ramadhan T . 2016 *Laporan Akhir Model Konservasi Terpadu di DAS Cicatih Menggunakan SWAT-ModFlow* (Bogor)
[8] Meinzer O E 1923 *The Occurrence of Ground Water in the United States* vol 489
[9] Richard W. Healy 2010 *Estimating Groundwater Recharge :* (United States: Cambridge University Press :)
[10] Nimmo J R, Horowitz C and Mitchell L 2015 Discrete-storm water-table fluctuation method to estimate episodic recharge *Groundwater* **53** 282–92
[11] United States Geological Survey 2017 Water-Table Fluctuation (WTF) Method

Sustainability of Slum Upgraded Area, Case of Mojosongo, Surakarta, Indonesia

S Sunarti[1]*, Nany Yuliastuti[2], Annisa Amellia Purwanto[3], and Kharunia Putri[3]
[1]Department of Urban and Regional Planning, Faculty of Engineering, Diponegoro University, Indonesia, *ID ORCID: https://orcid.org//0000-0003-3700-4016,
[2]Department of Urban and Regional Planning, Faculty of Engineering, Diponegoro University, Indonesia, ID ORCID: https://orcid.org//0000-0002-9700-0052
[3]Department of Urban and Regional Planning, Faculty of Engineering, Diponegoro University, Indonesia
[1]*sunarti@pwk.undip.ac.id, [2]n.yuliastuti@pwk.undip.ac.id, [3]aamellia30@gmail.com, [4]kharuniaputri13.kp@gmail.com

Abstract. Sustainable development is a challenge in urban development, especially in a slum area. Management of slum is not optimal, and several slums become slums again after upgrading due to the absence of environmental management, such as Mojosongo Housing, Surakarta City, Indonesia. This article aims to create policy recommendations for environmental quality improvement in slum. The research uses a quantitative method. Secondary data were obtained from related agencies regarding environmental conditions, population data, and policies related to slum management. Primary data was collected via observation and questionnaires to 30 respondents, consisting of natives and immigrants in Mojosongo Housing. The analysis technique uses a comparative analysis of data and literature review. The results show that social and economic aspects have been proven to cause a decrease in environmental quality. The challenge of slum management in strategic areas to be sustainable is via community empowerment of social and economic aspects of livelihoods. In addition, supervision and control of development for the public interest are recommended to be more effective by applying police power (an instrument of zoning regulation) and a taxation mechanism.

1. Introduction

Slums are an urban phenomenon often found in developing countries [1,2]. Slums are areas that decrease the quality of life due to the community's social and economic powerlessness, which causes environmental degradation [2]. Slums in urban areas reflect the failure of development strategies to accommodate urbanization and population growth [3]. Slums affect the city's image due to urban development's decline in environmental, economic, social, and cultural quality [4]. The negative impacts of slums forced the government to solve it through the slum upgrading as an implementation of SDGs (Sustainable Development Goals) number 11 namely cities inclusion without slums. Based on data from the Ministry of Public Works and Public Housing, the area of slum in Indonesia reached 86,562 Ha in 2020 [5], which was previously 37,407 Ha in 2014 [6]. The increase of slums shows that the management of slums has not been solved optimally.

Sustainable development is a challenge in urban development, especially in slum [7]. The management of slums often focuses on certain aspects, especially physical ones [8]. Management of slum needed a comprehensive approach [7, 9] and the involvement of all stakeholders [10, 11].

Therefore, slum upgrading should consider the integration of social, economic, and environmental aspects by the sustainable development concept to ensure the success of improving settlement quality.

Slum in Indonesia became slums again after upgrading, one of which is the Mojosongo Housing, Surakarta City. Mojosongo housing was successfully upgraded through the SSUDP (Semarang Surakarta Urban Development Program), urban renewal development action plan, and FED (final engineering design) program in 1999. However, after more than 20 years, Mojosongo Housing re-registered as a slum area in the Surakarta Mayor's Slum Decree Number 640/69.9 of 2020. Mojosongo housing, located in a strategic area as an economic growth area, which the Mojosongo Market supports, led to new activities requiring space. The space requirement uses other space functions that illegal. In addition, the transfer of house ownership, the low capacity of the community, and the absence of community groups have an impact on the decrease of environmental quality in Mojosongo Housing.

Previous researchers have explored slum management related to the physical, economic, and social conditions [3,12,13]. Discussions related to stakeholder involvement in slum management [14,15] and slum transformation [9,16-18] also have been analyzed in previous studies. The transformation of slums is a change from a state of physical and non-physical that requires changes in the community's culture, behavior, and mindset [19]. However, the debates on how local governments maintain and supervise to increase environmental quality are still little discussed by previous researchers. Hence, this article aims to create policy recommendations for improving the environmental condition in Mojosongo Housing.

2. Methods

The research used quantitative methods, deductively conducted via a synthesized formulation of theories [20]. The research was conducted for five months, from March to July 2022. Secondary data was collected from agencies related to slum management in Mojosongo Housing and government policies in slum management. Primary data was collected through observation to determine changes of environmental condition and questionnaire. The indicators in the questionnaire are (1) social (education, activities, organization), (2) economic (employment, income, land value), and (3) environmental (facilities and infrastructure). The research respondents are the community in Mojosongo Housing, consisting of native and immigrants. The number of samples used Slovin formula with a margin of error (e) 10% and the population (N) is 51 households. The detailed calculation of the respondents (Eq.1):

$$Sample\ (n) = N \div (1 + (N \times e^2)) = 51 \div \left(1 + (51 \times 10\%^2)\right) = 34\ respondents \qquad \textbf{(1)}$$

Sampling stopped at 30 respondents due to homogeneous population characteristics and consistent answer. The samples used has been considered representative, because the data results have been calculated for consistency. According to Sugiyono [21], the number of samples could be considered adequate if it has reached the level of "redundancy," which means that the data obtained are saturated.

3. Results and Discussion

The management of slums in Mojosongo Housing used land consolidation in RT 3 RW 1 (RT is the division of territory equal to a neighborhood unit, and RW is the division equal to a hamlet). The upgrading was conducted on 51 illegal housing units (24-200 sqm/unit). The SSUDP program certifies and arranges 51 housing units (35 sqm/unit). The program also built 10 kiosks (15 sqm) to support the economic activities and facilities (mosques, meeting halls, parks, and stages) to support the social activities. However, after 20 years, the environmental quality of Mojosongo Housing has deteriorated.

3.1. Environmental Degradation Factor in Mojosongo Housing

Mojosongo Housing as a strategic area become a magnet for trade activities and led to new activities that require space. The space requirements used existing public spaces and caused traffic congestion influencing the quality of the environment around. The economic growth in the Mojosongo area impacts the misused of the Tentara Pelajar Street sidewalk as a place to sell. The community built kiosks on the sidewalks and made canopies connected to their houses' roofs, so they seem to make the sidewalks an extension of their homes. These sidewalks caused reduced road capacity, congestion, disrupted cleanliness and reduced aesthetic value. In addition, the community also rents out some of these kiosks. This action is illegal because public space is government land and cannot be claimed for private interests.

Additional space requirements also occur in each housing unit. The increasing number of immigrants (64%) and the need for space due to increased family members have caused the community to expand their houses by utilizing public spaces. After the upgrading process, three agreements were made. There are: (a). the community is prohibited to sell the house, (b). the public space usage should fit its function, (c). the house's expansion can only add a second floor (mezzanine) without dismantling the roof. However, based on observations, there was a violation of the agreement due to the expansion of the house used an empty side and public spaces such as parks or places to dry clothes. 92% of houses have changed and only four houses maintained the original design after upgrading.

The survey results showed that the level of public trust in community leaders in Mojosongo is low due to the violation of public space usage by community leaders. Therefore, the Mojosongo Housing environment became disorganized and slum again. The violation also influenced by the majority of the population of Mojosongo Housing which immigrants dominate and small part is the 2nd generation of residents who are not fully aware of the agreement. Thus, the role of community leaders is necessary to be viable to ensure that the agreement is obeyed in Mojosongo Housing. In addition, public awareness of the maintenance of environmental quality is also needed. Sustainability of slum management can be achieved by considering environmental and ecological conditions [7,22]. The consensus after upgrading process was not written, it therefore sounds unlegalized. The researchers conclude that well-written operational regulations are needed to legalize sustainability of slum management after upgrading.

3.2. Analysis of Community Capacity Building Based on Social, Economic, and Environmental Aspects

The social aspect is one of the keys to sustainability of slum upgrading through community-based organizations to support the social identity, community empowerment, and sense belonging of community [18,22]. However, community organizations and social activities are not implemented in Mojosongo Housing. The absence of community to accommodates economic activities and the low level of community capacity made the community unable to take advantage of their area's potential optimally. The lack of the community's ability and education level that majority is intermediate has resulted in the unoptimal economic activity. The organization's formation could be done by reactivating the activities and associations by looking for new leaders that are trusted and well administered.

Based on the questionnaire, the existence of the Mojosongo Market has no significant effect. Most traders in the Mojosongo Market do not come from the Mojosongo Housing, and only 2% of the community sells in the market. Most communities are supplier in market (55%), 30% are parking attendants, and the rest are employees or civil. The average monthly income of the community is less than 2,000,000 (IDR). Moreover, several kiosks have been sold to outsider because of high land prices and economic needs. The community prefers to sell rather than develop the kiosk. Hence, it is necessary to provide community empowerment such as training and assistance to increase the knowledge and skills according to its characteristics. Suplliers in Mojosongo Housing also should be facilitated for training in increasing production and marketing. In addition, assistance could provide capital loans and network marketing formation by community organizations to increase economic turnover.

The development of the Mojosongo Housing also impacts the increasing of land value. Mojosongo housing is located in zone 281 Surakarta City with land value ranging from 2,000,000-5,000,000 (IDR) in 2020 [23]. In 1999, the community only paid 66,000 (IDR) monthly payments for 15 years. These trigger the occupants to sell their houses and move to other locations. Slums upgrading is successful if the community adapts and survives in that place [9,24,25]. Hence, community capacity building is needed to change the community's behavior and mindset to improving the quality of the environment.

Community capacity can also affect the control of environmental aspects. The misuse of public space for selling and expanding houses could be eliminated by empowerment. So, the community could be aware that environmental quality needs to be organized. First, by implementing locational and social relationships between community and environment [22]. Second, by implementing the maintenance to improve the quality of environment [26]. The success of the sustainability of slum management should be able to provide changes in the behavior and mindset of the community [27].

Community capacity building requires government assistance starting at the lowest local government [28] as activity driver, controller, and supervisor. Community capacity building also requires innovation, creativity, and technology to achieve a sustainable environment [22,29]. Increasing community capacity via empowerment is expected to facilitate community capacity building, access to productive resources, and participation actively in decision-making [26].

3.3. Challenges of Improving the Environmental Quality According to Sustainable Development Concept

As stated in SDGs number 11, improving the environmental quality should be supported by strengthening the social and economic aspects. Figure 1 is a diagram of the challenges of sustainable Mojosongo Housing development.

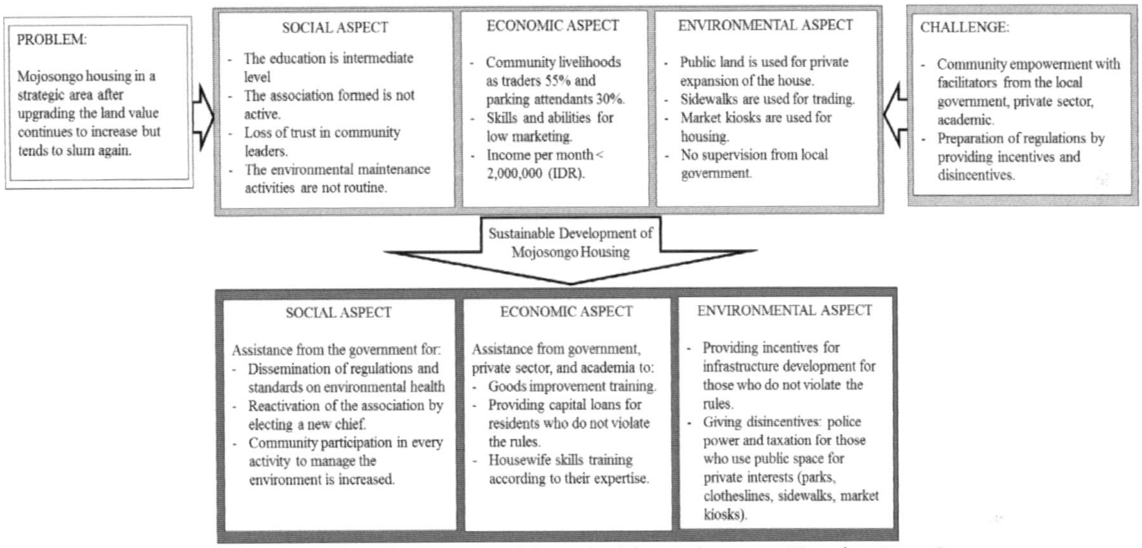

Figure 1. Diagram of The Challenges of Sustainable Mojosongo Housing Development.

The decreasing quality of the environment, the low level of education, and the community's economy in Mojosongo Housing are challenges for the local government to formulate directions and maintenance for sustainable development in slum [26,30]. The decreasing environmental quality in Mojosongo Housing is caused by sidewalks used for selling activities, and house expansion in public spaces. Misused of public space is caused by the absence of regulations related to land use restrictions. According to Vatter [31], the absence of land use restrictions resulted in the community being free to use land and rent out land according to their interests. The researchers believe it is necessary to implement police power (an instrument of zoning regulation) and taxation mechanisms by providing incentives and disincentives. Police power is the government's authority to control the community and property within its jurisdiction for the benefit of public security, health, safety, morals, and welfare [32]. By police power, the government could control operational and responsive development to reduce inappropriate land use [33]. Government intervention is needed via zoning regulations and financial or fiscal incentives for the community who had preserved the function of public space [34]. Furthermore, incentives and disincentives could be in subsidies, taxes, credits, property rights, capacity building, prohibitions, and obligations [35]. In this study, incentives are recommended through tax breaks and ease of development licensing. Meanwhile, disincentives can be given through the imposition of taxes and levies paid for misuse of lands, such as sidewalks for selling and public spaces for house expansion.

4. Conclusion

The research shows that the economic reasons caused Mojosongo Housing to become a slum again after being upgraded in 1999. The motive of surviving livelihood is one of the main obstacles to maintaining qualified housing conditions. Several misuses of claiming public land for individual interest are evidence of local government's low control and law enforcement. In this article, the combination of social and economic aspects has been proven to cause a decrease in environmental quality in Mojosongo. On the social aspect show that people's mindsets tend to focus on quick profits without thinking about long-term economic benefits. This is evident in the choice of selling houses and kiosks rather than developing a household economic business. The absence of a community that controls efforts to improve environmental quality exacerbates this condition. The economic aspect of Mojosongo Market also worsens the environmental quality caused by congestion and misused of sidewalks for selling.

The low level of community capacity should be supported by the active role of local government in creating control mechanisms. This article recommends policies that support the effective implementation for formulating regulations that can operationally control public space use through police power and taxation. Police power in the form of sanctions is not given legality to communities that use public space for personal interests, while traders who use public space are subject to high taxes and control. The government's authority is crucial to legalize justice, safety, and reduce negative impacts on Mojosongo's housing and surroundings. Collaboration of all stakeholders in the maintenance, supervision, and control to achieve sustainable slum development is essential. The research limitation is the limited number of respondents and the difficulty of finding respondents who work in the informal sector. Future research suggests collaborating with stakeholders in monitoring and controlling slum.

Acknowledgments

This research was financially supported by the Faculty of Engineering of Universitas Diponegoro, Indonesia through the 2022 Strategic Research Grant with a letter of assignment to conduct research number: 3178/S/pwk/4/UN7.5.3.2/PP/2022. Thank you to those who have helped in this research that are Development Planning Agency of Surakarta; residents of Mojosongo Berseri I Housing.

References

[1] Boonyabancha S 2009 Land for housing the poor-by the poor: Experiences from the baan mankong nationwide slum upgrading programme in thailand *Environment and Urbanization.* **21** 309-29
[2] Davis M 2013 Planet of slums *New Perspectives Quarterly.* **30** 11-2
[3] Degert I, Parikh P and Kabir R 2016 Sustainability assessment of a slum upgrading intervention in bangladesh *Cities.* **56** 63-73
[4] Fitri D A and Sulistinah 2021 Faktor-faktor penyebab munculnya permukiman kumuh daerah perkotaan di indonesia (sebuah studi literatur) *J. Swara Bhumi.* **1** 1-9
[5] Ministry of Public Works and Public Housing of the Republic of Indonesia 2020 *Data and resources kawasan kumuh di indonesia, (*Jakarta: Ministry of Public Works and Public Housing of the Republic of Indonesia)
[6] Ministry of Public Works and Public Housing of the Republic of Indonesia 2014 *Buletin ciptakarya: Lembaran baru penuntasan kawasan permukiman kumuh, (*Jakarta: Ministry of Public Works and Public Housing of the Republic of Indonesia)
[7] Un-Habitat 2012 *Sustainable housing for sustainable cities: A policy framework for developing countries, (*Nairobi: UNON)
[8] Clapham D 2005 *The meaning of housing: A pathways approach, (*United Kingdom: Policy Press)
[9] Sunarti S, Rahdriawan M, Dewi A and Widiarta J 2018 Hierarchy model of tambak lorok slum area transformation to support marine tourism kampong. *Proc. Int. Conf. on Cities 2017: Multi Perspectives on Peri-Urban Dynamics Towards Sustainable Development*, 2018 Surabaya, Indonesia: IOP Publishing) p 1-9

[10] Kotus J and Sowada T 2017 Behavioural model of collaborative urban management: Extending the concept of arnstein's ladder *Cities.* **65** 78-86
[11] Sunarti S, Yuliastuti N, Dewi L A and Bellia P S 2020 Sustainability development environment: Improving social economy in kalikondang village, demak regency *E3S Web of Conf.* p 1-9
[12] Doe B, Peprah C and Chidziwisano J R 2020 Sustainability of slum upgrading interventions: Perception of low-income households in malawi and ghana *Cities.* **107** 102946
[13] Sunarti S, Esariti L and Haulah L 2019 Spatial assessment of slum area in jepara city from physical and non-physical aspects perspective *Proc. Int. Conf. on the 3rd Geoplanning.* (Semarang: IOP Publishing) p 1-10
[14] Archer D 2012 Baan mankong participatory slum upgrading in bangkok, thailand: Community perceptions of outcomes and security of tenure *Habitat International.* **36** 178-84
[15] Purwanto E, Sugiri A and Novian R 2017 Determined slum upgrading: A challenge to participatory planning in nanga bulik, central kalimantan, indonesia *Sustainability.* **9** 1261
[16] Aduwo B, Ibem E O and Opoko P A 2013 Residents' transformation of dwelling units in public housing estates in lagos, nigeria: Implications for policy and practice *Int. J. of Education and Research.* **1** 5-20
[17] Avogo F A, Wedam E A and Opoku S M 2017 Housing transformation and livelihood outcomes in accra, ghana *Cities.* **68** 92-103
[18] Sunarti S, Syahbana J A and Manaf A 2019 Space transformation in a low-income housing community in danukusuman, surakarta *Int. J. of Housing Markets and Analysis.* **12** 265-80
[19] Poutiatine M I 2009 What is transformation?: Nine principles toward an understanding of the transformational process for transformational leadership *J. of Transformative Education.* **7** 189-208
[20] Benu F L and Benu A S 2019 *Metodologi penelitian kuantitatif, (*Jakarta: Prenada Media Group)
[21] Sugiyono 2016 *Metode penelitian kuantitatif, kualitatif dan r&d, (*Bandung: PT Alfabet)
[22] Smets P and Lindert P V 2016 Sustainable housing and the urban poor *Int. J. of Urban Sustainable Development.* **8** 1-9
[23] Ministry of Agrarian Affairs and Spatial Planning/Head of National Land Agency of the Republic of Indonesia 2022 *Online data base zona nilai tanah, (*Jakarta: ATRBPN)
[24] Sunarti 2016 *Transformasi kawasan kumuh di kota surakarta (studi kasus: Danukusuman dan mojosongo)* Disertasi Doktor, Universitas Diponegoro
[25] Sunarti S, Syahbana J A and Manaf A 2014 Slum upgrading without displacement at danukusuman sub-district surakarta city *Int. Trans. J. Eng. Manag. Appl. Sci. Technol.* **5** 213-25
[26] Sunarti S, Helmi M, Widjajanti R and Purwanto A A 2021 Modelling behavioural change from sea-based 'helicopter latrines' to land-based shared improved latrines in the demaan, jepara, indonesia *J. of Water, Sanitation and Hygiene for Development.* **11** 10-25
[27] Asa, Adekunle O, Lawal, Omotosho K, Morakinyo and Opeyemi K 2018 Informal transformation in public housing: A case study of abesan low income housing estate lagos, nigeria *Int. J. of Civil Engineering, Construction and Estate Management.* **6** 1-13
[28] Yuliastuti N, Wahyono H, Syafrudin S and Sariffuddin S 2017 Dimensions of community and local institutions' support: Towards an eco-village kelurahan in indonesia *Sustainability.* **9** 245
[29] Moulaert F, Maccallum D, Mehmood A and Hamdouch A 2013 *The international handbook on social innovation: Collective action, social learning and transdisciplinary research, (*United Kingdom: Edward Elgar Publishing) p 53-66
[30] Sudmeier-Rieux K, Paleo U F, Garschagen M, Estrella M, Renaud F G and Jaboyedoff M 2015 Opportunities, incentives and challenges to risk sensitive land use planning: Lessons from nepal, spain and vietnam *Int. J. of Disaster Risk Reduction.* **14** 205-24
[31] Vatter M H 2020 Stratified zoning in central cities *J. of Housing Economics.* **50** 101716
[32] Merriam-Webster.Com 2022 Police power. Dictionary, Merriam-Webster
[33] Gallent N and Kim K S 2001 Land zoning and local discretion in the korean planning system *Land Use Policy.* **18** 233-43

[34] Koomen E, Dekkers J and Van Dijk T 2008 Open-space preservation in the netherlands: Planning, practice and prospects *Land Use Policy.* **25** 361-77
[35] Undang-Undang Nomor 26 Tahun 2007 Tentang Penataan Ruang.

Environmental Aspects of City Participatory Budgeting in Poland

Brygida Klemens[1]*
[1] Department of Regional Policy and Labor Market, Faculty of Economics and Management, Opole University of Technology, Opole, 45-036, Poland, ORCID ID: 0000-0002-4606-163X
[1]*b.klemens@po.edu.pl

Abstract. Participatory budgets are an increasingly common form of co-decision of residents about the quality of life and development directions of urban space. In Poland, participatory budgets have been realized for a decade, and projects submitted to these budgets respond to needs such as e.g. leisure infrastructure, environmental conditions, recreation, health, and education. The study aims to analyze the phenomenon of civic participation and the use of participatory budgets as a tool of co-governance in the context of environmental issues and urban development. A desk research analysis of national and foreign literature data was performed, and the existing statistical data on participatory budgets in Poland were analyzed. In addition, the websites of selected cities in Poland were analyzed in the context of implemented projects and citizen's activity within the framework of participatory budgets. The results indicate that the initial enthusiasm of the citizen's implementation has decreased, but the COVID-19 pandemic has aroused renewed interest in the topic. The study's implication is recommendations, e.g.: creating a civic budget database featuring uniform categories or creating Green Civic Budgets in every city.

1. Introduction

The development of cities is determined by many factors, such as economic, legal, sociological, political, cultural, environmental, etc. Today, cities demonstrate the highest ability to create economic growth and play an increasingly important role in formulating and implementing development policy at the national and subnational levels [1]. If one understands the social and economic development of areas as the tempo and direction of changes in local communities that lead to better living conditions and quality and that the spatial structure resulting from this growth is a derivative of many circumstances, it is possible to classify them in several primary groups [2]: circumstances related to the location in geographic space; related to the position in economic structures; natural environment circumstances; human environment quality; technical infrastructure development; social infrastructure development; activity and effectiveness of local self-governments; entrepreneurship and innovativeness of the local community. Therefore, a territory's growth is affected by a combination of endogenous factors, which are only slightly dependent on the local community's activity, and exogenous factors in which the local population's features are critical. At the same time, the role of endogenous factors, such as technological innovations [3], government authority [4], cultural heritage, or social participation [5], is becoming increasingly emphasized.

More and more papers are devoted to the development of urban ecosystems that can be viewed as the real natural greenery and blue areas that provide a whole range of services, such as air purification, micro-climate regulation, noise, and vibration reduction, rainwater collection, aesthetic experience, recreational or educational services [6-10]. Moreover, it can be estimated that the benefits of an urban ecosystem can substantially reduce a city's operating costs, e.g., by cooling urban areas using trees [7,11]. The introduction of greenery, trees especially, to the urban environment is generally met with high approval and is a very welcome element of the environment [12,13]. It is currently emphasized that the availability of green areas is deemed as vital as essential services [14].

High quality of life in cities requires making decisions to benefit the environment, such as urban climate improvement, smog and car traffic reduction, expansion of urban green areas, and walking and recreational spaces [15]. All city users are satisfied when they can participate in their desired consumption [16-17]. Recognition of new processes or the changing role of development stakeholders can be a substantial developmental opportunity. Modern cities face many problems, such as urban land or apartment shortages, low communication availability, many vacant buildings that generate gaps in the urban fabric, and hindering city consolidation as a common good [18]. There are also problems of air pollution, noise, and little green and blue infrastructure [19], i.e., environmental aspects.

Regardless of the size of cities, their developmental issues require the involvement of various stakeholders, including competent authorities, inhabitants, entrepreneurs, or non-government organizations. To implement the city co-management model, all stakeholders should be able to act knowingly to benefit the city's present and future [20]. Moreover, external entity behavior can also contribute to the improvement of the city's operating conditions. It is crucial to skilfully determine the direction of changes, which is the basis for forecasting and planning processes. Viewing a city as a common good facilitates comprehension that collective property will be defended and transformed only when all stakeholders are involved in the process and considering all aspects and elements of the common good. More importantly, it fits into the territory and becomes a system in which nature, history, and heritage are intertwined [21-22].

Contemporarily, discussions over public management involve using the participative management model [23]. It is referred to as the interaction between the self-government administration and the community using consulting procedures [24] and involving all stakeholders in the decision-making process. The social participation phenomenon has a long-standing tradition, and a dynamic increase in literature on this topic can be observed from the 1960s [25]. Inhabitant participation is the foundation of a democratic state but applies to regional or local democracies. There are no contraindications against the inhabitants' use of urban public goods. However, getting involved in the decision-making on the manners of using such goods brings a new quality to this process and is becoming increasingly popular [26-27]. Political participation involves participating in elections, referendums, public consultation, political meetings, etc. In contrast, non-political participation focuses on meeting the needs of various social groups inhabiting a familiar territory and involves various activities unrelated to elections [28-29].

Inhabitant meetings play an essential role in building local social participation but mainly include rural meetings as meetings in other units are incidental [29]. Active local communities are more rarely in cities, and people live with the feeling of no daily interpersonal relations. However, technology and *social media* play an increasingly important role in taking participatory action by enabling the organization of large groups of inhabitants around a given topic in nearly real-time. The COVID-19 pandemic has substantially accelerated the transfer of citizen participation to virtual reality and ITC technologies.

Social participation concerns various topics, including the environment [30-31]. A manifestation of such activities in Poland is, for example, the smog alarm, which is a social movement aimed at reducing air pollutant emissions. The national Polish Smog Alarm was launched in 2015 and was a continuation of the urban project launched in 2012 in Kraków [32]. A feature characteristic of Poland is the inhabitant participation deficit, which is a result, among others, of the management process in self-government, which is (incorrectly) viewed as governance instead of acting in the inhabitants' interest in the course of public affairs management [33].

One manifestation of city co-management is the use of the participatory budget tool. It allows for involving the widest stakeholders circle possible in the decision-making process [34]. The inhabitants declare the projects they deem necessary and then vote to decide whether to use public funds to implement them. A participatory budget is a "decision-making process in which the inhabitants co-create the given city's budget, thereby making the common decision on the distribution of a specific amount of public funds" [35].

The first participatory budget, as a bottom-up initiative, was announced in 1989 in Porto Alegre, Brazil. In the 1990s, it was disseminated in Latin America and then in European states [36]. In Poland, the participatory budget (also known as the civic budget) does not have a long history, as the first budget of this kind was introduced in Sopot in 2011. However, it is becoming more popular as a tool for activating and mobilizing urban communities. The deficit of inhabitant participation in the widely understood social dialogue resulted in legislative changes which implemented the concept of civic budget in the legislation in 2018 [37]. The civic budget is becoming more popular in Poland.

Regarding environmental challenges, the local level (cities) plays a prominent role in the fight for clean air and implementing climate change adaptation activities. Inhabitants displeased with urban politics often decide to implement specific activities and projects favorable for the environment and people. Many initiatives aim to make cities more "green" [31]. It also turns out that the chance to implement "green projects" increased in cities where self-government units provided support in preparing project applications for the civic budget, as in the case of the "Kraków in green" initiative and the fact that every fifth project concerned the urban greenery in the 2017 edition in Kraków [38].

The participatory budget brings many benefits for the inhabitants, e.g., urban space arrangement and development, enriched infrastructure, small architecture, increasing the sense of inhabitants responsibility for the city, inhabitant activation, or the sense of improving the quality of life by providing the ability to take walks or other everyday activities in a friendly environment, air pollution reduction through green areas expansion or more sustainable transport, and many others [39, 7]. The study aims to analyze the phenomenon of civic participation and the use of participatory budgets as a tool of co-governance in the context of environmental issues and urban development.

The COVID-19 pandemic caused substantial changes in the operation of cities and the way of thinking of their inhabitants. The social movement called the ECG (*Economy for Common Good*) deemed the current epidemic crisis as an opportunity to change the paradigm of the dominant economic model that contributes to climate changes, biodiversity decline, and aggravating social and economic inequalities toward a better future [40, 41].

2. Methods

The paper is focused on a review of the literature concerning the concept of citizen participation in cities, mainly in terms of the use of the participatory budget tool. The general context of the deliberations included environmental aspects and city development. A desk research analysis of national and international literature data was conducted. The criteria of the literature used in this study were the research topic focused on cities and environmental context.

The statistical data were obtained from reports of the Institute of Urban and Regional Development (data from 2020 and 2021), an institution dealing in urban development, and from the selected cities' websites. The urbanization rate in Poland has been decreasing systematically for over a dozen years: in 2005, it was 61.39%, whereas, in 2020, it was 59.86% [42]. However, it is still the predominant share of the population. Given this data, the deliberations on the right to the city are justified and timely.
Two research hypotheses were defined:

H1: The civic budget is an increasingly popular form of participation used by city inhabitants.

Taking into account the impact of the COVID-19 pandemic on the changes in public services provision, a second research hypothesis was defined:

H2: The possibility of long-distance voting in the civic budget made inhabitants more eager to participate in public life in this form during the COVID-19 pandemic.

3. Results and Discussion

3.1. Urban Civic Budgets in Poland in 2021

In 2021, Poland's number of successful projects in urban civic budgets was 3,745. The Institute for Urban and Regional Development that conducted this analysis has divided the successful projects into 13 categories: urban furniture (985 projects, 26.3%); physical activity infrastructure (902 p., 24.1%); communication routes (772 p., 20.6%); lectures, events, actions (745 p., 19.9%); greenery (700 p., 18.7%); fixed asset purchases (16.3%); small architecture I: small objects (11.7%); networks and systems (7.6%); small architecture II: roof, walls, roofing (6.9%); buildings (3.9%); traffic organization (3.2%); art and heritage (3.2%); other (1.9%) [43]. The successful projects could have concerned various categories. Therefore, the information does not add up to 100%.

3.2. "Green" Category in Urban Civic Budgets in Poland in 2020 and 2021

Table 1 presents the number of projects in the “greenery” category. however, this category does not exhaust all projects that could be described as “environmental”. The data was presented for 2020 and 2021 for comparison purposes.

Table 1. Number (share) of projects in urban civic budgets in the “greenery” category in Poland in 2020 and 2021.

Activity	Year	Small cities (5-20 thousand)	Medium cities (20-50 thousand)	Medium cities (50-100 thousand)	Large cities (100-200 thousand)	Large cities (> 200 thousand)	Total
Trees	2021	10 (23.8)	31 (43.7)	27 (42.2)	41 (35.7)	188 (46.1)	297 (42.4)
	2020	6 (40)	19 (36)	12 (38)	30 (35)	191 (49)	258 (45)
Bushes	2021	13 (31)	24 (33.8)	21 (32.8)	42 (36.5)	160 (39.2)	260 (37.1)
	2020	3 (20)	22 (42)	8 (25)	27 (32)	142 (36)	202 (35)
Greenery (general)	2021	20 (47.6)	13 (18.3)	26 (40.6)	34 (29.6)	69 (16.9)	162 (23.1)
	2020	6 (40)	7 (13)	13 (41)	30 (35)	66 (17)	122 (21)
Flowers and decorative plants	2021	10 (23.8)	16 (22.5)	14 (21.9)	23 (20)	80 (19.6)	143 (20.4)
	2020	6 (40)	11 (21)	6 (19)	16 (19)	88 (23)	127 (22)
Lawns	2021	4 (9.5)	21 (29.6)	8 (12.5)	21 (18.3)	60 (14.7)	114 (16.3)
	2020	4 (27)	7 (13)	2 (6)	10 (12)	57 (15)	80 (14)
Outdoor maintenance	2021	2 (4.8)	6 (8.5)	2 (3.1)	9 (7.8)	37 (9.1)	56 (8)
	2020	0	0	1 (3)	0	10 (3)	11 (2)
Flowery meadows	2021	1 (2.4)	2 (2.8)	3 (4.7)	7 (6.1)	35 (8.6)	48 (6.9)
	2020	0	6 (11)	2 (6)	9 (11)	32 (8)	49 (9)
Rain gardens	2021	0	1 (1.4)	1 (1.6)	5 (4.3)	8 (2)	15 (2.1)
	2020	0	3 (6)	0	0	14 (4)	17 (3)
Green roofs	2021	1 (2.4)	1 (1.4)	0	1 (0.9)	20 (4.9)	23 (3.3)
	2020	0	2 (4)	0	1 (1)	7 (2)	10 (2)
Clearing (greenery removal)	2021	2 (4.8)	1 (1.4)	2 (3.1)	4 (3.5)	14 (3.4)	23 (3.3)
	2020	0	2 (4)	1 (3)	0	4 (1)	7 (1)
Other greenery types	2021	1 (2.4)	3 (4.2)	1 (1.6)	8 (7)	10 (2.5)	23 (3.3)
	2020	0	6 (11)	0	5 (6)	26 (7)	37 (6)
Projects with at least one activity in the category	2021	42 (100)	71 (100)	64 (100)	115 (100)	408 (100)	700 (100)
	2020	15 (100)	53 (100)	32 (100)	85 (100)	390 (100)	575 (100)

Source: [43, p. 69; 44, p. 51]

The overall number of projects in the “greenery” category of the Polish urban civic budgets increased from 575 to 700. The projects most implemented preferably in this field were classified into the following categories: trees; bushes; greenery (general); flowers and decorative plants; lawns. Less numerous projects concerned green roofs or rain gardens, among others.

3.3. Disadvantages of Civic Budgets

It is worth noticing that the civic budget is also criticized. The most common objections include the following: treating the budget as a trend, something temporary; using it as a political marketing tool; substantial limitation of inhabitant participation, e.g., in works over the budget's regulations by public officials; treating the budget as a form of the grant competition, etc. (more examples in [45]).

3.4. Civic Budget Voters in Some Polish cities

These factors contributed to the decline of overall interest in taking part in the civic budget. An analysis conducted for the six biggest cities in Poland (Warsaw, Kraków, Łódź, Wrocław, Poznań, and Gdańsk) demonstrated that voting declined systematically from 2014-2018 (Table. 2).

Table 2. Civic budget voters in the biggest Polish cities.

year	Voters
2014	654,307
2015	634,289
2016	503,289
2017	434,481
2018	411,273

Source: [46]

The period of COVID-19 pandemic had a positive impact on the interest in the participatory budget. When analyzing the data for three out of the six biggest Polish cities (no comparable data for all cities), it is possible to notice that the number of votes is systematically increasing (Table 3).

Table 3. Civic budget voters in Warsaw, Wrocław, and Kraków in 2019-2021

year	Voters
2019	229,938
2020	245,828
2021	253,020

Source: [47-49]

3.5. The COVID-19 Pandemic Impact on Civic Budgets in Polish cities

The COVID-19 pandemic and the related *lockdown* forced the inhabitants to remain locked up, occasionally in complete isolation. The inability to leave the apartment or the ability to leave it for a very short time was an everyday reality and a substantial challenge for city inhabitants. The COVID-19 pandemic caused an increase in the city inhabitants' interest in migrating to the countryside [50-51]. ICT technologies have substantially improved the inhabitants' access to public services and various forms of inhabitant participation [52]. The COVID-19 pandemic has accelerated technological changes and the implementation and use of electronic systems in an unprecedented way. A substantial number of administrative matters can be handled using remote technologies, meaning that access to administrative public services is easier, faster, and cheaper. Inhabitants of cities have better infrastructure that enables them to use ICT technologies. The case may be that these factors resulted in an increased interest in civic budgets during the pandemic.

3.6. Environmental Aspects in Civic Budgets in Polish Cities and Best Practices

The negative effects of climate change are becoming increasingly noticeable. There is an intensification of violent weather phenomena that takes on the nature of long-term temperature anomalies, water availability issues, high temperatures caused by the heat output of buildings, or weak green and blue infrastructure [53, 7]. Environmental aspects in city development are becoming an 6increasingly pressing issue [54]. Around the world, we have the so-called *best practices* on how to deal with environmental issues to make cities more sustainable and friendly to their inhabitants [55-57]. Some cities went a step further and implemented the so-called Green Civic Budgets. One example of this is Gdańsk which announced the enrolment for such a budget for the third time. In 2023, it will feature five topics: a) vegetation planting and care in urban greenery areas; b) new green areas; c) rain garden construction; d) restored and modernized development of existing green areas; e) green activities [58]. The implementation of environmental projects that would improve quality of life and make it possible to stay in green areas that are unrestricted by various regulations on using protective masks and social distance is a perfect fit for the civic budget's concept.

4. Conclusion

Did the possibility of long-distance voting in the civic budget made inhabitants more eager to participate in public life in this form during the COVID-19 pandemic? The analyzed data shows that the pandemic has contributed to the reversal of the attitudes of city inhabitants towards using ICT tools and taking greater care of environmental issues in 2019-2021. It can be stated that hypothesis 2 was verified correctly. In the case of hypothesis 1, the time before the COVID-19 pandemic suggests a rather negative verification.

Civic budgets can be great citizen participation tools provided that they are used to improve the inhabitants' quality of life and their involvement in the co-determination of the cities' future. It is necessary to focus on broader dissemination of the concept of creating Green Civic Budgets that would serve to determine the inhabitants' needs concerning the broadly understood environmental protection and ecology. A good practice would also be the establishment of workshops on the preparation of civic budget applications and even broader promotion of successful projects.

The analysis of civic budgets is made difficult due to the lacking a uniform method for their categorization and lacking a database, including a database of archive data. The methodology for assessing the implementation of civic budgets is also created individually for each budget. A good recommendation would be to create a civic budget database featuring uniform categories. Future research will cover a wider scale of cities, rural areas, and regions in Poland.

References

[1] Klemens B and Derlukiewicz N 2019 *Tourist cluster and their significance for the economy of cities,* (Madrid: IBIMA) pp 10331-10343

[2] Heffner K and Klemens B 2013 Koncepcje zmian i nowe procesy przestrzenne na obszarach wiejskich w Polsce *Kryzys finansowy a programowanie rozwoju jednostek przestrzennych*, ed S Korenik and A Mempel-Śnieżyk (Wrocław: PN UNI EKON) p 91

[3] Gorzelak G 2000 Zewnętrzna interwencja jako czynnik rozwoju lokalnego (na przykładzie Programu Inicjatyw Lokalnych) *Studia Regionalne i Lokalne* 3(3).

[4] Tuziak B 2016 Autorytet władz gminnych jako element kapitału społecznego a nierówności w rozwoju na poziomie lokalnym *Nierówności Społeczne a Wzrost Gospodarczy* 48 (4):347-359.

[5] Sowada T 2019 Ruchy miejskie w działaniu. Oblicza partycypacji *Studia i Prace z Geografii* 75, Poznań

[6] Bolund P and Hunhammar S 1999 Ecosystem Services in Urban Areas *Ecological Economics* 29(2) pp 293-301

[7] Jamontt M, Kociszewski K, Platje J, 2020 Wroclaw participatory budget as a tool to strenghten natural capital and the urban climate resilience in the years 2016-2018 *Studia Miejskie* 37

[8] Białobok S 1976 Ochrona zadrzewień i roślin w najbliższym otoczeniu człowieka *Ochrona przyrodniczego środowiska człowieka* ed. D Cichy and W Michajłow (Warszawa: PWN)
[9] Gill S et al. 2007 Adapting cities for climate change: the role of the green infrastructure *Built Environment* 33 (1) pp 115–133
[10] Bielecki A and Będkowski K 2020 The Issue of Green Area Accessibility - the Example of the City of Łódź *Studia Miejskie* 40
[11] Gómez-Baggethun E and Barton D N 2013 Classifying and Valuing Ecosystem Services for Urban Planning *Ecological Economics* 86 pp 235-245
[12] Szczepanowska H B 2012 Miejsce terenów zieleni w strukturze zintegrowanego projektowania, zarządzania i oceny ekologicznej inwestycji miejskich *Człowiek i Środowisko* 36 (1-2) pp 25-49
[13] Kendal D, Ordóñez C, Davern M, Fuller R A, Hochuli D F, van der Ree R, Livesley S J and Threlfall C G 2022 Public satisfaction with urban trees and their management in Australia: The roles of values, beliefs, knowledge, and trust *Urban Forestry & Urban Greening* Vol 73 127623
[14] Chrobak K and Kryczka P 2018 Wpływ wybranych rozwiązań urbanistycznych na dostępność przestrzenną terenów zieleni – przykład wrocławskiego Ołbina *Acta Scientiarum Polonorum Administratio Locorum* 17 (2) pp131-142
[15] Budner W and Gorynia M 2021 Czy pandemia zmieni nam miasta i regiony? *Obserwator Finansowy* 08.06.2021
[16] Czornik M 2012 *Konsumpcja miejska. Ekonomiczne refleksje nad ewoluowaniem funkcji miejskich* (Katowice: Wydawnictwo UE)
[17] Czornik M 2020 Collective Benefits as an Impulse for Urban Development *Studia Miejskie* 37
[18] Paes da Cunha C M and Gonçalves L M 2021 The city as a common good: ideas e ideals *Revista Nacional de Gerenciamento de Cidades* vol. 09 no 72
[19] Pauleit S, Andersson E, Anton B, Buijs A, Haase D, Hansen R, Kowarik I, Stahl Olafsson A and Van der Jagt S 2019 Urban green infrastructure – connecting people and nature for sustainable cities *Urban Forestry & Urban Greening* Vol 40
[20] Kawińska A 2013 Współodpowiedzialność za miasto *Odpowiedzialność - przestrzeń lokalnego społeczeństwa obywatelskiego, biznesu i polityki* ed. R Geilser (Opole: Wyd. UO) p 9
[21] Salzano E 2010 "L'habitat dell'uomo bene comune", in Cacciari P., a cura di, La societŕ dei beni comuni. Una rassegna (Roma: Ediesse)
[22] Garau C and Mistretta P 2014 The Territory and City as a Common Good *Quaderni di Careggi*
[23] Hajduk S 2018 Partycypacja obywatelska jako forma komunikacji społecznej w zarządzaniu rozwojem przestrzennym miasta *Handel Wewnętrzny* 1 (372) p 229
[24] Kuźnik F 2005 Stare i nowe koncepcje zarządzania publicznego w strukturach samorządu terytorialnego *Z teorii i praktyki zarządzania publicznego* ed. B Kożuch and T Markowski (Białystok: Fundacja Współczesne Zarządzanie)
[25] Huxley M 2013 Historicizing planning, problematizing participation *International Journal of Urban and Regional Research* 37(5)
[26] Polko A 2015 Strategie kolektywnego działania w miejskich przestrzeniach publicznych *Studia KPZK* (Warszawa: PAN) nr 161
[27] Szewczuk-Stępień M, Adamska M and Klemens B 2017 Strategie oddziaływania na interesariuszy rozwoju lokalnego kierowanego przez społeczność *Projekty lokalne i regionalne - interesariusze projektu.* ed. H Brandenburg and G Tobor (Katowice: Wyd UE)
[28] Kowalik J 2010 Aktywność obywateli w samorządzie gminnym z perspektywy władz lokalnych *Annales Universitatis Mariae Curie-Skłodowska Lublin – Polonia Sectio K* Vol. XVII issue 1
[29] Tybuchowska-Hartlińska K 2012 Partycypacja obywatelska na poziomie lokalnym *Preferencje Polityczne* nr 3
[30] van der Jagt A P N, Buijs A and Dobbs C. *et al.* 2022 An action framework for the participatory assessment of nature-based solutions in cities *Ambio*

[31] Legutko-Kobus P, Rzeńca A, Skubała P and Sobol A 2020 Miasta i ich mieszkańcy w obliczu wyzwań adaptacji do zmian klimatu (Warszawa: KPZK PAN).
[32] https://polskialarmsmogowy.pl/
[33] Sobol A 2018 Moje, nasze czyje? - Postrzeganie miasta jako dobra wspólnego *Studia KPZK PAN* Rozwój lokalny i regionalny. Teorie i zastosowania issue 184 p 278
[34] Sobol A and Rzeńca A 2018 Budżet partycypacyjny jako narzędzie polityki rozwoju małych miast województwa śląskiego i łódzkiego *Space - Society - Economy* 24
[35] Kębłowski W 2013 Budżet obywatelski. Krótka instrukcja obsługi (Warszawa: Instytut Obywatelski)
[36] Rachwał M 2013 Participatory Budgeting as a New Form of Co-decision in Local Financing *Przegląd Politologiczny* 4 pp 173-185
[37] Błaszak M 2019 Budżet obywatelski w Polsce i dylematy z nim związane *Ruch Prawniczy Ekonomiczny i Socjologiczny* 3 pp 203-220
[38] Mucha A 2019 Odmiany budżetów obywatelskich. Zieleń w budżetach obywatelskich *Raport o stanie polskich miast. Partycypacja publiczna*, ed. P Pistelok and B Martela (Warszawa-Kraków: IRMiR)
[39] Platje J, Paradowska M and Kociszewski K 2018 Ekonomika Transportu – teoria dla praktyki (Wrocław: Wyższa Szkoła Bankowa we Wrocławiu)
[40] Latemore G 2021 COVID and the Common Good *Philosophy of Management* vol 20
[41] https://www.ecogood.org/
[42] https://bdl.stat.gov.pl/bdl/start
[43] Martela B, Janik L and Bubak G 2022 Barometr budżetu obywatelskiego. Edycja 2021 (Kraków: IRMiR)
[44] Martela B, Bubak G and Janik L 2021 Barometr budżetu obywatelskiego. Edycja 2020 (Kraków: IRMiR)
[45] Pistelok P 2019 Dąbrowski Budżet Partycypacyjny - model 2.0 *Raport o stanie polskich miast. Partycypacja publiczna* ed P Pistelok and B Martela (Warszawa-Kraków: IRMiR) pp144-146
[46] https://www.miasto2077.pl/budzet-obywatelski-na-fali-spadkowej/
[47] https://bo.um.warszawa.pl/taskPropose/indexResults?regional=1
[48] https://budzet.krakow.pl/
[49] https://www.wroclaw.pl/rozmawia/wroclawski-budzet-obywatelski
[50] de Luca C, Tondelli S and Åberg H E 2020 The COVID-19 Pandemic Effects in Rural Areas *TeMA-Journal of Land Use, Mobility and Environment* (Naples: Laboratory of Land Use Mobility and Environmental)
[51] Gallagher S 2020 Escape to the Country: Will People Leave Cities Behind Post-Pandemic? *The Independent* 12.08.2020 *(Electronic Materials)*
[52] Klemens B and Solga B 2021 Pandemia COVID-19 i jej wpływ na sytuację społeczno-gospodarczą kraju i regionów *Studia Śląskie* vol. 88-89 pp 431-449
[53] van Vuuren, Detlef P et al. 2011 The Representative Concentration Pathways: An Overview *Climatic Change* 109(1) pp 5-31
[54] Klemens B, Solga B, Heffner K and Gibas P 2022 Environmental and Energy Conditions in Sustainable Regional Development *Energies* **2022** *15* 5758.
[55] Rowe P G and Hee L 2019 A City in Blue and Green. A Singapore Story (Springer Singapore)
[56] Lee H, Mayer H and Chen L 2016 Contribution of trees and grasslands to the mitigation of human heat stress in a residential district of Freiburg Southwest Germany *Landscape and Urban Planning* Vol 148 pp 37-50
[57] Westerhoff L, Sheppard SRJ, Iype DM, Cote S and Salter J 2018 Social mobilization on climate change and energy: An evaluation of research projects in British Columbia Canada *Energy Research & Social Science* Vol 46 pp 368-380
[58] https://www.gdansk.pl/budzet-obywatelski/Co-to-jest-Zielony-Budzet-Obywatelski,a,192102

Water Quality Status and Water Quality Index of Selulu Reservoir, East Kalimantan, Indonesia

Vita Pramaningsih[1*], Agus Dwi Purwanto[2], Ratna Yuliawati[3] and Deny Kurniawan[3]
[1]Environmental Health, Faculty of Public Health, Universitas Muhammadiyah Kalimantan Timur, Samarinda, East Kalimantan, 75124, Indonesia, ORCHID ID: 0000-0001-8104-0618
[2]Green Borneo Consultant, Samarinda, Kalimantan Timur, Indonesia
[3]Environmental Health, Faculty of Public Health, Universitas Muhammadiyah Kalimantan Timur, Samarinda, East Kalimantan, 75124, Indonesia
[1]*vp799@umkt.ac.id, [2]agusdwipurwanto81@gmail.com, [3]ry190@umkt.ac.id, [4]dk658@umkt.ac.id

Abstract. Selulu Reservoir is located in Waru District, North Panajam Paser Regency. It is one of the tourist destinations in supporting the tourism sector. Land use changes and urbanization have potential impacts on the environment. This study aimed to determine Water Quality Status and Water Quality Index (WQI) in Selulu Reservoir. Hopefully, the result will become the basis for future reservoir management to keep the water quality well. The method in this study used the survey method. These research steps include surveys, water samples, laboratory analysis, and data analysis. Parameters analyzed include temperature, pH, BOD, COD, and DO. The results showed that the reservoir's water quality in all samples still met the standard, except for pH at point 3. The difference fluctuating value found in COD at point 3, although it still meets the standard. This location is close to residential areas that dump their waste into the reservoir. The Water Quality Status of the reservoir still meets the standards, and WQI is in the excellent category. This is due to the dominant land use of forest and plantations with few settlements around the reservoir, domestic waste is small, and the reservoir capacity is still supportive. Management is needed to maintain reservoir water quality in the future.

1. Introduction

Water is a natural resource that people need; its function must be preserved and remains useful for human life and others [1]. This reservoir is in Subdistrict Waru, Panajam Paser Utara Regency. The total area of Subdistrict Waru is 553.88 km^2, with a percentage of 16.62% of the total district area [2]. The research location at 116°37'01.62" East Longitude and 01°23'22.7" North Latitude.

Selulu Reservoir has the potential to become a tourist destination if it is managed well, and the new capital city of Indonesia will be built in this district. It supports the tourism sector where the beauty and natural wealth are pretty abundant, especially types of fish. The livelihoods of the community around the Selulu Reservoir include fishing and fish cultivation in the form of cages. Most of the community employment in Panajam Paser Utara Regency is agriculture, plantation, and forestry, which fisheries by 40% and trade, restaurants, and accommodation services by 24%. Mining and quarrying sector by 2.34%, manufacturing industry by 10%, construction by 7%, community and individual services by 13%, and others 3.66% [2]. Communities around the Selulu Reservoir use the reservoir water for bathing, washing, and toileting then the wastewater flows directly into the reservoir. In addition, there is community fish cultivation carried out by making cages. These activities influence water quality decline if it continues without management. Domestic activities influence water pollution in the reservoir to impact the water quality, and some parameters are not following the standard, such as pH, DO, BOD, COD, and Fecal Colliform [3]. Uncontrolled community economic activities can affect the reservoir water quality decline showing the water quality status [4]. Selulu reservoir is used as a source of raw water for Water Treatment Plant (Drinking Water Company) and needs to be kept to the water quality.

Status of Water Quality and Water Quality Index (WQI) is used to understand the water quality and manage the resources in the reservoir and lakes [5]. WQI is adequate to calculate the lake water quality in terms of water landscape [6]. Based on several studies and conditions around the Selulu Reservoir, this study aims to determine the Status of Water Quality and WQI. It is done to know the reservoir water quality that supports welfare communities and supporters of the new capital city of Indonesia in the tourism sector. Considering development, the new capital city of Indonesia will increase urbanization and land use changes that impact the environment. The resulting study will be essential water quality information, especially for Selulu Reservoir.

2. Methods

The research method in this study used the survey method. The steps were carried out by the preparation stage, field survey, water sampling, laboratory analysis, and data analysis. It was organized for around 6 months, starting in January 2022. The laboratory analyzing water samples is the Regional Health Laboratory of Balikpapan City. The research location is shown in Figure 1 in Selulu Reservoir, Waru District, North Panajam Paser Regency, East Kalimantan.

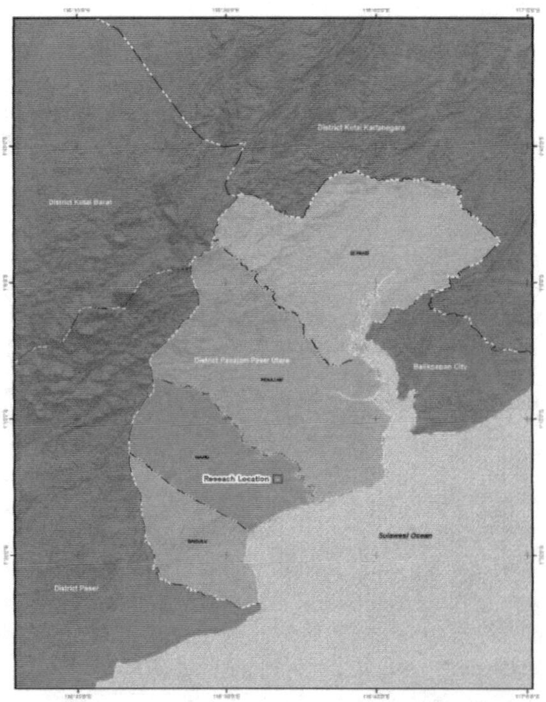

Figure 1. Research Location in Subdistrict Waru

Sampling uses the purposive sampling method by taking at 3 points on the right, middle and left side of the reservoir to consider that it can represent the condition of the reservoir water and accessibility of the location. The parameters analyzed were temperature, pH, BOD, COD, and DO. The Pollution Index (IP) calculation is used to determine the Water Quality Status, in accordance with the Decree of the Minister of the Environment No. 115 of 2003 concerning Guidelines for Determining the Water Quality Status, in the formula (1).

$$PI_j = \sqrt{\frac{\left(C_i/L_{ij}\right)_M^2 + \left(C_i/L_{ij}\right)_R^2}{2}} \tag{1}$$

Where PIj is the Pollution Index want to know, Ci is the concentration of the measured water quality parameter, Lij is the quality standard of the measured parameter, M is the maximum value or the highest value. At the same time, R is the average value that has been calculated. The results of calculating IP values and their categories [7] are presented in Table 1.

Table 1. River Water Pollution Index Values and Categories Water Quality Status

IP	Category
0≤IP≤1	Good (Meets Quality Standards)
1<IP≤5	Lightly Polluted
5<IP≤10	Medium Polluted
IP>10	Heavily Polluted

The Water Quality Index (WQI) is calculated based on the Regulation of the Minister of Environment and Forestry of the Republic of Indonesia No. 27 of 2021 concerning the Environmental Quality Index [8]. The determination of WQI is done by multiplying the quality of the index value by the percentage of meeting the quality standard obtained from the sum of the sample points that meet the quality standard by the number of samples in percent. Index value quality is presented in Table 2. WQI is obtained from the total number of Index Values per water quality and is matched with the vulnerable numbers in the Water Quality Index category in Table 3.

Table 2. Index Value Quality

Index Value Quality	Category
70	Good (Meets Quality Standards)
50	Lightly Polluted
30	Medium Polluted
10	Heavily Polluted

Table 3. Water Quality Index of Category

Index Value Quality	Category
90 ≤ x ≤ 100	Excellent
70 ≤ x < 90	Good
50 ≤ x < 70	Medium
25 ≤ x < 50	Unsuitable

IOP Conf. Series: Earth and Environmental Science 1111 (2022) 012039 doi:10.1088/1755-1315/1111/1/012039

$0 \leq x < 25$	Very Unsuitable

3. Results and Discussion

The results of laboratory analysis for the parameters of temperature, pH, BOD, COD, and DO from the Selulu Reservoir water are presented in Table 4. The analysis results were matched with the surface water quality standard class II following the East Kalimantan Regional Regulation No. 2 of 2011 concerning Water Quality Management and Water Pollution Control [1]. All parameters meet the Quality Standards except pH at Point 3, which is 5 with a quality standard of 6-9. The BOD and COD parameters have the same value at all points except for COD, which shows the difference in numbers at each point. The temperature is seen only at Point 1, which has a different but not very significant value. The highest COD value is at Point 3, with a value of 10 mg/L, and the lowest is at Point 1, with a value of 6 mg/L. Point 3 is close to a few residential areas, so the reservoir water still has the capacity for COD. COD values for all points still meet the standard. The COD value indicates that the waters are polluted with organic matter and become the primary pollutant in Lake Chaohu, China [9]. Dense community activities around the reservoir and waste disposal cause the pH, BOD, COD, and DO values to exceed the standard [3]. The water quality of the Selulu Reservoir for several parameters of temperature, pH, BOD, COD, and DO still meets the standard. This needs to be maintained by managing the environment around the reservoir.

Table 4. Laboratory analysis result of Water Quality Selulu Reservoir

No	Parameters	Point 1	Point 2	Point 3	Average	Standard	Method
1	Temperature	30	31	31	30.67	deviasi 3	Elektrometry
2	pH	6	6	5	5.67	6-9	Elektrometry
3	BOD	1	1	1	1.00	3	Oxydirect
4	COD	6	9	10	8.33	25	Spektrofotometry
5	DO	7	7	7	7.00	4	Electrometry

Source: Primary Data, 2022

Water quality status is the level of water quality conditions that indicate polluted or good conditions in a water source within a particular time by comparing it with the water quality standards [7]. Details of the Pollution Index (IP) calculation to determine the Water Quality Status of the Selulu Reservoir are presented in Table 5, Table 6, and Table 7. The pollutant Index value of the Selulu Reservoir is included in the category of good water quality status (meets quality standards).

Table 5. Pollution Index and Water Quality Status at Selulu Reservoir Point 1

Parameters	Result Analysis (Ci)	Standard (Lij)	Ci New	Ci/Lij	Ci/Lij New
PH	6	7	6	0.5	0.5
BOD	1	3	1	0.333333333	0.333333333
COD	6	25	6	0.24	0.24
DO	7	4	0	0	0
	Ci/Lij R				0.268333333
	Ci/Lij M				0.5
	Pollutant Index value (IP)				0.401249784
	Water Quality Status				Good (Meets Quality Standards)

Source: Primary Data, 2022

Table 6. Pollution Index and Water Quality Status at Selulu Reservoir Point 2

Parameters	Result Analysis (Ci)	Standard (Lij)	Ci New	Ci/Lij	Ci/Lij New
PH	6	7	6	0.5	0.5
BOD	1	3	1	0.333333333	0.333333333
COD	9	25	9	0.36	0.36
DO	7	4	0	0	0
	Ci/Lij R				0.298333333
	Ci/Lij M				0.5
	Pollutant Index value (IP)				0.411705464
	Water Quality Status				Good (Meets Quality Standards)

Source: Primary Data, 2022

Table 7. Pollution Index and Water Quality Status at Selulu Reservoir Point 3

Parameters	Result Analysis (Ci)	Standard (Lij)	Ci New	Ci/Lij	Ci/Lij New
PH	5	7	5	0.625	0.625
BOD	1	3	1	0.333333333	0.333333333
COD	10	25	10	0.4	0.4
DO	7	4	0	0	0
	Ci/Lij R				0.339583333
	Ci/Lij M				0.625
	Pollutant Index value (IP)				0.502962146
	Water Quality Status				Good (Meets Quality Standards)

Source: Primary Data, 2022

The highest IP value is in the measurement results of point 3, where the location is close to settlements. The pH value at point 3 does not meet the standard of 5 and is acidic. All chemical pollutants in the water will decrease along with the increase in the pH value of the water. In addition, a high pH will reduce the retention/absorption capacity of phosphorus in sediments which causes eutrophication [10]. pH value also affects the oxidation process of ammonia and nitrite. Low pH inhibits the oxidation process of ammonia and nitrite in waters where the optimal pH is 7 and 7.5 [11]. Nutrition and pH are also significant factors in eutrophication [12]. The highest COD result is at point 3 compared to all measurement points, although it still meets the standard. COD value is an indicator of organic matter pollution in the waters. Based on the study in Lake Chaohu, the COD parameter is the primary pollutant, and implemented an RCP (River Chief Policy) system to tackle the pollution of rivers and lakes in China [9].

The current Selulu Reservoir Water Quality Index (WQI), based on the results of water quality analysis, shows that it is in a suitable category at a value of 70; it is vulnerable to $70 \leq x < 90$, with the water quality of all samples meeting the standard.

Table 8. Selulu Reservoir Water Quality Index (WQI)

Water Quality Status	Number of Monitoring that meets water quality status	Percentage of Fulfillment of Water Quality	Index Value Quality	Index Value per Water Quality
Good (Meets Quality Standards)	3	100%	70	70
Lightly Polluted	0	0	50	0
Medium Polluted	0	0	30	0
Heavily Polluted	0	0	10	0
Total	3			WQI = 70
	Water Quality Index Category (WQI)			Good

Source: Primary Data, 2022

Calculating the Water Quality Index is a way to determine the water quality in the lake [13] and evaluate surface water quality, including rivers, lakes, reservoirs, and estuaries [5]. An innovation that links WQI and hydrological variables, urbanization provides a multidimensional evaluation of water quality [14]. The Selulu Reservoir Water Quality Index needs to be maintained in line with the development new capital city of Indonesia. This long-term development will change land use a lot, and major urbanization will occur when a new city has been built in North Panajam Paser Regency. Urbanization has an impact on the water quality in the city. The lake water quality is inferior due to the influence of urbanization and the growth of cities around the lake [15]. Selulu Reservoir is located in Waru District, which borders Panajam District, the new capital city of Indonesia. Changes in land use area in North Panajam Paser Regency in 2010-2016 were mostly plantations of 3.55% (from 21.96% to 25.51%); Forest by 1.42% (from 34.34% to 32.92%); Bushes by 1.36% (from 10.12% to 8.76% and HTI (Industrial Plantation Forest) by 1.19% (from 18.27% to 17.09%) [16]. Land use change in Panajam Subdistrict in the oil palm plantation sector is increasing, and the forest area is decreasing [17]. Selulu Reservoir has the potential as a tourist destination that must be managed properly, especially the water quality. The pollution load capacity in the reservoir still meets good water quality standards. Tourism is one of the factors that cause water pollution in surface water, such as rivers, lakes, and reservoirs. Deterioration of lake water quality due to grazing/livestock and tourism [18].

4. Conclusion

The water quality status of the Selulu Reservoir still meets the quality standards, and its WQI is in a suitable category. It means the reservoir can still accommodate the pollution load that enters. There are only a few settlements around the reservoir, and plantations dominate it. The water quality is still up to standard and does not exceed the capacity. Community activities, domestic activity, urbanization, and land use influence the WQI in the river, reservoir, and lake surface water. Therefore, it must be managed, monitored, and evaluated to sustain future water quality.

Acknowledgments

The author would like to thank the Regional Health Laboratory, Balikpapan, who helped analyze the study's water samples.

References

[1] Peraturan Daerah Provinsi Kalimantan Timur 2011 Peraturan Daerah Provinsi Kalimantan Timur No.02 Tahun 2011 tentang Pengelolaan Kualitas Air dan Pengendalian Pencemaran Air *Samarinda : Pemerintah Provinsi Kalimantan Timur*

[2] Yusuf M and Karim S A 2022 *Kabupaten Panajam Paser Utara dalam Angka 2022* (BPS Kabupaten Penajam Paser Utara)

[3] Pramaningsih V, Kurniawan D, Syamsir and Mardiana 2021 Water Quality Analysis Of Benanga Reservoir, In Samarinda, East Kalimantan, Indonesia *Procedia Environ. Sci. Eng. Manag.* **8** 353–61

[4] Goher M E, Mohamed E M M, Abdo M H, El Dars F M, Korium M A and Elsherif A A 2017 Water Quality Status and Pollution Indices of Wadi El-Rayan *Sustain. Water Resour. Manag.*

[5] Uddin M G, Nash S and Olbert A I 2021 A review of water quality index models and their use for assessing surface water quality *Ecol. Indic.* **122** 107218

[6] Chang N, Luo L, Wang X C, Song J, Han J and Ao D 2020 A novel index for assessing the water quality of urban landscape lakes based on water transparency *Sci. Total Environ.* **735** 139351

[7] Keputusan Menteri Negara Lingkungan Hidup 2003 Keputusan Menteri Negara Lingkungan Hidup Nomor 115 Tentang Pedoman Penentuan Status Mutu Air *Jakarta Menteri Negara Lingkung. Hidup* 1–15

[8] PermenLHK RI No. 27 2021 *Peraturan Menteri Lingkungan Hidup dan Kehutanan Republik Indonesia No. 27 Tahun 2021 tentang Indeks Kualitas Lingkungan Hidup* (Jakarta: Kementrian Lingkungan Hidup dan Kehutanan)

[9] Xu X, Wu F, Zhang L and Gao X 2020 Assessing the effect of the Chinese river chief policy for water pollution control under uncertainty—using chaohu lake as a case *Int. J. Environ. Res. Public Health* **17**

[10] Temporetti P, Beamud G, Nichela D, Baffico G and Pedrozo F 2019 The effect of pH on phosphorus sorbed from sediments in a river with a natural pH gradient *Chemosphere* **228** 287–99

[11] Le T T H, Fettig J and Meon G 2019 Kinetics and simulation of nitrification at various pH values of a polluted river in the tropics *Ecohydrol. Hydrobiol.* **19** 54–65

[12] Zhang H, Zong R, He H, Liu K, Yan M, Miao Y, Ma B and Huang X 2021 Biogeographic distribution patterns of algal community in different urban lakes in China: Insights into the dynamics and co-existence *J. Environ. Sci. (China)* **100** 216–27

[13] Vasistha P and Ganguly R 2020 Water quality assessment of natural lakes and its importance: An overview *Mater. Today Proc.* **32** 544–52

[14] Fadel A, Kanj M and Slim K 2021 Water Quality Index variations in a Mediterranean reservoir: a multivariate statistical analysis relating it to different variables over 8 years *Environ. Earth Sci.* **80** 1–13

[15] Birawat K K, Hymavathi T, C.Nachiyar M, Mayaja N A and Srinivasa C V. 2021 Impact of urbanisation on lakes—a study of Bengaluru lakes through water quality index (WQI) and overall index of pollution (OIP) *Environ. Monit. Assess.* **193** 1–20

[16] Widjayatnika B, Baskoro D P T and Pravitasari A E 2018 Analisis Perubahan Penggunaan Lahan dan Arahan Pemanfaatan Ruang untuk Pertanian di Kabupaten Penajam Paser Utara, Provinsi Kalimantan Timur *J. Reg. Rural Dev. Plan.* **1** 243

[17] Maulidya A, Damayanti A, Indra T L and Dimyati M 2021 Prediction of Land Change for Oil Palm Plantations in Penajam Subdistrict, Penajam Paser Utara Regency, East Kalimantan Province *J. Phys. Conf. Ser.*

[18] Roşca O M, Dippong T, Marian M, Mihali C, Mihalescu L, Hoaghia M A and Jelea M 2020 Impact of anthropogenic activities on water quality parameters of glacial lakes from Rodnei mountains, Romania *Environ. Res.* **182**

Environmental Assessment Using Integrated Risk Based Approach (IRBA) at Jabon Landfill, Sidoarjo

Arlini Dyah Radityaningrum[1], Talent Nia Pramestyawati[1*], Achmad Chusnun Ni'am[1], Eko Wahyudi[1], Mohamad Ferdaus Noor Aulady[2], and Nur Laila Hamidah[3]

[1]Environmental Engineering Department, Faculty of Civil Engineering and Planning, Institut Teknologi Adhi Tama Surabaya, Surabaya, 60117, Indonesia

[2]Civil Engineering Department, Faculty of Civil Engineering and Planning, Institut Teknologi Adhi Tama Surabaya, Surabaya, 60117, Indonesia

[3]Physics Engineering Department, Faculty of Industrial Technology and Systems Engineering, Institut Teknologi Sepuluh Nopember Surabaya, Indonesia

[1*]talentnia@gmail.com

Abstract. Jabon landfill is located above the land belonging to the local government of Sidoarjo city with 8 hectares. The study's objective is to evaluate the current state of Jabon landfill management using the Integrated Risk Based Approach (IRBA). Risk Index (RI) assessment results using IRBA obtained a 548,08 value with a category of moderate hazard evaluation so that the Jabon site can be gradually forwarded and rehabilitated into a sanitary landfill. Waste composition measured dominated by three types of waste biodegradable waste, non-biodegradable, and hazardous waste 69%, 29%, and 2%, respectively. The strategy needed for revitalizing the Jabon landfill is developing a system to minimize the impact of pollution on water, soil, and air so that it will be more environmentally friendly. Sanitary landfill should be considered to apply in Jabon landfill.

1. Introduction

The large population and the diversity of activities in big cities in Indonesia have resulted in service problems in urban infrastructures, such as solid waste management [1]. The rate of population growth is always directly proportional to the level of consumption and community activities [2]. Population growth in Sidoarjo Regency has increased from year to year. In 2010, the population of Sidoarjo was 1,949,595, and there was an increase of 16%; the population of Sidoarjo in 2019 was 2. 262,440 people [3], so the population growth in Sidoarjo Regency will be directly proportional to increasing the waste. Sidoarjo, as a big city, has various waste problems. Sidoajo has one landfill in Kupang Village, Jabon District, Sidoarjo Regency. Jabon landfill is large, about 8 hectares, and managed by the government. Solid waste generation in Jabon Landfill is about 500 tons/day [4].

Every landfill will produce leachate that contains various pollutants. Leachate can contaminate the wells of the residents around Jabon Landfill. A study about water quality from residents around Jabon landfill shows the pollution status of the 18 wells showing that two wells are moderately polluted, 15 are lightly polluted, and one is in good condition [5]. The wells used by the residents have contaminants and are not safe to drink. Leachate from landfill is a source of harmful pollution to the environment because it has a toxic effect. Leachate contains heavy metals and organic and inorganic compounds such as ammonia, sulfates, and metal cations. Leachate from the landfill has poisoned nearby water supplies

[6] because the Regional Drinking Water Company (PDAM) does not yet provide service to the Jabon landfill. The resident primarily relies on the use of dug wells.

A previous study about the implementation of solid waste management at Jabon Landfill in 2015 showed 65% [7]. The previous study determined solid waste management generally in Sidoarjo based on solid waste separation, collecting, transportation, treatment, and final dumping. Jabon landfill has operated for 14 years. The existing Jabon landfill has reached 92%, so the dumpsite for the landfill is insufficient. In other words, the landfill's capacity is overloaded [8]. The landfill area is expected to complete in early 2021. Based on that reason, the government must decide to close or rehabilitate the Jabon Landfill. The renewability of this research is to evaluate the Jabon landfill based on a risk index assessment to give the result to close or rehabilitate the Jabon Landfill.

Land expansion is needed for sustainable waste management at the Jabon landfill. Jabon landfill needs to be evaluated based on the amount of incoming waste, waste management policies at the landfill, and aspects of the surrounding environment. There are various methods to evaluate the waste generation at the landfill. One of them is the Integrated Risk Based Approach (IRBA). This study is to decide whether to close or rehabilitate the Jabon Landfill with IRBA methods. The purpose of this study is IRBA methods implemented in Jabon landfill from Ministerial Regulation of Public Works No 3, 2013. IRBA methods were also used to evaluate Cahaya Kencana Landfill in Banjar City. That method obtained a score of 524 that landfill needs to be rehabilitated must determine the strategy for modification of the unit design of leachate treatment [9].

2. Method

2.1 Time and Study area description

This research was conducted in January-July 2020. The location of the Jabon landfill is 33 km or can be taken in about 90 minutes via land from Juanda Airport, Sidoarjo. Jabon landfill has an area of 8 hectares. Jabon landfill is located in Tambak Kalisogo Village, Jabon District, Sidoarjo Regency, with coordinates: S=07°32'51,2" E=112°45'54.1".

2.2 *Risk Assessment*

The Risk Index (RI) can be utilized to evaluate landfill sites for closure or rehabilitation. Risk Index/ RI calculated with formula as below.

$$RI = \sum_{i=1}^{n} WiSi \quad (1)$$

Where:
Wi = Weightage of the with variable ranging from 0 – 1000
Si = Sensitive index of the with variable ranging from 0 – 1
RI = Risk Index variable from 0 – 1000

The first step is to evaluate the risk. The minimum sensitive index is zero, which shows little danger or negligible risk, and the maximum sensitive index is one denotes the most significant possible risk. The higher value indicates the greater the risk to human health and the urgency of the situation at the landfill site. The data is collected by sampling and environmental laboratory analysis. The next step is comparing the score with criteria for hazard evaluation based on that shown in Table 1. The following priority falls as the overall value rises. The lowest values represent poor environmental sensitivity and minimal effects.

Table 1. Criteria for hazard evaluation based on the risk index

No	Risk Index	Hazard Potential	Action recommendation
1.	750-1000	Very High	Close the dump with no more landfilling in the area. Take remedial action to mitigate the impacts
2.	600 –749	High	Close the dump with no more landfilling in the area. Remediation is optional
3.	450 – 599	Moderate	Immediate rehabilitation of the dumpsite into a Sustainable Landfill
4.	300 – 449	Low	Rehabilitate the dumpsite into a Sustainable Landfill in a phased manner
5.	< 300	Very Low	Potential site for future landfill

3. Results and Discussion

3.1 Landfill Jabon Condition

The main road to enter the Jabon landfill location from the highway is a dirt road with a length of 500 meters and a road width of 5 meters. Jabon landfill only has one weighbridge that is actively used every day. Jabon landfill has one cell with an area of 7 hectares. The leachate treatment plant at the Jabon landfill has an oxidation and deposition basin measuring 10 x 10 meters in length, then directly discharged to the outlet. At the Jabon landfill, clean water is provided from dug wells. Landfill capacity used with an area of 8 ha, it has an operational number of 5 excavators and five bulldozers. The waste transportation fleet at the Jabon landfill has 25 units in the form of 13 dump trucks, 6 arm roll trucks, three four-wheeled vehicles, and three three-wheeled vehicles.

3.2 Composition and Generation of Solid Waste

Waste composition measured dominated by three types of waste biodegradable waste 69%, non-biodegradable waste such as plastics, rubber, and metal 29%, and hazardous waste such as paint cans and batteries 2%. Figure 1 present the composition of the Jabon Landfill. The primary composition of solid waste in developing countries such as Indonesia is a biodegradable waste. It has different composition compared with solid waste composition in developed countries such as the united state primarily consists of paper 27%, food waste 15%, and plastics 13% [10]. The biodegradable waste consists of slowly biodegradable.

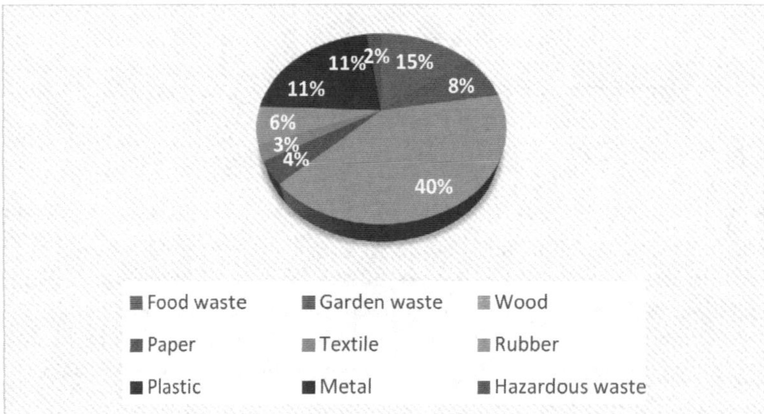

Figure 1. Jabon Landfill Composition

Solid waste generation in Jabon Landfill increases every year. In 2015, solid waste generation was 377.538,94 m^3 and increased by 407.872,2 m^3 in 2020. Figure 2 presents the solid waste generation at Jabon Landfill for five years. Separating food waste and plastic waste significantly reduces household waste, while source sorting of food waste changes the composition of residual household waste significantly [11].

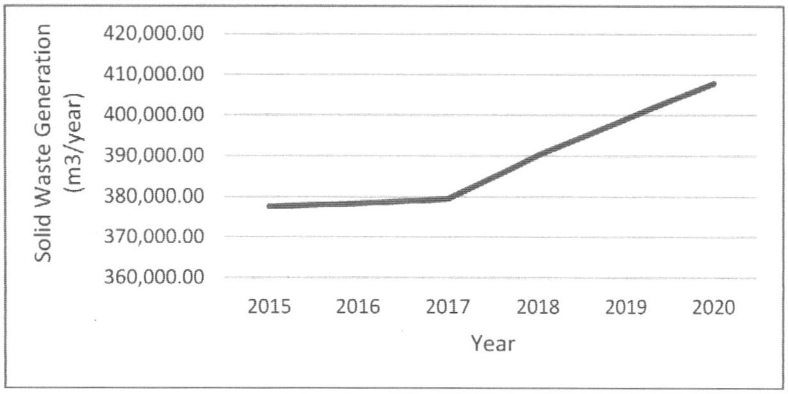

Figure 2. Solid waste generation of Jabon landfill

3.3 Laboratory result

The measurement of water quality consists of the quality of the groundwater and leachate. The results of groundwater measurements are compared with the quality standard of the Minister of Health Regulation No. 32 of 2017. As shown in Table 2, the results depicted that 4 out of 27 total parameters have higher results than the regulatory limit, viz. TDS, color, hardness, and manganese [12].

Accordint to the results study in Table 2, it can be concluded that groundwater quality can be consumed by adding treatment before the water is consumed, such as filtration. Another study assessing the risk index at five landfills presented that groundwater quality could not be consumed [13]. It is different from the current results, where groundwater can be consumed if there is no other source with pre-treatment.

Table 2. Comparison quality of groundwater and regulatory limit

Parameter	Unit	Laboratory Result	Regulatory Limit
TDS	mg/L	1250	100
Color	Pt/Co	75	50
Hardness	mg/L	621	500
Manganese	mg/L	0,6	0,5

One of the most popular methods for getting rid of municipal solid waste is to put it in landfills, but doing so might cause environmental leaks of landfill leachate, a type of wastewater with complicated compositions and high levels of toxins [14]. The measurement of leachate results for the BOD parameter have a value of 2880 mg/L, COD has a value of 8730 mg/L, and the TDS has a value of 10200 mg/L. Based on the regulatory limit from the Minister of Environmental Regulation about landfill leachate, it has a higher score than the regulatory limit [15]. In the previous study in 2018, Jabon Landfill leachate had a BOD value of 3864 mg/L and a COD value of 4120 mg/L. Compared with this research result, the BOD value is less than the limit, and the COD is higher than the limit [4]. Another study shows that two landfills from 14 landfills in different cities in Indonesia have less value than the regulatory limit [16]. Promoting sustainable municipal solid waste management can benefit from understanding the characteristics of landfill leachate [17].

3.4 Risk Index result

Risk Index (RI) assessment results use IRBA, as presented in Table 3, achieved a value of 548.08 with a category of moderate hazard evaluation so that the Jabon site can be forwarded and rehabilitated into a sanitary landfill gradually. A critical aspect of this rehabilitation landfill process is the leachate treatment unit in the Jabon landfill.

Table 3. Risk Index of Jabon landfill

No	Parameter	Attribute Weightage	Landfill Jabon measurement	Sensitivity Index	Score
I. Site-Specific Criteria					
1	Distance from nearest wáter supply source (m)	69	1000	0,5	34,5
2	Depth of filling of waste (m)	64	3	0,25	16
3	Area of the dumpsite (Ha)	61	8,5	0,40	24,15
4	Groundwater depth (m)	54	4	0,54	28,93
5	Permeability of soil (1 x 10-6 cm/s)	54	40	1	54
6	Groundwater quality	50	Potable if no alternative	0,5	25
7	Distance to critical hábitats such as wetlands and reserved forest (km)	46	20,4	0,42	19,47
8	Distance to the nearest airport (km)	46	34,4	0,00	0,00
9	Distance from surface water (m)	41	20	1,00	41,00
10	Type of underlying soil (% clay)	41	0,0004	0,75	30,78
11	Life of the site for future use (years)	36	1	0,05	1,80
12	Type of waste (Municipal Solid Waste (MSW)/ Household Waste (HW))	30	78 % MSW /22 % HW	0,25	7,50
13	The total quantity of waste at the site (ton)	30	1942306	1,00	30,00
14	Quantity of waste disposed of (ton/day)	24	641	0,57	13,69
15	Distance to the nearest village in the predominant wind (m	21	1000	0,50	10,50
16	Flood prone (flood period in years)	16	10	0,50	8,00
17	Annual rainfall at the site (cm/y)	11	1478,7	1,00	11,00
18	Distance from the city (km)	7	18,3	0,46	3,20
19	Public acceptance	7	No public concerns	0,00	0,00
20	Ambient air quality - CH_4 (%)	3	24.5 %	1,00	3,00
II. Related to characteristics of waste at dumpsite					
21	Hazardous contents in waste (%)	71	0,02	0,05	3,55
22	Biodegradable fraction of waste at the site (%)	66	69	0,81	53,21
23	Age of filling (years)	58	14	0,60	34,80
24	The moisture of waste at a site (%)	26	0,6	1,00	26,00
III. Related to leachate quality					
25	BOD of leachate (mg/L)	36	2880	1,00	36,00
26	COD of leachate (mg/L)	19	8730	1,00	19,00
27	TDS of leachate (mg/L)	13	10200	1,00	13,00
	Total				**548,08**

4. Conclusion

Risk Index (RI) assessment results using IRBA obtained 548,08. The value indicated that Jabon landfill has a moderate level of hazard evaluation. Jabon landfill consists of 14 zones, with one zone being nearly complete. Jabon landfill consists of biodegradable, non-biodegradable, and hazardous waste, which account for 69%, 29%, and 2%, respectively. To laboratory results, TDS, color, hardness, and manganese exceeded of quality standard. The following study should ensure that efforts to evaluate waste generation and reduction at the Jabon landfill are suitable. As a final point, we believe that the present study could be considered a guide for future studies on solid waste management in Jabon landfill. The limitation of is current study is the Covid-19 pandemic.

References

[1] Damanhuri E and Padmi T 2010 *Pengelolaan Sampah* (ITB Press)
[2] Talent Nia Pramestyawati and I. D. A. A. Warmadewanthi 2020 Non Residential Solid Waste Reduction in Krembangan District, North Surabaya *Jurnal Iptek* **24**
[3] Badan Pusat Statistik Kabupaten Sidoarjo 2019 *Kabupaten Sidoarjo dalam Angka*
[4] Saputra A F B and Mirwan M 2018 Evaluasi Pencemaran Lindi pada Air Sumur sekitar TPA Jabon *Envirotek* **10** 55–9
[5] Tangahu B, Kartika A A, Sambodho K, Marendra S M and Arliyani I 2021 Shallow Groundwater Pollution Index Around the Location of Griyo Mulyo Landfill (Jabon Landfill) in Jabon District, Sidoarjo Regency, East Java, Indonesia *J. Ecol. Eng.* **22** 199–210
[6] Achmad Chusnun Ni'am and IDAA. Warmadewanthi 2013 EFEKTIFITAS Typha angustifolia DAN Eichhornia crassipes DALAM MENGOLAH LEAHATE DENGAN SISTEM CONSTRUCTED WETLAND *Seminar Nasional Manajemen Teknologi XVIII* (Program Studi MMT-ITS)
[7] Tara P N D, Sunarko B and Kriswandana F 2015 Study Penerapan Pengelolaan Sampah di TPA Kupang, Kecamatan Jabon Kabupaten Sidoarjo *Gema Kesehatan Lingkungan* **13** 16–9
[8] Kementerian Pekerjaan Umum dan Perumahan Rakyat 2021 Kurangi Dampak Pencemaran, Kementerian PUPR Selesaikan Sistem Sanitary Landfill TPA Sampah Jabon di Sidoarjo Seluas 5,89 Hektar *Kementerian Pekerjaan Umum dan Perumahan Rakyat*
[9] Iman M S and Pandebesie E S 2020 Environmental Assesment Using Integrated Risk Based Approach (IRBA) at Cahaya Kencana Landfill Site *IPTEK Journal of Proceedings*
[10] I H, Shafy A and Mansour M S M 2018 Solid Waste Issue: Souces, Composition, Disposal, Recycling and Valorization *Egyptian Journal of Petroleum* **27** 1275–90
[11] Edjabou M E, Takou V, Boldrin A, Petersen C and Astrup T F 2021 The influence of recycling schemes on the composition and generation of municipal solid waste *Journal of Cleaner Production* **295** 126439
[12] Indonesian Minister of health 2017 *Minister of Health Regulation No. 32 of 2017 about Regulatory Limit for higiene sanitation, pool, solus per aqua and public bath*
[13] Panyaungan J Kajian Penerapan Penilaian Indeks Resiko Tempat Penimbunan Sampah Di Indonesia **5** 8
[14] Ma S, Zhou C, Pan J, Yang G and Sun C 2022 Leachate from municipal solid waste landfills in a global perspective: Characteristics, influential factors and environmental risks *Journal of Cleaner Production* **333**
[15] Indonesian Minister of Environmental *Minister of Environmental Regulation about Regulation Limit of Landfill Leachate Quality*
[16] Emalya N, Munawar E, Rinaldi W and Yunardi Y 2020 Landfill Leachate Management in Indonesia: A Review *IOP Conf. Ser.: Mater. Sci. Eng.* **845** 012032
[17] Vaccari M, Tudor T and Vinti G 2019 Characteristics of leachate from landfills and dumpsites in Asia, Africa and Latin America: an overview *Waste Management* **95** 416–31

Planning and Implementation of Hazardous Waste Management Emergency Response Program – A Case Study of PT. X Medical Hazardous Waste Treatment

M Candra Nugraha Deni[1]* and Aulia Shabirah[2]
[1]Environmental Engineering, Faculty of Civil Engineering and Planning, National Institute of Technology, Bandung, 40124, Indonesia, 0000-0002-5489-0083
[2]Environmental Engineering, Faculty of Civil Engineering and Planning, National Institute of Technology, Bandung, 40124, Indonesia
[1]*candranugraha@itenas.ac.id, [2]auliashabirah18@mhs.itenas.ac.id

Abstract Increasing the amount of medical waste can cause environmental and health problems if it is not managed properly. This is because medical waste is categorized as hazardous and toxic waste (B3), which can also create hazardous waste management emergencies situation. PT.X is a company that treats medical waste by using an incinerator. The purpose of this study is to plan and implement a medical waste management emergency response program at PT.X by determining the potential for hazards to occur in medical waste processing, implementing an emergency response system, and developing and implementing an emergency response program. The data collection method used is by direct observation and discussion with the company staff, as well as through the available data at the company. Risk assessment was conducted using the matrix of likelihood and consequence of potential risks, and the results showed 4 activities that have extreme risk. By planning and implementing a B3 waste emergency response program, PT. X has optimized its preparedness in dealing with emergency conditions so that in the event of an emergency it can be handled optimally.

1. Introduction

The increase in the number of hospitals will have an impact on the amount of medical waste, of which medical waste is categorized as hazardous and toxic waste (B3). B3 waste is the residue of a business and/or activity containing B3. Meanwhile, medical waste is solid waste consisting of infectious waste, pathological waste, sharp object waste, pharmaceutical waste, cytotoxic waste, chemical waste, radioactive waste, pressurized container waste, and waste containing high heavy metals [4].

The category of medical waste is based on the Minister of Health's decision No. 1204 of 2004, namely infectious, pathological, cytotoxic, sharp objects, pharmaceuticals, chemical substances, radioactive, high-pressure metals, and pressurized containers [6].

PT.X is a medical waste treatment company that uses incinerators to process the waste from their customers i.e., hospitals, clinics, drugs company, etc. Incineration is one of the technologies for treating solid waste by burning at a minimum temperature of 800 °C, which aims to process solid waste into gas and ash (bottom ash and fly ash) [7]. Pathogenic waste, such as waste originating from health care facilities, especially those with infectious characteristics, is best treated by the incineration method. This is because pathogenic microorganisms found in the infectious waste can be destroyed in incinerators at high temperatures [12]. At high temperatures, the waste undergoes oxidation and changes its phase from solid or liquid to gas, which is mainly in the form of CO_2 (carbon dioxide) and H_2O (water) [5].

The B3 waste needs to be managed in accordance with the provisions so as not to cause material losses, impacts on the environment or humans, or trigger the occurrence of B3 waste management emergencies condition. An emergency is a dangerous situation in which the situation can threaten human safety, cause pollution and/or damage to the environment, and require direct countermeasures to minimize the occurrence of more dangerous impacts [9].

Having considered the abovementioned, the objective of this research is to plan and develop an emergency response program for medical waste processing activities at PT. X by identifying potential hazards in the treatment area.

The B3 waste management emergency response program can be described as an emergency response system planning document that includes infrastructure (equipment, facilities, infrastructure provided) and response capacity [1]. This B3 waste management emergency response program contains a plan for the prevention, preparedness, and implementation of a waste management emergency program by taking into account the results of the risk assessment.

2. Methods

Primary data is obtained directly from The discussions with several staffs of the company, namely the general manager, HSE staff, and field workers in the incinerator section. Meanwhile, secondary data is obtained through company documents namely the currently available emergency response plan and standard operational procedure (SOP). The research was carried out within 3 months, from August 2021 to September 2021.

The preparation of the emergency response program goes through several stages, namely:

2.1. Hazardous Waste Emergency Risk Identification and Analysis

Emergency risk identification and analysis are determined by looking at the potential hazards of each activity in the processing area. Furthermore, the level of emergency risk (Table 1) is determined using the risk matrix method based on the level of likelihood (Table 2) and the consequence/impact (Table 3). Consequence describes the effects of the risk, meanwhile, the likelihood describes the tendency of the consequences and the frequency of the risk sources at each stage [8].

Table 1. Risk Matrix Based on Consequence/Impact and Likelihood [2].

Likelihood		Consequence/Impact				
		1	2	3	4	5
		Insignificant	Minor	Moderate	Mayor	Catastrophic
A	Almost Certain	H	H	E	E	E
B	Likely	M	H	H	E	E
C	Possible	L	M	H	E	E
D	Unlikely	L	L	M	H	E
E	Rare	L	L	M	H	H

Legend:
E : Extreme risk, immediate action required
H : High risk, senior management attention needed
M : Moderate risk, management responsibility must be specified
L : Low risk, managed by routine procedures

Table 2. Measure of Likelihood [2].

Level	Descriptor	Detail Description
A	Almost certain	Is expected to occur in most circumstances
B	Likely	Will probably occur in most circumstances
C	Possible	Might occur at some time
D	Unlikely	Could occur at some time
E	Rare	May occur only in exceptional circumstances

Table 3. Measure of Consequence/Impact [2].

Level	Descriptor	Detail Description
1	Insignificant	No injuries, low financial loss
2	Minor	First aid treatment required, on-site release immediately contained, medium financial loss
3	Moderate	Medical treatment required, on-site release contained with outside assistance, high financial loss
4	Mayor	Extensive injuries, loss of production capability, off-site release with no detrimental effects, major financial loss
5	Catastrophic	Death, toxic release off-site with detrimental effect, huge financial loss

2.2. *Emergency Program*

The preparation of the emergency response program includes a plan for the prevention, preparedness, and implementation of hazardous waste management emergency program by taking into account the results of the emergency risk assessment. The process of preparing the emergency response program involves various parties, both those who contribute and those who have the potential to receive the emergency impact. The preparation of the emergency response program refers to the structure contained in the Regulation of the Minister of the Environment No. 74 of 2019 about the Emergency Program for the Management of Hazardous and Toxic Materials and/or Hazardous and Toxic Waste [9].

3. Results and Discussion

3.1. *Hazardous Waste Emergency Risk Identification and Analysis*

The flow of medical waste management starts with the management in the sources by separating, collecting, and storing it. Furthermore, medical waste is transported and processed through incineration. The incineration process produces residue in the form of ash, which is then managed by a third party.

Hazard identification and emergency risk analysis are carried out in the processing area based on field observations, the company's Standard Operational Procedure documents, and interviews with the staffs. Then, a risk level can be determined based on hazard identification and risk analysis. A risk matrix is used to identify and analyse the emergency risk.

Table 4. Hazardous Waste Emergency Risk Identification and Analysis

Activity	Sub Activity	Potential Hazard	Risk	Risk Value		
				Consequence/ Impact	Probability	Risk Level
Unloading medical waste	Lowering medical waste from trucks	Unbalanced weight can lead to spilled bin and scattered waste	Waste spill can cause environmental pollution and when in contact with human will cause infection	3	E	M
		Stabbed by sharp object of medical waste	Stabbed by sharp object will cause injury and be infected	3	E	M
Medical waste combustion process	Put medical waste to a caster equipped with bin tippler	An unbalanced weight can lead to spilled bin and scattered waste	Waste spill can cause environmental pollution and when in contact with human will cause infection	3	E	M
		Waste stuck (does not go in completely into the incinerator) resulting in a need to be handled manually by the operator	Operator may fall into the incinerator and can be infected by the waste	5	C	E
	Primary combustion chamber	The temperature is not as it should be (800°C-1000°C)	Fire can spread from the engine hopper, produce harmful emissions	4	B	E

Activity	Sub Activity	Potential Hazard	Risk	Risk Value		
				Consequence/ Impact	Probability	Risk Level
	Ash scraper conveyor (the cooling process of the hot ash before poured into an ash container)	Exposed ash	If inhaled, it causes respiratory problems, contact with skin and eyes causes burning	3	C	H
		Operator error (failed to check the water level)	If the water level is too high, it will cause the water to overflow around the area so it can pollute the environment	3	C	H
	Secondary combustion chamber	The temperature is not as it should be (1000°C-1200°C)	Fire can spread from incinerators, produce harmful emissions	4	D	H
	Air pollution control system (maintenance)	The outside of the chimney has a porous part at the bottom. This creates a hole in the bottom of the chimney	Smoke comes out does not match with the supposed height (±20 m)	3	D	M
	Taking bottom ash from incinerator to the storage area	An unbalanced weight can lead to spilled ash bag	The spill can cause environmental pollution and when contact with humans causes health problems	3	A	E
	Taking fly ash from Air Pollution Control (APC) and move ash bag to the storage area	An unbalanced weight can lead to spilled ash bag	The spill can cause environmental pollution and when contact with humans causes health problems	3	A	E

At the processing area, three activities have a high level of risk, namely when the process of entering medical waste into the incinerator, the operational temperature, and the process of transferring bottom ash and fly ash residue from the incinerator to the temporary storage area. When medical waste is put into the incinerator chamber, the medical waste often gets stuck at the top of the hopper, and to overcome this, the operator needs to climb up to the top of the hopper and load the stuck waste without any tools. This can endanger the operator, who has the potential to fall and get injured or even become infected with medical waste. Another high-risk level is when transferring bottom ash and fly ash residues from the incinerator to a temporary disposal storage. If a jumbo bag containing ash falls, it can cause the ash to fly or fall to the floor, and if it spills, spreads and contacts with the humans, it will cause health problems. Bottom ash and fly ash are also harmful to the environment, so spilled ash has the potential to cause environmental pollution. This waste, when dropped on an unprotected surface, releases toxic compounds into the groundwater system of plant organisms [10]. Meanwhile, fly ash contains heavy metals, dioxins, and furans [3]. As a result, a program to avoid the emergencies situation must be developed and implemented.

3.2. Emergency Program

Based on the identification and analysis of emergency risks in the processing area, there are 3 activities that have a high risk and could potentially create a medical waste emergency. The resulting impact can be cumulative. Due to the magnitude of the impact that can be caused, it is necessary to make efforts to manage B3 waste in a comprehensive and sustainable manner [11].

The preparation of the emergency program refers to the guidelines from the government, is made up of infrastructure and response functions.

IOP Conf. Series: Earth and Environmental Science 1111 (2022) 012041 doi:10.1088/1755-1315/1111/1/012041

3.2.1 Infrastructure

a. Organization, which is in the form of an organizational structure that has the task of preparing, implementing, conducting internal evaluations, and revising the emergency response programs.
b. Coordination, which includes the flow of emergency communication and inter-team relationship systems as well as coordination procedures with other agencies. facilities and equipment, namely in the form of equipment needed to control emergency situations.
c. Countermeasures procedures, namely written procedures that include procedures for dealing with emergencies. This procedure refers to emergency response based on the characteristics of B3 waste found in the processing area.
d. Emergency Training and Rehearsal, namely emergency simulation activities involving all workers in the activity area and held at least once a year.

3.2.2 Countermeasure Function

The countermeasure function includes an implementation mechanism for handling that will be carried out when an emergency occurs, starting from the receipt of the initial report until the completion of emergency situation. This section describes the flow of the implementation of emergency response, as can be seen in Figure 2.

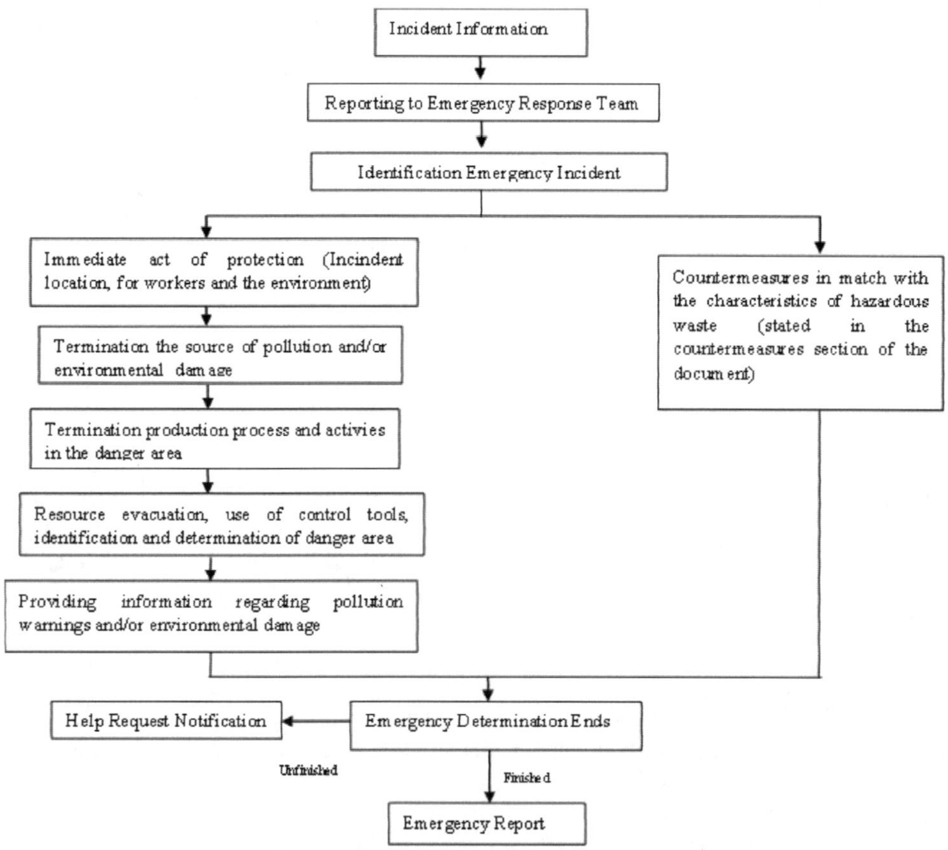

Figure 1. Emergency Response Flow

In the response flow, it is explained that after an emergency occurs, the emergency response team immediately identifies and takes immediate protective actions for workers and the environment. These actions can be carried out based on the characteristics of the hazardous waste contained in the processing area and following the mitigation procedures that have been established. If the emergency team cannot handle the emergency, then the relevant agency should be contacted for further action following the coordination flow that has been established. After the emergency condition is overcome, the coordinator of the emergency response team can make a written report to the company director.

4. Conclusion

The processing area has potential hazards and risks based on its activities. A risk assessment was conducted to identify the risk level from the sequences of unloading, burning, and ash handling activities.

The results of the risk level analysis were obtained. In the identification and analysis of emergency risks, three activities had an extreme level of risk, namely activities during the process of entering medical waste into incinerators, the operational temperature of burning, and activities during the process of moving bottom ash and fly ash from incinerators to the temporary storage area.

The emergency response program contains a plan for implementing prevention, preparedness, and implementation of B3 waste management emergency response involving various parties, contributing to and potentially receiving the impact of an emergency B3 waste management. This emergency program consists of infrastructure (organization, coordination, facilities and equipment, response procedures, emergency training, and rehearsals) and response functions. The emergency response program was developed based on the guidance from the regulations, and the emergency response rehearsal has already been conducted with the workers to understand the program.

References

[1] Alala Pratama 2022 Industrial Management in the Implementation of B3 Emergency and B3 Waste At PT. X *ENVITATS (Environmental Engineering Journal ITATS) 2(1): 1-10*

[2] Australian/New Zealand (AS/NZS) 4360 1999 Risk Management Guidelines (Strathfield: Standards Association of Australia) p 42

[3] Darmawan A 2022 A Review of Efforts to Manage Ash from Medical Waste Treatment with Stabilization and Solidification (S/S) in Indonesia: A Review *Jurnal Serambi Engineering 7(3)*

[4] Government Regulation Number 22 of 2021 Concerning the Implementation of Environmental Protection and Management

[5] Human Resources Development Agency 2018 Training Module for Thermal Incineration WtE Technology

[6] Hospital Environmental Health Requirements: A Decree of the Minister of Health of the Republic of Indonesia No. 1204 of 2004

[7] Latief A 2012 The Benefits and Impacts of Using Incinerators on the Environment *Semarang. Mechanical Engineering Department, Semarang State Polytechnic. Vol. 05: 20-22*

[8] Merpaung Ningsih 2016 Risk Analysis of Occupational Health and Safety in the Production Department of PT. Berkat Manunggal Jaya *Industrial Engineering Online Journal 5(1)*

[9] Regulation of the Minister of Environment and Forestry of the Republic of Indonesia Number 74 of 2019 concerning Emergency Response Program for Management of Hazardous and Toxic Materials and/or Hazardous and Toxic Waste

[10] Sabiha-Javied TM, Tufail M, Khalid S 2008 Heavy Metal Pollution From Medical Waste Incineration At Islamabad And Rawalpindi *Pakistan. Microchemical Journal 90: 77–81*

[11] Setiyono S 2001 The Legal Basis for Hazardous Waste Management *Environmental Engineering Journal 2(1)*

[12] Utami R.D 2011 Improving Incinerator Performance in Medical Waste Destruction at RSUD Dr. Soetomo Surabaya *Surabaya. Environmental Engineering Scientific Journal 7(2)*

The Effect of Chimney Size on Knock-Down Pyrolysis Tank Design as Rice Husk Burner Using CFD Simulation

Mohamad Ramadani Rudiatama[1], Hamdi Hamdi[2], Dianta Mustofa Kamal[3], Johny Wahyuadi M. Soedarsono[4], Ahmad Maksum[5]*
[1] D3 Mechanical Engineering, Dept of Mechanical Engineering, Politeknik Negeri Jakarta, Depok 16425, Indonesia. https://orcid.org/0000-0002-9235-6172
[2]D3 Mechanical Engineering, Dept of Mechanical Engineering, Politeknik Negeri Jakarta, Depok 16425, Indonesia
[3] Graduate Program, Dept of Mechanical Engineering, Politeknik Negeri Jakarta, Depok, 16425, Indonesia, https://orcid.org/0000-0001-9336-8936
[4] Prof Johny Wahyuadi Laboratory, Department of Metallurgical and Materials Engineering, Universitas Indonesia, Depok 16424, Indonesia, https://orcid.org/0000-0001-6051-2866
[5] Research Center for eco-friendly Technology, Politeknik Negeri Jakarta, Depok 16425, Indonesia, https://orcid.org/0000-0003-1800-9137
[1]abdirama625@gmail.com, [2]hamdi@mesin.pnj.ac.id,
[3]dianta.mustofakamal@mesin.pnj.ac.id
[4]jwsono@metal.ui.ac.id, [5]*ahmad.maksum@mesin.pnj.ac.id

Abstract. This study reports the simulation of rice husk pyrolysis units in various sizes of chimneys to investigate the effect of chimney diameter on temperature and heat distribution. The modeling design was carried out using Solidworks 2020 software with 3d modeling, and flow simulation features where the diameters of the chimneys were varied, namely 3, 4, 5, 6, and 8 inches. Simulation of the pyrolysis process with an input temperature of 450 °C and an average fluid velocity of 2.344 m/s for chimney diameters of 3, 4, 5, 6, and 8 inches has resulted in a condenser heat-temperature (TH) of 169.84 °C, 170.04 °C, 171.36 °C, 170.31 °C, and 168.05°C. It produces cold-temperature (TC) of 41.97 °C, 42.88 °C, 42.31 °C, 42.38 °C, and 45.35°C, respectively. The simulation results also show that the larger the chimney size of the pyrolysis tank, the greater the need for heat from the stove fire; conversely, the tank capacity of the husk becomes smaller.

1. Introduction

Rice husk waste can be processed into advanced materials such as silica, RHA (rice husk ash), or gas producers, which can also be used as a natural inhibitor [1,2]. The process of managing rice husks can be done by combining combustion, gasification, and pyrolysis [3,4]. The pyrolysis method was chosen because it has several advantages when applied to the combustion process in open spaces. When compared with the Combustion process based on research conducted by Holubcík (2020) which explained that compared to full combustion, the emissions resulting from the pyrolysis method are lower[5]. In addition, the pyrolysis method produces RH-Ash, Liquid smoke, and Producer gas, which will be further processed into silica and syngas, respectively [6]. In the village, the pyrolysis method is used because it only needs to reach a temperature of 500-700 °C if compared to the Gasification process, which requires a temperature of up to 700-1000 °C [7]. The pyrolysis tank needs to have an even combustion to obtain an optimal output.

Several studies related to the design of reactors using the pyrolysis system have been carried out before. S. A. Babajo (2021) has developed a laboratory-scale pyrolysis reactor system, but the design is

Made specifically for Jatropha Cake and Polystyrene materials [8]. B. L. Salvi (2020). developed a shell and tube condenser design in the fast pyrolysis process to process bio-oil made from wheat straw with bio-oil output, and a tank design gas producer does not use a chimney. M. R. Rudiatama (2022) designed a pyrolyzer tank as a producer of husk charcoal with a chimney concept in the middle of the chamber[9]. In the literature of previous studies, no predictions or calculations were found regarding the effectiveness of the chimney size on heat distribution in the pyrolysis tank, especially those carried out using flow simulations.

Therefore, it is necessary to study the analysis and simulation of heat distribution in the pyrolysis tank due to the difference in the size of the chimney to reduce the risk of fabrication losses due to the wrong design of the pyrolysis tank.

2. Materials and Methods

2.1 *Materials*

The material used in this study is in the form of rice husks obtained from waste from rice milling. Rice husks were chosen as the material for the husk charcoal to be taken, namely Rice Husk Ash (RHA). RHA will be used as the primary material for making nano silica after combustion through a pyrolysis reactor.

i. Rice Husk Waste: Rice husks are selected as the primary material for making RHA through a pyrolysis tank. Rice husks are washed and pre-treated first before entering the biomass pyrolyzer unit.
ii. Pyrolysis tank: On the inside of the pyrolysis tank, the material used is 304 stainless steel. Making pyrolysis tank by rolling and jointing using welding.
iii. Insulator: Insulator material in the form of lightweight brick and cement. The selection of insulators is carried out by considering the ease of repair and maintenance.

2.2 *Methods*

In the analysis of the pyrolysis tank reactor unit, based on the previous study by M. R. Rudiatama (2022), the chimney part of the tank was modified by changing the size to several variations. After that, an overall flow simulation is made on the pyrolysis unit to see heat distribution. The design modification and simulation process using Solidworks 2020 software with 3d modeling and flow simulation features. The data taken are in the form of mass from the tank, temperature-heat output (TH out) on pyrolysis, and temperature-cold (TC out) output on the condenser, also shown flow visualization results. The reaction time on the flow simulation is assumed to be as long as 1 hour.

2.3 *Modification of pyrolysis tank design*

Flow simulation experiments on different tank models are based on changes in the diameter of chimneys with sizes of 3 inches in Figure 1(a), 4 inches in Figure 1(b), 5 inches in Figure 1(c), 6 inches in Figure 1(d), and 8 inches on Figure 1(e). Model differences in design and modification can be seen in Figure 1. This change is used to determine the impact of heat distribution in pyrolysis tanks that use chimneys.

To find out the capacity obtained and the weight of the tank, calculations are needed to analyze the mass and optimum capacity of each model made. The density of rice husks uses $125 \frac{kg}{m^3}$[9], so that the size of each pyrolysis tank can be known.

Calculations on the volume of the tank with the result in the unit of m^3.

$$V = \frac{1}{4}\pi D^2 t \quad \textbf{(1)}$$

Where V is the volume of the tank, D is the diameter of the chimney size, t is the height of the tank design, and $t = 1\ meter$.

The formula for determining the weight of rice husks in kg.

$$m = V \, x \, \rho \quad \textbf{(2)}$$

Where m is the mass of the tank, V is the volume of the tank, and ρ is the density of rice husk.

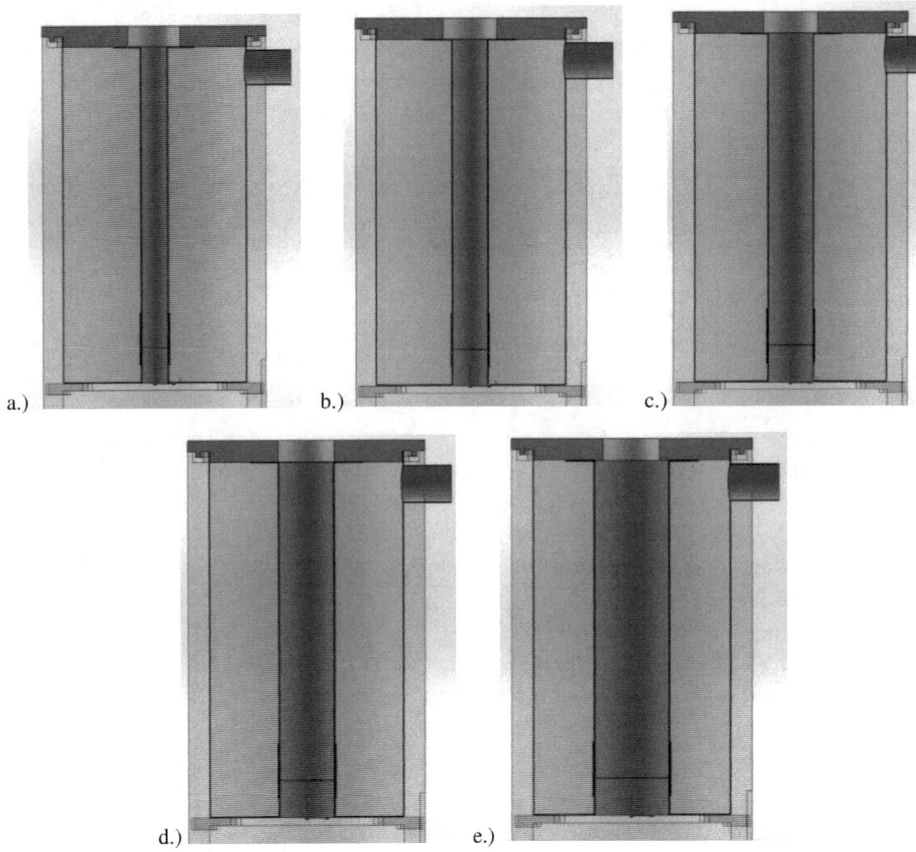

Figure 1. Detailed view of pyrolysis tank: a.) 3 inches; b.) 4 inches; c.) 5 inches; d.) 6 inches; e.) 8 inches.

2.4 Flow simulation inside the tank
Flow simulation analysis is used to analyze the distribution of heat in the fluid in the pyrolysis tank with the following parameters used;

- Pyrolysis uses temperatures that range from 450°C, based on previous studies [5,7,10].
- The velocity of the fluids in the reactor tank assumed 3 m/s.
- The velocity of the cold fluids in the condenser is assumed as 10 m/s.
- Mass flow rate uses 1 L/min [11].

3. Results and Discussion
The pyrolysis tank uses AISI 304 material, with the tested fluid assumed to be steam mixed with carbon dioxide. The simulation was tested using Solidworks flow simulation 2020 software. In the simulation process, the fluid is in the pyrolysis tank, then undergoes cooling through the condenser. The parameters were tested to contain visualizations of heat transfer and distribution in each reactor design and the temperature of the fluid produced from the stove from under the reactor.

Changes in design specifications can be seen in Table.1, which shows differences in the mass of pyrolysis tanks in each chimney size. At maximum capacity, it is measured when the condition of the rice husk is complete, but for the optimum capacity it is calculated using the following equation;

$$Capacity_{opt} = Capacity_{max}\ x\ 75\%$$

Where $Capacity_{opt}$ is an optimum capacity of the tank, and $Capacity_{max}$ is a maximum capacity that resulted from the theoretical calculation, and t is the height of the tank design and $t = 1\ meter$. 75% is the assumed value that the optimum tank capacity is filled.

3.1 Specification of modified pyrolysis tank unit

Simplify the results of the calculations and specifications in each modified design on the reactor tank. It can be seen based on Table 1. In the previous study, the size variable used was an optimum capacity of 20 kg, whereas M.R. Rudiatama used a chimney size of 4 inches.

Table 1. Design and specification of modified pyrolysis tank

Size [Inch]	Inner (d1) [mm]	Outer (d2) [mm]	D1 [mm]	D2 [mm]	t [mm]	V[m^3]	Mass[kg]	max Capacity [kg]	Optimum Capacity [kg]
3”	76,2	82,2	550	544	1000	0,227	54,9349	28,3757	21,28
4”	101,6	107,6	550	544	1000	0,223	57,1422	27,9026	20,92
5”	127	133	550	544	1000	0,218	59,3251	27,3029	20,47
6”	152,4	158,4	550	544	1000	0,212	61,2808	26,5767	19,93
8”	203,2	209,2	550	544	1000	0,197	65,728	24,7443	18,55

3.2 Visualization of flow simulation inside the tank

The simulation results are produced by cutting the top plane with an offset of 0.3 m from the bottom of the tank. Once simulated, it is seen in Figure 7(a) that reactors with a diameter of 3 inches have a maximum room temperature of 450 °C and a minimum temperature of 448.2 °C. The maximum temperature occurs in the area around the chimney. In Figure 7(b), a reactor with a diameter of 4 inches has a minimum temperature of 447.85 °C with maximum temperature conditions limited to around the chimney but has a broader spread radius. In Figure 7(c), a reactor with a diameter of 5 inches has a minimum temperature of 446.67°C. The maximum temperature condition has a wider radius and is more evenly distributed. In Figure 7(d), a reactor with a diameter of 6 inches has a minimum temperature of 446.49 °C with maximum temperature conditions having a larger radius and visually more solid. In Figure 7(e), a reactor with a diameter of 8 inches has a minimum temperature of 447.43 °C. The maximum temperature conditions occur around the chimney, visually having a solid color but not too massive as the previous size. Based on (Vershinina et al., 2022), the overall design variations meet rice husk material's pyrolysis burning method temperature conditions [7].

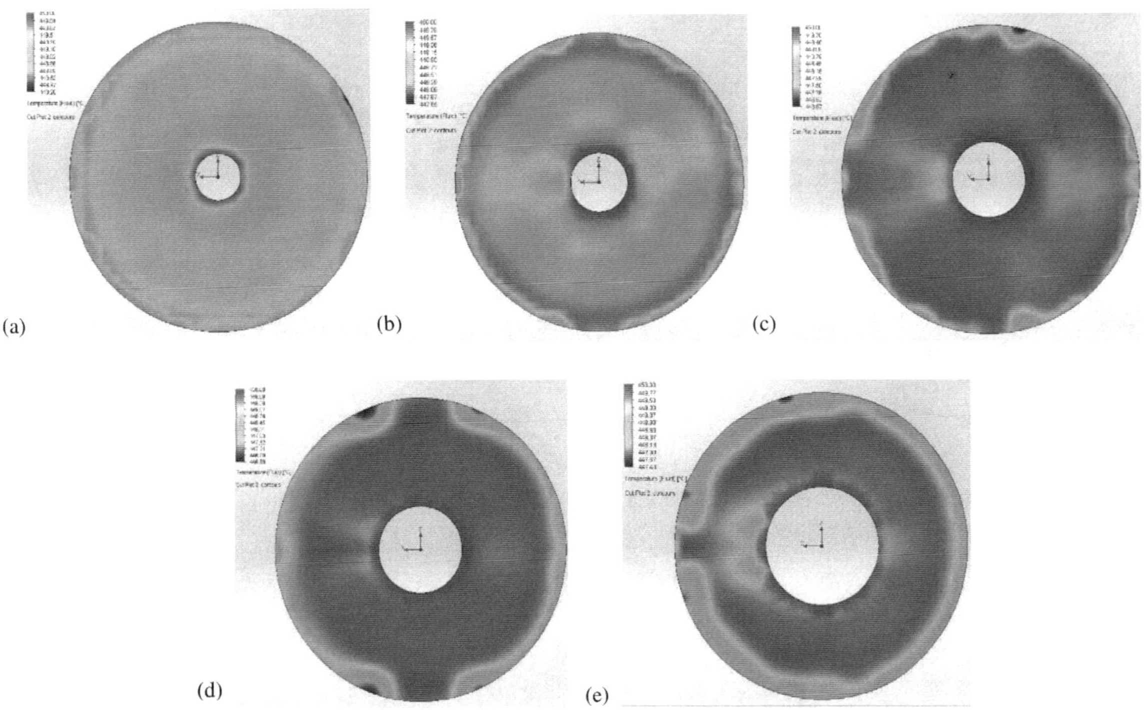

Figure 2. Intersection view of the distribution of fluid flow distribution in the pyrolysis tank

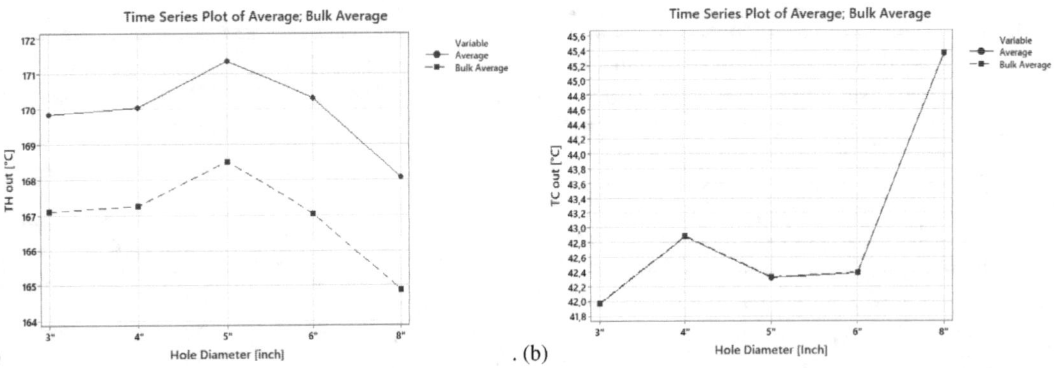

(a) . (b)

Figure 3. Average output temperature: (a) TH out; (b) TC out

The test results using flow simulation obtained results in the form of Temperature Heat (TH out) from the fluid flowing from the conducted heater to the pyrolysis tank, as in Figure 3(a). 3a, the average TH out was 169.84°C, 170.08°C, 171.36°C, 170.31°C, and 168.05°C, the fluid that came out had the highest condition in the 5 inches size chimney, which experienced a temperature drop of 61.91%, then decreased in the 8 inches diameter chimney with heat loss reaching 62.65%. Meanwhile, in fluid conditions in the form of cold temperature water (TC out), as in Fig Figure 3(b), the condenser has an average temperature of 41.97°C, 42.88°C, 42.31°C, 42.38°C, 45.35°C. There was a significant increase in temperature in the tank with a diameter of 8 inches with a heat gain of 81. 40°C.

The results of the modification of the chimney also affect the tank's weight and the capacity of the tank space. In Figure 4(a), it can be seen that the larger the diameter of the chimney, the tank space increases in mass, but in Figure 4(b), it has decreased in volume. In Fig Figure 5, it can be seen that the tank is experiencing a decrease in capacity, increasing the size of the chimney. In this study, it can be a solution to the research of M.R. Rudiatama; with a specific temperature, the use of a chimney of one size does not meet the optimal temperature conditions in the combustion of the pyrolysis method [9].

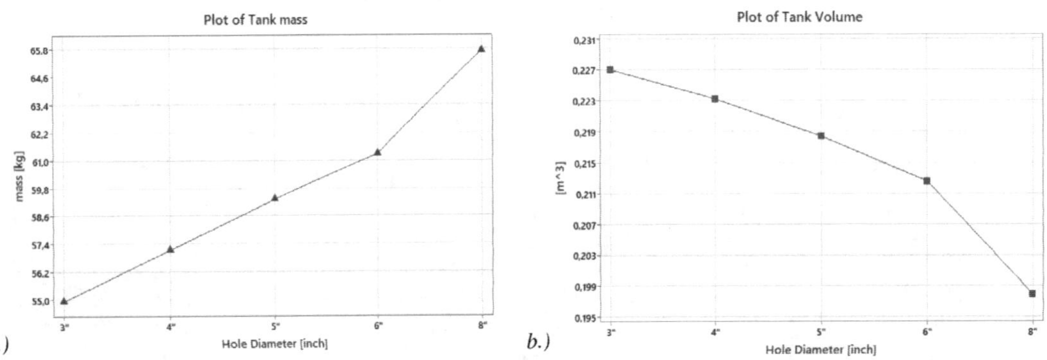

a.) *b.)*

Figure 4. Comparison of spesifications on the result of design modifications : a.) tank mass; b.)tank volume

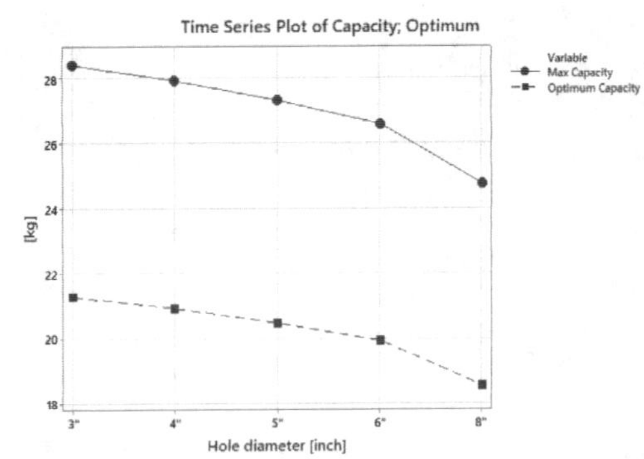

Figure 5. Comparison between maximum capacity and optimum capacity.

4. Conclusions

The design of a pyrolysis tank using a chimney in the middle of the tank can help accelerate the combustion process of husk charcoal. The simulation results of 3 to 8 inches in diameter show that the larger the chimney size, the larger the stove fire. On the other hand, the larger the diameter of the chimney, the smaller the storage capacity.

Moreover, the lowest temperature-heat (TH) of the fluid leaving the pyrolysis tank is achieved at a 5 inches flue, where the heat loss is up to 61.91%. Meanwhile, the highest cold temperature (TC) of the fluid coming out of the pyrolysis tank, which is 81.40 °C, is obtained at a chimney size of 8 inches.

In further research in this study, with the lack of experimental data, it is necessary to measure the actual temperature variable in the reactor design that has been made. Furthermore, it is necessary to analyze the effectiveness of the pyrolysis biomass design based on temperature-heat (TH) out and temperature-cold (TC) out of the pyrolysis system and condenser system accelerating.

Acknowledgments

The authors would like to thank the facilities provided by the Research Center for eco-friendly Technology and financial support from the Politeknik Negeri Jakarta through the Indexed International Publication Grant with Contract number: B.263/ PL3.18/PT.00.06/2022, Fiscal Year 2022.

References

[1] Kaban A P S, Ridhova A, Priyotomo G, Elya B, Maksum A, Sadeli Y, Sutopo, Aditiyawarman T, Riastuti R and Soedarsono J W 2021 Development of White Tea Extract as Green Corrosion Inhibitor in Mild Steel Under 1 M Hydrochloric Acid Solution *Eastern-European Journal of Enterprise Technologies* **2** 6–20

[2] Maksum A, Rustandi A, Permana S and Soedarsono J W 2019 Roasting-Quenching Pretreatment in the Calcination Process to Improve the Purity of Rice Husk Bio-Silica *JP Journal of Heat and Mass Transfer* **16** 313–26

[3] Hameed S, Sharma A, Pareek V, Wu H and Yu Y 2019 A Review on Biomass Pyrolysis Models: Kinetic, Network and Mechanistic Models *Biomass and Bioenergy* **123** 104–22

[4] Bakari R, Kivevele T, Huang X and Jande Y A C 2020 Simulation and optimisation of the pyrolysis of rice husk: Preliminary assessment for gasification applications *Journal of Analytical and Applied Pyrolysis* **150**

[5] Holubčík M, Klačková I and Ďurčanský P 2020 Pyrolysis Conversion of Polymer Wastes to Noble Fuels in Conditions of the Slovak Republic *Energies (Basel)* **13**

[6] Li W, Ge P, Chen M, Tang J, Cao M, Cui Y, Hu K and Nie D 2021 Tracers from Biomass Burning Emissions and Identification of Biomass Burning *Atmosphere (Basel)* **12**

[7] Vershinina K, Nyashina G and Strizhak P 2022 Combustion, Pyrolysis, and Gasification of Waste-Derived Fuel Slurries, Low-Grade Liquids, and High-Moisture Waste: Review *Applied Sciences (Switzerland)* **12**

[8] Babajo S A, Enaburekhan J S and Rufa'i I A 2021 Design, Fabrication and Performance Study of Co-Pyrolysis System for Production of Liquid Fuel from Jatropha Cake with Polystyrene Waste *Journal of Applied Sciences and Environmental Management* **25** 407–14

[9] Ramadani Rudiatama M, Irza Ramadhan F, Indra Kusumo P, Syakuro A, Ricardo Marulitua Tambun D, Hamdi and Ahmad Maksum 2022 Perancangan Tangki Reaktor Pyrolyzer Kapasitas 20 kg sebagai Penghasil Rice Husk Ash (RHA) *Seminar Nasional Inovasi Vokasi* vol 1 (Jakarta)

[10] Huang Y, Li B, Liu D, Xie X, Zhang H, Sun H, Hu X and Zhang S 2020 Fundamental Advances in Biomass Autothermal/Oxidative Pyrolysis: A Review *ACS Sustainable Chemistry and Engineering* **8** 11888–905Aladin A, Modding B, Syarif T and Dewi F C 2021 Effect of Nitrogen Gas Flowing Continuously into the Pyrolysis Reactor for Simultaneous Production of Charcoal and Liquid Smoke *Journal of Physics: Conference Series* vol 1763 (IOP Publishing Ltd)

IOP Conf. Series: Earth and Environmental Science 1111 (2022) 012043 doi:10.1088/1755-1315/1111/1/012043

Implementation of REDD+ Policies in Indonesia to Reducing Deforestation Amid Rising Earth's Temperature

Galih W. Pradana[1*], Hamsinah Djaing[2], Badrudin Kurniawan[1] and Alycia Putri Febriani[1]

[1]Public Administration, Faculty Social and Law, State University of Surabaya, Surabaya, 60231, Indonesia, [1]0000-0001-6875-381X, [3]0000-0003-0885-8362

[2]Public Administration, Universitas Hasanuddin, Makassar, 90245, Indonesia, 0000-0003-0265-9055

[1*]galihpradana@unesa.ac.id, [2]hamsinahdjaing@gmail.com

[3]badrudinkurniawan@unesa.ac.id, [4]alyciaputri.20002@mhs.unesa.ac.id

Abstract. The existence of forests that function as a temperature cooler and a CO2 absorber on earth is seriously threatened due to deforestation that is not decreasing in number but increasing. Many factors cause deforestation, ranging from industrial and new land clearing for plantations. Even without realizing it, massive deforestation is very influential on the global temperature increase, which threatens the safety of residents and other living things on earth. In Indonesia, there is a REDD+ policy prepared by the Ministry of the Environment, which is a step to reduce the effects of greenhouse gases due to deforestation. This study aims to analyze in depth the implementation of REDD+ policies in Indonesia and examine how these policies develop against the reality of forest conditions in Indonesia today. This research is a descriptive study using the literature study method, where data collection regarding current REDD+ and deforestation policies in Indonesia is then carried out by searching for data and information through documents, photos, pictures, written, and unwritten documents that can support writing. This study finds that the implementation of REDD+ policies has a significant impact on reducing deforestation and forest sustainability in Indonesia.

1. Introduction

NASA scientists staged a large protest before the JPMorgan Chase facility in Los Angeles not long ago. This protest is based on scientists' claims that the globe has suffered for several decades due to global warming. According to the most recent IPCC report, current and planned fossil energy infrastructure emissions are already more than double the amount that will cause the earth's temperature to rise by at least 1.5 degrees Celsius, resulting in a variety of natural disasters such as heat, fire, storms, floods, and droughts [1]. As the proportion of carbon dioxide gas (CO2) and other gases in the earth's atmosphere increases, the atmosphere must store more heat from the sun. This is what is causing global warming. Table 1 illustrates this. According to the Ministry of Environment, various factors directly impact the rise in geothermal energy, including greenhouse gases, motor vehicle fumes, industrial smoke, and deforestation, accounting for 60% of Indonesia's emissions [2].

The central topic of this study is deforestation as one of the reasons for an increase in the earth's temperature, where deforestation has at least a substantial enough influence on the global warming phenomena. This is due to forests' role as absorbers and more excellent CO2 on Earth. Therefore their disappearance results in the non-absorption of CO2, which then concentrates in the atmosphere, making it thinner and contributing to the earth's increasingly heated state. Deforestation is a state in which forests diminish on a large scale due to land conventions for infrastructure, towns, agriculture, mining, and plantations. This shift in forest conditions contributes to global warming [3]. Indonesia has the world's second highest deforestation and forest degradation rate, estimated at approximately 2 million hectares per year. It is one of the top 5 global GHG emitters amongst developed and developing countries, as seen in table 2 [4]. Referring to data released by the Indonesian Central Bureau of Statistics in 2014, several areas in Indonesia that experienced the most considerable deforestation were Riau, with a deforestation rate of 201,939.4 hectares per year, followed by Central Kalimantan with a deforestation rate of 58,835.6 hectares per year, Kalimantan East and North of 38,339.4 hectares per year, and West Kalimantan of 29,919.6 hectares per year [5]. The data did experience a significant decline in 2019, but it seems that this has not overcome the problem of increasing temperatures in Indonesia. Based on a report from the Indonesian Meteorology, Climatology, and Geophysics Agency, there was a temperature anomaly wherein 2016 was the hottest year, with an anomaly value of 0.8 degrees Celsius throughout the observation period from 1981 to 2020. With an anomaly value of 0.4 degrees Celsius, 2021 ranks as the eighth hottest year, while 2020 and 2019 rank second and third with anomaly values of 0.7 degrees Celsius and 0.6 degrees Celsius, respectively. In comparison, the global average temperature information released by the World Meteorological Organization (WMO) in its last report in early December 2020 also puts 2016 as the hottest year (first rank) [6].

Table 1. Increase in Earth's Surface Temperature (2011-2021)

No.	Year	Degrees Celcius
1	2011	0,61
2	2012	0,65
3	2013	0,68
4	2014	0,75
5	2015	0,9
6	2016	1,02
7	2017	0,92
8	2018	0,85
9	2019	0,98
10	2020	1,02
11	2021	0,85

Table 2. 10 Countries with the Most Primary Forest Land Loss (2020)

No.	Country	Thousand hectare
1	Brazil	1.704
2	Rep. Congo	491

3	Bolivia	277
4	Indonesia	270
5	Peru	190
6	Colombia	166
7	Cameroon	100
8	Laos	89,7
9	Malaysia	73
10	Mexico	68,4

The global climate initiative Reducing Emissions from Deforestation and Forest Degradation (REDD+) initially included carbon markets as one potential policy mechanism to finance various forestry projects in developing countries [4]. Indonesia was an early leader in REDD+, and processes to engage and seek consent from constituencies and communities were widely applied in developing programs at various scales [9]. However, the Indonesian institutional setting also features several constraining factors, including tax breaks and other incentives that stimulate deforestation and forest degradation, regulations that conflict with conservation aims, and poor implementation of pro-REDD+ reform [10]. A vital debate related to REDD+ is permanence. While avoiding deforestation or forest degradation can reduce carbon emissions and mitigate climate change, this effect is challenged because forest carbon can be released when protected forests are harvested or converted to non-forest land [11].

2. Methods

This research uses a literature study approach [12]. The documents used are from 2006-2020. The basic steps and essential choices in conducting a literature study in this research use four phases [13] (1) designing a review, which includes the following activities: a) determining whether this review is necessary and what the contribution of conducting a REDD+ policy review is; b) assessing the potential audience of the REDD+ policy; c) analyzing the specific objectives and research questions that will be discussed in this review, and d) analyzing the appropriate methods to be used to address the specific objectives of this review. e) Examining the REDD+ policy review search strategy (including search terms, databases, inclusion and exclusion criteria, etcetera). In Phase (2) of conducting the review, the activities carried out in this phase, namely: a) Assessing whether the search plan developed in phase one succeeded in producing suitable samples or needed adjustments; b) Assessing what the practical plan for selecting articles is; c) Assessing how the search and selection process is documented; d) Assessing how the quality of the search and selection process will be assessed. Phase (3) this phase includes the following activities: a) determining what type of information needs to be abstracted in order to meet the objectives of the specific review; b) determining what type of information is required to carry out a specific analysis; c) determining how reviewers will be trained to ensure the quality of this process; and d) determining how this process will be documented and reported. Phase (4) is writing a review, while the activities carried out in this phase are a) analyzing motivation and the need for this review to be communicated. b) Considering what reporting standards are appropriate for this specific review; c) Writing down what information needs to be included in the review; d) Ensuring and writing down the level of information provided is sufficient and appropriate to allow transparency so that readers can judge the quality of the review; e) Results are presented and clearly explained; f) review contributions are communicated.

3. Results and Discussion

3.1. Deforestation reduction strategies in Indonesia through the implementation of REDD+ policies

In addressing the rate of deforestation in Indonesia, the government implements various policies, such as selective logging, followed by reforestation or reforestation efforts. Another policy that supports efforts to reduce deforestation rates is the REDD+ (Reducing Emissions from Deforestation and Forest Degradation) policy, which is designed to use financial incentives to reduce emissions of greenhouse gases caused by deforestation. Indonesia changed the original idea of REDD+ from a market-based instrument to a government-funded one. Community involvement and non-monetary benefits are increasingly being raised in the REDD+ discourse [14]. (REDD+) plays a crucial role in the Indonesian climate policy [15]. REDD+ not only aims to reduce the increase in greenhouse gases but also includes the roles of conservation, sustainable forest management, and increasing forest carbon stocks [16]. This scheme also helps in reducing poverty and sustainably improving the economy. This policy involves stakeholders in the implementation process. However, what is more, important is that the voices of the community, indigenous peoples, and traditional communities must also be taken into account so that the rights of those who live in the forest are guaranteed. The REDD+ policy is also a financial incentive program for countries willing to protect their forests as the world's lungs. This program has a target, especially for countries with extensive forests but facing deforestation problems, such as Indonesia and Brazil [17].

The REDD+ policy in Indonesia has existed and has been implemented since 2009, to be exact, in October. At that time, President Susilo Bambang Yudhoyono committed to reducing CO2 emissions in Indonesia by 62%. This has received support from the Government of Norway, which approved the signing of a Letter of Intent on May 26, 2010, in which the Norwegian government agreed to support Indonesia through contributions based on emission reductions in line with the REDD+ scheme. The contribution in question is channeled to the Indonesian financial mechanism, namely the Indonesian Environmental Fund [16]. Indonesia also adopted a moratorium on issuing exploitation licenses, the 'One Map' policy, and stricter monitoring practices. As a result, in 2020, Indonesia was awarded its first payment by Norway and the GCF for reducing emissions in 2017 [18].

3.2. The development of REDD+ policies in Indonesia is based on the reality of current forest conditions.

Indonesia's success in dealing with deforestation received recognition from NICFI (National Climate and Forest Initiative Norway). This recognition was conveyed through a press release where NICFI acknowledged that Indonesia was able to reduce the rate of deforestation and mass conversion of peatlands. This is clear evidence that Indonesia is committed to contributing to global climate change mitigation as well as protecting biodiversity through the REDD+ policies that have been implemented. This slick cooperation between Indonesia and Norway regarding the REDD+ policy did not continue because the Indonesian government officially terminated cooperation in reducing greenhouse gas emissions from deforestation—to be precise, on Friday, September 10, 2021. This is related to Indonesia's commitment to reduce greenhouse gas emissions by 26% by 2020 and received a response from the Norwegian government to contribute in the form of Result-Based Payment funds, with a scheme agreement of US$ 5 per ton. However, this agreement did not result in the disbursement of funds until the end of 2020, when the Ministry of Foreign Affairs finally decided to stop the cooperation but did not change Indonesia's commitment to reducing greenhouse gas emissions [17].

Another finding states that at the end of August 2020, Indonesia received official funding from the Green Climate Fund for REDD+ policies of USD 103.8 million, which will be managed by the Environmental Fund Management Agency (BPDLH), and Indonesia is the largest recipient of funding beyond Brazil. This financial support will strengthen coordination, implementation, and the overall REDD+ architecture [19]. In addition, based on the Ministry of Foreign Affairs records, Indonesia has made excellent progress thanks to the cooperation agreement with Paris (Paris Agreement). The achievements that have been obtained are that the deforestation rate has decreased to the lowest in 20 years, namely in 2020, and the decrease in the area of forest fires in Indonesia [20]. In the 2019/2020 period, the deforestation rate decreased by 75%. This is a tangible manifestation and not just an assumption resulting from the government's cooperation and commitment to dealing with deforestation [7]. Changes in information exchange and collaboration networks and, perhaps most importantly, how REDD+ has been institutionalized" over time seem to reflect a process of objectification rather than a strive to implement significant reform across sectors and interests to keep forests standing in Indonesia [21].

4. Conclusion

Global warming has experienced an increasing trend lately, one of which is caused by deforestation or deforestation. The role of forests is vital in absorbing CO2, so their existence is essential and must be maintained. The government has made various efforts to reduce the rate of deforestation in Indonesia, one of which is REDD+ cooperation with the Norwegian government through financial support, which significantly reduces carbon emissions caused by greenhouse gas effects and deforestation. However, this collaboration could not continue due to no further commitments related to funding in 2020. However, the Indonesian government remains committed to reducing carbon emissions, which is also supported by the GCF and the Paris Agreement, which also significantly impact reducing the rate of deforestation in Indonesia.

References

[1] Irach H 2022 Ilmuwan NASA Demo Besar-besaran, Peringati Dampak Pemanasan Global dalam Waktu Dekat *Fimela*
[2] Lhkppi D Pemanasan Global
[3] Wahyuni H and Suranto S 2021 Dampak Deforestasi Hutan Skala Besar terhadap Pemanasan Global di Indonesia *JIIP J. Ilm. Ilmu Pemerintah.* **6** 148–62
[4] Boer H J 2018 The role of government in operationalizing markets for REDD + in Indonesia *For. Policy Econ.* **86** 4–12
[5] Badan Pusat Statistik 2017 Angka Deforestasi Netto Indonesia Di Dalam Dan Di Luar Kawasan Hutan Tahun 2013-2020 (Ha/Th) 335–58
[6] BMKG 2022 Ekstrem Perubahan Iklim Di Indonesia | BMKG *BMKG*
[7] Katadata 2021 Laju Deforestasi Hutan Primer Indonesia Peringkat 4 di Dunia
[8] NASA 2022 Global Temperature | Vital Signs – Climate Change: Vital Signs of the Planet
[9] Boer H J 2019 Deliberative engagement and REDD+ in Indonesia *Geoforum* **104** 170–80
[10] Kaisa K K, Maria B, Efrain M, Sirkku J, Moira M, Cynthia M and Bimo D 2017 Analyzing REDD+ as an experiment of transformative climate governance: Insights from Indonesia *Environ. Sci. Policy* **73** 61–70
[11] Li X, Zhang X and Yang H 2020 Estimating the opportunity costs of avoiding oil palm-based deforestation in Indonesia: Implications for REDD+ *Chinese J. Popul. Resour. Environ.* **18** 9–15
[12] Snyder H 2019 Literature review as a research methodology: An overview and guidelines *J. Bus. Res.* **104** 333–9
[13] Kraus S, Breier M and Dasí-Rodríguez S 2020 The art of crafting a systematic literature review in entrepreneurship research *Int. Entrep. Manag. J.* 16
[14] Ekawati S, Subaru, Budiningsih K, Sari G K and Muttaqin M Z 2019 Policies affecting the implementation of REDD+ in Indonesia (cases in Papua, Riau and Central Kalimantan) *For. Policy Econ.* **108** 101939
[15] Rakatama A, Pandit R, Iftekhar S and Ma C 2020 Policy forum: Improving the acceptability of REDD+ projects among local households in Indonesia *For. Policy Econ.* **116** 102172

[16] KLHK D P- Pertanyaan Seputar REDD+ dan Implementasi REDD+ di Indonesia
[17] Fajrian H 2022 Pemerintah Setop Kerjasama REDD+, Norwegia Tetap Ingin Dukung RI Artikel ini telah tayang di Katadata.co.id dengan judul "Pemerintah Setop Kerjasama REDD+, Norwegia Tetap Ingin Dukung RI" , https://katadata.co.id/happyfajrian/berita/613caeb9d6cb3/pemerint *Kata Data*
[18] Pham T T, Moeliono M, Yuwono J, Dwisatrio B and Gallo P 2021 REDD+ finance in Brazil, Indonesia and Vietnam: Stakeholder perspectives between 2009-2019 *Glob. Environ. Chang.* 70 102330
[19] Kemenkeu INDONESIA MAKIN DIPERCAYA: Green Climate Fund Setujui REDD+ Results-Based Payment USD 103,8 juta
[20] Prabowo D 2021 Indonesia Akhiri Kerja Sama Terkait Pengurangan Emisi dan Deforestasi dengan Norwegia *Kompas.com*
[21] Moeliono M, Brockhaus M, Gallemore C, Dwisatrio B, Maharani C D, Muharram E and Pham T T 2020 REDD+ in Indonesia: A new mode of governance or just another project? *For. Policy Econ.* 121 102316

Independent Smallholders Farmers Group Sustain Biodiversity, Social and Economy After a Roundtable on Sustainable Palm Oil Certification: Case Study Sites in Jambi Province, Indonesia

Rosyani [1*], D C Zemp[2], B Irawan[3], Ermadani[4], and A Nizori[5]

[1] Department of Social of Agricultural Economics, Faculty of Agriculture, Jambi University, Indonesia

[2] Department of Biodiversity, Macroecology, and Biogeography, University of Goettingen, Göttingen, Germany

[3] Department of Forestry Faculty of Agriculture, Jambi University, Indonesia

[4] Department of Agro Eco-Technology Faculty of Agriculture, Jambi University, Indonesia

[5] Department of Technology of Agricultural Products, Faculty of Agriculture, Jambi University, Indonesia

[1*]rosyani_pertanian@unja.ac.id

Abstract. The purpose of this research is to analyze environmental impacts by assessing the number of forest crops planted among oil palm plantations, comparing soil qualities between certified and non-certified areas, analyze economic impacts by calculating the group's income, and analyze the strategies used by strengths to be balanced biodiversity and economic development. The data collection uses observing, interviewing, and sampling the soil in the oil palm plantation and control site. The economic analysis was analyzed by using cost and benefit analysis. There has been a very significant change as the location of the river border has been planted with several types of forest plants. The chemical properties of soil in Gapoktan and non-Gapoktan farming areas were low pH or acid soil reaction. However, organic C, total N, and available P in Gapoktan soils were higher than in the non-Gapoktan soils. Smallholders' strategies can maintain the sustainable management of oil palm plantations. It can be concluded that plot scales in smallholder plantations can support the biodiversity of oil palm plantations; so that runoff water can be restrained, and the biodiversity enrichment in oil palm can generate synergies between economic, social, and ecological functions.

1. Introduction

Globally, pressure demands increased palm oil to be produced sustainably, especially according to the roundtable on sustainable palm oil (RSPO) standard. To date, adopting sustainable palm oil standards is voluntary for many countries. Also, it is unclear whether early adopters achieve better financial positions than those who do not practice the standards [1]. Land conversion to produce oil palm has been recognized as one of the main drivers of tropical deforestation [2]. Smallholder oil palm plantations have increased from 3,125 ha in 1979 to more than 6 million hectares in 2019, accounting for about 41% of Indonesia's total oil palm plantations [3].

Reducing tropical deforestation caused by expanding agricultural and other commodities, especially palm oil, requires good interventions for engaging small-scale farmers. However, smallholders identify their main challenges in achieving certification [4]. Furthermore, independent oil palm smallholders are grouped according to many categories, based on landholding size, the farmer's ethnicity, local or migrant farmers, and the farm location of palm oil plantation (peat or minerals soil). The largest categories of farmers in their study owned, on average, 49.6 ha [5]

The black campaign *Elaeis guineensis* associated with oil palm's reputation as the deforestation driver has driven many buyers from Europe and North America to demand sustainability-produced palm oil. Industry groups created voluntary certification systems, such as the Roundtable on Sustainable Palm Oil (RSPO). Although the overall RSPO-certified area is small compared to the total oil palm area, there have been some reports of the limited success of RSPO in reducing deforestation [6]. Are there any studies about the RSPO impacts on biodiversity? However, smallholder participation in this certification program has been limited until now due to their limited capacity to organize and comply with the complicated certification process [7]. More literature reviews are here for smallholders in advancing RSPO monoculture [8].

Many studies argue that oil palm plantations in Indonesia have low biodiversity values being a monoculture system [9]. However, limited empirical studies have measured the impacts of smallholder RSPO certification on livelihood and biodiversity. This integrated assessment can help us understand the effectiveness of RSPO in meeting competing needs.

In general, smallholder palm oil production in Indonesia can be divided into two groups: independent smallholders and scheme smallholders [10]. Independent smallholders of palm oil manage their plantations individually or in a group (Gapoktan) and are not totally helped by the government [11].

This study focuses on one independent smallholder farmer's group (Gapoktan) in Jambi Province, Indonesia, Gapoktan of Tanjung Sehati which obtained RSPO certification and implemented environmentally friendly agricultural practices. This RSPO positively impacted their environment, economy, social and institutional contexts [12].

The result is significant for this study. Therefore the purpose of this research is to analyze environmental impacts by assessing the number of forest crops planted among oil palm plantations and comparing soil qualities between certified and non-certified areas, to analyze economic impacts by calculating the group's income, and to analyze the strategies used by strengths to be balanced biodiversity and economic development. Hopefully, the results of this study will help generate strategies to protect the social systems of oil palm by planting other plants sustainably.

2. Methods

2.1. *Study Area*

Some communities in Jambi Province of Sumatra, Indonesia, embraced RSPO certification early on. More than half of the total oil palm plantations in Jambi Province are cultivated by smallholders as of 2020, 716,328 ha out of the total is 1.074.600 Ha [13]. One early adopter group of smallholders is Gapoktan Tanjung Sehati, a farmer group certified by RSPO in the year 2014 and maintained its certification until now. Many farmers of Gapoktan Tanjung Sehati have been planting trees within the palm oil plantations and allowing pioneer plants to grow in the border area of the river in their garden area planted with forest crops. Some of the sustainable practices, such as protection of river borders and protection in the basin location, are required by RSPO certification. The Smallholders group, which obtained RSPO certification, must comply with the regulation of Indonesia's law[14].

The research activities were conducted in smallholder oil palm landscapes in Merangin District, Jambi Province, Indonesia. The location of the smallholder (Gapoktan), which received RSPO certification in 2014, is shown in Figure 1.

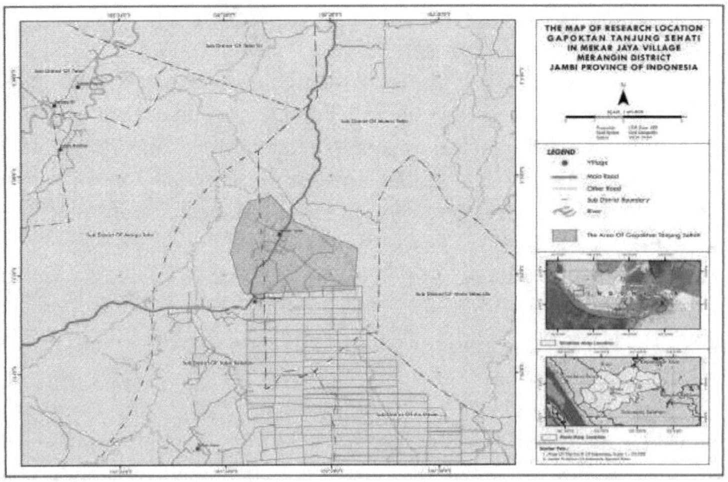

Figure 1. The location of Gapoktan Tanjung sehati

2.2. *Materials, data collection, and analysis*

Materials of this study were collected from primary data in the field, such as data on plants and areas, while maps of river locations in the plantation and collected by the survey from the local independent smallholder farmers. Biodiversity data were collected through field observations using EFForTSBiodiversity Enrichment Experiment (BEE) method from January to December 2020. Soil samples were taken at four locations in Gapoktan Tanjung Sehati, and two sample points were taken at Sahut Village, which was a monoculture plantation planted only with an oil palm plantation around Gapoktan Tanjung Sehati. Then the data were analyzed in the laboratory. Soil samples were taken at four locations of Gapoktan Tanjung Sehati and two sample points at nearby Sahut village (control Area without Certification).

3. Result and Discussion

3.1. *Ecosystem Biodiversity*

The results of interviews with farmers Gapoktan Tanjung Sehati revealed that between the years 2000 and 2007, their plantation lands were found still monoculture. Furthermore, the local NGO, Setara Foundation, came to the village and empowered the farmers' community about the management of environmentally sound plantations, and the community began to practice it. In the vicinity of the plantation, there is a small river with a length of approximately 3000 meters, named Sungai Batuah, and the wide plantation area was about 346,57 hectares. There has been a very significant change at the river border area as the location of the river border has been planted with several types of forest plants, as *Swietenia macrophylla, Entada phaseoloides, Nauclea orientalis, Macaranga triloba, Alstonia scholaris, Fagraea fragrans, Artocarpus integer, Durio zibethinus, Tectona grandis, Phoebe hunanensis,* and *Aquilaria malaccensi.*

Based on topographic maps and field observations, topography in the study area consists of level (0 - 3%), undulating (3 - 8%), and rolling (8-15%). The dominant topographic conditions are level and rolling. The soil types found in the Gapoktan location are Ultisol and Inseptisol. Ultisol is on the topography of undulating and rolling, while Inseptisol is on the level and undulating topography. The physical and chemical properties of the soil were determined based on the analysis of soil samples taken at four sites on Gapoktan, namely site 1 (TG0), site 2 (TG1), site 3 (TG2), and site 4 (TG3 with RSPO Certification) fields and two sites in control areas which have no implementation of RSPO certification (control 1 and control 2). The results of soil's physical and chemical properties are presented in Table 1.

Table 1. Soil quality in the Gapoktan study area and the control villages

No.	Parameters	Unit	Laboratory result					
			TG 0	**TG 1**	**TG 2**	**TG 3**	**Control 1**	**Control 2**
1	pH							
	- H_2O	-	4,6	4,8	4,7	4,6	4,5	4,8
	- KCl	-	3,8	3,7	3,7	3,6	3,7	4,1
2	C- Organic	%	1,6	2,21	1,81	1,76	0,98	1,13
3	N- Total	%	0,2	0,24	0,15	0,16	0,11	0,12
4	C/N ratio		10	9,2	12,1	11	8,9	9,4
5	P- available	Ppm	4,6	4,1	3,8	3,5	2,3	1,8
6	Exchangeable bases							
	Ca, Mg, K, Na	me/100 g	3,5	5,14	4,24	4,48	6,65	5,16
	(CEC)	me/100 g	12	16,5	14,2	14,6	18,6	16,2
	Base Saturation	%	29	31,1	29,8	30,7	35,7	31,8
7	Acidity							
	- Exchangeable Al	me/100 g	2,1	1,94	1,77	2,41	1,92	2,68
	- Exchangeable H	me/100 g	1,8	0,96	1,14	1,35	1,06	1,47
8	Texture		Clay	Clay	Clay	Clay	Clay	Clay
	Coarse sand	%	1,6	1,6	0,9	2,3	0,6	1,1
	Very fine sand	%	4,8	2,6	2,4	2,7	1,2	2,6
	Dust	%	30	23,3	32,9	29,6	26,9	41,6
	Clay	%	64	72,6	63,8	65,4	71,3	54,7
9	Permeability	cm/hour	1,3	0,98	1,35	2,04	1,39	2,36
10	Bulk Volume	g/cm^3	1,3	1,14	1,28	1,3	1,15	1,35

Note: Control 1 and control 2 were the location of samples taken without RSPO Certification TG0, TG1, TG2, and TG3, which were the location of the sample taken with RSPO Certification

Based on soil analysis in Table 1, it can be seen that the physical properties of soil in Gapoktan and non-Gapoktan areas were clay texture and rather slow to medium permeability. The chemical properties of soil in Gapoktan and non-Gapoktan farming areas were low pH or acid soil reaction. However, organic C, total N, and available P in Gapoktan soils were higher than in the non-Gapoktan shown in Figures 2.

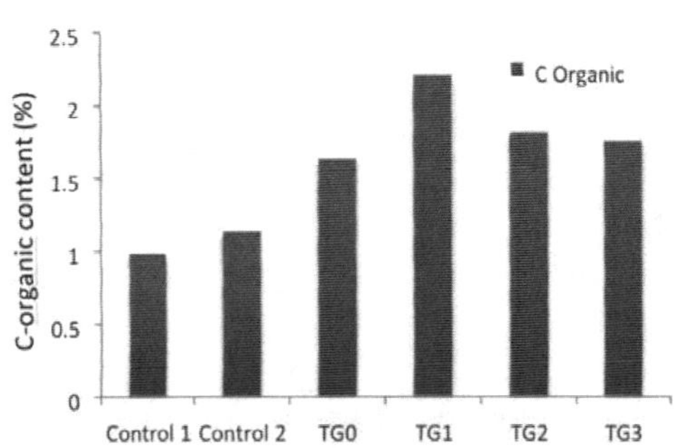

(a) Analysis of C-organic content from soil samples taken at four sites of Gapoktan (RSPO Certification) and two sites in control areas

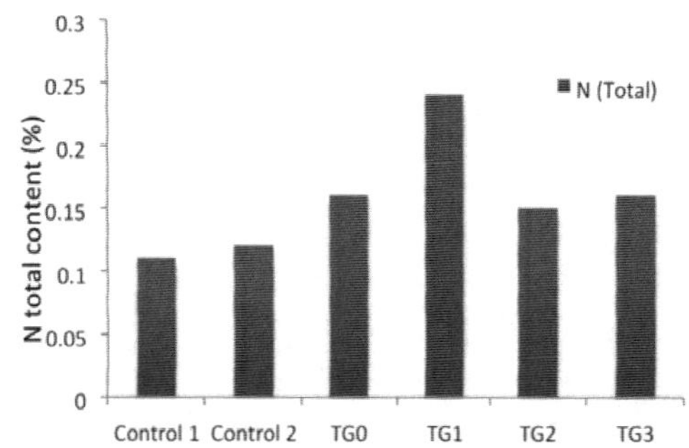

(b) Analysis of N-total content from soil samples taken at four sites of Gapoktan (with RSPO Certification) and two sites in control areas (without RSPO Certification.)

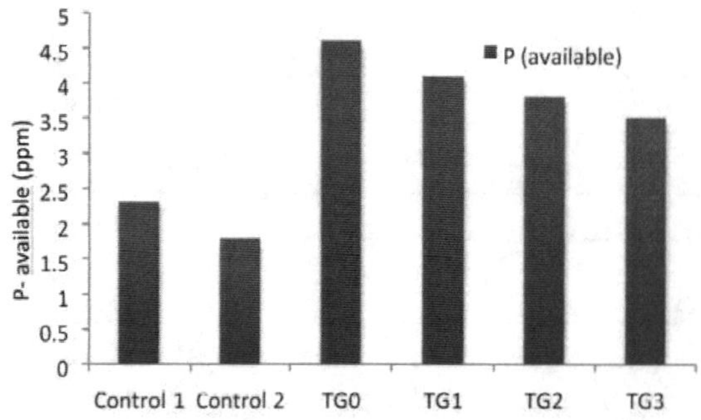

(c) Analysis of P-available from soil samples taken at four sites of Gapoktan (with RSPO Certification) and two sites in control areas without RSPO Certification.

Figure 2. Analysis of soil samples

It is estimated that the higher nutrient contents result from Gapoktan soils because organic fertilizer (cow manure) was applied in the Gapoktan land for many years as the source of macronutrients such as Nitrogen (N) and Phosphorous (P available). The values of soil cation exchange capacity (CEC) and base saturation in Gapoktan and non-Gapoktan are low and have relatively similar results. Inseptisol and Ultisol are soiled with very acid-to-acid soil reactions and low to deficient organic matter and nutrient content [15]. Therefore, organic fertilizers such as cow manure can be an essential source of organic matter to improve soil quality through increasing soil organic matter and nutrients.

According to previous researchers Teuscher et al. [7], this experiment expects a combination of island size, tree diversity levels, and above-average productive and cost-effective composition to achieve high returns in ecosystem function. This involves identifying the best performing tree species in their most productive composition under oil palm plantation conditions, which do not have a negative impact on oil palm yields. The early positive responses of birds and invertebrates to biodiversity enrichment treatments are promising and suggest that tree islands can be a suitable measure for increasing biodiversity in poor landscapes.

3.2 Analysis of Economy impact:

Another term for PTN (Rupiah given by farmers to Gapoktan/kg = Rp. 230) means the agreement of Gapoktan that must be paid per farmer. The farmer has to pay Rp 230/kg for selling.The results of the economic benefit obtained by group smallholders In Managing Oil Palm Plantation (Located in RSPO certified) are 4.882.995 Kg Total of production, The currency Rp.230.- and GCB (GAPOKTAN Cooperation Benefit) Rp.1.123.088.850

The production analysis and the amount of money of farmers set aside for Gapoktan Tanjung Sehati produced rapid economic growth for Gapoktan Tanjung Sehati. Gapoktan Tanjung Sehati cooperative obtained Rp.1,123,088,850/year savings in 2020. The funds are utilized for the needs of cooperative economic activity, meaning transportation fund marketing of Fresh Fruit Bunch. From the fund, the cooperative has purchased two trucks for their group necessary. The funds are also used to finance local Islamic education school for farmer's children, build the water towers for Gapoktan plantation, and build cattle farming. The organic waste from cattle farming is still used for organic fertilizer by Farmer's group. Whereby the expenditure will return as the farmer's benefit, and now it has been realized as the farmer's benefit.

3.3. *Analysis of Social Impact*

At the research location, 6 farmer groups who are members of the Tanjung Sehati Gapoktan have joined. The close relationship between the groups that has been built has enabled farmers to run oil palm farming businesses. Starting from processing to harvesting and marketing are carried out together. This means that socially, the psychological strength among farmers has been built. Farmers can trust in finance and build new knowledge about plant biodiversity. Relating to social analysis [16]. The motivation of independent smallholders to participate in certification schemes seems to be a proactive risk-reduction approach rather than a reactive one that could help mitigate current challenges smallholders face or improve sustainability. This means that smallholders become strong and can share knowledge to understand the meaning of the RSPO certificate. After understanding the farmers together with psychology, who can implement the principles and criteria of the RSPO?

3.4. *Sustainability Strategy*

Based on the research undertaken by smallholders in the plantation areas, the right strategy is to maintain other tree crops and oil palms. This is because the monoculture of oil palm will accelerate the water runoff process, causing rainwater containing various nutrients to be released to the sea's surface. Accordance to field results, Oil palms have been planted with other tree crops plants well. In the plantation areas, oil palms are planted by intercropping (plot scale) systems. Previous researchers, Gerard et al. [9], mentioned that past events that biodiversity enrichment in oil palm could result in a trade-off between economic and ecological functions under certain conditions. In its initial phase, EFForTS-BEE significantly increases yields per oil palm, which compensates for yield losses from oil palm removal at the plot scale. In this analysis of net yield changes, the researchers found an overall neutral effect on outcomes for small plots and varied widely but generally had a significant positive effect for large plots. An early sign of the negative influence of tall trees on oil palm yields, this positive effect may change in the long term. However, this study's results encourage further research to identify more diverse management strategies for oil palm plantations that harmonize ecological and economic benefits.

Moreover, previous researchers also made similar statements [17]. There is a need to encourage wider adoption of the RSPO P&C scheme by oil palm smallholders for better agricultural management, social justice, environmental protection, and smallholder welfare. Such strategies can contribute to rehabilitating ecosystem functions and increasing the resilience of local livelihoods in the face of fluctuations in world market prices and environmental changes. Future results of this experiment will shed more light on whether the area is suitable for diversifying oil palm and, if so, which management strategies are specific and of concern to all stakeholders. The following strategies that farmers apply are maintaining the riparian areas by planting pioneer trees on the left and right sides to maintain soil fertility and moisture.

The CRC-EFForTS Biodiversity Enrichment study is planned to answer questions about the potential of enriched oil palm landscapes to maintain or enhance biodiversity and ecosystem functions and services while minimizing economic losses. The expected outcome of this experiment is a combination of island size, tree diversity level, and above-average productive and cost-effective composition to achieve high benefits in ecosystem goals [7]. Furthermore, the authors claim that Biodiversity in oil palm plantations can synergize between economic and ecological functions under certain conditions. For example, the tree species are Sungkai (*Peronema canescens*), Petai (*Parkia speciosa*), and Jengkol (*Archidendron pauciflorum*), which show the best growth performance in terms of tree diameter, as well as have high survival rates when they coexist with oil palm plantations.

The Partnering experience is one of the important points for smallholders in applying RSPO principles and criteria. In line with previous research conducted by Gatto [18]. A general policy implication of our results from Jambi is that the contract schemes between village communities and palm oil companies contributed positively to economic development. However, it is stated that not all independent smallholders who get an RSPO certificate can improve their economy through the market price. This could RSPO failure implementation [7]. Meanwhile, the interviews of the smallholders showed that the farmers' trust in certification schemes did not work, which was supposed to be a precondition for good governance [19]. A few smallholders even accused the certification process of being unfair because all smallholders had to wait many years to get their certificates because the audit failed. This finding contrasts with Rametsteiner et al. [20] previous research, which claimed that independent audits incentivize improving management practices. Another smallholder reported that members of an already certified farmer group did not receive the premium price as promised for their fresh fruit brunch price agreement, which led to disappointment and mistrust among the smallholder farmers.

4. Conclusion

Based on the results, it can be concluded that oil palm plantations' biodiversity has many benefits. Moreover, biodiversity and economic ecosystem management strategies are compatible and have benefitted the sustainability of smallholders' plantations and economic welfare. Implementing RSPO certification has resulted in positive changes in the environmental knowledge of independent smallholder farmers. The farmers also have a positive impact on their economy. Independent smallholders can reduce sales transportation costs because they have arranged transportation through group collaboration. From a social perspective, independent smallholders have gained the power to market access, knowledge about sustainable plantation management, and RSPO principles and criteria. There has been a very significant change as the location of the river border has been planted with several types of forest plants. The chemical properties of soil in Gapoktan and non-Gapoktan farming areas were low pH or acid soil reaction. However, organic C, total N, and available P in Gapoktan soils were higher than in the non-Gapoktan soils.

References

[1] Ayompe, L. M., Schaafsma, M., & Egoh, B. N. (2021). Towards sustainable palm oil production: The positive and negative impacts on ecosystem services and human wellbeing. *Journal of Cleaner Production*, *278*, 123914. https://doi.org/https://doi.org/10.1016/j.jclepro.2020.123914

[2] Keraf, A. Sonny. 2010. Etika Lingkungan Hidup. Jakarta: PT Kompas Media Nusantara [3] Directorate of General Estate Crops. (2020). Tree Crop Estate Statistics of Indonesia 2018-2020 [Online] Available at: http://ditjenbun.pertanian.go.id/pojok-media/publikasi/.

[4] Watts, J. D., Pasaribu, K., Irawan, S., Tacconi, L., Martanila, H., Wiratama, C. G. W., Musthofa, F. K., Sugiarto, B. S., & Manvi, U. P., 2021. Challenges faced by smallholders in achieving sustainable palm oil certification in Indonesia. World Development, 146, 105565. https://doi.org/https://doi.org/10.1016/j.worlddev.2021.105565

[5] Jelsma, I., Slingerland, M., Giller, K. E., & Bijman, J. (2017). Collective action in a smallholder oil palm production system in Indonesia: The key to sustainable and inclusive smallholder palm oil? The Journal of Rural Studies, 54, pp. 198–210. https://doi.org/10.1016/j.jrurstud.2017.06.005.

[6] Carson, J. B., Tesluk, P. E., and Marrone, J. A. (2018). Shared leadership in teams: an investigation of antecedent conditions and performance. Acad. Manag. J. 50, 1217–1234. doi: 10.5465/amj.2007.20159921

[7] Teuscher, M., Gérard, A., Brose, U., Buchori, D., Clough, Y., Ehbrecht, M., Hölscher, D., Irawan, B., Sundawati, L., Wollni, M., & Kreft, H. (2016). Experimental Biodiversity Enrichment in Oil-Palm-Dominated Landscapes in Indonesia. *Frontiers in plant science*, *7*, 1538. https://doi.org/10.3389/fpls.2016.01538

IOP Conf. Series: Earth and Environmental Science 1111 (2022) 012044 doi:10.1088/1755-1315/1111/1/012044

[8] Beckert, C and M Varandas eds., tica as e Politicas Ambientais (Environmental Ethics and Environmental Policies), (Lisbon, Portugal: Centro de Filosofia da Universidade de Lisboa, 2004), pp. 133- 15 1.

[9] Gerard, A., Wollini, M., Hölscher, D., Irawan, B.,, Sundawati. L., Teuscher, M., Kreft, H. (2017). Oil-palm yields in diversified plantations: Initial results from a biodiversity enrichment experiment in Sumatra, Indonesia.Agriculture, Ecosystems and Environment, Page; 240-259.

[10] Brandi, C., Cabani, T., Hosang, C., Schirmbeck, S., Westermann, L., Wiese, H. (2013). Sustainability Certification in the Indonesian Palm Oil Sector: Bene ts and Challenges for Smallholders. Available at: https://www.die-gdi.de/uploads/medi a/Studies_74.pdf.

[11] Brandi, C., Cabani, T., Hosang, C., Schirmbeck, S., Westermann, L., Wiese, H. (2015). Sustainability standards for palm oil. J. Environ. Dev. 24 (3), 292–314.

[12] Rosyani, Rukayah and Heiko Faust. (2018). Strategies to Sustain RSPO Certification a Case Study of Oil Palm Smollholder Group Association Tanjung Sehati Merangin, Jambi Province, Sumatera Indonesia. Proceeding. International Conference on Oil Palm & Environment Embracing sustainable Palm Oil; Solution for Local Production and Global Change" https://www.infosawit.com/news/7971/rspo-berikan-peluang-petani-kelapa-sawitberkembang Bali ICOP 2018.

[13] Plantation office of Jambi Province, (2020), Jambi in Numeral Data 2019. Cooperation Among of Development Planning Agency.

[14] Suharto, R, KarimHusein, Sartono, Desi Kusumadewi, Asril Darussamin, Dhiny Nedyasari, Djaka Riksanto, Hariyadi, Abdul Rahman, Tomoyuki Uno, Piers Gillespie, Chandra Arianto, dan Rauf Prasodjo, 2015. Studi Bersama ISPO dan RSPO: Kerjasama Kementerian Pertanian RI- United Nations Development Programme (UNDP)- Sustainable Palm Oil Initiative (SPOI), (www.rspo.org, www.id.undp.org dan www.inpop.id)

[15] Hardjowigeno. (2003). Soil Classification and Pedogenesis, Jakarta, Akademia, Pressindo

[16] Martens, K., Kunz, Y., Rosyani, I., & Faust, H. (2020). Environmental Governance Meets Reality: A Micro-Scale Perspective on Sustainability Certification Schemes for Oil Palm Smallholders in Jambi, Sumatra. *Society & Natural Resources*, *33*(5), 634–650. https://doi.org/10.1080/08941920.2019.1674436

[17] Napitupulu, Rosyani, Ernawati HD, Rukaiyah Rofiq. (2017). RSPO Certification Impact on Oil Palm Smallholders' Welfare In Jambi Province. Journal: Agricultural Policy Analysis, vol 15 (2).

[18] Gatto, M, Wollini, M, Rosyani Asnawi, and Matin Qaimb. (2017). Oil Palm Boom, Contract Farming, and Rural Economic Development: Village-Level Evidence from Indonesia. World Development Journal. Vol. 95, pp. 127–144. http://www.sciencedirect.com/science/article/pii/S0305750X16301541

[19] Vatn, A., 2015. Environmental governance. Institutions, policies and actions. Cheltenham Glos: Elgar

[20] Rametsteiner, Ewald, and Markku Simula. (2003). Forest certification instrument to promote sustainable forest management? Journal of Environmental Management, 67 (1):87–98.

Self-Reported Pharmaceutical Storage, Use, and Improper Disposal to The Environment Among Urban Parents in Indonesia

Anindrya Nastiti[1], Arlieza Raudhah Riyanto[2], Ade Supriatin[3], Dwina Roosmini[4], Siska WD Kusumah[5], Rumaisha Milhan[2*], Gertjan Geerling[6,7]
[1]Environmental Management Technology Research Group, Faculty of Civil and Environmental Engineering, Institut Teknologi Bandung, Bandung, 40132, Indonesia, 0000-0003-4857-5287
[2]Faculty of Civil and Environmental Engineering, Institut Teknologi Bandung, Bandung, 40132, Indonesia
[3]Industrial Hygiene and Technology Laboratory, Faculty of Civil and Environmental Engineering, Institut Teknologi Bandung, Bandung, 40312, Indonesia, 0000-0002-9211-9644
[4]Center for Environmental Studies, Institut Teknologi Bandung, Bandung, 40312, Indonesia, 0000-0003-3420-4581
[5] Faculty of Civil and Environmental Engineering, Institut Teknologi Bandung, Bandung, 40132, Indonesia, 0000-0002-7588-5216
[6]Radboud University, the Netherlands
[7]DELTARES, the Netherlands
[1]anindrya@itb.ac.id, [2]arliezaraudhah@gmail.com, [3]adeesupriatin2@gmail.com, [4]droosminire@gmail.com, [5]siskarius@gmail.com, [6*]rumaisha09@gmail.com
[7]gertjan.geerling@ru.nl,

Abstract. Pharmaceuticals are increasingly used to treat and prevent disease globally. However, this trend has the potential to increase pharmaceutical discharge into the environment, especially from households. Pharmaceuticals released into the environment can harm human health, e.g., antimicrobial resistance. Thus, it is important to understand the drivers of pharmaceutical use, storage, and disposal among the community. The main objectives of this study are twofold: (1) to identify the commonly stored and used pharmaceuticals based on its classification; and (2) to identify the disposal method of unused pharmaceuticals at home. In October 2018, an exploratory online survey was conducted on 262 urban parents in the Jabodetabek area and Bandung with children ≤5 years old. This online survey revealed that the most stored medicine at home was acetaminophen. Irrational use of medicines was more common in the treatment-seeking behavior of parents rather than children. Doctor's prescription is the most factor affecting buying decisions of pharmaceuticals at home, whether for themselves or their children. In most cases, pharmaceutical waste was treated as domestic waste, which was improper disposal. To reduce improper use and disposal, we suggest a targeted campaign to raise awareness of pharmaceutical waste's unintended health and environmental impacts.

1. Introduction

Pharmaceuticals are increasingly used to treat and prevent disease globally, thus also prompting an increase in unused pharmaceuticals discharged into the environment. Household is one of the sources of pharmaceutical waste; people often dispose of expired, leftover, or unused pharmaceuticals from home

[1]. Pharmaceuticals are often found in wastewater, surface water, groundwater, and even drinking water in low concentrations ranging from micrograms to nanograms per liter [2]. The presence of drugs in natural water systems and the entry of drugs into drinking water facilities may affect water quality and public health [3]. Antibiotic resistance in some bacterial strains develops as a serious consequence of the accumulation of antibiotics in the environment. Antibiotics need to be given special attention among pharmaceuticals since their environmental contamination facilitates the development and spread of antibiotic resistance among bacteria, including pathogens [4]. Pharmaceutical waste in households is commonly disposed of as domestic or household waste mixed with other waste. This study is focused on urban parents since parents are responsible for handling the medicines at home, including their children's medicines. Urban areas (Jakarta, Bogor, Depok, Tangerang, Bekasi (Jabodetabek) area and Bandung) were chosen since they have the densest population in Indonesia. The main objectives of this study are twofold: (i) to identify the most used and stored medicines at home; and (ii) to identify the disposal practice of unused medicine at home.

2. Methods

In October 2018, an exploratory online survey using a snowball sampling approach. This method, snowball sampling using virtual networks, is useful for studying "hard-to-reach" populations. The number of response rates is higher than the traditional snowball technique [5]. Data were collected from parents with young children (≤5 years of age) living in urban Jakarta, Bogor, Depok, Tangerang, Bekasi (collectively known as *Jabodetabek*, the largest megapolitan area in Indonesia), and Bandung, the capital city of West Java. Those areas are the most populated province in Indonesia. Parents with young children were chosen because usually, parents hold the responsibility of handling pharmaceuticals at home, and young children tend to get sick, so they need pharmaceuticals more than adults. To achieve 90% confidence and a margin error of 0.05 with a population of more than 1,000,000 people, our total number of online survey respondents was 262.

Respondents were asked to list all brand names of pharmaceuticals stored in their medicine cabinet, including those they had consumed within the last month. After identifying trade names or brands of pharmaceutical products, they were classified according to *Daftar Obat Esensial Nasional* 2017 (List of National Essential Medicine 2017; DOEN; Indonesian Ministry of Health). Other questions in the online survey included factors affecting the buying decisions of medicine for children and themselves (e.g., based on doctor's prescription, recommendations from relatives, etc.), disposal methods of unused medicine, and participants' exposure to RUM materials. (See Appendix A for the complete list of questions). Then, the top pharmaceuticals used and stored by participants were presented and classified by drug class and active ingredients. This research obtained ethical approval from the Research Ethic Committee of Universitas Padjajaran, Bandung, Indonesia.

3. Results and Discussion

Most respondents were female (91%) and aged between 30 and 34 (47%). As many as 27% of the respondents were from Jakarta, while 40% were from Bogor, Depok, Tangerang, Bekasi, and 32% from Bandung. About half of them (53%) were not exposed to the Rational Use of Medicine (RUM) educational material. On average, each participant stored three to four brands of medicine, up to 24. In total, the medicine the respondents stored at home contained 167 active ingredients belonging to 53 drug classes. **Figure 1** shows that paracetamol (i.e., acetaminophen), antipyretic, and analgesic drug were the most abundant active ingredients stored at home, consistent with the observation that these were the most stored drug classes. Furthermore, a total of 12 drugs classified as antibiotics were stored at home across the entire population surveyed.

Respondents' children consumed medicine containing 76 active ingredients across 29 drug classes in one month (**Figure 2**). The five most common were paracetamol, pseudoephedrine HCl (temporarily relieves bronchial asthma symptoms), glyceryl guaiacolate (an expectorant for cough relief from minor respiratory infections), chlorpheniramine maleate, and cetirizine HCl (both antihistamines). Only three out of the 262 respondents indicated that their children had taken antibiotics within the last month. The most frequently observed drugs taken were analgesics and antipyretics, antihistamines, vitamins and minerals, herbs, decongestants, and expectorants. In comparison, the medicine taken by the adult respondents within the last month contained 103 active ingredients across 38 drug classes (**Figure 2**), and only five had taken antibiotics.

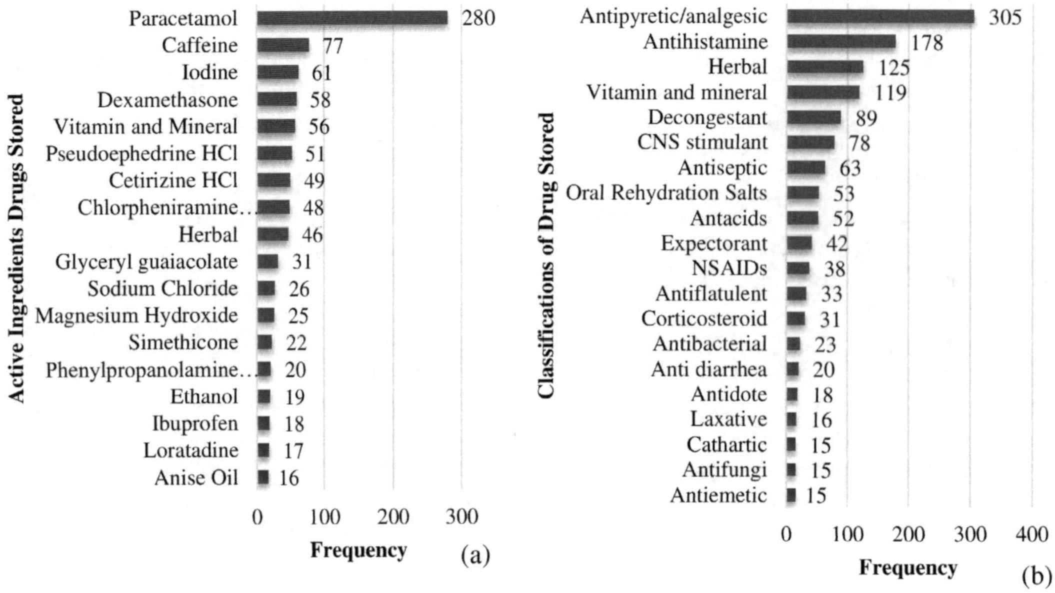

Figure 1. Medicine stored in respondents' house; (a) by active ingredients, and (b) by drugs classification

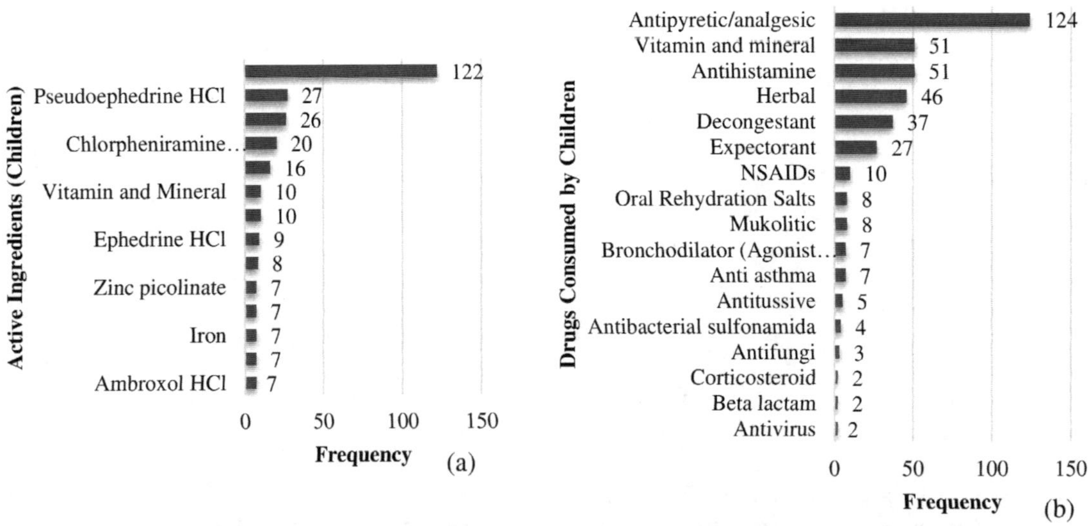

Figure 2. Medicine(s) consumed by respondents' children within the last month; (a) by active ingredient, and (b) by drug classification

Respondents themselves consumed medicine containing 15 active ingredients across 19 drugs classification. Paracetamol (35.49%) was the most active ingredient consumed by parents, followed by pseudoephedrine HCl (17.17%) and dextromethorphan Hbr (15.64%), as shown in **Figure 3**. The first two active ingredients were the same as their children's most consumed medicine. The most consumed drugs classification by respondents was antipyretic and analgesic (39.3%), followed by decongestant

(23.66%). In this study, the average number of types of drugs stored per respondent (approximately three to four brands), and specific types of drugs stored, were not much different from the results of a national-scale study, with an average of approximately four to five drugs, and a maximum of 18 brands [5]. Most ailments start with complaints of pain, a signal that the body gives to one's consciousness that cell damage exists. In this study, the drug class that parents mostly kept at home was OTC analgesics—antipyretics. This implies that most respondents were concerned more with alleviating the symptoms rather than looking for and treating the cause.

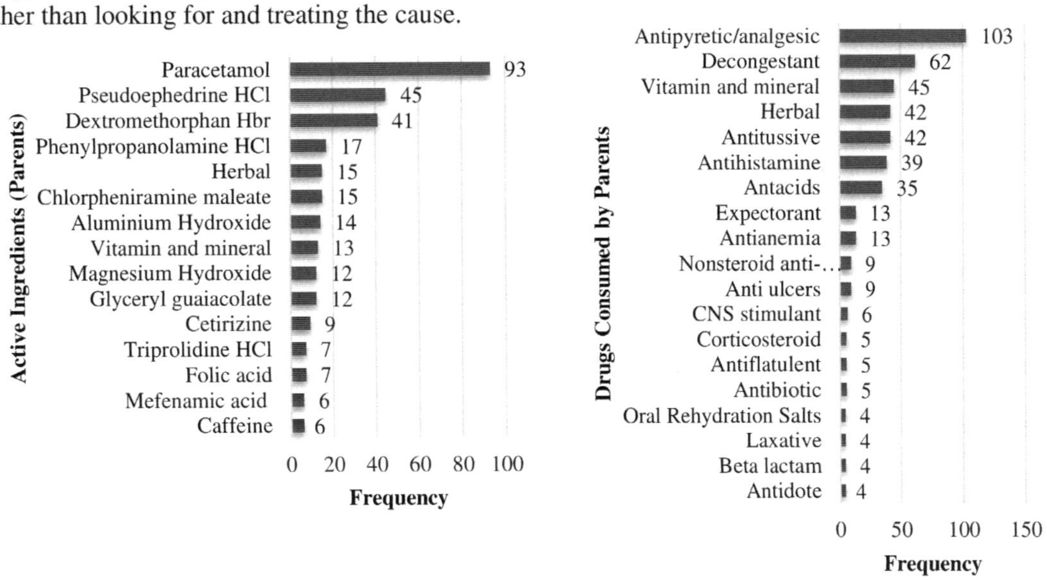

Figure 3. Medicine(s) consumed by respondents' themselves within the last month; (a) by active ingredient, and (b) by drug classification

The availability of analgesic and antipyretic drugs at home has become a modern necessity for first aid kits, even when no one is sick [6]. Additional medicines kept at home are sourced from previous periods of illness, where medication was not taken because the symptoms had subsided, or supplies were guarded for future use if the symptoms reappeared [7]. The numbers were insignificant regarding medicines for non-communicable diseases (NCDs), such as anti-hypertensive, diabetes, and heart medication. Although NCDs in Indonesia are included in the top 10 diseases based on regional health profiles in West Java, in this study, the relative age of parental respondents was still quite young [8]. Thus, the at-risk population was largely unaccounted for.

In developing countries, the prevalence of self-medication can be affected by difficulty in accessing treatment (both geographically and financially), as well as experience with previous treatment [9]. In addition, marketing exposure is also an important factor, such as in a study in Batu City, Indonesia, where the pattern of self-diagnosis, self-medication, and purchases of cough medicine brands was influenced by circulating advertisements [10]. Even though medically, cough subsidence does not necessarily indicate the disease has been cured. Thus, the pattern of self-medication in this study area affected the provision of OTC drugs repeatedly and continuously by the community, even when not needed, increasing the potential for drug waste generation. Such knowledge is essential for formulating a source-oriented pharmaceutical waste management policy.

With respect to drug types, the present study did not report any drugs categorized as very dangerous that are included in the U- or P-List of the Resource Conservation and Recovery Act [11], such as chemotherapy drugs and blood thinners. However, drugs such as non-steroidal anti-inflammatory drugs (NSAIDs), GI tract medications, nerve stimulants, and respiratory medications can have adverse effects on the environment, especially on the morphology and physiology of aquatic animals [12]. Furthermore, the intrusion of antibiotics also raises concerns about the possibility of multi-resistant microbes forming in surface water [13].

We also asked what factors affect the decisions to buy medicine for parents and children. Most respondents indicated that prescriptions from doctors and pediatricians were the primary motivating factors behind buying specific medicine brands for their children (64.57%) and themselves (40.38%). Furthermore, 9.27% and 66.2% of the respondents bought medicine for their children based on the recommendations of relatives and self-treatment based on symptoms, respectively. Doctors' prescriptions also served as the most common driver for the respondents' decision to buy medicine for themselves, followed by the influence of relatives. This implies that when ill, children are more likely to see a doctor than parents, and self-treatment based on symptoms, internet searches, previous experience, and promotional media also contributed to the rationale for buying certain brands of medicine in both children and parents.

Most respondents (78.6%) treated unused or expired medicine as domestic waste. This practical method requires the least time and effort. The remaining respondents disposed of medicine in sinks or toilets buried it, gave it to a third party, or followed other disposal mechanisms. As discussed earlier, participants are generally more aware of the direct health risks rather than the environmental ones associated with the misuse of pharmaceuticals; thus, people may feel that disposing of pharmaceutical waste through the sewer systems, or household garbage is not of concern, particularly when these pharmaceuticals are perceived as non-hazardous. Another possible factor driving the improper disposal of pharmaceutical waste is that participants may perceive household disposal as a less significant source of environmental contamination than wastes and effluents generated by hospitals and industries. Unsurprisingly, garbage and sewer disposal are quite common in other countries as well, including developing countries whose practices are influenced by the absence of clear regulations, inadequate facilities, and a lack of supervision or education [14, 15, 16]. According to Government Regulation No. 101/2014, pharmaceutical waste in Indonesia is categorized as Category 2 hazardous waste, which is considered to have a delayed effect on human health and the environment, with sub-chronic or chronic toxicity. Although this regulation has been further expanded into the Ministry of Environment and Forestry Regulation No. 56/2015, it pertains only to the pharmaceutical waste generated by formal healthcare facilities. Pharmaceutical waste produced by household activities is yet undefined by any national regulations and/or guidelines [17]. Efforts to take back unused medicine have been implemented in New Zealand, Australia, the US, Ireland, and Sweden. However, this service is scarce in Indonesia.

Studies have shown that raising public awareness is vital when it comes to the proper disposal of unused household pharmaceuticals [1, 16, 18, 19]. Finally, as the availability of safe disposal options is a key predictor of safe disposal behaviors [20], the government is responsible for establishing the legal framework, providing guidelines, and managing safe disposals of household pharmaceutical waste. The primary limitation of this study was the use of snowball sampling for online questionnaires; hence, there was a tendency to recruit respondents within the social circle of the researcher.

4. Conclusion

The study aimed to identify the most stored and used pharmaceutical products as well as the disposal method of pharmaceutical waste by urban parents in Indonesia via online surveys. Acetaminophen was cited as the most stored medicine at home and a small percentage of respondents stored antibiotics at home. Most of the respondents do improper pharmaceutical waste disposal by treating it as domestic solid waste. In the future, the government should establish a comprehensive law on safe handling of household pharmaceutical waste and raise public awareness.

Acknowledgments

We thank Sabrina Farah Salsabilla for her assistance in data pre-processing. This research was funded by the International Research Program of LPPM ITB and part of the *Thriving Waters, Healthy Society* research activity of the Alliance for Water, Health, and Development—a consortium between Indonesia and the Netherlands.

References

[1] Shaaban H, Alghamdi H, Alhamed N, Alziadi A, Mostafa A. *Environmental Contamination by Pharmaceutical Waste: Assessing Patterns of Disposing Unwanted Medications and Investigating the Factors Influencing Personal Disposal Choices*. Journal of Pharmacology and Pharmaceutical Research. 2018; p. 1.

[2] Shaaban H, Gorecki T. *Current trends in green liquid chromatography for the analysis of pharmaceutically active compounds in the environmental water compartments*. Talanta. 2015; 132: p. 739-752.

[3] Sharma K, Kaushik G. *NSAIDS in the Environment: From Emerging Problem to Green Solution*. Annals of Pharmacology and Pharmaceutics. 2017; **2**(14): p. 1077.

[4] Andersson DI, Hughes D. *Microbiological effects of sublethal levels of antibiotics*. Nature Reviews Microbiology volume. 2014 May; **12**(7).

[5] Baltar F, Brunet I. *Social research 2.0: Virtual snowball sampling method using Facebook*. Internet Research. 2012; **22**(1).

[6] Gitawati R. P*attern of Household Drug Storage*. National Public Health Journal. 2014 August; **9**(1).

[7] Kulkarni P, Garg A, Ajmera A, Mahajan A, Gadekar S, Kulkarni P. *Global Scenario of Self Medication: A Review of Literature*. Indian Journal of Commerce and Management Studies. 2018 January; **9**(1).

[8] Kristina SA. *A Survey on Medicine Disposal Practice among Households in Yogyakarta*. Asian Journal of Pharmaceutics. 2018 November; **12**(3).

[9] Indonesian Health Ministry. *Laporan Nasional Riskesdas 2018*. Jakarta: Indonesian Health Ministry, Health Research and Development Agency; 2018.

[10] Al-Worafi Y. *Drug Safety in Developing Countries: Achievements and Challenges.* Academic Press of Elsevier; 2020.

[11] Titani M, Hidayat R, Hidayati IR. *The Correlation of Cough Medicine Advertisement In Television On Community Self-Medication In Krajan Sae Beji Village Junjero Batu City (Based on BPOM Head Regulation Number 8 in 2017)*. Farmasains: Jurnal Ilmu Farmasi dan Kesehatan. 2019 Januari; 4(2).

[12] US EPA. GovInfo. [Online]; 2020. Available from: https://www.govinfo.gov/app/collection/uscode/2020/.

[13] Akash MSH, Rehman K, Sabir S, Gul J, Hussain I. *Potential Risk Assessment of Pharmaceutical Waste: Critical Review and Analysis*. Pak J Sci Ind Res Ser A. 2020: p. 209-219.

[14] Richa B, Kishore SR. *The antibiotics, waste and treatment plants: presenting a possibility of antibiotic resistance in bacteria of surface waters - A Review*. Research Journal of Chemistry and Environment. 2020 August; **24**(8).

[15] Kotchen M, Kallaos J, Wheeler K, Wong C, Zahller M. *Pharmaceuticals in wastewater: Behavior, preferences, and willingness to pay for a disposal program*. Journal of Environmental Management. 2009 March; **90**(3).

[16] Constantino VM, Fregonesi BM, Tonani KAdA, Zagui GS, Toninato APC, Nonose ERdS. *Storage and disposal of pharmaceuticals at home: a systematic review*. Ciencia e Saude Coletiva. 2020; **25**(2).

[17] Bashaar M, Thawani V, Hassali MA, Saleem F. *Disposal practices of unused and expired pharmaceuticals among general public in Kabul*. BMC Public Health. 2017; **17**(45).

[18] Kusumah. Personal Communication. 2021..

[19] Rogowska J, Zimmermann A, Muszynska A, Ratajczyk W, Wolska L. *Pharmaceutical Household Waste Practices: Preliminary Findings from a Case Study in Poland*. Environmental Management. 2019; 64: p. 97-106.

[20] Bungau S, Tit DM, Fodor K, Cioca G, Agop M, Iovan C, et al. *Aspects Regarding the Pharmaceutical Waste Management in Romania. Sustainability*. 2018 August 7; **10**(2788).

Land Conservation and The Potential Goal for Food Security in Urban City

Ketut Sumantra[1]* and Ni Gst.Ag. Gde Eka Martiningsih[2]
[1] Department of Agrotechnology, Faculty of Agriculture and Business, University of Mahasaraswati Denpasar, 80231 Indonesia, ORCID ID 0000-0003-0669-7745
[2] Department of Agribusiness, Faculty of Agriculture and Business, University of Mahasaraswati Denpasar, 80231 Indonesia, ORCID ID 000-0001-5534-4968
[1]*ketut.sumantra@unmas.ac.id, [2]ekamartini@gmail.com

Abstract. Population growth and increasing demand for land for urban development have led to the conversion of agricultural land. Land conversion harms the reduction of agricultural areas and indirectly affects food availability, so efforts are needed to preserve it. The study aims to obtain the availability of paddy fields and the conservation strategies. Evaluation of land use change and availability was analyzed by high-resolution SPOT 6/7 satellite imagery data, slope maps, soil type maps, sub-district administrative boundaries maps, raw rice field maps, spatial pattern planning maps, climate data, and criteria for land suitability for food crops. The strategy of conservation of paddy fields based on the wisdom of *Subak* (Balinese cultural base irrigation system for paddy fields) was analyzed by SWOT. The analysis showed that the converted land areas were 78.5 Ha. The available paddy fields that needed to be conserved were 895.02 Ha, consisting of a moderate suitability class of 556.74 Ha and a very suitable suitability level of 338.28 Ha. The strategy for conservation of paddy fields based on *Subak* wisdom must be implemented through 1) Developing tourist attractions based on *Subak* agricultural activities and rituals, 2) Developing a network of cooperation with the government and business people, 3) Implementing strict *Subak* rules *(awig-awig)*, 4) Encouraging *Subak* to be more business-oriented through plant diversification that has market prospects.

1. Introduction

In recent decades, many urban areas have experienced dramatic growth. The high rate of urban population growth has become a global issue that requires comprehensive attention and management because it impacts various aspects of development [1].

Urban agriculture is defined as agricultural activities carried out within the city (intra-urban) and the suburbs (peri-urban) to produce, process, and distribute various food and non-food products by utilizing humans, materials, products, and services in urban areas [2]. Aside from being a food provider [2–4], other benefits of urban agriculture are a green corridor, playing a role in environmental, economic, social, health, and education aspects, and tourists [5,6]. Urban agricultural land use often competes with other sectors such as industry, settlement, trade, and tourism [7]. This competition impacts the emergence of land-use change activities [8]. The land-use change can have a negative impact if not adequately controlled.

Decreasing agricultural land area indirectly affects food availability [3], increasing dependence on food imports, decreasing biodiversity, reducing green open space, and damaging the environment and local culture [9]. To provide such food, it is necessary to preserve sustainable food cropland in urban areas through approaches that can synergize social, cultural, environmental, and economic aspects [10], revise the Urban Spatial Planning to protect the rice fields [11], and increase the contribution of research on land use by various divisions adopted by local wisdom [8]. Population growth and increasing land requirements for development activities have influenced sustainable land use. In Indonesia, the area of paddy fields from year to year continues to decline. In 2018, the land area remained at 7.1 million hectares, but in 2017 it decreased to 7.75 million hectares [12]. Similarly, in Bali Province, the area of rice fields from 2015 to 2017 has decreased by around 2,077 hectares [13].

Denpasar City has an area of 12,778 Ha (2.27% of the total area of Bali Province) consisting of 2,509 Ha of paddy field, 505 Ha of non-paddy agriculture land, and the remaining 9,764 Ha is non-agricultural land, such as roads, settlements, offices, and river [13]. On the other hand, the area of paddy fields in Denpasar-Bali has decreased by an average of about 1.28% per year [13].

Land conversion occurs due to foreign investment entry, labor availability, and population [14]. The impact of the development of shopping centers, settlements, business places such as restaurants, and other infrastructure also triggers land use change [15] in addition to the social and cultural conditions of the local community in land use [16].

The law on the protection of Sustainable Food Agricultural Land and the regulation of the Bali Provincial Spatial Plan has mandated the protection of agricultural land. However, the implementation has not shown encouraging results. Instead, the rate of function of paddy's field was 800 Ha/year during 1998-2018, which exceeded the tolerance limit of 540 Ha/year [11], suggesting that the revision of regional regulations has urgency on protecting agricultural land and Subak as a potential cultural heritage.

Sustainable agriculture aims to balance economic, environmental, and social aspects and create a resilient agricultural system in the long run [17]. This concept has not given equal attention to cultural and spiritual services, let alone integrated into environmental management. Integrating socio-economic, environmental, cultural, and spiritual values is Balinese local wisdom used as a Subak philosophy called Tri Hita Karana [18]. Tri Hita Karana is the concept of a balanced and harmonious relationship of three elements, namely the relationship between humans and God (*parahyangan*), the relationship between humans and humans (*pawongan*), and the relationship between humans and the environment *(palemahan)* [18]. Subak is not just water management but a religious and cultural fortress against globalization and other threats to the local wisdom, including land-use change [19]. Therefore, studying spirituality as the core of the dynamic balance of ecological, economic, and social pillars can significantly contribute to sustainable agricultural development by national cultural identity.

Many studies on the evaluation and determination of sustainable agricultural land have been carried out, as reported by [7,20,21], but the research has not linked the role of society in it. This research, in addition to determining the availability of land, also determines the strategies that must be carried out to preserve rice fields in urban areas with a sustainable *subak* system approach.

Subak plays a vital role in the agricultural system in Bali; for that reason, the Denpasar City Government set a pilot by applying local wisdom through the *Subak Lestari* Program, located in North Denpasar and East Denpasar Districts. The priorities of the Subak Lestari program are protecting and improving farmers' welfare, conservation and promotion of ecosystem services, cultural preservation, and tourism development [22]. The problem is whether, in the era of 4.0, Subak was able to be the guarantor of land sustainability and physical security. This study aims to obtain the existing area of rice fields and evaluate the suitability of land for food crops, as well as its conservation strategy based on local wisdom on Subak sustainability in Denpasar-Bali.

2. Methods

This study is located in the Subak Lestari area, in North Denpasar and East Denpasar Districts, which extends between 1090 47"- 1100 8' 20" E longitude and 70 32'- 70 28' 54 "S latitude with a total area of about 1,034.82 km2. The research materials and tools include (1) SPOT 6/7 high-resolution satellite imagery with a scale of 1: 25,000, (2) slope maps, (3) soil types maps, (4) sub-district boundaries maps, (5) zoning plan maps, (7) temperature data from 2011-2018, (8) annual rainfall data, and (9) drainage data. The tools in the study include the Global Positioning System (GPS), soil sampling equipment, table of quality/characteristics of irrigated fields, interview guides, voice recorders, and ArcGIS 10.3 software.

To generate the agricultural suitability map, first, the high-resolution satellite data (SPOT 6/7) for the year 2018 was processed using ArcMap for mapping the existing urban agricultural land by using an on-screen digitizing procedure and overlaid by the boundary of agricultural land of the zoning plan map as a shape file. Second, the various potential land characteristic map essential for the growth of various agricultural plants (slope, temperature, soil type, rainfall, and drainage maps) were prepared. The score and weight for each thematic map were calculated based on the provision of FAO [23]. After that, the geographical information system modeling was performed by integrating all thematic maps with their calculated weightage using the weighted overlay analysis method. The agricultural suitability map was finally obtained based on their surface values, and they are further classified into four agricultural suitability levels. Land suitability levels refer to R&D Ministry of Agriculture, 2016 consist of (S1-highly suitable; S2- moderately suitable; S3-marginally suitable; and NS-not suitable). Suitability of land agroclimatic on commodity paddy and corn as in Table 1 and Table 2.

Table 1. Rice plant agro-climatic requirements

No	Requirements Usage / Characteristics Land	Class Suitability Land				Weight
		S1	S2	S3	N	
1	Temperature mean (°C)	24-29	22-24	18-22	<18	5
			29-32	32-35	>35	
2	Rooting media (RC)					
	Drainage	Rather obstructed, moderate	Delayed, fine	Very hampered, a bit fast	Fast	5
	Soil texture	Smooth, a bit fine	Currently	Rather Rough	Rough	5
3	Slope (%)	< 3	3-5	5-8	> 8	5
	Danger erosion	Very low	Low	Currently	Heavy	5

Table 2. Corn crop agro-climatic requirements

No	Requirements Usage / Characteristics Land	Class Suitability Land				Weight
		S1	S2	S3	N	
1	Temperature mean (°C)	20-26	-	16-20	<16	5
			26-30	30-32	>32	
2	Rainfall (mm)	500-1,200	1,200-1,600	>1,600		3
			400-500	300-400	<300	
3	Drainage	Good, a bit hampered	Rather fast, medium	Hampered	Very hampered, hurry	4

No	Requirements Usage / Characteristics Land	Class Suitability Land				Weight
		S1	S2	S3	N	
4	Rooting Media Texture	Smooth, a bit smooth, medium	-	Rather Rough	Rough	4
5	Slope (%)	< 8	8-16	16-30	> 30	4

The SWOT analysis tool was used as a strategic planning approach to formulating strategies for preserving urban agricultural land in Denpasar City, Indonesia. SWOT analysis is a continuation of the situation analysis of internal-external factors, where 'strengths' apply to current forces associated with preserving urban agricultural land, whereas 'opportunities' refers to what actions could be taken to enhance this preservation, and 'weakness' refers to current problems whereas 'threats' are problems waiting to happen. First, face-to-face structured interviews with stakeholders were conducted to conduct the SWOT analysis. A group workshop and personal interviews were held among 42 member participants of pakaseh (Subak leader) as the person who knows the characteristic of the Subak. Second, the open interviews and focus group discussion results were used to find the strengths, weaknesses, opportunities, and threats. Finally, formulate a strategic plan for preserving agricultural land in an urban area based on local wisdom. Figure 1 shows the flowchart of this study.

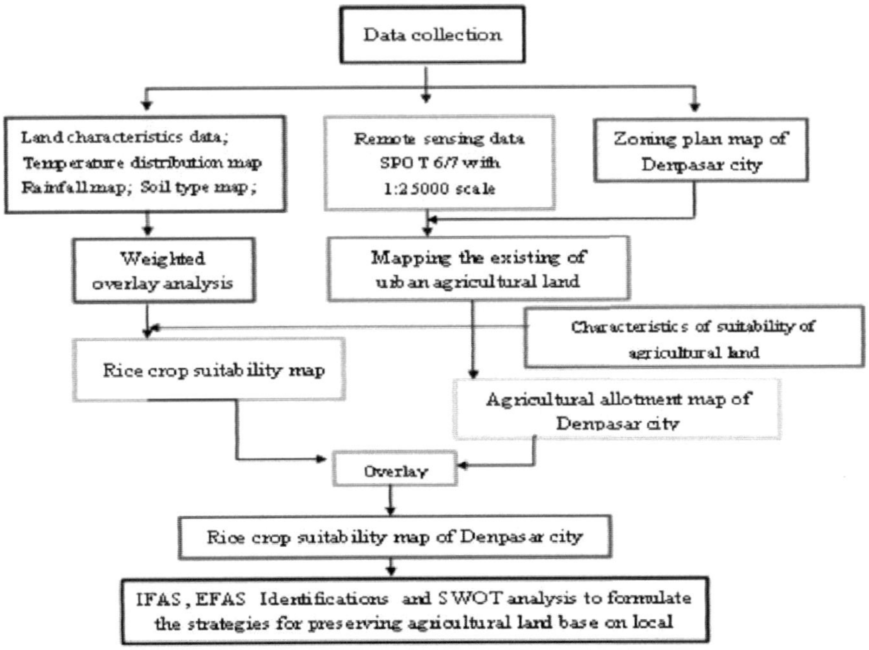

Figure 1. Evaluation flowchart of the suitability of agricultural land

3. Results and Discussion

3.1 Land Suitability for Crops

Based on the zoning plan map of Denpasar City (Figure 2), the North Denpasar and East Denpasar Districts are dominated by residential areas, and some are used as agricultural land. Figure 3 shows the agricultural allotment map in East Denpasar and North Denpasar Districts. Based on the figure, the total area of agricultural land in East Denpasar is 461.51 Ha, and in North, Denpasar is 433.49 Ha.

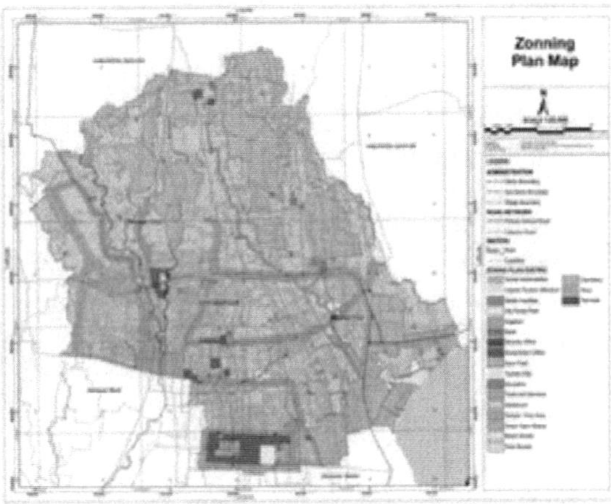

Figure 2. Zoning plan map of North Denpasar and East Denpasar Districts in 2011-2031

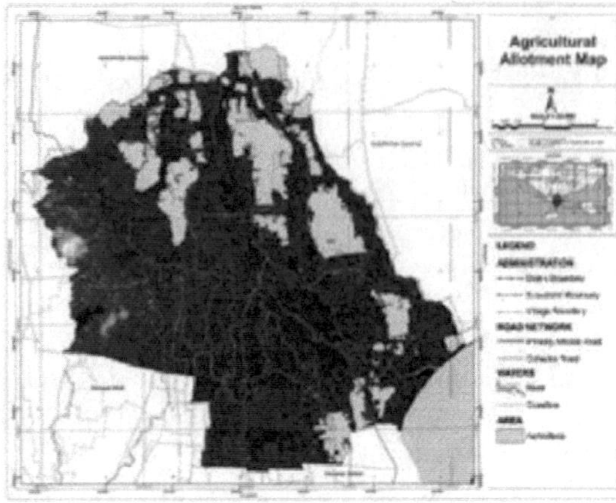

Figure 3. Agricultural allotment map of North Denpasar and East Denpasar Districts in 2019

Figure 2 and Figure 3 show the results of the analysis of the suitability of agricultural land, with the zoning plan map showing a decrease in land area. The area of paddy fields in 2019 was reduced by 78.5 Ha. The most significant land conversion occurred in East Denpasar District, reaching 59.72 Ha, while the rice field conversion in the North Denpasar sub-district reached 18.78 Ha. Agricultural land conversion in the two sub-districts was used for settlements, trade, and services. The remaining crop fields supporting food availability in the two sub-districts are 895.03 Ha (Figure 3). The land available for suitability for rice and corn plants is in categories S1 and S2, as shown in Figure 4 and Figure 5.

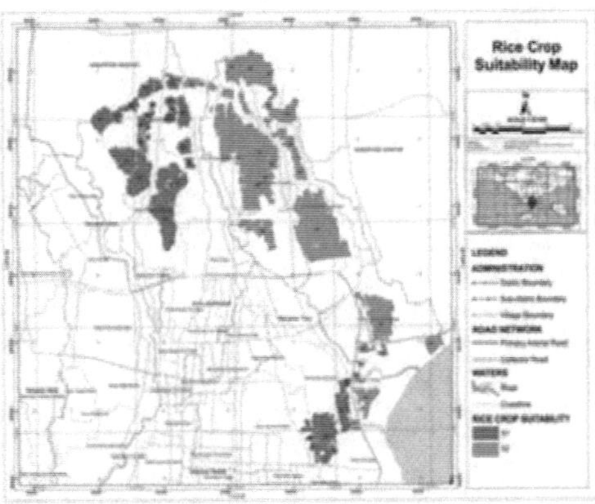

Figure 4. Rice crop suitability map of North Denpasar and East Denpasar Districts

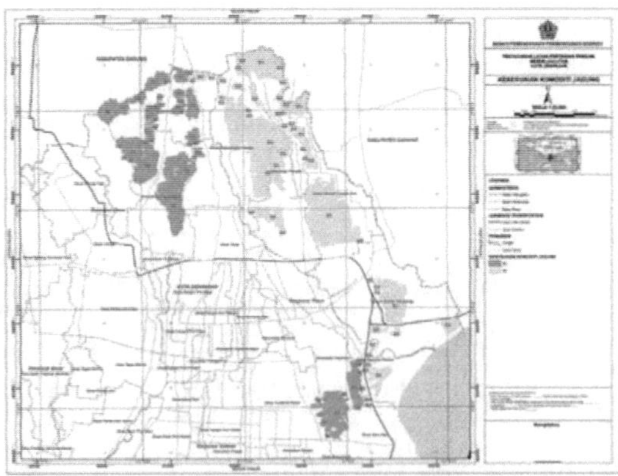

Figure 5. Corn crop suitability map of North Denpasar and East Denpasar Districts

Based on Figure 4 and Figure 5, the area of rice and corn crops with the category of moderately suitable (S2 = 556.74 Ha) is larger than the area of rice crops with the highly suitable (S1=338.28 Ha). The results of agricultural land suitability for each District are shown in Table 3.

Table 3. Rice and Corn crop suitability classes in each village in North Denpasar and East Denpasar

No	Sub-District	Village	Area (Ha) S1	Area (Ha) S2	Total (Ha)
1	North Denpasar	Peguyangan Kaja	112.52	7,43	119.94
2	North Denpasar	Peguyangan Kangin	4.11	169.33	173.44
3	North Denpasar	Tonja	-	0.01	0.01
4	North Denpasar	Ubung Kaja	18.45	-	18.45
5	North Denpasar	Peguyangan	121.65	-	121.65
Total Sub-District North Denpasar			256.73	169.34	433.5
6	East Denpasar	Kesiman Kertalangu	-	71.61	71.61
7	East Denpasar	Kesiman Petilan	2.11	21.81	23.92
8	East Denpasar	Penatih Dangin Puri	-	168.38	168.38
9	East Denpasar	Sumerta Kelod	8.76	-	8.76
10	East Denpasar	Kesiman	70.68	0.38	71.06
11	East Denpasar	Penatih	-	117.79	117.79
Total Sub-District East Denpasar			81.55	379.97	461.52
Total			338,28	549,31	895,02

Based on Table 3. in North Denpasar District, the potential for food crops in the very suitable category (S1) is broader than the land in the appropriate category (S2), respectively 256.73 Ha and 176.77 Ha. The most dominant limiting factor in East Denpasar District is the soil permeability because the Regosol soil type dominates the soil.

3.2. Strategy for Sustainable Rice Field Conservation in Urban Areas to Support Food Security

UNESCO has recognized Subak in Bali as a World Cultural Heritage [24]. This recognition reflects several things, including the recognition of a) the existence of Subak institutions, b) the Subak system that applies the Tri Hita Karana (THK) concept, and c) the Subak paddy landscape that contains cultural activities. The importance of the role of Subak in the agricultural system in Bali, the Denpasar City Government, set a pilot by utilizing local wisdom through the *Subak Lestari* Program, with priority programs being the protection and improvement of farmers' welfare, preservation and promotion of ecosystem services, cultural preservation, and tourism development [22]. From 895.02 hectares of rice fields in North Denpasar and East Denpasar, conservation needs to be done with Balinese local wisdom.

A SWOT analysis procedure is needed to formulate a strategy for conserving Subak based on local wisdom. The analysis was performed to maximize the strengths and opportunities while minimizing weaknesses and threats simultaneously. SWOT matrices and proposed strategies for Subak land conservation are presented in Tables 4 and 5, respectively.

Table 4. IFAS Identification of Subak local wisdom in conserving rice fields in Denpasar City

No	Strength	Weight	Rating	Score
1	Subak has a solid philosophical foundation and clear objectives.	0.09	4	0.36
2	The average production that farmers can achieve is between 5-7 tons per hectare of the dry grain harvest	0.06	3	0.18
3	Subak is the only organization farmers, so if empowered will have a robust set (bargaining position) outside the ordinary.	0.06	4	0.24

4	Organizations run by democratic have regulations (awig-awig)	0.07	4	0.28
5	Government support in the form of laws on Subak, policy, and funding	0.07	4	0.28
6	Subak landscape and panorama attractive	0.06	3	0.18
7	Strategic location and close to the provincial capital	0.08	4	0.32
8	Member subak have mutual feelings of respect and mutual belief	0.06	3	0.18
9	All land subak already uses irrigation technical.	0.03	3	0.09
No	**Weaknesses**	**Weight**	**Rating**	**Score**
1	Subak has a dominant social culture compared to business culture,	0.06	3	0.18
2	Moderate rainfall and relatively limited water sources	0.05	3	0.15
3	Production facilities still feel difficult and expensive	0.05	2	0.10
4	Transforming labor from the agricultural sector to non-agriculture	0.05	3	0.15
5	Lack of ability Subak in provide facility	0.05	2	0.10
6	Conversion of paddy fields	0.06	3	0.18
7	Lack of Accessed Street in the Subak area	0.05	2	0.10
8	Lack of administration system and finance,	0.05	2	0.10
	Total internal factors	1.00		3.17

Based on IFAS analysis (Table 4), the total score strength is more significant than the weakness, with an IFAS value of 3.17. Four dominant parameters namely: 1) Has strong philosophy and concrete goals (score 0.36), 2), Strategic location and close to the provincial capital (score 0.32); 3) The organization is run democratically because it has rules (awig-awig), and 4) Government support in the form of laws on Subak, policy, and funding (score 0.28). Whereas dominant factor weakness in preservation paddy field based subak: 1) Subak has a dominant social culture compared to business culture (score 0.18), 2) Transforming labor from the agricultural sector to non-agriculture (score 0.15), 3) Moderate rainfall and relatively limited water sources (score 0.15), and 4)) Conversion of paddy's field (score 0.18).

Table 5. EFAS Identification of Subak local wisdom in conserving rice fields in Denpasar City

No	Opportunities	Weight	Rating	Score
1	Market opportunities for agriculture product	0.08	3	0.24
2	Diversification opportunities for plants that have high selling points	0.08	4	0.32
3	Availability of business loans from the government	0.06	3	0.18
4	Supporting from the government because in line with the vision development tourist culture	0.08	4	0.32
5	There is funding from the government for Subak conservation.	0.10	3	0.30
6	Strategic location and close to the provincial capital	0.08	3	0.24
7	Subak opportunity as a tourist attraction	0.09	4	0.36
8	Synergy Among agriculture, animal husbandry, and fisheries must be encouraged to realize CLS (Crop, Life Stock System).	0.07	3	0.21
No	**Threats**	**Weight**	**Rating**	**Score**
1	Shifting the conditions of socio-agrarian society towards industrial society caused a decrease in interest in work in agriculture,	0.06	3	0.18

2	Taxes for agricultural land are still high, so they need to be subsidized	0.05	2	0.10
3	Population growth and conversion of paddy fields.	0.06	3	0.18
4	The growth of private entrepreneurs engaged in agribusiness and managed professionally can be a competitor of the Subak community	0.07	3	0.21
5	More many worker effort farmers were taken by power work outside subak	0.06	2	0.12
6	Constraint pests and diseases	0.05	1	0.05
	Total external factors	1.00		3.01

Analysis EFAS (Table 5) shows that the opportunity factor value is more significant than the threat factor value, with a total EFAS score of 3.01. Four factors dominant as opportunity subak local wisdom in conserve paddy fields, namely 1) Subak opportunity as a tourist attraction (score 0.36), 2) Diversification opportunities for plants that have high selling points (Score 0.32), 3) There is funding from the government for Subak conservation (score 0.30), 3) Strategic location and market opportunities score 0.24). Whereas the most dominant threat faced are 1) The growth of private entrepreneurs engaged in agribusiness and managed professionally can be a competitor of the Subak community (score 0.21), 2). Shifting the conditions of socio-agrarian society towards industrial society, causing a decrease in interest in work in agriculture, 3) Population growth and conversion of paddy fields, and 4) More farm workers are taken from outsiders.

The SWOT analysis results indicate that eight strategies can be carried out in preserving paddy's field based on the local wisdom of *Subak* (Table 6). The eight strategies are 1). Increase the value of ritual and spirituality. 2). Collaborating with the government in funding, fostering, and developing Subak as a tourist attraction. 3). Encourage Subak to be more business-oriented through diversification of plants that are adaptive to the environment and have market prospects. 4). Create business units, and 5). Strengthening and implementing Subak rules both for supporting businesses and protecting paddy fields. 6). Empower the Subak members in plant and unique business management. 7. Establishing the Subak economic business institutions. 8) Selective and rigorous in developing facilities to reduce land-use change.

Table 6. SWOT matrix and alternative strategies in preservation paddy field

Internal Factor / **External Factor**		**Strength (S)**		**Weakness (W)**	
		Factor	Score	Factor	Score
		Subak has a solid philosophical foundation and clear objectives	0.36	Subak has a dominant social culture compared to the business culture	0.18
		The organization is run democratically because it has rules (*awig-awig*)	0.28	Transforming labor from the agricultural sector to non-agriculture	0.15
		Government support in the form of laws on Subak, policy, and funding	0.28	Moderate rainfall and relatively limited water sources	0.15
		Strategic location and close to the provincial capital	0.32	Conversion of paddy fields	0.18
Opportunity (O)		**SO Strategy**		**WO Strategy**	
Factor	**Score**				
Diversification opportunities for plants that have high selling points	0.32	1. Increase the value of ritual and spirituality by the Subak's foundations and goals and apply the *Subak awig-awig* rules both		3. Encourage Subak to be more business-oriented through diversification of plants that are adaptive to the environment and	
Support from the government for developing cultural tourism	0.32				

Factor	Score	SO strategy (cont.)	WO strategy (cont.)
There is funding from the government for Subak conservation	0.30	for soil protection and business development diversification of potential market-integrated plants with the tourism sector (S1, S2, S3, O1, O2, and O4) 2. Collaborating with the government in funding, fostering, and developing Subak as a tourist attraction. (S1, S2, S4, O2, O3, O4)	have market prospects and create tourist attractions so that agriculture can prevent labor transformation and suppress the conversion of agricultural land (W1, W2, W3, O1, O2, O4) 4. Subak community, together with the government, creates business units, reinforces the rules for both protecting land and business units owned, and opens new business fields (W3, W4, O2, O4)
Subak opportunity as a tourist attraction	0.36		
Threat (T)		**ST strategy**	**WT Strategy**
Factor	**Score**		
Shifting the conditions of socio-agrarian society towards industrial society caused a decrease in interest in work in agriculture.	0.18	5. Strengthening and implementing Subak rules both for supporting businesses and protecting paddy fields (S1, S3, T1, T2) 6. Empowering the Subak members both in the field of plant management and unique business management based on the potential of Subak local wisdom to minimize business competition and labor competition from outside (S2, S4, T3, T4)	7. Collaborate with the government in strengthening the Subak institutions by establishing the Subak economic business institutions (W1, W2, T1, T3). 8. Selective and rigorous in developing facilities to reduce land-use change (W2, W4, T2, T3)
Population growth and conversion of paddy fields	0.15		
The growth of private entrepreneurs engaged in agribusiness and managed professionally can be a competitor of the Subak community	0.21		
More farmworkers are taken from outsiders	0.12		

Increasing the values of ritual and spirituality according to the *Tri Hita Karana* philosophy and applying the *Subak* rules (*awig-awig*) are necessary for protecting the paddy field [21] and developing business units of *Suba*k based on plant diversification. Preservation of paddy fields is the essence of the relationship between ritual, spirituality, and sustainable development. Sustainable development with three main pillars (ecology, economy, and social) recognizes the existence of ecosystem services in environmental management [25]. Still, cultural and spiritual services do not seem to have received equal attention. Subak is not only water management but is a stronghold of religion and culture against globalization and various other threats to local wisdom [19]. Therefore, the study of aspects of spirituality as the core of the dynamic balance of the ecological, *Subak* activities, *Subak* philosophy, and *Subak* rituals can be used as tourist attractions [11]. In addition, *Subak*, which has beautiful natural scenery and fresh air conditions, can be developed as an agro-ecotourism area [25,26]. To preserve and enhance the bargaining position and protection of agricultural land, *Subak* must assume a more strategic role and function. The technology that needs to be offered to the local farmers is appropriate technology that is relevant to meet the reality of needs and becomes a solution to the community's real problems. Agricultural intensification is essential to ensure food security and slow the expansion of agricultural and cultural land. If the reality of the problem is the low productivity of crops, then the appropriate technology to increase production needs to be developed. If the problem is limited agricultural land, farmers can focus on producing high-value food products through crop diversification and health products through the application of organic agriculture [27,28]. If the real problem for farmers is finding a market for the raw crops, the technology that needs to be developed is processing and packaging technology and designing applications to help local farmers find markets for raw crops. Not all of the various processing, distribution, and marketing technologies can be done by farmers themselves. Still, it would be perfect if done together in the form of agricultural economic business institutions formed by Subak.

4. Conclusion

Availability of raw rice fields in North Denpasar and East Denpasar is 895.02 Ha, with a moderately suitable level (S2) broader than land with a very suitable level (S1) with an area of 556.74 Ha (S2) and 338.28 Ha (S1), respectively. The concept of preservation must be implemented because of several opportunities and strengths that can be developed as a land conservation strategy: 1) The development of tourist attractions is based on agricultural activities and rituals. 2) Developing a network of cooperation with the government and entrepreneurs. 3) Implementation of customary rules that are strict for controlling land-use change. 4) Encouraging the traditional conservation to be more business-oriented through diversification of plants that are adaptive to the environment and have market prospects. More personal views might reflect the investigated phenomena' factual descriptions to avoid subjective findings. Further study is recommended in a broader area of Bali to generate generalizability of findings.

Acknowledgments
The author thanks the Rector, the Chairman of the Institute for Research and the Community Services of Universitas Mahasaraswati Denpasar, and the Denpasar City Planning Agency for the approval and guidance in this study.

References

[1] Liu Y 2018 Introduction to Land Use and Rural Sustainability in China *Land use policy* **74** 1–4
[2] Fresco L, Huizing H, Van Keulen H, Luning H and Schipper R 1992 *Land Evaluation and Farming Systems Analysis for Land Use Planning FAO/ITC*
[3] Sukiman 2015 Dinamika Sosial Ekonomi Petani Jagung Kuning di Desa Tonrorita Kecamatan Biringbulu Kabupaten Gowa 1985-2005 *J. Rihlah* **II** 81–100
[4] Martin M and Molin E 2019 Environmental Assessment of an Urban Vertical Hydroponic Farming System in Sweden *Sustain.* **21** 1–14
[5] Santoso E B, Aulia B and Widya R R 2018 Measuring Performance of Urban Farming for Sustainable Urban Development in the City of Surabaya, Indonesia EasyChair Preprint Measuring Performance of Urban Farming for Sustainable Urban Development in the City of Surabaya, Indonesia Eko Budi Santoso, *13th Int. Congr. Asian Plan. Sch. Assoc.* 12–4
[6] Bisaga I, Parikh P and Loggia C 2019 Challenges and Opportunities for Sustainable Urban Farming in South African Low-Income Settlements: A Case Study in Durban *Sustain.* **11**
[7] Lanya I, Subadiyasa N and Adi G P . 2017 Impact of Spatial Plan on the Conversion of Subak Rice Fields and Food Security in Bandung and Gianyar Regencies Bali Province *IOP Conference Series: Earth and Environmental Science* (IOP Publishing) pp 1–9
[8] Liu Y, Li J and Yang Y 2018 Strategic Adjustment of Land Use Policy Under the Economic Transformation *Land use policy* **74** 5–14
[9] Primdahl J, Kristensen L S and Busck A G 2013 The Farmer and Landscape Management: Different Roles, Different Policy Approaches *Geogr. Compass* **7** 300–14
[10] Battisti L, Gullino P and Larcher F 2018 Using The Ecosystem Services' Approach for Addressing Peri-Urban Farming in Turin Metropolitan Area *Acta Hortic.* **1215** 427–32
[11] Norken I N, Suputra I K and Arsana I G N K 2015 Water Resources Management of Subak Irrigation System in Bali *Appl. Mech. Mater.* **776** 139–44
[12] Badan Pusat Statistik 2018 *Statistik Indonesia Tahun 2017* (Jakarta)
[13] Badan Pusat Statistik 2010 *Bali Dalam Angka 2010* (Denpasar)
[14] Gao J, Wei Y D, Chen W and Yenneti K 2015 Urban Land Expansion and Structural Change in the Yangtze River Delta, China *Sustain.* **7** 10281–307
[15] Pondaag C ., Pakasi C B D and Kumaat R M 2018 Faktor-Faktor yang Mempengaruhi

Terjadinya Alih Fungsi Lahan Pertanian di Kabupaten Minahasa Tenggara *Agri-Sosioekonomi* **14** 151
[16] Adnyana I M, Puja I N and Arthagama I D M 2017 Evaluasi dan Penetapan Lahan Pertanian Berkelanjutan untuk Pencegahan Alih Fungsi Lahan di Kabupaten Bangli *ECOTROPHIC J. Ilmu Lingkung. (Journal Environ. Sci.* **11** 76–80
[17] Rose D C, Sutherland W J, Barnes A P, Borthwick F, Ffoulkes C, Hall C, Moorby J M, Nicholas-Davies P, Twining S and Dicks L V. 2019 Integrated Farm Management for Sustainable Agriculture: Lessons for Knowledge Exchange and Policy *Land use policy* **81** 834–42
[18] Sardiana I K 2018 The Study of Development of Urban Farming Agrotourism Subak-Irrigation-Based in Sanur Tourism Area, Denpasar City, Bali *J. Indones. Tour. Dev. Stud.* **6** 33–40
[19] Roth D 2014 Environmental Sustainability and Legal Plurality in Irrigation: The Balinese subak *Curr. Opin. Environ. Sustain.* **11** 1–9
[20] Eghdami H, Azhdari G, Lebailly P and Azadi H 2019 Impact of Land Use Changes on Soil and Vegetation Characteristics in Fereydan, Iran *Agric.* **9** 1–17
[21] Martiningsih E 2015 The Spirit of Ritual in Community Engagement: A Case Study in Subak Asia Pacific *J. Sustain. Agric. Food Energy* **3** 10–3
[22] Putri N L A, Yuliantari M, Suryawardani and Diarta I K S 2019 Strategi Perekayasaan Subak Sembung sebagai Daya Tarik Wisata di Perkotaan **9** 29–38
[23] Food and Agriculture Organization 1999 Land Evaluation and Farming System Analysis for Land Use Planning.
[24] Yamashita S 2013 The Balinese Subak as World Cultural Heritage: In the Context of Tourism *J. Kaji. Bali* **3** 39–68
[25] Lanya I, Subadiyasa N, Sardiana K and Adi G 2018 Planning of Agro-Tourism Development Specific Location in Green Open Space Sarbagita Area Bali Province *IOP Conference Series: Earth and Environmental Science* (IOP Publishing) pp 1–8
[26] Diarta I K S and Sarjana I M 2018 Strategi Pengembangan Subak Padanggalak sebagai Daya Tarik Wisata di Kota Denpasar Bali *Media Konserv.* **23** 281–92
[27] Ramankutty N, Ricciardi V, Mehrabi Z and Seufert V 2019 Trade-Offs in the Performance of Alternative Farming Systems *Agric. Econ. (United Kingdom)* **50** 97–105
[28] Alexander I 2019 Perception of Organic Farming among Urban Vegetable Farmers in Delta Central Agricultural Zone *J. Agric. Ext.* **23** 76–83

IOP Conf. Series: Earth and Environmental Science 1111 (2022) 012047 doi:10.1088/1755-1315/1111/1/012047

Effect of Permanent Magnet Structure on The Performance of LSPMSM with a Power of 22 kW and 3000 rpm

NY Do[1], TA Le[2], XC Ngo[3*]
[1] Center for Mining, Electro-Mechanical research, Hanoi University of Mining and Geology, Hanoi, 100000, Viet Nam, ORCID ID: 0000-0002-5661-0268
[2] Faculty of Electrical Engineering, Hanoi University of Industry, Hanoi, 100000, Viet Nam
[3] Department of Industrial Engineering and Management, Hue University, Thua Thien Hue, 530000, Viet Nam
[1]donhuy.humg@gmail.com, [2]leanhtuan0985@gmail.com,
[3*]cuongngoxuan@gmail.com

Abstract. Energy efficiency and sustainability are important goals for countries all over the world. With the advancement of material technology, many new generations of high-efficiency motors are being researched and developed to replace induction motors (IM) to save energy. Line-start permanent-magnet synchronous motors (LSPMSM) are high-performance and ultra-high-performance motors that are emerging as an alternative to squirrel-rotor induction motors. Constructing the permanent magnets in the rotor is critically important in ensuring LSPMSM performance. The study uses the finite element method to analyze the characteristics of the motor. The study presented the influence of the permanent magnet structure in the rotor of the LSPMSM with a power of 22 kW and a speed of 3000 rpm on the electromagnetic field distribution in the motor, as well as the motor's working characteristics to optimize the permanent magnet structure to optimize the performance of the LSPMSM. The analysis results show that the proposed LSPMSM has a 92.6% higher efficiency than the same-type induction motor. This is a suitable replacement for high-power and high-speed IM, which contributes to energy efficiency, so using fewer fossil fuels to reduce environmental impact is a sustainable development solution.

1. Introduction

Energy conservation is currently a significant concern for all nations in the world. One of the most crucial conditions for sustainable growth, addressing the risk of fossil fuel depletion and the negative consequences of environmental contamination, is the economic and effective use of energy [1].

Economies are overgrowing, and the major global energy challenge is to secure energy supplies to meet growing demand. The International Energy Agency (IEA) estimates that world energy demand will increase by more than 50% between now and 2030. This means that emissions of carbon dioxide (CO_2), the main greenhouse gas, will increase by 52%. The cumulative investment required to meet projected demand is estimated at some 17 trillion USD. Combining growing energy demand with environmental protection is today's major challenge [2].

Many developing nations have been forced to "save energy and reduce emissions" to move toward a more "balanced and sustainable" output, which has decreased energy consumption per unit of GDP [3]. This is due to the intense pressure to limit carbon emissions and rising energy demand.

Energy, the environment, and sustainable development are closely related. However, since all energy sources have some environmental impact, there is a strong correlation between energy efficiency and environmental impact. Increasing energy efficiency is frequently accompanied by decreased resource use and pollution [4]. The use of electricity for electric motors represents the most considerable energy use in industrial facilities. To reduce the need for additional new power sources, reduce CO2 and SO2 emissions, and have a less detrimental effect on the environment, one solution to the sustainable and economical use of energy is to increase the efficiency of power use for motors.

Among electrical appliances, the electric motor consumes the largest amount of electricity, accounting for about 70% of the entire electricity grid. Currently, induction motors (IM) are widely used in the market, but improving IM efficiency to IE3 or IE4 according to IEC60034-30 is difficult. In the last two decades, the rapid development of rare earth permanent-magnet technology with high magnetic field density and significant energy products, such as neodymium magnet types, has promoted the application of these types of magnets in the field of electric machines to produce motors with high energy efficiency. Among the permanent-magnet motors, the line-start permanent-magnet synchronous motor (LSPMSM) emerges as an alternative to IM with a squirrel-rotor to save energy [5, 6].

Although LSPMSM is claimed to be a motor with high efficiency, high power factor, and high torque density, the operating parameters may not be guaranteed if the design is not good. They may not even meet the standards of IM [7]. Besides, LSPMSM often researches and manufactures with a small capacity and low speed below 1500 rpm. There are very few types of LSPMSM with high power and speed of 3000 rpm [5–9]. But high-power and high-speed motors are often used in many places, such as mining, oil, and gas. Optimizing performance for high-power LSPMSM up to 22kW at a high speed of 3000rpm was a significant challenge in the design process. In the study [10], a hybrid rotor structure is proposed in motor design to enhance starting and motor synchronization for high-power LSPMSM to reduce harmonic amplitude, reduce torque ripple, and reduce losses to improve motor efficiency further. Research work [11] studies the influence of the stator winding type of the LSPMSM 6/8 on the synchronization process during starting, thereby improving the operating efficiency of the LSPMSM.

In the study [12], an optometric analysis was performed to find the best configuration of the air gap length, magnet thickness, magnet width, and the number of conductors per slot, along with modifications of the rotor slot. Simultaneous steady-state and transient performance optimization of a 4-kW LSPMSM under efficiency maximization and starting torque maximization is examined in a study [13]. A commercial IM-based LSPMSM design has been carried out to meet IE4 efficiency standards using a hybrid rotor with the same stator and windings as the IM. Optimized LSPMSM based on a genetic algorithm performs better in the transient and steady-state than initial LSPMSM [14].

The study [15] presents a new analytical method that models the rotor to optimize the electromagnetic force distribution in the air gap to optimize the torque ripple of the LSPMSM has been presented. A study [16] has shown that using a permanent magnet with a curved shape will give more power factor and good synchronization ability. It is suitable for large-power LSPMSMs, which are suitable for high loads, such as fans, pumps, and compressor drives, especially in the mining industry. The study [17] proposes a design process for LSPMSM to optimize torque parameters and power factors in the steady working and start modes. Study [18] analyzed rotor structure with auxiliary poles and magnetization direction to improve the performance of LSPMSM.

Thus, the analysis of design parameters and the design parameters to optimize the performance of the LSPMSM has been and is of great interest. This paper studies the influence of permanent magnet structure on the performance of LSPMSM with a high power of 22 kW, a high speed of 3000 rpm, and a rated voltage of 690/380VAC. From the obtained results, the paper will evaluate and draw some conclusions for application in high-performance LSPMSM to replace IM used in mining.

2. Influence of design parameters on LSPMSM

2.1. LSPSMSM

The LSPMSM is an improvement over the permanent magnet synchronous motor and the IM. In other words, it is a hybrid between IM and synchronous motors by placing permanent magnets on the rotor of the IM. Therefore, LSPMSM not only possesses the advantages of a permanent magnet synchronous motor but also has the advantage of IM, which is the ability to start directly.

The stator of a typical LSPMSM motor has the same stator construction as the IM. The LSPMSM rotor is similar to the IM's rotor in that it has a squirrel cage, but the LSPMSM is different from the IM in that this motor has additional permanent-magnet rods in the rotor steel core. The article will analyze the design parameters of the LSPMSM based on the original IM with a power of 22 kW and a speed of 3000 rpm with the shape shown in Figure 1 and the parameters in Table 1.

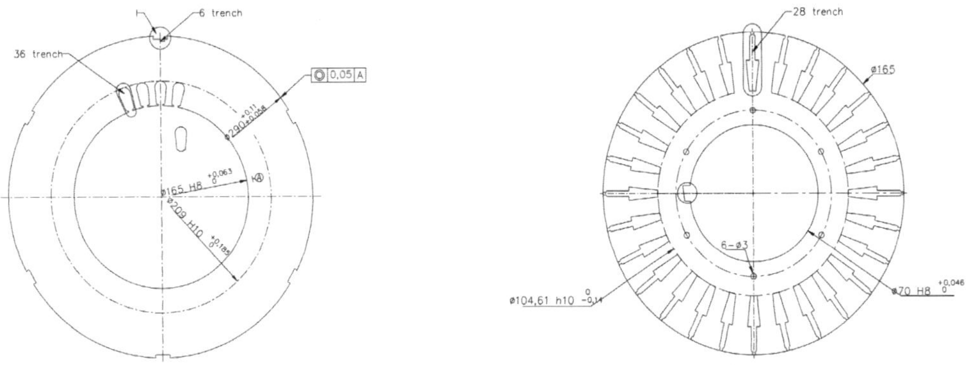

a) b)

Figure 1. Stator (a) and rotor (b) of IM with a power of 22 kW and speed of 3000 rpm.

Table 1. IM parameters

Parameters	Symbol	Value	Unit
Rated power	P	22	kW
Rated voltage Y/ Δ	U_n	690/380	V
Speed	n_n	3000	rpm
Number of stator slots		36	
Number of rotor slots		28	
Air gap length	δ	0.4	mm
Power frequency	f	50	Hz
Performance	η	89	%

2.2. Analysis of the effect of permanent magnet structure on the operation of LSPMSM

The LSPMSM rotor is a squirrel cage with additional permanent magnet rods in the rotor steel core. When calculating the design of LSPMSM, the most complicated part is the calculation of permanent magnets, and it is significantly more difficult for high-speed motors of 3000 rpm.

The calculation of permanent magnets is not reasonable, leading to intricate machining, low motor efficiency, failure to start, and long starting time. In addition to the high cost of permanent magnets in LSPMSM, the non-optimal calculation makes it difficult to commercialize this type of motor in practice [1,17].

The permanent magnets used in the LSPMSM are rare-earth NdFeB magnets with a residual magnetic density (B_r), maximum magnetic field strength (H_{cJ}), minimum magnetic field strength (H_{cB}), the maximum working temperature of magnet (T_w) and maximum energy storage (BH_{max}). The formula determines the volume of a permanent magnet V_m:

$$V_m = \frac{2.k_{of}k_{fd}(1+K_{EC})P_n}{\pi^2\xi.2.p.f.B_r.H_C} \quad \textbf{(1)}$$

where, k_{of} - the overload factor, select $k_{of} = 2$; k_{fd} - Magnetic form factor, preliminary select $k_{fd} = 1$; K_{EC} - Electromotive force factor $(0.6 \div 0.95)$, select $K_{EC} = 0.9$; P_n - Rated power; ξ - Magnet utilization factor $(0.3 \div 0.7)$, select $\xi = 0.5$; B_r - residual magnetic density; H_c- maximum magnetic field strength.

Typically, to ensure optimal design, fabrication, and assembly of permanent magnets in LSPMSM, the thickness of permanent magnets is usually in the range of values from 5mm÷6mm. From that, determine the basic parameters of permanent magnets, in which the size parameter of permanent magnets has the corresponding value: h= 5.5mm; b=40mm

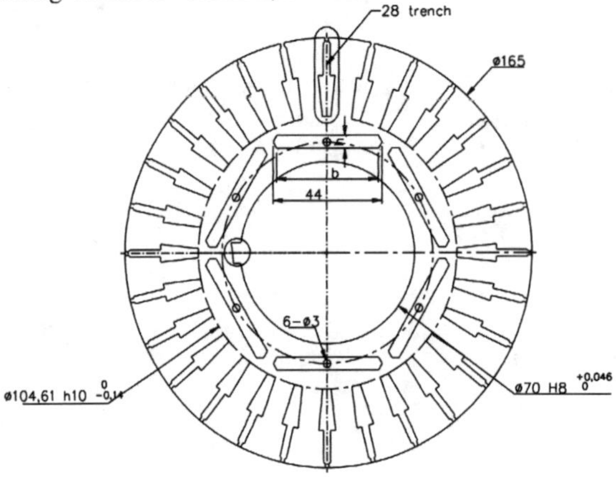

Figure 2. Parameters of permanent magnets in proposed LSPMSM.

Usually, LSPMSM is an improvement over the IM. Where the tooth, groove, and stator winding configurations remain the same, the rotor is calibrated by adding permanent magnet rods.

3. *Finite element method for motor analysis*

Evaluation of design parameters affecting the working process of LSPMSM based on the finite element method (FEM). FEM is used to solve the problem of magnetic field distribution in the magnetic circuit and the space around the electromagnetic device, which is the solution of the Poisson equation written for the electromagnetic field of the electric motor model [1].

This model is built based on Maxwell-Ampere law. According to the Maxwell - Faraday equation written for the case of an electric motor at a steady state as follows [5,10]:

$$\nabla \times \vec{H} = \vec{J} \tag{2}$$

where $\vec{J}$ density of direct current flowing in an electromagnet coil, A/m^2; $\vec{H}$- magnetic field strength, H/m.

Magnetic field strength $\vec{H}$ is related to flux density $\vec{B}$ as follows:

$$\vec{B} = \mu_0 \mu_r \vec{H} \tag{3}$$

where μ_0 - vacuum permeability; μ_r- relative magnetic permeability.

In the electromagnetic field, $\vec{B}$ calculated over the magnetic vector potential $\vec{A}$ as follows:

$$\vec{B} = \nabla \times \vec{A} \tag{4}$$

Substituting (2) and (3) into (1), we get the equation:

$$\nabla \times \left(\frac{1}{\mu_0 \mu_r} \nabla \times \vec{A}\right) = \vec{J} \tag{5}$$

Equation (4) has a general form of Poisson's equation, which can be interpreted in the analytical model corresponding to the Oxyz coordinate system as follows:

$$\frac{1}{\mu_0\mu_r}\left(\frac{\partial^2 A}{\partial x^2}+\frac{\partial^2 A}{\partial y^2}+\frac{\partial^2 A}{\partial z^2}\right)+J=0 \quad \textbf{(6)}$$

Solve (5), find $\vec{A}$ then based on (2) and (3) to calculate the magnetic flux density B and the magnetic field strength H as follows

$$\vec{B}=B_x\vec{i}+B_y\vec{j}+B_z\vec{k}=\left(\frac{\partial A_z}{\partial y}-\frac{\partial A_y}{\partial z}\right)\vec{i}+\left(\frac{\partial A_x}{\partial z}-\frac{\partial A_z}{\partial x}\right)\vec{j}+\left(\frac{\partial A_y}{\partial x}-\frac{\partial A_x}{\partial y}\right)\vec{k} \quad \textbf{(7)}$$

With the voltage applied to the coil terminals and the DC resistance of the coil, the current density can be determined from the equations:

$$U=R_{dc}i+L\frac{di}{dt};i_{dc}=\frac{U}{R_{dc}} \quad \textbf{(8)}$$

Where U - voltage applied to the two coil terminals; R_{dc} - resistance of the coil; L- coil inductance; I_{dc} - current in the coil; S_{dq} - coil cross-section.

FEM is a technique for solving equation (5) to determine the magnetic vector potential $\vec{A}$, then calculate the flux density $\vec{B}$ and magnetic field strength $\vec{H}$ according to formulas (3) and (2), thereby determining the magnetic field distribution in space with high accuracy. From there, it will be of great help to the designer and operator in adjusting the parameters of the machine's electromagnetic mechanism to optimize the distribution of the magnetic field in space to improve the machine's performance. FEM consists of 4 basic steps that discrete the analysis domain into elements. Elements are linked together to form a mesh, choose the membership function, approximate the solution on each element, match all the elements in the analytic domain to get the system matrix and solve the system matrix by an iterative method.

4. Results and Discussion

Parameters of the proposed LSPMSM with a power of 22 kW speed of 3000 rpm used in the simulation have parameters as in Table 2.

Table 2. LSPMSM parameters

Parameters	Symbol	Value	Unit
Types of permanent magnets		NdFeN35	
Dimension of permanent magnets DxRxC		29x5x115	mm
Rated load torque	M_n	17,5	N.m

Calculating according to FEM requires the support of digital computers and software programs written based on FEM mathematics.

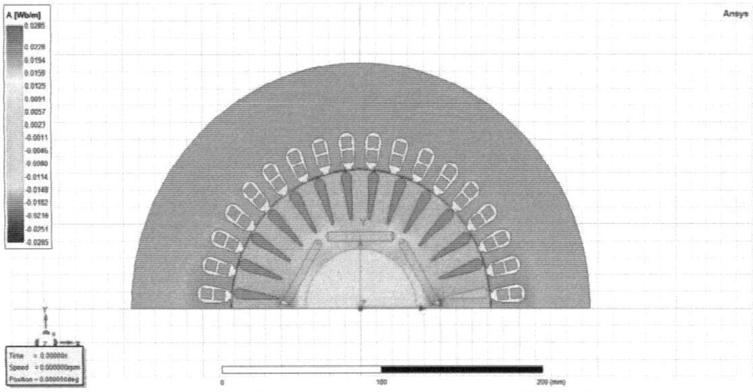

Figure 3. Simulate flux of proposed LSPMSM.

The results of the electromagnetic field analysis of the proposed LSPMSM are shown in Figure 3. It is clear that the magnetic field is intensely concentrated at the permanent magnet ends with the U-shaped electromagnet structure. This can cause saturation of the magnetic circuit of the steel foils at the ends of the permanent magnets. This uneven magnetic field distribution leads to reduced motor efficiency, difficulty starting the motor, and difficulty in synchronizing.

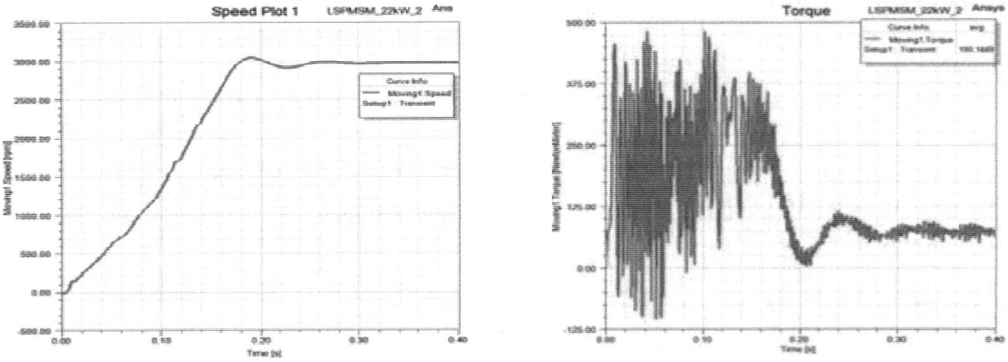

a) b)

Figure 4. Starting characteristics with a rated load of proposed LSPMSM. a) Speed; b) Torque.

Figure 4 shows starting characteristics with a rated load of LSPMSM with a capacity of 22kW, 3000 rpm, and a U-shaped permanent magnet structure. The motor can be started with a rated load. With this configuration, when rated load, the motor reaches its maximum speed after 0.26 s. Besides, it only needs to go through 5 "bumps" segments to achieve the maximum speed characteristic. However, the results in Figure 4.b show that the working torque of the motor fluctuates strongly, causing ripple and creating stability for the motor during operation.

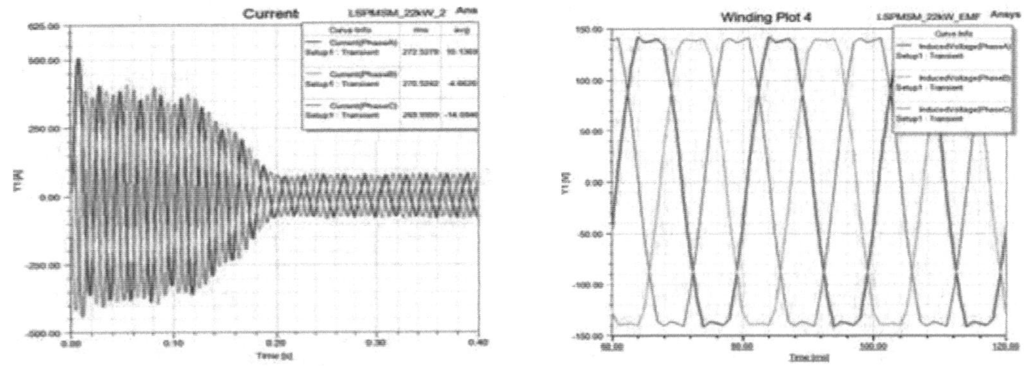

a) b)

Figure 5. Starting current with a rated load (a) and electromotive force (b) of proposed LSPMSM.

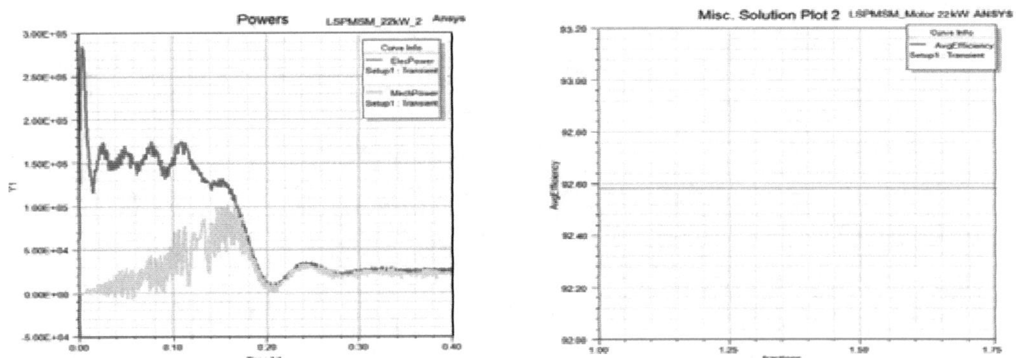

a) b)

Figure 6. Electromagnetic and mechanical power characteristics (a) and performance (b) of proposed LSPMSM

Figure 5 show that the current characteristic waveforms of the LSPMSM with the U-configuration in steady-state mode are non-sinusoidal. All configurations of LSPMSM have current harmonics. The main reason is the different structures of permanent magnets, affecting the magnetic field distribution in the magnetic circuit, and the air gap. The generated harmonics will reduce the efficiency of the motor.

In a steady state, Ansys/Maxwell2D software is also used to evaluate the performance of the LSPMSM. Two power characteristic curves (electromagnetic and mechanical power) are shown in Figure 6.a. Performance calculation results in a steady state of LSPMSM with a capacity of 22 kW, a speed of 3000 rpm, and a performance of 92.6% (IE3) in Figure 6.b.

Thus, with the proposed magnet structure, motor performance can improve from 89% for IM of the same type (in Table 1) to 92.6% (Figure 6.b) for LSPMSM. In addition, the above simulations also show that with this type of structure, the LSPMSM can be completely self-started and synchronous. This is a promising solution to replace high-power and high-speed IM.

Research results show that combining new materials and suitable electromagnet structure design will create a new generation LSPMSM with super high efficiency (IE4) to replace traditional low-efficiency IM (IE1). Correspondingly, the motor efficiency increased by about 3.6% (up from 89% of the IM to 92.6% of the LSPMSM). According to the International Energy Agency (IEA), electric motors consume 40% of total energy, which accounts for 75% of electricity consumption. For example, in Vietnam, in 2021, the total electricity consumption will reach 225.3 x109 kWh [19]. If we replace about 20% of low-power IM (IE1) with high-efficiency LSPMSM (IE4), the equivalent amount of energy saved in one year is about 486.648 x 106 kWh.

With the total electric energy saved as above, it will reduce about 5% of the total capacity of a coal-fired power plant or 16% of the total capacity of an oil-fired power plant. Thereby reducing the use of fossil energy sources, reducing CO_2, SO_2 emissions, and dust emissions, and ensuring economic development with environmental protection.

5. Conclusion

According to the analysis results obtained after simulating an LSPMSM with a capacity of 22 kW, a speed of 3000 rpm, and a U-shaped structure, the LSPMSM with this structure is simple and easy to manufacture, the materials used to make electromagnets are few, and the cost of electromagnets is low. The proposed LSPMSM can be started and synchronized with the rated load. However, the magnetic field distribution in the motor still has many points that may become saturated soon, the motor is noisy when operating, and the motor efficiency is only at IE3. To improve the performance of the LSPMSM, the structure of the permanent magnet can be optimized, but this increases the fabrication cost of the LSPMSM.

The primary objectives of many nations are the production of energy, the efficient use of energy, and the resolution of environmental issues. The industrial equipment that uses the most energy is the electric motor. The development of replacement super and high-efficiency motors for IM, such as the LSPMSM discussed in the paper, has helped make energy use more efficient and less harmful to the environment. This is a strategy for the industry's sustainable growth.

Acknowledgments

This paper is supported with research funding by Project code DT.CNKK.QG.012/21

References

[1] Do N Y, Ngo T T, Ngo X C and Le A T 2021 Numerical simulation method application in the design of a line-start permanent magnet synchronous motor In national Conference on Mechanical, Electrical *Electr. Autom. Eng. MEAE2021* 2–7

[2] European Commission 2006 Environment fact sheet: sustainable development *Environment fact sheet: energy for sustainable development* pp 1–4

[3] Wang Q and Chen Y 2010 Energy saving and emission reduction revolutionizing China's environmental protection *Renew. Sustain. Energy Rev.* **14** 535–9

[4] Dincer I and Rosen M A 1999 Energy, environment and sustainable development *Appl. Energy* **64** 427–40
[5] Melfi M J, Umans S D and Atem J E 2013 Viability of highly-efficient multi-horsepower line-start permanent-magnet motors *Annu. Pet. Chem. Ind. Conf.* **1** 1–10
[6] Isfahani A H, Vaez-Zadeh S and Rahman M A 2011 Evaluation of synchronization capability in line start permanent magnet synchronous motors *IEEE International Electric Machines and Drives Conference, IEMDC* pp 1346–50
[7] Elistratova V 2015 *Optimal design of line-start permanent magnet synchronous motors of high efficiency* (Ecole Centrale de Lille)
[8] Stoia D, Chirilă O, Cernat M, Hameyer K and Ban D 2010 The behaviour of the LSPMSM in asynchronous operation *International Power Electronics and Motion Control Conference* pp 45–50
[9] Takahashi A, Kikuchi S, Mikami H, Ide K and Binder A 2012 D-q Space vector analysis for line-starting permanent magnet synchronous motors *International Conference on Electrical Machines, ICEM 2012* pp 136–42
[10] Yan B, Yang Y and Wang X 2021 Design of a Large Capacity Line-Start Permanent Magnet Synchronous Motor Equipped with Hybrid Salient Rotor *IEEE Trans. Ind. Electron.* **68** 6662–71
[11] Zhao W, Tian M, Wang X and Sun Y 2020 Analysis of the Synchronization Process and the Synchronization Capability for a Novel 6/8-Pole Changing LSPMSM *IEEE Trans. Magn.* **56**
[12] Sarac V, Minovski D and Janiga P 2022 Parametric Analysis for Performance Optimization of Line-Start Synchronous Motor with Interior Asymmetric Permanent Magnet Array Rotor Topology *Electronics* **11** 1–22
[13] Faramarzi Palangar M A M, Kahourzade S and Soong W L 2021 Simultaneous efficiency and starting torque optimization of a line-start permanent-magnet synchronous motor using two different optimization approaches *Arab. J. Sci. Eng.* **46** 9953–64
[14] Palangar M F, Mahmoudi A, Kahourzade S and Soong W L 2020 Optimum Design of Line-Start Permanent-Magnet Synchronous Motor Using Mathematical Method *IEEE Energy Conversion Congress and Exposition* pp 2064–71
[15] Jing L, Gong J and Lin Y 2019 Analysis and reduction of cogging torque of line-start permanent magnet motors *Prog. Electromagn. Res. M* **78** 115–24
[16] Lyskawinski W, Jedryczka C and Szelag W 2017 Influence of magnet and cage shape on properties of the line start synchronous motor with powder hybrid rotor *International Symposium on Electrical Machines, SME*
[17] Sarani E and Vaez-Zadeh S 2017 Design Procedure and Optimal Guidelines for Overall Enhancement of Steady-State and Transient Performances of Line Start Permanent Magnet Motors *IEEE Trans. Energy Convers.* **32** 885–94
[18] Ugale R T and Chaudhari B N 2021 Performance Enhancement of Line Start Permanent Magnet Synchronous Motor with a Special Consequent Pole Rotor *IEEE Trans. Energy Convers.* **36** 1972–82
[19] Nguyen T S 2022 Vietnam Electricity in 2021 - Results through challenges

Comparison Performances of Microfiltration and Rapid Sand Filter Operated in Water Treatment Plant

B D Marsono[1*], A. Yuniarto[2], A Purnomo[3], ES. Soedjono[3]
[1]Faculty of Civil, Planning, and GeoEngineering, Sepuluh Nopember Institute of Technology, Surabaya, 60111, Indonesia, 0000-0003-2164-7092
[2]Department of Environmental Engineering, Institut Teknologi Sepuluh Nopember, Surabaya 60111, Indonesia, 0000-0002-7492-6901
[3]Department of Environmental Engineering, Institut Teknologi Sepuluh Nopember, Surabaya
[1]*bowodjok@gmail.com

Abstract. Water treatment of river water is generally carried out conventionally, a rapid sand filter consisting of a coagulation unit, flocculation, sedimentation, and sand filtration. One of the disadvantages of sand filters is that it requires a large volume of water for washing, inconsistent quality, and escape of microorganisms less than 0.01 mm in size. Membrane technology can solve this problem with the advantages of including a more compact unit with a larger surface area per volume. The immersed membrane has another advantage that it is possible to integrate with sedimentation units so that it can save the land. This paper will study a comparison of the performance of the microfiltration membrane with sand filters in two drinking water treatment plants. The comparison method is based on technical specification, capacity, and quality performance. The results show that the quality of the microfiltration membrane filtrate is better than the sand filter. Overall immersed membrane microfiltration with pore size 0.05 µm had removal efficiency of E. Coli was 100%, turbidity 93 %, TSS 100%. Organic permanganate value 37%, and Microplastics 81.5%, whereas pore size 0.07 µm removal E. Coli was 99.8%, turbidity 52.9 %, Organic permanganate value 17.6%, and Microplastics 37.1%. The productivity of microfiltration membranes is more significant than rapid sand filters. The removal efficiency of WWTP Ngagel was better than Siwalan Panji.

1. Introduction

The water treatment system of rapid sand filters generally includes coagulation units, flocculation, sedimentation, sand filter, and disinfection. If the turbidity of the raw water is high, a pre sedimentation unit can be added as a preliminary treatment to settle discrete particles. WTP (Water treatment plants) produce water with quality that meets the requirements according to the drinking water quality standards used in Indonesia, namely the Regulation of the Minister of Health of the Republic of Indonesia No. 492 / MENKES/PER/IV/2010, which includes physical, chemical, bacteriological, and radiological requirements. Membrane technology is distinguished according to pore size from large to small, respectively, MF (Microfiltration), UF (Ultrafiltration), NF (Nanofiltration), and RO (Reverse Osmosis). The smaller the membrane pores, the more significant pressure is required. Likewise, microfiltration materials can remove suspended particles, ultrafiltration removes macromolecules, nanofiltration can separate valence ions two or more, and Reverse Osmosis can remove valence ions 1. Membranes are generally made of inorganic materials such as ceramics, carbon, and metals and have the advantage of stability against heat and pressure [1].

Microfiltration is a process with a driving force of different pressures where colloids are suspended, and microporous membranes can hold particles of 0.1-20 μm. MF is usually operated at a relatively low TMP (Trans Membrane Pressure) (< 50 psi or 3.4 bar or 0.35 Mpa), and the permeate flux is very high [2]. Microfiltration is a membrane with an asymmetric shaft thickness of 10-150 μm. With a pore size of 0.05-10 μm, the allowable driving force is < 2 bar [3].

Membrane technology has developed rapidly to treat wastewater and drinking water. Biological wastewater treatment using membranes is known as membrane *side stream* and membrane immersed or submerged. A membrane side stream is a membrane that is operated separately from the biological processing unit while the immersed membrane is installed inside the biological reactor. The MF process is generally used for water treatment systems. After the sand filter, MF can be applied to groundwater and surface waters. MF pore size usually varies from 0.1 to 10 microns [4].

With pore sizes of 0.1 and 0.01 μm, respectively, microfiltration and Ultrafiltration membranes are designed to remove particulate matter such as turbidity and microorganisms through filtration mechanisms, resulting in water quality superior to conventional treatment processes. For example, MF and UF can consistently produce filtrates with turbidity below 0.1 NTU, independent of feed water quality. In addition, protozoan microorganisms such as *Giardia* (5-15 μm) and *Cryptosporidium* (3-5 μm) can be separated by MF or UF membranes [5].

According to [6], in raw water at the ADWTP (Advanced Drinking Water Treatment Plant), there are 6614 ± 1132 microplastics/L, while in treated water, there are 930 ± 71 microplastics/L, so the efficiency of microplastic removal is 82.1 - 88.6%. Microplastic removal using membranes has been widely studied due to the efficiency of the membrane, which can adequately remove substances in water. The removal efficiency of microplastic using membranes is up to 99.9%, but this depends on the durability, size, and concentration of microplastics [7].

One of WTP in Sidoarjo is in Siwalan Panji. There are five water treatment installations in Siwalanpanji's WTP, consisting of 4 conventional packages and one installation with ultrafiltration, interconnected, and production capacity up to 165 L/s. For the process and operation of the ultrafiltration unit at the Siwalanpanji's WTP, the processing unit consists of intakes, collecting wells, pre-sedimentation, and collector tanks.

The Water treatment system at Ngagel III Surabaya is a conventional treatment consisting of flash mixings, slow mixings, sedimentation, and filtration. WTP Ngagel III takes raw water from the Surabaya River with a treatment capacity of up to 1750 L/s. Filter units total 16 units with a discharge of 110 L/s per unit. The filter uses anthracite media and silica sand. The filter has 8.32 m in length and 3.57 m in width. The filter operates with a filtration rate of 9.8 $m^3/m^2/h$ and a backwash rate of 31 $m^3/m^2/h$. This article will evaluate the MF/UF membrane and rapid sand filter in the two WTP for performance comparison.

2. Methods

A comparison of the performance of membranes with rapid sand filters was carried out in two WTP. Both installations use raw river water upstream from the Brantas River, East Java. The membrane unit at WTP Siwalan Panji is called an ultrafiltration unit with a side stream module, as in WTP Ngagel, using an immersed membrane microfiltration, is a research pilot placed in a sedimentation unit. Rapid sand filter in WTP Siwalan Panji single media while Ngagel III dual media.

Performance comparison includes specifications, capacity, and quality of the filtrate produced. The compared specifications include pore size, operating value, filtration rate, and material type. The capacity compared includes production volume, ration volume backwash to production, and flow rate. The quality comparison includes the parameters of Turbidity, Organic matter, microplastics, and E. Coli. At the same time, the removal efficiency is calculated based on feed water and effluent of membrane and sand filter unit. Water sampling takes feed water and filtrates it on membrane units and rapid sand filters.

3. Results and Discussion

3.1. Technical Specifications

The technical specifications between the MF membrane and the rapid sand filter in the two WTP show that the filtration rate of the rapid sand filter is greater than that of the MF. The pore size membrane is smaller than the rapid sand filter, which is 0.05-0.07 μm and four μm, respectively. Pore size is essential in removing water substances, especially turbidity, particles, and microorganisms. The operating pressure on the microfiltration membrane of 0.9 bar at WTP Ngagel is due to the immersed membrane type, which requires a maximum suction pressure of 1 bar. UF on Siwalan Panji operates on a side stream with a pressure of 2 bar. While the Rapid sand filter operates by gravity with a pressure of 0.2 bar. The duration of membrane operation is more than 24 hours, while the rapid sand filter averages 12 hours. Table 1 compares the technical specifications of membranes and rapid sand filters.

Table 1. Comparison of Membrane and Rapid sand filtration specifications

No	Parameters	Units	Membrane MF/UF		Rapid Sand Filter	
			Pilot WTP Ngagel	WTP Siwalan Panji	WTP Ngagel	WTP Siwalan Panji
1	Membrane type		Immersed MF	Side Stream UF	Dual Media	Single Media
2	Filtration rate	m^3/m^2/hour	6.5	1.5	9.9	7.6
3	pore size	μm	.05	0.07	4	4
4	Operating pressure	bar	0.9	2	0.2	0.2
5	Surface area per unit	m^2	0.156	0.5	29.8	3.96
6	Long Operation	hours	> 48	24	12	12

The comparison of capacity can refer to the filtration rate, the suction pressure, and the operation's duration. Another parameter that indirectly affects the capacity is the pore size, while the surface area to volume ratio will affect the unit's effectiveness. The filtration rate or flux in the immersed membrane of WTP Ngagel operates at a suction pressure of 0.9 to 0.7 bar, reaching 6.5 $m^3/m^2/h$ lower than the sand filter which reaches 9.9 $m^3/m^2/h$ and 7.6 $m^3/m^2/h$ at WTP Siwalan Panji. Even though the membrane filtration rate is lower, the membrane operation duration is longer and can reach more than one and a half times the length of operation of the grained media filter. Thus, membranes are more competitive in productivity than filters with grained media. A comparison of immersed membrane and filter specifications can be seen in Table 1.

The parameter of the ratio of the surface area of the membrane in the form of a sheet plate is greater than that of a grained media filter. This type of membrane has a surface on two sides that plays a role in filtering, while the grained media filter has only one surface at the top. Thus, membrane units are more compact/compact than filter units.

MF Membrane Siwalan Panji operates side stream in a module, while the rapid sand filter has six units. WTP Ngagel III, under research, used an MF membrane module immersed in a sedimentation tank while the rapid sand filter operated 16 units with a capacity of 110 L/s per filter unit.

3.2. *Capacity*

The suction pressure of 0.8 to 0.6 bar of immersed membrane shows that flux decreased from 178 $L/m^2/h$ in the first hour to 68 $L/m^2/h$ at the 35th hour and 40 $L/m^2/h$ at the 75th hour. Flux decreases with a decrease in suction pressure due to the fouling of the membrane surface. Turbidity throughout membrane operation is relatively stable at 0.2 – 0.6 NTU.

Performance comparisons from the quantity aspect show that immersed membrane microfiltration is much better than dual media filters. If we compare the dual media filter operated in the Ngagel III IPA and the Microfiltration Membrane Immersed Unit to treat 110 L/s water, the study showed that the membrane filter surface area was twice that of the dual media filter with the number of membrane units 398 units versus 1. The need for the land area is one-fifth more efficient.

The energy requirements for pumping are relatively comparable. The microfiltration membrane requires a relatively small head, a 10 m suction head, and a suction head according to the reservoir elevation. In contrast, the dual media filter in Ngagel from the channel after the sedimentation bath to the filter requires a head of about 10 meters.

An important parameter for comparison is the flow rate per surface area. In this case, the immersed membrane, which has two sides on the surface, shows a capacity of 16.3 $L/s/m^2$, while the rapid sand filter is 3.72 and 3.96 $L/s/m^2$. This shows that immersed membranes are more compact and effective in treating water. This capacity far exceeds the comparison stated by [9] namely for MF and rapid sand filter 0.02 – 0.2 $L/s/m^2$, 0.6 – 2 $L/s/m^2$ respectively.

The water requirement for washing on dual media filters with a backwash speed of 31 m/hour for 20 minutes is 308 m^3, while the membrane with a total of 398 units requires 20 m^3 of washing water only. Water produced with a dual media filter operating duration reaches 4,752 m^3, while the membrane reaches 9,504 m^3. Thus, the percentage of washing water to the production volume for dual media filters reaches 6.5%, while the microfiltration membrane reaches 0.2%. This will benefit from the aspect of productivity and the wastewater or sludge produced. The comparison of membrane and rapid sand filter capacities is described in Table 2 below.

Table 2. Capacity Comparison of Membrane and Rapid Sand Filter

No	Parameters	Units	Membrane MF/UF		Rapid Sand Filter	
			Pilot WTP Ngagel	WTP Siwalan Panji	WTP Ngagel	WTP Siwalan Panji
1	Capacity per unit	L/s	110	30	110	8.3
2	Surface effective	m^2	60.9	70.0	29.8	3.96
3	number unit	unit	398.2	36 modules	16	6
5	Land requirement	m^2	6.0	-	29.7	3.96
6	Flow rate per surface area	$L/s/m^2$	18.3	1.57	3.72	2.10
7	Backwash rate	m/hour	3.9	3.0	31	28
8	The volume of water backwash	m^3/day	20	24	308	28
9	Volume of production	m^3/day	9,504	2,592	4,752	852.4
10	Ratio water backwash to production	%	0.2	0.9	6.5	3.29

3.3. *Filtrate quality*

Quality comparisons, especially the parameters studied, namely turbidity, organic substances, Total Suspended Solid, Microplastics, and *E. Coli*. The measurement results showed that overall the quality parameters of membrane filtrates were better than conventional filters except for organic substances parameters. The turbidity quality of the filtrate produced by the membrane reaches 0.39 NTU while the filter reaches 2.6 NTU. The organic substances produced by the filter are better than the membrane, reaching 9.48 mg/L and 11.85 mg/L, respectively. On the parameters of *E. Coli*, microfiltration membranes and rapid sand filter Ngagel III WTP have filtrate 0 MPN/100 ml and 3 MPN/100 ml, respectively. This result is much better than Siwalan Panji WTP filters have *E. Coli* 410 MPN/100 ml dan 0 MPN/100 ml, respectively. The results of the filtrate quality comparison can be seen in Table 3.

Table 3. Comparison of The Filtrate Quality of Membrane and Rapid Sand Filter

Parameters	Membrane MF/UF		Rapid Sand Filter	
	Ngagel III	Siwan Panji	Ngagel III	Siwan Panji
Turbidity (NTU)	0.39	1.13	0.6	0.45
Organics (mg/L KMnO4)	11.85	14.75	9.48	17.4
TSS (mg/L)	0	-	0	-
E. Coli (MPN/100 ml)	0	3	410	0
Microplastic (particle/L)	5	22	5	38

The filtrate quality in this study was better than the membrane with a pore of 0.07 μm operated at the Siwalan Panji PDAM Sidoarjo, where it reached 1.13 NTU, organic substances 14.75 mg/L, and *E. Coli* 2.7 MPN/100 ml [8]. A sand sieve with a pore size of 4 μm can separate *protozoan* microorganisms (4–100 μm), while the type of bacteria (0.2–1.0 μm) and viruses (0.01–0.1 μm) will pass in the filtrate. However, water treatment systems must affix disinfectants to kill these bacteria and viruses. On the other hand, microfiltration membranes with pores less than 0.2 μm can remove bacteria of the type *E. Coli, Salmonella, and Cholera* [9]. The size of *E. Coli* is larger than the membrane's pore size. *E. Coli* bacteria are rod-shaped with a size of 1.0-1.5 μm x 2.0-6.0 μm [10], while the pore size of the microfiltration membrane used is 0.05-0.07 μm.

Performance comparisons between membrane microfiltration and Rapid sand filter in removing turbidity, *E. Coli*, and microplastics show that microfiltration membranes are better than rapid sand filters. While the organic parameter is sometimes a rapid sand filter better than the MF membrane, this is due to the adsorption process in the cake and TSS (Total Suspended Solid) on the membrane and filter media. Removal turbidity efficiency in both Ngagel and Siwalan Panji WTP in membrane units above 90% as reported by [9, 10, 11] using MF hollow fiber and flat sheet reached 92-99%. Figure 1 compares the MF membrane's removal efficiency to the Rapid sand filter in two WTP units.

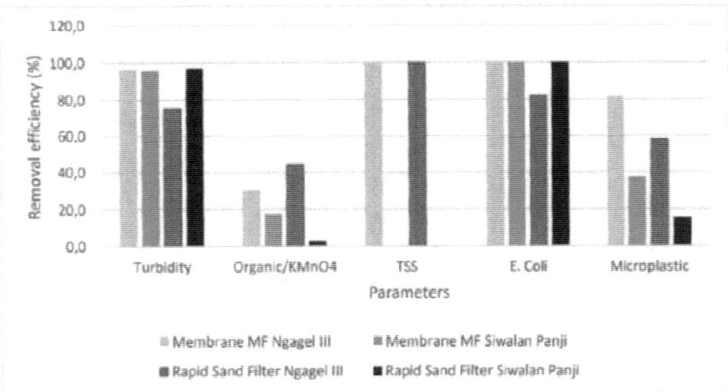

Figure 1. Comparison of Removal efficiency

Microplastic parameters cannot be removed entirely by MF pore 0.05μm due to the microplastic size, which is elongated fiber but smaller than the membrane pore. Microplastic efficiency removal with MF reached 81.5% in WTP Ngagel III, while in Siwalan Panji, it was 58.3%. The form of the microplastic examined, as shown below, can escape the membrane pore. The performance removal efficiency of microplastics reaches 86 – 99% reported by [6,11].

4. Conclusion

The quality of the microfiltration membrane filtrate gives better and more consistent results than dual media filters, particularly turbidity and *E. Coli* parameters. The productivity of microfiltration membrane filtrate is also greater than rapid sand filters, and the water requirement for backwash is less than rapid sand filters. Overall MF have removal efficiency of *E. Coli* parameters reach 100%, turbidity 53 - 96%, TSS 47 - 100%, Organic substances 17 - 31%, and Microplastics 37 - 81%. The removal efficiency of *E. Coli* and turbidity is relatively stable. Future research could be continued on energy needs, operating and maintenance costs, and membrane shortages if operated for a long time.

Acknowledgment

The author would like to thank the Environmental Engineering Department of ITS for the final project and all parties who have assisted in completing this research. Besides that, the author would like to thank the Director of the PDAM Surabaya and PDAM Sidoarjo, the Head of the Production Section, the Head of the Laboratory Section, and the PDAM Operator for providing the data and facilities needed during this research period.

References

[1] Crittenden JC, Trussell RR, Hand DW, Howe KJ T 2012 Water Treatment Principles and Design. MWHs Water Treatment. John Wiley & Son, Inc. 12-819 *John Wiley Son, Inc* **12** 2012
[2] Scott K, Mahmood A J, Jachuck R J and Hu B 2000 Intensified membrane filtration with corrugated membranes *J. Memb. Sci.* **173** 1–16
[3] Mulder M 2021 *Basic Principles of Membrane Technology*
[4] Koyuncu 2015 Advances in Water Treatment by Microfiltration, Ultrafiltration, and Nanofiltration *Elsevier* 2015
[5] Subcommittee on Periodical Publications of The 2008 Microfiltration and ultrafiltration membranes for drinking water **12** 84–97
[6] Wang Z, Lin T and Chen W 2020 Occurrence and removal of microplastics in an advanced drinking water treatment plant (ADWTP) *Sci. Total Environ.* **700** 134520
[7] Padervand M, Lichtfouse E, Robert D and Wang C 2020 Removal of microplastics from the environment. A review *Environ. Chem. Lett.* **18** 807–28
[8] Rachmawati F, Marsono B D, Masduqi A and Purnomo A 2021 Technical evaluation of ultrafiltration unit in Siwalanpaji's water treatment plant, PDAM Sidoarjo *IOP Conf. Ser. Earth Environ. Sci.* **896**
[9] Gray N F 2013 *Filtration Methods* (Elsevier)
[10] Rahayu WP, Nurjanah S. K E 2018 *Escheria Coli: Patogeneitas, Analisis dan Kajian Resiko*
[11] Lebeau T, Lelievre C, Buisson H, Cleret D, Venter LWV C P 1998 Immersed membrane filtration for the production of drinking water: combination with PAC for NOM and SOCs removal. Desalination *J. Am. Water Work. Assoc.* **117** 219–31

Isotherm Coefficient for Benzene and Toluene Adsorption on Granular Activated Carbon for Preparation of Biobarrier

Galih Rineksa[1], Achmad Bustomy[1], Don-Hee Park[2], and Misri Gozan[1]*
[1] Biorefinery Lab., Department of Chemical Engineering, Faculty of Engineering, Universitas Indonesia, Kampus Baru UI, Depok, 16424, Indonesia
[2] Department of Biotechnology and Bioengineering, Chonnam National University, Gwangju, 500757, Korea
[1]galih.rineksa11@ui.ac.id, [2*]mrgozan@gmail.com

Abstract. Biobarrier combines the adsorption and biodegradation processes which are carried out simultaneously. This method can increase the rate of contaminant reduction, reduce contaminant toxicity and extend the life of activated carbon. This study aims to evaluate the characteristics of benzene and toluene adsorption on granular activated carbon (GAC). The ratio of adsorption in the expanded bed to the fixed bed determines the operating conditions in the benzene biodegradation process experiment. The results showed that the GAC was distributed in the micropore and mesopore regions with a pore diameter of 15 Å to 23 Å. The molecular diameters for benzene and toluene are 5.285 Å and 5.72 Å, respectively, which fit the pores. The adsorption capacity of activated carbon can be represented well using the Freundlich equation. Freundlich's constants for benzene and toluene are Kf = 17.87 (1/n = 0.46) and Kf = 25.88 (1/n = 0.45).

1. Introduction

Water treatment processes have been widely developed to produce drinking water, especially to remove pollutants in surface water and groundwater. Methods include magnetic treatment [1], lighting-aeration bioreactor [2], advanced oxidative [3], and bio-barrier [4], among others. The adsorption process plays an important role in advanced wastewater treatment. Activated carbon used for adsorption commonly comes in powdered and granular forms [5]. Hydrocarbons are the most common pollutants in water, especially benzene and toluene [6,7]. Benzene and toluene are contaminants from solvents or intermediates in various chemical and manufacturing industries.

The function of bio-barrier is to adsorb pollutants, which generally include toluene and benzene, which are hazardous contaminants due to their carcinogenic properties. They are commonly found in soils polluted by nearby petrochemical industries. Activated carbon is among the most effective adsorbent for such pollutants owing to its high porosity, which helps immobilize the microbes. Activated carbon also poses no contaminants to the water [8].

This research is part of the design of a bioregenerator for applying the adsorption and degradation of pollutants containing hydrocarbons. This study tests the adsorption properties of benzene and toluene isotherms on granular activated carbon. This study aims to determine the adsorption capacity of available activated carbon to be used as a part of bio-barrier research. To this day, the research on adsorption on GAC using benzene and toluene as part of the bio-barrier is still relatively limited. Biobarrier plays an essential role in bioremediation.

2. Methods

2.1 Materials

Granular activated carbon was obtained from 2 different companies, PT. Daiti (labelled "PTD") and Bratachem (labelled "BTC"). Benzene and toluene of analytical grades were purchased from Merck, Indonesia.

2.2 Experiment Procedure

The carbon preparation stage is divided into two steps: the separation of GAC based on the mesh and the cleaning and drying. This study uses GAC with a filtered size of 12 and a homogeneous mesh. GAC was first cleaned to remove impurities. Next, GAC was steamed for 120 minutes in an autoclave and then in an oven for 12 hours (110℃). Storage was then carried out in a desiccator to prevent moisture.

The adsorption capacity test of benzene and toluene in water on activated carbon was carried out by making benzene solutions with concentrations of 100 ppm, 200 ppm, 400 ppm, 600 ppm, 800 ppm, and 1000 ppm, as well as toluene solutions with concentrations of 100 ppm, 200 ppm, 400 ppm, and 500 ppm. A total of 0.25 g of GAC for each concentration, benzene, and toluene, was put into a 100 ml Erlenmeyer flask for various concentrations. The Erlenmeyer flasks were tightly sealed with aluminum foil and plastic. The solutions were put into an auto-shaker for two days at a shaker speed of 27 rpm. The arrangement of experiments is illustrated in Figure 1.

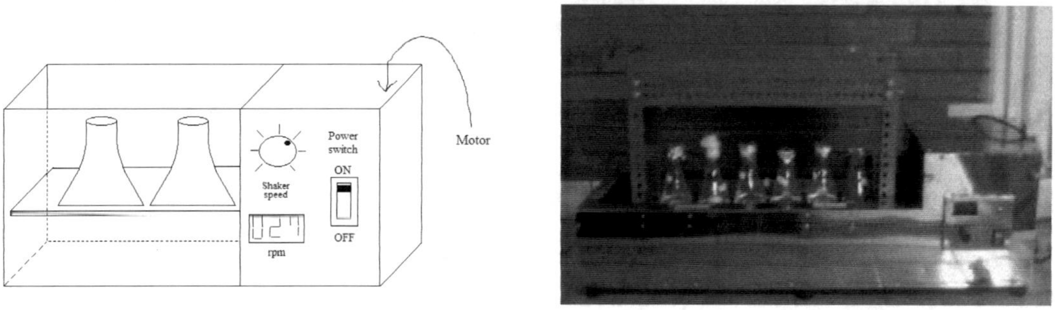

Figure 1. Isothermal adsorption experiment equipment diagram (left) and setup (right)

After two days, the absorbance was measured. The absorbance values obtained for each concentration were plotted in a graph with log Ce as the abscissa and log Qe as the ordinate. The values of the Freundlich constant and 1/n of the GAC were obtained using linear regression.

2.3 Standard Calibration of Pollutant Models with UV Spectrophotometer

UV spectrophotometer calibration needs to be done because the output data from this tool is in the form of absorbance data. For each compound, there is a specific wavelength. This wavelength can be obtained from literature data or scanned for the wavelength used for benzene and toluene. Based on analysis by spectrophotometer, the wavelengths used are 254 nm for benzene and 260 nm for toluene. Calibration was done by making pollutant samples with a single component with varying concentrations. Calibration curves for benzene and toluene were obtained from the absorbance measurement results.

2.4 Analysis

A UV spectrophotometer was used to determine standard pollutant models with known concentrations to obtain a correlation between pollutant concentrations in solution and adsorption. The absorbance values of benzene at 254 nm and toluene at 260 nm were determined from the solutions made in a particular concentration using a UV spectrometer. The values of the void, as well as the densities of GAC and the materials, were calculated using equations (1), (2), and (3).

$$GAC\ Density = \frac{weight\ of\ GAC}{GAC\ volume} \quad (1)$$

$$Material\ Density = \frac{weight\ of\ dry\ GAC}{GAC\ wet\ volume - \frac{the\ weight\ of\ wet\ GAC - the\ weight\ of\ dry\ GAC}{water\ density}} \quad (2)$$

$$Void\ volume = \frac{picnometric\ volume - volume\ of\ wet\ GAC}{picnometric\ volume} \quad (3)$$

Autosorb - 6B Gas Sorption System (Quantachrome) was used with a sample of 0.5 grams of GAC to determine the GAC surface area, pore diameter, and pore volume. The contents of benzene and toluene were analyzed in the liquid using GC. GC calibration was carried out to determine the area of benzene and toluene. The retention times of CS_2, benzene, and toluene were previously determined. Benzene and toluene concentrations measured in this calibration were 25 ppm, 50 ppm, 100 ppm, and 200 ppm. The data obtained from this calibration were the graphs between the concentrations of benzene and toluene plotted against the area. The column used was Silicon SE-30, with a column temperature of 70℃, injection temperature of 100℃, attenuation of 7, nitrogen pressure of 1.0 kgf/cm^2 with a flow rate of 30 ml/min, hydrogen pressure of 0.6 kgf/cm^2 and air pressure of 0.5 kgf/cm^2.

3. Results and Discussion

3.1 GAC Characterization

The characterization of GAC aims to determine the type of activated carbon used. Activated carbon can be divided into gas-phase adsorbent carbon and liquid-phase adsorbent carbon based on its use. Based on pore size, activated carbon can be divided into micropore, mesopore, and macropore. The results of GAC characterization are shown in Table 1.

Table 1. Characterization of Granular Activated Carbon (GAC)

Properties	Value	
Wet carbon density in bed	1.20*	g/cm^3
Void of the bed	0.48*	--
Activated carbon surface area	1065**	m^2/g
Pore volume	0.314 – 0.488	ml/g
Pore diameter	15 - 23	Å

*Measured in wet conditions
**Measured in dry conditions

The density of GAC in the wet state is 1.2 g/cm^3, indicating that GAC has a heavier density than water. The bed has a void fraction of 0.48, indicating that the GAC in the bed has a reasonably close distance between particles. Density and invalid data are used in the theoretical calculation of the fluidized minimum velocity. The surface area of the analyzed GAC is 1065 m^2/g. The pore diameter analysis showed that the pore diameters of GAC range from 15 to 23. This put the GAC in both micropore (< 20 Å) and mesopore (20 – 500 Å) regions. In comparison, the molecular diameters of benzene and toluene are 5.85 Å and 5.72 Å, respectively, which are significantly smaller than the pore diameter of GAC, making the GAC feasible for use in the adsorption process of benzene and toluene.

The pore volume of the GAC ranged from 0.314 ml/g to 0.488 ml/g. The pore volume of the GAC largely determines the adsorption capacity. The larger the volume of the GAC, the greater the adsorption capacity. Another thing that affects the adsorption capacity is the diameter of the molecule. With a smaller molecule diameter, more molecules are being adsorbed per unit mass. Thus the adsorption capacity is higher.

Table 2. Properties of PTD and BTC Activated Carbon

Properties	PTD	BTC
Surface area (m^2/g)	1065	370
Pore volume (ml/g)	0.314 – 0.488	0.03 – 0.26
Pore diameter (Å)	15 - 23	7 - 15

From Table 2, it can be seen that PTD GAC has a larger surface area, pore volume, and pore diameter. The larger the pore volume, the greater the adsorption capacity of GAC.

3.2 Isothermal Adsorption

The isothermal adsorption experiment was carried out for two days to achieve equilibrium, where the GAC had fully absorbed the pollutants in the sample solutions. The results are the equilibrium concentration values, which were analyzed by a UV spectrophotometer and plotted in Figure 2. Various models can be used to represent isothermal adsorption. The Freundlich and Langmuir models are the simplest and most commonly used adsorption models. In most cases, the Freundlich equation is more suitable for explaining adsorption in the liquid phase.

The Langmuir equation uses specific absorption mechanisms with a finite number of absorption nuclei. The Freundlich equation modifies the Langmuir equation, i.e., if the absorption nucleus is enormous or is assumed to be infinite relative to the number of solute molecules. In high solute concentrations, the concentration of the adsorbed substance will not approach a constant value. The values of the equilibrium concentration of the liquid phase (C_e) are converted to log C_e and plotted against the adsorption capacity of the GAC (log Q_e). Equation (4) shows the Freundlich model equation linearized by taking the logarithms of both sides, and Equations (5) and (6) show how to get the K_f and 1/n from the intercept and slope values from the linear regression.

$$\log Q_e = \log K_f + \frac{1}{n} * \log C_e \quad \textbf{(4)}$$

Thus,

$$K_f = 10^{intercept} \quad \textbf{(5)}$$

And

$$\frac{1}{n} = slope \quad \textbf{(6)}$$

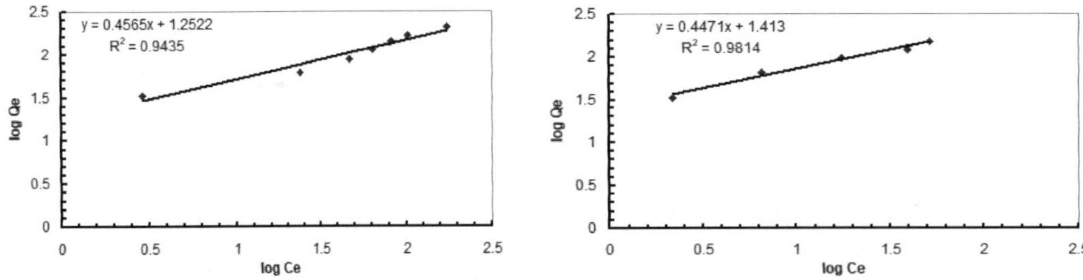

(a) (b)

Figure 2. Adsorption Isotherm for (a) benzene and (b) toluene, with Freundlich model.

The adsorption test results for benzene solution samples with PTD and BTC GAC were 17.87 and 16.59, respectively. The K_f value is almost the same as the K_f in the literature [9]. The value of K_f indicates that the maximum adsorption ability of benzene on GAC is quite good. The value of 1/n for

PTD GAC from the experiment is almost the same as the literature, which is 0.46. However, the value of 1/n for BTC GAC differs from the literature.

The adsorption test results for toluene solution samples with PTD and BTC GAC were 25.88 and 22.59, respectively. The K_f value differs from the K_f in the literature [9]. This difference can be attributed to differences in the pore size of the GAC. The value of K_f indicates that the maximum adsorption ability of toluene on GAC is not good enough. The 1/n value for PTD GAC from the experiment is almost the same as the literature, which is 0.45. Like the benzene adsorption experiment, the value of 1/n for BTC GAC from toluene adsorption differs from the literature.

The measurement of K_f is necessary since adsorption using GAC can cause zonation, which stems from differences in K_f values. Such zonation results in the zoning of contaminant-degrading microbes. By knowing the microbe zones, a special treatment for each species or group of microbes can be carried out. This is very beneficial for further research, namely bio-barrier design.

4. Conclusion

The results showed that the GAC used was distributed in micropore and mesopore regions with pore diameters ranging from 15 to 23 Å, significantly larger than benzene's molecular diameter (5.285 Å) and toluene and (5.72 Å). The experimental results of benzene adsorption with the GAC showed quite good K_f and 1/n values compared to the literature. On the other hand, the experimental results of toluene adsorption with the GAC showed lower K_f and 1/n values compared to the literature. However, the GAC in the experiment was still able to absorb toluene quite well. The results of this research can be continued with experiments or the application of bio-barrier absorption of hydrocarbon substances in a larger bioreactor.

Acknowledgement

This research was supported by The Ministry of Research and Higher Education of Indo- 415 nesia's financial support through the research grant "Hibah Penelitian Dasar Unggulan Perguruan 416 Tinggi" (PDUPT) 2021 (NKB-854/UN2.RST/HKP.05.00/2022). The authors also gratefully acknowl edged the financial support for publication by Universitas Indonesia through the scheme of "Hibah Publikasi Terindeks Internasional" (PUTI) Q1 2022 (NKB-1151/UN2.RST/HKP.05.00/2022).

References

[1] Saksono N, Gozan M, Bismo S, Krisanti E, Widaningrum R and Song S K 2008 Effects of magnetic field on calcium carbonate precipitation: Ionic and particle mechanisms *Korean J. Chem. Eng.* **25** 1145–50

[2] Ramadhan M Y A, Harahap A F P, Hartono H and Gozan M 2020 The Effect of Lighting, Aeration and Stirring on Industrial Synthetic Resin Wastewater Model Treatment by Using Tubular Bioreactor *IOP Conf. Ser. Earth Environ. Sci.* **520**

[3] A Tiehm, M Gozan, A Müller, H Schell, H Lorbeer P W 2002 *Sequential anaerobic/aerobic biodegradation of chlorinated hydrocarbons in activated carbon barriers* (IWA Publishing)

[4] Bustillo-Lecompte C F, Kakar D and Mehrvar M 2018 Photochemical treatment of benzene, toluene, ethylbenzene, and xylenes (BTEX) in aqueous solutions using advanced oxidation processes: Towards a cleaner production in the petroleum refining and petrochemical industries *J. Clean. Prod.* **186** 609–17

[5] Yang S, Zhang Q, Lei Z, Wen W, Huang X and Chen R 2019 Comparing powdered and granular activated carbon addition on membrane fouling control through evaluating the impacts on mixed liquor and cake layer properties in anaerobic membrane bioreactors *Bioresour. Technol.* **294** 122137

[6] Pascale R, Bianco G, Calace S, Masi S, Mancini I M, Mazzone G, and Caniani D 2018 Method development and optimization for the determination of benzene, toluene, ethylbenzene and xylenes in water at trace levels by static headspace extraction coupled to gas chromatography–barrier ionization discharge detection *J. Chromatogr. A* **1548** 10–8

IOP Conf. Series: Earth and Environmental Science 1111 (2022) 012049 doi:10.1088/1755-1315/1111/1/012049

[7] Mohammadi L, Bazrafshan E, Noroozifar M, Ansari-Moghaddam A, Barahuie F and Balarak D 2017 Adsorptive Removal of Benzene and Toluene from Aqueous Environments by Cupric Oxide Nanoparticles: Kinetics and Isotherm Studies *J. Chem.* **2017**
[8] Liu S J, Zhao Z Y, Li J, Wang J and Qi Y 2013 An anaerobic two-layer permeable reactive biobarrier for the remediation of nitrate-contaminated groundwater *Water Res.* **47** 5977–85
[9] H. Sontheimer, John C. Crittenden R S S 1988 *Activated Carbon for Water Treatment* (Universitas Michigan: DVGW-Forschungsstelle, Engler-Bunte-Institut, Universitat Karlsruhe)

Effect of Voltage Unbalances on the Performance of a Three-phase Transformer

Ngo Xuan Cuong[1], Le Anh Tuan[2], Do Nhu Y[3*]
[1] School of Engineering and Technology, Hue University, Thua Thien Hue, 530000, Viet Nam, ORCID ID: 0000-0002-0571-2168
[2] Faculty of Electrical Engineering, Hanoi University of Industry, Hanoi, 100000, Viet Nam
[3] Department of Electrification, Hanoi University of Mining and Geology, Hanoi, 100000, Viet Nam, ORCID ID: 0000-0001-6395-2875
[1]ngoxuancuong@hueuni.edu.vn, [2]leanhtuan0985@gmail.com, [3*]donhuy@humg.edu.vn.

Abstract. Mitigation and adaptation to climate change have been major driving forces in modernizing electric power infrastructure. Low power quality has a significant impact on electrical equipment efficiency. Voltage unbalances one of the issues with low power quality. This study used a simulation method for a power supply with a voltage unbalance to evaluate the effect of voltage unbalance on the performance of 3-phase transformers. The research results show that when voltage unbalance occurs, theworking point with the maximum efficiency shifts and transformer efficiency is reduced in all load cases. When there is an increase in VUF, the efficiency of the transformer is reduced.With a small VUF, the drop is insignificant, but when the VUF is greater than 3%, the drop starts to increase. In the case of voltage unbalance, when the positive sequence voltage is greater than the phase voltage, both the no-load loss and the load loss increase significantly. Both these losses are reduced when the positive sequence voltage is less than the phase voltage. Due to phase shift, both these losses are almost unchanged. The research results are the starting point for general studies on voltage unbalance and power quality.

1. Introduction

In recent years, climate change has often come to the fore in all research fields, especially in environmental policy. Unreasonable use of electricity and improper operation of electrical equipment also partly cause adverse impacts on the environment and give rise to global warming. Energy efficiency simply means using less energy to perform the same task – eliminating energy waste. Power quality is a term used to describe electrical energy efficiency in electric energy. This means that if the equipment is operating correctly due to the power it receives, thepower quality is good. Unwanted situations can arise in the electrical network. Electrically powered equipment may malfunction, fail prematurely, or shut down unexpectedly when disturbances occur. Poor power quality comes at a cost. Power outages can cost money, but damaged assets can cost even more, including potential losses due to downtime in a manufacturing process. Damaged assets caused by power quality events that caused increased heat will almost certainly reduce equipment life. Common power quality issues are classified into voltage unbalanced and harmonic distortion [1].

Voltage unbalance is defined as the deviation of individual phase voltage magnitudes from the rated values, with the individual phase voltages not being equal in magnitude or phase displacement. Inequality among phase voltages can be attributed to unsymmetrical transmission and distribution line impedances, unbalanced or unstable power utilities, unbalanced three-phase loads, open delta transformer connections, blown fuses of three-phase capacitor banks, uneven distribution of single-phase loads across the three phases, single-phase traction loads, weak rural power electric systems with long transmission lines, or even unidentifiable causes [2].

Voltage can impact customer equipment; for example, three-phase induction motors can experience temperature rises due to currents in the windings caused by negative sequence voltage. Reduced torque and lower full-load speeds can also result from voltage unbalance, potentially preventing the motor from performing its function adequately. Other effects include increased noise and vibration [3].

A voltage unbalance can also have an impact on network and equipment performance. It can cause an increased voltage drop on conductors, necessitating the use of more prominent conductors. It can result in further losses on each phase of transformers, increasing hot spot temperature. Unbalance can cause the generation of triplen harmonics to increase, potentially overloading harmonic filters and capacitor banks [4].

Voltage can be caused by the load [5,6]or the supply [7]. A study [5] investigated the effect of three-phase transformer connection type on voltage, and current unbalance propagation in power networks. Analytical results show that wye/wye and delta/wye (with neutral wire at secondary) connections have the most excellent voltage and current unbalance propagation in an unbalanced load. Furthermore, it has been determined that increasing the no-load secondary line-to-line voltage in step-up transformers increases the VUF value at first. It then decreases, whereas in step-down transformers, increasing the no-load secondary line-to-line voltage decreases the VUFvalue monotonously [5].

A study of three-phase unbalanced voltage improvement on the secondary side of a distribution transformer was conducted in the study [6]. The proposed method involves compensating sources that introduce three different single-phase currents into the connection point of the grid. For the delta-delta connected transformer, the worst unbalanced voltages occurred at phase loads of 95%, 65%, and 80% in phases a, b, and c, respectively. This resulted in a voltage unbalance factor of 10.56% and 0.34% for zero-sequence and negative-sequence, respectively. On the contrary, both voltages' unbalance factors were zero if the transformer supplied the balanced load. The zero-sequence voltage is theessential factor. The simulation results also show that unbalanced voltage, even when the load isthree-phase balanced, is a common problem and complex in practice [6].

The study [7] presents a study of the operation of three-phase distribution transformers under unstable voltage conditions. This paper presents a detailed numerical study of a distribution transformer with a primary delta connection and a secondary star connection with a solidly grounded neutral. Three-phase distribution transformer's operating characteristics will be inferior to those of a balanced case, but most customers are unaware that unbalance can result in a higher power factor. Voltage imbalance can reduce transformer efficiency and increase transformer losses [7].

The study presents a simulation for comparison of some transformer connections commonly used in railways to supply AC traction loads for voltage unbalance mitigation [8]. Voltage unbalance is caused primarily by unbalanced currents at points of standard coupling drawn by unevenly distributed loads. The power system components will suffer from overheating, additional losses of lines and transformers, interference with communication systems, and other adverse effects resulting from a significant amount of negative sequence current injected into the system.

The above analysis has shown that additional studies on the effect of voltage unbalance on 3-phase distribution transformers are needed. This paper presents the effects of voltage unbalance on the efficiency and power loss of a 3-phase distribution transformer. The study also suggests various phase difference cases for the research system.

Section 2 of the paper builds a research model. Section 3 presents the analysis of simulation results of a 3-phase transformer operating under various unbalanced conditions. Finally, conclusions and future research are presented in Section 4.

2. Methods

The study simulated a 3-phase transformer working with a phase unbalance source. Standard voltage unbalance cases are described and used as case studies to evaluate the effects of voltage unbalance on the performance of a 3-phase distribution transformer.

2.1. Simulation model for a three-phase transformer

The proposed effects are studied using the "Three-Phase Transformer (Two Windings)" model on MATLAB/ Simulink software. Three single-phase transformers are coupled in the Three-Phase Transformer (Two Windings) model. The paper uses the manufacturer's parameters to build simulation parameters for a three-phase transformer based on a linear transformer combined with hysteresis characteristics on MATLAB/Simulink software. A typical transformer commonly used is the 630 kVA three-phase transformer. It is used in the experiment in this study with its initial parameters from manual [9,10] given in Table 1.

Table 1. Parameters of three-phase transformer 630 kVA from manual.

Parameters	Symbol	Value	Unit
Rated power	P	630	kVA
HV No-load voltage	V_{NL1}	6000	V
LV No-load voltage	V_{NL2}	1200	V
HV Rated current	I_{1n}	60.6	A
LV Rated current	I_{2n}	304.3	A
Short circuit voltage,	$V_{sc\%}$	3.5	%
No-load current	$I_{NL}\%$	3	%
No-load	P_{NL}	2800	W
Short circuit losses	P_{sc}	4700	W

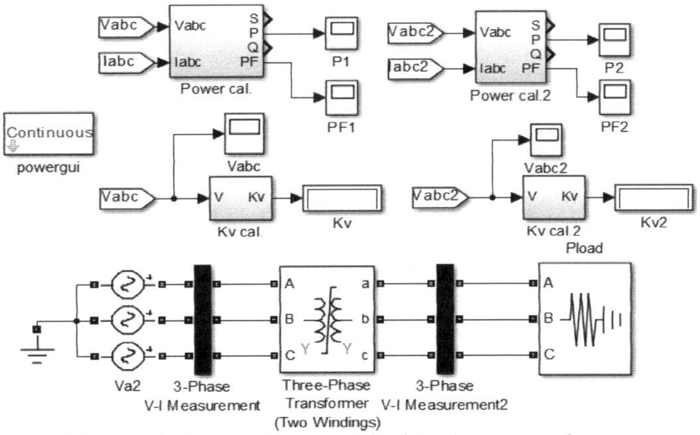

Figure 1. Simulation model of 3-phase transformer.

The transformer uses a Y/Y connection. In the model, the resistances are constant, which means that the skin effect is ignored. The simulation model of the 3-phase transformer on Matlab-Simulink is shown in figure 1.

The losses in the transformer include load and no-load losses and are calculated by:

$$P_T = P_1 - P_2 - P_L + P_{NL} \tag{1}$$

P_1 - 3-phase high-voltage input power, P_2 - low voltage 3-phase power.

The no-load loss is independent of the load and is caused by the induced voltage in the core. It consists of two components: hysteresis loss (P_H) and eddies current loss (P_{FE}). Furthermore, this loss is calculated by the following formula:

$$NL = P_{Fe} + P_H \tag{2}$$

Eddy's current loss in the proposed model is calculated by:

$$P_{Fe} = 3R_m I_m^2 \tag{3}$$

Where Rm - magnetization resistance, ohm; Im - true RMS value of magnetizing current, A. The load loss in the proposed model is the copper loss, which is proportional to the square of thetrue root mean square (RMS) value of the load current increased by the current harmonic components.

$$P_L = 3R_1 I_{1.r}^2 + 3R_2 I_{2,r}^2 \tag{4}$$

Where $R1$ - high voltage phase resistor; $R2$ - low voltage phase resistance; $I1,r$ - true RMS value of highvoltage load current; $I2,r$ - true RMS value of low voltage load current.
Power Factor (*PF*) is calculated according to the following relationship:

$$PF = \frac{P}{\sqrt{P^2+Q^2}} \tag{5}$$

where P - the active input power, W; Q - the reactive input power, Var.
The efficiency of a transformer is the ratio of the 3-phase low-voltage output power P_2 to the 3-phase high-voltage input power P_1:

$$eff = \frac{P_2}{P_1} 100 \tag{6}$$

2.2. Voltage unbalance
The method of measuring the unbalanced voltage level in most practical studies is either the percent voltage unbalance defined by the National Electrical Manufactures Association or the voltage unbalance factor (VUF) defined by the International Electrotechnical Commission (IEC). Both are positive real quantities that can indicate the degree of voltage unbalance. The percent voltage unbalances a method for calculating the ratio of the maximum deviation of the average voltage to the average of the three voltages using sampled data. The VUF is the negative sequence voltagedivided by the positive sequence voltage [11].

In this paper, the IEC or the "true" definition is used to define the degree of voltage unbalance, as the VUF definition is adequate in measuring asymmetry. A voltage unbalance factor is the ratio of the negative-sequence voltage component (V_{neg}) to the positive-sequence voltage component (V_{pos}). This definition is consistent with IEC standard 61000-4-30 [12]:

$$K_V = \frac{V_{neg}}{V_{pos}} 100 \tag{7}$$

V_{neg} is the negative-sequence voltage component; V_{pos} is the positive-sequence voltage component; K_V is the voltage unbalance factor, %.

A supply system may have various unbalances, and voltage variations above and below therated value are possible. Voltage unbalance can thus be divided into two types: overvoltage unbalance (OVU) and Undervoltage unbalance (UVU). OVU is a condition in which the three-phase voltages are not equal, and the positive sequence component is greater than the rated value. In contrast, UVU is a condition in which the three-phase voltages are not equal, and the positive sequence component is less than the rated value [7].

We analyzed the effects of the unbalance regime on VUF from 0% to 6%. Because the magnitude of the positive-sequence component affects the motor behaviour differently for the same VUF, several cases can occur for the exact value of VUF. We considered four different conditions for each value of

VUF. The phase voltage unbalance cases considered for system simulation are presented in Table 2. Table 2 presents the modules and displacement angles of the three phase-to-neutral voltages.

Table 2. Cases of phase voltage unbalance are considered for system simulation.

Case	Unbalance conditions	K_V (%)	V_{a}, V	V_{b}, V	V_{c}, V	V_{pos}, V	V_{neg}, V
1	Different amplitudes, equal phase shifts, and positive-sequence voltages are less than phase voltages	0	4900∠0	4900∠-120°	4900∠120°	4900	0
		1	4754.4∠0°	4900∠-120°	4900∠120°	4851	48.53
		2	4611.8∠0°	4900∠-120°	4900∠120°	4804	96.07
		3	4471.8∠0°	4900∠-120°	4900∠120°	4757	142.7
		4	4334.6∠0°	4900∠-120°	4900∠120°	4712	188.5
		5	4200∠0°	4900∠-120°	4900∠120°	4667	233.3
		6	4192.4∠0°	4050∠-120°	4900∠120°	4381	262.8
2	Different amplitudes, equal phase shifts, and positive-sequence voltages are greater than phase voltages	0	4900∠0	4900∠-120°	4900∠120°	4900	0
		1	5048.5∠0°	4900∠-120°	4900∠120°	4949	49.5
		2	5200∠0°	4900∠-120°	4900∠120°	5000	100
		3	5354.6∠0°	4900∠-120°	4900∠120°	5052	151.5
		4	5512.5∠0°	4900∠-120°	4900∠120°	5104	204.2
		5	5673.7∠0°	4900∠-120°	4900∠120°	5158	257.9
		6	5777.8∠0°	5150∠-120°	4800∠120°	5209	312.5
3	Equal amplitudes, different phase shift, and positive-sequence voltage equals phase voltage	0	4900∠0	4900∠-120°	4900∠120°	4900	0
		1	4900∠1.719°	4900∠-120°	4900∠120°	4900	49
		2	4900∠3.437°	4900∠-120°	4900∠120°	4898	97.96
		3	4900∠5.154°	4900∠-120°	4900∠120°	4896	146.9
		4	4900∠6.869°	4900∠-120°	4900∠120°	4892	195.7
		5	4900∠8.581°	4900∠-120°	4900∠120°	4888	244.4
		6	4900∠10.29°	4900∠-120°	4900∠120°	4882	292.9

3. Results and Discussion

In this section, simulation results are presented and discussed. In order to evaluate the overall efficiency of the transformer, we analyze the effects of voltage deviation on the efficiency in different load cases from 60% to 100% of the rated value. The results are shown in Figure 2, in which the cases are balanced voltage cases (Normal), case 1 with VUF of 5% (Case 1), and case 2 with VUF of 5% (Case 2).

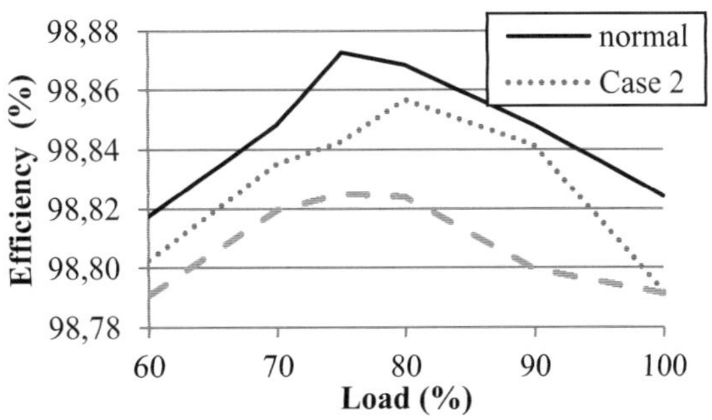

Figure 2. Dependence of transformer efficiency on load in case of voltage unbalance.

It is clear that, in the usual case, the transformer efficiency peaks when the load reaches 75%. However, when there is a voltage unbalance, the peak point shifts. In the case of an increase in the positive-sequence voltage, the efficiency reaches its maximum when the load reaches 80%. In all load cases, when voltage unbalances occur, transformer efficiency is degraded.

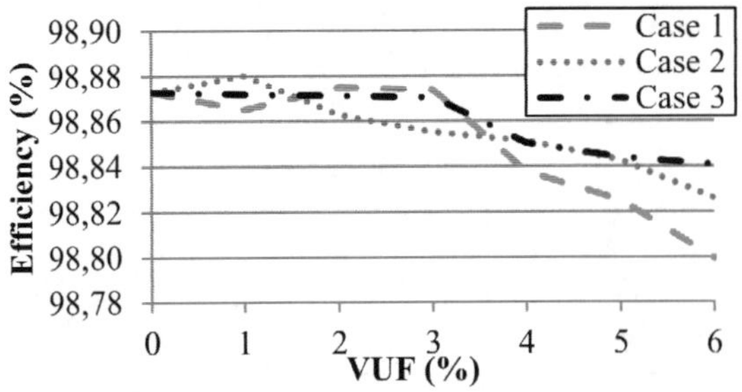

Figure 3. Dependence of transformer efficiency on VUF.

For a more detailed analysis of the impact of VUF on performance, the study was conducted on a transformer with a load of 75%, and 3 cases of voltage unbalance are presented in Table 2. Figure 3 shows the dependence of transformer efficiency on VUF. In general, when there is an increase in VUF, the efficiency of the transformer is reduced. With a small VUF, the decrease is insignificant, but when the VUF is greater than 3%, the decrease begins to increase. In the case study, the maximum efficiency drop was 0.08% with case 1, when positive-sequence voltages are less than phase voltages.

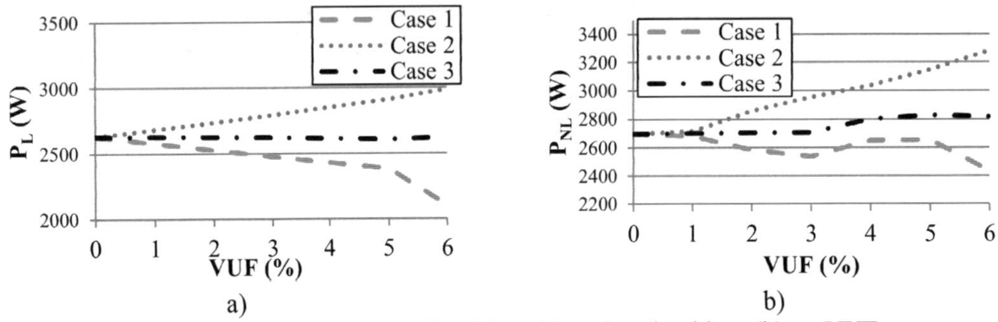

Figure 4. Dependence of load loss (a) and no-load loss (b) on VUF.

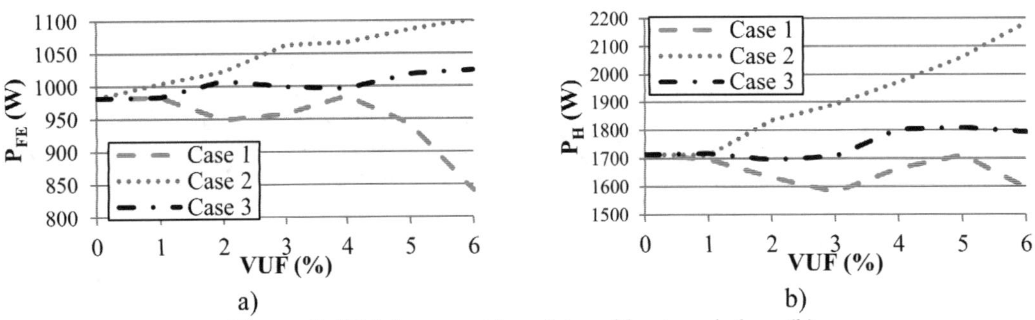

Figure 5. Eddy's current loss (a) and hysteresis loss (b).

Assessment of transformer losses due to voltage can be done through no-load and load losses. In Figure 4, the load and no-load losses in VUF are shown. It is clear that in the case ofvoltage unbalance, when the positive sequence voltage increases above the phase voltage (Case 2), both the no-load loss and the load loss increase significantly. In the case of phase unbalance, when the positive sequence voltage decreases more than the phase voltage (Case 1), both these losses decrease. Both these losses are almost unchanged in the case of voltage phase difference when the positive sequence voltage does not increase (Case 3).

The no-load loss consists of two components: hysteresis loss (P_H) and eddy current loss (P_{FE}). Figure 5 shows hysteresis and eddy current loss according to VUF in different voltage unbalance cases.

Obviously, in the case of voltage unbalance, when the positive sequence voltage increases more than the phase voltage (Case 2), both hysteresis loss and eddy current loss increasewith the increase of VUF, in which hysteresis loss increases most significantly. In the case of voltage unbalance, when the positive sequence voltage is lower than the phase voltage (Case 1), both these losses tend to decrease with the increase of VUF.

4. Conclusion

This paper builds a model to study the effect of voltage unbalance on the performance of a 3-phase transformer. When voltage unbalances occur, the operating point with the maximum efficiency of the transformer shifts, and the transformer efficiency is reduced in all load cases. When there is an increase in VUF, the efficiency of the transformer is reduced. With a small VUF, the performancedrop is negligible, but when the VUF is greater than 3%, the drop starts to increase. In the case of voltage unbalance, when the positive sequence voltage is greater than the phase voltage, both the no-load loss and the load loss increase significantly. In the case of voltage unbalance, when the positive sequence voltage is less than the phase voltage, both these losses are reduced. Both these losses are almost unchanged in the case of voltage unbalance due to phase shift. The research results are the starting point for general studies on voltage unbalance and power quality. Improving equipment performance and power quality contributes significantly to energy saving in particular and environmental problems in general. Therefore, further experimental studiesof other types of electrical machines and systems are needed to understand power quality's effects on electrical engineering systems fully.

5. References

[1] Chattopadhyay S, Mitra M and Sengupta S 2011 *Electric power quality* (Dordrecht: Springer)
[2] Kini P G, Bansal R C and Aithal R S 2007 A novel approach toward interpretation and application of voltage unbalance factor *IEEE Trans. Ind. Electron.* **54** 2315–22
[3] Y N D and Cuong N X 2022 Impact of Voltage Unbalance and Harmonics on Induction Motor in Operation Mode *D.C., Vu, N.P., Long, B.T., Puta, H., Sattler, KU. (eds) Advances in Engineering Research and Application ICERA 2021. Lecture Notes in Networks and Systems* (In: Nguyen: (Springer, Cham) 366.)
[4] Von Jouanne A and Banerjee B 2001 Assessment of voltage unbalance *IEEE Trans. Power Deliv.* **16** 782–90
[5] Kolagar A D, Hamedani P and Shoulaie A 2012 The effects of transformer connection type on voltage and current unbalance propagation *2012 3rd Power Electronics and Drive Systems Technology (PEDSTC) (IEEE)* pp 308–14
[6] Chang C, Cheng S and Boyanapalli B 2022 Three-Phase Unbalance Improvement for Distribution Systems Based on the Particle Swarm Current Injection Algorithm *Energies* **15**(9) p 3460
[7] Edan M 2013 Effect of Unbalance Voltage on the Operation and Performance of a Three Phase Distribution Transformers Journal of Babylon University, Engineering Sciences **5** pp 2-5
[8] Firat G, Yang G and Al-Ali H A H 2015 A comparative study of different transformer connections for railway power supply- mitigation of voltage unbalance
[9] Cuong NGO X and Y DO N 2021 Influence of Harmonics on the Working Efficiency of a 6 / 1.2 kV Transformer in a Pit Mine *Inżynieria Mineralna* **2** 149–56
[10] Дзюбан В С, Ширнин И Г, Ванеев Б Н, Гостищев В М 2001 Справочник энергетика угольной шахты *Донецк «Юго-Восток, Лтд»* **2** 447
[11] Zheng G and Xu Y 2010 An intelligent three-phase voltage unbalance measuring instrument based on the ATT7022C *2010 IEEE International Conference on Automation and Logistics (IEEE)* pp 674–8
[12] Santoso S, McGranaghan M F, Dugan R C and Beaty H W 2012 *Electrical power systems quality* (McGraw-Hill Education)

Detection of Sea Level Rise Monitored from Surabaya Tidal Station Data from 2013 to 2021

Dina Fitriana[1], Mufti Petala Patria[2]* and Eko Kusratmoko[3]
[1] Master Program of Marine Science, Faculty of Mathematics and Natural Sciences, Universitas Indonesia
[2] Departemen of Biology, Faculty of Mathematics and Natural Sciences, Universitas Indonesia
[3] Departemen of Geography, Faculty of Mathematics and Natural Sciences, Universitas Indonesia
[1]dina.fitriana01@ui.ac.id1, [2]*mpatria@sci.ui.ac.id, [3]eko.kusratmoko@sci.ui.ac.id

Abstract. Sea level rise is a certainty because of changes in human life that use machines so that it produces pollution that increases global warming which can make the polar ice caps melt so that the volume of sea water increases. The impact of sea level rise includes changes in coastline, tidal flooding and other impacts that can reduce the quality of life of the community. Surabaya is the second big city on the island of Java after Jakarta which has a population of 2,878,314 people based on the results of the 2020 BPS population census. Of course, this automatically makes the city nicknamed the city of heroes an important city in Indonesia. Therefore, the Geospatial Information Agency as one of the agencies providing tidal data has placed one of its observation stations in the city of Surabaya, located in Ujung Village, Semamir District, Surabaya City, at coordinates 07° 12' 0.1" South Latitude 112° 44' 26.1 " East longitude. The results of data processing using the smallest smoothing method resulted in the LAT tidal datum values, namely -0.636 m, MSL 1.131 m and HAT 2.850 m. The tidal type at the Surabaya tidal station is a mixed mixed tide type which tends to double daily, with an average tidal harmonic constant value of O1 = 0.265 m, S2 = 0.230 m, K1 = 0.476 m, and M2 = 0.377 m. The most dominant tidal harmonic constant is M2, and the lowest harmonic constant is M4 while sea level rise in the city of Surabaya based on the analysis of tidal observation data for 9 (nine) years is 0.199 mm/month.

1. Introduction

The sea is a source of life that has a strong influence on human civilization. Many facts show that many of the world's big cities are in coastal areas which are the meeting area between sea and land, one of which is the city of Surabaya. The city of Surabaya is the capital of East Java Province, located between 07° 9' to 07° 21' south latitude and 112° 36' to 112° 54' east longitude. The area is low land with an altitude of 3-6 meters above sea level, except in the south an altitude of 25-50 meters above sea level. The population in the city of Surabaya is 2,878,314 people based on the results of the 2021 population census [1].

The Geospatial Information Agency manages tidal stations throughout Indonesia (Figure 1), one of which is in the city of Surabaya, the location of the station is in Ujung Village, Semamir District, Surabaya City, at coordinates 07° 12' 0.1" South Latitude 112° 44' 26.1" East Longitude. The Surabaya tide station operates 24 hours a week with minute observation data that can be monitored online so that incoming data can be monitored from the office. The Surabaya tide station has 3 sensors that make observations together. Tidal station data Each year the tides are processed to produce tidal datums, namely LAT (Lowest Astronomical Tide), MSL (Mean Sea Level) and HAT (Highest Astronomical Tide). Tidal station observation data has been stored since 2011 and is still operating. This long data is certainly very good for detecting sea level rise because the observations are directly in situ, so that it can calculate the actual condition of sea level rise in the city of Surabaya.

Sea level rise is a change in sea level in a certain area within a certain period. The development of industry up to 5.0 of course also has an impact on global warming, thus making the earth's condition warmer than its original condition, this right is one of the causes of sea level rise, because the increase in earth's temperature makes the polar ice caps melt so that it automatically increases the volume of sea water. [2] reports that the average global surface temperature has increased by 0.3 - 0.60C since the end of the 19th century and until 2100 the earth's temperature is expected to increase by about 1.4 - 5.8^{0}C, since from 1961 to 1993, the world's sea level has increased at an average rate of 1.8 mm/year (1.3 – 2.3 mm/year). Research in 2010 [3] reported that from 1993 to 2003 sea level rise averaged 3.1 mm/year (2.4 – 3.8 mm/year). The global mean sea level observed by satellite altimetry has a trend of 2.9 ± 0.4 mm/year between January 1993 and December 2008. The sea level rise rate in Indonesia is +4.6 mm/year. This positive velocity value indicates that the sea level in Indonesia will continue to rise, so it is predicted that there will be submergence in several areas in Indonesia [4]. The prediction of sea level rise in the Surabaya area is 4.8 mm/year, using tidal data in 2010 from the tidal station of the Tanjung Perak Meteorology Geophysics Agency, Surabaya.

In this study, data was collected from the Surabaya tidal station belonging to the Geospatial Information Agency with an observation duration of 9 (nine) years from 2013 to 2021. the precision of the resulting data. The purpose of this study was to detect sea level rise at the Surabaya tidal station. So it is hoped that the coastal community of Surabaya can prepare themselves for the impact of sea level rise, especially on the occurrence of shoreline changes, and tidal flooding. With this research, it can provide information about the value of sea level rise at the urabaya tidal station and can be used as a reference for research on the same topic, especially in the marine sector.

2. Methods

The stages carried out in this research are starting from data collection, data processing, data analysis, and report writing. The data used in this study is tidal observation data from 2011 to 2021 which was observed from the Surabaya tidal station, one of the stations managed by the Geospatial Information Agency. The data processing method used in this study is the smallest smoothing method which is one of the most widely used tidal data processing methods. After the calculation process is complete, an analysis of the results of data processing is carried out so that it can be presented in this research report.

2.1. Surabaya tidal station

The Surabaya tide station is located at the pier of PT. PAL so that to get to the location a permit is needed from the relevant institution before visiting the Surabaya tidal station. Currently, the Surabaya station is a station that can be accessed remotely from the Geospatial Information Agency office, so that data collection is carried out directly at the Geospatial Information Agency office. The data used in this study is data from 2013 to 2021.

The best data for 2013 uses sensor 2 because the data is better. Although the data is not full because it is empty, the void can be ignored because it can still be processed. The best data for 2014 uses sensor 1 because the data is better. Visible all available data with no broken graphs. Although the data is not full but the void can be ignored because it can still be processed. The best 2015 data uses sensor 1 because the data is better. Visible all available data with no broken graphs. The best 2016 data uses sensor 1 because the data is better. Visible all available data with no broken graphs. The best 2017 data uses sensor 3 because the data is better. Visible all available data with no broken graphs. The best 2018 data uses sensor 3 because the data is better. Visible all available data with no broken graphs. The best 2019 data uses sensor 3 because the data is better. Visible all available data with no broken graphs.

The best 2020 data use sensor 3 because the data is better. Visible all available data with no broken graphs. Lack of data is found in September and December. The best 2021 data use sensor 3 because the data is better. Visible all available data with no broken graphs. There is a lack of data in January, which is between the 5th and 6th of January. Then there's the data shortage in May, and October

2.2. *Tidal data processing*

Tidal data processing using software in accordance with the technical instructions for processing tidal data. The first stage is validation, namely cleaning of data from blunders and systematic data so that validated data is produced. Then the second stage is processing the data with the smallest smoothing method so that the tidal datum and tidal harmonic components are obtained.

2.3. *Harmonic component*

Tidal constants are harmonic with time, so they are called tidal harmonics. Data processing in this study resulted in 9 (nine) tidal harmonic constants as follows:

Table 1. Tidal Harmonic Component

No	Code	Harmonic Constituents
1	M2	The largest lunar constituent
2	K1	Declination Luni-Solar
3	S2	The largest solar constituent
4	O1	Principal Lunar
5	P1	Principal Solar
6	N2	Elliptical Lunar
7	K2	Declination Luni-Solar
8	M4	Twice the angular velocity M2
9	MS4	M2 and S2 Combination

Of the 9 main tidal harmonic components, there are 4 harmonic constants commonly used in determining the tidal type, namely M2, S2, K1, and O1. Tidal type classification is determined based on the comparison between the total amplitude of the single harmonic constants A(K1), A(O1), and the total amplitudes of the double harmonic constants, namely A(M2), A(S2). This comparison is known as "Formzahl" the equation is:

$$F = \frac{A_{K1}+A_{O1}}{A_{M2}+A_{S2}} \quad (1)$$

Description:

F : Formzahl number

A_{K1} : amplitude K1

A_{O1} : amplitude O1

A_{M2} : amplitude M2

A_{S2} : amplitude S2

2.4. *Tidal datum*

A tidal datum is a local datum determined based on a tidal phase. There are 3 (three) calculations for the tidal datum, namely LAT (Lowest Astronomical Tide), MSL (Mean Sea Level) and HAT (Highest Astronomical Tide).

2.5. Analysis of sea level rise
After obtaining the tidal harmonic components, linear regression analysis was carried out to obtain the value of sea level rise at the Surabaya tidal station based on existing research data.

$$Y=a\ X+b \qquad (2)$$

Description:
Y = dependent variable
X = independent variable
a and b = regression parameters

3. Results and Discussion
Location, Surabaya City is located between 07° 9' and 07° 21" South Latitudes and between 112° 36" and 112° 54" East Longitudes. Most of the regions are lowland, which is around 3 - 6 m above the sea level. Except on the south region, the elevation is up to 25 - 50 m above the sea level [1]. The increase in sea level elevation can have a bad impact not only on human life, but also on the ecosystem of coastal areas including the tidal flood that occurs, namely the entry of water into the land during high tides at sea, even though there was no flood in residential areas around the coast, but because there is an increase in sea level the area becomes flooded, in the long term if sea level rise continues it will cause erosion so that it can change the coastline and even change the function of the land, for example areas that were previously residential into water areas due to continuous erosion due to sea level rise. Because coastal areas are very vulnerable to the impacts caused by sea level rise, it is necessary to detect the rise in coastal areas, one of which is the city of Surabaya. The results of the research will be described in detail in the following table, pictures, and descriptions.

3.1. Tidal harmonic constant
Data processing is carried out monthly so that monthly tidal harmonic constants are obtained so that when graphically depicted, tidal harmonic constants are obtained (Figure 1).

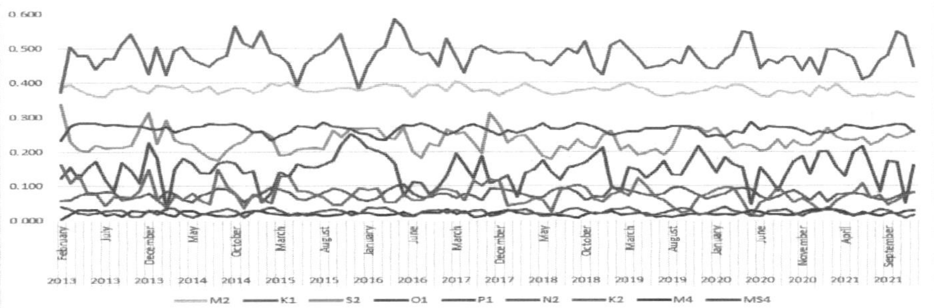

Figure 1. Tidal Harmonic Components of Surabaya Station

Based on the analysis of the results of the Surabaya tidal station data processing, it is found that the most dominant tidal harmonic constant is K1 which is the combined harmonic component of the moon and sun. The component whose value has the least effect at the Surabaya tidal station is M4, which is the lunar component. Formal calculations prove that the tidal type from the Surabaya station is a mixed mixed tidal type, which tends to double daily (Table 2).

Table 2. Tidal Harmonic Components for 9 Years Period

	M2	K1	S2	O1	P1	N2	K2	M4	F
MAX	0.403	0.585	0.337	0.285	0.251	0.104	0.163	0.032	0.039
MIN	0.356	0.371	0.175	0.232	0.032	0.052	0.010	0.005	0.006
AVERAGE	0.377	0.476	0.230	0.265	0.147	0.075	0.073	0.019	0.024

IOP Conf. Series: Earth and Environmental Science 1111 (2022) 012051 doi:10.1088/1755-1315/1111/1/012051

3.2. Tidal datum

The results of the processing of the tidal datum at the Surabaya station are as follows (Figure 2):

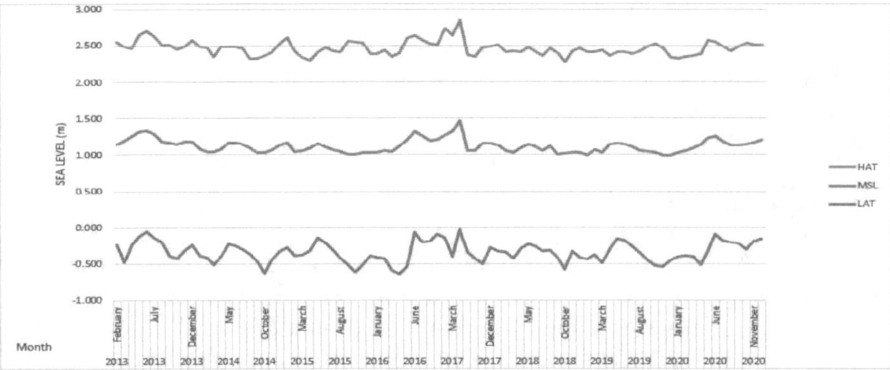

Figure 2. Tidal Datum for Surabaya Station

The tidal datum for Surabaya station is local at Surabaya station, the LAT, MSL, and HAT values are as follows (Table 3):

Table 3. The Tidal Datum of the Surabaya Station

	LAT	MSL	HAT
MAX	-0.029	1.464	2.850
MIN	-0.636	0.998	2.276
AVERAGE	-0.332	1.131	2.478

LAT is the lowest water level during the observation data period, which is -0.636 m. MSL is the average sea level during the observation period which is 1,131 m and the HAT value which is the highest sea level value during the observation is 2,850 m.

3.3. Sea level rise

Changes in sea level are a certainty because from time to time the sea level continues to move up and down to form tides, with linear regression calculations the value of R2 = 1, with the equation y = 0.0002x + 1.1228. Thus the first month or x = 1, then the value of y = 1123 mm, so that the 99th month x = 99 then the value of y = 1142,600 mm. So that the upward trend in MSL from 2013 to 2019 is 0.198 mm, if the MSL trend graph is made as follows in Figure 3.

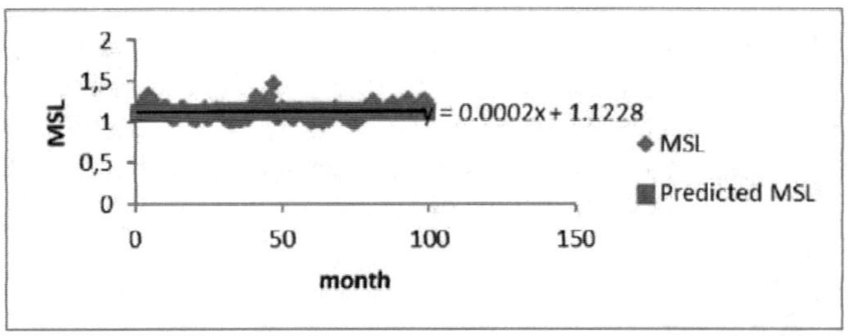

Figure 3. MSL Trend

4. Conclusion

Detection of sea level rise using tidal data at Surabaya station produces some information, namely the harmonic component that dominates in Surabaya waters is a mixture of moon and sun components, namely M2, so that the type of tide in Surabaya is also mixed with the average formal value for 99 months is 0.024 mixed mixed tidal type tends to double daily. The tidal datum for Surabaya LAT station is -0.636 m, MSL 1.131 m and HAT 2.850 m. Sea level rise at the Surabaya station from 2013 to 2021 was detected at 0.198 mm/month which was calculated for 99 months or equal to 2.376 mm/year.

References

[1] Badan Pusat Statistik 2021 *Surabaya Municipality in Figures 2021* (Surabaya: Badan Pusat Statistik)
[2] Internationall Panel Climate Change 2007 *The Physical Science Basis* (Cambridge: Cambridge University Press)
[3] Fenoglio-Marc L and Tel E 2010 Coastal and global sea level change *J. Geodyn.* **49** 151–60
[4] Handoko E Y, Yuwono and Ariani R 2019 Indonesian Sea Rise Ananlysis 1993-2018 Using Altimetry *Geoid* **15** 1

IOP Conf. Series: Earth and Environmental Science 1111 (2022) 012052 doi:10.1088/1755-1315/1111/1/012052

Detection of Red-Eared Slider (*Trachemys scripta elegans*) using Environmental DNA with Cytochrome b Primer

Shafa Talitha Risti[1], Andi Eko Maryanto[2*], Noviar Andayani[1]
[1]Department of Biology, Faculty of Mathematics and Natural Sciences, Universitas Indonesia, Depok, Indonesia 16424
[2]Department of Biology, Faculty of Mathematics and Natural Sciences, Universitas Indonesia, Depok, Indonesia 16424, 0000-0003-0624-0753
[1]shafa.talitha@ui.ac.id, [2]*andi.maryanto@ui.ac.id, [3]andayani@ui.ac.id

Abstract. Biomonitoring surveys for the presence of invasive species need effective and efficient methods. Molecular methods based on the detection of genetic material in environmental samples or environmental DNA (eDNA) provide better resolution for detection, especially in elusive species. The eDNA-based method is an approach that has been widely used for conservation and business management because it is a non-invasive method. Cytochrome b primer is a specific primer that can be used to amplify Red-Eared Slider eDNA from water samples, eDNA was amplified using qPCR. The results obtained after performing qPCR are in the form of Ct values that can be used to estimate the biomass of Red-Eared Slider.

1. Introduction

Environmental DNA (eDNA) is a method for biomonitoring, such as surveying animal or plant communities in an area and detecting the presence of invasive species [1]. An eDNA-based method is an approach that has been widely used for conservation and management because it is a non-invasive method, where the required DNA samples are taken from the environment in the form of cell lysis or excretion from living organisms [2]. eDNA-based biomonitoring differs from conventional biomonitoring, which requires more effort for sampling and is also invasive. Conventional biomonitoring methods rarely show all species in the environment, so all specimens must be obtained, and species identified [1]. Detecting the presence of species is also problematic in environments with low population density, so eDNA-based research is needed. In addition to detecting invasive or endangered aquatic species, eDNA-based biomonitoring can also examine species diversity and biomass [1].

The Red-eared slider (*Trachemys scripta elegans*) was used as a source of eDNA. This native species from Southeast America is included in the IUCN 100 of the world's worst invasive list and is considered a significant threat to aquatic flora and fauna [3]. This species is widely traded as pets, and many individuals have been dumped into the wild outside their original place of origin, including in Asian countries [3]. Trade in this species has been reported in many countries. In Indonesia, according to The Asian Turtles Conservation Network, red-eared sliders are widely seen on the islands of Sumatra, Java, Kalimantan, Sulawesi, and Irian [3]

The cytochrome b gene is often used as a target gene because the mutation rate of the Cyt b gene is different at the first codon and the second codon position has a low substitution rate. It does not vary between individuals [4], meaning that the same gene contains regions specific to a particular organism.

Some regions are similar to each type of organism. Thus, it can be used as a genetic marker to identify specific species [5]. This work has aimed to determine the character, ability, and efficiency of CYTB primers in detecting turtle eDNA and estimating their biomass.

2. Methods

2.1. Container Set Up

There were a total of six containers used in this experiment, three smaller containers were used to put one individual in each container, and three bigger containers were used to put three individuals in each container. Each smaller container was filled with 300 mL of water, and the bigger containers were filled with 500 mL. All individuals were kept in the containers for three days prior to sampling.

2.2. Sample Collection

Water samples were taken from each container once every week for four weeks and stored in a plastic bottle disinfected with 70% alcohol to avoid unwanted contamination. In addition, water samples for negative control were taken before the turtles were put into the containers. The plastic bottles that contained water samples were labeled, the bottles which contained water from smaller containers were labeled with K1, K2, and K3, and the bottles which contained water from bigger containers were labeled with KG1, KG2, and KG3.

2.3. Sample Processing

After all the samples were taken, the filtration process began using a filtration device consisting of a container to put the water samples, a container to collect the filtrate, and a hose to connect the filtration machine and the containers. The water and negative samples went through a 0,45 μm pore size filter until it was clogged. Filters were then cut into two pieces then stored in a 1.5 mL tube containing 1 mL Longmire buffer. Once the filtration process finished, eDNA from each filter was extracted using the Phenol: Chloroform: Isoamyl (25:24:1) method and ethanol precipitation method [6] with modification. Modifications were made by adding 2-mercapto and beads to help destroy the filters with the help of a microtube homogenizer. Positive controls were made by extracting tail tissue from Red-Eared Slider using Gen Elute-E Single Spin Tissue DNA kit. After all the samples were ready, the samples were quantified using spectrophotometers to see the DNA concentration and purity of DNA (A260/280). The accepted ratio of A260/280 ranges between 1,800 – 2,000. After all the samples were ready, Quantitative PCR was conducted using the species-specific primers cytochrome b (CYTB) Cytb -Tse-F (5'- CCTCCAACATCTCTGCTTGA-3') and Cytb-Tse-R (5'-ATTGTACGTCTCGGGTGATG-3'). Each plate contained negative control, positive control, dan three replicates for each sample. The qPCR was carried out in a 20 μl reaction containing 10 μl bio line master mix, 0,8 μl of each primer, 7,4 μl Nuclease Free Water, and 1 μl extracted eDNA sample following these conditions: 45 cycles, 95 °C for 5s, 53 °C for 10s, 72 °C for 15 s, and 95 °C for 120 s for an additional denaturation step. The standard curve was made beforehand in the same condition.

2.4. Statistical Analyses

The data was analyzed using the Shapiro-Wilk Goodness of Fit Test to test the normality of the distribution. The relationship between turtle biomass and Ct value was examined using Scatterplots. Deviation from the linearity test confirmed the linear relationship between turtle biomass and Ct value. The tests were performed using IBM SPSS Statistics ver. 23.

3. Results and Discussion

3.1. eDNA Amplification using qPCR and Cytochrome b primer

qPCR was performed to get a standard curve to verify the specificity of the cytochrome b primer for the red-eared slider. To make a standard curve, a five-step serial dilution was conducted on extracted tail tissue DNA with an initial concentration of 16 μl. The results showed that all samples were successfully amplified to the lowest concentration. Based on Figure 1. the highest concentration had a

Ct value of 18.22, and the lowest concentration had a Ct value of 22.89. The amplified negative control showed a Ct value of 37.14. The standard curve showed 82% efficiency with a slope of -3.8324 and an R^2 value of 0.9987.

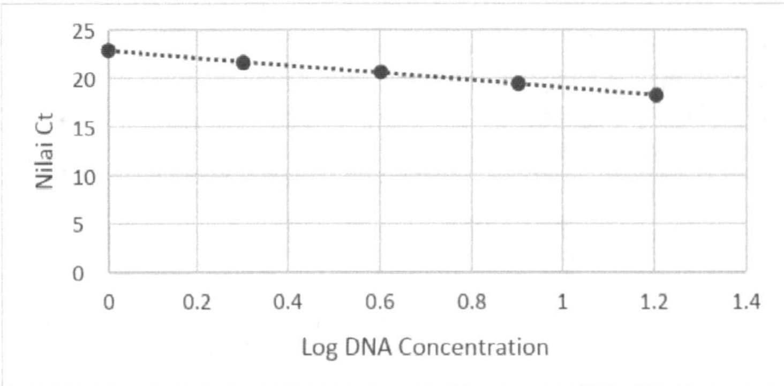

Figure 1. Standard Curve for *Trachemys scripta elegans*

A Ct value below 29 indicates a positive reaction indicating many nucleic acid targets in the sample. At the same time, a Ct value in the range 30-37 also indicates a positive reaction with moderate nucleic acid targets in the sample, and a Ct value in the range 38-40 indicates a weak reaction. The number of nucleic acid targets in the sample is very small and can indicate the presence of contamination [7].

The results obtained after performing qPCR using cytochrome B primers on water samples showed that all samples were successfully amplified. Based on Table 1. the container contained one individual from weeks 1 to 4 (K1, K2, K3) had an average Ct value of 24.24. In contrast, a container containing three individuals from weeks 1 to 4 (KG1, KG2, KG3) had an average Ct value of 22.40.

Table 1. Ct Values for Water Samples

	Samples	Average Ct Value (3 Replicates)
Week 1	K1	28.35384488
	K2	27.23093838
	K3	23.90026444
	KG1	25.71647741
	KG2	23.8962325
	KG3	25.28255649
Week 2	K1	21.61917987
	K2	23.53166667
	K3	26.59266667
	KG1	18.23736482
	KG2	22.78233333
	KG3	24.67566667
Week 3	K1	22.15133333
	K2	20.58966667
	K3	23.443
	KG1	20.48033333
	KG2	24.108
	KG3	20.99333333
Week 4	K1	23.021

K2	19.38733333
K3	31.12066667
KG1	21.77166667
KG2	21.67066667
KG3	19.21566667

3.2. eDNA for Estimating Biomass of Red-eared slider

eDNA was obtained from water samples in containers containing one and three individuals, and containers containing one and three had different biomass. The biomass of each container can be seen in Table 2.

Table 2. Ct Values for Water Samples

	Samples	Biomass (g)
Week 1	K1	10
	K2	5
	K3	5
	KG1	20
	KG2	20
	KG3	20
Week 2	K1	7
	K2	8
	K3	12
	KG1	33
	KG2	23
	KG3	22
Week 3	K1	13
	K2	8
	K3	7,2
	KG1	24
	KG2	25
	KG3	29
Week 4	K1	8,3
	K2	8,6
	K3	8,2
	KG1	28,3
	KG2	33,4
	KG3	24,6

After qPCR was performed, the Ct values were obtained, which can be seen in Table 1. then to see the relationship between the Ct value obtained and the biomass, a Scatter-Plot Linearity test was carried out, which previously had a normality test that stated that the sample was usually distributed. After that, the deviation from linearity significance test was carried out based on decision making if the value of sig. If a deviation from linearity is more significant than 0.05 (Sig. > 0.05), then a linear relationship exists between variables.

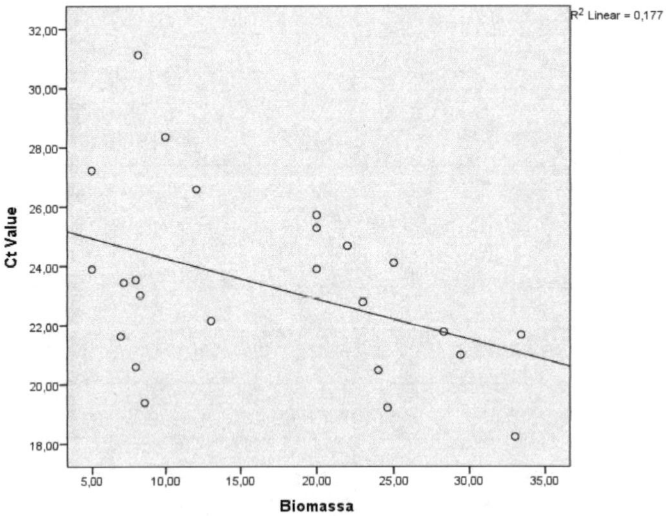

Figure 2. Relationship between Ct Value and Red-eared slider Biomass

Based on Figure 2, the results showed negative linearity; the more significant the biomass, the smaller the Ct value. Sig. test results Deviation from linearity shows a value of 0.139, ensuring a linear relationship. The value of Ct represents the polymerization cycle where the fluorescence signal crosses the threshold. The higher the target concentration, the faster the signal passes through the threshold, thus giving a lower Ct value. If the DNA concentration decreases, the Ct value will also increase [8]. Based on the statistical analysis results, it can be concluded that larger biomass tends to produce smaller Ct values, and small Ct values indicate higher target DNA concentrations [8]. This result is similar to a study conducted by [9], where eDNA copies were higher in an aquarium containing three individuals compared to an aquarium containing one individual. There was a significant positive correlation between eDNA copies and biomass. Therefore, eDNA can be used to estimate biomass [9].

4. Conclusion

eDNA method for biomonitoring could replace conventional biomonitoring. Cytochrome b primer is a practical primer for amplifying *Trachemys scripta elegans* eDNA in water samples. The eDNA method can be used to estimate the biomass of the target organism.

Acknowledgments

This study was financially supported by the Faculty of Mathematics and Natural Sciences, Department of Biology, Universitas Indonesia.

References

[1] Valdez-Moreno M, Ivanova N V., Elías-Gutiérrez M, Pedersen S L, Bessonov K and Hebert P D N 2019 Using eDNA to Biomonitor the Fish Community in a Tropical Oligotrophic Lake *PLoS One* **14** 1–22

[2] Dejean T, Valentini A, Duparc A, Pellier-Cuit S, Pompanon F, Taberlet P and Miaud C 2011 Persistence of Environmental DNA in Freshwater Ecosystems *PLoS One* **6** 1–4

[3] Ramsay N F, O'Riordan R M and Chou L M 2007 The Red-Eared Slider (Trachemys Scripta Elegans) in Asia: A Review *Biol. invaders Inl. waters Profiles, Distrib. Threat.* 161–74

[4] Singkam A ., Solihin D . and Affandi R 2018 Keragaman Morfometrik dan Gen Cytochrome B DNA Mitokondria Ompok Hypopthalmus di DAS Batang Hari. *Biospecies* **11** 89–97

[5] Kurniah A N, Puspitasari R L and Perdana A T 2020 Detection of Cytochrome B Pork in Packaged Meatball from Traditional Markets Around Jakarta *Artik. Pemakalah Paralel* 685–9

IOP Conf. Series: Earth and Environmental Science 1111 (2022) 012052 doi:10.1088/1755-1315/1111/1/012052

[6] Renshaw M A, Olds B P, Jerde C L, Mcveigh M M and Lodge D M 2015 The Room Temperature Preservation of Filtered Environmental DNA Samples and Assimilation into A Phenol-Chloroform-Isoamyl Alcohol DNA Extraction *Mol. Ecol. Resour.* **15** 168–76
[7] Mishra B, Ranjan J, Purushotham P, Saha S, Payal P, Kar P, Das S and Deshmukh V 2022 High Proportion of Low Cycle Threshold Value as an Early Indicator of Covid-19 Surge *J. Med. Virol.* **94** 240–5
[8] Troutman M 2015 ThermoFisher Scientific
[9] Takahara T, Minamoto T, Yamanaka H, Dol H and Kawabata Z . 2012 Estimation of Fish Biomass Using Environmental DNA *PLoS One* **7** 1–8

Evaluation of Slope Stability at Interface using Thin Soil Material Model in Finite Element Software

Fathiyah Hakim Sagitaningrum[1]*, Samira Albati Kamaruddin[2], Ramli Nazir[3], Budi Susilo Soepandji[4] and Idrus Muhammad Alatas[5]
[1] Ph.D. Candidate, Razak Faculty of Technology and Informatics, Universiti Teknologi Malaysia, Kuala Lumpur, 54100, Malaysia
[2] Razak Faculty of Technology and Informatics, Universiti Teknologi Malaysia, Kuala Lumpur, 54100, Malaysia
[3] Centre of Tropical Engineering, Universiti Teknologi Malaysia, Johor Bahru, 81310, Malaysia
[4] Civil Engineering Department, Faculty of Engineering, Universitas Indonesia, Depok, 16424, Indonesia
[5] National Science and Technology Institute, Jakarta, 12630, Indonesia
[1]*fathiyah.hakim@gmail.com, [2]samira.kl@utm.my, [3]ramlinazir@utm.my, [4]budisus@eng.ui.ac.id, [5]hb_idrus@yahoo.com

Abstract. Cases of landslides on clay shale slopes have been an intriguing study in Indonesia. Most of the slopes failed due to the weathering of the clay shale rock. As studies focused on the properties of clay shale, several FEM analyses indicated a unique translational type of failure between the interface of clay shale and its overburden. However, the modeling of the interface in a case study was still limited. This study aims to evaluate a landslide using a thin-soil interface model using FEM software. The model used two conditions, extreme groundwater level conditions. There were several iterations of the shear strength parameters of the interface layer conducted. Results indicated that the thin-soil interface could portray the interface condition at the landslide. Compared to laboratory interface results, the interface would fail at a high degree of saturation due to water infiltration at the surface. Also, granular overburden with high permeability had a higher chance of failure than cohesive overburden. From the results, this study opened the path to incorporating the interface layer in clay shale slope analysis. However, further studies are needed using the thin-soil interface model in other clay shale landslide cases to synthesize its repeatability.

1. Introduction

Landslides have been one of the most common causes of geotechnical engineering failures on natural and manufactured slopes. In Indonesia, landslides were affected by various factors such as soil type, rainfall intensity, and geological terrain. One of the problematic soil types which initiated landslides is the clay shale soil.

Many studies of clay shale in Indonesia focused on its properties, including its unique weathering property. Due to weathering, the initial properties of clay shale would change through the degradation of the structure [1]. The degradation of clay shale would further degrade its shear strength, which led to slope instability [2,3]. Various clay shale slope analyses indicated a typical translational failure plane. However, these analyses only incorporated the degraded shear strength parameters to the clay shale layers [4–6]. From these analyses, a landslide in Semarang during the toll road construction indicated a unique behavior. The study argued that the translational failure might be due to the interface failure between the clay shale rock and its overburden [6].

From the previous issue, it was clear that studies of soil-soil interfaces are needed. The recent trend of the interface shear strength also indicated that the research had shifted from the soil-solid interface to soil-soil and soil-rock interfaces [7]. Other than experimental studies, the studies of interface modeling were also developing. Most of the models used the Discrete Element Method (DEM) and Finite Element Method (FEM) to model the interface behavior in its element size [8–10]. However, recent back-analyses of clay shale slope demanded that the interface be included in a full-scale model analysis [4–6]. From the problems, the developing interface modeling could not answer the needs of landslide back-analyses. Thus, it is essential to explore the possibility of including the interface effect in landslide back-analyses.

One of the efforts is introducing the thin-soil material as the interface model. A previous study explored the thin-soil material using the FEM software using back-analysis of a clay shale slope [10]. However, the study was limited to the natural water table, which could not give a wide range of interface conditions at failure. Thus, this study would like to evaluate the thin-soil material model for different water table conditions. At the end of this study, interface shear strength laboratory results would be plotted simultaneously with the FEM results. Thus, an analysis of interface properties that led to slope instability can be deducted.

2. Methods

Previous research on the interface model focused on the element-size interface behavior [8–10]. Some of these models used ABAQUS and FLAC3D, FEM analysis software commonly used on the element-size model. On the contrary, the most common FEM software used for the back-analysis of a landslide is Plaxis. In Plaxis itself, the interface analysis could use the provided interface function taken from the properties of the surrounding soil. However, a previous Semarang-Bawen toll road slope analysis showed that the interface function had several limitations [2]. In this study, the thin-soil material model was used to replace the interface function of Plaxis.

For this study, Plaxis version 8.6 was used for the thin-soil interface model iteration since it is commercially available and still widely used in several practices. The back-analysis used the Semarang-Bawen toll road slope failure in 2015, which had already been used in previous interface slope stability research [2,10]. The slope was taken as the case study due to its known interface failure possibility from the landslide report [2].

The thin-soil layer was limited to one meter tall, which is very small compared to the overall slope geometry. The model used for the soils was the Mohr-Coulomb model. Mohr-Coulomb was selected due to the limited data available on the slope, which only provided the boring log, cone penetration test, index properties, and triaxial tests [2]. Thus, using other models in the Plaxis software would lead to inaccurate results.

Another variation to assure that the range of interface shear strength accommodated any critical condition is the groundwater level. Two groundwater level combinations were used in this study: low-groundwater-level and high-groundwater-level. In the model, the thin-soil interface layer was kept saturated. The condition was necessary to ensure that the model could depict the slope situation at failure and the interface laboratory testing assumption.

For the Plaxis 8.6 analysis steps, two analyses were used in this study. The analyses were the total multipliers step and the Phi-C reduction analysis. The total multipliers analysis was used to calculate the weight of the entire slope due to its gravity. Afterward, the Phi-C reduction analysis was used to calculate the safety factor. The limit of the factor of safety ranged from 1.00 to 1.10. The translational failure plane deducted from the landslide analysis was also used as a control for the model. If any results did not comply with the range of a safety and failure plane factor, the iteration of the shear strength parameters should be repeated for another variation. Overall, the flowchart of the thin-soil interface model iteration in this study can be seen in Figure 1. Around five iterations were used in this study to ensure a linear trend for each groundwater level variation.

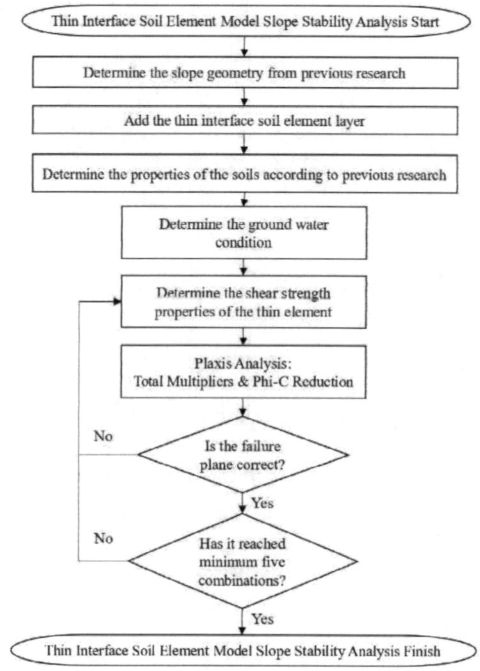

Figure 1. Thin-soil interface model analysis flowchart

Lastly, the laboratory test results using the Progressive Reversal Interface Direct Shear Test (PRIDST) were plotted into the interface model results. The interface shear strength experiment was conducted using undisturbed clay shale and laterite-sand overburden variations according to the tuff breccia on the Semarang-Bawen toll road slope failure [1]. The variations of laterite-sand overburden included: 100% sand: 75% sand: 25% laterite, 50% sand: 50% laterite, 25% sand: 75% laterite, and 100% laterite. A range of interface cohesion and friction angle from the laboratory results would be able to be analyzed from the plotting.

3. Results and Discussion

3.1. Interface Shear Strength Parameters Results at the Thin-Soil Interface Material

As stated in the previous section, the thin-interface model used previous research geometry of the Semarang-Bawen toll road landslide on KM 483+200 in Figure 1 [2]. The soil properties input used the Mohr-coulomb material type and the clay shale interface soil layer (as seen in Table 1). However, the interface shear strength properties were determined using the iteration in this study.

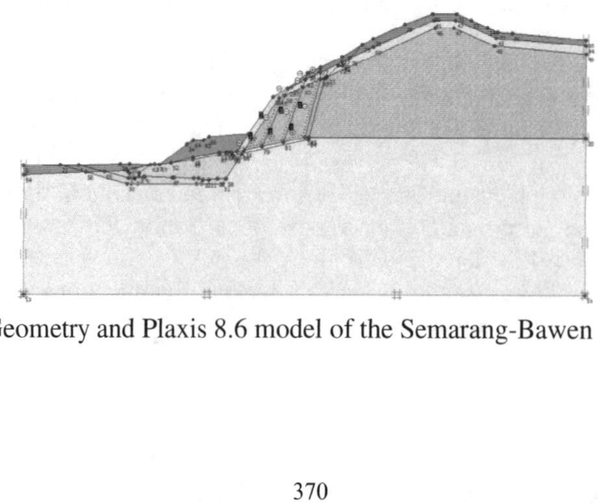

Figure 2. Geometry and Plaxis 8.6 model of the Semarang-Bawen toll road slope

Table 1. Soil properties input for Plaxis 8.6

Soil Properties	Tuff Breccia	Weathered Tuff Breccia	Silty Clay	Fresh Clay Shale	Interface Clay Shale
Color	Purple	Slight Green	Dark Green	Slight Brown, Dark Yellow	Light Yellow
Unsaturated Unit Weight (kN/m³)	17	17	15	18	18
Saturated Unit Weight (kN/m³)	18	18	17	18	18
Elastic Modulus (kN/m²)	150000	60000	10000	100000	100000
Poisson Ratio	0.25	0.25	0.30	0.30	0.30
Shear Modulus (kN/m²)	60000	24000	3846	38460	38460
Cohesion (kN/m²)	100	50	15	250	
Friction Angle (°)	38	30	30	41	

From the geometry of the slope, two groundwater level conditions were determined using the phreatic line determination. The results for the low-groundwater-level (LWL) and high-groundwater-level (HWL) input into the software can be seen in Figures 3 and Figure 4.

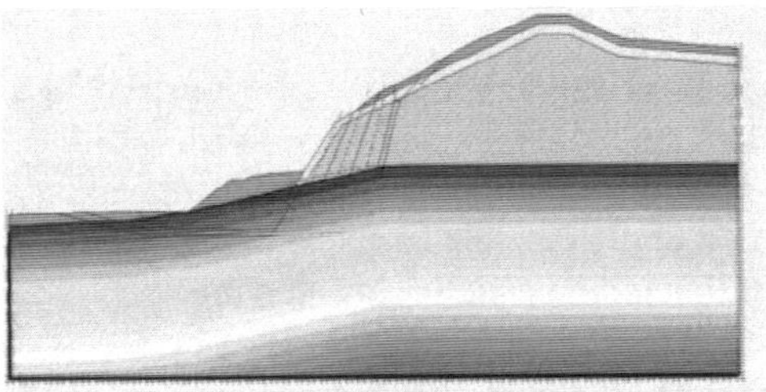

Figure 3. Low-Groundwater-Level (LWL) model for Plaxis 8.6 thin-soil interface model

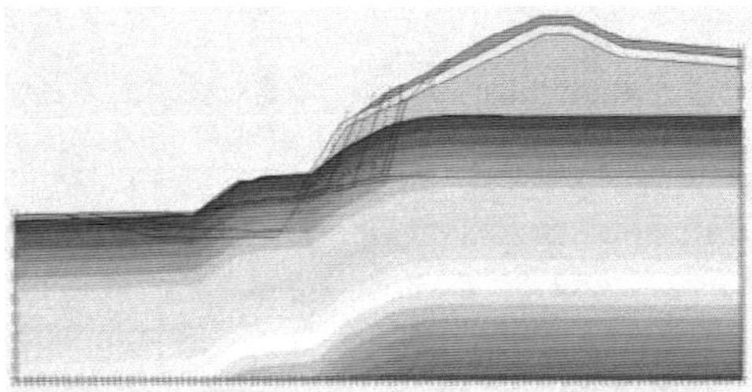

Figure 4. High-Groundwater-Level (HWL) model for Plaxis 8.6 thin-soil interface model

Lastly, the iteration was controlled using the factor of safety range and the failure plane during the landslide (as seen in Figure 5). As previously mentioned, the limitation was needed to ensure that the interface contributed to the landslide force. Thus, the range of the results in this study would be limited to only interface cohesion and interface friction angle that would initiate landslide in two extreme groundwater conditions. Thus, the limitation and the minimum number of iterations taken into this study would limit the bias of the results data.

Figure 5. Failure plane reference for the Semarang-Bawen toll road slope failure

3.2. *Interface Shear Strength Parameters Results at the Thin-Soil Interface Material*

After several iterations of the slope stability analysis using Plaxis 8.6 for both groundwater level conditions, a series of interface cohesion and friction angle with its Factor of Safety (FS) can be deducted. The results can be seen in Table 2.

Table 2. Interface residual shear strength parameters at failure condition

Trial	LWL			HWL		
	Cohesion (kN/m^2)	Friction Angle (°)	FS	Cohesion (kN/m^2)	Friction Angle (°)	FS
1	1	12	1.023	1	31	1.004
2	5	11	1.066	4	29.75	1.006
3	9	9.5	1.036	7	28.75	1.018
4	13	8.5	1.059	10	27.75	1.021
5	15	7.5	1.041	12	26.5	1.003
6	17	6.5	1.008	14	25	1.009
7	19	6	1.018	17	23.5	1.011
8	21	5.5	1.03	21	23	1.003

The cohesion for both the LWL and HWL was capped at 1 kPa and 21 kPa. The lowest 1 kPa was used due to the limitation of the Plaxis 8.6 software, which needed a positive cohesion value higher than zero. At the same time, the highest 21 kPa was selected according to the highest cohesion values from the laboratory results. Lastly, the FS ranged between 1.000 and 1.060. The ones nearing 1.10 were seen at low cohesion due to the difficulty of having the exact failure plane which controlled the back-analyses.

3.3. *Thin-Soil Interface Soil Model Results Plotted Against Laboratory Results*

The results from both groundwater level conditions were plotted in one graph. The graph consisted of the interface cohesion as the y-axis and the interface friction angle as the x-axis. Afterward, previous laboratory results were added to the graph. Thus, a range of cohesion and friction angle which would initiate failure could be acquired. The properties of the laboratory results which would be considered in the analysis are its degree of saturation and its laterite-sand ratio. The results can be seen in Figure 6.

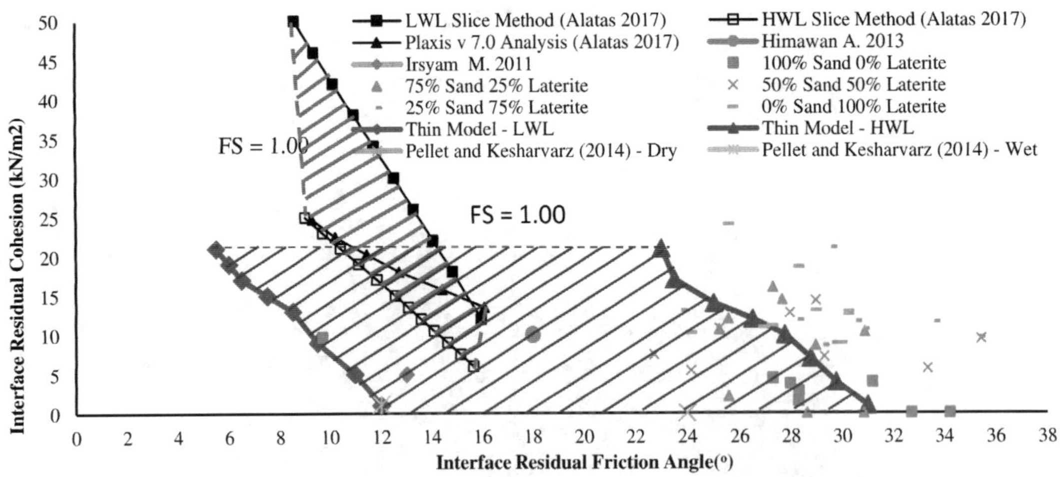

Figure 6. Comparison of interface shear strength thin-soil interface model with laterite-sand overburden variations

For comparison, the results of interface shear strength between clay shale with other soils were also included in the graph [11,12]. Furthermore, previous back-analyses were conducted on the same slope using the limit equilibrium slice method and finite element Plaxis ver. Seven were also added [13].

The previous slope stability analysis of the Semarang-Bawen landslide was seen to have higher cohesion and a narrower range of a factor of safety. Several reasons were known for the different results between this study and the previous study [13]. First, the previous study did not include any interface function in the analysis. Second, the cohesion values were capped according to the residual shear strength of the fresh clay shale. Third, the estimation of the groundwater level was different from this study. Lastly, the Plaxis analysis only included the existing groundwater level condition at failure.

This study argued that the thin-soil interface method would give a much similar situation to the landslide with a broader range than the previous analysis [13]. From Figure 6, it was clear that the laterite-sand variations, the residual clay shale properties [11], and the interface shear strength of steel-clay shale rock [12] mostly fell into the range of FS = 1.00. Thus, the thin-soil interface method could sufficiently depict the interface shear strength behavior in the FEM software. However, the thin-soil interface had several limitations, such as low repeatability and a highly user-dependent tendency. The reason was that this method was developed according to the landslide of the Semarang-Bawen toll road, but it might not be repeatable in other landslides. Several considerations, such as the depth of the interface and the landslide cases, should also be considered when using the model.

Table 3. Overburden variations of laterite-sand combinations at FS = 1.00 zone

Number	Laterite-Sand Overburden Variations	Degree of Saturation (%)
1	100% Sand	18.89%, 25.30%, 32.67%, 40.42%, 59.04%, 103.06%
2	75% Sand: 25% Laterite	38.66%, 45.91%, 53.07%, 60.97%, 101.68%
3	50% Sand: 50% Laterite	47.23%, 53.27%, 98.85%
4	25% Sand: 75% Laterite	48.85%, 100.59%
5	100% Laterite	36.29%, 50.32%, 100.52%

Lastly, a trend for the degree of saturation between various overburden conditions can be deducted from Table 3. The higher the degree of saturation, the more it is susceptible to failure. However, the 100% sand variation indicated a different trend where some lower degrees of saturation also initiated failure according to the thin-soil interface model. The reason might be due to the high permeability of sand and its granular texture. Both reasons lead to the low adhesion of the sand overburden to the wet undisturbed clay shale at the interface. Previous research indicated that higher clay content would give higher interface shear strength until a maximum value [14]. Other research also stated that clay had higher interface shear strength than sandy clay [15]. Thus, the results of this study are comparable with previous studies of the interface shear strength, where a higher clay ratio of the overburden would give higher interface shear strength.

From Table 3, the results also indicated that water had a significant role in determining the interface failure, where a higher degree of saturation would lower the interface shear strength. Although [16] argued that a higher degree of saturation would give higher interface shear strength of clay interface with concrete, other studies agreed with the current study's results [1–3]. Thus, water could be concluded to reduce the interface shear strength dominantly.

4. Conclusion

The study evaluated the thin-soil interface model using FEM software for interface slope stability back-analyses. In addition, laboratory results of interface shear strength tests were included in the analysis. Several conclusions can be deducted from the results and analysis of the thin-soil interface model.

First, the thin-soil interface model could be used to conduct a back-analyses slope stability case. The conclusion was made from the comparison to previous studies, which fell inside the FS = 1.00 zone, indicating failures of the slopes. Second, a comparison with laboratory results showed that the thin-soil interface model indicated that a higher degree of saturation at the interface would lead to failure regardless of the overburden variation. Attention should be given to 100% sand, which is purely granular, and failed at a low degree of saturation.

Lastly, further studies are needed to explore the possibility of the thin-soil interface model at other slopes. Clearly, the method was developed limitedly for the Semarang-Bawen toll road landslide. Thus, it needed several iterations on other cases to understand its repeatability and errors.

Acknowledgments

The first author is a Ph.D. international student partially funded by the International Doctoral Fellowship (IDF) from Universiti Teknologi Malaysia (UTM).

References

[1] Sadisun I A, Shimada H, Ichinose M and Matsui K 2005 Study on The Physical Disintegration Characteristics of Subang Claystone Subjected to A Modified Slaking Index Test *Geotech. Geol. Eng.* **23** 199–218

[2] Alatas I M, Kamaruddin S A, Nazir R and Irsyam M 2016 Effect of Weathering On Disintegration and Shear Strength Reduction of Clay Shale *J. Teknol.* **78** 93–9

[3] Oktaviani R, Rahardjo P P and Sadisun I A 2018 Landslides Induced by Slaking of Geomaterial *MATEC Web Conf.* **229** 1–6

[4] Irsyam M, Susila E and Himawan A 2007 Slop Failure of An Embankment on Clay Shale at Clay Shale at Km 97+500 of The Cipularang Toll Road and The Selected Solution *Int. Symp. Geotech. Eng. Gr. Improv. Geosynth. Hum. Secur. Environ. Preserv. Bangkok, Thail.* 531–40

[5] Pratama I T 2021 Studi Stabilitas Lereng Clay Shale di Kalimantan dengan Menggunakan Metode Kesetimbangan Batas dan Pendekatan Probabilistik dan Deterministik *J. Rekayasa Konstr. Mek. Sipil* **04** 103–13

[6] Alatas I M, Kamaruddin S A, Nazir R, Irsyam M and Himawan A 2015 Shear Strength Degradation of Semarang Bawen Clay Shale Due to Weathering Process *J. Teknol.* **77** 109–18

[7] Sagitaningrum F H, Kamaruddin S A, Nazir R, Soepandji B S and Alatas I M 2020 Soil-Solid Interface Shear Strength Review and Its Possibility on Interlayer Slope Stability Analysis *IOP Conf. Ser. Earth Environ. Sci.* **426**

[8] Li Y-K, Han X-L, Ji J, Fu D-L, Qiu Y-K, Dai B-C and Lin C 2015 Behavior of Interfaces between Granular Soil and Structure: A State-of-the-art Review *Open Civ. Eng. J.* **9** 213–23

[9] Jebali H, Frikha W and Bouassida M 2018 Application of Vacuum Consolidation for the Improvement of Tunis Soft Soil *Sustain. Civ. Infrastructures* 100–7

[10] Sagitaningrum F H, Kamaruddin S A and Alatas I M 2021 *Analysis of Interface Slope Stability Using Thin Soil Materials in Finite Element Software* vol 15 (Indonesian Society for Geotechnical Engineering)

[11] A H 2013 *Alternatif Mitigasi Kelongsoran dan Perkuatan Lereng Galian Tower SUTT STA 19+255 Pembangunan Jalan Tol Semarang - Solo Tahap I Ruas Ungaran - Bawen Paket V Tinalun - Lemah Ireng STA 18+000 - 21+825 (in Indonesian)* (Bandung: PT. LAPI ITB)

[12] Pellet F L and Keshavarz M 2014 Shear Behavior of The Interface Between Drilling Equipments and Shale Rocks *J. Pet. Explor. Prod. Technol.* **4** 245–54

[13] Razak I M A 2017 *Kesan Luluhawa terhadap Kekuatan Ricih Syal Lempung dalam Penentuan Parameter Kestabilan Cerun (in Malaysian)* (UTM)

[14] Kim D, Nam B H and Youn H 2018 Effect of Clay Content on The Shear Strength of Clay–Sand Mixture *Int. J. Geo-Engineering* **9**

[15] Shakir R R and Zhu J-G 2008 Mechanical Behavior of Soil and Concrete Interface *ICEM 2008 Int. Conf. Exp. Mech. 2008* **7375**

[16] Shakir R R and Zhu J 2009 Nonlinear Elastic Model for Compacted Clay Concrete Interface *Front. Archit. Civ. Eng. China* **3** 187–94

Heat Effectiveness Analysis of Knock Down Smoke LCS (Liquid Collection System) Design with a Mixture of Ethylene Glycol and Water as Cooling Media

Faiz Irza Ramadhan[1], Johny Wahyuadi M. Soedarsono[2], Rini Riastuti[3], Ahmad Maksum[4]*

[1] D4 Powerplant Engineering, Dept of Mechanical Engineering, Politeknik Negeri Jakarta, Depok 16425, Indonesia, https://orcid.org/0000-0002-9217-6999
[2] Prof. Johny Wahyuadi Laboratory, Department of Metallurgical and Materials Engineering, Universitas Indonesia, Depok, 16424, Indonesia, https://orcid.org/0000-0001-6051-2866
[3] Department of Metallurgical and Materials Engineering, Universitas Indonesia, Depok, 16424, Indonesia, https://orcid.org/0000-0003-3431-0413
[4] Research Center for eco-friendly Technology, Politeknik Negeri Jakarta, Depok 16425, Indonesia, https://orcid.org/0000-0003-1800-9137

[1]faizirza@gmail.com, [2]jwsono@metal.ui.ac.id, [3]riastuti@metal.ui.ac.id, [4]*amad.maksum@mesin.pnj.ac.id

Abstract. One of the methods commonly used to treat biomass waste, including rice husks, is pyrolysis and gasification, both of which produce smoke that can disturb the surrounding environment. This study aims to optimize the conversion process of smoke released in the pyrolysis process of rice husks into two products at once, namely liquid smoke and producer gas. This study designed a Liquid Collection System (LCS) with a shell and tube type knock-down condenser mechanism where the hot smoke is cooled so that the components of the condensed smoke (become liquid smoke) are separated from the non-condensable components (become producer gas). The composition of the LCS coolant, a mixture of Ethylene Glycol and Water, is simulated with the help of Computational Fluid Dynamics (CFD) software to optimize the heat transfer effectiveness value obtained. The simulation results show that the heat transfer efficiency of the volume variation of Ethylene Glycol in the mixture with the volume ratio of Ethylene Glycol and Water is 0:100, 10:90, 20:80, and 30:70 is 9.43%, 13.21%, 15.32%, and 15.43%, respectively. The highest efficiency value was obtained at the highest Ethylene Glycol content, which was a ratio of 30:70.

1. Introduction

The processing and utilization of organic waste or biomass also developed along with the increase in technology, especially in Indonesia as an agricultural country with agricultural commodities that are favored with a rice harvest area of 1.9 million km^2. In 2021 rice yields reached 54.42 million tons of GKG (Dry Milled Grain) [1]. Based on the research of Udjianto, D., rice husks amount to about 25% of the total rice harvested, concluded that the harvest from West Sumatra Province amounted to 1.5 million tons of GKG with a by-product of rice husks of around 370 thousand tons [2]. Rice husk processed by pyrolysis method will become rice husk char. This char, when further processed, can become silica synthesis with a certain degree of purity[3]. Liquid smoke is suitable as a natural inhibitor for steel based on rice husk charcoal composition [4,5]. In addition, there is producer gas (syngas) by non-condensable gases that have other selling points, such as those that can also be used as fuel in further processing[6].

LCS is a liquid collection system using a shell and tube condenser in a compact square shape to maximize the production place, especially for the small business enterprise or home business. Analysis related to LCS (Liquid Collection System) with output in the form of grade 3 liquid smoke has been carried out before. Research by Luo, G. et al. conducted lab-scale production trials to manufacture wood-based bio-oil with a 3-stage LCS scheme with stage 2 condenser cooling media using a 50/50 water-propylene glycol mixture[7]. Keklikcioglu, O. and Ozceyhan, V. researched heat transfer research on a mixture of ethylene glycol and water using conical divergent, convergent, divergent-convergent pipe cables with turbulence parameters in the pipe [8]. Shankara, R, P. et al performed a heat transfer analysis of variations in the mixture of ethylene glycol and water as a coolant on the radiator[9].

This study was conducted to simulate the smoke flow through the condenser to become liquid smoke. To find an effective method for changing the phase of pyrolysis emission, this study focuses on the investigation of the coolant that is effectively used, considering the cost and maintenance of the condenser (LCS). Heat transfer is one of the workflows in this condensation process. The cooling system is a component that has a significant role in the effectiveness of the condenser by researching the coolant with optimal ethylene glycol composition. The process is passed from the beginning to the output, as in figure 1 below.

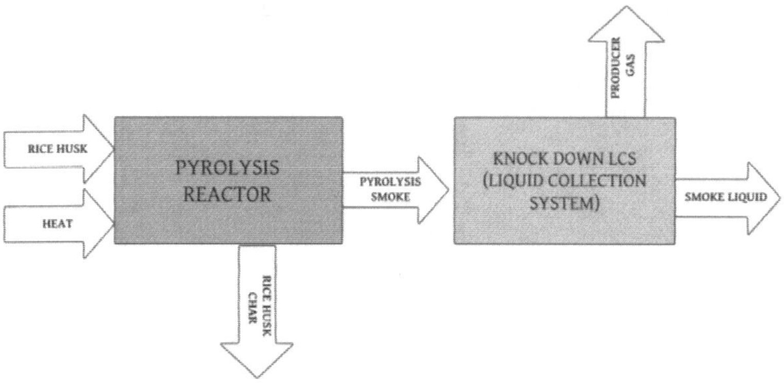

Figure 1. Rice Husk Pyrolysis Workflow Chart

2. Methods and Materials

The reactor, in the form of pyrolysis, uses a process of thermal degradation of materials with minimal oxygen conditions. Pyrolysis will break the long chain of hydrocarbon elements so that when viewed macro, there will be changes in the physical properties of materials processed into gaseous and liquid phases. The pyrolysis method is considered to be able to reduce solids emissions derived from the reaction process slowly. The condenser is a component of collecting liquid smoke as part of an effort to maximize side effects or emissions in the form of combustion solids that cause pollution in the liquid phase.

2.1. Methods of Simulation

The LCS or Condenser is conducted in shell and tube type condenser with a capacity of 124.2 liters. Based on the research of Ramadhan, Faiz Irza et al., rice husks are used as a material for making grade 3 liquid smoke that has been pre-treated[10]. The components in figure 2. consist of a pyrolizer, connecting pipe, and LCS (Liquid Collection System) condenser. The biomass or rice husk is inserted into the feedstock at 20kg. Then the feedstock is closed, and a pyrolysis process is carried out with a temperature between 400°C-500°C for 1 hour. Then the smoke from the rice husk flew through the connecting pipe and entered the condenser. The cooling medium is circulated from the reservoir towards the condenser and back to the reservoir (closed cycle) at room temperature. The flow rate of the cooling media is a maximum of 70 L/min circulated using a pump with a discharge head specification of 6 m. The emission of pyrolysis is smoke, and non-condensable gases flow through the condensation process resulting in grade 3 liquid smoke and producer gas.

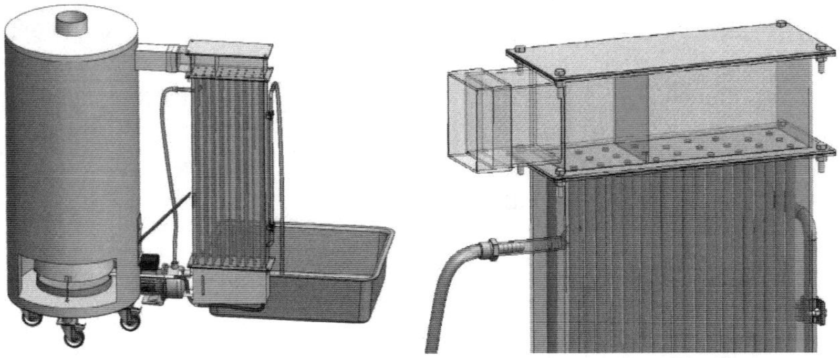

Figure 2. Knock Down LCS Placement and Inlet Design Configure

2.2. *Materials LCS (Liquid Collection System)*

This study focused on determining the value of effective heat transfer in heat absorption and release by several variables of cooling media as a reference for condenser performance. The variables of the cooling medium are pure water and also variations in ethylene glycol-water mixture with a ratio of Ethylene Glycol:Water (EG:W) 0:100, 10:90, 20:80, 30:70. Heat transfer analysis tested using CFD (Computational Fluid Dynamics) software. It is repeated up to 200 times in iterations to be taken evenly from each ratio of the cooling medium used. Table 1. Consists of thermophysical properties of ethylene glycol-water [11].

Table 1. Thermophysical Properties of Ethylene Glycol-Water

Material	EG:W 0:100	EG:W 10:90	EG:W 20:80	EG:W 30:70
Density, ρ (kg/m³)	997.1	1012.56	1027.93	1042.77
Specific Heat, Cp (J/kg K)	4108	4091	3960	3845
Thermal Conductivity, k (W/m K)	0.613	0.555	0.498	0.4405
Dynamic Viscosity (mPa/s)	0.891	1.1755	1.46	1.7445
Thermal Difussity, α.10⁷ (m²/s)	1.47	1.368	1.266	1.164
Expansion Coefficient, β (K⁻¹)	0.00021	0.00025	0.00028	0.00032
Prandtl Number (Pr)	6.07	8.64	11.21	13.78

3. Results and Discussion

The liquid phase will go down and accommodate the storage area under the condenser. The level of effectiveness in the condenser is influenced by the condenser material used, the surface area of the inner and outer walls of the pipe, the cooling media, and the variable temperature.

A decrease in the temperature of the heated fluid through the transfer of convection heat on the condenser wall will result in condensation on the pipe wall, where later smoke and other solid particles pass through and are mixed in the water and become liquid phase. In this study, some points will be concerned with heat transfer rate (Q), overall heat transfer coefficient (U), the difference in the average logarithm temperature (LMTD), and the effectiveness value (ε), which is commonly used to determine the performance of heat transfer.

3.1. *Heat Flow Simulation*

After running simulations of 200 iterations of each ethylene glycol variable, the construction of the input and output heat energy flow generation model in the heat supply system has been analyzed. Table 2. Summarize data obtained by the simulation. In the performance test, the pyrolysis condenser functions as a place to convert the smoke from pyrolysis waste rice husks into liquid. The smoke channeled by the connecting pipe will enter the condenser pipe.

The first smoke that came into the inlet started at a temperature of 350ºC with composition from the subdomain data consist air, carbon dioxide, and water. The coolant subdomain consists volume of water and ethylene glycol. The condenser model is knocked down, with the bottom and top assembly being a particular part and using a bolt to assemble. The material is aluminum AISI 302 with conductivity thermal 16.2 W/m.K. Heat transfer increases the water inlet to 1ºC based on simulation and turns the smoke temperature to 319-299ºC, making the condenser run well. The surface heat transfer area to 25 tubes is 1.06 m^2, with the highest total heat transfer of 7667.23 W.

Table 2. Summary of Coolant Heat Transfer Simulation

Data	EG:W 0:100	EG:W 10:90	EG:W 20:80	EG:W 30:70
Temp. Hot in (°C)	350	350	350	350
Temp. Hot out (°C)	319.36	308.05	300.20	299.84
Temp. Cool in (°C)	25	25	25	25s
Temp. Cool out (°C)	26.2	26.71	26.89	27.18
Heat Transfer average, Q (W)	4220.5	6014.21	6647.28	7667.23
Heat Transfer Max, Q max (W)	44772.27	45525.66	41028.86	45617.63
Heat Transfer Coef, U (W/m^2 .C)	3440.34	2071.61	2538.89	2372.64
Log Mean Temp. The difference, LMTD (°C)	308.81	302.64	298.41	298.06
Correction Factor (F)	0.99	0.99	0.99	0.99
Specific Heat Minimum, Cmin ($C_{p.c}$)	137.76	140.08	126.24	140.36
Effectiveness, ε (%)	0.0943	0.1321	0.1532	0.1543
Number of Transfer Units, NTU	0.613	0.65	0.884	0.743

The tube structure is aligned and crossed as the first design test and uses small dimensions to expand the area of motion of the coolant to accelerate the heat transfer rate. According to the law of heat transfer, the rate of such heat dissipation and a small surface area produces a small heat transfer. The displacement coefficient is obtained through the convection of the tube wall, which has a thickness of 2mm, thus affecting the value of the magnitude Q. The tube was made small to avoid losses on inner walls, but after being considered and analyzed during the simulation to build up a scale of the mass carried by the smoke. The material used considers the weld so that leakage does not occur. Compared to Turgut, Mert Sinan, et al., and Saputro, the diameter of the tube will have a significant result through the heat transfer dissipation and making the broader separation by the tube fin design [12,13].

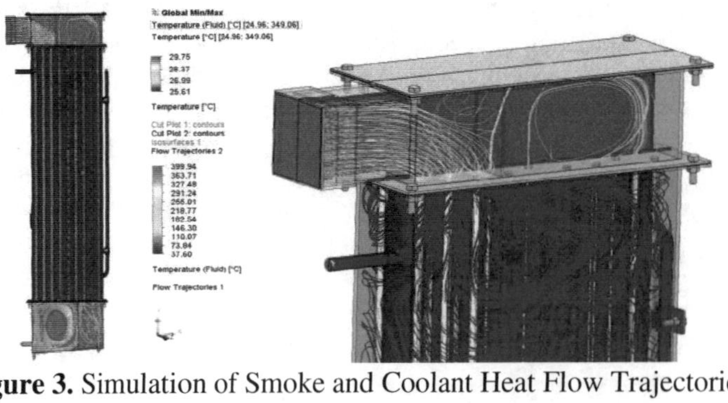

Figure 3. Simulation of Smoke and Coolant Heat Flow Trajectories

3.2. Material and Volume Content

Figure 3. is AISI Aluminium 302 material affecting the reaction and composition in the liquid, is iron that is not easy to rust and is used daily for consumption. It is obtained that the temperature of the liquid smoke begins to decrease in the simulation and is collected in the liquid collection chamber below, and the producer will be separated from the liquid smoke. The tube laminated by the hydrophobic film becomes condensed smoke or steam higher than the tube hydrophilic [14]. The wall roughness becomes an obstacle in optimizing the potential displacement provided by the wall tube.

Based on Figure 4. The condenser efficiency with ethylene glycol, 30% of the volume, is the highest, with a total of 15.43%, resulting from calculating the average displacement over 200 iterations against its maximum heat transfer. The more the value of the average Q, the more efficiency will also increase because the average value is close to the maximum value, so the heat transfer process is higher between smoke and water. As the Prandtl number increase, it will reduce the thermal boundary layer thickness because the Prandtl number is the ratio of diffusivity momentums to thermal diffusion, and heat transfer will occur by fluid momentums [14].

a.

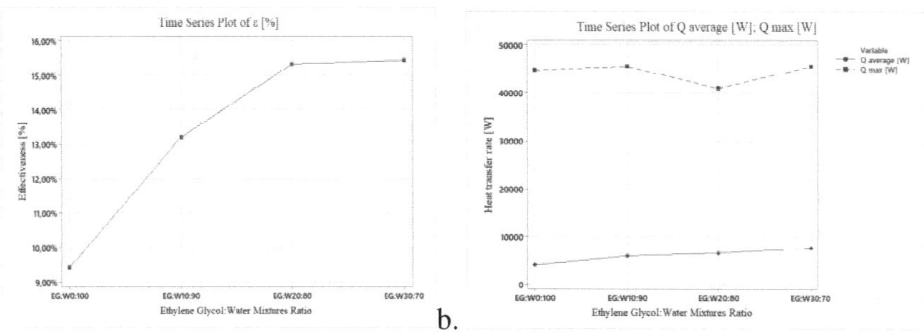

Figure 4. a. Effectiveness Graph; b. Heat Transfer Average and Maximum Line Plot

With a result of 15%, it is considered sufficient to prove that the condenser is running correctly and there are no failures in software analysis, either welding leaks or other connections. The condenser can be adequately reassembled without any obstacles, so the design and circulation of the first stage are already running well.

4. Conclusion

In this study, simulation of heat transfer flow in LCS (Liquid Collection System) has been successfully carried out, where the coolant with the highest volume addition of 30% ethylene glycol produces a higher heat transfer value than the others. The addition of ethylene glycol additives has been shown to increase the heat transfer rate, which increases with an increase in dynamic viscosity. Based on the simulation results, it can be strongly assumed that adding ethylene glycol to the coolant will significantly increase the efficiency of the LCS unit. However, further research is still needed with the actual temperature data on the LCS unit so that the effectiveness of the condensation process on the LCS unit can be known more accurately.

Acknowledgment

The authors would like to thank the facilities provided by the Research Center for Eco-friendly Technology and financial support from the Politeknik Negeri Jakarta through the Indexed International Publication Grant (PIT 2022) with Contract number: B.263/ PL3.18/PT.00.06/2022, Fiscal Year 2022.

References

[1] BPS 2022 *Statistical Yearbook of Indonesia 2022* vol 1101001
[2] Udjianto T, Sasono T and Manunggal B P 2021 Potensi Sekam Padi Sebagai Bahan Bakar Alternatif PLTBm di Sumatra Barat *Jurnal Teknik Energi* **11** 11–8
[3] Aminudin M R and Amaria A 2021 Artikel Review : Synthesis and Characterization of Nanosilica from 17–33

[4] Maksum A, Rustandi A, Permana S and Soedarsono J W 2019 Roasting-quenching Pretreatment in The Calcination Process To Improve The Purity of Rice Husk Bio-silica *JP Journal of Heat and Mass Transfer* **16** 313–26

[5] Paul Setiawan Kaban A, Mayangsari W, Syaiful Anwar M, Maksum A, Riastuti R, Aditiyawarman T and Wahyuadi Soedarsono J 2022 Experimental and Modelling Waste Rice Husk Ash as a Novel Green Corrosion Inhibitor Under Acidic Environment *Mater Today Proc* **62** 4225–34

[6] Abdullah N A, Putra N, Hakim I I and Raldi A. K 2017 A Review of Improvements to The Liquid Collection System Used in The Pyrolysis Process for Producing Liquid Smoke

[7] Luo G, Chandler D S, Anjos L C A, Eng R J, Jia P and Resende F L P 2017 Pyrolysis of Whole Wood Chips and Rods in a Novel Ablative Reactor *Fuel* **194** 229–38

[8] Keklikcioglu O and Ozceyhan V 2022 Heat Transfer Augmentation in a Tube with Conical Wire Coils Using a Mixture of Ethylene Glycol/water as a Fluid *International Journal of Thermal Sciences* **171** 107204

[9] Prasanna Shankara R, Banapurmath N R, D'Souza A, Sajjan A M, Ayachit N H, Yunus Khan T M, Badruddin I A and Kamangar S 2022 An Insight into The Performance of Radiator System Using Ethylene Glycol-Water Based Graphene Oxide Nanofluid *Alexandria Engineering Journal* **61** 5155–67

[10] Ramadhan F I, Rudiatama M R, Kusumo P I, Syakuro A, Ricardo D, Tambun M, Maksum A, Mesin J T and Jakarta P N 2022 Analisis Efektivitas Kondensor Aligned Tube Model Knock Down pada Sistem Rice Husk Slow Pyrolisis *Seminar Nasional Inovasi Vokasi* vol 1 (Jakarta) pp 177–84

[11] Ogut E B and Kahveci K 2016 Mixed Convection Characteristics of Ethylene Glycol and Water Mixture Based Al2O3 Nanofluids *Proceedings of the 2nd World Congress on Mechanical, Chemical, and Material Engineering (MCM'16)* 1–9

[12] Turgut M S, Turgut O E and Abualigah L 2022 Chaotic quasi-oppositional arithmetic optimization algorithm for thermo-economic design of a shell and tube condenser running with different refrigerant mixture pairs *Neural Comput Appl* **34** 8103–35

[13] Ayun Q, Kurniawan S and Saputro W A 2020 Perkembangan Konversi Lahan Pertanian Di Bagian Negara Agraris *Vigor: Jurnal Ilmu Pertanian Tropika Dan Subtropika* **5** 38–44

[14] Li H, You Z and Zhang H 2022 Experimental investigation on the heat transfer enhancement of steam condensation on tube with hydrophilic-hydrophobic hybrid surface *J Phys Conf Ser* **2280** 012059

The Effect of The Combination of Zeolite and Activated Biochar-Based Sugarcane Bagasse on The Increase of CH_4 in Biogas with A Different Time Variation

Adinda Putri Amalia[1], Ambar Pertiwiningrum[1*] and Nanung Agus Fitriyanto[1]
[1]Departement of Animals Product Technology, Animal Science Faculty, Gadjah Mada University, Yogyakarta Indonesia
[1]adindaputri00@mail.ugm.ac.id , [2*]artiwi@mail.ugm.ac.id,
[3]nanungagusftriyanto@ugm.ac.id

Abstract. This study aimed to determine the effect of variation in biogas purification time on methane (CH_4) content, heating value, and combustion efficiency. The statistical analysis used was a one-way ANOVA analysis. The purification used activated bagasse biochar and zeolite. The variations of purification time were divided into three treatments, which are 10 minutes, 20 minutes, and 30 minutes. The contents of CH_4 in biogas with purification times, respectively, were 281201,45 ppm, 313563,05 ppm, and 339112,92 ppm. The biogas field heat value results with purification times were 242.79 kJ, 248.37 kJ, and 255.35 kJ. The biogas' heat value with purification time was 515.60 kJ, 890.07 kJ, and 849.50 kJ. The results of combustion efficiency with purification times of 10 minutes, 20 minutes, and 30 minutes, respectively, were 47,95%, 28,20%, and 30,84%. Based on the results, it can be concluded that biogas purification using a combination of activated bagasse biochar and zeolite absorbent with different purification times has better values than biogas purification using an adsorbent not activated. It can increase the CH_4 content in biogas and biogas field heat value, but it reduces the heat value and biogas combustion efficiency due to incomplete combustion.

1. Introduction

Biogas is one of the renewable energy sources. Biogas is produced from organic wastes, such as biomass, human, livestock, etcetera. The process of making biogas is through anaerobic digestion, and the results can be used as energy. The production of biogas from livestock manure, especially cows, has great potential and many advantages. The energy produced by biogas is more environmentally friendly. Besides that, using biogas also reduces the greenhouse effect and other harmful emissions [1].

The absorbents used in this study are a combination of biochar, sugarcane bagasse, and zeolite. Sugarcane bagasse is commonly used as boiler fuel to produce energy for sugar manufacturing. Sugarcane bagasse has cellulose, which contains an active carboxyl group, and lignin which contains a phenolic group [2]. Sugarcane bagasse can use as an ingredient for making activated carbon or biochar. Sugarcane bagasse is used as biochar because it has high carbon content [3].

Zeolite is a crystalline solid and has pores of microporous size. Zeolites are tetrahedral alumina (AlO_4) and silica (SiO_4). The use of zeolites depends on their ability to carry out ion exchange, absorption, or become a catalyst [4]. Zeolite has a well-organized crystalline form with interconnected pores in all directions. This makes zeolites have a large surface area. The surface area and the associated pores can make zeolite one of the best absorbents [5].

IOP Conf. Series: Earth and Environmental Science 1111 (2022) 012055 doi:10.1088/1755-1315/1111/1/012055

The sugarcane bagasse used for this study will be converted into biochar sugarcane bagasse by a pyrolysis process. Biochar, also known as activated carbon, is the result of combustion without oxygen, producing a solid, black, and pore residue. Biochar is produced by decomposing organic matter by removing the water content and volatile components [6]. Wastes in the agricultural sector, such as sugarcane bagasse, rice husks, corn cobs, and coconut husks, are potential materials for making activated carbon [7]. The biochar sugarcane bagasse and zeolite used for purification will be activated first with potassium hydroxide (KOH) and sodium hydroxide (NaOH). Activation is a process of enlarging the pores of an absorbent so that the absorbent will change physically and chemically. The surface area of the absorbent will increase so that it will have a higher absorption power [8].

The contents of biogas are methane (CH_4), carbon dioxide (CO_2), hydrogen sulfide (H_2S), and many more. CH_4 is the main content of biogas that can be used as energy, but biogas also has many impurity gases that can disrupt the combustion process. Biogas needs to be purified first before it can be used as energy. Biogas purification has the primary function of increasing the calorific value of biogas so it can become an alternative energy source used by humans. Several methods for the biogas purification process have been applied, such as biogas purification using zeolite, calcium hydroxide, NaOH, and other absorbent materials [9]. The primary process of biogas purification is absorption. Biochar, sugarcane bagasse, and zeolite will absorb the impurity gases, mainly CO_2.

In the previous study, a combination of zeolite and biochar-based sugarcane bagasse had the best result in the combination of 50:50 in % volume. However, the absorbent was not yet activated, so in this study, the renewability is used as the activated absorbent [10]. This study investigated the utilization of combination biochar sugarcane bagasse activated with zeolite activated in biogas purification performance with different times of biogas purification. Biogas used in this study is from cattle manure fermentation. There are three different times of biogas, 10 minutes, 20 minutes, and 30 minutes. Biogas will flow through the absorbents for a predetermined time by the continuous method. This study is to find the best time for biogas purification that will produce the best CH_4 composition, field heat value, biogas heat value, and the best efficiency of combustion.

2. Methods

This study was conducted from December 2021 to March 2022 at the Leather, Waste, and By-Products Laboratory, Faculty of Animal Science UGM, The Agrotechnology Innovation Center of Universitas Gadjah Mada, and Recycling Innovation Room in The Agrotechnology Innovation Center of Universitas Gadjah Mada. Biogas composition was used to investigate the content of CH_4 in biogas after the purification process. Biogas derived from cattle manure fermentation at The Agrotechnology Innovation Center. For analyzing CH_4 contents in biogas using an analysis named gas chromatography to analyze.

2.1 Biochar production

Biochar sugarcane bagasse is made by collecting sugarcane bagasse from ice sugarcane sellers. The sugarcane bagasse is still wet when it is collected, so it must be dried under sunlight first. After it is dry, sugarcane bagasse is cut into smaller pieces. The cut of sugarcane bagasse was pyrolyzed using the pyrolysis reactor at a temperature of 225°C for 3 hours until it turned black and became biochar. The biochar sugarcane bagasse is then cooled down in a pyrolysis reactor for 24 hours until it cools down and can be collected in a gunny sack.

2.2 Biochar and zeolite activation

The biochar sugarcane bagasse is then activated using an activator. The activator that is used is a 20% KOH activator. Biochar sugarcane bagasse was mixed with KOH until it drowned and left for 4 hours. After 4 hours, the biochar sugarcane bagasse was repeatedly washed using distilled water. Then the biochar sugarcane bagasse was dried in the oven at 110°C for 24 hours. The zeolite was activated with a 5% NaOH activator. Zeolite was mixed with NaOH, then put the mixture in the oven at 80°C for 4 hours. After that, the mixture was repeatedly washed using distilled water. The zeolite was also dried in the oven at 110°C for 24 hours.

2.3 Biogas purification
Biogas purification is done by the absorption method. Biogas will pass through the purification column, which has been filled with a combination of biochar, sugarcane bagasse, and zeolite. Biogas will go through the compressor and pass through the purification column. Purified biogas will collect in a biogas balloon holder. Biogas balloon holder that contains purified biogas then connected to compressor and gas cylinder also will take samples in Venoject for the gas chromatography analysis. The combination of absorbents uses 50% of zeolite and 50% of biochar sugarcane bagasse. The treatments were distinguished by differences in purification time, namely 10 minutes, 20 minutes, and 30 minutes, with the continuous method. Each treatment was repeated three times. The results obtained were tested by gas chromatography analysis and the calorific value of biogas.

3. Results and Discussion
Biogas purification was performed by absorption of impurity gases of biogas through the absorption column. Biogas purification uses the combination of biochar, sugarcane bagasse, and zeolite with a ratio of 50:50 in % volume. The purification times were divided into three times, 10 minutes, 20 minutes, and 30 minutes, using a continuous method that did not change the absorbent until the end. The increase in CH_4 is presented in Figure 1. Figure 1 shows that CH_4 in biogas increases within the time of purification. The increase of CH_4 in biogas is due to the presence of the absorbent that absorbs impurity gases from biogas. From the data in Figure 1, the highest CH_4 content in biogas is in 30 minutes of purification. It means that the longer the time of purification, will make the better results of CH_4 content because the more biogas passing through the purification column. The content of CH_4 in biogas after purification with activated carbon and gel silica produces the highest content in 30 minutes of purification [11].

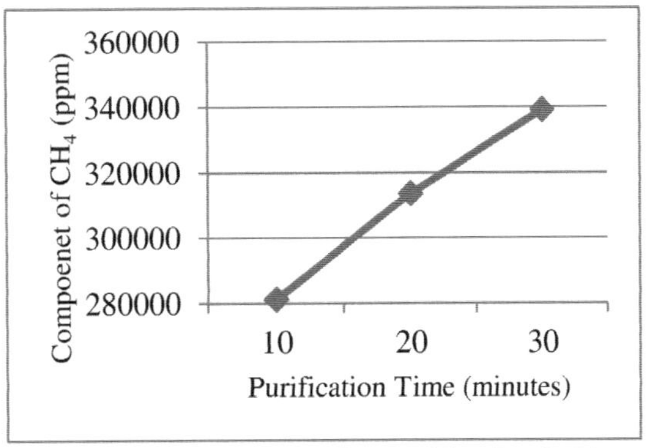

Figure 1. Component of CH_4 after purification in biogas.

The field heat values analyze by doing a combustion process with purified biogas. Purified biogas was placed in the gas cylinder at 3 bar of pressure. Then the gas cylinder will be connected to the stove. Prepare 1 L of distilled water and put it in a saucepan. Put the saucepan on the stove, turn on the stove and press the timer. Every 30 seconds, write down the temperature and gas cylinder's pressure. Does it until the flame turns off. After that, turn off the stove and calculate the difference in temperature (ΔT). After having ΔT, then calculate the field heat values. The increase in field heat value is presented in Figure 2. Figure 2 shows that the field heat value in biogas increases during purification. Biogas' increased field heat values are due to the absorbent that absorbs impurity gases from biogas. From the data in Figure 2, the highest field heat value is in 30 minutes of purification. It means that the longer the purification time, the better the field heat value results because the more biogas passing through the purification column. The previous study stated that the field heat value decreased during a purification time of 25 minutes, and the highest field heat value was at 20 minutes. Furthermore, in the previous study, the field heat value of the non-activated purification has a lower value than the activated one [10]. It can be done because maybe there is a saturation in the absorbent used in purification. The more it is used, the more it will be saturated and lose its ability to absorb the impurity gases in biogas, decreasing in 25 minutes. The ability absorption of zeolite can decrease if continuously used because it will produce a film layer on zeolite's surface. The film layer in zeolite's surface will make zeolite challenging to absorb the impurity gases in biogas [12].

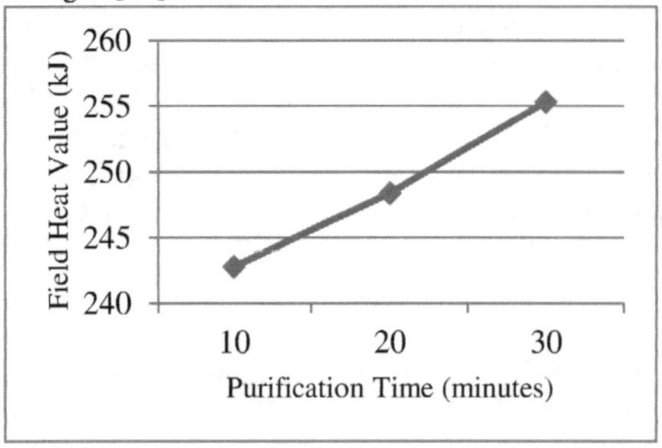

Figure 2. The experiment on the biogas heat value

The increase in biogas heat value is presented in Figure 3. Figure 3 shows that the biogas heat value in biogas increases within the time of purification. However, then decreased purification time by 30 minutes. It can be done because of the incomplete combustion of biogas. Some impurity gases are left in biogas cylinders that can affect the combustion. Because of the incomplete combustion will affect the biogas heat value itself. From the data in Figure 3, the highest field heat value is in 20 minutes of purification. Biogas produces gases, especially impurity gases. The impurity gases are water vapor and CO_2. Water vapor contents in biogas can decrease the biogas heat value [13]. The highest biogas heat value is in 10 minutes of purification, and in the next time purification, the biogas heat value decreasesing. Although the biogas heat was decreased compared with the previous study, the research with activated absorbent has a better result than the non-activated one [10].

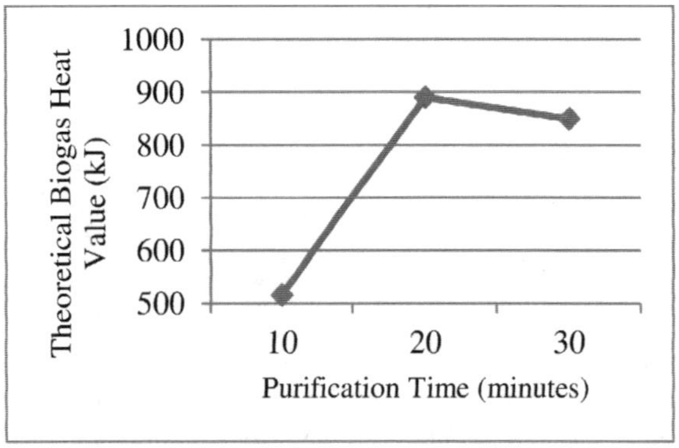

Figure 3. The theoretical biogas heat value

The efficiency of biogas combustion is presented in Figure 4. Figure 4 shows that the highest efficiency of biogas combustion value is in 10 minutes of purification. Then in 20 minutes, it decreases and increases again in 30 minutes of purification. The efficiency of biogas combustion is affected by field heat value, and biogas heat value, which means the combustion process and heat value analysis affected the results. Heat value analysis does through a combustion process, an incomplete combustion process where some impurity gases are left in the biogas cylinder that can affect the combustion. Because incomplete combustion will affect the efficiency of biogas combustion itself, some impurity gases and gases cannot be desorbed due to the balanced pressure between the environment and the biogas system. So, there are still gases remaining in the activated carbon or absorbent. One example of a gas that cannot be desorbed is CO_2. The remaining CO_2 gases in the tank can reduce the efficiency of biogas combustion. The previous study stated that the highest efficiency of biogas combustion is in 20 minutes of purification. The biogas value with non-activated absorbent has a lower result than the activated one [10].

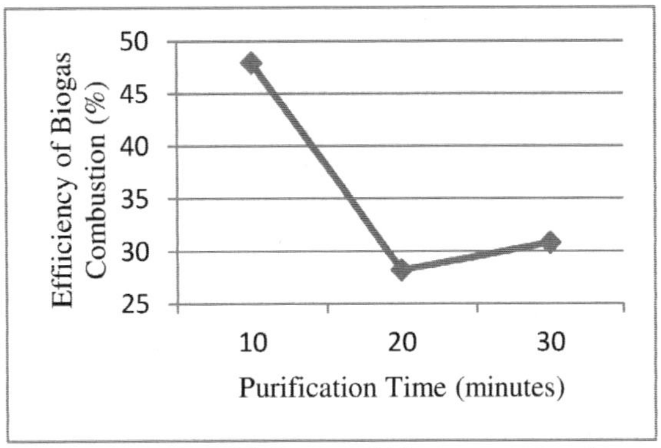

Figure 4. The efficiency of biogas combustion

4. Conclusion

Based on this study that has been done, it can be concluded that the longer the purification time can increase the CH_4 content and the field heat value. At the same time, there is a decrease in biogas heat value and efficiency of biogas combustion. This can happen because some mass of gases are left in the absorbent or biogas cylinder and causes incomplete combustion, thereby reducing the biogas heat value and efficiency of biogas combustion. Based on the study that has been carried out using activated absorbents produces better results than a study using non-activated absorbents. This research was a limitation to biogas amount, so it was challenging to run the research with more extended purification. The suggestions for the following research were enlarging the reservoir of the biogas so it can increase the amount of biogas, so the following research can use the longer purification time and know when the absorbent is saturated. It is necessary to check the micropores on the activated absorbents, so it can find out whether the activation is successful. The following research can vary the absorbents used in research, such as using livestock and agricultural wastes that are rarely processed or utilized.

Acknowledgments

The research was funded by the PTUPT program of the Universitas Gadjah Mada. The authors would like to thank students from the Faculty of Animal Science Universitas Gadjah Mada who helped with data collection.

References

[1] Putri D A, Saputro R R and Budiyono 2012 Biogas production from cow manure *Int. J. Renew. Energy Dev.* **1** 61–4

[2] Yoseva P L, Muchtar A and Sophia H 2022 Pemanfaatan limbah ampas tebu sebagai adsorben untuk peningkatan kualitas air gambut *JJurnal JOM FMIPA* **2** 56–63

[3] Pratama B S, Aldriana P, Bambang I and Saptati A . D 2018 Konversi Ampas Tebu Menjadi Biochar dan Karbon Aktif untuk Penyisihan Cr (VI) *J. Rekayasa Bahan Alam dan Energi Berkelanjutan* **2** 7–12

[4] Nugroho K A 2017 Pemanfaatan Zeolit ZSM-5 dalam Proses Peningkatan Mutu Kualitas Air *Dep. Chem.* **1** 1–6

[5] Anggara P A, Wahyuni S and Prasetya A T 2013 Optimalisasi zeolit alam Wonosari dengan proses aktivasi secara fisis dan kimia *J. Rekayasa Bahan Alam dan Energi Berkelanjutan* **2** 72–7

[6] Syauqiah I, Amalia M and Kartini H A 2011 Analisis Variasi Waktu Dan Kecepatan Pengaduk Pada Proses Adsorpsi Limbah Logam Berat Dengan Arang Aktif *Info Tek.* **12** 11–20

[7] Imani A, Sukwika T and Febrina L 2021 Karbon Aktif Ampas Tebu sebagai Adsorben Penurun

Kadar Besi dan Mangan Limbah Air Asam Tambang *J. Teknol.* **13** 33–42
[8] Amanda I R, Putri K, Artika N, Julia R R D and Sanjaya A S 2019 Aktivasi Biochar Dari Kayu Macaranga Gigantea Menggunakan Zncl2 (Activation Of Biochar Frommacaranga Gigantea Wood Using Zncl2) *J. Kim. Mulawarman* **17** 6–10
[9] Rajagukguk K and Satria Arysca Wisnu 2019 Design of Biogas Purification To Reduce Carbon Dioxide (Co2) and Hydrogen Sulfide (H2S) *J. Tek. Mesin Univ. Bandar Lampung* **7** 3–8
[10] Besari R N 2020 *PENINGKATAN NILAI KALOR BIOGAS DENGAN METODE VARIASI RASIO DAN WAKTU PURIFIKASI MENGGUNAKAN ADSORBEN ZEOLIT DAN ARANG AMPAS TEBU RACHMAD NUR BESARI, Prof. Ir. Ambar Pertiwiningrum, M.Si., Ph.D., IPM., ASEAN Eng.* (UGM)
[11] Ritonga A M, Masrukhi and Kusmayadi R P 2021 Permunian biogas metode adsopsi menggunakan down-up purifier dengan arang aktif dan silika gel sebagai adsorben *J. Agric. Biosyst. Eng. Res.* **1** 72–80
[12] Hamidi N, Wardana I N G and Widhiyanuriyawan D 2011 Peningkatan Kualitas Bahan Bakar Biogas Melalui Proses Pemurnian Dengan Zeolit Alam *Rekayasa Mesin* **2** 227–31
[13] Septyana G A 2015 *Energi Alternatif Biogas*

Capturing Carbon Dioxide from Human-Driven Vehicles by Green Structures for Carbon Neutrality

Tetiana Tkachenko[1] and Viktor Mileikovskyi[2*]
[1] Department of Environment Protection Technologies and Labor Safety, Faculty of Engineering Systems and Ecology, Kyiv National University of Construction and Architecture, Kyiv, 03037, Ukraine, ORCID ID 0000-0003-2105-5951
[2] Department of Heat and Gas Supply and Ventilation, Faculty of Engineering Systems and Ecology, Kyiv National University of Construction and Architecture, Kyiv, 03037, Ukraine, ORCID ID 0000-0001-8543-1800
[1]tkachenkoknuba@gmail.com, [2*]mileikovskyi@gmail.com

Abstract. For carbon neutrality of cities, we should compensate carbon dioxide gains from vehicles. Human-driven ones are not carbon-free because cyclists perform hard work. Green roofs are an adequate carbon-capturing facility, especially for the dense building. We used the previous author's data for carbon dioxide sequestration of them. The carbon dioxide gained from a cyclist is calculated using aggregated indicators for hard work and rest. Also, the experimental data for a cyclist on an exercise bike is used as an alternative. For recumbent bicycles, drag is 0.2÷0.3 of drag for classical ones. This allows estimation of the carbon dioxide gain for this promising technology. The results show that road bicycles require only 4.4 m^2 of extensive and 1.37 m^2 of intensive green roofs per cyclist. Recumbent bicycles need two-time fewer areas. Electric public conveyances and cars require one-two order of magnitude more area.

1. Introduction

A carbon-neutral city is one of the modern concepts for decreasing the technogenic impact on the environment and throttling climate changes [1–3]. Zero energy consumption using renewable energy accompanied by energy-efficient consumption is one of the main principles of a carbon-neutral city [1]. Nevertheless, one of the powerful sources of carbon dioxide is transport [4–24], which is one of the critical parts of city infrastructure [4]. It emits approx. 1/5 of the total carbon dioxide in the European Union [17]. A 15-minute city is the perspective concept to decrease the transport need [25]. Nevertheless, some objects can not be distributed evenly by a city. There are sights, staff, etcetera. Thus, the minimum of transport is always necessary. Production is also transport-dependent.

Most vehicles are not carbon-neutral. There are much data on the release of carbon dioxide by fuel and electrical-driven devices [26]. There is an opinion that human-powered vehicles are carbon-neutral [27]. There is an analysis of the emissions-safety factor of road junctions, but carbon dioxide emissions are computed only from cars. Thus, we did not find any research data on carbon dioxide emission in cities from this vehicle type and measures for its compensation. This causes improper designs of carbon-neutral cities because the opinion above is wrong. Humans on such vehicles release more carbon dioxide because of more frequent breathing. For modern carbon-neutral bicycle cities, taking into account the release is crucial. The work is devoted to estimating the release from different human-driven vehicle types.

Human-driven vehicles are recumbent bicycles, fully suspended bicycles (velomobiles), etcetera. Practically, most human-driven vehicles are bicycles instead of the highest drag of a vertically sitting person. It is because of the best visibility of the surrounding for safety. Other vehicles require special traffic laws and enough-quality roads to avoid danger due to blind spots in front of the legs and low weight. However, these measures will be reasonable if we show a significant advantage in carbon dioxide release. Using the averaged reference data in a ventilation handbook for carbon dioxide release during hard work is enough to estimate one for a cyclist. Also, we can use experimental data for bicycles. The main problem is the minimal amount of data about recumbent bicycles in cities. We can use the relation between drags of classic and recumbent bicycles in the hypothesis that they are primary loads. After that, it is necessary to recalculate the data per year and person using speed data for such vehicle types and average yearly path. Comparing the results with cars and public conveyances is essential to show the effect.

If we confirm a significant carbon dioxide release, we should find environment-friendly facilities to absorb it. Landscaping with outdoor greening is the most natural and accessible one [28,29]. Green structures [30] are an effective alternative, which can compensate for the lack of ground areas. Our researches show that green structures and indoor and outdoor greening can effectively solve different technical problems [31–37]. To sequestrate carbon dioxide emissions enough, we should properly design them using carbon dioxide sequestration. Such data is present in the author's work [37] for different green roof types. The carbon dioxide sequestration on 1 hm^2 of extensive green roofs with *Lolium perenne* is S = 17 t/(year·hm^2) [37]. The intensive green roof on the Royal Tower building in Kyiv sequestrates S = 54.61 t/(year·hm^2). The estimation data will allow us to find the necessary green roof area per cyclist for carbon-nature bicycle city planning.

2. Methods

We will try to get the universal estimates applicable to almost any place. A person on a human-driven vehicle performs hard work. We can use two approaches to calculate carbon dioxide gains: aggregating indicators and specific gains for the corresponding metabolism. The first approach uses the carbon dioxide emission for hard work [38] – g_{CO2} = 90 g/(h·person). If a person is not riding, he releases carbon dioxide – $g_{CO2,0}$ = 45 g/(h·person). The second approach requires experimental data, for example, in the framework of VEGA European Union's project for the operation of a stationary exercise bike for the speed of 25 km/h – 172.8 and 32.4 g/(h·person). This data is obtained closer to real ones but for some people with fixed physical conditions and current health states. The result has no advantage compared to the data [38].

The average velocity [39] on a straight road using a road bicycle is v = 25÷30 km/h. Using a mountain bicycle allows the speed of v = 20 km/h. Thus, the total and over the rest value specific carbon dioxide gain from a straight road is, accordingly, [g/(km·person)]

$$G_{CO2} = \frac{g_{CO2}}{v};\ \Delta G_{CO2} = (g_{CO2} - g_{CO2,0})/v. \quad \textbf{(1)}$$

A recumbent bicycle gives the drag of 20...30 % of the bicycle [40]. The drag by [41] is [N]

$$F_d = (c_d A)\rho\, v^2/2, \quad \textbf{(2)}$$

where c_d – drag coefficient; A – the area of the midsection [m^2]; ρ – air density [kg/m^3]; v – the velocity of air attack (for calm, it is equal to the vehicle speed).

The rolling resistance has been estimated by the data in [69-71]. By [69] at a normal force N for a pair "wheel-surface" with rolling resistance coefficient C_{rr}, the resistance [N]

$$F_{rr} = C_{rr}\, N. \quad \textbf{(3)}$$

In this case, N is the weight of a bicycle and the cyclist on it. The average mass of people [42] is m_p = 62 kg. The average mass of bicycles [43] is 17 lbs or m_b = 7,71 kg for road bicycle and 25 lbs or m_b = 11,3 kg. The total weight for the average acceleration of gravity g = 9,80665 m/s^2 is [N]

$$N = g\,(m_p + m_b). \tag{4}$$

For road bicycles, by equation (4), the weight N = 684 N, for mountain one – 719 N. For production bicycle tires [44], C_{rr} = 0.0022÷0.0050. We will accept 0.0022 for road bicycles and 0.0050 – for mountain ones. By equation (3), for road bicycles, F_{rr} = 1,5 N and for mountain ones, F_{rr} = 3,6 N. The force is negligible and less than the weight of demi-season clothes.

We will assume that the power applied to pedals is used to compensate for the drag. If at the same velocity, the drag for a recumbent bicycle falls to 0,2÷0,3 of the value for a classic one, then the complex ($c_d A$) for the recumbent bicycle is equal to 0,2÷0,3 of the value for a classic one. By equation (2), if the cyclist applies the same force, the recumbent bicycle will ride with $1/(0{,}2÷0{,}3)^{1/2}$ = 1,83÷2,24 times more incredible velocity.

For cars, we can use the data of [45]. European Union statistics says that each person drives the average distance per day d = 5.6÷19.0 km/day [46] or D = 365,25 d = 2045.4 6939.75 km/year. The most driving country is Germany, the less one is Greece. The median is corresponding to Italy – d = 11.4 km/day or D = 4163.85 km/year. We can recalculate carbon dioxide gain per km to the same one per year [g/(year·person)]:

$$\Gamma_{CO2} = G_{CO2}\,D. \tag{5}$$

3. Results and Discussion

Let us analyze the results by equations (1) and (5) of carbon dioxide gains (Figure 1). The figure shows that human-driven vehicles are very close to but not fully carbon-free. They are at least 2.33 times less carbon-emissive than electric-driven public conveyances. Cars are the most carbon-intensive personal vehicles, even for four persons per one. They surpass even public conveyances except for five persons per car, but this is not a usual option. Recumbent bicycles provide more than two times fewer gains. Cyclists on mountain bicycles release 25÷30 % greater than on-road ones. Thus, bicycles should be used only if the way contains poor-quality or very inclined roads.

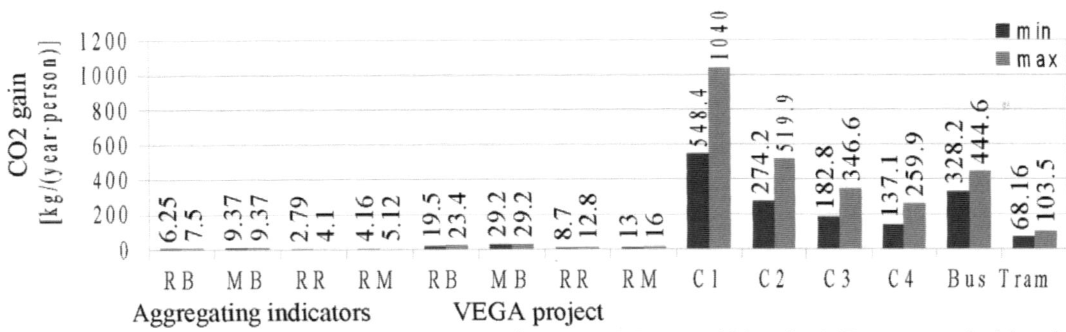

Figure 1. Carbon dioxide gains Γ_{CO2} per year and person: RB – road bicycle; MB – mountain bicycle; RR – a recumbent bicycle with road wheels; RM – the same mountain ones; C1-C4 – car, 1-4 persons.

Let us use the median distance to obtain more probable results to calculate the possibilities of green goofs to capture the carbon dioxide gains (Figure 2). The results show that if we accept the aggregating indicators and median path, we need the maximum 10000/2270 = 4,4 m² of an extensive green roof or 1,37 m² of the intensive one for carbon neutrality of road bicycles.

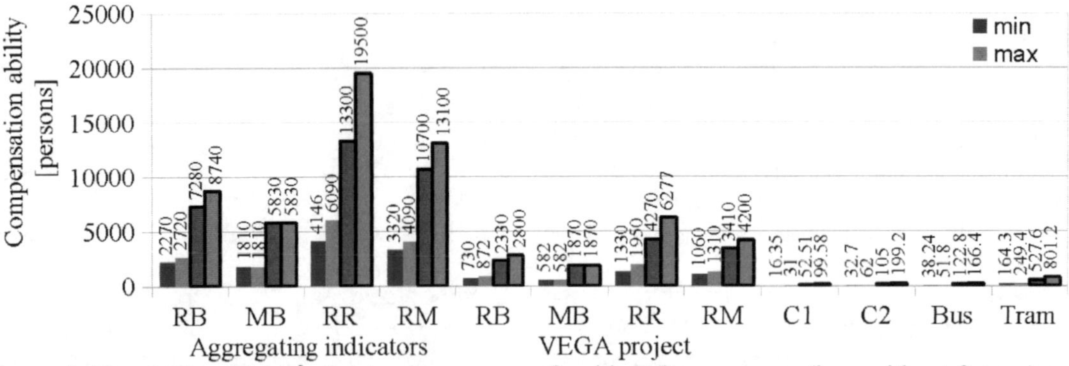

Figure 2. The ability of 1 hm² of extensive green roofs with *Lolium perenne* (bars without frames) and intensive green roofs like Royal Tower, Kyiv (bars with frames) to compensate for carbon dioxide emissions from vehicles, median average daily distance, abbreviations see Figure 1.

In the same conditions, for cars with two persons per car, we require a minimum of 161,3 m² of extensive green roofs or 50 m² of intensive ones. For trams, the corresponding values are 40,1 m² and 12,48 m². Therefore, switching from car city to bicycle city concept accompanied by green structures implementation can provide carbon neutrality of transport. Electric public conveyances require more than nine times greater area of green roofs than road bicycles. The introduction of recumbent bicycles can approximately two times decrease the necessary green roof area to release the areas for compensation of other gains.

4. Conclusion

Human-driven vehicles (bicycles, recumbent bicycles, velomobiles) are near but not carbon-neutral. A riding cyclist releases more carbon dioxide than a car passenger. It is possible to estimate the additional carbon dioxide release using aggregating indicators of hard work or experimental data for exercise bikes. The first data is averaged by many people. The second one is closer to the actual load but not averaged for different people with different physical conditions. The lowest carbon dioxide release of transport corresponds to human-driven vehicles, especially recumbent bicycles. Cyclists on bicycles release approximately two-time more carbon dioxide due to higher drag. For the most carbon neutrality, mountain bicycles should be used only when necessary: on very inclined or on poor-quality roads. For road bicycles, 4,4 m² of extensive and 1,37 m² of intensive green roofs can compensate the carbon dioxide emission of each cyclist. For cars with approximately two persons, we need 161,3 and 50 m². Electric public conveyances release significantly greater carbon dioxide than cyclists, which requires 40,1 and 12,48 m² of green roofs.

Acknowledgments

This work is performed in the framework of the following grant programs, state grant program by the Ministry of Education and Science of Ukraine "Creating perspective technologies of forming the safe building environment combining "green structures," photo design, and engineering systems," 2022 – 2023, registration number 0122U001197. "Multilevel Local, Nation- and Regionwide Education and Training in Climate Services, Climate Change Adaptation, and Mitigation – ClimEd," 15.11.2020 – 14.11.2023, registration number 619285-EPP-1-2020-1-FI-EPPKA2-CBHE-JP.

References

[1] Komninos N 2022 Net Zero Energy Districts : Connected Intelligence for Carbon-Neutral Cities *Land* **11**
[2] Pinto G, Abdollahi E, Capozzoli A, Savoldi L, and Lahdelma R 2019 Optimization and

Multicriteria Evaluation of Carbon-neutral Technologies for District Heating *Energies* **12**
[3] Huang B, Xing K, Pullen S and Liao L 2020 Exploring Carbon Neutral Potential in Urban Densification : A Precinct Perspective and Scenario Analysis *Sustainability* **12** 4814
[4] P S, T F, W W, Y M, and K P 2022 *Energy Report 8*
[5] P S, W W, T F, Y M, and K P 2022 *Energy Report 8*
[6] Kebede L, Segni G and Tama R 2022 Diesel-Fueled Public Transport Vehicles and Air Pollution in Addis Ababa, Ethiopia : Effects of Vehicle Size, Age and Kilometers Travelled *Atmos. Environ. X* **13** 100144
[7] M M H and S-T L 2022 The International Conference on Energy and Green Computing ICEGC'2021 (Meknes) *E3S Web of Conferences 336*
[8] Hakkim H, Kumar A, Sinha B and Sinha V 2022 Air Pollution Scenario Analyses of Fleet Replacement Strategies to Accomplish Reductions in Criteria Air Pollutants and 74 VOCs Over India *Atmos. Environ. X* **13** 100150
[9] Wei N, Zhang Q, Zhang Y, Jin J, Chang J, Yang Z, Ma C, Jia Z, Ren C, Wu L, Peng J and Mao H 2022 Super-Learner Model Realizes the Transient Prediction of CO 2 and NOx of Diesel Trucks : Model Development, Evaluation and Interpretation *Environ. Int.* **158** 106977
[10] Herfurth D 2022 Kfz-Steuer: Statt CO2-Preis und Klimaprämie *Wirtschaftsdienst* **2021**
[11] Pusztai Z, Kőrös P and Friedler F 2021 Vehicle Model for Driving Strategy Optimization of Energy Efficient Lightweight Vehicle **88** 385–90
[12] Lyu P, Slade P, Liu Y and Wang Y 2021 Science Direct Review of The Studies on Emission Evaluation Approaches for Operating Vehicles *J. Traffic Transp. Eng. (English Ed.* **8** 493–509
[13] Adame L F, Rodriguez-martínez A, Cerezo J and Rosenberg J 2021 A Sustainable Development Assessment to 2050 of the Transport Sector in Mexico *Chem. Eng. Trans.* **86** 469–74
[14] Jnr B A 2021 Integrating Electric Vehicles to Achieve Sustainable Energy as a Service Business Model in Smart Cities *Front. Sustain. Cities* **3** 1–12
[15] Wang Y, Yuan H, Martinez A, Hong P, Xu H and Bockmiller F R 2021 Polymer Electrolyte Membrane Fuel Cell and Hydrogen Station Networks for Automobiles : Status , Technology , and Perspectives *Adv. Appl. Energy* **2** 100011
[16] Mendoza P, Demuynck J, Bosteels D, Wilkes T, Mueller V and Recker P 2022 Future-Proof Heavy-Duty Truck Achieving Ultra-Low Pollutant Emissions with a Close-Coupled Emission Control System Including Active Thermal Management *Transp. Eng.* **9** 100125
[17] Peppas A, Kollias K, Dragatogiannis D A and Charitidis C A 2021 Sustainability Analysis of Aluminium Hot Forming and Quenching Technology for Lightweight Vehicles Manufacturing *Int. J. Thermofluids* **10** 100082
[18] Kapoor P, Costall A W, Sakellaridis N, Lammers R, Buonpane A, Limited C E, Street P and Eca L 2022 Multi-objective Optimization of Turbocharger Turbines for Low Carbon Vehicles Using Meanline and Neural Network Models *Energy Convers. Manag.* **15**
[19] Sitati C N, Oludhe C, Oyake L and Mbandi A M 2022 A Street-Level Assessment of greenhouse gas emissions associated with traffic congestion in the city of Nairobi , Kenya The problem : Nairobi City with Uhuru Highway as a case study *Clean Air J.* **32** 1–12
[20] Liu J, Li M, Zeng Y, Yin M and Zhang X 2022 An Improved Methodology for Evaluating Energy Service Demand for China's Passenger Transport Sector *Adv. Clim. Chang. Res.* **13** 290–300
[21] Joubert J W and Gräbe R J 2022 Real driving emissions data : Isuzu FTR850 AMT *Data Br.* **41** 107975
[22] Li Y, Wu Q, Zhang Y, Huang G and Jin S 2022 Mapping Highway Mobile Carbon Source Emissions Using Traf fi c Flow Big Data : A Case Study of Guangdong *Front. Energy* **10** 1–13
[23] Dioha M, Ruggles T H and Dioha M O 2022 Exploring the role of electric vehicles in Africa ' s energy transition : A Nigerian case study *ISCIENCE* **25** 103926
[24] Dandan X, Lin L, Tengteng L, Xiaojun J and Dongzhi G 2022 7th International Conference on Energy Science and Applied Technology ESAT 2022 (Yichang) *E3S Web of Conferences 352* p 03034

[25] Allam Z, Bibri S E, Chabaud D and Moreno C 2022 The Theoretical , Practical , and Technological Foundations of the 15-Minute City Model : Proximity and Its Environmental , Social and Economic Benefits for Sustainability *Energies* **15** 6042
[26] Department for Business Energy & Industrial Strategy 2022 *2022 G overnment Greenhouse Gas Conversion Factors for Company Reporting Methodology Paper for Conversion factors*
[27] Anon Calculation of Emissions
[28] Mallick J, Almesfer M K, Alsubih M, Ahmed M and Kahla N Ben 2022 Estimating Carbon Stocks and Sequestration With Their Valuation Under a Changing Land Use Scenario : A Multi-Temporal Research in Abha City , Saudi Arabia *Front. Energy Evol.* **10** 1–15
[29] Muthalib A A, Ngii E and Putera A 2022 The Ability of Green Open Spaces in Greenhouse Gas Control to Achieve Green Cities in Kendari City *Int. J. Energy Econ. Policy* **12** 327–31
[30] Kuronuma T, Watanabe H, Ishihara T and Kou D 2018 CO2 Payoff of Extensive Green Roofs with Different Vegetation Species *Sustainability* **10** 1–12
[31] Deeb A S, Tkachenko T and Mileikovskyi V 2021 Environmental Assessment of Relationships and Mutual Influences in the System " Protective Forest Plantations – Anthropogenic Landscapes " *IOP Conf. Ser. Earth Environ. Sci.* **940**
[32] Tkachenko T and Voloshkina O 2018 The Role of " Green Structures " in Reducing the Environmental Footprint of Urbocenoses *Int. J. Eng. Technol.* **7** 214–20
[33] Tkchenko T, Mileikovskyi V, Dziubenko V and Tkachenko O 2020 Improvement of the Safety of Multi-floor Housing *IOP Conf. Ser. Mater. Sci. Eng.* **907**
[34] Tkachenko T and Mileikovskyi V 2019 Solution of Sick Building Syndrome Problem using Indor Plants *Environ. Sci. Eng. Manag.* **3** 405–11
[35] Tetiana T and Viktor M 2020 Increasing Indoor Air Quality by a Natural Sanitizing Interior *E3S Web Conf.* **211** 1–8
[36] T T and V M 2020 Advances in Intelligent Systems and Computing *Proc. of the 19th Int. Conf. on Geometry and Graphics. ICGG 2021 (Sao Paulo)* pp 139–51
[37] Tkachenko T and Mileikovskyi V 2020 Methodology of Thermal Resistance and Cooling Effect Testing of Green Roofs *Songklanakarin J. Sci. Technol.* **42** 50–6
[38] Markus T 2018 Handbook on Marine Environment Protection *Handbook on Marine Environment Protection* ed M Salomon and T Markus (Cham: Springer International Publishing) pp 545–62
[39] Anon 2022 The speed achieved by a cyclist
[40] Anon How Recumbent Bikes are Designed: Design Pluses And Minuses
[41] Edgar J 2020 *Car Aerodynamic Testing for Road and Track* (Dalton New South Wales: Edgar Media)
[42] Anon How much do people weigh on average around the world?
[43] Anon How Much Does a Bike Weigh? (Every Bike Type)
[44] Anon Rolling Resistance
[45] G T 2022 2022 Government greenhouse gas conversion factors for company reporting. Methodology Paper for Conversion factors. Draft Report (Department for Business, Energy & Industrial Strategy)
[46] Anon Eurostat Statistics explained. File: Average distance per person per day (kilometres) v3.png

Velcro Product Design with Biomimicry Approaches

Filia Nauli Hapsari[1], Ratna Purwaningsih[1]*, Faradhina Azzahra[1], and Diana Puspita Sari[1]
[1]Industrial Engineering, Faculty of Engineering, Diponegoro University
Jl. Prof. H. Soedarto, S. H. Tembalang, Semarang 50275
[1]filianauli@students.undip.ac.id, [2*]ratna.purwaningsih@ft.undip.ac.id,
[3]faradhinaazzahra@lecturer.undip.ac.id, [4]dianapuspitasari@lecturer.undip.ac.id

Abstract. Nature has experienced so many adjustments throughout its life cycle, such as evolutions, climate change, and natural selection, and it still manages to exist today. Therefore, it will be an excellent example to be imitated into a product or process since it is most likely sustainable and comply human needs. Imitating nature's process, design, or system is called Biomimicry Approach. In this research, researchers develop a new design for Velcro. Velcro is a fastener that consists of 2 parts: hook and loop. Both parts are made of nylon thread and will be fastened together. Current Velcro products are not durable because of the design and also materials. It can be easily wrinkled and damage the fabric it is glued. This study aims to develop Velcro design to have a sustainable design using the technical steps in each biomimicry design spiral methodology stage. Researchers used Biomimicry Approach to find the best development for the Velcro product and generate a conceptual design for Velcro. The natural model used in the conceptual design was a *Lentinula edodes* (shitake mushroom). The design assessment indicator used were von mises stress, safety factory, and displacement. The results showed that the development design of Velcro with the Biomimicry Design Spiral Approach was acceptable and safe.

1. Introduction

Sustainability is a design process that produces a product development framework or changes in people with a design that meets the sustainable development goals [1]. It is related to the practice of design, education, and research that contribute to sustainable development. In other words, sustainability is a process to reach sustainable development [2]. The concept of sustainable development was found after the Brundtland Report was held. The three fundamental critical areas of sustainable development are social, economic, and environmental. Sustainability can be reached only if it contributes to economic growth and social progress and promotes environmental sustainability [3].

To attain one of those pillars, environmental sustainability, paying attention to product design sustainability is essential. Sustainable design seeks to reduce any negative impacts on the environment. Therefore, as an instrument of sustainable development [4], sustainable design should meet the needs of today's consumers without sacrificing future generations to meet their needs [5]. One way that can be used to obtain sustainable design is by imitating nature's design. This term is called Biomimicry. Biomimicry emerged in 1982 and was popularized by a scientist and writer named Janine Benyus in her book "Biomimicry: Innovation Inspired by Nature". Biomimicry is defined by Janine Benyus (1997) as a new science that learns models from nature and then imitates or takes inspiration from it in terms of designs and processes to solve human problems". Biomimicry consists of two words; Bios means life, and mimesis means imitate. It concludes that Biomimicry is a concept that imitates nature's process, form, and models [6]. The Biomimetic Concept makes nature an inspiration [2]. In this research, Velcro design development uses the Biomimicry concept as one of the sustainable design solutions.

Nature is an excellent example of an exemplary process since it does not produce waste everything is recycled, reused, and has no destructive impact on the environment. The concept of Biomimicry encourages innovators to look creatively at nature to generate a design that will reduce the harmful impact on the environment and gives inherent sustainability strategies. Besides, everything inside nature (animals, plants, organisms) also has experienced lots of adjustment, such as evolutions, natural selection, and the existence of climate change that causes a long-term shift in temperature and weather patterns. However, nature still manages to exist today. It gives a more significant possibility to make a better design.

To assist the Biomimicry design process, a design spiral methodology was formed to give methods to further applied Biomimicry in the design process. Biomimicry Design Spiral Methodology was developed by Carl Hastrich, an industrial designer in 2005. It is a step-by-step process of transforming nature's strategies into innovative and sustainable design solutions [6]. There are several ways to integrate biology into a design: redefining the design problem, retrieving inspiration from nature, and establishing guideline criteria for product engineering. The Biomimicry Design Spiral Methodology describes steps to translate human problems and identify natural solutions. The spiral shape is inspired by the nautilus shell's shape that consists of 5 activities: Identify/Define/Distill, Interpret/Biologize/Translate, Discover, Emulate, and Evaluate. Figure 1 below shows the step of the Biomimicry Design Spiral Methodology.

Biomimicry approaches have been widely applied worldwide; for example, the design for Shinkansen Train in Japan is inspired by the Kingfisher bird, the black box concept in an airplane is inspired by the woodpecker bird, the shark skin inspires a swimsuit and so many more. Several researchers in Indonesia utilize the Biomimicry concept to design their innovative products. In the field of building architecture, Almatisha et al. (2021) applied biomimicry architecture approaches to design Marine Research Centre and Oceanarium [8], Nasution et al. (2020) also reviewed Biomimicry architecture approaches to design Rokan Hulu Butterfly Park and its Conservation Center [9]. In the field of product design, several researchers used biomimicry approaches to design swimming hand gloves [10,11], a playing tool for children [12], and also to design Search and Rescue (SAR) ship propellers [13].

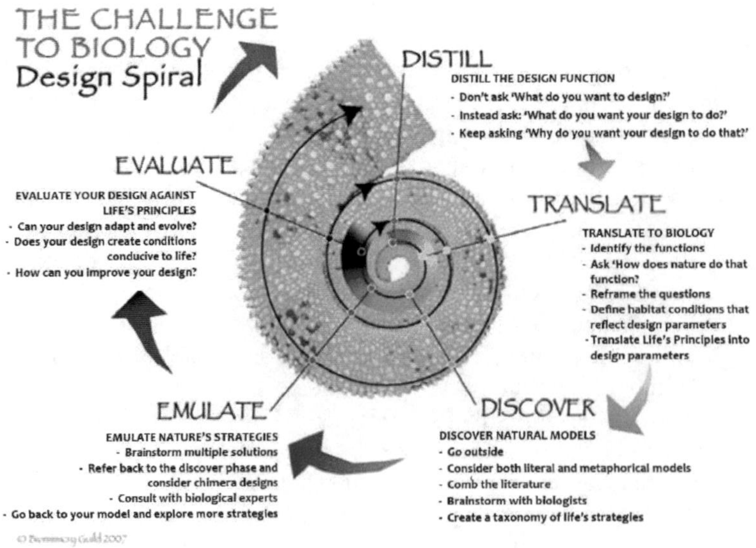

Figure 1. Biomimicry Design Spiral [7].

This research identifies a sustainable design for Velcro. Geroge de Mestral first invented Velcro in Swiss in 1941. Velcro is a fastener used to fasten two sides of the fabric. It is divided into two parts: hook and loop, as shown in Figure 2. The hook is the rougher part made from rougher nylon thread, whereas loops are the softer, more acceptable part and made from softer nylon thread. The cocklebur plant inspires this product design in North America. This plant height is between 50 – 120 cm and belongs to the Asteraceae family or the miserly family. So, cocklebur is still related to chrysanthemum, sunflower, and dandelion [14].

Velcro has so many purposes. It is widely used in health areas, for example, joint bracers as a fastener and adjustable. Besides that, Velcro is used for shoe tighteners in the area of the garment. In military areas, Velcro is used in military uniforms to fasten many patches so they will not fall off. In the electronic area, Velcro is helpful to fasten cable and make them looks tidy. The Velcro helps to tighten the cable, so it will not tangle. Last, Velcro is used to tidy up car stuff in the transportation area. There are still many other velcro functions in real life [15]. On the original website the Velcro, it is said that with a particular strength of Velcro, it can hang a picture without nails because it provides super strong holding power even on rough surfaces and in extreme conditions.

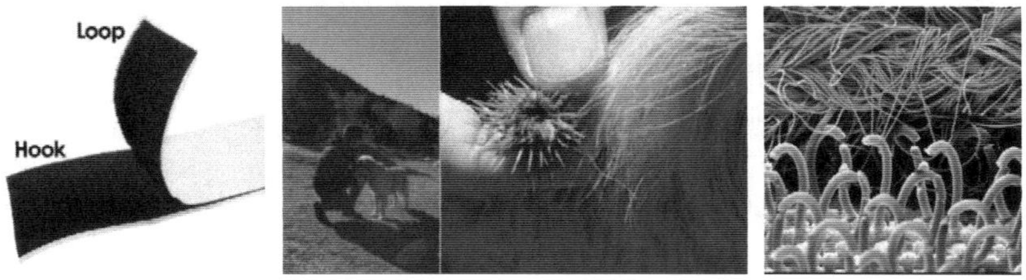

Figure 2. Velcro [14].

Besides Velcro's practical uses, Velcro is still made of polypropylene with customized backings (polypropylene, polyester, PVC) that is thought to take anywhere from 20-200 years to decompose [22]. In long-term use, the Velcro will damage the fabric it is glued. The loop part can be ragged and not as tight as before. Then, it will be thrown away and not be used anymore. In addition, it stated that the hook on the Velcro, originally designed to resemble the Arctium minus, lacks reusability and physical size [21]. Velcro is unlike other things that can be directly recycled when it does not perform as well as the first time. Based on the existing conditions, we need to develop Velcro design to be more sustainable.

Figure 3. Lentinula edodes [16].

In this research, we aim to make a new fastener design with Biomimicry Design Spiral Methodology inspired by mushrooms. The kind of mushroom that we look into is specifically shitake mushroom. Figure 3. shows the morphology of the Shitake Mushroom. This research aims to imitate the shape of pileus and mushroom stem, to be further translated into the form of a new fastener design. Pileus has a blunt edge, reducing the possibility of damaging fabric that directly sticks to the fastener. The stem will embed to the surface of the fastener. Also, the material will be further discussed to obtain an environmental-friendly Velcro.

IOP Conf. Series: Earth and Environmental Science 1111 (2022) 012057 doi:10.1088/1755-1315/1111/1/012057

2. Methods

To design Velcro with a biomimicry approach, the researchers spend three months, from October 2021 until December 2021, executing several steps according to the design spiral methods methodology. The first step is to *distill*, where the researcher needs to filter the desired product function by answering the question, "what should the design be able to do?". The current best fastener is Velcro, inspired by cocklebur, but it leaves damage if attached to a fabric. The researcher wants to design a better fastener that is easy to use, durable, does not leave damage, and also sustainable. The second step is *translat*ion, where the researchers discover how the concept of nature solves the problem and translate the solutions into product ideas. There are two activities in this step: construction studies to develop an idea of nature to the fastener and material studies to discover the material that has the same characteristics as the inspiration from nature. One of the concepts of nature that can be implemented to have desired product function is the shape of *Lentinula edodes,* as shown in Figure 3. Meanwhile, the material that is most suitable for fasteners nowadays is nylon. Nylon has several advantages, such as being lightweight, not easy to deform, and resistant to water and heat.

The third step is to *emulate.* The researchers discover and explore the natural principles of shitake mushrooms to refine the initial concept. Then, several alternative designs for fasteners emerged using the computer-aided design (CAD) software CATIA. The fourth step is to *observe.* The researchers select the best alternative to apply to the product design by assessing the conceptual design with a tensile test using ANSYS. The last is to *evaluate.* After the assessment, it will be known whether the design has fulfilled the desired function or not. In this step, the advantages and disadvantages of the design can be identified to get a better design. The conclusion of this research method can be seen in Figure 4.

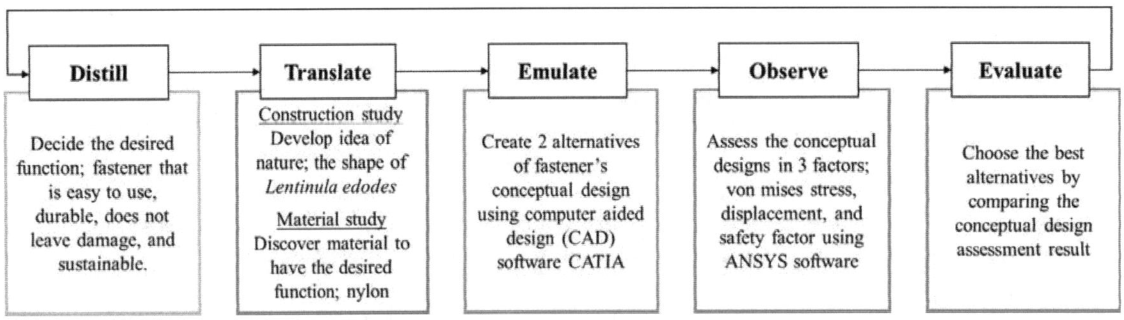

Figure 4. Biomimicry Design Spiral of Velcro Design

3. Results and Discussion

The Velcro was designed using CATIA software. Two alternative conceptual designs vary the space between the stems for the hook part. Both have a height of 0,4 mm, a stem diameter of 0,3 mm, and a pileus diameter of 0,5 mm. But, the space between the stems is 0.75 mm apart for the 1st design (as shown in Figures 5 a, b, and c) and 0,5 mm apart for the 2nd design (as shown in Figures 5 d, e, and f). It is shaped like a mushroom, the material is cut at an angle, and the ends are formed to look like caps or umbrella-like tops. The velcro loop adjusts the hook's specification so it can fit well. The design material also has been adjusted to the real one, nylon.

After that, the design was tested using ANSYS Software. The assessment simulates the velcro material to lift a 5 kg load. The researchers conduct three testing indicators to analyze the product's stress: Von Mises Stress, Safety Factor, and Displacement. These indicators are selected based on research conducted by Gustomo et al. (2020) [17]. *Von mises stress* is a stress that causes a material failure if the material accepts triaxial stress that produces energy. Failure happens when the considerable strain energy of triaxial stress equals the strain energy of the standard tensile test of the material when it starts to dissolve. With this indicator, the researchers try to find the maximum von mises stress value from every design [18]. The material is safe if the *von mises stress* is smaller than the yield strength. *Von mises stress* unit is Pascal. The second indicator is a safety factor. This factor is a safety indicator of the product used. The safety is counted from the dissolved stress from the material divided by the tensile stress. A safety factor of less than 1 shows there will be a permanent failure. In ANSYS Software, the greater the safety factor, the better. It can be seen from the product's color in the ANSYS analysis. The last factor is the displacement factor. Displacement is a shape transformation caused by the movement of part of the product because loads are loaded to both sides of the product [19].

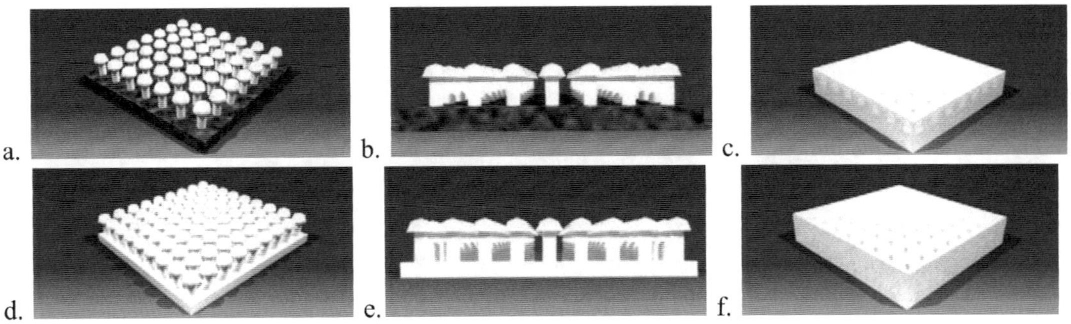

Figure 5. (a) Isometric View of Hook on the 1st design, (b) Isometric View of Hook on the 1st design, (c) Isometric View of Loop on the 1st design, (d) Isometric View of Hook on the 2nd design, (e) Isometric View of Hook on the 2nd design, (f) Isometric View of Loop on the 2nd design

From the simulation conducted for the first design (as shown in Figure 6a), the maximum value of Von mises stress that occurs under load conditions is 0.00572 Mpa, while the minimum value is 0 Mpa. This value is far below the yield strength value as the standard of nylon 645 material of 53.77 Mpa [20]. Most of the distribution stress analysis results on the Velcro design are still blue, so we can conclude that the von mises stress in velcro design is within the safe limits. For the displacement indicator (as shown in Figure 6b), we obtained the maximum value at the load condition is 0,00005764 mm, whereas the minimum value is 0 mm. The displacement is still located in the elastic area, the existence of load is not giving any significant change in shape, and still acceptable. In the safety factor (as shown in Figure 6c), every part of the Velcro is still blue, which means that Velcro has a safety factor above the allowed load, and the design is safe. From the three factors that are being analyzed from the new velcro design, it can be concluded that the second design of Velcro is excellent and safe.

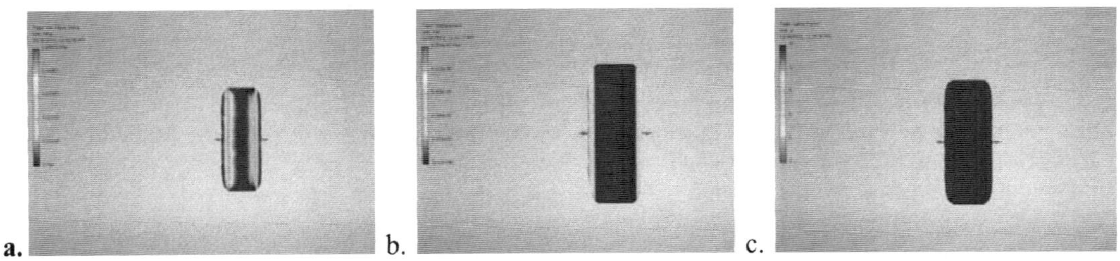

Figure 6. (a) Von mises stress test result on the 1st design, (b) Displacement test result on the 1st design, (c) Safety factor test result on the 1st design.

For the second design, the *von mises stress* indicator test has a maximum value of 0.1105 Mpa and a minimum value of 0 Mpa. The visualization of stress distribution in Figure 7a shows that the second design is bluer than the first one. As for the displacement indicator (Figure 7b), the second design has a maximum value of 0,0000771 mm, while the minimum value is 0 mm, indicating that lifting the load does not significantly change the design. For the safety factor (Figure 7c), all parts are blue, so it can be said that nylon, as the material for Velcro is declared safe to lift 5kg loads. Other research done by

Trilock Velcro Anchor (2018) revealed that Velcro with the same material also passed the safety factor test when lifting 10kg loads [5].

a. 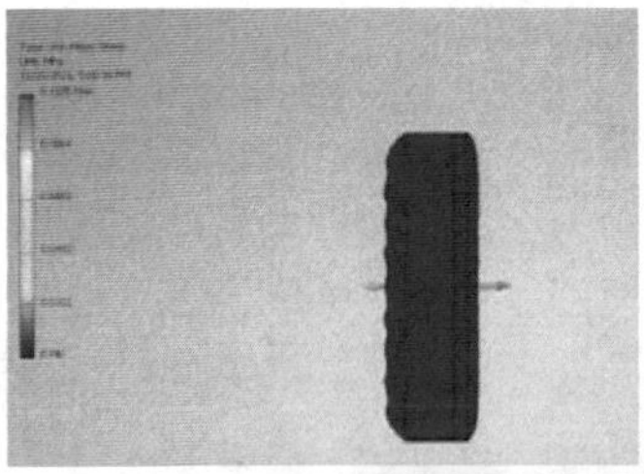

b. 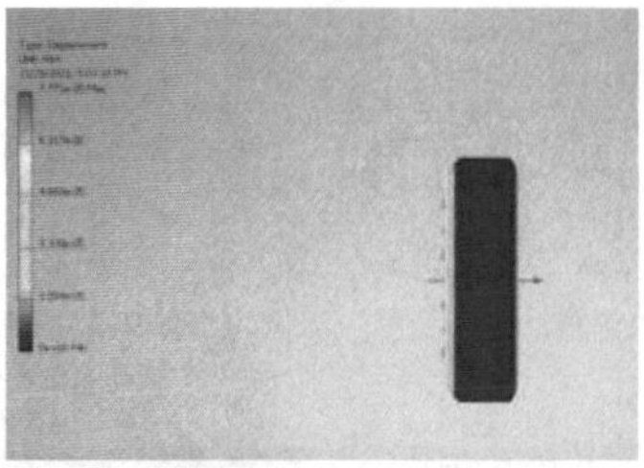

c. 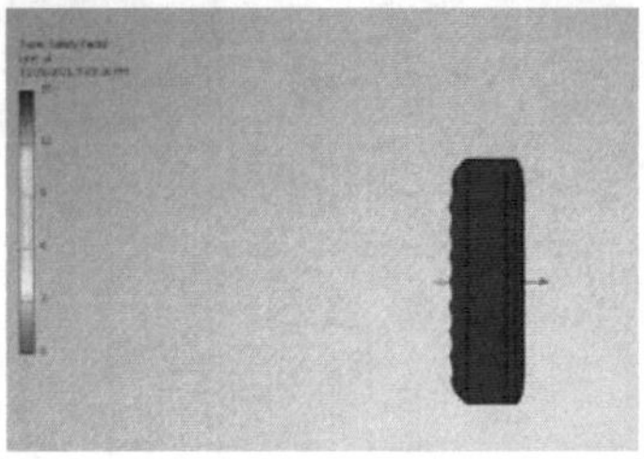

Figure 7. (a) Von mises stress test result on the 2nd design, (b) Displacement test result on the 2nd design, (c) Safety factor test result on the 2nd design.

The comparison of the two alternatives is shown in Table 1. For the von mises stress indicator, the second design is more able to feel the pressure because the second design's maximum value is higher than the first design. With the stress loaded, the first design feels more thorough all over the surface rather than the second design. The displacement indicator of the second design has a higher maximum value than the first. It indicates that the second design experiences more remarkable shape change when given a load. As for the safety factor indicator, both design result is equal. So based on the indicators above, the researcher chose the first design to be applied and used as a fastener that is more durable than the existing Velcro.

Table 1. Comparison between alternatives based on indicators

Indicators	1st design	2nd design
Von mises stress	√	
Displacement	√	
Safety factor	√	√

4. Conclusion

Nature can give lots of inspiration to make and design a product. It has experienced many adjustments, so it will be a perfect imitation if we imitate its shape, process, or system. The stage or phase of the Biomimicry design spiral can be generated into several steps to help design the product. This experiment using the Biomimicry Design Spiral Methodology resulted in a design of a new fastener design that is inspired by nature, specifically *Lentinula edodes* or also well known as shitake mushroom. The von misses stress, displacement, and safety factor test were obtained. From those three tests, the first design with 0,75 mm space between the stems is acceptable and safe.

For future research, it will be better if this sustainable design is equipped with also a sustainable material. The researchers have found that econyl is more sustainable compared to nylon. It is regenerated nylon made from waste such as industrial plastic, waste fabrics, and fishing nets from oceans that is the same quality as virgin nylon. The vision is to make the world better by delivering a sustainable product with a closed-loop regeneration process. In addition, the assessment can also be done in other ways such as "Kric Krac" or peel test to determine velcro resistance.

Acknowledgment
We thank Dr. Ratna Purwaningsih, ST., MT as our lecturer on the Design for Sustainability Subject, who has encouraged us to join this conference held by Universitas Indonesia. We also appreciate and thank our beloved friends who helped write and revise this paper, namely Skolastika Devi Kusuma C and Mikhael Sinaga. Without their help, this paper would not have been published well.

References
[1] Benyus J M 1997 *Biomimicry: Innovation inspired by nature* (New York: William Morrow Paperbacks)
[2] Steadman P 2008 *The evolution of designs: Biological analogy in architecture and the applied arts* (New York: Routledge)
[3] Fiorentino C 2013 Design for Sustainability vs. Design for Resilience: A Time Scale Problem? *Int. J. Sustain. Educ.* **8** 29–35
[4] Vezzoli C, Ceschin F, Osanjo L, M'Rithaa M K, Moalosi R, Nakazibwe V and Diehl J C 2018 Design for sustainability: An introduction *Designing Sustainable Energy for All* (Springer) pp 103–24
[5] NLG 2018 NLG TriLock Velcro Anchor Product code: 101454 (United Kingdom: Never Let Go)
[6] The Biomimicry Institute 2016 The power of the Biomimicry Design Spiral
[7] Peters T 2011 Experimental green strategies: Redefining ecological design research *Archit. Des.* **81** 14–9
[8] Almatisha S F, Dermawati and Pusatarini R A 2019 Implementasi pendekatan arsitektur biomimikri melalui penggunaan self-cleaning concrete pada perancangan marine research centre dan oceanarium *Seminar Inteletual Muda #2, Peningkatan Kualitas Hidup dan Peradaban dalam Konteks IPTEKSEN* pp 280–5
[9] Nasution F A, Aldy P and Susilawaty M D 2020 Kajian arsitektur biomimikri dalam perancangan rokan hulu butterfly park and conservation center *J. Arsit. Zo.* **3** 322–37
[10] Purwaningsih R, Susanto N, Prastawa H, Susanty A and WP S N 2020 Biomimicry class perancangan produk dengan biomimicry desain spiral pada sarung tangan renang *J. Pasopati* **2** 83–90
[11] Purwaningsih R, Wicaksono P A and Saptadi S 2020 The bio-mimicry method in creative process of new product design inspired by nature solution *AIP Conference Proceeding*
[12] Haslina R 2014 Pengembangan desain alat bermain untuk anak TK berbentuk modular menggunakan konsep biomimicry berdasarkan lokomosi ulat Manduca Sexta *J. Tingkat Sarj. Senirupa dan Desain* **3** 1–6
[13] Joyodiharjo B J, Piliang Y A and Larasati D 2019 Pendekatan biomimicry pada desain sarana search and rescue dengan teknologi additive manufacturing *Seni, Teknol. dan Masy.* **2** 23–7
[14] Velcro Brand 2020 Which velcro brand fastener is the strongest?
[15] Han G 2008 The 20 best household uses for Velcro *Apartm. Ther.*
[16] Wurth M 2018 Pilze selbst anbauen (Austria: Biorama)
[17] Gustomo G and Anis S 2020 Analisis kekuatan rangka bodi bus listrik MD12E Perseroan Terbatas Mobil Anak Bangsa dengan metode elemen hingga *J. Mech. Eng.* **54** 1–5
[18] Bhandari V B 1994 *Design of machine elements* (New Delhi: Tata McGraw-Hill)
[19] Ali M A 2018 *Perbaikan kekuatan meja mesin cedu CNC 3 axis* (Universitas Islam Indonesia)
[20] Farani I, Singh N, Colangelo F, Luciano R, Conazzi F, Fraternali F 2019 High-Performance Nylon-6 Sustainable Filaments for Additive Manufacturing *Materials* **12** 39-55
[21] Saunders B E 2019 A biomimetic study of natural attachment mechanisms - Arctium minus part 1 *Roboritics and Biomimetics* (Springer)
[22] Stanes E R, Gibson C R 2017 Materials that linger: An embodied geography of polyester clothes *Geoforum* **85** 27-36

Research of The Reasons for Decreasing the Degree of Desliming of The Sylvinite Ore of the Tyubegatan Deposit

Murodjon Samadiy[1*], Jasur Makhmayorov[1], Bakhodir Abdullayev[1], Shokir Rakhimkulov[1], Ilkham Usmanov[1]

[1]Yangiyer branch of Tashkent Chemical Technological Institute, Department of Chemical Technology of Inorganic Substances, Yangiyer, Uzbekistan.
[1*]samadiy@inbox.ru

Abstract. The purpose of the work was to analyze the work of the first stage of the UE "Dekhkanabad plant of potassium fertilizers" to find out the reasons for the decrease in the efficiency of the process of desliming sylvinite ore of the Tyubegatan deposit. It has been established that the decrease in the efficiency of the desliming process is due to a decrease in the content of potassium chloride and an increase in the insoluble residue in the ore, an increase in the proportion of weakly sludge carbonate and sulfate minerals in the insoluble residue, compared with silicate minerals, a decrease in the differentiated distribution of the insoluble residue in coarse and fine-grained fractions to 1,76, while the design data is 4.22.

1. Introduction

In the world, the production and use of mineral fertilizers are growing every year. This is due to an increase in the yield and quality of agricultural products, providing the earth's growing population with food. For average growth and development of plants, nitrogen, phosphorus, and potash fertilizers are primarily needed [1–3].

In recent years, as a result of specific large-scale measures, high results have been achieved in scientific research in obtaining phosphorus and potash fertilizers based on local raw materials and providing agriculture with high-quality mineral fertilizers. In this regard, the production of flotation potassium chloride from the local sylvinite raw materials of the Tyubegatan deposit should be specially noted [4–7].

The rapid growth of the economy of Uzbekistan provides for the rise of industry to a qualitatively new level, the further intensification of the production of finished products based on the deep processing of local raw materials, and the development of the production of new types of products and technologies. In this regard, a further increase in the capacity for the extraction and enrichment of sylvinites from the Tyubegatan deposit, and their processing into potash fertilizers, is of particular importance. However, during the operation of the processing plant of UE "Dekhkanabad Plant of

Potash Fertilizers", insufficient efficiency of ore desliming processes and, as a result, a decrease in the selectivity of the subsequent sylvite flotation were revealed.

Analysis of the work of the plant showed: - that the content of insoluble residues (i.r.) in the mined ore ranges from 1% to 10%, and to ensure the design content of i.r. in the ore (~ 3%) at the mine, blending and homogenization of the ore sent to the factory is carried out. At the same time, a significant amount of ore with a potassium chloride content below 32% and i.r. Above 3.25% is stockpiled at the mine. Desliming is a pre-treatment operation consisting of removing the most finely dispersed part of crushed ores (sludge) from the pulp to improve the quality of the concentrate [8]. The use of one or another method of desliming sylvinite ores depends on the content of n.d [9]. At factories operating only according to the flotation desliming scheme, the content of n.d. in ore does not exceed 2.5% [10].

Potash ores also contain water-insoluble minerals (up to a few percent in sylvinites), which appear in the flotation circuits as sludge. Sludges also adsorb amines, and the sludge must be removed from the ore before flotation. Since some sludge remains in the sylvinite, the desliming operation is followed by the so-called blinding of the remaining sludge. The fatty amines used as a flocculant are adsorbed on the sludge's minerals, preventing the amine collector's adsorption on these water-insoluble fine particles [11, 12]. In the scientific, technical, and patent literature, many methods are given to improve the efficiency of the separation of i.r. from sylvinite ore. Improving the flotation desliming of sylvinite ores is relevant in all countries producing KCI by the flotation method.

2. Methods

The objects of study were sylvinites of the Tyubegatan deposit located in the Dekhkanabad district of the Republic of Uzbekistan. The productivity of extraction of the original ore is 2100 thousand tons per year, and the annual productivity of potash fertilizers of the unitary enterprise "Dekhkanabad Plant of Potash Fertilizers" is 600 thousand tons per year. The product quality corresponds to the current GOST 4568-95 Potassium chloride. Potassium salt reserves of the Tyubegatan deposit in Uzbekistan: category (A + B + C1) - 215,278 thousand tons with KCl content of 35.95%, category C2 - 60,923 thousand tons with a content of 35.39%. Raw ores are transported by road to the processing complex of the unitary enterprise "Dekhkanabad Plant of Potash Fertilizers".

Research spent with sylvinite ore of the Tyubegatan deposit containing 20 % KCI and 3 % i.r. Deslurrying of low-grade sylvinite ores was spent on the modeling installation simulating industrial conditions and in industrial conditions of first turn UE "Dekhkanabad Plant of Potash Fertilizers" [13]. The essence of deslurrying consists of processing crushed to a fraction less than 1 mm of low-grade sylvinite in the sated solution of sylvinite, branch of the suspension containing particles of 100-200 microns in first and 30-40 microns in the second hydro cyclones. The chemical analysis of initial, intermediate, and end-products spent by known methods [14–17]. X-ray analysis of samples of insoluble impurity has been spent on x-ray diffractometer DRON-6, equipped with a tube with Co-anode and graphite monochromator with the length of a wave λ =1.79021 Å in a range from 5 to 76 degrees on 2 θ.

3. Results and discussion

In order to clarify the reasons for the low degree of deslurrying, the mineralogical composition of the i.r. Sylvinite ores of the Tyubegatan deposit four samples with different content of i.r. sample no. 1 contains 30.4% KCl and 3.25% i.r., sample no. 2 25% KCl and 4.00% i.r., sample no. 3 23.2% KCl and 4.95% i.r. and sample no. 4 20.1% KCl and 5.30% i.r. Studies have shown that i.r. According to chemical analysis, Sylvinite ores of the Tyubegatan deposit contain 65.10-72.50% non-silicate minerals and are mainly represented by carbonate and sulfate minerals (Table 1).

Table 1. Mineralogical contents of water-insoluble remains of sylvinites from the Tyubegatan deposit

Name of minerals	Mineralogical content of insoluble impurities, wt. %				Average for ore samples
	1	2	3	4	
Sulfates	20,92	20,10	16,90	18,50	19,10
Carbonates	51,50	45,00	55,60	49,30	50,35
Total nonsilicate minerals	72,42	65,10	72,50	67,80	69,45
Silicates	27,58	34,90	27,50	32,20	30,54

The analysis showed that i.r. weakly sludge carbonate and sulfate minerals mainly represent the potash ore of the Tyubegatan deposit. Based on the data obtained, the decrease in the efficiency of hydromechanical deslurrying of ore in hydro cyclones, in comparison with the design indicators, can also be explained by a decrease in the content in i.r. Easily sludged silicate and clay minerals. According to X-ray diffraction analysis, insoluble impurities are mainly represented by carbonate minerals, ranging from 38 to 61%. Carbonate minerals are represented by dolomite – 6-9% and magnesite 10- 54%. The content of silicate minerals varies from 32 to 56%, of which clay minerals range from 7 to 29%. Unlike i.r. Source ores in "free" i.r. The content of silicate minerals is 79-86%, of which clay minerals make up 43-56% and non-clay minerals - 29-36%, which is another confirmation of the decrease in the efficiency of hydromechanical deslurrying due to a decrease in the composition of i.r. Ores of the Tyubegatan deposit silicate minerals, which are easily slimed. To confirm the salt composition of i.r. X-ray diffraction patterns and IR spectra were taken (Fig. 1 and 2). The X-ray pattern has diffraction peaks related to calcium and magnesium carbonates to two-water and anhydrous calcium sulfate. Peaks 5.40; 4.56; 3.16 and 2.47 Å belong to calcium sulfate, 3.40; 1.875; 1.626 Å belong to calcite, and peaks 3.38; 2.75; 2.44; 1.818; 1.434 Å - silicates.

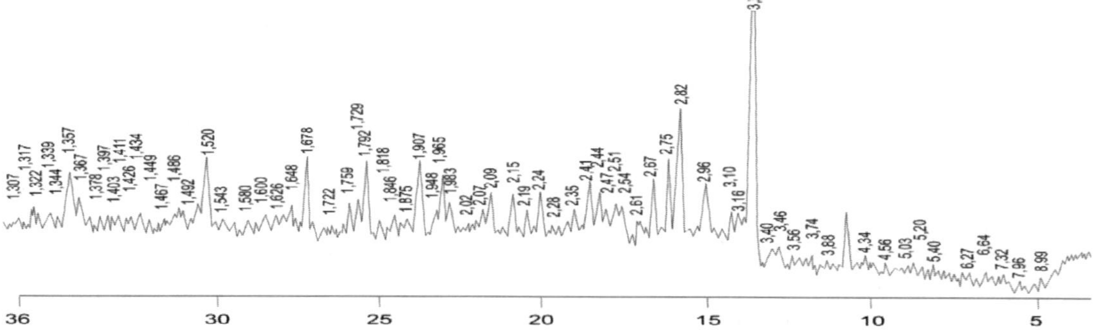

Figure. 1. X-ray pattern of water-insoluble residue

The IR spectra show absorption bands at 1008.44 cm^{-1} related to sulfate groups. Absorption bands 3557.42; 3430.55 cm^{-1} refer to gypsum hemihydrate and dihydrate, and absorption bands 878.32; 467.21 cm^{-1} – refer to silicates. These data confirm the results of chemical and X-ray phase analyses.

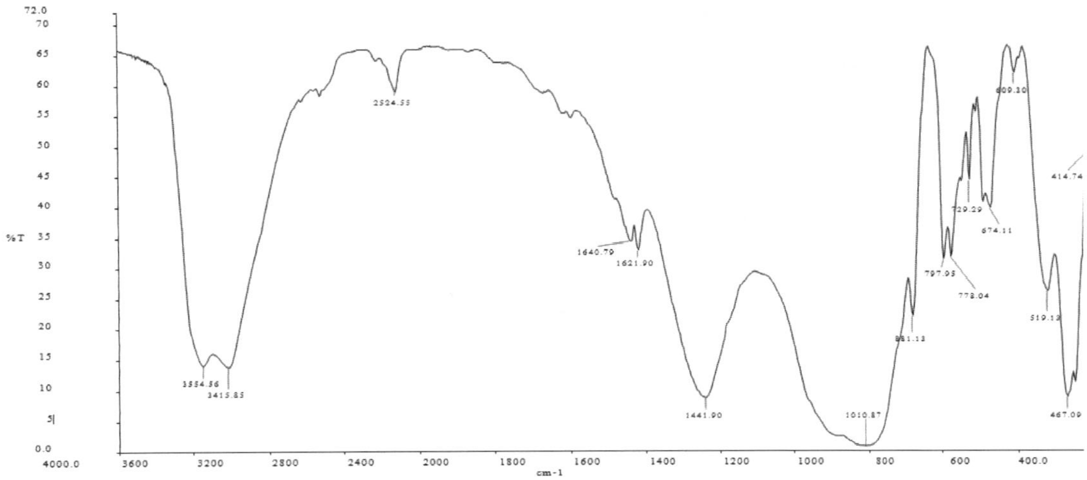

Figure 2. IR spectrum of the water-insoluble residue

Figure 3 shows micrographs of the original sylvinite ore and i.r. microscopic images were taken with an NSZ- 810 stereomicroscope at 50x magnification.

a b

Figure 3. Microscopic image of the water-insoluble residue and sylvinite of the Tyubegatan deposit: a – sylvinite, b – i.r.

Figure 3-a shows a micrograph of the original sylvinite. The bright part in the photo is sodium chloride, potassium chloride, and carbonates. It is dark blotches, i.r. Figure 3-b shows a micrograph of i.r. It has more dark spots due to the sulfate and silicate components of the i.r. This is another confirmation of the composition of the i.r. sylvinite ore of the Tyubegatan deposit.

Distribution studies, i.r. according to ore fineness fractions, four samples of the above composition were carried out on low-grade sylvinite ore. The samples were sequentially crushed to a particle size of less than 1 mm in a snap, and a hammer crusher and wet sieving were carried out. After wet sieving, the content of i.r. It was analyzed in fractions +0.1 mm, -0.1 mm, and in “free” dispersed sludge. In the ore samples taken for wet screening, the ratio of the -0.1 mm fraction to the +0.1 mm fraction was also determined (Table 2).

Content i.r. in the +0.1 mm fraction increases from 2.42% to 4.68%, while the content of i.r. in the -0.1 mm fraction remains at the same level of 8.33-8.90%. With a decrease in potassium chloride content in the ore, the indicators of the differentiated distribution of i.r. Also, decrease in coarse and fine-grained fractions of crushed ore. The ratio of the -0.1 mm fraction to the +0.1 mm fraction decreases from 3.68 to 2.35, while the content of i.r. in the fine fraction was 4.39 times more than in the coarse fraction. This, in turn, leads to a decrease in the degree of extraction of i.r. in the -0.1 mm fraction from 35.2% to 20.02%. This explains another reason for the decrease in the efficiency of hydromechanical deslurrying of ore.

Due to the low efficiency of the processes of deslurrying sylvinite ore and the consequent decrease in the selectivity of subsequent flotation, testing of deslurrying cycles was carried out to determine ways to improve the existing technology and flow diagram.

Table 2. The results of the analysis of the fractional content of sylvinite ores of the Tyubegatan deposit

Indicators	Samples			
	1	2	3	5
Content i.r. in the particle size fraction +0.1 mm, %	2,42	3,40	4,51	4,68
Content i.r. in thc particle size fraction -0.1 mm, %	8,90	8,61	8,33	11,01
Fraction ratio -0.1 mm to fraction +0.1 mm	3,68	2,53	1,85	2,35
Extraction i.r. into the particle size fraction -0.1 mm, %	35,20	21,35	20,65	20,02
Extraction i.r. into freely dispersed sludge, %	9,20	6,77	6,69	5,73

Testing of deslurrying cycles was carried out with the determination of the particle size distribution of ore deslurrying products, distribution of KCl, and i.r. by size classes in crushed ore and ore deslurrying products. The main test results are shown in Table 3.

Table 3. The main results of testing the unitary enterprise "Dekhkanabad Plant of Potash Fertilizers"

Indicators		Sylvanite samples				
		1	2	3	4	The average
Source ore						
Content i.r., %		3,19	3,17	3,20	3,20	3,20
Feed deslurrying ore						
Size fraction content -0.3 ÷ +0 mm, %		77,90	24,10	39,00	45,20	46,55
Content i.r., %	in general deslurrying nutrition	2,77	2,39	2,43	2,73	2,63
	in fraction -0.15 ÷ +0 mm	2,60	4,09	4,68	3,70	3,77
	in fraction +0.15 mm	3,01	1,83	1,90	2,50	2,31
	The ratio of the contents of i.r. in the fraction -0.15 ÷ +0 mm and in the fraction +0.15 mm	1,86	2,23	2,46	1,48	1,76
Sands of hydrocyclones of the 1st stage of deslurrying						
i.r. content, %		1,83	1,76	1,88	1,78	1,81
Sands of hydro cyclones 2nd stages of deslurrying						
i.r. content, %		2,74	2,78	2,82	2,69	2,76
Feeding the main sylvinite flotation						
i.r. content, %		2,03	2,06	1,94	1,95	2,05

An analysis of the data in Table 3 shows that at a content of i.r. 3.2%, the content of the fineness fraction -0.3 mm varies from 24.10% to 77.90%, with an average content of 46.55%. Content i.r. in the general diet, deslurrying averages 2.63%, in the -0.15 mm fraction 3.77%, in the +0.15 mm fraction 2.31%. The ratio of the content of i.r. in the -0.15 mm fraction to the +0.15 mm fraction is 1.76, while the ratio of 4.22 was assumed during the design.

4. Conclusion

Thus, the research conducted on the study of the composition of sylvinite ores with different contents of potassium chloride and insoluble residue, mineralogical composition, and insoluble residue showed that the insoluble residues of ores of the Tyubegatan deposit are represented by slightly sludge carbonate and sulfate minerals, with a decrease in the content of potassium chloride and an increase in insoluble residue, they decrease and indicators of differentiated distribution of insoluble residue in coarse and fine-grained fractions of crushed ore, which together leads to a decrease in the efficiency of the ore deslurrying process.

References

[1] Kurbanov E and Kuziev R 2001 The Current State of Soil Fertility in Uzbekistan and Some Ways to Improve It *Min. Bull. Uzb.* 94–6
[2] Samadiy M A, Mirzakulov K C and Yorboboev R.Ch., Boynazarov B. 2014 *Analysis of technology of deslurrying potassium ores of Tyubegatan* (Minsk)
[3] Samadiy M A, Mirzakulov K C, Melikulova G E, Boynazarov B T and Rakhmatov K B 2015 Investigation process of deslurrying sylvinite ores of the Tyubegatan deposit *Chem. Chem. Technol.* 57–62
[4] Gordeev A and Chernyakov B 2001 Some aspects of the food problem of the world *Quest. Econ.* 50–9
[5] Mineev V G and Bychkova L . 2003 State and prospects for the use of mineral fertilizers in world and domestic agriculture *Agrochem. Mon. Sci. J.* 5–11
[6] ICED 1989 *Report of the International Commission on Environment and Development (ICED)*
[7] Zakharenko V A 2000 World trends in the production and use of mineral fertilizers *Agrochemistry* 14–5
[8] Samadiy M A, Lutfullaev S L and Mirzakulov K C 2016 Physical and chemical characteristics of the insoluble rests in water and their influence on process deslurrying sylvinites of Tyubegatan *Austrian J. Tech. Nat. Sci.* 87–92
[9] Kolos 1982 *Fertilizers, their properties and methods of use*
[10] B.M. B, B.S Z, Namazov S S and Tukhtaev S 2013 The state and prospects of innovative developments in the field of technology of inorganic substances and chemicalization of rural production *Republican Scientific and Technical Conference* (Tashkent) pp 3–21
[11] Arsentiev V A and Leja J 1977 *No Title*
[12] Arsentiev V A, Dendyuk T V, Gorlovsky S I and Forssberg K E 1988 No Title *Proceedings 16th International Mineral Processing Congress,* (Amsterdam) pp 1439–1449.
[13] Samady M ., Turaev K A and Mirzakulov K C 2014 Problems of introduction of innovative ideas, technologies and projects in manufactures *Republican Scientific and Technical Conference* pp. 357–8
[14] M.M V and L.N E 1975 *Methods of the analysis of complex fertilizers*
[15] GOST 20851 1995 *Fertilizer minerals. Methods of definition of a mass fraction of potassium* (M: Standards Publishing House)
[16] 24024. G 1981 *Phosphorus and inorganic connections of phosphorus. Methods of definition of sulphates* (M: Standards Publishing House)
[17] Burrriel M F and Rameres M X 1972 *Photometry of flame* (Mir)

Optimization Model of Cirata Reservoir Management Using Discrete Markov and ARIMA Discharge Forecasting Methods

Dhita Septiani Yogo Sri Waluyo[1]* and Mariana Marselina[1]
[1]Environmental Engineering, Faculty of Civil and Environmental Engineering, Bandung Institute of Technology, Bandung, 40132, Indonesia
[1]*dhita.sysw@gmail.com, [2]mariana.marselina@yahoo.com

Abstract. Even though the management operation guidelines have been determined, the optimal management of the Citarum Cascade Reservoir has not yet been achieved. Optimal management of cascade reservoirs in an integrated manner is difficult to be carried out due to complex operating procedures and high uncertainty of hydrological components. So, research on the optimal management of each reservoir considering the discharge forecasting model as input discharge (Qin) was proposed. The objective of this study was to optimize Cirata Reservoir's operational management. Hence all raw water demands are met without any water wasted through the spillway, as indicated by the correlation between inflow discharge (historical and model) and the correlation between trajectory (guideline and actual). In this case, discharge forecasting methods used were Discrete Markov and ARIMA. The continuous with five years return period (R5 continuous), continuous with ten years return period (R10 continuous), 3-classes Discrete Markov, and 5-classes Discrete Markov methods were applied for guideline trajectories. Based on the results, the correlation coefficients between inflow discharge were 0.63 (3-Classes Discrete Markov); 0.78 (5-Classes Discrete Markov); and 0.68 (ARIMA (1,0,0)(1,0,1) & ARIMA (0,0,2)(1,0,1)). Optimization simulations were carried out for 16 scenarios with combinations of four discharge models and four guideline trajectories during the 2016-2020 period. Based on the research, the scenario by the 5-Classes Discrete Markov discharge model and the R10 continuous guideline trajectory is the most optimal management of the Cirata Reservoir, with a correlation coefficient between the trajectory was 0.91.

1. Introduction

Water resources are renewable natural resources that important for human survival. The necessity for water will rise along with human population growth. However, water availability is difficult to manage and predict accurately due to high spatial and temporal variability [1]. Nevertheless, land use conversion, climate change, and water pollution could also affect water availability. Therefore, a multipurpose reservoir utilization optimization model was proposed to reduce water use problems and environmental damage [2]. The reservoir will hold water during the rainy season and expect to reduce the potential for flooding. In the dry season, water scarcity would be overcome by releasing reservoir water stored previously during the rainy season.

In 2018, Citarum Cascade Reservoir, one of the multipurpose reservoirs in Indonesia, experienced problems including discharge decrement and water pollution [3]. The River Basin

Research Station (BBWS) Citarum report shows that the Citarum Cascade Reservoir will fulfill only 32% of raw water demands in Bandung City in 2020. Even though a reservoir management operation plan in an integrated manner has been determined from January to December, the water supply for electricity purposes is minimal, along with the agriculture and raw water demands [4,5]. Management of cascade reservoirs in an integrated manner is difficult to be carried out due to complex operating procedures and the high uncertainty of hydrological components [6]. Thus, the management of each reservoir is needed to conduct. So, Cirata Reservoir, in the middle of Citarum Cascade Reservoir, could be optimized to anticipate the downstream demand.

Cirata Reservoir management refers to PP 37 of 2010 article 45, which states that the reservoir operation pattern includes dry, normal, and wet year operation patterns. There are two ways to manage reservoir operations. Those are through deterministic and optimal ways. Reservoir management, in a deterministic way, only uses one operating pattern, such as only a dry operation pattern. However, deterministic management results in excess water conditions in the dry and normal operating patterns, and water shortage conditions in the wet operating pattern, so the reservoir objective cannot be fulfilled. Meanwhile, optimal reservoir management uses dry, normal, and wet operating patterns, which are expected to comply with the objectives and constraints [7].

The hydrological component that needs to be considered in managing reservoirs is the discharge. Discharge has a stochastic nature, so it is not easy to predict the amount of discharge at any time. Moreover, climate change is also a factor that could affect the continuity of raw water sources. A discharge forecasting model is necessary to foresee the discharge uncertainty and the annual classification for anticipating climate randomness [8]. Based on those problems, this study's goal was to optimize the operational management of the Cirata Reservoir so that all raw water needs are fulfilled without any water wasted through the spillway. The research of optimization using several discharge forecasting models was conducted in Saguling and Kedung Ombo Reservoirs, where both types of research used spatial correlation and discrete Markov methods to forecast discharge value. Optimal reservoir management was carried out by two scenarios and six scenarios for Saguling and Kedung Ombo Reservoirs research, respectively [9,10]. Meanwhile, this study uses 16 scenarios combined with four model guideline trajectories and four forecasting discharge methods.

2. Methods

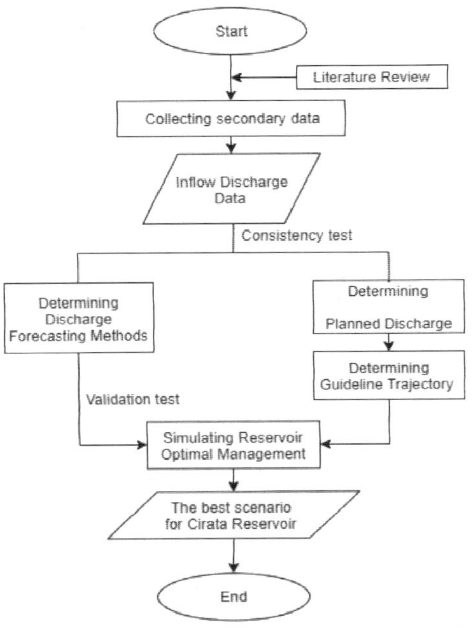

Figure 1. Research method flowchart

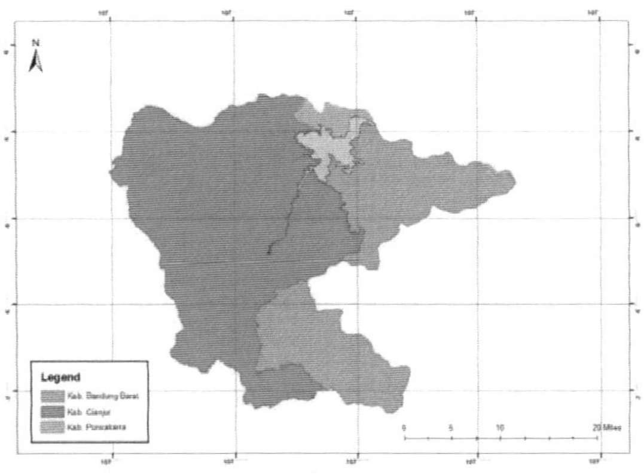

Figure 2. Middle Citarum Watershed

The research location for the Cirata Reservoir is in 3 regencies. Those are Purwakarta Regency, Cianjur Regency, and West Bandung Regency in West Java Province, Indonesia. In addition, Cirata Reservoir is also part of Citarum Cascade, where the Citarum River flows. Figure 2. gives information about the middle Citarum watershed and the location of Cirata Reservoir. Hydrological component data used for this research is monthly average discharge data from 2000-2020, obtained from discharge recording posts in the Cirata Reservoir watershed owned by PT Pembangkit Jawa Bali UP Cirata. Based on Figure 1, the research method was divided into two stages of analysis: discharge analysis consisting of determining planned discharge, discharge forecasting model, and guideline trajectory, and then optimal reservoir management, which was carried out by 16 combined scenarios.

2.1. Consistency Test

The consistency test was also conducted to recognize the truth of the field data, which could be influenced by factors such as shifting in the specifications of the measuring instrument, displacement in the measuring instrument, and the environmental change around the measuring instrument [11]. In this case, the consistency test was carried out using the RAPS (rescaled adjusted partial sums) method. The method's advantage is avoiding the impacts of different data units and random errors in the analysis [12].

2.2. Planned Discharge

The planned discharge is a discharge arranged to occur at a specific return period that is expected to pass through a river. The purpose of determining the planned discharge is to predict the allocation of water required downstream for DMI (Domestic, Municipal, and Industries) and irrigation. Two methods are applied to attain planned discharge. Those are continuous and discrete Markov. The continuous (historical statistic) method used Weibull probability distribution as equation (1). Meanwhile, the discrete Markov method divided ordered discharge into 3 or 5 classes.

$$P(X \geq x) = \frac{m}{n+1} 100\% \quad \textbf{(1)}$$

The guideline trajectory generally aims to predict the reservoir volume at a certain time using the ripple method. The concept of this method is the cumulative curve of reservoir inflow against time which can be mathematically expressed as follow [13]:

$$V(t) = \int_0^t Q(t)dt \quad \textbf{(2)}$$

Guideline trajectory was determined by categorizing each year as super dry/dry/normal/wet/super wet (5-classes) or dry/normal/wet (3-classes) using discrete Markov and continuous planned discharge as input resulting in four guideline trajectory models. Classification for three classes employed continuous planned discharge with five years return period (R5 continuous) and discrete Markov 3-classes planned discharge. On the other hand, continuous planned discharge with ten years return period (R10 continuous) and discrete Markov 5-classes planned discharge were utilized for the classification of five classes.

2.3. ARIMA Discharge Forecasting Method

ARIMA is a combination of models (p, d, q), where AR (p) is the order of Autoregressive components, I (d) Integrated is the degree of difference involved (non-seasonal difference), and MA (q) is the order of Moving Average components [14,15]. In this study, the interpretation of Autoregressive is the discharge value at the time which is influenced by the previous period until the p-period discharge value. Meanwhile, Moving Average means that the value of the discharge amount would be affected by its error, whereas Integrated is the difference based on the data. Therefore, the requisite conditions that must be met for ARIMA methods are static data. If the data is stationary at the same level, then the order is 0, but if the data is stationary at the first difference, the order is 1, and so on [16]. The seasonal pattern is constantly repeated and determined by the autocorrelation coefficient two or three-time-lag identification. In this case, it showed that the autocorrelation was different from zero, so autocorrelation indicated a pattern in a data set. ARIMA method with the seasonal component was notated as $ARIMA\ (p, d, q)(P, D, Q)m$, where m is the amount of period per season and (P, D, Q) were part of the seasonal from a model.

2.4. *Discrete Markov Forecasting Method*

The Discrete Markov method is a general stochastic process with the concept that the probability of a discharge at a specific time (t) depends only on the condition of the discharge at the previous time $(t - 1)$. This method consists of two pulls, which are determining the condition (state) and the quantity (value) [17]. First, to determine the discharge's condition (state), a discharge probability set needs to be summarized into a transition matrix. This was done by dividing the ordered discharge data into classes based on the cumulative distribution curve to attain the probability of each condition from $(t - 1)$ time to (t) time, then putting each probability into a transition matrix. For Discrete Markov 3 classes, the data were divided into three dry (R0), normal (R1), and wet (R2).

On the other hand, the five classes categorized the data as very dry (R0), dry (R1), normal (R2), wet (R3), and very wet (R4). Afterward, the transition matrix was transformed into discharge classification. The second step was determining the quantity (value) by averaging the discharge data in each class to earn a specific discharge value. The discharge classification was then transformed to its precise value for each class, so the discharge model during the period was acquired.

2.5. *Validation Test*

In order to find a representative model, a validation test is needed. Through a statistical test, historical discharge was compared with the model discharge. This study used the correlation, Root Mean Square Error (RMSE), and Mean Absolute Percentage Error (MAPE) as validation tests. The correlation test aims to know the relation degree between two variables, while RMSE and MAPE were applied to detect the accuracy of each model [18].

2.6. *Reservoir Optimal Management*

Reservoir operation is defined as regulating how much water is drawn from a source (e.g., river) to be further transferred between reservoirs and to be released from the reservoir for consumption (e.g., for DMI) or utility (e.g., hydropower production) [19]. Reservoirs operate the management according to their necessity. In its operation, reservoir management has several obstacles, including the law of water mass balance, where there is a limit to the volume of water in the reservoir following the mass balance; reservoir storage volume limitation; reservoir water level elevation; limit of inlet and outlet discharge, where the inlet discharge is less than the reservoir capacity by releasing the excess into the spillway, while the outflow adjusts to the downstream water demand; and planned discharge as an input discharge.

Meanwhile, the operational constraint includes the maximum turbine discharge allowed. Furthermore, optimal reservoir management using a discharge forecasting model and annual classification is encouraged to apprehend the discharge and climate uncertainties [8]. The objective of this optimal Cirata Reservoir management is to fulfill all raw water demand downstream without any water being wasted through the spillway.

3. Results and Discussion

This research used monthly average inflow discharge data for discharge analysis during 2000-2020. Figure 3. illustrates that the inflow of Cirata Reservoir consists of outflow from Saguling Reservoir and water resources of the local Cirata watershed. In this case, the discharge data need to do a consistency test, as mentioned before, so by applying the RAPS method. The data are proven to be consistent for all resources. Afterward, predicting the discharge availability for raw water demand is compulsory, in this case, using the Weibull method to attain the inflow reliability of Cirata Reservoir, as shown by the graph in Figure 4. The figure demonstrates the water allocation downstream was 114.45 m^3/s (R5 dry, the reliability discharge with a probability of 80%), of which 87.98 m^3/s is for DMI demand (R10 dry, the reliability discharge with a probability of 90%) and 26.47 m^3/s for irrigation need (R5 dry – R10 dry).

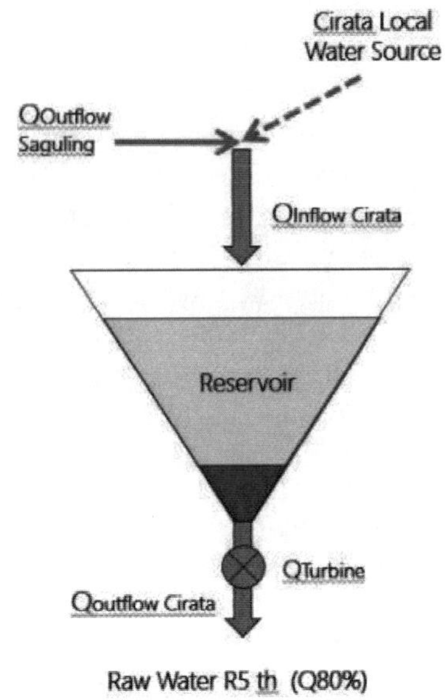

Figure 3. Discharge flow scheme

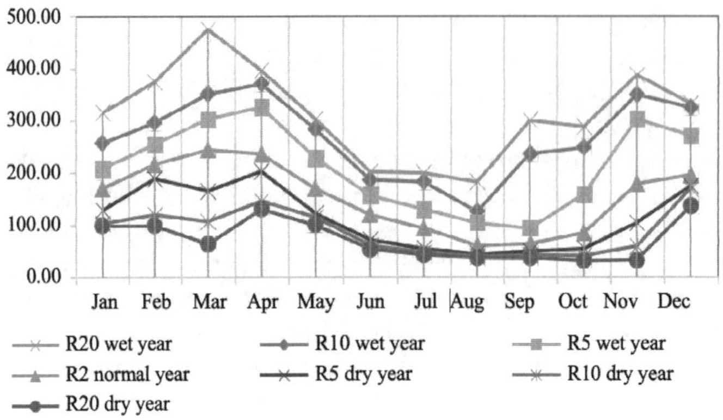

Figure 4. Graph of Cirata Reservoir inflow reliability discharge during 2000-2020

The planned discharge as input for the ripple method will produce guideline trajectories for each method which are displayed in figures 5-8. The R5 continuous guideline trajectory complies with the planned discharge of R5 dry (the reliability discharge with a probability of 80%), R2 normal (the reliability discharge with a probability of 50%), and R5 wet (the reliability discharge with a probability of 20%). The discrete Markov method used the discrete Markov planned discharge R0, R1, and R2 (3- classes). Based on Figure 5, the R5 continuous method for the dry year experiences excess water in April and May. Compared to Figure 6, there was no such event in the discrete Markov 3 class method. However, both methods had overflow water in normal and wet years from April to June.

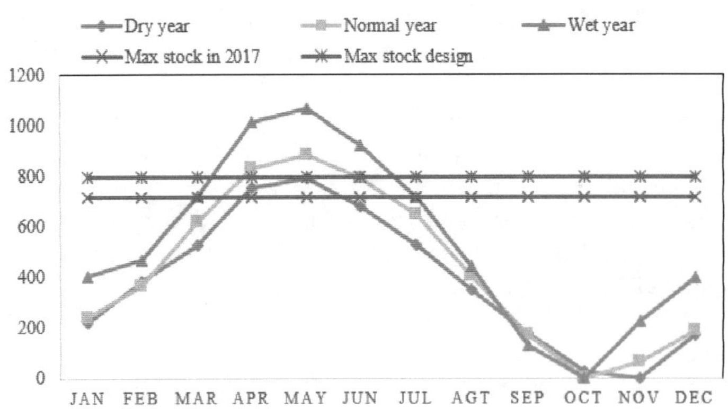

Figure 5. Guideline trajectories R5 Continuous (in Mio m^3)

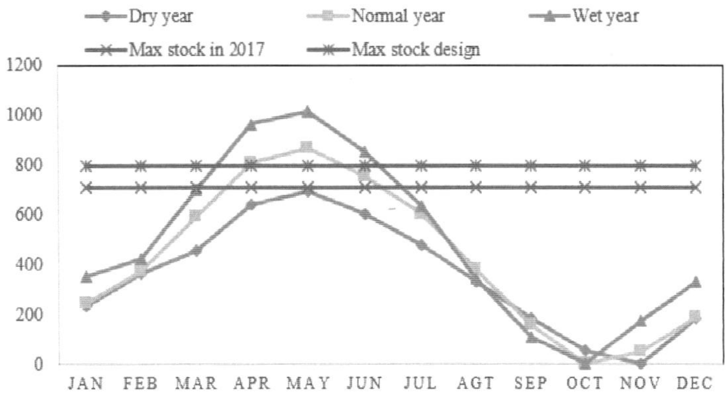

Figure 6. Guideline trajectories 3-classes Discrete Markov (in Mio m^3)

The R10 continuous guideline trajectory used the planned discharge of R10 dry (the reliability discharge with a probability of 90%), R5 dry (the reliability discharge with a probability of 80%), R2 normal (the reliability discharge with a probability of 50%), R5 wet (the reliability discharge with a probability of 20%), and R10 wet (the reliability discharge with a probability of 10%). Whereas the discrete Markov method was applied, the Discrete Markov planned discharge R0, R1, R2, R3, and R4 (5- classes). Figures 7 and 8 show that both guideline trajectory methods for dry years suffered from an overspill of water in April and May, while in normal, wet, and very wet years, the overflow water occurred from April to June.

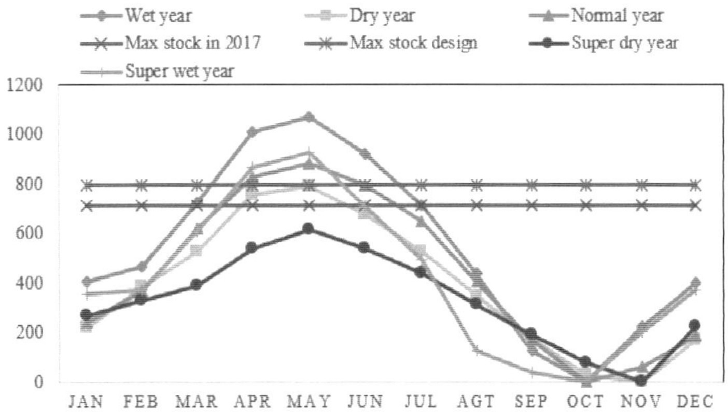

Figure 7. Guideline trajectories R10 Continuous (in Mio m^3)

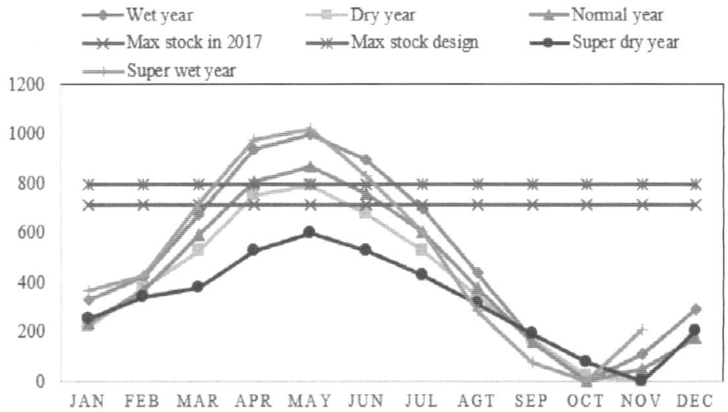

Figure 8. Guideline trajectories 5-classes Discrete Markov (in Mio m^3)

ARIMA models need statistical software to be acquired. Through R software, three discharge ARIMA models qualified for the discharge forecasting method in Cirata Reservoir for 2016-2020. However, only two models had the most significant correlation coefficient. Those are ARIMA (1,0,0)(1,0,1) and ARIMA (0,0,2)(1,0,1), with the amount of 0.68 for both models. The ARIMA models were then compared to the historical discharge illustrated in Figures 9 and 10. In the meantime, consecutively, the 3-classes and 5-classes Discrete Markov models were demonstrated in Figures 10 and 11. In this case, the correlation coefficient 3-classes model was close to 0.64. On the contrary, the 5-classes model had a greater correlation coefficient than the 3-classes (0.78), so bigger classes will give a better correlation.

Table 1. describes the validation test result for each discharge model. According to the table, the 5- classes discrete Markov model had the best correlation (0.78). However, the accuracy model was the worst compared to the others, as shown in the MAPE value, while the RMSE test tells that the 3-classes Discrete Markov model is the minimum accurate model. Even so, all the models will exert for reservoir optimization simulation.

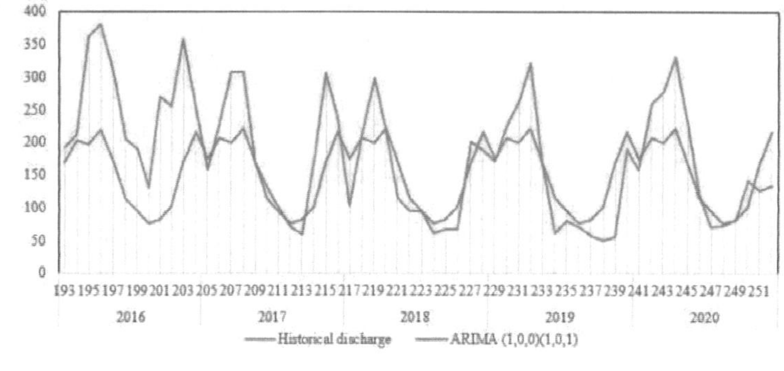

Figure 9. Comparison of Historical Discharge and ARIMA (1,0,0)(1,0,1) model (in m^3/s)

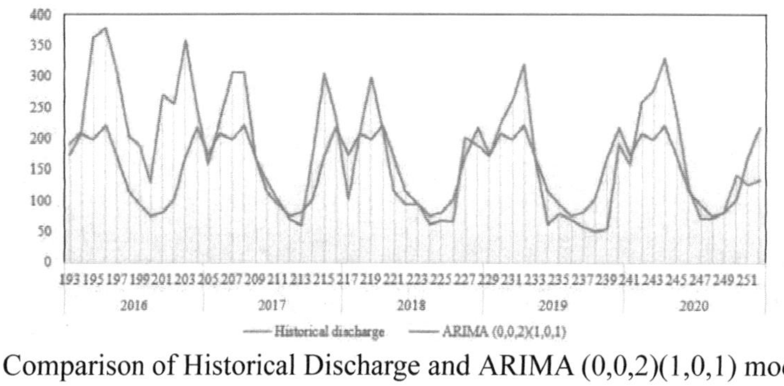

Figure 10. Comparison of Historical Discharge and ARIMA (0,0,2)(1,0,1) model (in m^3/s)

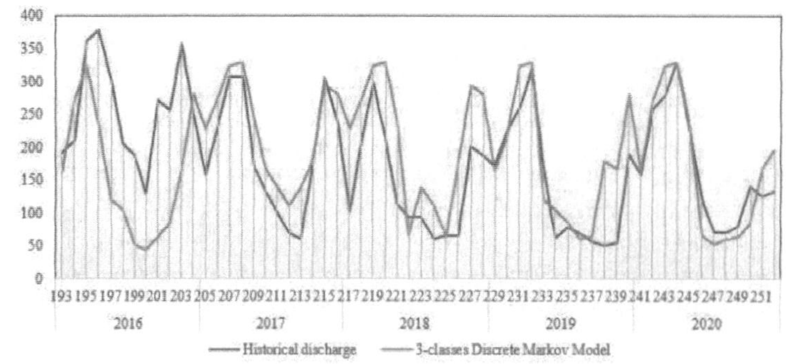

Figure 11. Comparison of Historical Discharge and 3-classes Discrete Markov model (in m^3/s)

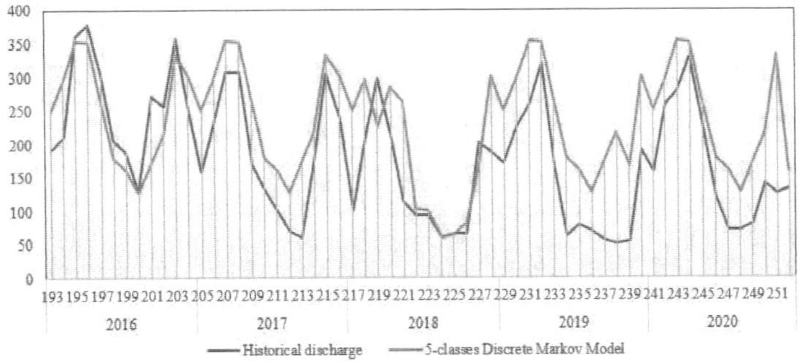

Figure 12. Comparison of Historical Discharge and 5-classes Discrete Markov model (in m^3/s)

Table 1. Validation test results for each discharge forecasting model

Discharge Forecasting Model	Validation Test		
	Correlation	MAPE	RMSE
ARIMA (0,0,2)(1,0,1)	0.68	31%	73.24
ARIMA (1,0,0)(1,0,1)	0.68	31%	73.50
3-Classes Discrete Markov	0.63	44%	79.46
5-Classes Discrete Markov	0.78	56%	78.37

Table 2. Annual discharge classification

Year	3-Classes	5-Classes
2016	Wet	Normal
2017	Wet	Super Wet
2018	Wet	Wet
2019	Normal	Normal
2020	Wet	Normal

The annual classification for 3-classes and 5-classes during 2016-2020 is provided in Table 2. The table shows the discharge tends to have a wet year in 3-classes and an average year in 5-classes during the period. Before optimizing the reservoir, information on the existing condition of the Cirata Reservoir economic environment was needed. The latest research informed that the maximum volume at normal water levels measured in 2017 dropped to 714 million m^3 [20]. So, the constraints for optimizing Cirata Reservoir consist of R5 dry year reliability discharge (114.45 m^3/s) as the minimum and turbine discharge as the maximum discharge required, and then the volume in 2017 (714 million m^3) as the maximum stock limit.

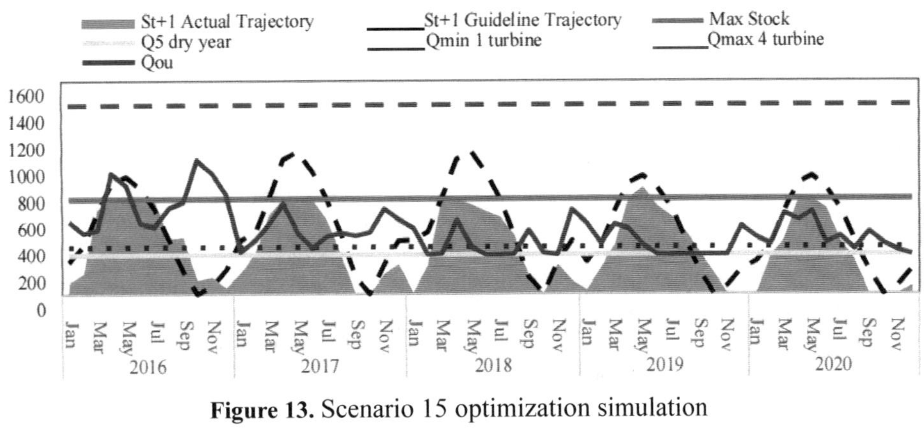

Figure 13. Scenario 15 optimization simulation

Table 3. Scenario Correlation Summary

Scenario	Discharge Forecast Method	Guideline Trajectory	R Discharge	R Trajectory
1	ARIMA (1,0,0)(1,0,1)	R5 Continuous	0.68	0.83
2	ARIMA (1,0,0)(1,0,1)	3-Classes Discrete Markov	0.68	0.84
3	ARIMA (1,0,0)(1,0,1)	R10 Continuous	0.68	0.85
4	ARIMA (1,0,0)(1,0,1)	5-Classes Discrete Markov	0.68	0.85
5	ARIMA (0,0,2)(1,0,1)	R5 Continuous	0.68	0.83
6	ARIMA (0,0,2)(1,0,1)	3-Classes Discrete Markov	0.68	0.84
7	ARIMA (0,0,2)(1,0,1)	R10 Continuous	0.68	0.85
8	ARIMA (0,0,2)(1,0,1)	5-Classes Discrete Markov	0.68	0.85
9	3-Classes Discrete Markov	R5 Continuous	0.63	0.81
10	3-Classes Discrete Markov	3-Classes Discrete Markov	0.63	0.82
11	3-Classes Discrete Markov	R10 Continuous	0.63	0.83
12	3-Classes Discrete Markov	5-Classes Discrete Markov	0.63	0.82
13	5-Classes Discrete Markov	R5 Continuous	0.78	0.89
14	5-Classes Discrete Markov	3-Classes Discrete Markov	0.78	0.89
15	5-Classes Discrete Markov	R10 Continuous	0.78	0.91
16	5-Classes Discrete Markov	5-Classes Discrete Markov	0.78	0.90

Table 3 explains scenario 15 as the most optimal scenario, with the best correlation, which is not only the correlation between discharge model and historical data (R Discharge) but also the correlation between guideline and actual trajectory (R Trajectory) by 0.91. Since both ARIMA models had the same R Discharge, the R Trajectory for each method was also applied to them. Compared to research in Saguling Reservoir with the same scenario (scenario 10), the R Discharge and R Trajectory in this research were minor [9]. The simulation optimization for scenario 15 is illustrated in Figure 13. During the period, there was an overflow in June 2019 of 84 million m^3 which was gradually caused by a decline in maximum stock volume.

4. Conclusion

In conclusion, the 5-classes Discrete Markov discharge model is the best model for Cirata Reservoir, despite its accuracy level, compared to the others. Meanwhile, the R10 continuous method is more suitable for guideline trajectory in Cirata Reservoir during the period. For all scenarios, the discharge in the reservoir could meet the water demand. Unfortunately, they still had water wasted through the spillway in June 2019, with the 15th scenario being the least overspill. The results also show that the correlation coefficient of the discharge model and historical data inclines with the correlation coefficient of guideline and actual trajectory, proving a better scenario for optimizing the Cirata Reservoir. As a recommendation, this study needs to compare with another optimization model to fulfill the other objectives of Cirata Reservoir. Therefore, in addition to all raw water demands achieved, the profit from electricity can also be maximized.

Acknowledgments
This research was supported by the ITB Research Fund of 2022.

References

[1] Sitompul M and Efrida R 2018 Evaluasi Ketersediaan Air DAS Deli Terhadap Kebutuhan Air (Water Balanced) *J. Rekayasa Sipil* **14** 121
[2] Xu X, Bin L, Pan C, Ding A and Chen D 2014 Optimal Reoperation of Multi-Reservoirs for Integrated Watershed Management with Multiple Benefits *Water (Switzerland)* **6** 796–812
[3] Pembangkitan Jawa Bali 2018 *Menghadirkan Nilai Tambah dan Harapan Baru bagi Pemangku Kepentingan* (PT Pembangkitan Jawa Bali)
[4] Suwarno R I and M C M 2020 Profil BBWS Citarum (Bandung)
[5] Badan Pengkajian dan Penerapan Teknologi 2021 Teknologi Modifikasi Cuaca Atasi Defisit

Level Air Waduk Citarum (Bandung)
[6] Thuillier K and Qm B 2019 To cite this version : Linear Programming For Metabolic **126** 118–27
[7] Corsita 2014 Pengelolaan Waduk Kaskade Citarum Untuk Pengembangan Air Baku Infrastruktur Air Minum DKI Jakarta (Bandung: Institut Teknologi Bandung)
[8] Sabar A 2010 *Tren Global Pembangunan Infrastruktur Sumber Daya Air yang Berkelanjutan* (Bandung: Fakultas Teknik Sipil & Lingkungan - ITB)
[9] M. Marselina, A. Sabar, I. R. S. Salami D M 2015 Model Prakiraan Debit Air Dalam Rangka Optimalisasi Pengelolaan Waduk Kedung Ombo *J. Teh. Lingkung.* **21** 77–86
[10] Hutagalung S E M and Sabar A 2015 Model Prakiraan Debit Air Dalam Rangka Optimalisasi Pengelolaan Waduk Kedung Ombo *J. Teh. Lingkung.* **21** 77–86
[11] Kamiana I 2011 *Teknik Perhitungan Debit Rencana Bangunan Air* (Jakarta: Graha Ilmu)
[12] You Q, Jiang H, Liu Y, Liu Z and Guan Z 2019 Probability Analysis and Control of River Runoff-Sediment Characteristics Based on Pair-Copula Functions: The Case of The Weihe River and Jinghe River *Water (Switzerland)* **11**
[13] Takal K M, Sorgul A R and Balakarzai A T 2017 Estimation of Reservoir Storage Capacity and Maximum Potential Head for Hydro-Power Generation of Propose Gizab Reservoir, Afghanistan, Using Mass Curve Method *Int. J. Adv. Eng. Res. Sci.* **4** 98–104
[14] Fashae O A, Olusola A O N I and U C G 2019 Comparing ANN and ARIMA model in predicting the discharge of River Opeki from 2010 to 2020 *Sustain.* **35** 69–77
[15] Ghimire B N 2017 Application of ARIMA Model for River Discharges Analysis *J. Nepal Phys. Soc.* **4** 27
[16] Wahab A 2020 Tracing The Influence of Islamic Feast Days on The Main Commodity Inflation According to The Expenditure Group **9** 47–59
[17] Wulandari S, Sabar A and Marselina M 2017 Comparison of Hydrology Component Spatial Correlation Model (Continuous Model) and Markov Chain (Case Study of Cikapundung Watershed , Saguling Reservoir and Cipanunjang Reservoir) **12** 13779–87
[18] Marselina M and Sabar A 2021 A Discharge Estimation Model Using a Spatial Correlation Method in Cipanunjang-Cileunca Reservoir, Bandung District, West Java Province *IOP Conf. Ser. Earth Environ. Sci.* **716**
[19] B. Dobson, T. Wagener F P 2019 An Argument-Driven Classification and Comparison of Reservoir Operation Optimization Methods *Adv. Water Resour.* **1** 74–86
[20] Salawangi W K 2020 *Model Integrasi Pengelolaan Sumber Air Waduk Kaskade Citarumdalam Memenuhi Kebutuhan Air Baku di Hilir 1105/S2-TL/TML/2020* (Bandung: Institut Teknologi Bandung)

IOP Conf. Series: Earth and Environmental Science 1111 (2022) 012060 doi:10.1088/1755-1315/1111/1/012060

The Study of Biopore Infiltration Holes (BIH) Implementation to Reduce Waterlogging in President University Campus Area

Alifah Shafina Ashabie[1]*, Yunita Ismail Masjud[2]
[1] Environmental Engineering Department, Faculty of Engineering, President University, Bekasi City, 17530, Indonesia, [1]000-0003-0239-7921
[2] Environmental Engineering Department, Faculty of Engineering, President University, Bekasi City, 17530, Indonesia, 0000-0002-3297-8850
[1]*alifah.ashabie@student.president.ac.id, [2]yunitaismail@president.ac.id

Abstract. The continuous infrastructure development results in the lack of green open space as water catchment areas into the ground. The reduced water infiltration zona will cause runoff. The soil on President University's campus is fine-grained, with high plasticity due to clay mineral particles. One factor that affects rainwater utilization is the soil type. Clay types have slower permeability, resulting in reduced rainwater infiltration areas. One of the technologies to reduce water runoff is the Biopore Infiltration Hole (BIH). To find out how much water can infiltrate into the soil to reduce waterlogging and to plan the total amount of Biopore Infiltration Hole (BIH) that can be applied at President University. This study uses quantitative analysis methods to identify the ideal Biopore Infiltration Hole (BIH) requirement on President University's campus and determine the BIH. The findings of this study indicate that the total planned implementation of BIH that can be used on the campus of President University is 1711.8 pcs in an open space zona of 13,600 m^2 and, based on the annual infiltration calculation, is 0.97 m^3/min that significant enough to reduce waterlogging in the President University Campus Area.

1. Introduction

Infrastructure is the driver of economic growth. Infrastructure is capital goods or social overhead capital, which is the basis or essential means of meeting social needs, which are then indirectly used to produce or increase production. Every year, there is an increase in infrastructure development, which improves the community's welfare because infrastructure development is focused on connecting with the production centers of the people's economy. Infrastructure development continues to be carried out so that a large land area has been covered by cement [1]. Asphalt has covered many roads, so rainwater cannot infiltrate the ground properly and cause waterlogging. Even if there is high rainfall, it will cause flooding. The regulation regulates the allocation of 30% RTH, which aims at maintaining urban balance ecosystems, hydrological systems, and micro-climates, as well as other clean air that has the potential to increase the availability of community ecosystems [2].

Rainwater that falls to the earth cannot seep directly into the ground and will become runoff. So the lack of green open space for catchment areas is that they cannot withstand the flow of water caused by high rainfall, resulting in waterlogging and even flooding [3]. Floods are a significant problem that can occur annually in flood-prone environments, so these areas need special attention. Another thing that can cause flooding is that the rainwater that falls on the earth cannot infiltrate the ground properly because many lands are already covered with cement which can cause puddles also cause flooding. Bekasi Regency is the location where the President University campus is located. Geographically Bekasi Regency is located between 60 10'53 "-60 30" 6 "S and 1060 48" 28 "-1070 27" 29 "East Longitude. This location makes Bekasi Regency in the western part from north to south across West Java. Bekasi covers an area of 127,388 hectares, one of which area is the Cikarang Utara District. The temperature range of Bekasi Regency is between 28°C -32°C [4]. The President University campus area has three train stations: WTP-2, Cibarusah, and BA-2. Rainfall data from 2015 to 2019 were utilized to calculate rainfall intensity. The greatest rainfall recorded at the WTP-2 rain station in the last five years has been 67 mm, 129 mm, 120 mm, 75 mm, and 120 mm. The greatest rainfall at the Cibarusah rain station in the previous five years was 88 mm, 135 mm, 114 mm, 90 mm, and 95 mm. The greatest rainfall at the BA-2 rain station in the previous five years was 61 mm, 110 mm, 89 mm, 72 mm, and 80 mm in the most recent year, 2019.

The President University area's groundwater condition is of clay type, meaning that it has certain particles that will produce plasticity in the soil when mixed with water [5]. Therefore, by using biopore technology, this research will use small land in the President's University area. By storing some of the surface water runoff in the ground, the biopore infiltration hole (BIH) is a concept that can be used to save water or increase groundwater reserves. The biopore infiltration hole is built into the ground as a perpendicular cylinder hole with a diameter of 10 centimeters and a height of 100 centimeters. The distance between the biopore infiltration holes is 100 centimeters [6]. The use of this biopore technology uses PVC pipe as the primary material. Several holes in the biopore infiltration hole are also made to accelerate water absorption into the soil. In an earlier study, biopore incorporated in organic waste of fruit peel had a significant infiltration outcome every week. This method takes four weeks and has varied values since rain does not fall optimally until the measurements are taken. The low groundwater level ensures that trash does not stagnate and that decomposition proceeds as efficiently as possible. This is due to an increase in the rate of infiltration. Similarly, prolonged rain might reduce the pace of biopore infiltration, causing the infiltration process to fail [7].

2. Methods

This research was conducted experimentally by field test in several locations. The first experiment was carried out near the campus prayer room or at coordinates -6.284919°, 107.170460°, and the second location for data collection was near the campus B building area, or at coordinates -6.285779°, 107.170460°. The field testing process was carried out in a grass area, and two trials were conducted at each location.

2.1 Experimental method

The experimental method is (a) to immerse the ring vertically into the ground to a depth of 5 centimeters using a wooden block and a hammer. Make sure the double ring infiltrometer can stand firmly, (b) avoid leakage around the ring wall by compacting the part of the soil that is in contact with the ring wall, (c) install the ruler vertically on the inner ring, (d) flood the measuring ring with a constant depth and measure the rate at which water enters the soil, (e) install a float to ensure the water level is appropriate, (f) queasy steady-state flow is assumed to occur when the rate of decrease of water in the ring becomes constant. Then the measurement is stopped.

2.2 Data analysis method

2.2.1 Hypothesis

H0: $(\mu 1-\mu 2) = 0$, which means that all data follow a specific distribution

H1: $(\mu 1-\mu 2) \neq 0$ means that all data do not follow a particular distribution

Microsoft Excel was used to perform the Smirnov-Kolmogorov analysis. The statistical test's goal is to see if the independent variables significantly affect the dependent variable. If the p-value is less than alpha 0.05 (5%), the null hypothesis is accepted, and no difference between treatments exists.

IOP Conf. Series: Earth and Environmental Science 1111 (2022) 012060 doi:10.1088/1755-1315/1111/1/012060

2.2.2 *Calculation of Infiltration*
The maximum rate of water that can enter the soil at one time is referred to as infiltration capacity. The formula for infiltration is shown below [8].

$$f = FC + (f_0 + FC).\ e^{-kt} \quad \textbf{(1)}$$

2.2.3 *Statistical Testing Using ANOVA*
H0: (μ1-μ2) = 0, which means that all population data is the same, and there is no influence from the location factor.
H1: (μ1-μ2) ≠ 0 means that all population data are not the same, and there is an influence from the location factor.

The analysis of variance (ANOVA) test was run in Microsoft Excel. The variance test aims to determine whether the independent variable has a significant impact on the dependent variable. If the p-value is less than alpha 0.05 (5%), the null hypothesis is accepted, and no effect between locations is observed.

3 Results and Discussion
In order to plan for waterlogging control, planners must have as much information as possible about the situation. These are commonly known as hydrological data.

3.1. *Hydrological Data Analysis*
The rainfall data used in this study came from the three stations closest to the research site. The rain stations visited were WTP-2, Cibarusah, and BA-2. The most significant rainfall recorded at the WTP-2 rain station in the last five years has been 67 mm, 129 mm, 120 mm, 75 mm, and 120 mm. The most significant rainfall at the Cibarusah rain station in the previous five years was 88 mm, 135 mm, 114 mm, 90 mm, and 95 mm. The most significant rainfall at the BA-2 rain station in the previous five years was 61 mm, 110 mm, 89 mm, 72 mm, and 80 mm in the most recent year, 2019, as shown in the table below.

Table 1. Rainfall in the area and planned area

Year	Rh Max WTP-2	Rh Max Cibarusah	Rh Max BA-2	Rh Plan
2010	0	111	68	59.67
2011	0	75	110	61.67
2012	109.5	74	74	85.83
2013	105	79	174	119.33
2014	184	142	213	170.67
2015	67	88	61	72.00
2016	129	135	110	124.67
2017	120	114	89	107.67
2018	75	90	72	79.00
2019	120	95	80	98.33

The projected rainfall is used to determine the first computation of rainfall intensity towards runoff. The anticipated rainfall frequency study comprises calculating statistical parameters (dispersion measurement), establishing the kind of distribution, and evaluating the distribution alignment. The dispersion measurement resulted in average rainfall of 99 mm and a total maximum rainfall plan of 987.83 mm. After obtaining the dispersion calculation results, the distribution must be calculated, as shown in the table below:

Table 2. Distribution using Gumbel Type I Method

No	Period	*X*	Sd	Sn (n=10)	Yn (n=10)	Yt	Xt
1	2	99	36.334	0.9496	0.4952	0.3668	93.8705
2	5	99	36.334	0.9496	0.4952	1.5004	137.4614
3	10	99	36.334	0.9496	0.4952	2.251	166.1812
4	25	99	36.334	0.9496	0.4952	3.1999	202.4654
5	50	99	36.334	0.9496	0.4952	3.9028	299.383
6	100	99	36.334	0.9496	0.4952	4.6012	256.1055

The following table shows the results table for calculating the frequency distribution and distribution using the Log Pearson Type III method as shown in the table below:

Table 3. Distribution using log Pearson type III method by period

No	Period	Log(x)	Sd log(x)	Cs	k	Y=log(x)+ k(Sd log (x))	X=10Y
1	2	1.9706	0.1498	0.6770	-0.099	1.955789	90.321
2	5	1.9706	0.1498	0.6770	0.8	2.090430	123.15
3	10	1.9706	0.1498	0.6770	1.328	2.169507	147.74
4	25	1.9706	0.1498	0.6770	1.030	2.261016	182.40
5	50	1.9706	0.1498	0.6770	2.359	2.323918	210.82
6	100	1.9706	0.1498	0.6770	2.755	2.383226	241.67

Methods for determining the type of distribution must be used following the requirements of each method. The following is a table of conformity requirements, along with calculations for each method as shown in the table below:

Table 4. Type of Distribution

No	Type of Distribution	Terms	Result	Description
1	Gumbel Method	Ck = 5.402	Ck = 1.659	Not meet the criteria
		Cs = 1.139	Cs = 1.088	Not meet the criteria
2	Log Normal Method	Ck = 5.383	Ck = 5.426	Meet the criteria
		Cs = 1.137	Cs = 1.221	Meet the criteria
3	Log Pearson Method Type III	Cs ≠ 0	Cs = 0.6770	Meet the criteria

According to the table above, the two methods that meet both requirements (Ck and Cs) are the Log Normal Method and the Log Pearson Method type III. Log Normal method with Cs = 1.221 and a requirement of Cs = 1.137, and Ck = 5.426 and a requirement of Ck = 5.383. Log Pearson Method Type III with a Cs value of 0.6770 and a Cs requirement of 0.

The distribution alignment test, also known as the fit test, is used to determine whether or not a distribution is acceptable. The distribution alignment test in this analysis is Smirnov Kolmogorov. The Smirnov-Kolmogorov test is more appropriate for the alignment test method with the chosen distribution Log Pearson Type III. Based on the comparison table below, it can be seen that Dmax < D0 (0.202 < 0.409) is critical, so the distribution method tested is acceptable Dmax < Dcritical. The following table shows the results of the Smirnov-Kolmogorov test calculation:

Table 5. Smirnov-Kolmogorov calculation

1	2	3	4 = 1-3	5	6	7 = 1-6	8 = 4-7
Xi	M	P(x)=M/(n+1)	P(x<)	F(t)=(Xi-Xrt)/Sd	P'(x)=M/(n-1)	P'(x<)	D
50.67	1	0.091	0.576	-3.54	0.111	0.556	0.020
61.67	2	0.182	0.485	-3.38	0.222	0.444	0.041
72.00	3	0.273	0.727	-.244	0.333	0.667	0.060
79.00	4	0.364	0.636	-1.81	0.444	0.556	0.080
85.83	5	0.455	0.379	-1.19	0.556	0.278	0.101
98.33	6	0.545	0.788	-0.06	0.667	0.667	0.121
107.67	7	0.636	0.03	0.78	0.778	0.889	-0.859
119.33	8	0.727	0.606	1.84	0.889	0.444	0.162
124.67	9	0.818	0.848	2.32	1.000	0.667	0.181
179.67	10	0.909	0.909	7.30	1.111	0.556	0.202
						Dmax=	0.202

Rainfall intensity is the quantity of rainfall in a unit of time represented as a percentage. The intensity of rainfall is obtained from the slope of the curve (the tangent of the curve) recorded by the automatic rainfall meter. The unit of rainfall intensity is mm/hour. The Mononobe method is used to calculate rainfall intensity on this analyst. The intensity of Mononobe rain is calculated as follows [8].

$$I = \frac{R_{24}}{24} x \left| \frac{24}{t} \right|^{\frac{2}{3}} \quad \textbf{(2)}$$

The following table displays the results of the rainfall intensity calculation:

Table 6. Rainfall intensity calculation

Return Period	Intensity (I)		
	2	5	10
R24 (mm)	121.33	506.17	987.83
t	(mm/h)	(mm/h)	(mm/h)
1	175.478	175.478	342.463
2	26.499	110.544	315.738
3	20.222	84.361	164.639
4	16.293	69.639	135.906
5	14.386	60.013	117.121
6	12.793	53.144	103.716
7	11.495	47.954	93.587
8	10.516	43.870	85.616
9	9.722	40.557	79.150
10	9.062	37.806	73.781
11	8.504	35.478	69.239
12	8.025	33.479	65.337
13	7.608	31.739	61.942
14	7.241	30.209	58.956
15	6.916	28.851	65.306
16	6.625	27.636	53.934
17	6.362	26.541	51.798
18	6.124	25.549	49.861
19	5.908	24.645	48.097
20	5.709	23.816	46.479
21	5.526	23.054	44.992

22	5.357	22.350	43.618
23	5.201	21.697	42.344
24	5.056	21.090	41.160
Total	396.976	1099.499	2145.780

The following is the rational method's runoff discharge formula [8].

$$Q = 0.278\, x\, C\, x\, I\, x\, A \tag{3}$$

The flow coefficient uses a coefficient (C) of 0.7, considering that the President University Campus is included in the industrial environment. The calculation of runoff discharge on campus is known to have a rainfall intensity (I) of 0.6 m/h and 13,600 m^2 of the total catchment area (A). The total runoff discharge in a 13,600 m^2 open space at President University is 1050 m^3/h.

3.2. Infiltration Analysis

An infiltrometer is used to measure the rate of infiltration. A double ring infiltrometer is used in this test. The following graph results of the infiltration rate are known from the experimental treatment carried out at two points, as shown in the graphic below.

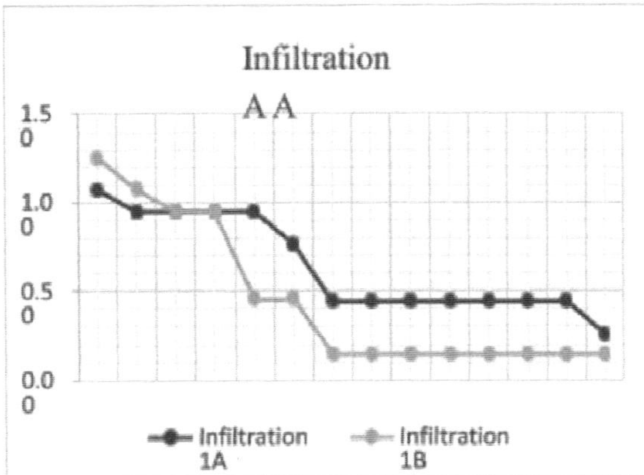

Figure 1. Graphic infiltration A

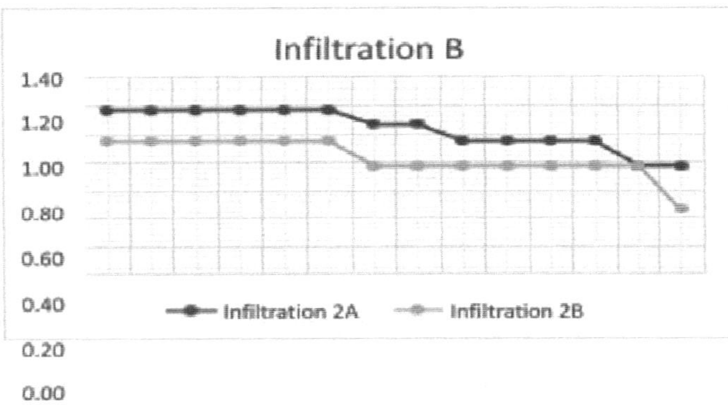

Figure 2. Graphic Infiltration B

Table 7. Calculating the rate of infiltration 1A

No	Δt	Δh	ft (cm/h)		FC ft-FC	log (ft-FC)
1	2	0.4	12	0.2	11.8	1.07
2	2	0.3	9	0.2	8.8	0.94
3	2	0.3	9	0.2	8.8	0.94
4	2	0.3	9	0.2	8.8	0.94
5	2	0.3	9	0.2	8.8	0.94
6	4	0.4	6	0.2	5.8	0.76
7	4	0.2	3	0.2	2.8	0.45
8	4	0.2	3	0.2	2.8	0.45
9	4	0	3	0.2	2.8	0.45
10	4	0	3	0.2	2.8	0.45
11	4	0	3	0.2	2.8	0.45
12	4	0	3	0.2	2.8	0.45
13	4	0	3	0.2	2.8	0.45
14	4	0	3	1.2	1.8	0.26

Table 8. Calculating the rate of infiltration 1B

No	Δt	Δh	ft (cm/h)		FC ft-FC	log (ft-FC)
1	2	0.6	18	0.1	17.9	1.25
2	2	0.4	12	0.1	11.9	1.08
3	2	0.3	9	0.1	8.9	0.95
4	2	0.3	9	0.1	8.9	0.95
5	2	0.1	3	0.1	2.9	0.46
6	4	0.2	3	0.1	2.9	0.46
7	4	0.1	1.5	0.1	1.4	0.15
8	4	0.1	1.5	0.1	1.4	0.15
9	4	0.1	1.5	0.1	1.4	0.15
10	4	0.1	1.5	0.1	1.4	0.15
11	4	0.1	1.5	0.1	1.4	0.15
12	4	0.1	1.5	0.1	1.4	0.15
13	4	0.1	1.5	0.1	1.4	0.15
14	4	0.1	1.5	0.1	1.4	0.15

Table 9. Calculating the rate of infiltration 2A

No	Δt	Δh	ft (cm/h)		FC ft-FC	log (ft-FC)
1	2	0.5	15	0.2	14.8	1.17
2	2	0.5	15	0.2	14.8	1.17
3	2	0.5	15	0.2	14.8	1.17
4	2	0.5	15	0.2	14.8	1.17
5	2	0.5	15	0.2	14.8	1.17
6	2	0.5	15	0.2	14.8	1.17
7	2	0.4	12	0.2	11.8	1.07

8	2	0.4	12	0.2	11.8	1.07
9	2	0.3	9	0.2	8.8	0.94
10	2	0.3	9	0.2	8.8	0.94
11	2	0.3	9	0.2	8.8	0.94
12	2	0.3	9	0.2	8.8	0.94
13	2	0.2	6	0.2	5.8	0.76
14	2	0.2	6	0.2	5.8	0.76

Table 10. Calculating the rate of infiltration 2B

No	Δt	Δh	ft (cm/h)		FC ft-FC	log (ft-FC)
1	2	0.3	9	0.1	8.9	0.95
2	2	0.3	9	0.1	8.9	0.95
3	2	0.3	9	0.1	8.9	0.95
4	2	0.3	9	0.1	8.9	0.95
5	2	0.3	9	0.1	8.9	0.95
6	2	0.3	9	0.1	8.9	0.95
7	2	0.2	6	0.1	5.9	0.77
8	2	0.2	6	0.1	5.9	0.77
9	2	0.2	6	0.1	5.9	0.77
10	2	0.2	6	0.1	5.9	0.77
11	2	0.2	6	0.1	5.9	0.77
12	2	0.2	6	0.1	5.9	0.77
13	2	0.2	6	0.1	5.9	0.77
14	2	0.1	3	0.1	2.9	0.46

There are differences in infiltration rates at the first and second points based on monitoring data from two points. The first point's infiltration rate can absorb water faster and more efficiently. Based on the time difference for the first ten minutes of two minutes and four minutes of the next minute until it reaches a constant. Because it is well known in this area that the soil used is clay backfill. Meanwhile, water infiltration into the soil takes longer at the second location. In calculating the infiltration rate using the Horton method, the infiltration rate of this study is 0.97 m^3/min.

The results of the principal parameter analysis were used to test the significance of statistical tests in this study. A statistical test aims to determine the significance of the influence of the independent dependent variable. The ANOVA test null hypothesis indicates that all populations' mean is the same. Another possibility is that the population means differ from one another. If P-value< 0.005, the null hypothesis is rejected. The statistical test presented above demonstrates significant infiltration results from the two locations used to calculate water infiltration into the soil. The P-Value is less than 0.05, indicating that the infiltration treatment from two locations has a significant effect on each other.

3.3 Biopore Infiltration Hole (BIH) Planning

The current lack of vacant land in the President University area makes implementing Biopore infiltration holes (BIH) challenging. The installation of paving blocks on existing vacant land has resulted in a lack of vacant land. Determining the biopore infiltration hole at President University is difficult due to the current state of vacant land, which has much paving. According to a site survey, President University has a total area of 13,600 m^2 with a permeability area of 6,365 m^2. Biopore infiltration holes are planted with a maximum distance of 100 centimeters between one hole and another. With a diameter of each hole of 10 centimeters.

Based on these provisions and the consideration of Regulation No. 12 of 2009 of the State Department of Environmental Affairs Regarding Rainwater Utilization, the ideal maximum amount of biopore infiltration is 50 holes for a land area of 100 m^2. Planning for the amount of BIH uses rain intensity data at the 2-year return period. At the same time, the water infiltration rate data uses the highest field infiltration test data from the sampling results, which is 0.97 m^3/min. The number of biopore infiltration holes and the maximum required BIH is calculated using the following formula [8].

$$Amount\ of\ BIH = \frac{Rain\ Intensity\ (\frac{mm}{h})\ x\ Watertight\ Area\ (m^2)}{Infiltration\ Rate\ per\ Hole\ (\frac{m^3}{h})} \quad (4)$$

$$Amount\ of\ Max\ BIH = \frac{Open\ Available\ Area\ (m^2)x\ Ideal\ number\ of\ holes\ (pcs)}{Ideal\ land\ area\ (m^2)} \quad (5)$$

The number of BIH that can be used in the President University area, which has a surface area of 13,600 m^2, is 1711.8 pcs. According to State Minister of Environment Regulation No. 12 of 2009, the maximum number of BIH utilized in the President University region is 3182.5 pcs.

4. Conclusion

The use of biopore effectively reduces the impact of waterlogging in clay soil, particularly in the President University Campus Area, Jababeka Cikarang Utara. Quantitative information has been used to validate the research, which was then carried out by calculating biopore data. The total runoff discharge in the open space of the President University campus area of 13,600 m^2 is reported to be 1050 m^3/hour. The infiltration rate measured from two places in the campus area examined is 0.97 m^3/minute. With a total area of 13,600 m^2, this significantly reduces waterlogging in the President University campus area. Based on the completed findings and discussions, as you can see: (a) Water infiltration into the soil in the President's University area is known to take 0.97 m^3/min that significant enough to reduce waterlogging in the President University Campus Area, (b) The President University campus has 1711.8 biopore infiltration holes planning which could be implemented to cover 13,600 m^2 of green open space.

Acknowledgments

The author would like to thank Allah SWT for giving the strength, wisdom, support, and knowledge to complete this study. Also, the author would like to express their gratitude to the Environmental Engineering of President University lecturers for guiding and supporting us.

References

[1] Adharani Y 2017 Penaatan dan Penegakan Hukum Lingkungan pada Pembangunan Infrastruktur dalam Mewujudkan Pembangunan Berkelanjutan (Studi Kasus Pembangunan PLTU II di Kecamatan Mundu Kabupaten Cirebon) *PADJADJARAN J. Ilmu Huk. (Journal Law)* **4** 61–83

[2] Presiden Republik Indonesia 2007 Undang-Undang Republik Indonesia Nomor 26 Tahun 2007 Tentang Penataan Ruang *Undang Undang Republik Indonesia* vol 9 p 10

[3] Prakoso P and Herdiansyah H 2019 Analisis Implementasi 30% Ruang Terbuka Hijau Di Dki Jakarta *Maj. Ilm. Globe* **21** 17

[4] Cahyanto A D, Pramudya B and Noor E 2019 Model Pengembangan Kebijakan Integrasi Sistem Manajemen Mutu dan Lingkungan (Studi Kasus Kawasan Industri Jababeka, Cikarang, Bekasi) *J. pengelolaan Sumberd. alam dan Lingkung.* **6** 31–44

[5] Hafizhan A 2020 *Analisis Faktor-Faktor Penyebab Banjir di Kota Bekasi* (Universitas

Muhammadiyah Surakarta)
[6] Menteri Negara Lingkungan Hidup 2009 Peraturan Menteri Negara Lingkungan Hidup Nomor 12 Tahun 2009 Tentang Pemanfaatan Air Hujan pp 1–10
[7] Juliandari M 2013 Efektivitas Lubang Resapan Biopori Terhadap Laju Resapan (Infiltrasi) *J. Teknol. Lingkung. Lahan Basah* **1** 1–10
[8] Syarifudin A 2017 *Hidrologi Terapan* (Yogyakarta: ANDI)

IOP Conf. Series: Earth and Environmental Science 1111 (2022) 012061 doi:10.1088/1755-1315/1111/1/012061

Potential Distribution of Skipjack Tuna Using MODIS Satellite Imagery in The Southern Waters of West Java - Banten

Mamat Suhermat[1,2], Muhammad Dimyati[1], Supriatna Supriatna[1*] and Martono[2]
[1]Faculty Mathematics and Natural Sciences, Universitas Indonesia
[2]Center of Atmospheric Research and Technology (BRIN), Bandung 40173, Indonesia
[1]msuhermat@gmail.com, [2]muh.dimyati@yahoo.com, [3*]ysupris123@gmail.com,
[2]martono_lapan@yahoo.com

Abstract. The West Java southern waters - Banten is among the skipjack tuna Katsiuwonus Pelamis distribution areas. It is well known that the skipjack tuna potential distribution is influenced by the environmental circumstances of the waters. This study was performed to establish the potential distribution of skipjack tuna in the West Java southern waters– Banten to support the potential management program of marine fishery resources. The data used consisted of skipjack tuna catch position, sea surface temperature, and chlorophyll-a concentrations from June 2019 to August 2019. Descriptive analysis and maximum entropy (maxent) model were applied to analyze skipjack tuna's potential distribution. The results revealed that the potential area of skipjack tuna is scattered in parts of the southern waters of West Java - Banten, with a validation test of the model performance of 0.92 (excellent).

1. Introduction

As one of the largest archipelagic countries, Indonesia has high fishery resources reserves both offshore and coastal areas. Indonesian Exclusive Economic Zone (EEZ) has the sustainable potential of Indonesia's marine fish resources of 12.54 million tons per year. Skipjack is one of the dominant fishery potentials in Indonesian waters[1]. Skipjack tuna has a high economic value, so it has become the main export commodity of Indonesian fisheries [2]. The export value of skipjack tuna reached USD 903.30 million in 2020 [3].

Oceanographic environmental circumstances strongly affect the skipjack tuna's distribution and abundance[4,5]. Therefore, skipjack tuna always migrate to a suitable place with their habitat [6–9]. One of the Indonesian waters that have a high potential for skipjack tuna is the southern waters of West Java – Banten, which is in the Fisheries Management Area of the Republic of Indonesia 573 [10,11]. Skipjack tuna production at PPN Palabuhanratu reached 1,230,124 kg in 2020 [12].

Several studies have been conducted to predict fish catch areas using remote sensing satellite imagery [13,14]. Remote sensing data benefit from the extensive spatial coverage and continuous time periods allowing us to draw better conclusions regarding the variation of habitat associated with surface/subsurface conditions built upon a great number of integrated and reliable data [14,15]. Sea surface temperature (SST), taken from MODIS satellite imagery, has been applied to explain the main habitat characteristics of tuna species in the oceans globally [16,17].

The species distribution model is an empirical model which relates field observations to statistically derived environmental predictor variables [18]. The theory of the species distribution model asserts that the model could foresee the potential spatial distribution of a range of species by connecting the points of occurrence to the predictor variables [19]. Physical parameters are modeled, and certain fish food chains have been carried out using the Generalized Additive Model (GAM) and Empirical Cumulative Distribution Function (ECDF) models [20,21]. The maximum entropy model (Maxent) [22] is one of the most widely applied machine learning algorithms to infer species distributions. In various types of research, the maximum entropy system has been discussed for both terrestrial [23] and marine [24].

Research on the potential distribution of skipjack tuna in the West Java southern waters– Banten is important to do in an effort to support the potential management program of marine fishery resources. The utilization of modeling compiled with remote sensing technology is a good choice to study and determine the potential distribution of skipjack tuna in these waters.

2. Methods

The studied area in this study is the eastern Equatorial Indian Ocean waters covering the area of 105°–109° East Longitude and 7°– 10° South Latitude, as shown in Figure 1.

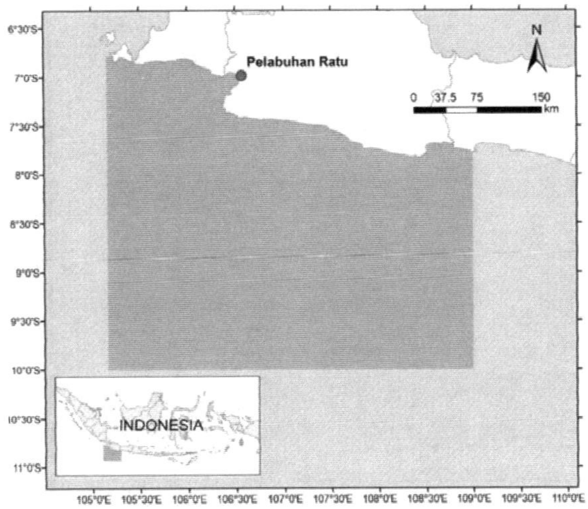

Figure 1. Research location

The data applied include oceanographic satellite data in the form of sea surface temperature (SST), chlorophyll-a concentration (from the site http://oceandata.sci.gsfc.nasa.gov//), and the catch data of skipjack tuna [12]. The satellite image used is Aqua MODIS Level 3 for 2019, which has been radiometrically and geometrically corrected with a spatial resolution of 4 km x 4 km. The satellite imagery was downloaded using the Google Earth Engine (GEE) cloud-based platform to display oceanographic data like SST and chlorophyll-a [25-27]. Furthermore, the image is cropped according to the research location. The cropped image is then Reprojected into a global projection system using the World Geodetic System 1984 (WGS 84) datum and saved in GeoTIFF format using ArcGIS software. Data that is already in GeoTIFF format is then converted into ASCII format (.asc) using ArcGIS software to be ready for use in analysis using Maxent software.

Spatial analysis was used with the Maximum Entropy Model (Maxent) method developed by [22]. In this study, the maxent model was run 500 iterations with a grid of 4 km X 4 km. The data were randomly divided into 2 categories: training data and test data. Model validation was carried out using 25% of the test data.

Maxent's main concept is to assess the target probability distribution by looking for a geographically uniform distribution that has maximum entropy based on environmental factors at the site of occurrence of a species [28,22]. The entropy value (H) of the probability distribution P in the X region is specified as:

$$H(x) = \quad -\sum P(x) \ln ln\, P(x) \tag{1}$$

To determine the target probability distribution, a known distribution is needed. In Species Distribution Modeling (SDM), the known distribution is a collection of species occurrence data. The complementary log-log transformation (clog log) is used to estimate the probability of species occurrence as follows:

$$probability\ of\ occurrence\ =\ 1\ -\ exp(-\ exp(H(x)P(x)) \tag{2}$$

3. Results and discussion

The results of determining the distribution of sea surface temperature and chlorophyll-a concentration from Aqua MODIS are shown in figure 2.

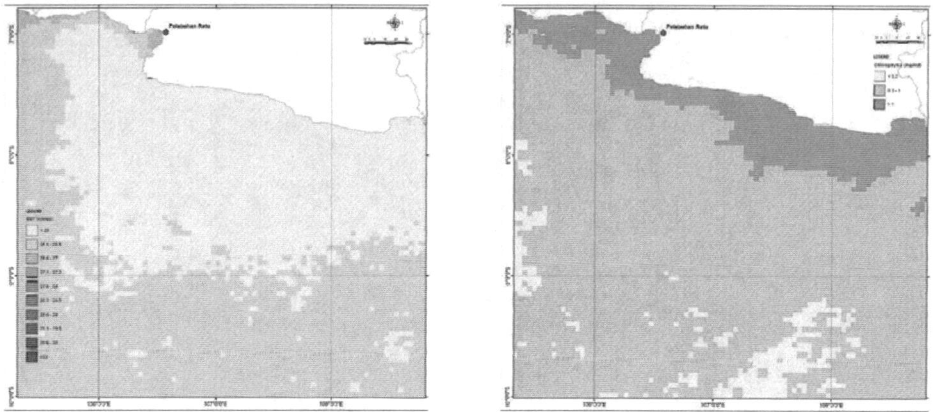

Figure 2. Spatial Distribution of SST and chlorophyll-a concentration

The distribution of sea surface temperatures in 2019 varies between 26.0 – 28.6 °C. The water area is dominated by sea surface temperatures of 26.6 – 27.0 °C. The average chlorophyll-a concentration during 2019 was 0.8 mg/m3. Based on the results of the classification of chlorophyll-a concentration values, skipjack catch locations were found in chlorophyll-a less than 0.3-0.1. The potential distribution model of skipjack tuna in 2019 using the Maximum Entropy Model (Maxent) is displayed in figure 3.

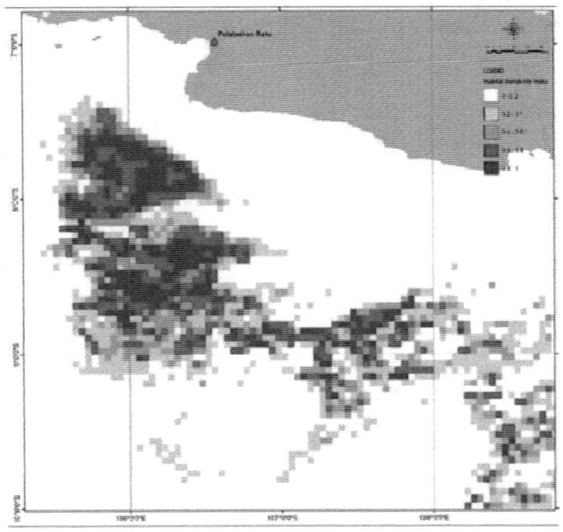

Figure 3. Potential distribution model of skipjack tuna in 2019

The results showed that the model that had been built had an Area Under Curve (AUC) of 0.92. The model with an AUC value of 0.9 – 1 is included in the excellent classification [29].

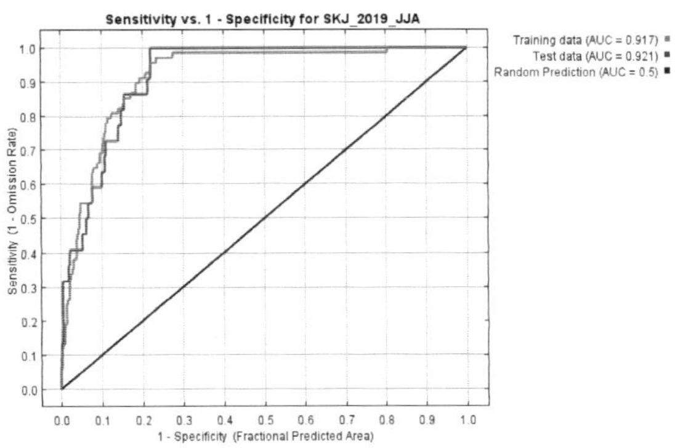

Figure 4. Receiver Operating Characteristic (ROC) Curve

Environmental variables that affect the model results are shown by the response curve. The x-axis shows the value of the distribution of parameters. In contrast, the relationship between the index value of the occurrence of skipjack tuna to the parameters is shown by the y-axis. This study shows that SST affects the model in the range of 24.0 – 28.5 °C, with the highest influence in the range of 25.5 – 26.2 °C. The model has the effect of chlorophyll-a concentration in the range of 0 – 0.1 mg/m3. The results of the SST response curve are in accordance with the study [30,31] that the skipjack habitat is in an area with an SST of 21.6 – 30 °C. The results of the response curve of chlorophyll-a are in accordance with research [16] that the skipjack habitat is in an area with a concentration of chlorophyll-a 0.08 – 0.18 mg/m3.

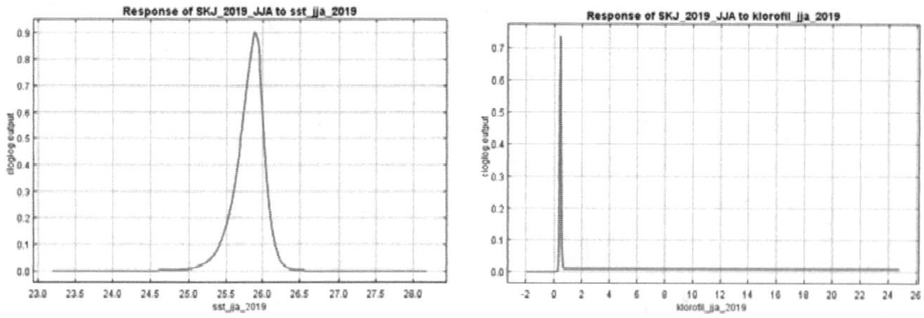

Figure 5. Response Curve

4. Conclusion

SST and chlorophyll-a are substantial predictors in foreseeing the skipjack tuna habitat. The results of Maximum Entropy (Maxent) modeling in this study have a high confidence level with a model performance validation test of 0.92 (excellent). The potential area for skipjack tuna is concentrated at 105.5°- 109°E. These results can provide new information for skipjack tuna resources management.

Acknowledgments

The authors thanks to National Research and Innovation Agency (BRIN), which has supported and funded degree by research program. This research also supported by Center of Atmospheric Research and Technology, National Research and Innovation Agency and Department of Geography, Faculty of Mathematics and Natural Sciences, Universitas Indonesia.

References

[1] Apriliani I M, Rizal A, Hamdani H and Denika A 2019 Catch of Skipjack Tuna (Euthynnus sp.) in National Fisheries Port Pengambengan, Bali, Indonesia *World Scientific News* **120** 144–53

[2] Azzahra A N, Permata C A D and Akhirta N 2017 Pemetaan Potensi Penangkapan Ikan Cakalang di Perairan Sulawesi *Seminar Nasional Penginderaan Jauh* pp 331–9

[3] MMAF 2020 *Ministry of Marine Affairs and Fisheries Performance Report*

[4] Muhling B, Liu Y, Lee S-K, Lamkin J, Roffer M, Muller-Karger F and Walter J 2015 Potential Impact of Climate Change on the Intra-Americas Sea: Part 2. Implications for Atlantic bluefin tuna and skipjack tuna adult and larval habitats *Journal of Marine Systems* **148** 1–13

[5] Townhill B, Couce E, Bell J, Reeves S and Yates O 2021 Climate Change Impacts on Atlantic Oceanic Island Tuna Fisheries *Frontiers in Marine Science* **8** 634280

[6] Lehodey P, Senina I, Calmettes B, Hampton J and Nicol S 2013 Modelling the impact of climate change on Pacific skipjack tuna population and fisheries *Climatic Change* **119** 95–109

[7] Dueri S, Bopp L and Maury O 2013 Projecting the impacts of climate change on skipjack tuna abundance and spatial distribution *Global change biology* **20**

[8] Johnson J and Pacific O 2018 *Effects of Climate Change on Ocean Fisheries Relevant to the Pacific Islands*

[9] Mediodia H J, Kahui V and Noy I 2020 *Sea Surface Temperature and Tuna Catch in the Eastern Pacific Ocean under Climate Change* (Munich)

[10] Jatmiko I, Zedta R, Agustina M and Setyadji B 2019 Genetic Diversity and Demography of Skipjack Tuna (Katsuwonus pelamis) In Southern and Western Part of Indonesian Waters *ILMU KELAUTAN: Indonesian Journal of Marine Sciences* **24** 61–8

[11] Firdaus M 2018 Profil Perikanan Tuna Dan Cakalang di Indonesia *MARINA* **4** 23–32

[12] KKP 2020 *Laporan Data Log book Perikanan Tangkap Tahun 2020* (Kementerian Kelautan dan Perikanan)

[13] Coletto J, de Pinho M and Madureira L 2018 Operational oceanography applied to skipjack tuna (Katsuwonus pelamis) habitat monitoring and fishing in south-western Atlantic *Fisheries Oceanography* **28**
[14] Yen kuo-wei, Lu H-J, Chang Y and Lee M-A 2012 Using remote-sensing data to detect habitat suitability for yellowfin tuna in the Western and Central Pacific Ocean *International Journal of Remote Sensing - INT J REMOTE SENS* **33** 7507–22
[15] Lan K-W, Shimada T, Lee M-A, Su N-J and Chang Y 2017 Using Remote-Sensing Environmental and Fishery Data to Map Potential Yellowfin Tuna Habitats in the Tropical Pacific Ocean *Remote Sensing* **9**
[16] Mugo R, Saitoh S-I, NIHIRA A and KUROYAMA T 2010 Habitat characteristics of skipjack tuna (Katsuwonus pelamis) in the western North Pacific: a remote sensing perspective *Fisheries Oceanography* **19** 382–96
[17] Zainuddin M, Farhum A, Safruddin S, Selamat M B, Sudirman S, Nurdin N, Syamsuddin M, Ridwan M and Saitoh S I 2017 Detection of pelagic habitat hotspots for skipjack tuna in the Gulf of Bone-Flores Sea, southwestern Coral Triangle tuna, Indonesia *PLoS ONE* **12**
[18] Guisan A and Thuiller W 2005 Predicting species distribution: Offering more than simple habitat models *Ecology Letters* **8** 993–1009
[19] Fournier A, Barbet-Massin M, Rome Q and Courchamp F 2017 Predicting species distribution combining multi-scale drivers *Global Ecology and Conservation* **12** 215–26
[20] Yanti Siregar E, Siregar V and Agus S 2018 FISHING GROUND ANALYSIS OF YELLOWFIN TUNA (Thunnus albacares) IN WEST-SUMATERA WATERS BASED ON GAM MODEL *Jurnal Ilmu dan Teknologi Kelautan Tropis* **10** 501
[21] Sukresno B, Hartoko A, Sulistyo B and Subiyanto 2015 Empirical Cumulative Distribution Function (ECDF) Analysis of Thunnus.sp Using ARGO Float Sub-surface Multilayer Temperature Data in Indian Ocean South of Java *Procedia Environmental Sciences* **23** 358–67
[22] Phillips S J, Anderson R P and Schapire R E 2006 Maximum entropy modeling of species geographic distributions *Ecological Modelling* **190** 231–59
[23] Peterson A, Papes M and Eaton M 2007 Transferability and model evaluation in ecological niche modeling: A comparison of GARP and Maxent *Ecography* **30** 550–60
[24] Alabia I, Saitoh S-I, Mugo R, Igarashi H, Ishikawa Y, Usui N, Masafumi K, Awaji T and Seito M 2015 Seasonal potential fishing ground prediction of neon flying squid (Ommastrephes bartramii) in the western and central North Pacific *Fisheries Oceanography* **24** 190–203
[25] F Firdaus, T G Pin and Supriatna. 2020. Distribution of phytoplankton using remote sensing in Cimandiri Estuary, Sukabumi, West Java. IOP Conf. Series: Earth and Environmental Science 481 (2020) 012066 IOP Publishing.
[26] I Annisa & S. Supriatna. 2021. Mapping chlorophyll-a and Total Suspended Solid (TSS) distribution in the waters of Ciletuh Bay. IOP Conf. Series: Earth and Environmental. Science 623 (2021) 012087 IOP Publishing.
[27] H I Akbar, S Supriatna and M P Patria. 2020. The relationship between total suspended matter and sea current for the spread of Chlorophyll-a and fishing area in Jakarta Bay. IOP Conf. Series: Earth and Environmental Science, 561, (2020) 012052 IOP Publishing.
[28] Phillips S J, Anderson R P, Dudík M, Schapire R E and Blair M E 2017 Opening the black box: an open-source release of Maxent *Ecography* **40** 887–93
[29] Hosmer D W and Lemeshow S 2010 *Applied Logistic Regression Second Edition* (New York: John Wiley & Sons, Inc.)
[30] Druon J-N, Chassot E, Murua H and Soto M 2016 Preferred feeding habitat of skipjack tuna in the eastern central Atlantic and western Indian Oceans: relations with carrying capacity and vulnerability to purse seine fishing
[31] Druon J-N, Chassot E, Murua H and Lopez J 2017 Skipjack Tuna Availability for Purse Seine Fisheries Is Driven by Suitable Feeding Habitat Dynamics in the Atlantic and Indian Oceans *Frontiers in Marine Science* **4** 315

IOP Conf. Series: Earth and Environmental Science 1111 (2022) 012062 doi:10.1088/1755-1315/1111/1/012062

The Effect of Washing and Air Scouring on The Performance of Immersed Membrane Microfiltration

Bowo Djoko Marsono[1] and Virgina Jane Ujiane[1]
[1] Environmental Engineering, Faculty of Civil, Planning, and Geo-Engineering, Sepuluh Nopember Institut of Technology, Surabaya, 60111, Indonesia.
[1]bowodjoko@enviro.its.ac.id, [2]virginajane22@gmail.com,

Abstract. The quality of raw water in Surabaya is decreasing, and conventional treatment units are no longer efficient in treating raw water to become drinking water. Microfiltration membranes applied in sedimentation units can produce better quality results on limited land that uses a small number of chemicals. The material will be deposited on the membrane's surface throughout the filtration process. The deposition of these materials can cause fouling, leading to a decrease in membrane productivity. This study involved a detailed investigation into the membrane washing method and air scouring implication in the microfiltration process to understand the most efficient technique to increase membrane productivity. Washing was done manually with a gentle scrubbing method and backwashing with time variation for 20 minutes and 10 minutes. Air Scouring was done with air pressure variations of 0,3 bar and 0,6 bar. The study showed that water flux was affected by the washing and air scouring method. Permeate quality was also found to be affected by washing and air scouring variation. Permeate quality analyses in this study are turbidity, TSS, micro-plastics, E-Coli, and organic substances.

1. Introduction

The turbidity of raw water processed by the PDAM Ngagel during the rainy season can reach 1000 NTU [1]. To reduce turbidity in its raw water, PDAM Ngagel still uses chemicals in the form of aluminum sulfate ($Al_2(SO_4)_3.14H_2O$) as a coagulant and conventional treatment units. The high value of turbidity causes increasing in coagulant necessity and water treatment cost. The use of sedimentation tanks in conventional drinking water treatment plants is also less optimal because it can only treat Total Suspended Solid (TSS) in treated water between 65-70%, so it requires a filter unit that needs a large area [2]. The solution to this problem is to apply membrane technology. Membrane technology installed in the sedimentation tank can reduce organic and inorganic compounds in the water without using chemicals [3]. Microfiltration membrane is a membrane technology that applies in driving force of about 1 bar and is used to separate particles in size between 0.1 to 10 μm [4]. The problem that usually emerges in the membrane filtration process is the deposition of material or particles on the membrane surface that can cause fouling. Fouling will reduce the rate of mass or volume through the membrane (flux) so that it affects the decrease in product accumulation (permeate). Although the occurrence of these impurities will increase the separation (rejection) in microfiltration, the decreasing of flux is a disadvantage in a microfiltration separation process, so impurities must be controlled to increase the efficiency of the microfiltration process, either by increasing the anti-impurity properties of the membrane material or by optimizing the membrane filtration process [4]. One of the optimization processes that can be done is by air scouring. The presence of fouling also causes the membrane to be washed periodically to remove the components that cause fouling from the surface and membrane structure. This study investigated the effect of the membrane washing method and air scouring in the microfiltration process to understand the most efficient technique to increase membrane productivity.

2. Methods

2.1. Preparation

Preparing the reactor, which will be placed in the IPAM Ngagel III Surabaya's sedimentation pond. The reactor is made of an iron and acrylic frame with dimensions of length x width x height, namely 60 cm x 15 cm x 50 cm. The type of membrane used in this study is a ceramic membrane flat sheet with a membrane pore size of 0,00005 mm and membrane area of 0,15 m^2. Membranes are connected to a pump by a hose and a series of pipes.

2.2. Implementation

The research was conducted by immersing the membrane reactor in a sedimentation pond. Treated water from the flocculation pond flowed through the sedimentation tank. Then, the treated water was processed by a microfiltration membrane. Membrane washing was done after permeate flux decreased. There are three variations of washing, namely manual washing using a gently scrubbing method with the addition of NaOCl, backwash in variations of 20 minutes, and backwash for 10 minutes. Backwash was applied with a pressure of 1 bar using 5% NaOCl. Air scouring is carried out for fouling minimization with two variations: air scouring with an air pressure of 0.3 bar and 0.6 bar. Water flux was measured, and permeate quality was analyzed to understand better the most efficient membrane washing method and air scouring technique to increase membrane productivity. Permeate quality analyses in this study are turbidity, TSS, micro-plastics, E-Coli, and organic substances. Those parameters were analyzed to compare the quality of the permeate before and after the implementation of washing and the quality of the permeate with air scouring implication and without air scouring implication in the filtration.

2.3. Data Analysis

Membrane performance is determined by two parameters, namely permeability, and selectivity. Qualitative analysis was carried out by knowing the membrane's selectivity, expressed by the rejection coefficient (R). The calculation of the rejection coefficient is carried out using the following equation.

$$R = 1 - \frac{Cp}{Cf} \times 100\% \quad \textbf{(1)}$$

Quantitative analysis was conducted to determine the effect of air scouring by comparing the flux produced when filtration using air scouring and without air scouring. Flux is calculated by dividing the volume of permeate (V) that passes through the area of membrane (A) in a particular unit of time (t). Flux calculation using the following equation.

$$J = \frac{V}{(Axt)} \quad \textbf{(2)}$$

Qualitative analysis was conducted to determine the effect of washing by testing the quality of feed water and permeate. Then for quantitative analysis, water flux recovery (WFR) is calculated by comparing the flux after washing (Jc) with the flux before washing (Jo) as in the following equation.

$$WFR = \frac{Jc}{Jo} \quad \textbf{(3)}$$

3. Results and Discussion

3.1. Effect of Manual Washing on Permeate

Manual washing was carried out by brushing the membrane surface with 5% NaOCl as a cleaning agent. NaOCl is very effective in killing bacteria and protozoa [5]. The flux decreases until 10% of the initial flux is an indicator to stop the filtration due to fouling. When fouling happens, membrane washing needs to be done to improve the performance of the membrane [6]. The appearance of the membrane in fouling conditions and after manual washing can be seen in Figure 1. The total production water before washing was 264.24 L. After washing was applied, the operating time became longer. This is indicated by a significant flux decrease that occurred only at the 79th hour. After washing, the total production water in the filtration is as much as 846.72 L. This shows that manual washing can prolong the operation time and increase water production by 69%. The comparison of the quantity test before and after manual washing using the gently scrubbing method can be seen in the graph in Figure 2.

Figure 1. Membrane visual before and after manual washing.

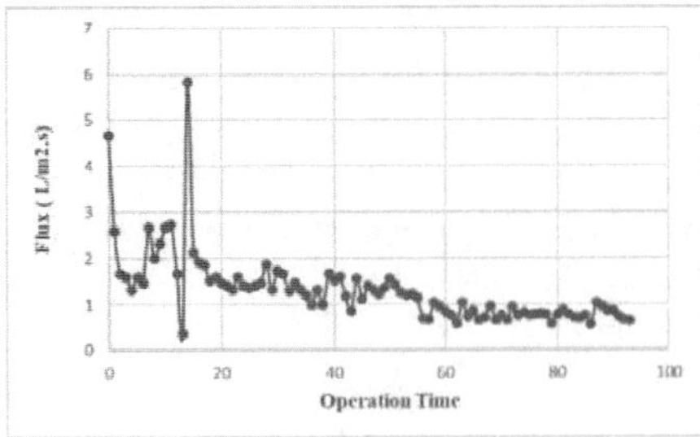

Figure 2. Manual Washing Flux Comparison with Time

Before washing, there was a decrease in flux at the 13th hour by 92.28%. of the initial flux. The initial flux when the membrane was operated was 4.67 L/m^2.s. Then there was a decrease in flux at the 13th hour to 0.36 L/m^2.s. The initial filtration pressure was 0.9 bar and then increased to 0.8 bar. A decrease in flux and an increase in pressure indicate the need for washing to improve membrane performance. After washing, the flux recovery value increased by 100%. Manual washing can also increase the pressure until it returns to the initial filtration pressure, from 0.8 pressure when fouling occurs to 0.9 bar. A comparison of the rejection coefficient before and after washing was carried out to determine the effect of manual washing on the quality of the permeate. The comparison of the membrane rejection coefficient before and after manual washing can be seen in Table 1. Based on the comparison of the rejection coefficients, it can be known that manual washing can increase the rejection coefficient of the turbidity parameter by 3%, the E.coli parameter by 1%, and the organic matter parameter by 17%.

Table 1. Comparison of Membrane Rejection Coefficient Before and After Manual Washing

Parameter	Before		After		Unit	R	
	Feed	Permeate	Feed	Permeate		Before	After
Turbidity	6,01	0,41	10,7	0,38	NTU	93%	96%
TSS	17	6	30	16	mg/L	65%	47%
Organic Matter	12,96	10,74	21,48	14,22	mg/L	17%	34%
E-Coli	1600	7,8	410	0	MPN/100mL	99%	100%
Microplastic	39	0	25	9	Particle/L	100%	64%

3.2. *Effect of Backwash on Permeate*

The backwash is done by pumping back the produced water, which has been added with a 5% NaOCl cleaning agent. NaOCl at a dose of 2-5 mg/L prevents fouling of the microfiltration membrane and can extend the washing interval to 3 months [7]. NaOCl is added during backwashing to control bacterial growth and degrade organic material on the membrane and in the hose connecting the pump circuit to the membrane [8]. The appearance of the membrane before washing and after washing can be seen in Figure 3. Backwash with a pressure of 1 bar for 10 minutes is less effective for cleaning foulants on the membrane surface. This is because the floc formed not only forms a cake on the surface of the membrane but has entered the membrane's pores. This process is called irreversible fouling, when solute moves from the membrane surface into the membrane material, narrowing and clogging pores [9]. Physical cleaning, such as backwash, is less effective when used to restore membrane flux that has undergone irreversible fouling. Irreversible fouling is usually overcome by chemical washing [10]. In the backwash variation for 10 minutes, before washing, the membrane was operated for 79 hours with total production water of 846.72 L. Filtration was stopped because there had been a decrease in flux. The total production water after washing is 735.18 L. This shows that backwashing for 10 minutes can only extend the operating time but cannot increase the amount of production water. The comparison of the results of the quantity test before and after the backwash for 10 minutes can be seen in the graph in Figure 4.

Figure 3. Membrane Visual Before and After 10 Minutes of Backwash.

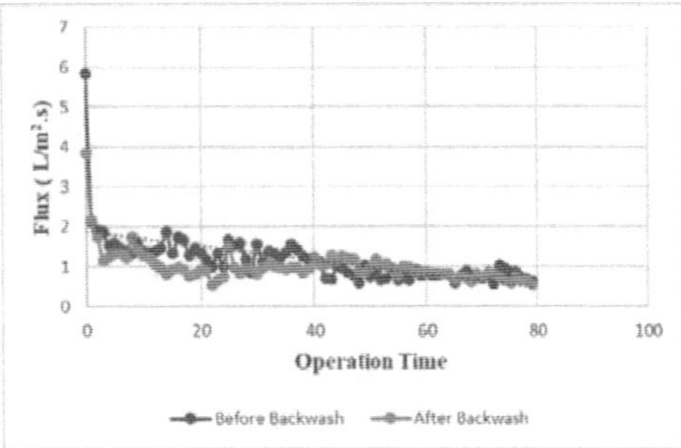

Figure 4. Flux Comparison with Time Before and After 10 Minutes of Backwash

Before washing, there was a decrease in flux at the 79th hour by 89.14% from the initial flux. The initial flux when the membrane was operated was 5.83 L/m^2.s. Then the flux decreased at 79 hours to 0.63 L/m^2.s. The initial flux when the membrane was operated after backwashing for 10 minutes was 3.83 L/m^2.s. Then the flux decreased at 80 hours to 0.51 L/m^2.s. The flux recovery value calculation results are only 65.71%; this is because the membrane washing interval is not paid attention to, so it affects the membrane condition. In the backwash variation for 20 minutes, it can be seen in Figure 5 that the washing results are less effective for cleaning foulants on the membrane surface.

The top of the membrane is cleaner than the bottom of the membrane. However, the upper membrane is still unclear when a backwash has been carried out. Because the particle diameter is smaller than the pore diameter, one of the fouling mechanisms will occur, namely standard blocking. Standard blocking occurs due to the adsorption of impurity molecules on the pore walls, causing pore narrowing [4]. Filtration before washing was carried out for 80 hours with total production water of 735.18 L. After washing, the operating time could be extended. This is indicated by the absence of a significant decrease in flux at the 80th hour. The total production water in the filtration after washing is 672.9 L. This shows that backwashing for 20 minutes can only extend the operating time but cannot increase the amount of production water. The comparison of the results of the quantity test before and after the backwash for 20 minutes can be seen in the graph in Figure 5,6. The result of the backwash flux recovery for 20 minutes of backwash is 69.56%. The flux recovery for this variation is higher than the backwash for 10 minutes. This proves that the washing time can affect the flux increase.

Figure 5. Membrane Visual Before and After 20 Minutes of Backwash.

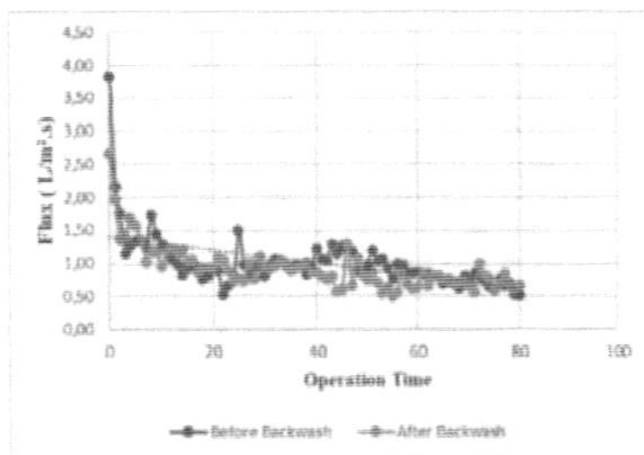

Figure 6. Flux Comparison with Time Before and After 20 Minutes of Backwash

The effect of backwash on permeate quality can be known by comparing the rejection coefficient before and after backwash. The calculation of the membrane rejection coefficient before and after 10 minutes of backwashing can be seen in Table 2.

Table 2. Comparison of Membrane Rejection Coefficient Before and After for 10 minutes Backwashing

Parameter	Before		After		Unit	R	
	Feed	Permeate	Feed	Permeate		Before	After
Turbidity	10,7	0,38	5,3	0,4	NTU	96%	92%
TSS	30	16	67	35	mg/L	47%	48%
Organic Matter	21,48	14,22	12,96	5,69	mg/L	34%	56%
E-Coli	410	0	160	0	MPN/100mL	100%	100%

Microplastic	25	9	40	21	Particle/L	64%	48%

Based on the comparison of the rejection coefficients, it can be seen that backwashing for 10 minutes can increase the rejection coefficient value of the TSS parameter by 1% and the parameter of organic matter by 22%. The effect of 20 minutes of backwashing on permeate quality is determined by comparison of the membrane rejection coefficient before and after backwashing for 20 minutes, which can be seen in Table 3. It can be seen that backwashing for 20 minutes can increase the rejection coefficient of the turbidity parameters by 6%, TSS by 52%, and microplastics by 8%. The result showed that backwash for 20 minutes affected improving membrane performance in removing turbidity, TSS, and microplastic parameters.

Table 3. Comparison of Membrane Rejection Coefficient Before and After 20 Minutes of Backwashing

Parameter	Before		After		Unit	R	
	Feed	Permeate	Feed	Permeate		Before	After
Turbidity	5,3	0,4	11,3	0,18	NTU	92%	98%
TSS	67	35	11	0	mg/L	48%	100%
Organic Matter	12,96	5,69	14,85	9,16	mg/L	56%	38%
E-Coli	160	0	1299	0	MPN/100mL	100%	100%
Microplastic	40	21	9	4	Particle/L	48%	56%

3.3. *Effect of Air Scouring on Permeate*

Figure 7. shows the membrane without air scouring. The cake formed is thicker. While in the image of the membrane appearance carried out by 0.3 bar air scouring, the cake formed was thinner. This is because, on the membrane, with the application of air scouring with an air pressure of 0.3 bar, air bubbles can create a two-phase flow that can prevent the formation of an impurity layer [11]. In the variation of air scouring with an air pressure of 0.3 bar, filtration without air scouring was carried out for 80 hours with total production water of 737.88 L. After applying air scouring with an air pressure of 0.3 bar, the operating time could be extended. This is indicated by the absence of a significant decrease in flux at the 80th hour. After applying air scouring with an air pressure of 0.3 bar, the total water production in the filtration is 816.42 L. This shows that the application of air scouring can extend the operating time and can increase the amount of production water. The comparison of the results of the filtration quantity test without air scouring and with the application of air scouring with an air pressure of 0.3 bar can be seen in Figure 8

Figure 7. Membrane Visual Without and With Air Scouring.

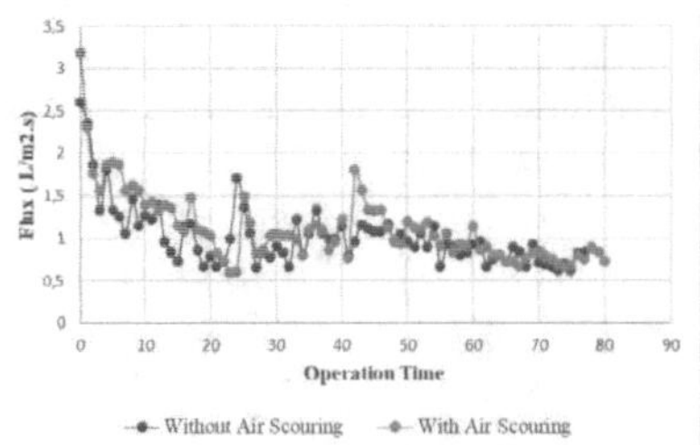

Figure 8. Flux Comparison with Time Without and With Air Scouring

Filtration operations with air scouring have a higher average flux but are not much different. This is thought to be because the applied bubbles are refined, so they are not optimal in preventing dirt deposits on the membrane surface. The increase in membrane flux is also related to the speed of airflow that flows into the diffuser, so the more significant the airflow velocity, the greater the flux increase [11]. In the variation of air scouring with an air pressure of 0.6 bar, the membrane without air scouring had a thicker and uneven cake. While in the image of the membrane appearance carried out by air scouring with the air pressure of 0.6 bar, the cake formed is thinner and evenly distributed over the entire surface of the membrane. This is because air bubbles can carry foulants evenly across the membrane surface. The flow of water or air moving along the membrane surface will cause an increase in shear stress on the membrane surface which can reduce the formation of a foulant layer that forms on the membrane surface so that after water scouring is applied, the cake formed looks thinner [12].

Filtration without air scouring was carried out for 67 hours with total production water of 436.02 L. After applying air scouring with an air pressure of 0.6 bar, the operating time could be extended. This is indicated by a reasonably stable flux and no significant decrease in flux until the 80th hour. The total production of water in the filtration after applying air scouring with an air pressure of 0.6 bar is 705.84 L. This indicates that the application of air scouring can extend the operating time and increase water production by 38.2%.

Filtration operation with 0.6 bar air pressure air scouring favors higher final and average fluxes. The greater the air pressure can form a counter flow that causes local mixing. The local mixing can increase the shear stress of the membrane to reduce fouling [13]. The effect of air scouring on permeate quality can be known by comparing the rejection coefficient in filtration without and implying air scouring. The calculation of the membrane rejection coefficient in filtration without and with implying air scouring in air pressure variation of 0.3 bar can be seen in Table 4.

Table 4. Comparison of Membrane Rejection Coefficient Before and After 20 Minutes of Backwashing

Parameter	Without AS		With AS		Unit	R	
	Feed	Permeate	Feed	Permeate		Without AS	With AS
Turbidity	5	.3	5	.3	NTU	94%	94%
TSS	67	51	67	43	mg/L	24%	36%
Organic Matter	12	7.9	12	4.4	mg/L	34%	63%
E-Coli	160	0	160	0	MPN/100mL	100%	100%
Microplastic	40	17	40	27	Particle/L	58%	33%

Based on the comparison of the rejection coefficients, it can be seen that air scouring with an air pressure of 0.3 bar can increase the rejection coefficient value of TSS parameters by 12% and organic substances by 29%. In infiltration that applies air scouring in air pressure variation of 0.6 bar, the calculation of the membrane rejection coefficient can be seen in Table 5.

Table 5. Comparison of Membrane Rejection Coefficient Before and After 20 Minutes of Backwashing

Parameter	Without AS		With AS		Unit	R	
	Feed	Permeate	Feed	Permeate		Without AS	With AS
Turbidity	7,96	0,54	11,3	0,21	NTU	93%	98%
TSS	4	1	11	0	mg/L	75%	100%
Organic Matter	10,1	7,6	14,8	8,5	mg/L	25%	43%
E-Coli	160	0	129	0	MPN/100mL	100%	100%
Microplastic	5	3	9	5	Particle/L	40%	44%

Based on the comparison of the rejection coefficients, it can be seen that air scouring with an air pressure of 0.6 bar can increase the rejection coefficient values for turbidity parameters by 5%, TSS by 25%, organic substances by 18%, and microplastics by 4%.

4. Conclusion

The results of this study conclude several things, including (1) washing can increase the membrane performance measured by flux recovery and membrane rejection coefficient with the most effective washing variation, namely manual washing. The resulting flux recovery after applied manual washing is 100% and can increase the rejection coefficient on the parameters of turbidity and organic matter (2) air Scouring can extend the operating time, increase the amount of production water, and improve the membrane rejection ability. In air scouring with an air pressure of 0.6 bar, the total water production can be increased by 38.2%. It can increase the rejection coefficient on turbidity, TSS, organic matter, and microplastics parameters. The more effective variation of air scouring is air scouring with an air pressure of 0.6 bar.

References

[1] Patriasani E 2010 Uji Kinerja Media Batu pada Bak Prasedimentasi *Prosiding Seminar Nasional Manajemen Teknologi XVI Program Studi MMT-ITS* pp 1–8

[2] Setiono B 2018 *Optimasi Bak Sedimentasi Dengan Penambahan Sekat Vertikal Untuk Pengolahan Air Bersih* (ITN Malang)

[3] Wenten L 1999 *Teknologi Membran Industrial* (Bandung: Institut Teknologi Bandung)

[4] Shalahuddin I and Wibisono Y 2019 Mekanisme Fouling pada Membran Mikrofiltrasi Mode Aliran Searah dan Silang *J. Rekayasa Proses* **13** 6

[5] Qin L, Shu Q, Wang Z, Shang C, Zhu S, Xu J, Li R, Zhu L and Yuan Z 2014 Cultivation of chlorella Vulgaris in dairy wastewater pretreated by UV irradiation and sodium hypochlorite *Appl. Biochem. Biotechnol.* **172** 1121–30

[6] Avlonitis S A, Kouroumbas K and Vlachakis N 2003 Energy consumption and membrane replacement cost for seawater RO desalination plants *Desalination* **157** 151–8

[7] Chang H, Liang H, Qu F, Liu B, Yu H, Du X, Li G and Snyder S A 2017 Hydraulic backwashing for low-pressure membranes in drinking water treatment: A review *J. Memo. Sci.* **540** 362–80

[8] Chae SR, Yamamura H, Ikeda K and Watanabe Y 2008 Comparison of fouling characteristics of two different poly-vinylidene fluoride microfiltration membranes in a pilot-scale drinking water treatment system using pre-coagulation/sedimentation, sand filtration, and chlorination *Water Res.* **42** 2029–42

[9] Mahmud R N 2005 Kinetika Fouling Membran Ultrafiltrasi (UF) Pada Pengolahan Air Berwarna : Pengaruh Interval dan Lamanya Pencucian Balik (Backwashing) Membran *Info Tek.* **6** 62–9

[10] Gruskevica K and Mezule L 2021 Cleaning methods for ceramic ultrafiltration membranes affected by organic fouling *Membranes (Basel).* **11** 1–15

[11] Ducom G, Puech F P and Cabassud C 2002 Air sparging with flat sheet nanofiltration: A link between wall shear stresses and flux enhancement *Desalination* **145** 97–102

[12] Cheryan M 1998 *Ultrafiltration and Microfiltration Handbook* (CRC Press)

[13] Fouladitajar A, Zokaee Ashtiani F, Rezaei H, Haghmoradi A and Kargari A 2014 Gas sparging to enhance permeate flux and reduce fouling resistances in cross-flow microfiltration *J. Ind. Eng. Chem.* **20** 624–32

Pressure And Spacer Effect On The Performance Of Immersed Microfiltration Membrane

Bowo Djoko Marsono[1]*, Inayah Agustina[2], Adhi Yuniarto[2], Joni Hermana[2]
[1]Department of Environmental Engineering, Faculty of Civil, Planning and Geo Engineering, Institut Teknologi Sepuluh Nopember, Surabaya, 60111, Indonesia, https://orcid.org/0000-0003-2164-7092
[2]Department of Environmental Engineering, Faculty of Civil, Planning and Geo Engineering, Institut Teknologi Sepuluh Nopember, Surabaya, 60111, Indonesia
[1*]bowodjok@enviro.its.ac.id

Abstract. The water treatment applied in PDAM Surabaya City to treat Kali Surabaya still uses a conventional process where the treatment process will get even harder with decreasing quality of raw water. Immersed microfiltration membrane is a reliable technology in water treatment compared to conventional technology. Nevertheless, the performance of the membrane decreases due to deposits that occur on the membrane surface and lead to fouling. This study aims to examine the effect of pressure and spacers on the performance of microfiltration ceramic membrane measured based on quality and quantity parameters. The research parameters used were turbidity, TSS, organic matter, total *coliform*, *E. Coli* and the number of microplastic while the quantity parameters included flux, amount of water production and the duration of membrane operation before fouling. This study tested immersed flat-sheet microfiltration ceramic membranes in sedimentation unit of PDAM Ngagel III Surabaya City. Membrane operation used cross-flow with variable pressure (0.6 bar, 0.7 bar, 0.9 bar) and spacer slopes (75°, 90°, 105° and without spacers). Based on the results of the study, it was found that the effect of optimal pressure occurred at 0.7 bar with the efficiency of eliminating turbidity, TSS, organic matter, microplastic, total *coliform*, and *E. Coli* parameters respectively, 94.66%, 96.88%, 24.29%, 62.27%, 100% and 100%. The average flux value at pressure 0.7 bar is 64,33 L/m^2.hour with operating duration of 55 hours and permeate production volume 530.70 L. The optimal spacer effect was achieved at slope of 90° with the efficiency of eliminating turbidity, TSS, organic matter, microplastic, total *coliform* and *E. Coli* parameters respectively 95.12%, 64.58%, 28.75%, 84.74%, 99.99% and 100%. The average flux value was 66,78 L/m^2.hour with an operating duration of 90 hours and permeate production volume 901.55 L.

1. Introduction

The level of pollution in Kali Surabaya in 2015-2017 is categorized as moderate to severe [1]. Kali Surabaya is a source of raw water for the city which is managed by PDAM Surabaya. Water treatment applied to PDAM Surabaya City is still using the conventional process by chemical precipitation. The decrease in raw water quality causes operational and maintenance costs to increase, the need for land area increases, operations and maintenance become complicated to be able to produce water that is in accordance with the drinking water quality requirements in Permenkes No. 492 of 2010 [2]. This encourages the thought to modify the existing drinking water treatment with new technology.

Membrane technology is a very reliable technology in water and wastewater treatment [3]. MF (Microfiltration) and immersed UF (Ultrafiltration) membranes have been widely applied in the water treatment sector [4] and wastewater [5] which can be combined with conventional treatment [6] as well as in the pharmaceutical [7] and food industries [8]. Membrane technology is a clean technology that is environmentally friendly because it doesn't add chemicals (coagulants, flocculants) as in conventional water treatment [9]. However, membrane performance can change over time due to deposition of impurities that occur on the membrane surface called fouling [3] Fouling occurs due to accumulation, deposition and physico-chemical interactions between the content of the feed water and the membrane surface [10], [11]. The fouling phenomenon can reduce membrane productivity due to an increase in membrane retention of the permeate (filtrate) due to decreased permeability and membrane flux. Therefore, it is necessary to have an effective strategy to control fouling and maintain flux in order to minimize energy consumption and operational costs [12], [13].

Mass transfer in the microfiltration membrane occurs due to pressure as a driving force given in the feed. The optimal operating pressure helps the removal of the product (permeate) through the membrane surface while the concentrate (retentate) will be rejected by the membrane [14]. However, the decrease in flux that occurs due to fouling causes a decrease in pressure so that the water quality will be worse [15]. One strategy that has been successfully applied to keep the flux stable is by placing a spacer on the membrane [16]. The efficiency of fouling control using spacers is influenced by the morphology, material, distance, geometry, thickness and internal angle of the spacer [15], [17]. The use of finned spacers with a 45° fin angle configuration can increase the membrane permeability up to 87% [16]. The angular orientation of the spacers is also able to increase the mass transfer in the membrane [18].

This study tested the performance of immersed microfiltration ceramic membranes at the IPAM sedimentation unit of PDAM Ngagel III Surabaya with cross-flow to reduce deposits on the membrane surface [19]. Performance of the membrane tested to determine their quantity and quality parameters. The main parameters of membrane performance efficiency are influenced by permeability, selectivity, polarization concentration and fouling phenomena [11].

1.1. Permeability
Permeability is strongly influenced by the number of pores, pore size, operating pressure and membrane thickness. In measurement, membrane permeability is expressed in flux. Flux is the amount of permeate volume that has passed unit area of the membrane in a certain time [11]. The flux rate will decrease with time (irreversible) due to concentration polarization, adsorption, compaction and fouling.

$$J = \frac{V}{A\,x\,t} \quad \textbf{(1)}$$

Where:
J = Flux ($L/m^2.h$)
V = Permeate volume (mL)
A = Area of membrane (m^2)
t = Time (h)

1.2. Selectivity
Selectivity is the ability of the membrane to distinguish components in a flow and separate them from one phase to another. In general, the flux phenomenon is inversely related to selectivity. The selectivity of the membrane can be measured by the rejection coefficient. The rejection coefficient is a measure of how reliable a membrane is in removing certain components (solutes) so that a good quality permeate can be produced. The greater the value of the rejection coefficient, the better the performance of the membrane [11].

$$R = 1 - \frac{Cp}{Cf}\, x\, 100\% \quad \textbf{(2)}$$

Where:
R = Rejection Coefficient (%)
Cp = Permeate concentration
Cf = Feed Concentration

2. Methods

2.1. *Membrane Preparation*

The study used a microfiltration membrane with a pore size of 0.05 μm made of ceramic (α-Al_2O_3) as the object being tested. The membrane configuration used is in the form of a flat sheet with an effective area of 0.15 m^2. The implementation of this research requires 4 flat-sheet membranes and 1 as a backup. Prior to data collection, an initial check of the membrane flux and pressure was carried out. Membrane checked to ensure that there were no defects in the membrane product and to ensure that the membrane could function properly.

2.2. *Reactor Preparation*

The membrane reactor in this study was designed to facilitate researchers in obtaining data related to the effect of membrane pressure and spacer slope on the performance of immersed membrane microfiltration when operated in the field.

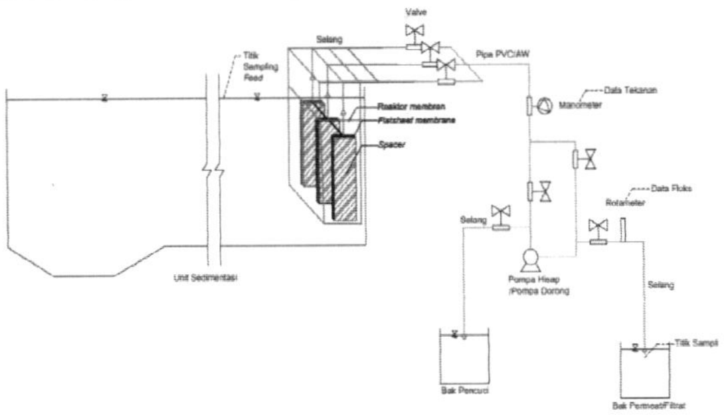

Figure 1. Schematic of a microfiltration immersed membrane reactor in a sedimentation tank

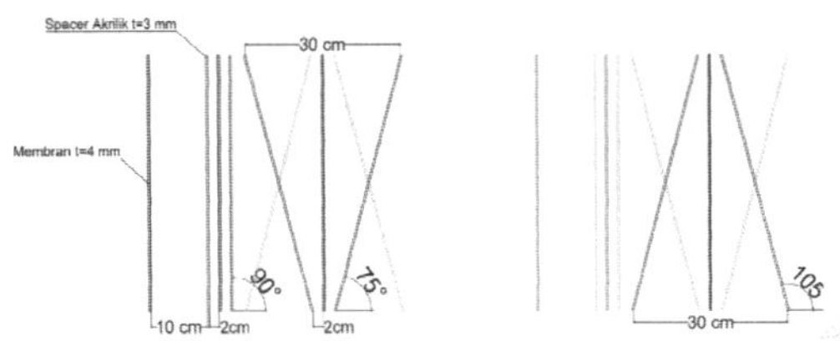

Figure 2. Schematic of the slope of the spacer in a membrane reactor

a. Immersed Reactor
   The reactor is rectangular in shape made of acrylic material with a thickness of 3 mm as the reactor wall and combined with an iron frame so that the reactor can be placed hanging on the wall of the sedimentation tank. On the inside of the reactor, a slot system is used to accommodate 3 flat-sheet membranes and 4 spacers made of acrylic with a thickness of 3 mm.
b. Non-immersed Reactor
   The part of the reactor that is not immersed is the flow and pressure control which is connected to the membrane outlet to take samples of the processing results. This reactor is placed on the operating path between the outlet of the sedimentation tank and canal III.

2.3 *Operational Preparation*

3 Flat-sheet microfiltration ceramic membranes were placed in the immersed reactor. Spacers are installed in slots according to the research conditions. The reactor is immersed in a sedimentation tank. Membrane operations were carried out with constant pressure for each variation and controlled using a manometer. During the initial operation of the reactor, clean water was prepared in the wash basin as the initial start-up of the pump. The negative pressure setting under the research conditions was carried out naturally based on a decrease in the pressure driven force on the membrane by initial set-up the maximum pressure that could be achieved by the membrane (0.9 bar) then operated to the lowest pressure at the research conditions and analyzed. Changes in permeate flux were observed every hour using a rotameter and measured manually using a measuring cup for 1 minute. The permeate effluent is temporarily accommodated in the permeate/filtrate bath to be reused as a solvent or washing water. The permeate effluent that is not analyzed will be flowed back to the outlet of the sedimentation tank to maintain the water level in the sedimentation tank. After the filtration process, the membrane is backwashed using washing water which is pumped into the membrane. The membrane performance test is carried out until there is a decrease in pressure and or if the resulting effluent discharge is ±15-20%

of the initial discharge or <100 mL. This research was conducted intermittently on/off the pump. This is done to maintain the performance of the pump used during operation to remain stable.

3. Results and Discussion

3.1. Analysis of the Effect of Pressure on Quality Parameters

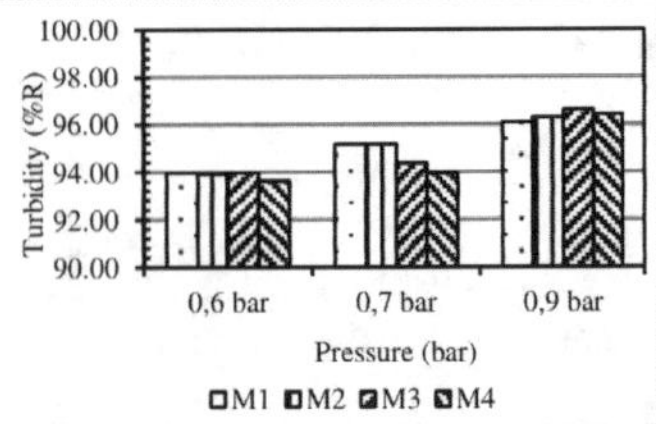

Figure 3. Pressure effect on turbidity

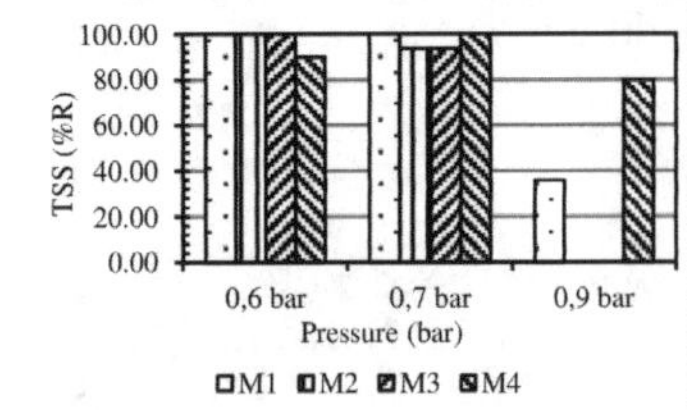

Figure 4. Pressure effect on TSS

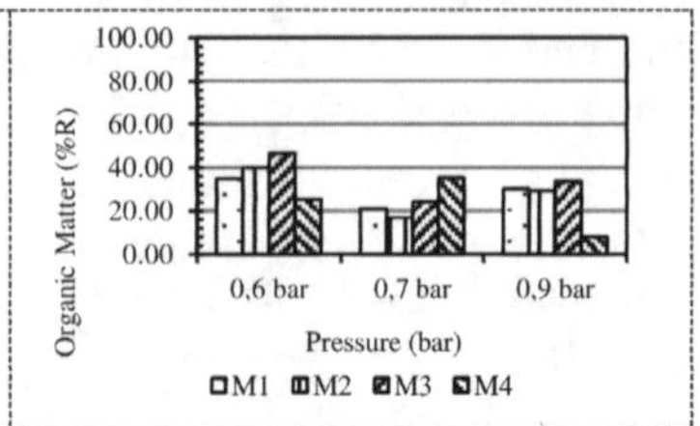

Figure 5. Pressure effect on organic matter

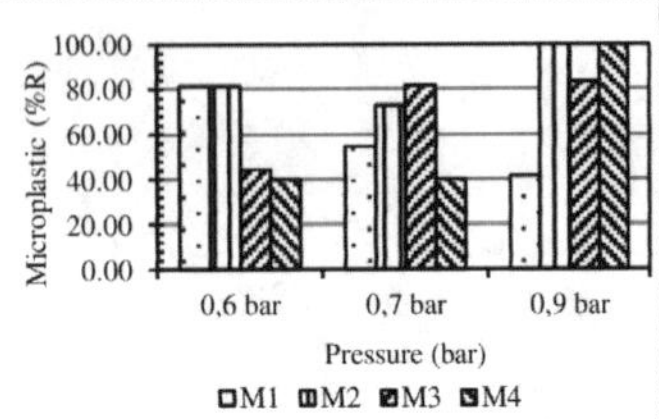

Figure 6. Pressure effect on microplastic

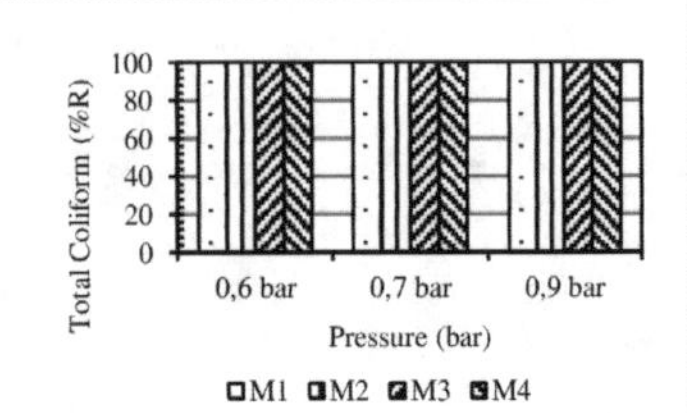

Figure 7. Pressure effect on total *coliform*

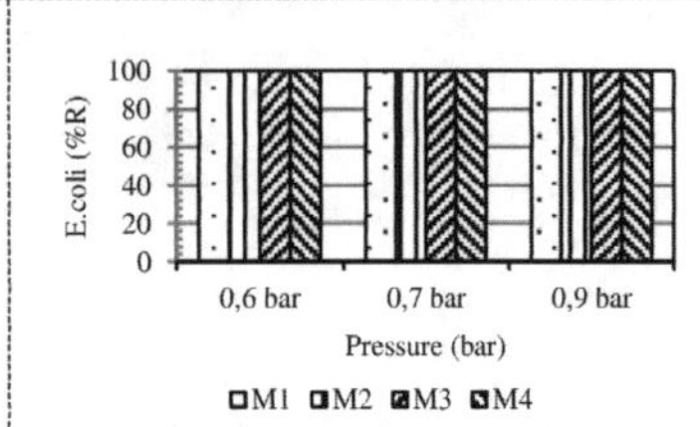

Figure 8. Pressure effect on *E. coli*

Based on the figure, the value of turbidity rejection coefficient (%R) in this study reached a maximum at a pressure of 0.9 bar with an average rejection of 96.34%. This indicates that the higher the operating pressure used, the higher the value of the resulting rejection coefficient for the turbidity parameter. The value of the rejection coefficient (%R) that can be produced by the membrane is also influenced by the ability of the membrane selectivity to the characteristics of the treated feed water [11]. In addition, the permeability of the solvent is higher than the permeability of the solute, so that the higher pressure has a higher solvent content [20]. The higher pressure allows the microfiltration membrane to achieve higher rejection efficiencies.

Fouling caused by suspended particles is classified as a type of fouling that is difficult to physically remove (irreversible fouling) [21]. The highest rejection coefficient (%R) for the TSS parameter occurred at a pressure of 0.6 bar with an average rejection of 97.50%. However, at a pressure of 0.9 bar there was a decrease in rejection until the average rejection value (%R) at a pressure of 0.9 bar was 28.93%. This indicates that the higher the set pressure, the greater the concentration of TSS that passes through the membrane. The accumulation of deposits at a pressure of 0.6 bar is thicker so that the pore size of the membrane decreases and the ability of rejection to TSS increases [22]. The suitability of the operating pressure applied is also influenced by the specifications of the type of membrane, the location of the implementation and the characteristics of the treated feed water [5], [19].

The findings in the data in Figure 5. prove that the removal process on the membrane is influenced by the pore size of the membrane and the size distribution of the removed particles. At low pressures, the pore size of the membrane tends to shrink due to the accumulation of organic matter that has occurred on the membrane surface, which is called concentration polarization [23]. Membrane permeability will only be able to be passed by dissolved organic matter with a very small size, while organic substances that are not dissolved will be retained on the membrane surface due to increased selectivity by the membrane [11], [22]. When the accumulation of organic matter has reached its maximum, the increase in pressure that occurs at 0.7 bar conditions will cause the solids that clog the membrane pores to escape because they get a pressure driven force to be able to carry out mass transport across the membrane [19].

This causes the membrane selectivity to decrease so that the value of the rejection coefficient produced at a pressure of 0.7 bar is lower.

The findings obtained based on microplastic particle analysis showed that the highest efficiency of microplastic removal was achieved at a pressure of 0.9 bar with an average rejection coefficient (%R) of 81.25%, while the lowest efficiency occurred at a pressure of 0.6 bar with a value of 0.6 bar. the average rejection coefficient (%R) is 61.85%. This indicates that the microplastic content is able to escape at lower pressure conditions. at low pressures the pore size of the membrane shrinks, so that microplastic particles that have a smaller size have the potential to clog the membrane pores more quickly [11]. The smaller the size of the microplastic in water, the more difficult it is to remove only physically [24]. Based on the results of the analysis of microplastics on the membrane effluent, it was possible that the size of the microplastics that passed was >0.05 µm. This can occur due to contamination from reactor connections or membrane frames that have been fragmented during operation.

The highest efficiency of total *coliform* removal was achieved at a pressure of 0.7 bar with an average rejection coefficient (%R) of 100%, while the lowest efficiency occurred at a pressure of 0.6 bar with an average rejection coefficient (%R) of 99.93%. When compared with the initial content in the feed water, the overall selectivity of the membrane to the total *coliform* parameter was high. This can happen because most of the total *coliform* bacteria have a size between 0.5 x 0.3 µm [24]. The rejection value achieved for each variation of membrane pressure on the *E. Coli* parameter has an average of 100% as shown in Figure 8. This proves that the *E. Coli* bacteria contained in the feed water have a size between 1-1.5 µm x 2-6 µm so that it can be retained by the pores of the microfiltration membrane at every pressure variation [24]. Elimination of *coliform* and *E. Coli* bacteria is strongly influenced by the pore size of the membrane used [25]. This is because the specification of the microfiltration ceramic membrane used has a very small pore size (0.05 µm) so it is very reliable in retaining thc retention of total *coliform* and *E. Coli* content in feed water with varying pressure conditions [26].

3.2. *Analysis of the Effect of Pressure on Quantity Parameters*

The analysis of the effect of pressure on the quantity parameter is carried out using data that refers to the condition of the M3 membrane, namely a membrane with a spacer slope of 75°. The operation of the membrane under conditions of initial pressure variation of 0.9 bar until the lowest pressure reached 0.55 bar occurred for 93 hours as shown in Figure 9. The change in pressure conditions can be seen from the increase in flux on the graph after a significant decrease. The increase in flux is due to the initial setting of the pump again after the intermittent process. During 90 hours of operating pressure, it can be concluded that the higher the operating pressure used, the greater the flux value produced. But the flux value will decrease rapidly over time due to the fouling phenomenon [11], [23]. Optimal pressure conditions with stable flux values and long operating time occur at a pressure of 0.7 bar.

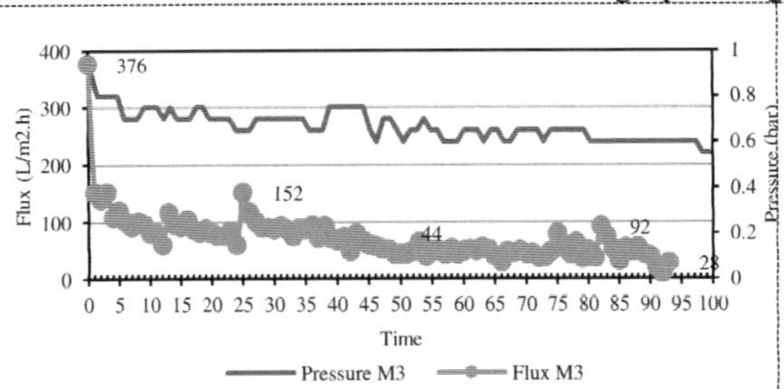

Figure 9. Pressure effect on flux

Table 1. Pressure effect on permeate volume

Pressure (bar)	Permeate volume (L)	Operation Time (h)	Average flux (L/m^2.h)
0,9	404,70	24	112,42
0,7	530,70	55	64,33
0,6	82,14	11	49,78

3.3. Analysis of the Effect of Spacer on Quality Parameters

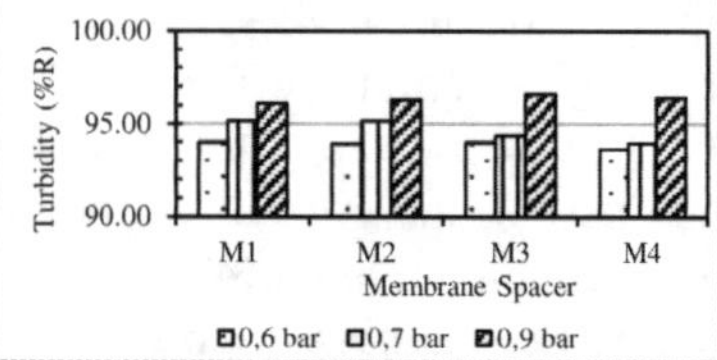

Figure 10. Spacer effect on turbidity

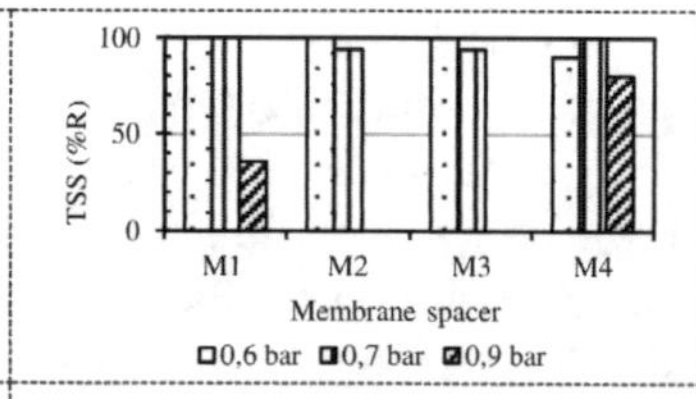

Figure 11. Spacer effect on TSS

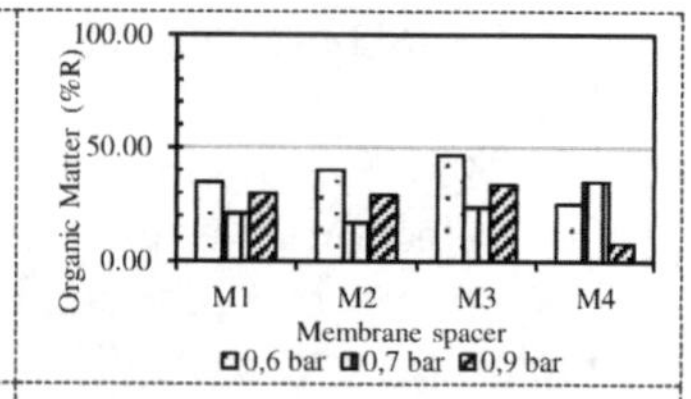

Figure 12. Spacer effect on organic matter

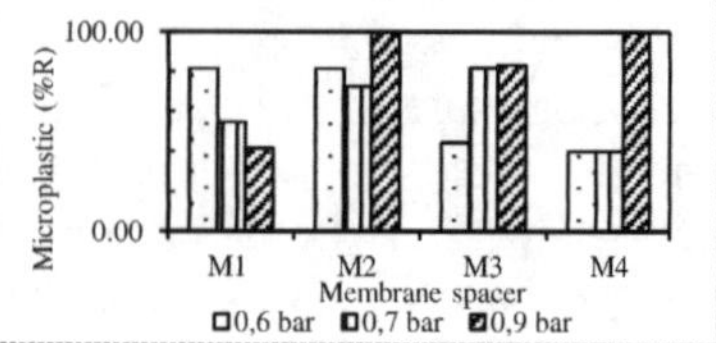

Figure 13. Spacer effect on microplastic

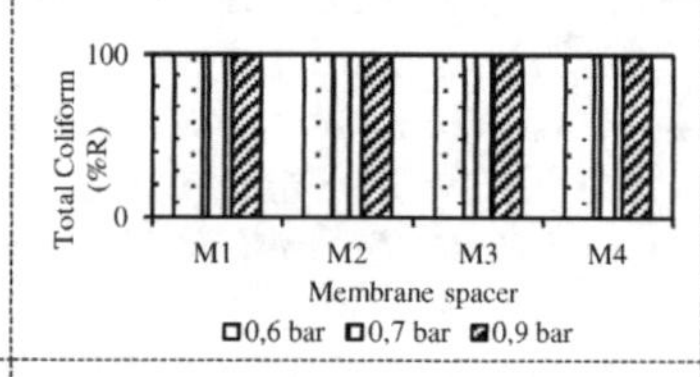

Figure 14. Spacer effect on total *Coliform*

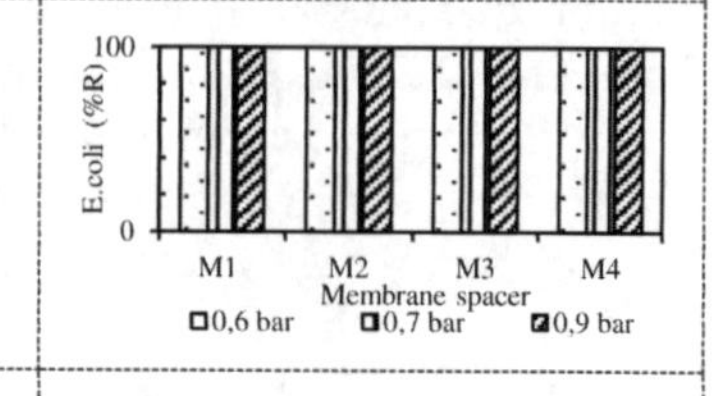

Figure 15. Spacer effect on *E. Coli*

The findings obtained that the average rejection coefficient (%R) of the turbidity parameter reached maximum on the M2 membrane with a spacer slope of 90° with an average rejection coefficient value of 95.12%. This proves that at 90° spacer tilt angle with cross-flow, it is able to produce mass transfer and concentration mixing on all parts of the membrane surface which is more evenly distributed, thereby reducing the formation of concentration polarization [27].

Based on the results of the analysis on the variation of the slope of the spacer used, the highest average rejection coefficient (%R) was achieved to exclude TSS in the feed water by 90% on M4 membrane conditions with a spacer slope of 105°. This proves that the selectivity of the membrane reaches its optimum at M4 membrane conditions due to the accumulation of high levels of impurities on the membrane surface causing the membrane pore size to decrease. Therefore, the TSS particles that are able to pass through the membrane have smaller and fewer sizes [19].

The value of the rejection coefficient (%R) of the parameters of organic substances as a whole is still low. This is because the microfiltration and ultrafiltration membranes are unable to remove dissolved organic and inorganic particles before they are converted into suspended particles [25]. In addition, the concentration of dissolved organic matter which is higher than that of undissolved organic matter >0.5 μm will affect the removal efficiency produced by the submerged microfiltration membrane [26]. Based on the results of the analysis on the variation of the slope of the spacer used, the highest average rejection coefficient (%R) achieved is 34.78% on M3 tires with a slope of 75°

The highest average rejection coefficient (%R) achieved for removing microplastics in feed water was 84.74% under M2 membrane conditions with a spacer slope of 90°, while the lowest average rejection value (%R) was 59.23%. obtained in the condition of the M1 membrane with a configuration without a spacer. This proves that the selectivity of the membrane reaches its optimum at M2 membrane conditions due to the 90° spacer tilt creating a fairly close distance to the membrane. The potential for more evenly distributed pollutant contact and increased accumulation of other pollutants is very high. One of the accumulated pollutants is microorganisms. The high efficiency of the membrane in removing microorganisms indicates that the concentration of microorganisms on the membrane surface is quite high. In the presence of microorganisms, degradation of plastic polymers can occur by bio fragmentation because microorganisms are able to produce cutinase enzymes that can break polymer bonds in plastics [27], [2], [28].

The results of the membrane selectivity analysis on the variation of the spacer slope obtained that the highest average rejection coefficient (%R) was achieved at 100% under M3 tires with a slope of 75°. The increase in total *coliform* rejection on the M3 membrane can occur because at a slope of 75° with the spacer forming the letter "V", the *coliform* concentration that enters through the bottom of the reactor will be pushed with a high cross-flow flow rate and retained in the spacer. The narrow distance at the bottom of the spacer can increase the turbulence of the incoming feed water so that the accumulated *coliform* concentration becomes even and reduces the polarization of the *coliform* concentration on the membrane [23], [17]. The ability of the microfiltration membrane to remove *coliform* bacteria ranges from 99.99%. From the results of the membrane effluent analysis on the *E. Coli* parameters, it can be seen that the *E. Coli* content in each spacer slope variation has met the applicable quality standard of 0 MPN/100 mL. This result is in line with the membrane research that has been carried out by [5] and [2] on the removal of *E. Coli* in which all *E. Coli* content can be removed 100%.

3.4. Examples Analysis of the Effect of Spacer on Quantity Parameters

Flux decrease in all four membranes occurred for 90 hours. At the 1st hour the highest flux value was reached at 160 L/m^2.hour on the M2 membrane with a slope of 90°, while the lowest flux value at the 1st hour of 128 L/m^2.hour occurred on the M1 membrane with no spacer condition. At the 90th hour, the highest flux value occurred at M2 membrane conditions of 34 L/m^2.h and the lowest at M1 membrane conditions at 24 L/m^2. h. This proves that the addition of spacers is able to increase the membrane permeability so that the resulting flux value increases [27].

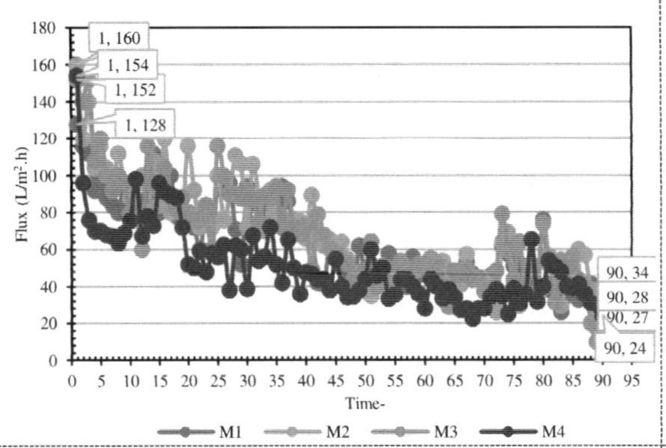

Figure 16. Spacer effect on flux

Table 2. Pressure effect on permeate volume

Spacer	Permeate volume (L)	Operation Time (h)	Average flux (L/m^2.h)
M1 (without spacer)	839,79	90	62,21
M2 (90)	901,55	90	66,78
M3 (75)	895,19	90	66,31
M4(105)	712,95	90	52,81

4. Conclusion

The best performance of immersed membrane microfiltration with optimal pressure effect occurs at pressure of 0.7 bar. The removal efficiency on the parameters of turbidity, TSS, organic matter, microplastic, total *coliform* and *E. Coli* for each variation was quite high but the quantity parameter produced at a pressure of 0.7 bar was able to maintain the flux value. The effect of spacer provision on optimal quality and quantity parameters occurs at 90° spacer slope. The provision of quality parameters for each spacer variation gives an insignificant difference but at 90° spacer slope produces the highest permeability with a long operating time

References

[1] N. I. Said dan D. R. K. Hartaja, "Kondisi Kualitas Air Baku Instalasi Ngagel Perusahaan Daerah Air Minum (PDAM) Kota Surabaya," Jurnal Air Indonesia, Vol. 10, No. 2 (2018) 52-66.

[2] N. S. Mahardini dan F. H. Kusuma, "Pengolahan Air Baku Menjadi Air Minum Dengan Teknologi Membran Mikrofiltrasi dan Ultrafiltrasi," PMKP, Vol. 1, No. 10 (2002) 1-13.

[3] R. Rahmawati, M. R. Bilad, A. M. Laziz, N. A. H. M. Nordin, N. Jusoh, Z. A. Putra, T. M. I. Mahlia dan J. Jaafar, "Finned spacer for efficient membrane fouling control in produced water filtration," Journal of Environmental Management, Vol. 249 (2019) 109359

[4] T. Lebeau, C. Lelièvre, H. Buisson, D. Cléret, L. W. Van De Venter, & P. Côté, "Immersed membrane filtration fo the production of drinking water: combination with PAC for NOM and SOCs removal," Desalination, Vol. 117, No. 1–3 (1998) 219–231.

[5] S. B. Al-Shammari, Bou-Hamad, and M. Al-Tabtabaei, "Comparative performance evaluation of microfiltration submerged and pressurized membrane treatment of wastewater," Desalination and Water Treatment, Vol. 49, No. 1-3 (2012) 26-33.

[6] S. J. Judd and P. Hillis, "Optimisation of combined coagulation and microfiltration for water treatment," Water Research, Vol. 35, No. 12 (2001) 2895-2904.

[7] R. W. Baker, "Membrane Technology and Applications," In Membrane Technology and Research, 2nd ed. California: McGraw-Hill (2004)

[8] D. Novianto, Y. Dinarianasari dan A. Prasetyaningrum, "Pembuatan Refined Carrageenan Dari Rumput Laut Jenis Euchema cottonii," Jurnal Teknologi Kimia dan Industri, Vol. 2, No. 3 (2013) 109-114

[9] I. G. Wenten, "Teknologi Membran dalam Pengolahan Air Limbah dan Industri. Studi Kasus: Pemanfaatan Ultrafiltrasi untuk Pengolahan Air Tambak", Conference Paper (2004, Jan) 1-22.

[10] V. Chen, A. G. Fane, S. Madaeni and I. G. Wenten, "Particle deposition during membrane filtration of colloids: Transition between concentration polarization and cake formation," Journal of Membrane Science, Vol. 125, No. 1 (1997) 109–122.

[11] M. Mulder, "Basic Principles of Membrane Technology," in Kluwer Academic Publisher, 2nd ed. (1996)

[12] R. W. Field, D. Wu, J. A. Howell and B. B. Gupta, "Critical Flux Concept for Microfiltration Fouling," Journal of Membrane Science, Vol. 100, No. 3 (1995) 259–272.

[13] H. Yamamura, K. Kimura and Y. Watanabe, "Mechanism involved in the evolution of physically irreversible fouling in microfiltration and ultrafiltration membranes used for drinking water treatment," Environmental Science and Technology, Vol. 41, No. 19 (2007) 6789–6794.

[14] I. G. Wenten, "Teknologi Membran: Prospek dan Tantangannya di Indonesia," In Forum Guru Besar ITB, Bandung: ITB, (2016) 40-56

[15] S. Kerdi, A. Qamar, A. Alpatova, J. S. Vrouwenvelder, and N. Ghaffour, "Membrane filtration performance enhancement and biofouling mitigation using symmetric spacers with helical filaments," Desalination, Vol. 484 (2020, Mar) 114454.

[16] R. Rahmawati, M. R. Bilad, N. I. M. Nawi, Y. Wibisono, H. Suhaimi, N. Shamsuddin dan N. Arahman, "Engineered spacers for fouling mitigation in pressure driven membrane processes: Progress and projection," Journal of Environmental Chemical Engineering, Vol. 9, No. 5 (2021) 106285.

[17] B. Wu, Y. Zhang, Z. Mao, W. S. Tan, Y. Z. Tan, J. W. Chew, T. H. Chong, and A. G. Fane, "Spacer vibration for fouling control of submerged flat sheet membranes," Separation and Purification Technology, Vol. 210 (2019) 719–728.

[18] Y. Taamneh and K. Bataineh, "Improving the performance of direct contact membrane distillation utilizing spacer-filled channel," Desalination, Vol. 408, (2017) 25–35.

[19] L. K. Wang, J. P. Chen, Y. T. Hung, and N. K. Shammas, "Membranes and Desalination Technologies," in Handbook of Environmental Engineering, Vol. 13, New York: Humana Press (2011)

[20] R. Riduan, "Analisa Pengaruh Kandungan Suspended Solid (SS) Air Baku terhadap Kinerja Membran Mikrofiltrasi," Info-Teknik, Vol. 4, No. 1 (2003) 27–33.

[21] F. C. Kramer, R. Shang, L. C. Rietveld, dan S. J. G. Heijman, "Fouling control in ceramic nanofiltration membranes during municipal sewage treatment," Separation and Purification Technology, Vol. 237 (2020) 116373.

[22] S. Notodarmojo dan A. Deniva, "Penurunan Zat Organik dan Kekeruhan Menggunakan Teknologi Membran Ultrafiltrasi dengan Sistem Aliran Dead-End (Studi Kasus: Waduk Saguling, Padalarang)," ITB Journal of Sciences, Vol. 36, No. 1 (2004) 63–82.

[23] S. Redjeki, "Proses Desalinasi Dengan Membran," Direktorat Penelitian Dan Pengabdian Kepada Masyarakat (DP2M), (2011)

[24] R. Nurtsani, "Analisis Bakteri Patogen Escherichia coli Pada Tiram (Crassotrea sp.) yang Berasal dari Perairan Laut Kecamatan Baru," Tugas Akhir, Dept. Ilmu Kelautan, FIKP, Universitas Hasanuddin (2018)

[25] N. F. Gray, "Filtration Methods. In Microbiology of Waterborne Diseases: Microbiological Aspects and Risks" 2nd ed., Chapter 35, Elsevier (2014).
[26] K. Guerra, and J. Pellegrino, "Development of a Techno-Economic Model to Compare Ceramic and Polymeric Membranes," Separation Science and Technology, Vol. 48, No. 1 (2013) 51–65.
[27] P. S. Goh, W. J. Lau, M. H. D. Othman, and A. F. Ismail, "Membrane fouling in desalination and its mitigation strategies," Desalination, Vol. 425, (2017, Agt) 130–155.
[28] F. Zamani, H. J. Tanudjaja, E. Akhondi, W. B. Krantz, A. G. Fane, dan J. W. Chew, "Flow-field mitigation of membrane fouling (FMMF) via manipulation of the convective flow in cross-flow membrane applications," Journal of Membrane Science, Vol. 526, (2017) 377–386.

IOP Conf. Series: Earth and Environmental Science 1111 (2022) 012064 doi:10.1088/1755-1315/1111/1/012064

The Influence of Storage on Structural Damage of Vegetable-Tanned Goat Skin by Microbes

Muhammad Iwan Alwan[1], Arnieda Asfarina Parameswari[1], Ambar Pertiwiningrum[2*], Nanung Agus Fitriyanto[1], Viagian Pastawan[1]
[1]Department of Technology and Livestock Products, Universitas Gadjah Mada, Yogyakarta, 55281, Indonesia
[2]Department of Technology and Livestock Products, Universitas Gadjah Mada, Yogyakarta, 55281, Indonesia, 0000-0001-7592-2542
[1]muhammad.iwan.alwan@mail.ugm.ac.id,[2]arniedaap@mail.ugm.ac.id,
[3]*artiwi@mail.ugm.ac.id, [4]nanungagusfitriyanto@mail.ugm.ac.id,
[5]virgian.pastawan@ugm.ac.id

Abstract. This study aimed to determine the influence of storage on the structural damage of vegetable-tanned goat skin by microbes. The storage time was divided into three treatments, namely treatment P0 (0 months, storage), treatment P1 (6 months, storage), and treatment P2 (12 months, storage). The observation parameters for investigation included microbe contamination, pH levels, and levels of oil/fat. This study was conducted with Analysis Variance Completely Randomized Design (ANOVA) and Duncan Multiple Range Test (DMRT) to determine the differences between treatments. In contrast, analysis regression was used to obtain the optimum level. The research found that the highest amount of microbial contamination on the leather occurred at month 12, with the amount of microbial contamination reaching 4.46x103 CFU. The best storage for pH levels occurred at month 14 and pH 5.16. The skin with the best oil content is in the third month, with an oil/fat content of 7.97%.

1. Introduction

Before, the goat was multi-purpose livestock. Goats are exploited for all parts, such as for food and clothing. Goats can adapt to various environments and be raised to produce meat, dairy, and hide [1]. The potential of goat production encourages Indonesian farmers to rear them. For example, based on the Central Bureau of Statistics [2] data, the goat population in the region of Yogyakarta had increased from 416.400 to 421.287 in 2021.

The rising goat population also increases the by-product, which includes the skin. Skin can be utilized to make various products with high economic value. One of the methods used to process goat skin is tanning. Tanning is the process of treating raw skin to make it stable and unputrefied [3]. The purpose of tanning is to turn raw skin into tanned leather, which is more resistant to chemical reactions, physical damage, and microbial activity.

Generally, skin tanning is divided into three types based on the substance being used: vegetable tanning (counter current and rapid tanned methods), mineral tanning (chrome tanning, aluminum tanning, and fat tanning), and synthetic tanning [4]. Vegetable tanning is turning raw skin into tanned leather using plants' tanning substances (tannin). Tannin is the essential substance in plants found in the wood, fruits, leaves, bark of trees, and roots, which is formed naturally [5]. The characteristics of tannins are: tightening (astringent), amorphous, preserving the skin from microbes, and providing color to tanned skin. The principle of vegetable tanning is inserting tannins into the skin tissue, such as collagen, so that a chemical bond is formed to prevent microorganisms, chemical and physical damage, and eventually turning the skin into leather.

However, the skin is tanned, and the opportunity of tanned skin getting damaged remains exists. Among the factors affecting structural damage is storage. The more extended storage will increase the possibility the skin to be damaged. Damage can be characterized by the number of microbes and the skin's chemical quality, including pH and oil/fat levels.

The microbes found on the skin will affect its structure. The longer storage time will lead to microbe contamination on the skin. This research aimed to define the influence of storage on the structural damage of vegetable-tanned goat skin by microbes. The research was conducted using different storage duration treatments so that it is expected to be able to determine the factors that cause structural damage that occurs in vegetable-tanned goat skin. The skin is a by-product of slaughtering livestock which is easily damaged physically, chemically, and biologically, so there is a need for handling that can prevent skin damage [6].

2. Methods

This study was carried out from February 16, 2021, to February 21, 2022. The examination for pH levels, water content, and level of oil/fat was held at the Center for Leather, Rubber, and Plastics (BBKKP) Jalan Sukonandi No. 9 Yogyakarta 55166. At the same time, the microbial contamination was examined in the Laboratory of Leather Technology, Animal By-Product, and Waste, Animal Science Faculty, Universitas Gadjah Mada. The study utilized one sheet of vegetable-tanned goat skin for different storage times (0 months, 6 months, 12 months). Subsequently, the goat skin was investigated for total microbes (TPC), pH level, and oil/fat test.

2.1 Total Plate Count

Firstly, the vegetable-tanned goat skin that had been stored was cut into small pieces and weighed as much as 5 grams. Subsequently, the sample was put into Erlenmeyer and shaken at 100 rpm. It was then poured into the reaction tube (T1) and filtered with filter paper. The growth medium was made with 1.75 grams of PCA for 100 ml of distilled water that was then shaken and sterilized with an autoclave at 121^0 C for 15 minutes. The result in the reaction tube (T1) was taken as much as 1 ml and poured into the reaction tube (T2) that had been filled with 9 ml of distilled water. After being mixed with vortex until homogeneous, it was then taken as much as 0.1 ml to be incubated at 37 ^{0} C for 48 hours.

2.2 pH Test

One gram of adhesive sample was dissolved in newly boiled water, then added with water until it reached 10 ml. The solution was left relaxed at room temperature and then measured for pH using a pH meter.

2.3 Oil/fat Content

Fat content in the skin can be determined by the Soxhlet method. In Soxhlet, the extracting solvent in the Soxhlet flask was heated to its boiling point to evaporate. This solvent vapor rose through the return cooling pipe, condensing and dripping on the extracted material. This solvent soaked the material, and when the height exceeded the height of the solvent flow pipe, the extract flowed into the Soxhlet flask. The collected extract was heated again so that the solvent would evaporate and the fat would be left on the flask.

2.4 Data Analysis
Data analysis was carried out by analyzing all the data obtained using ANOVA, which was compiled based on a Completely Randomized Design. The analysis results that had significant differences were then tested using Duncan with Microsoft Excel software. The regression analysis was used to obtain the optimum test results.

3. Results and Discussion
3.1 Total Microbes
The amount of microbial contamination is the number of colonies found on goat skin expressed in CFU (Colony Forming Units). The result of this study is identical to the data from SNI 01-6366-2000, where the maximum microbial is $1x10^4$.

Table 1. The results of microbial contamination (CFU) of vegetable-tanned goat skin

Replication	P0 (10^3 CFU)	P1 (10^3 CFU)	P2 (10^3 CFU)
1	1,0	2,9	3,4
2	1,2	3,9	4,6
3	1,3	4,1	5,5
Average	1,167	3,633	4,467

The analysis results in Table 1 showed a significant difference ($P < 0.05$) in the relationship between the storage and the amount of microbial contamination in vegetable-tanned goat skin. Treatment P2 had the highest number of microbes compared to P0 and P1. The regression analysis showed that the maximum microbial contamination occurred in month 12, with the amount of microbial contamination reaching $4,46x10^3$ CFU. Based on these results, there was a significant difference ($P < 0.05$) in the relationship between the type of storage and the amount of microbial contamination of vegetable-tanned goat skin. These results have difference from the study conducted [7], where the number of bacterial colonies on goat skin was 12.51 ± 1.64

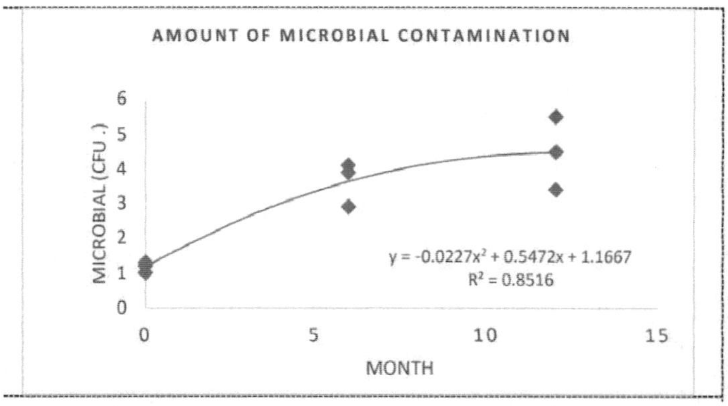

Figure 1. Graph of the Relationship Between Storage Time and the Amount of Microbial Contamination of Vegetable-Tanned Goat Skin.

From the graph above (Figure 1), it can be seen that there was an upward trend in the amount of microbial contamination. This means that the storage time is directly proportional to the amount of microbial contamination. The longer storage time will affect the number of microbes in the skin and, therefore, will damage the skin.

Two factors that cause the increasing number of microbes in the skin are temperature and environment. Room temperature is perfect for microbes to grow because most microbes can grow at room temperature, which is 37°C. Research from [8] states that cold and frozen storage can destroy spoilage microbes. This means that the skin is stored in a place with a cold temperature, and the number of microbes on the skin is likely to be reduced. In addition, a non-sterile environment will make it easier for microbes to grow. An unsterile environment will make it easy for microbes to breed. Microbes on the skin can be bacteria, molds, and yeasts. The bacteria present on the skin of livestock include *Staphylococcus sp.*, *Streptococcus* sp., and *Corynebacterium sp.* Meanwhile, the types of molds include *Microsporum sp.*, *Trichophyton sp.*, and *Epidermophyton sp.* The yeasts present on the skin are *Malassezia* sp. and *Pityrosporum sp* [9].

Another factor that causes an increase in the total number of bacteria is RH (Relative Humidity). The average humidity of D.I. Yogyakarta is around 80-92% [10]. According to [11], RH above 65% causes easy mold growth. Mold growth will form white, gray, greenish, and slightly blackish clusters. The growth of mold can damage the structure of the skin and make the skin vulnerable to damage.

3.2 pH Test

This study showed data in accordance with SNI 4593:2011, with the pH ranging between 3,5- 7.

Table 2. pH test results of vegetable-tanned goat skin

Replication	P0	P1	P2
1	2,41	4,23	5,11
2	2,47	4,31	5,14
3	2,45	4,37	5,13
average	2,44 ± 0,0031[a]	4,30 ± 0,070[b]	5,13 ± 0,153[c]

The results of the ANOVA analysis in Table 2 showed a significant difference ($P>0.05$) in the relationship between the length of storage time and the pH level of vegetable-tanned goat skin. The treatment P2 had a higher pH level than P0 and P1. The regression analysis results stated that there was a significant difference ($P>0.05$) in the relationship between pH levels and storage time of vegetable-tanned goat skin. The best storage was in month 14 and at pH 5,16.

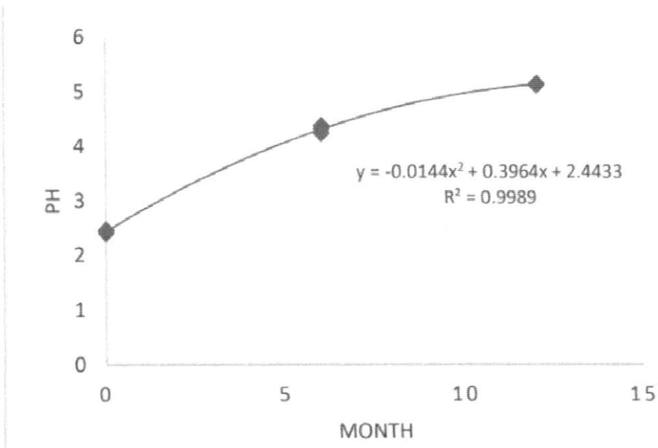

Figure 2. Graph the relationship between storage time and pH levels of vegetable-tanned goat skin.

From the graph above (Figure 2), it can be seen that an increase in pH levels often takes a long time of storage. The longer the skin storage time, the higher the pH level. The skin type and tanning material can also influence the skin's pH level. The skin of different animals will have different skin compositions, affecting the skin's pH. In addition, skin that is tanned using vegetable tanners will be easier to have a raise or decrease in pH level.

3.3 Oil/Fat Level Test

This study showed the suitability of the results with data from SNI 4593:2011, where the oil/fat content ranged from 8.0% to 15%.

Table 3. The Test Result of Oil/Fat Content of Vegetable-Tanned Goat Skin (%)

Replication	P0	P1	P2
1	7.64	9.47	12.85
2	7.71	9.45	12.91
3	7.78	9.49	12.92
average	7.71 ± 0.070 [a]	9.47 ± 0.020 [b]	12,89 ± 0.038[c]

The result of the analysis (Table 3) showed that there was a fundamental difference ($P>0.05$) in the relationship between the length of storage time and the level of oil/fat in vegetable-tanned goat skin. Treatment P2 had the highest level of oil/fat compared to P0 and P1. The result of regression analysis stated that there was a significant difference ($P>0.05$) in the relationship between oil/fat level and storage time of vegetable-tanned goat skin. The best oil level occurred in the third month, with oil/fat levels of 7,97%.

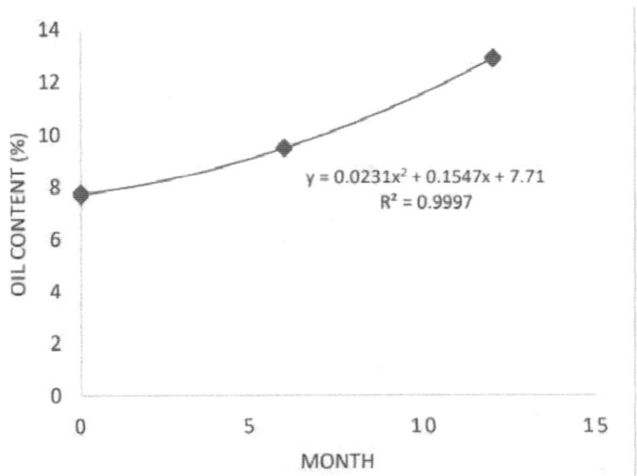

Figure 3. Relationship between storage time and oil/fat level of vegetable-tanned goat skin

Based on the chart (Figure 3), it can be concluded that more extended storage will increase the oil/fat level of the skin. It can be influenced by several factors, including temperature and skin type. The higher the storage temperature, the higher the oil/fat level. The skin from different goats will also have different oil/fat levels.

IOP Conf. Series: Earth and Environmental Science 1111 (2022) 012064 doi:10.1088/1755-1315/1111/1/012064

4. Conclusion

Based on the research, it can be concluded that the length of storage time will significantly impact the amount of microbial contamination, pH, and oil/fat content that will also damage the structure of vegetable-tanned goat skin. The massive amount of microbial contamination occurred in 12 months with a microbial count of $4{,}46 \times 10^3$ CFU. The best storage for pH level was in the 14th month at a pH concentration of 5,16.

Acknowledgments

The research was funded by the RTA program of the Universitas Gadjah Mada with assignment letter number 3550/UN1.P.III/Dit-Lit/PT.01.05/2022. The authors would like to thank students from the Faculty of Animal Science Universitas Gadjah Mada who helped with data collection.

References

[1] Kurniasih N N, Fuah A M and Priyanto R 2013 Karakteristik Reproduksi dan Perkembangan Populasi Kambing Peranakan Etawah di Lahan Pasca Galian Pasir *J. Ilmu Produksi dan Teknol. Peternak.* **1** 132–7

[2] Badan Pusat Statistik 2022 *Populasi Kambing Menurut Provinsi* (Indonesia)

[3] Mustakim, Aris S W and Kurniawan A P 2015 Perbedaan Kualitas Kulit Kambing Peranakan Etawa (PE) dan Peranakan Boor (PB) yang Disamak Krom *J. Ternak Trop.* **11** 38–50

[4] Sutyasmi S, Widowati T P and Setyadewi N M 2016 Pengaruh mimosa pada penyamakan kulit jaket domba samak nabati menggunakan sistem C-RFP, ditinjau dari sifat organoleptis, fisis, dan morfologi kulit *Maj. Kulit, Karet, dan Plast.* **32** 31–8

[5] Kasmudjiastuti E, Sutyasmi S and Widowati T P 2015 Pemanfaatan tanin dari kulit kayu tingi (Ceriops tagal) sebagai bahan penyamak nabati: pengaruh penambahan alum dan mimosa *Maj. Kulit, Karet, dan Plast.* **31** 45–54

[6] Anggara D F, Sutardjo D S and Suradi K 2013 Pengaruh Penggunaan Jenis Asam pada Proses Pickle terhadap Kualitas Kimia Kulit Kelinci Peranakan New Zealand White *J. Livest. Sci.* **2** 1–12

[7] Melia S, Juliyarsi I and Hayatuddin M 2014 Karakterisitik Kimia dan Total Koloni Bakteri Gelatin dari Beberapa Jenis Kulit Ternak *J. Peternak. Indones. (Indonesian J. Anim. Sci.* **16** 188–92

[8] Siburian E T P, Dewi P and Kariada N 2012 Pengaruh suhu dan waktu penyimpanan terhadap pertumbuhan bakteri dan fungi ikan bandeng *J. Life Sci.* **1** 101–5

[9] Pasaribu F H, Pribadi E S and Sunardi 1998 Temua Bakteri dan Kapang pada Kulit Sapi Mentah Impor dan Lokal : Suatu Studi Pendahuluan *Media Vet.* **5** 17–21

[10] Badan Pusat Statistik 2019 *Rata Rata Suhu Udara dan Kelembaban Udara Setiap Bulan di Yogyakarta* (Yogyakarta)

[11] Dignard C and Mason J 2018 *Caring for Leather, Skin, and Fur* (Ottawa, Canada)

IOP Conf. Series: Earth and Environmental Science 1111 (2022) 012065 doi:10.1088/1755-1315/1111/1/012065

An Overview of Creative Cities and Ecotourism Development in Jepara District, Indonesia

Lala Hucadinota Ainul Amri [1], Nova Amalia Sakina [2]*, Noor Azly Mohammed Ali [3] and Rusmadiah Anwar [4]

[1]Business Administration, Madiun State Polytechnic, Madiun, 63133, Indonesia, ORCID ID: 0000-0002-6087-5576

[2]*School of Environmental Science, Universitas Indonesia, Jakarta, 10430, Indonesia, ORCID ID: 0000-0002-0672-7986

[3]Print Technology, College of Creative Arts, Universiti Teknologi MARA, Shah Alam, 40450, Selangor, Malaysia, ORCID ID: 0000-0001-5711-7276

[4]National Design Centre, College of Creative Arts, Universiti Teknologi Mara, Shah Alam, 40450, Selangor, Malaysia, ORCID ID: 0000-0001-6803-6972

[1]lalahuca@pnm.ac.id, [2]*nova.amalia01@ui.ac.id, [3]noora568@uitm.ed.my, [4]rusma935@uitm.edu.muy

Abstract. The city today is conceptualized using terms of making cities inclusive, safe, resilient, and sustainable. The concept of creative cities began to be discussed when the Indonesia Creative Cities Network (ICCN) was formed as part of the creative industry development program. The ecotourism concept is one aspect that must be implemented in creative cities. The implementation of ecotourism must be concerned with a sustainable environment and involve society as a tourist. According to the direction of the Indonesian president that environmental sustainability is the primary goal of Indonesia's development. Jepara, as a developing district in Indonesia, has the potential to be developed as a strategic and policy framework as a benchmark for other districts/cities in Indonesia. This research aims to serve as an overview of a conceptual framework that must be considered to develop creative cities and ecotourism in Indonesia. The results show that the conceptual issue correlated with creative city and ecotourism must be handled and accommodated to facilitate the development of creative city and ecotourism. This framework can support further research in this developing concept and can be implemented in other developing districts or cities.

1. Introduction

All cities in Indonesia have an attraction that, if developed, can become an attractive characteristic and bring in tourists. In Indonesia, several regions have already received recognition as creative cities [1]. It is good this thing continues to be improved in quality and quantity. Efforts to start a network of Indonesian creative cities began to be seriously discussed in creative economy workshops organized by the government. Indonesia Creative Cities Network (ICCN) is part of the creative industry development program by presenting superior works and products in every city in Indonesia to prepare themselves for the ASEAN Economic Community program [2]. Ideas and initiatives to develop creative cities emphasize the process of creation, innovation, and individual talent [3].

The policy direction and strategy of the Ministry of Tourism and Creative Economy (Kemenparekraf) or Agency for Tourism and Creative Economy (Baparekraf) describes the spirit and efforts made to change the current condition into the desired condition. The ultimate goal is to increase the contribution of the tourism sector and the creative economy to the resilience of the national economy. This can be realized through the development of 5 strategic pillars of tourism and the creative economy, namely the Pillar of Tourism Destinations and Creative Economy Products, the Pillar of Tourism Marketing and the Creative Economy, the Pillar of the Tourism Industry and Creative Economy, the Pillar of Human Resources and Tourism and Creative Economy Institutions and the Pillar of Creativity [4].

The Kemenparekraf/Baparekraf targets growth in investment and access to finance in the tourism sector and the creative economy. The ratio of standardized and certified tourism and creative economy businesses is targeted to increase from 10% in 2020 to 30% in 2024. Total investment in tourism and the creative economy is targeted to increase from US$ 2 billion in 2020 to US$ 3 billion in 2024. The ratio of tourism and creative economy businesses that have access to finance to the total tourism and creative economy businesses is targeted to increase from 1.8% in 2020 to 4.6% in 2024 [4] [5].

Based on optimizing the creative economy, tourism development in Indonesia in the last five years has been very rapid because each region is trying to optimize its regional potential to boost the tourism industry. This is supported by the existence of regional autonomy, which allows regions to maximize the utilization of their potential to increase Regional Original Income (PAD). This potential development is carried out concerning sustainable tourism development [6]. As one of the efforts to develop the tourism sector in Jepara, the Government of Jepara prepares a tourism development plan which includes product development (tourism objects and attractions, tourism support facilities, accessibility) to support the concept of developing creative cities, creative industries, and creative economy [7] [8]. The development of Jepara's creative cities has its problems; there is no standard reference for developing a creative city that is environmentally friendly. For this reason, an in-depth study is needed so that it can be used as a benchmark for other cities.

2. Methods

This research was carried out using a qualitative approach, in principle wanting to examine, explain, critically describe, or describe a phenomenon, an event, or an event of social interaction in society to seek and find meaning in a real context (natural settings). Therefore, this type of qualitative research is descriptive by collecting soft data, not complex data to be processed with statistics. As in research with a quantitative approach, data collection with a qualitative approach is in the form of field research, and there is also library research [9]. The literature used in this research is documentation of previous research and reports published by the government. Researchers have several key questions that were formulated previously. In this study, the researcher provides an in-depth descriptive study of the tourism development document of Panjang Island, located in Jepara, Indonesia. The site was chosen because it has a development plan published on the government's website. It is assumed to have a development direction supporting creative cities, industries, and the creative economy. For this reason, this research will discuss more deeply the direction of development and the conceptual issues related to creative cities and industries that must be handled and accommodated to facilitate the development of creative cities.

3. Results and Discussion

3.1. Indonesia as a Prospective Country Investment Destination

Prospects and investment opportunities in the tourism sector in Indonesia show that the continued improvement in national business and economic conditions after the global economic crisis has strengthened international confidence in Indonesia, making Indonesia a prospective investment destination. The World Investment Prospects Survey 2014-2016 reports that from 159 executive responses of Transnational Companies (TNCs) from developed and developing countries, the prospect of investing in Indonesia is ranked third or quiet perspective [4].

Of course, this is related to sustainability, where Indonesia's assessment of sustainable competitiveness is very important for investment prospects, given that investment is one component of GDP and its growth [5]. In general, there are 3 (three) main things that entrepreneurs always consider when making investments: - Political stability and security that provide business certainty. - A flexible and proactive bureaucracy so that it can serve the wishes of entrepreneurs but still be within the corridor of applicable Laws and regulations. - Able to provide a conducive climate for business so that entrepreneurs can benefit because the company is not a social entity. However, tourism investment is currently relatively not optimal to move the tourism industry more evenly in various provinces and tourism destinations in Indonesia. Currently, most of the investment activities in Indonesia are concentrated in Bali, Java, and Batam, with the dominance of business types in the hotel, restaurant, and transportation sectors.

3.2. Creative Economy in Indonesia

In the Ministry of Tourism and Creative Economy's Strategic Plan document, creativity means the capacity or effort to produce or create something unique and new and create a solution to a problem or do something different. While the Creative Economy is the creation of added value based on creativity. Creative Economy Enterprises are business entities, both legal entities and non-legal entities, that transform and utilize creativity to produce goods and services and are recognized as having intellectual property rights, both registered and attached [4]. On the other hand, Creative has various definitions. Creative industries have multiple definitions and meanings, and the debate continues on the inherent linkages between creative and cultural industries [10].

In Presidential Regulation Number 72 of 2015, creative economy products are classified into 16 sub-sectors which are then detailed by the Central Statistics Agency (BPS) into 206 five-digit KLBI or Indonesian Standard Business Field Classifications (BEKRAF, 2017). The details of the sixteen creative economy sub-sectors in a row according to the KBLI sequence are as follows: (a) architecture, (b) interior design, (c) visual communication design, (d) product design, (e) movies, animations, and videos, (f) photography, (g) crafts, (h) culinary, (i) music, (j) fashion, (k) applications and game developers, (l) publishing and printing, (m) advertising, (n) television and radio, (o) performing arts, and (p) fine arts.

3.3. Strengthening Ecotourism Organizations in the Regions

Through decentralization, local governments are required to manage regional finances accountable and transparently. With the existing normative policies, local governments can change policies and regional financial management systems. However, this paradigm has not become a national perception that is evenly distributed at all levels, and the absence of a national and holistic spatial plan that is used as the basis for the development of economic resources, especially for the tourism sector, which has a direct impact on regional economic development [4]. Tourism is still considered a sector of choice. It has not been considered a strategic sector that contributes significantly to regional development and the community's welfare, so the strengthening of organizations in charge of tourism development has not been evenly distributed in various regions. As a result, cross-regional coordination in the integrated handling of cross-regional tourism assets often encounters obstacles and obstacles. On the other hand, the weak understanding of tourism often positions tourism as a complementary sector that does not have a strategic position in the organizational structure of regional development.

3.4. Potential of Jepara, Indonesia

Jepara is a city located in the province of Central Java, where the western and northern parts are bordered by the sea, while the eastern part is a mountainous area. The geographical position of Jepara Regency is located in the northern part of Central Java Province with coordinates 110 9' 48.02"-100 58'37.40" East Longitude and 5 43'20.67" – 6 47'25.83" South Latitude with boundaries the area includes the West: Java Sea, North: Java Sea, East: Pati and Kudus Regencies, South: Demak Regency. Jepara Regency has an area of 1,004,13,2 km^2, including 16 Districts, 184 Villages, and 11 Special Villages, with a sea area of 2,112,836 km^2. Jepara Regency is a coastal district with a beach length of more than 82 km, so it has the potential for marine tourism that can be developed to create jobs and improve the regional economy. The tourism potential in Jepara Regency includes nature tourism, marine/beach tourism, religious tourism, historical tourism, cultural tourism, craft centers, and culinary tourism. Other tourism potentials developing in Jepara Regency are Nature Tourism, namely Bandengan Beach, Kartini Beach, Karimunjawa Islands, Ombak Mati Bondo Beach, and Tempur Tourism Village (Wind.Temple), Plajan Tourism Village. Cultural Tourism (Mantingan Tomb, Kartini Museum, Portuguese Fort, Ari-Ari Monument, Jepara Regency Pendopo, Hian Thian Siang Tee Temple), Artificial Tourism (Jepara Ourland Park and Turtle Ocean Park), Shopping Tour (Center for Carving and Sculpture) Mulyoharjo, Troso Ikat Weaving Center, Monel Craft Center). The tourism potential of the Jepara Regency is expected to become integrated tourism in Jepara Regency and can support the potential for the development of Panjang Island Tourism [7].

Jepara Regency is very supportive in terms of tourism development. This can be seen in the Regional Medium-Term Development Plan (RPJMD) and several tourism studies that have been carried out. The potential of Jepara for tourism development is enormous, with a market share of not only local tourists but well-known domestic tourists and even foreign tourists as a buffer for the Karimunjawa Islands, which has been designated as a National Strategic Area. To encourage tourism growth, Jepara Regency offers the development of Panjang Island tourism potential to investors with an investment of IDR 30,807,605,000.- investors will get development and management rights by considering environmental aspects and cooperating with the surrounding community as a contribution and reduction of social impacts (Bakar & Zainon, 2019) so that the Aqua Edu Culture Park - Panjang Island can be developed which is highly competitive and provides a different experience for tourists [8].

3.5. Creative City Development

The first parameter of a creative city is that the city can develop the potential of a creative economy. Second, there is a role and synergy between the government, academia, business actors, and the community. Third, sustainable strategies and innovations create creative activities oriented to regional economic development [1]. The principle of tourism development on Panjang Island has a development principle that supports creative industries by considering tourism development that does not disturb the environment but can preserve and utilize the environment. Development of tourism with tourist attractions that involve tourists in conserving biodiversity on small islands, as well as being disability friendly by using Braille, which is effective in helping people with disabilities [11].

The development of supporting infrastructure for tourism development on Panjang Island has the concept of organizing a tourism system on land and the coast with an emergency evacuation system but made with high creativity, with relatively low wave conditions on Long Island with the highest wave of ±2 meters in December, the level of security on the island safe length with data on an average rainfall of 311mm/year, highest rainfall of 1122 mm/month in December, and lowest rainfall of 5 mm/month in October [7].

Cooperation with higher education in terms of supporting the provision of information centers and natural laboratories, providing essential public facilities that are environmentally friendly and do not change the landscape, building elements from natural materials, and environmentally friendly solar cell energy. Strengthening the characteristics of Long Island tourism, including the lumban tradition, pilgrimage to the guardian's tomb, and dance performances. Improve the local economy, such as crab products taken from local fishermen.

The development of Panjang Island is intended to build an integrated tourism system with other tourism in Jepara Regency, which focuses on marine tourism, beaches, education, and environmentally friendly culture. In addition, another goal is to create jobs in the tourism sector, especially for the people of the Jepara district. Creating competitive tourism and creative industry investment climate for investors. Creating an increase in the creative tourism economy for the community. Introducing and preserving culture and facilitating research and conservation of marine ecosystems.

3.6. Development Concept Analysis

The concept of developing Panjang Island to support it as a creative city can be seen from the development concept made by the team from the Jepara Steam Power Plant (PLTU), which can be seen in Figure 1. below:

IOP Conf. Series: Earth and Environmental Science 1111 (2022) 012065 doi:10.1088/1755-1315/1111/1/012065

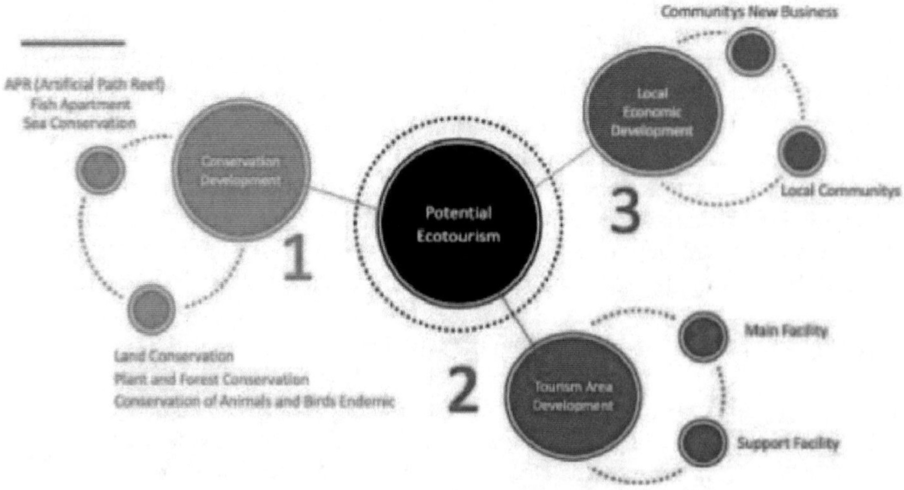

Figure 1. The concept of developing Panjang Island

The development of Panjang Island is planned to have buildings and facilities, which can be seen in Figure 2:

Figure 2. The development planning of Panjang Island

The creative aspect can be seen in the art of architecture and product design. Architectural aspects can be seen from the arrangement of building plans, applying the concept of creativity. In product design, landscapes, the status of places, and the existing diversity can be uniquely combined. This proves the use of the concept of creativity. Other creative economy subsectors, such as music and performing arts, can be optimized by providing unique performance venues. The place is designed to be open and environmentally friendly, following the principles of development, not disturbing the environment, preserving nature, and making good use of the environment. The sub-sectors, crafts, culinary, fashion, and fine arts, have unique booths. This is considered because of the needs of tourists. In addition, it raises the creative industry production sector in the surrounding area, such as the creative batik industry, in which all types have their patterns and uniqueness [12]. This is where the role of the creative industry sub-sector becomes crucial. This creative industry sub-sector has its uniqueness and different types of creativity. The creative city uses this difference to create a city that fully supports the creative industry and the creative economy.

4. Conclusion

Developing creative cities in Indonesia requires an integrated approach to regulatory aspects, natural resources, and the condition of the surrounding community. This requires flexibility in responding to the challenges of creative city development in finding innovative solutions, involving local communities in planning, decision-making, and delivering economic benefits without harming nature. To get to a creative city, it is necessary to formulate missions and ideas that are prepared by taking into account the strategic environmental factors, both external and internal that influence it as well as the strengths, weaknesses, opportunities, and challenges that exist in its goal as a creative city.

It considers the strategic environmental factors, both external and internal, that influence it and the strengths, weaknesses, opportunities, and challenges in its goal as a creative city. The principle of developing a creative city in the district of Jepara is quite good. However, despite its potential, everything has not been optimized to support 'creative cities, creative industries, and the creative economy. While the opportunity to introduce a community-based management approach, namely the involvement of local stakeholders (especially local communities) in the management system, is quite well implemented. Because the primary supporting force for the creative economy is the local community which is empowered and well trained, and other indicators following the SDGs are environmental sustainability that other districts/cities can follow.

Acknowledgment
We want to acknowledge The Ministry of Higher Education Malaysia for financial support under the FRGS-RACER grant scheme (RACER/1/2019/SSI07/UITM/1). We also thank Universiti Teknologi MARA, particularly the Department of Printing Technology, College of Creative Arts, UiTM.

References
[1] Ardhanariswari K A and Probosari N 2020 The Quadruple Helix Communication Model in Yogyakarta's Creative City Development Model Komunikasi Quadro Helix dalam Mengembangkan Yogyakarta sebagai Kota Kreatif **5** 294–306
[2] Rahadi R A 2016 Opportunities and Challenges for Micro-Small and Medium Business in Indonesia Facing ASEAN Economic *J.M.D.K* **18** 45–53
[3] Kemenparekraf & B 2015 Siaran Pers Indonesia Creative Cities Conference Merajut Jejaring Kota Kreatif Indonesia 2015
[4] Kementerian Pariwisata dan Ekonomi Kreatif RI 2014 Ekonomi Kreatif: Kekuatan Baru Menuju 2025 *Www.Kemenpar.Go.Id* **1** 5–24
[5] Badan Pusat Statistik 2021 *Statistik Indonesia 2021*
[6] Jepara P K 2020 Rencana Kerja Pemerintah Daerah Kabupaten Jepara Tahun 2020
[7] Bappeda 2018 Investment Profile Aqua Edu Culture Park Panjang Island.
[8] Central Java Governance Development of Panjang Island – Central Java 17
[9] Yusuf M 2013 Yusuf M. METODE PENELITIAN Quantitative, Qualitative, and Combined Research.2014..pdf
[10] Leng K S, Badarulzaman N, Samat N, Siratsharifah M and Dawood R S 2015 Creative cities research in Penang, Malaysia: A review of conceptual and methodological framework *Plan. Malaysia* **13** 1–18
[11] Mohamad Rozy S N, Vermont V V, Zainal Abidin S and Anwar R 2021 Ways and Challenges of Product Labelling for the Visually Impaired *Environ. Proc. J.* **6** 79–84
[12] Rosman R, Khairi H, Legino R and Naser F L 2021 Overview of Malaysian Traditional Batik Authenticity *Environ. Proc. J.* **6** 125–9

Sustainable Development in the Context of the Architecture of Environmental Friendly Medical Centers in Rural Areas (Case for Ukraine)

Iryna Bulakh[1*,2]
[1] Department of Design and Urban Development, Faculty of Architecture, Technical University of Darmstadt, Darmstadt, 64287, Germany, ORCID ID 0000-0002-3264-2505
[2] Department of Design Architectural Environment, Faculty of Architecture, Kyiv National University of Construction and Architecture, Kyiv, 03037, Ukraine, ORCID ID 0000-0002-3264-2505
[1]*bulakh@stadt.tu-darmstadt.de, [2]bulakh.iv@knuba.edu.ua

Abstract. The study of ways to improve medical care in rural areas is relevant for many countries, including Ukraine, whose territory continues to be destroyed for five months (at the time of writing) due to Russian military aggression. The article is based on the results of research by various scientists, primarily related to the sustainable development of architecture and urban planning, as well as the author's research in recent years. The article's purpose is to search for new architectural forms of organizing medical care in rural settlements that would contribute to the influx of young people and the development of medicine. An analytical method, a review of information sources, a full-scale survey with photo fixation, a survey of rural residents, and conceptual and exploratory design as part of the educational process at the university were used. The result of the study is a proposal for the design of new types of buildings with two main functions ("work" + "housing").

1. Introduction

Access of people to medical care in rural areas is a problem for many countries, including Ukraine. This is due to various factors, including the state priority for developing medicine in large cities, the small number and low population density in rural areas, and the lack of funding opportunities. Urbanization has a particular and weighty influence, which contributes to the excessive growth of cities and the simultaneous "extinction" of villages and small towns. As a result, rural youth seek higher education in cities and no more extended plan to return to their places of origin. But we understand that for any country's normal and successful development, a balance must be found between urban and rural areas. The entire population of the country cannot live exclusively in cities! This will lead to various negative consequences, including degradation of agriculture, transport collapse, and increased emergence and spread of pandemics due to excessive population density in cities. But how do they motivate young people to return or move to the countryside? How do we provide the necessary medical care to rural residents? This article attempts to answer these questions.

The theoretical basis for the preparation of this article is research related to various issues of sustainable design, architecture, and urban planning. In particular, the latest articles by Rudolph-Cleff A. and co-authors [1–3], Shebek N., Timokhin V., Tretiak Y., Kolmakov I. and Olkhovets O. [4, 5],

Pleshkanovska A. [6–9], Kovalska G. and co-authors [10–12], Semerikov S. and co-authors [13–15], Linda S. and co-authors [16, 17], Chukharev S. and co-authors [18, 19], Dyomin M. [20, 21], Adeyeye K. and co-authors [22–30], Ivashko Yu. and co-authors [31], Chernyshev D. and co-authors [32, 33], Kovalchuk O. and co-authors [34–36]. Also, some of the author's preliminary studies related to this topic served as the basis of the article [37–46]. Issues related to the problems of environmental and sustainability were studied based on a study by I. M. Lami and B. Mecca [47].

The purpose of this article is to attempt to find a new way to develop medical care in rural areas from the point of view of the architectural aspect. The study is relevant in connection with the permanent state of this problem, which was especially aggravated after the start of Russia's military invasion of the territory of Ukraine. Because of the war, a significant territory of Ukraine is almost completely erased, destroyed, and scorched. This is the territory that is or was in occupation and the one on which shelling is periodically carried out. In Ukraine, the medical infrastructure suffered greatly. According to the data of the Ministry of Health of Ukraine, as of the end of June, rockets, and projectiles of the Russian occupiers damaged 628 objects of medical institutions and destroyed another 118 without the possibility of their further restoration [48]. Thus, already today, it is necessary to understand how it is possible to restore one of the most critical sectors of the life of a country destroyed by war. How do we build medicine in rural areas so that its new level of development becomes a "magnet" for young families? However, it should be noted that all the arguments and the proposed solution in this article are based on the normal situation. In contrast, the actual military situation does not provide an accurate and stable basis for solutions other than straightforward and temporary measures (mobile medical centers, tents, etc.). Thus, all the suggestions in this article can be applied in areas that are not directly a combat zone in the countryside of Ukraine or after the end of the active phase of the war.

2. Methods

To conduct this study, the following methods were used: analytical (analysis of the existing approach to providing access to medical services for residents of rural areas); review of information sources (government programs, reports, scientific articles, reports at conferences, etc.); field survey with photo fixation, interviewing rural residents, conceptual and search as part of the educational process at the university.

3. Results and Discussion

3.1. Analysis of the architecture of medical institutions in rural areas of Ukraine (before the war)

In most rural settlements of Ukraine, access to medical services is low. This conclusion was obtained based on the author's preliminary and more comprehensive study [49]. The main reasons for this are similar to other countries, which allows for universal results that will be useful in other regional contexts.

The main reasons include:

- The insufficient overall financing level in rural areas, particularly in the medical sector.
- Lack of modernization and reconstruction of the existing medical infrastructure with its gradual physical deterioration and unsuitability for use.
- Prescription and referral of residents of rural areas to receive medical care in enlarged medical centers and hospitals in nearby cities.
- A complex stereotype of the "unpopularity" of living in rural areas among young people is their unwillingness to return to the villages after receiving higher education.
- Low population density over long distances in rural areas.
- Lack of state, local or private funding and motivation programs that stimulate the influx of the young population into the countryside (including doctors and medical personnel).

As a result, those few medical institutions located in rural areas of Ukraine have an inappropriate appearance (Figure 1, Figure 2). In recent years, new medical centers have appeared in some

Ukrainian villages, which in most cases were designed and built as part of the personal initiatives of communities and patrons (Figure 3, Figure 4). But against the background of the total number of Ukrainian villages, this indicates isolated examples and not a pattern.

Figure 1. Ambulatory care clinic in the village of Bezlyudivka, Kharkiv region, Ukraine.

Figure 2. Family Outpatient Clinic of Tsiblivska Rural Medical District, Kyiv region, Ukraine.

Figure 3. Outpatient clinic for the general practice of family medicine in the village of Belogorodka, Kyiv region, Ukraine.

Figure 4. Outpatient clinic for the general practice of family medicine in the village of Schastlivoe, Boryspil district, Kyiv region, Ukraine.

3.2. *Medical infrastructure of rural areas in Ukraine at present (military) moment*

The war in Ukraine is a tragedy that destroys the lives, health, and the possibility of the everyday existence of about 42 million Ukrainians, as well as many other countries that are helping in every possible way. Now it is impossible to talk about the exact consequences and scale since active hostilities continue, and there is no way to predict anything. The only thing that can be stated for sure is that the Russian military is purposefully destroying the medical infrastructure in cities and rural areas. According to the data of the Ministry of Health of Ukraine, as of the end of June, rockets, and projectiles of the Russian occupiers damaged 628 objects of medical institutions and destroyed another 118, without the possibility of their further restoration (Figure 5, Figure 6) [48].

Figure 5. The destroyed hospital in Volnovas, Ukraine.

Figure 6. The maternity hospital in Mariupol after the airstrike, Ukraine.

3.3. Proposals for the organization of medical care in rural areas

The main idea is that for the attractiveness of the work of doctors and medical personnel in rural areas, it is necessary to design and build dual-functional facilities that combine "work" (medical care) + "housing". Let's talk about very compact villages with a small number of inhabitants. It is advisable to build a cottage for the residence of a doctor and his family on the ground floor. There should be an office for receiving patients and a small area for visitors (doctor's assistant area, guest toilet, waiting area). In the case of a large village or a cluster of several villages close to each other, it is proposed to design a small medical center for the work of 3-5 doctors and nurses. Close to the medical center, a residential building should adjoin, which will act as a "magnet", especially for young professionals who usually face the problem of providing housing for their families. This idea was implemented and illustrated as part of an educational conceptual design among students of the Faculty of Architecture of the Kyiv National University of Construction and Architecture (Figure 7, Figure 8).

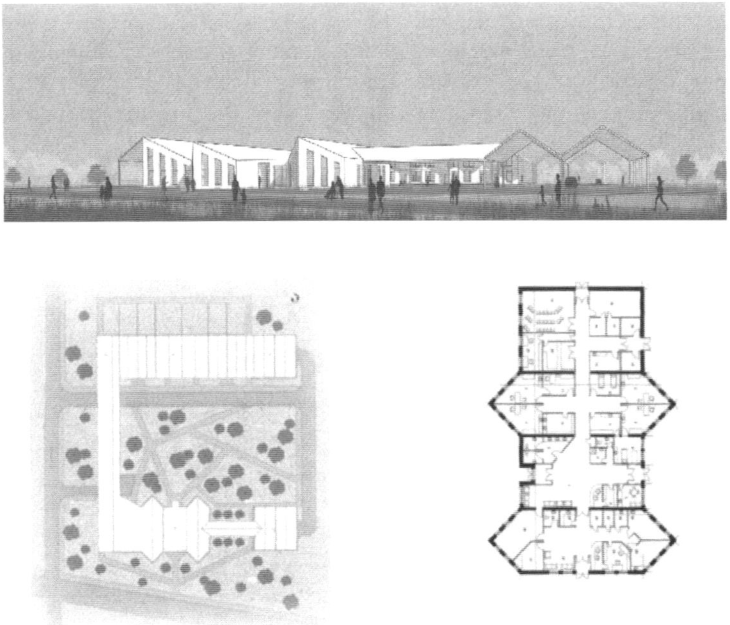

Figure 7. The Ambulatory clinic for family medicine for 25 visits per day in rural areas (the authors Maslovets O., prof. Bulakh I.).

Figure 8. The Ambulatory clinic for family medicine for 25 visits per day in rural areas (the authors Levitska M., prof. Bulakh I.).

A significant advantage of the ideas presented (as well as others that were not included as illustrations in this article) is the organization of various functional parts-buildings as a single architectural ensemble, which provides convenient access for visitors and staff of the medical center and at the same time, there is the isolation of the residential block. In both examples, the connection of two functional units is proposed using a covered passage (for ease of use in the cold and rainy seasons). Also, special attention in the projects was paid to the improvement of the territory of the medical center, which is a kind of "visiting card," demonstrating an attentive attitude to nature and its conservation.

4. Conclusion

To develop medicine in rural areas, it is necessary to create conditions for comfortable living and work for young doctors, nurses, and their families. For this, it is proposed to design new types of two-function buildings: cottages (for small villages) and medical centers (for large villages or several villages located nearby), which combine "work" and "housing". Every rural community is interested in providing access to therapeutic medical care, and such local and compact proposals can be quite feasible. This proposal does not pretend to be an exclusive and unambiguous statement. Still, it is a preliminary presentation of a new Ukraine architectural form of organizing the sustainable development of medicine in rural areas, waiting for feedback from the international scientific community.

Acknowledgments
The author of the article thanks Prof. Dr.-Ing. Annette Rudolph-Cleff, the Department of Design and Urban Planning of the Faculty of Architecture, and fund "Ingenium" of the Darmstadt University of Technology for their support and assistance.

References

[1] Rudolph-Cleff A, Knodt M, Schulze J and Engel A 2022 International Journal of Disaster Risk Reduction 72 102856
[2] Huck A, Monstadt J, Driessen P P J and Rudolph-Cleff A 2021 Journal of Contingencies and Crisis Management 29(1) 12–22
[3] Knöll M, Neuheuser K, Cleff T and Rudolph-Cleff A 2018 Environment and Planning B: Urban Analytics and City Science 45(4) 797–813
[4] Shebek N, Timokhin V, Tretiak Y, Kolmakov I and Olkhovets O 2020 E3S Web of Conferences 166 09001
[5] Shebek N, Timokhin V, Tretiak Y, Kolmakov I and Olkhovets O 2021 E3S Web of Conferences 280 04009
[6] Pleshkanovska A 2021 E3S Web of Conferences 280 04012
[7] Pleshkanovska A and Biriuk S 2021 Journal of Urban and Regional Analysis 13(2) 257−280
[8] Pleshkanovska A 2020 Bulletin of Geography 48(48) 155−164
[9] Pleshkanovska A 2021 Science and Innovation 17(6) 110−122
[10] Kovalska G and Smilka V 2019 International Journal of Innovative Technology and Exploring Engineering 9(1) 1593−1600
[11] Kashchenko O, Kovalska G, Gnatiuk L 2020 Wiadomosci Konserwatorskie 2020(61) 31−34
[12] Kovalska G, Merylova I and Bulakh I 2019 International Journal of Innovative Technology and Exploring Engineering 8(12) 1765−70
[13] Semerikov S, Chukharev S, Sakhno S, Striuk A, Osadchyi V, Solovieva V and Danylchuk H 2020 E3S Web of Conferences 166 00001
[14] Semerikov S, Chukharev S, Sakhno S, Striuk A, Iatsyshyn A, Klimov S, Osadchyi V, Vakaliuk T, Nechypurenko P, Bondarenko O and Danylchuk H 2021 E3S Web of Conferences 280 00001
[15] Semerikov S, Teplytskyi I, Soloviev V, Amelina S, Tarasenko R 2021 Journal of Physics: Conference Series 1840(1) 012036
[16] Cherkes B and Linda S 2019 IOP Conf. Ser.: Mater. Sci. Eng. 471 072019

[17] Linda S, Mykhaylyshyn O, Olena R and Mariia V 2020 IOP Conf. Ser.: Mater. Sci. Eng. 960(3) 032061
[18] Pysmennyi S, Fedko M, Shvaher N and Chukharev S 2020 E3S Web of Conferences 201 01022
[19] Sakhno S, Yanova L, Pischikova O and Chukharev S 2020 E3S Web of Conferences 166 06002
[20] Dyomin M, Dmytrenko A, Chernyshev D, Ivashko O 2020 Lecture Notes in Civil Engineering 73 365-73
[21] Dyomin M, Chernyshev D, Dmytrenko A, Kuzmenko T and Chang P 2020 Landscape Architecture and Art 16(16) 70-9
[22] Adeyeye K, Gallagher J, McNabola A, Ramos H M and Coughlan P 2021 Energies 14(14) 4222
[23] Hart J, Albadra D, Paszkiewicz N, Adeyeye K and Copping A 2022 Design Studies 80 101107
[24] Ramos H M et al. 2022 Water (Switzerland) 14(8) 1304
[25] Baïri A et al. 2022 International Journal of Modern Physics C 33(1) 2250002
[26] de Castro M et al. 2021 Journal of Building Engineering 44 102579
[27] Baïri A et al. 2021 Energies 14(22) 7479
[28] Devereux C, Coscia J, Adeyeye K and Gallagher J 2021 Sustainable Energy Technologies and Assessments 47 101377
[29] Mitrovic D et al. 2021 Water (Switzerland) 13(7) 899
[30] Coley D et al. 2021 International Journal of Architectonic, Spatial, and Environmental Design 16(1) 25–49
[31] Baiandin S, Ivashko Yu, Dmytrenko A, Bulakh I and Hryniewicz M 2022 International journal of conservation science 13(2) 381−94
[32] Chernyshev D, Ivashko Y, Kuśnierz-Krupa D and Dmytrenko A 2020 Landscape Architecture and Art 17(17) 13−21
[33] Chernyshev D, Ryzhakov D, Dikiy O, Khomenko O and Petrukha S 2020 International Journal of Emerging Trends in Engineering Research 8(9) 4967−73
[34] Kovalchuk O, Kochetov G and Samchenko D 2019 IOP Conference Series: Materials Science and Engineering 708(1) 012087
[35] Kochetov G, Prikhna T, Samchenko D and Kovalchuk O 2019 Eastern-European Journal of Enterprise Technologies 6(10−102) 6−14
[36] Krivenko P and Kovalchuk O 2020 Key Engineering Materials 864 180−88
[37] Bulakh I, Didichenko M, Kozakova O and Chala O 2020 E3S Web of Conferences 166 08001
[38] Bulakh I, Chala O and Divak V 2020 Civil Engineering and Architecture 8(4) 586−98
[39] Bulakh I 2019 Sci. innov. 15(5) 57−66
[40] Bulakh I, Kozakova O and Didichenko M 2019 International Journal of Innovative Technology and Exploring Engineering 9(1) 317−23
[41] Bulakh I and Merylova I 2020 Civil Engineering and Architecture 8(5) 1127−35
[42] Bulakh I, Kozakova L, Didichenko M and Chala O 2021 Civil Engineering and Architecture 9(7) 2476−91
[43] Bulakh I, Kashchenko T, Harbar M, Praslova V, Riabets Y and Divak V 2022 Civil Engineering and Architecture 10(3) 874–887
[44] Bulakh I 2022 IOP Conf. Series: Earth and Environmental Science 987 012011
[45] Bulakh I, Semyroz N, Kysil S, Bulhakova T, Mezhenna N, Abyzov V, Zymina S and Bulakh V 2022 Civil Engineering and Architecture 10(5) 1840−53
[46] Bulakh I 2020 Journal of Regional and City Planning 31(1) 82−96
[47] Lami I M and Mecca B 2021 Sustainability 13 142
[48] Press Service of the Ministry of Health of Ukraine. Information available: https://moz.gov.ua/article/news/evropa-doluchatimetsja-do-vidnovlennja-zrujnovanih-likaren-v-ukraini
[49] Bulakh I 2021 Urban Planning Basics of the Spatial Development of the System of Children's Medical Complexes (Kyiv: Kyiv National University of Construction and Architecture)

The Urban Environment and Public Health: Associations Between COVID-19 Cases and Urban Factors in Semarang City, Central Java, Indonesia

Wewin Wira Cornelis Wahid[1]* and Jawoto Sih Setyono[2]
[1]Department of Urban and Regional Planning, Diponegoro University, Semarang City, Central Java, 50275, Indonesia, 0000-0003-3919-0596
[2]Department of Urban and Regional Planning, Diponegoro University, Semarang City, Central Java, 50275, Indonesia, 0000-0003-4360-1917.
[1]*wewinwira@gmail.org, [2]jawoto@pwk.undip.ac.id

Abstract. UN-Habitat noted that COVID-19 cases were disproportionately found in urban areas. A question arises as to what urban factors are related and how they are related. Therefore, this study aims to analyze the association between urban factors and cumulative cases of COVID-19 in Semarang City, the fifth largest Indonesian city with diverse urban characteristics. Through literature review, twelve urban factors were selected and tested against cumulative COVID-19 cases. The study was conducted cross-sectionally during the new normal period until just before the delta variant of COVID-19 emerged. The association was analyzed using robust regression because the OLS regression assumptions were not met. This study found a strong association based on a high R2 (0.725) value. This study also found six significant variables and found that; (1) elderlies are a vulnerable group that needs to be cared for, (2) planned densities are associated with lower COVID-19 cases, and (3) COVID-19 spread among the poor is presumably undetected, (4) bus transportation and (5) commercial facilities have higher COVID-19 transmission risk. Finally, the study outlines implications for future planning to promote planned densities, light active mobility, reduce poverty, and increase healthcare capacity.

1. Introduction

UN-Habitat found that 95% of COVID-19 cases recorded in 2020 were in urban areas [1]. This calls for researchers to understand the relationship between urban areas and COVID-19. The COVID-19 pandemic also provides critical reflections on the importance of cities and could rally support for future transformative actions [2]. However, to know what changes are needed, understanding which urban characteristics or factors play an effect and how they have such an effect is crucial. One research suggested examining the physical and non-physical urban factors [3]. Hence, a literature review was done to find various urban factors that were thought to be related.

Non-physical factors include the socio-demographical and socio-economic urban factors. Research on population density found differing results. Some studies found a positive relationship [3,4], some found a negative relationship [5,6], and some found no significant association [8,9]. Some conclude that this is due to the opposing roles that population density has [5,7]. On the one hand, high population density means higher interactions and higher COVID-19 transmission [8]. However, density is related to; higher levels of service and accessibility, shorter trips, and higher levels of prosperity (thus able to work from home), all of which support COVID-19 transmission prevention [9–11].

COVID-19 morbidity has a demographic bias towards elderlies [4,7,12] due to inherent factors and their activity profile. Inherent factors include lowered immunity [13] and the prevalence of underlying diseases [14]. On the other hand, the elderly tend to visit hospitals more frequently, and they face a risk of nosocomial or cross-infections from COVID-19 silent spreaders in the hospital [12]. This is highly likely, as many infected persons tend to be asymptomatic [14]. Socio-economically, poor people and slum dwellers are more at risk of COVID-19. Poor people find it difficult to socially distance themselves due to low-level informal jobs requiring on-site work if they want to survive [15].

Moreover, most poor people live in slums characterized by many dwellers per house [16], resulting in an inability to self-isolate if a family member is infected. Communal water and sanitation utilities [17] also allows COVID-19 to spread through fomites. Behavioral aspects such as open defecation allow COVID-19 to spread fecal-orally via flies, mice, and cockroaches, or day-to-day river usage [18].

Physical factors include the built-up area and how it is configured. Humans spend most of their time in the built environment, while COVID-19 can be transmitted indoors through direct human contact, fomites, and aerosol [19]. Consequently, higher built-up density (coverage) and intensity (floor area ratio) were related to higher COVID-19 transmission. Meanwhile, the literature found differing results, as research on built-up density found both significant positive [4] and negative relationships [5,20]. Similarly, research on building intensity also found significant positive [12] and significant negative relationships [4]. However, researchers (on both density and intensity) that found negative relationships noted that the crowding factor most likely explains the differing results.

Research on certain facilities found that commercial facilities, health facilities, and transportation facilities were associated with higher COVID-19 cases [12,20–22]. Commercial and transportation facilities are generally associated with higher rates of activities, while health facilities allow for nosocomial and cross-infections. Thus, these facilities should be examined further.

Few have studied the relationship between land-use mixture and COVID-19 transmission. However, one research found different relationships at different locations [12]. They explained that there was a positive relationship in locations that depend on public transportation and a negative relationship in locations where basic needs were obtained within walking distance. There are indications that the mode of travel is more influential than a land-use mixture. However, highly diverse areas are generally presumed to have high human interactions, which might cause more COVID-19 infections.

Based on the literature examined, it seems that; a clear picture of the relationship between urban factors and COVID-19 cases has not been formed due to the different relationships found between some of the urban factors. This requires the experiment to be replicated and refined to paint a clearer picture of the issue. Based on the literature reviewed, 12 urban factors should be investigated for their relationship to higher or lower COVID-19 cases. Those factors are (1) elderly population, (2) poor population, (3) slum settlements, (4) sanitation utilities (availability of private toilets), (5) water utilities facilities (availability of household water access), (6) population density, (7) built-up area density, (8) building intensity, (9) healthcare facilities, (10) commercial facilities, (11) transportation facilities, and (12) land-use mixture. These 12 factors will be detailed and statistically analyzed for relationships with COVID-19 cases. To do this, a good location will be needed.

Semarang City, Indonesia, should provide an excellent setting to understand the relationship between urban factors and COVID-19 cases. The research needs to analyze differing degrees of each urban factor (for example, what a higher percentage of the built-up area would do? And what a lower percentage would do?). Consequently, a diverse location would be needed. Semarang has various degrees of urban factors. It is the fifth biggest metropolitan in Indonesia but is not as uniformly urbanized or dense as other bigger Indonesian cities (namely Jakarta, Surabaya, and other cities). Semarang still has areas with high and low populations and built-up density. It also has areas of sprawl in the urban fringe [23] and mixed-use areas in the urban core [24]. Semarang still has 431.5 hectares of slums (in 2021), and 4.34% of the population is still poor (in 2020). Regarding COVID-19 cases, Semarang has one of the highest Covid-19 cases in Indonesia. Overall, the differing degree of each factor is good enough to achieve clear distinctions of what a high or low value of each factor would do to COVID-19 cases.

To understand how cities should be planned and managed in the future, it is necessary to understand the relationship between urban factors and COVID-19 cases. Consequently, a research question was formulated, asking how urban factors and COVID-19 cases are associated, particularly in Semarang City. A city with sufficiently available and diverse urban factors. This study hopes to add to the knowledge of what healthy and resilient cities should look like.

2. Methods

This study was conducted with a cross-sectional design, using COVID-19 data from June 22, 2020, to May 20, 2021, the new normal period before the emergence of the delta variant. The period before that is excluded because it was a period of strict restrictions from the Semarang City Government. Strict restrictions caused less activity and behavioral changes (work from home, closed facilities, less travel), thus negating the expected influence of urban factors. The period after the delta variant of COVID-19 emerged was also excluded because the differences in transmission rates could reduce the study's accuracy. The delta variant was at least twice as transmissive as the initial variant [25].

Table 1. Information on research data

Variable	Data	Year
COVID-19 Cases	Number of Cumulative COVID-19 Cases	2020-2021
Elderly Population	Number of Elderly Persons	2019
Slum Settlements	Area of Slum Settlements	2021
Poor Population	Number of Poor Households	2020
Toilet Utilities	Percentage of Households with Permanent Toilets	2020
Water Utilities	Percentage of Households with Access to Clean Water	2019
Built-up Area Density	Percentage of Built-up Area	2021
Population Density	Population Density (Relative to Built-up Area)	2019
Building Intensity	Average Floor Area Ratio	2018
Healthcare Facilities	Area of Healthcare Facilities (Hospitals, Clinics)	2018
Commercial Facilities	Area of Commercial Use	
Transportation Facilities	Area of Transportation Facilities (Airports, Train Stations, Sea Ports, and Bus Stops)	
Land-use Mixture	Land-use Mixture (Obtained by Calculating Shannon Entropy Index)	2018

The data used are explained in Table 1; sub-districts or "kelurahan" are used as the data unit, amounting to 177 data points (all sub-districts in Semarang). The data will be analyzed statistically through multivariate regression analysis, a standard method of investigating multivariate associations. Either one of the OLS (Ordinary Least Square) multiple linear regression or robust regression will be used, depending on which method fits best. The data will first be processed using multiple linear regression (OLS) due to general assumptions that parametric statistics are more accurate when all assumptions are met, namely assumptions of; linearity, normality of errors, independence of errors, non-multicollinearity, and homoscedasticity [26,27]. However, if the assumptions are not met, outliers may exist within the data, and there is a risk of erroneously estimating the significance of relationships between variables [26,27]. If this problem arises, the data will be processed using robust regression.

Robust regression is a non-parametric regression analysis used as an alternative to parametric regressions [28]. Many robust estimators can be used. However, the three most popular estimators are; M-estimator, S-estimator, and MM-estimator [28,29]. Studies performed estimator comparisons to determine the best estimator [28,29]. Two statistical results used as indicators are; the R^2 value (how well the model explains the data) and the standard error (how accurate the model fits with the data). The results from the best estimator will then be interpreted by examining the R^2 (simultaneous relationship), p-value (relationship significance), coefficient (positive or negative relationship), and standard error (predictive accuracy) of each variable.

3. Results and Discussion

Statistical outputs from all the regressions were obtained, and the results were analyzed. This section consists of three parts. First is the part that determines the best regression model. Second is the interpretation of statistical outputs. Third, discussions on why each relationship occurred the way it did by referring to relevant literature.

3.1. *Best Regression Model*

Based on Table 2, the assumptions that were not met are; linearity, normality of errors, and homoscedasticity. This is presumably due to outliers. Outliers were found in the seven independent variables as well as in the dependent variable. Henceforth, the data was processed using robust regression, which was non-parametric and could manage outliers.

Table 2. Multiple linear regression assumption test results (OLS)

Assumption	Statistical Test		Results	Conclusions
	Function	Requirement		
Linearity	Deviation from Linearity	Each variable has p-value > 0.05	9/12 had a linear relationship	Assumption not met
Normality of errors	Kolmogorov-Smirnov Test	p-value > 0.05	p-value < 0.05	Assumption not met
Independence of errors	Durbin-Watson Test (DW)	DW value in-between dU and 4dU	Result 1.942) was within 1.9133-6.3972	Assumption met
Non-Multicollinearity	Variance Inflation Factor (VIF)	VIF values in-between 0.10-10	Each variable had VIF 0.1-10	Assumption met
Homoscedasticity	Glejser Test, Breusch Pagan Test.	p-value > 0.05	Both tests had p-value < 0.05	Assumption not met

Based on Table 3, the M-estimator is the best estimator. M-estimator had the highest R^2 value and lowest standard error amongst the robust estimators. Consequently, the output of the M-estimator will be used in this study. Examining the OLS output closer, it had fairly good accuracy based on R^2 and standard error. However, there is reason to believe that this occurs due to overestimation. Outliers can draw the regression line away from the best fit line, thus resulting in a wider margin of error [26].

Table 3. Comparison of regression outputs

Model		R^2	Standard Error of Regression	Number of Significant Variables
OLS		0.838	59.075	6
Robust Regression	M-Estimator	0.725	59.594	6
	S-Estimator	0.536	73.535	5
	MM-Estimator	0.582	61.199	5

3.2. *Interpretation of Statistical Output*

The R^2 value of 0.725 indicates a strong relationship between urban factors and the number of COVID-19 cases in Semarang City. Based on the p-value and coefficient of each variable in Table 4, it was found that six urban factors were significantly associated with COVID-19 cases. Suppose the predictive power (based on standard error) was rank ordered. The number of elderly residents was the most predictive, followed by (2) population density, (3) number of poor households, (4) area of transportation use, (5) density of the built-up area, and (6) area of commercial use.

Table 4. M-Estimator regression output

Variable	Coefficient	Standard Error	p-value
Number of Elderly Population	0.074211	0.003110	<0.0001
Number of Poor Households	-0.283480	0.072435	0.0001
% of Household Toilet Ownership	-0.218281	0.413374	0.5975
% of Households with Access to Clean Water	0.405870	0.599068	0.4981
Area of Slum Settlements	-0.601285	0.739487	0.4162
Population Density (Relative to Built-up Area)	-0.290641	0.067549	<0.0001
Built-up Area Density	0.574457	0.288638	0.0466
Area of Healthcare Use	1.107514	1.513290	0.4643
Area of Commercial Use	1.029473	0.374600	0.0060
Area of Transportation Use	-0.468439	0.183124	0.0105
Average Floor Area Ratio	-15.948350	17.667650	0.3667
Land-use Mixture	-0.267685	0.360286	0.4575
C	10.121900	65.619340	0.8774

3.3. Discussions on the Association between Urban Factors and COVID-19 Cases Examples
The relationship of each significant urban factor is discussed in this section. Various possibilities that most likely caused this relationship are explored by referring to relevant literature and further examining research data. Furthermore, as previously ranked in section (3.2), the significant urban factors will be discussed from the most predictive variable to the least predictive.

3.3.1 The Elderly is a Vulnerable Group That Needs to be Cared For
The statistical output found that the elderly population was positively associated with the number of COVID-19 cases. This means that sub-districts with a very high number of elderlies had a high to the very high number of COVID-19 cases and vice versa. This finding was in line with the existing literature. The vulnerability profile of elderlies was relatively high, considering their low immunity level [13], their tendency to have underlying diseases that contribute to comorbidities [14], and their risk of infection via frequent visits to the hospital for check-ups or treatment [15]. Elderlies may experience higher activity and social interactions during the new normal period. Combined with their high susceptibility to COVID-19 infection, the slightest exposure most likely caused morbidity.

3.3.2 Planned Densities are Associated with Lower COVID-19 Cases
The statistical output found that population density was negatively associated with the number of COVID-19 cases. On the other hand, built-up area density was positively associated with COVID-19 cases. This indicates that population density was not quite identical to built-up density. By examining data, it was found that; out of all the sub-districts with high built-up density (74), only a few (24) are categorized as high to very high population density. Meanwhile, the other (50) sub-districts (with high to very high built-up density) either have very low (3), low (24), or medium (23) population densities.

Differing urban development phenomena likely caused the relationship between these two variables in locations where each variable had an effect. Based on Figure 1, sub-districts with high to very high population densities are mainly in the urban core (29 out of 30 sub-districts). On the other hand, sub-districts with high to very high built-up densities were more spread into the urban fringe. Previous research has found the urban core to be more compact and mixed-use [24]. This is supported by the floor area ratio and land-use mixture data acquired. Meanwhile, the urban fringe zone was found to be experiencing sprawl [23].

Furthermore, the sub-districts with medium to very high COVID-19 cases were located in the Ngaliyan, Pedurungan, Semarang Barat, and Tembalang districts, all of which have been specifically found to experience urban sprawl [23]. In summary, most sub-districts with higher population density are located in mixed-use areas, while sub-districts with higher built-up density are located in sprawl areas. Therefore, looking at the differing transmission risk in mixed-use and sprawl areas might help understand this finding.

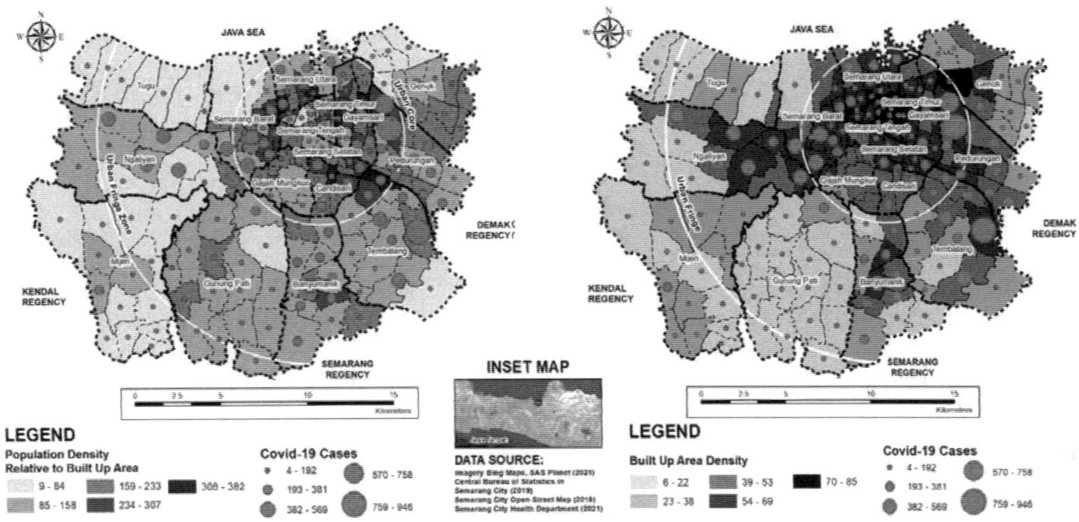

(a) Population Density (b) Built-up Area Density

Figure 1. Population and Built-up Densities Distribution and COVID-19 Cases in Semarang City

As hinted in the first section, several studies explained that crowding (or overcrowding) rather than density was the main factor that drove higher COVID-19 transmission [5,7]. Crowding occurs when high population density is not accommodated by adequate services, living space, and circulation areas [11]. There is a risk of crowding in locations with limited facilities. The situation in which many people shop in the same place because of limited shopping centers comes to mind. However, crowding can also occur on public transportation [11]. Limited services in a certain sub-district would encourage its residents to travel to the city center, and if that trip is taken by public transport (such as buses), there is a high risk of contracting COVID-19 [20].

Regarding the effects of crowding, it was recommended to promote compact development, mixed land use, and planned density [10,11,20]. Planned density is associated with adequate facilities and services for its residents [11]. The high availability of facilities and services makes the population more dispersed due to the many places they can choose to go to. Compact development with mixed land use allows workplaces and services to be closer to residential areas [20]. This allows residents to have shorter trips (to shops, pharmacies, and stations) [10]. Shorter trips might play a significant role, as prolonged exposure in confined spaces is generally considered risky.

In the case of Semarang City, sub-districts with higher population densities were indicated to be in planned areas characterized by mixed-use and compact development. Meanwhile, sub-districts with medium to high built-up densities were sprawl areas. Moreover, several sub-districts in the urban fringe zone have relatively lower service availability than the city center. This results in crowding at the nearby facilities and the possible use of public transportation to the city center.

3.3.3 COVID-19 Spread Among the Poor is Presumably Undetected

The statistical output found a negative relationship between the number of poor households and the number of COVID-19 cases. The data shows that out of all the sub-districts with very high number of poor people, only one has a high number of COVID-19 cases. Meanwhile, medium to very high COVID-

19 cases tend to be located in sub-districts with low poverty. Similar research on COVID-19 spread and the poor concluded that COVID-19 spread might be undetected among the poor [30]. This is in line with various literature which explains how the poor's physical environment, social life, and economic capacity promote COVID-19 spread [15–17,31]. Despite high transmission risks among the poor, there is a lower number of screenings due to the high cost of COVID-19 tests [30]. In Indonesia (during the observation period), COVID-19 PCR and Antigen tests were IDR 800,000 and IDR 240,000. Assuming a general poverty line by absolute and relative standards, the costs of COVID-19 tests are too expensive.

The World Bank sets the absolute poverty standard at US$ 1.90 per person per day, or around IDR 815,100 per month (with an exchange rate of IDR 14,300). Meanwhile, the March 2021 relative poverty line was IDR 472,525 per capita per month [32]. Consequently, a poor person has to spend 30%-50% of monthly income to pay for an Antigen test and 98%-169% for a PCR test. This expenditure is considered too large, resulting in reluctance among the poor to take a COVID-19 test. Thus COVID-19 detection among the poor is much lower.

3.3.4 Bus Transportation Likely to Have Higher Transmission Risk

Areas of transportation facilities were found to be negatively associated with the number of COVID-19 cases. This result means that more COVID-19 cases tend to be related to lower transportation facilities. Based on the research data sub-districts with more extensive transportation facilities (airports, seaports, and train stations) have fewer COVID-19 cases than smaller facilities (bus stops). This indicates that factors related to large and small transportation facilities cause differing levels of COVID-19 cases.

The preventive measures were less effective in small transportation facilities such as buses. Inter-city and inter-province bus users were only required to include negative COVID-19 test results when departing from and to Jakarta. Meanwhile, Semarang's Rapid Transit Bus (intra-city) only required passengers to wear masks and adjust their sitting distances. These efforts were less effective than those carried out in extensive transport facilities. As larger transportation facilities would require a negative COVID-19 test result, those facilities had significantly less transmission risk.

Studies have found a relationship between bus travel and higher risks of contracting COVID-19, especially in long durations of travel [20,22]. COVID-19 could transmit through the air via aerosols or fomites the carrier touched. Furthermore, bus passengers tend to get on and off at bus stops near their homes. Bus stops and the passenger's residence could be potentially connected if a passenger had been infected on the bus and transmitted COVID-19 to family members and neighbors. Therefore, the placement of bus stops might be a helpful indicator for identifying locations with higher risks.

3.3.5 Commercial Facilities Have Higher COVID-19 Transmission Risk

Areas of commercial use were found to be positively associated with COVID-19 cases. The data shows that there were lower COVID-19 cases in sub-districts with a lower concentration of commercial area. Interestingly, sub-districts in the urban core also tend to have lower COVID-19 cases, even though the commercial area is quite dense in that zone. This might show an inconsistency, but as said in the previous section (3.2), this variable has the lowest predictive accuracy out of all the other significant variables. Furthermore, this anomaly in the urban core area is still in line with the previously planned density sub-section and its advantages in preventing COVID-19 transmission.

The positive association between commercial facilities and COVID-19 cases should be examined by examining their activities. One is the screening mechanism via temperature checks used in commercial facilities during the new normal period. Although practical (cheap and easy), checking body temperatures is ineffective in detecting infected people [33]. People infected with COVID-19 sometimes show no symptoms (asymptomatic) [14]. Moreover, the accuracy of the temperature-checking tools varies depending on product quality and the technique used by the checker. Besides poor screening methods, specific characteristics in commercial facilities were also risk factors. For example, restaurants were found to spread COVID-19 through air conditioning [19] and contaminated food [34]. Consequently, commercial facilities have a high COVID-19 transmission risk profile.

4. Conclusion

This research was done to understand how urban characteristics or factors affected COVID-19 transmission by examining COVID-19 cumulative cases in Semarang City. Through statistical processing, it was found that six urban factors were associated with COVID-19 cases. The findings from this study provide implications for future planning. First, the number of elderly persons was the best predictor of COVID-19 cases. Consequently, ensuring the availability of hospital beds is a must, as inadequate hospital beds in some countries resulted in higher elderly mortality. Second, planned density seems to fare better against COVID-19 morbidity, which implies that planners should continue to promote mixed and compact development. Third, COVID-19 spread is likely undetected among the poor due to the high costs of COVID-19 tests. It is worth mentioning that Indonesia has lowered its COVID-19 test costs. However, long-term solutions to increase resilience should focus on reducing poverty. Fourth, bus transportation might have higher risks of COVID-19 transmission. This provides an opportunity to promote light active mobility. Initiatives in this aspect would also promote planned density by creating a transit-oriented neighborhood with adequate, accessible (walkable and bikeable) access and localized activities. Fifth, commercial facilities were found to have higher transmission risks due to poor prevention measures and the facility's activity profile. Therefore, in the rise of future COVID-19 waves, restricting commercial activities to curb COVID-19 spread is likely beneficial. Limitations and Future Research: first, researchers have not included connectivity and mobility factors due to the unavailability of mobility and connectivity data. Second, the Shannon-entropy formula purely measures diversity. Other methods that specifically weights more critical land use in mixed-used development might provide different results. Lastly, more non-parametric regression models could be used.

References

[1] UN-Habitat 2020 COVID-19 Response Plan *UN-Habitat* 1–16
[2] Sharifi A and Khavarian-Garmsir A R 2020 The COVID-19 pandemic: Impacts on cities and major lessons for urban planning, design, and management *Sci. Total Environ.* **749** 1–3
[3] Afrin S, Chowdhury F J and Rahman M M . 2021 COVID-19 Pandemic: Rethinking Strategies for Resilient Urban Design, Perceptions, and Planning *Front. Sustain. Cities* **3**
[4] You H, Wu X and Guo X 2020 Distribution of COVID-19 Morbidity Rate in Association with Social and Economic Factors in Wuhan, China: Implications for Urban Development 17 3417 *Int. J. Environ. Res. Public. Heal.* **14** 17
[5] Khavarian-Garmsir A R, Sharifi A and Moradpour N 2021 Are high-density districts more vulnerable to the COVID-19 pandemic? *Sustain. Cities Soc.* **70** 102911
[6] Liu L 2020 Emerging study on the transmission of the Novel Coronavirus (COVID-19) from urban perspective: evidence from China *Cities* **103** 102759
[7] Almagro M and Orange-Hutchinson A 2022 JUE Insight: The determinants of the differential exposure to COVID-19 in New York City and their evolution over time *J. Urban Econ.* **127** 103293
[8] Sy K T L, White L F and Nichols B E 2021 Population density and basic reproductive number of COVID-19 across United States counties *PLoS One* **16** 1–11
[9] Hamidi S and Zandiatashbar A 2021 Compact development and adherence to stay-at-home order during the COVID-19 pandemic: A longitudinal investigation in the United States *Landsc. Urban Plan.* **205** 103952
[10] Hamidi S, Sabouri S and Ewing R 2020 Does Density Aggravate the COVID-19 Pandemic?: Early Findings and Lessons for Planners *J. Am. Plan. Assoc.* **86** 495–509
[11] UN-Habitat 2021 *Cities and pandemics: Towards a more just, green and healthy future*
[12] Li X, Zhou L, Jia T, Peng R, Fu X and Zou Y 2020 Associating COVID-19 severity with urban factors: A case study of Wuhan *Int. J. Environ. Res. Public Health* **17** 1–20
[13] Bajaj V, Gadi N, Spielman A P, Wu S C, Choi C H and Moulton V R 2021 Aging, Immunity, and COVID-19: How Age Influences the Host Immune Response to Coronavirus Infections?

Front. Physiol. **11**
[14] Oran D P and Topol E J 2020 Prevalence of asymptomatic SARS-CoV-2 infection. A narrative review *Ann. Intern. Med.* **173** 362–8
[15] Wasdani K P and Prasad A 2020 The impossibility of social distancing among the urban poor: the case of an Indian slum in the times of COVID-19 *Local Environ.* **25** 414–8
[16] Friesen J and Pelz P F 2020 COVID-19 and Slums: A Pandemic Highlights Gaps in Knowledge About Urban Poverty *JMIR Public Heal. Surveill* **6**
[17] Ghosh S, Seth P and Tiwary H 2020 How does Covid-19 aggravate the multidimensional vulnerability of slums in India? A Commentary *Soc. Sci. Humanit. Open* **2** 100068
[18] Gwen W, Muzava M, Mapanda F and Tauro T P 2016 Comparative short-term effects of sewage sludge and its biochar on soil properties, maize growth and uptake of nutrients on a tropical clay soil in Zimbabwe *J. Integr. Agric.* **15** 1395–406
[19] Dietz L, Horve P F, Coil D A, Fretz M, Eisen J A and Wymelenberg V D K 2020 Coronavirus (COVID-19) Pandemic: Built Environment Considerations To Reduce Transmission ed *mSystems* **5**
[20] Yang Y, Li Y, Kral K, Hupert N and Dogan T 2021 Urban design attributes and resilience: COVID-19 evidence from New York City *Build. Cities* **2** 618
[21] Yip T L, Huang Y and Liang C 2021 Built environment and the metropolitan pandemic: Analysis of the COVID-19 spread in Hong Kong Build *Environ* **188**
[22] Ponte C, Carmona H A, Oliveira E A, Caminha C, Lima A S, Andrade J S and Furtado V 2021 Tracing contacts to evaluate the transmission of COVID-19 from highly exposed individuals in public transportation *Sci. Rep.* **11** 1–11
[23] Aprilia Y and Pigawati B 2018 Urban Sprawl Typology in Semarang City *Forum Geogr.* **32** 131–45
[24] Segnita A and Handayani W 2017 Pengukuran Penggunaan Lahan Campuran (Mixed Use) dengan Indeks Entropy di Kota Semarang *J. RIPTEK* **II** 135–50
[25] CDC 2021 Delta variant: What we know about the science *Centers Dis. Control Prev.*
[26] Casson R J and Farmer L D M 2014 Understanding and checking the assumptions of linear regression: A primer for medical researchers *Clin. Exp. Ophthalmol.* **42** 590–6
[27] Williams M N, Grajales C A G and Kurkiewicz D 2013 Assumptions of multiple regression: Correcting two misconceptions *Pract. Assessment, Res. Eval.* **18** 1–14
[28] Alma Ö G 2011 Comparison of Robust Regression Methods in Linear Regression *Int. J. Contemp. Math. Sci.* **6** 409–21
[29] Susan Y, Pratiwi H, Sulistijowati H and Liana T 2014 M Estimation, S Estimation, and MM Estimation in robust regression *Int. J. Pure Appl. Math.* 349–60
[30] Ruppel A, Halim M I, Kikon R, Mohamed N S and Saebipour M R 2021 Could COVID-19 be contained in poor populations by herd immunity rather than by strategies designed for affluent societies or potential vaccine(s)? *Glob. Heal. Action* **14**
[31] Gwen W 2021 Leaving no stone unturned in light of the COVID-19 faecal-oral hypothesis? A water, sanitation and hygiene (WASH) perspective targeting low-income countries *Sci. Total Environ.* **753** 141751
[32] Badan Pusat Statistik 2021 *Profil Kemiskinan di Indonesia Maret 2021* (Jakarta: Badan Pusat Statistik)
[33] Facente S N, Hunter L A, Packel L J, Li Y, Harte A, Nicolette G, McDevitt S, Petersen M and Reingold A L 2021 Feasibility and effectiveness of daily temperature screening to detect COVID-19 in a prospective cohort at a large public university *BMC Public Health* **21**
[34] Han J, Zhang X, He S and Jia P 2021 Can the coronavirus disease be transmitted from food? A review of evidence, risks, policies and knowledge gaps *Environ. Chem. Lett.* **19** 5–16

The Performance of Companies in Assisting the Environment By Issuing Green Bond

Pandior Benget Namora Tua[1*], Rofikoh Rokhim[2]
[1]Departement of Management, Faculty of Economics and Business, Universitas Indonesia, Jakarta, 16424, Indonesia
[2]Departement of Management, Faculty of Economics and Business, Universitas Indonesia, Jakarta, 16424, Indonesia, 0000-0002-2847-0552
[1]*pandior.benget@ui.ac.id, [2]rofikoh.rokhim@ui.ac.id

Abstract. Besides as well as a way for companies to raise money and for investors to differentiate their portfolios, as should be one of the purposes of the issuance of Green Bonds to mitigate a decrease in carbon emissions does not accompany climate change, the expansion in the issuance of Green Bonds. The emergence of green bonds evidences investor interest in financial and environmental instruments. The business must continue to operate well to satisfy investor expectations. Using the Purposive Sampling technique, this study examines 89 companies for the impact of several financial parameters on the success of businesses that issue green bonds using panel data regression and multiple linear regression techniques. The findings demonstrated that the factors Equity Multiplier, Total Asset Turnover, and Profit Margin significantly impacted the company's performance as measured by the ROE variable. The findings also demonstrate that issuers of Green Bonds frequently use their business activities to boost their performance.

1. Introduction

Climate change is a global problem that we are currently experiencing. According to the Intergovernmental Panel on Climate Change (IPCC), global temperatures will continue to rise by 0.2°C every ten years 14]. If this situation doesn't improve, it could endanger human existence as we know it on a worldwide scale. According to data published by World Research in 2018, ten nations are responsible for the majority of global carbon emissions. by causing 11,705.81 MtCO2e of emissions in just 2018 alone. The largest source of carbon emissions is industry, which is also responsible for emissions from other sources like forestry, agriculture, and household appliances.

One of the innovations and types of attempts in the financial sector to address this global climate catastrophe is the Green Bond, which was first used and published in 2008 by the World Bank in partnership with the Swedish Bank (SEB) [12]. The Green Bond's potential as a financial instrument is also seen favorably by many parties due to its capacity to distribute a sizable sum of money for projects that will later reduce carbon emissions or preserve the environment.

Like bonds in general, the government or corporation offers debt instruments to investors to finance a project or development in their business and pledges to pay back the principal debt combined with the interest coupon on the due date of payment. A green bond, in this instance, is a bond (Bond) that is only allocated to support ecologically friendly initiatives or developments (Green).

The fact that there are more issuers and investors worldwide every year shows that Green Bonds have attracted investors worldwide since their launch in 2008. This is encouraging for the environment and the climate's improvement since it shows that many businesses are trying to focus their operations on environmentally friendly activities.

Flammer claims that the issue of green bonds can be a way for businesses to signal their support for environmental betterment [7]. This process is known as signaling. Additionally, the goal of signaling is to make the information that the public and investors receive more broadly applicable to utilize it to make judgments. This signaling may also uniquely appeal to investors who value the environment highly. Signaling can be abused when a corporation merely wants to get the attention of investors and sympathy or interest from the public for its dedication to the environment. Yet, in reality the company is not doing what investors and the public would expect it to be doing to improve the environment. This practice is known as "greenwashing." However, [7] identifies instances where a company's public statement of environmental commitment positively affects the expansion of its shareholder base. This demonstrates that there are expectations for the company from the standpoint of the environment and from a financial one, where investors naturally also want a return on their investment. Companies must continue to operate well and stay out of bankruptcy to achieve these expectations [11] . Thus, this study aims to identify the variables that influence corporate performance, particularly that of firms that issue green bonds.

2. Methods

Financial ratios are universal indicators used to measure company performance so that company performance can also be compared. [5,13] . Du Pont analysis method divides the company's performance into three categories; The Profit Margin (PM) ratio, which serves as a proxy for Operating Efficiency, will demonstrate how the business makes money from its sales; Financial Leverage, which is proxied by the ratio of Equity Multiplier (EM), will explain how companies use debt and assets as sources of funding for their businesses. Asset Use Efficiency, proxied by the ratio of Total Asset Turnover (TAT), will explain the company's efficiency in using its assets to generate income.

Additionally, the three components of the company's performance measures will be evaluated, which the Return represents on Equity ratio (ROE). The approach is to compare performance before and after issuing the Green Bond by looking at the many performance-influencing variables, including Profit Margin, Total Asset Turnover, and Equity Multiplier. The company being observed must meet several preset criteria, such as having issued a Green Bond at least once and providing complete financial data for every year during which it is being observed. Then, the Thomson Reuters software allows the researcher to pinpoint 89 companies as the sample. Due to its advantages in the analysis of cross-sectional and time-series data, the regression approach employed in this study is the Panel Data Regression method. The researchers used two models—one before the issuance of Green Bonds, which was categorized as pre-issuance, and one after the issuance of Green Bonds, in the post-issuance category, with different variables to analyze one variable—to analyze the influence of the factors that can affect the performance of Green Bond issuing companies. Dependent (ROE). The following regression model is created using the [7,9] model with modifications:

$$Y_{it}(ROE) = \beta_1 + \beta_2 EM_{it} + \beta_3 TAT_{it} + \beta_4 PM_{it} + \beta_5 Size_{it} + u_{it} \quad (1)$$

The profitability ratio, typically denoted by the Return on Equity (ROE) or Return on Assets (ROA) variable, is the dependent variable in this study. Return on Assets is a ratio utilized by [14] to assess a company's performance and contrast it with other ratios [7]. However, ROE is employed to show how the company makes money from the investment it receives [1]. Additionally, ROA is used in Du Pont's research to calculate the critical ratio, ROE. Referring to (Ross, 2010), ROE, which is also the ratio used to measure corporate profitability and ROA, will be employed in this study as the dependent variable.

H1: The performance of the Green Bond issuing corporation is impacted by the operating efficiency factor.

The operating efficiency factor impacts the success of the corporation that issues green bonds. This is based on [6] research, which looks at how the Green Bond affects the company; a related variable is the profit margin ratio.

H2: The performance of the corporation that issues green bonds is impacted by the asset use efficiency factor.

The asset use efficiency factor impacts the success of the corporation that issues green bonds. This is based on the study done by [10], who looks at how Green Bonds affect the company. Total Assets, a factor in the Total Asset Turnover ratio, is a comparable variable.

H3: Financial leverage influences how well green bond issuing companies operate.
The financial leverage element impacts the success of the corporation that issues green bonds. This is based on the study by [9], who looks at how the Greenium label affects the company's stock price that issues Green Bonds. Total equity, a factor in the Equity Multiplier ratio, is a comparable variable.

H4: The performance of the Green Bond issuing corporation is impacted by the strong size element.
The company size factor impacts the success of the company that issues green bonds. This is based on studies by [7] and [8], which also used the control variable of firm size. The natural logarithm of the company's total assets is the same variable.

3. Result and Discussion

Table 1. Multicollinearity Test

Corporate Performance Variable	*Pre-Issuance* VIF	*Pre-Issuance* 1/VIF	*Post-Issuance* VIF	*Post-Issuance* 1/VIF
PM	1,74	0,58	1,82	0,55
TAT	1,62	0,62	1,39	0,71
EM	1,22	0,82	1,44	0,69
SIZE	1,02	0,98	1,91	0,52
Mean VIF	1,40		1,64	

The VIF value in both the pre-issuance model and the post-issuance model shows that the mean value of the variables employed in table 1 is less than 10, which is evident from the VIF value. This suggests that the two models' variables are free of multicollinearity symptoms and suitable for use in research.

Table 2. Heteroscedasticity Test

Corporate Performance Variable	*Pre-Issuance* Prob.	*Post-Issuance* Prob.
EM	0,57	0,01
TAT	0,28	0,32
PM	0,17	0,34
SIZE	0,41	0,01

Table 2 demonstrates heteroscedasticity in the post-issuance model, particularly in the PM and SIZE variables. This indicates a significant degree of similarity in the residuals, which might cause errors in interpreting the data processing results. To mitigate the symptoms of heteroscedasticity, the Seemingly Unrelated Regression (SUR) approach is applied to models [3]

Table 3. Normality Test

Model	Sig.
Pre-Issuance	0,16
Post-Issuance	0,01

Based on Table 3, it can be seen that the Post-Issuance model exhibits signs of abnormal data, but [2] claims that the normality test is unnecessary because Panel Data Regression was used to process the Post-Issuance data, so the results of the normality test on the Post-Issuance model can be disregarded.

Table 4. Model Estimation Test

Model	**Chow Test Prob > F**	**Hausman Test Prob > Chi 2**	**Lagrange Test P > Chibar2**	**Model Used**
Post Issuance	0,0000	0,0034	-	FEM

The FEM model was determined to be the best model based on the findings in table 4 for the first and second years of the Post-Issuance model. This is because the Hausman test result value is also less than the alpha value, and the Chow test result value is less than the alpha value of 0.05. Therefore, the third test is not required to be performed, and the FEM model will be used instead.

3.1 Hypothesis Test

Table 5. Partial Effect Test (T-Test)[a]

T-Test	***Pre-Issuance***		***Post-Issuance***	
	Coef.	**Sig.**	**Coef.**	**Sig.**
Constanta	-0,263	0,076**	-0,191	0,000***
PM	0,617	0,000***	0,763	0,000***
TAT	0,216	0,000***	0,287	0,000***
EM	0,004	0,102*	0,007	0,000***
SIZE	0,007	0,254	0,007	0,001***

[a]The coefficient of significance parameter at the level of 15%, 10%, 5% is indicated by the asterisk symbol (*) respectively: *,**,***

According to table 5, the factors EM, TAT, PM, and SIZE significantly impact ROE both before and after issuing the green bond.

Table 6. Difference in Effect between Pre-issuance model and post-Issuance model

Variable	***Pre-Issuance***	***Post-issuance***	**Difference**
PM	0,617	0,763	0,146
TAT	0,216	0,287	0,070
EM	0,004	0,007	0,003
SIZE	0,007	0,007	0,000

Table 6 shows a shift in the influence of the variables PM, TAT, EM, and SIZE, which is evident from the distinction between the two models. The difference frequently increases or is positive. PM can impact ROE because rising sales will result in rising net income. According to Table 5, the Coefficient value shows that the variable PM dominates other variables. This outcome shows that the corporation has a propensity for the performance of major companies to be influenced by the sales of its products. A healthy corporation is indicated by an increase in ROE value, followed by an increase in PM value (Ross, 2010), because of the improved performance indicated by the ROE variable, which is significantly influenced by higher sales (PM). It could also imply that businesses won't be exposed to the risks associated with using debt instruments. Because sales are the components that make up TAT and affect its value, which is a component of Net Income and a component of ROE, TAT can also impact ROE. A corporation that effectively uses its resources has a good TAT value. Because EM's rising value suggests decreased reliance on debt instruments for financing, which could either mean less equity or higher income, EM can impact ROE. A high EM number suggests that the business tends to finance itself through debt.

According to a study [7], companies that demonstrate a dedication to the environment have a rise in shareholders. The data processing results that show beneficial company performance is consistent with this. The expectation shareholders have for the company is good or positive performance for investors to continue holding the shares or make new purchases after these expectations are met.

4. Conclusion

According to the research method, the size of the company, which is proxied by the SIZE variable, has a significant impact. The research results on the 89 companies show that Operating Efficiency is proxied on the Profit Margin ratio, followed by Asset Use Efficiency, which is proxied on the Total Asset Turnover ratio.

A large amount of influence of the Profit Margin and Total Asset Turnover variables on Return on Assets, which is quite large, is an example of how Green Bond issuing companies tend to take advantage of operational activities or their primary business activities to maintain the company's performance.

According to the study's findings, the issue of the Green Bond has enhanced the impact of elements that influence a company's success. The growing impact of the parameters examined from before the Green Bond issuance to after the Green Bond issuance provides evidence of these results. Positive business performance shouldn't hinder a company from implementing green projects financed by green bonds. Then, the emissions that didn't decrease might be due to other factors.

To better understand the impact of factors that affect a company's performance, it is necessary to access more recent financial data on the companies that issue green bonds, classify the sample of green bond issuers into corporate sectors, and include non-financial factors, particularly those that can have an impact on the company's performance on the environment.

References

[1] Alarussi A P A and Alhaderi S 2018 Factors affecting profitability in Malaysia. *Journal of Economic Studies, 45*, 00-00. doi:10.1108/JES-05-2017-0124
[2] Desy A 2022 Behavior Analysis of Savings Between Pre-Covid-19 and During Covid-19 Era.
[3] Dwiningsih N 2020 Analisa Penggunaan Metode Penelitian Regresi Panel.
[4] Everuss L and Lever-Tracy C 2020 Global Warming. In.
[5] Farza K, Ftiti Z, Hlioui Z, Louhichi W and Omri A 2021 Does it pay to go green? The environmental innovation effect on corporate financial performance. *J Environ Manage, 300*, 113695. doi:10.1016/j.jenvman.2021.113695
[6] Flammer C 2012 Does Corporate Social Responsibility Lead to Superior Financial Performance? A Regression Discontinuity Approach. *SSRN Electronic Journal, 61*. doi:10.2139/ssrn.2146282
[7] Flammer C 2021 Corporate green bonds. *Journal of Financial Economics, 142*. doi:10.1016/j.jfineco.2021.01.010
[8] Ibrahim S A, Roslen S, and Yee L 2017 Green Bond and shareholders' wealth: a multi-country event study. *International Journal of Globalisation and Small Business, 9*, 61. doi:10.1504/IJGSB.2017.10005758
[9] Larcker D and Watts E 2020 Where's the Greenium? *Journal of Accounting and Economics, 69*, 101312. doi:10.1016/j.jacceco.2020.101312
[10] Lebelle, Jarjir S and Sassi S 2020 Corporate Green Bond Issuances: An International Evidence. *Journal of Risk and Financial Management, 13*, 25. doi:10.3390/jrfm13020025
[11] Liang D, Lu C, Tsai C F and Shih G A 2016 Financial Ratios and Corporate Governance Indicators in Bankruptcy Prediction: A Comprehensive Study. *European Journal of Operational Research, 252*. doi:10.1016/j.ejor.2016.01.012
[12] Maltais A, and Nykvist B 2020 Understanding the role of green bonds in advancing sustainability. *Journal of Sustainable Finance and Investment*, 1-20. doi:10.1080/20430795.2020.1724864

[13] Ross, S. A. W. R. J. J. F 2010 *Corporate finance*.
[14] Zhou X, and Cui 2019 Green Bonds, Corporate Performance, and Corporate Social Responsibility. *Sustainability, 11*, 6881. doi:10.3390/su11236881

The Role of Indonesian Investors in Achieving SDGs Through Green Bonds

Hallira Husin Haddad[1*], **Rofikoh Rokhim**[1]
[1] Department of Management, Faculty of Economics and Business, Universitas Indonesia, Jakarta, 16424, Indonesia
[1*]hallira.husin01@ui.ac.id, [2]rofikoh.rokhim@ui.ac.id

Abstract. With the ambition to achieve the Sustainable Development Goals (SDGs) agenda in 2030, many countries are increasing the implementation of green financing through green bonds. The question is, how appealing are green bonds for investors? This question is relevant as green bonds have been applied in some countries to help them to achieve sustainable development goals as part of sustainable finance. This paper compares the green bond yield, and conventional bonds yield to see if there is a significant difference between them. This becomes important as the environmental aspect is not the only consideration for investors in choosing green bonds. However, they also pay attention to the return value. The analysis was carried out on the performance of green bonds against conventional bonds in Indonesia from 2018 to 2021 and compared green bonds' yields with conventional bond yields with an independent sample t-test. The results show that Indonesian green bonds have an increased performance every year and green bond yields have no significant difference from conventional bond yields, where green bonds have slightly higher yields.

1. Introduction

Many studies tried to find the impact of green bonds on the environment. However, not much academic research to date is unable to provide a consensual view of the relationship between investor preference and its effect on green bonds' interest. This is important because green bonds are one of the effective ways to improve the environment, especially in reducing global warming. According to the Ministry of Finance [1], green investors purchased green bonds as much as 29 percent, while conventional investors purchased the remaining 71 percent. For this reason, based on the buyer's preferences, if it is related to the project being carried out, Indonesian green bonds have not been able to function following their nature as green bonds.

This shows that investors' appetite for green bonds is relatively low and is also influenced by risk preference, which is essential in deciding to invest. When referring to the performance of green bonds, a study [2] shows that the results of green bonds have a positive return announcement, an increase in long-term value and operating performance, an increase in environmental performance, an increase in green innovation, and an increase in long-term ownership for investors. Furthermore, another study [3]states that institutional investors have now realized the importance of the company's role in the environment for clients and beneficiaries of green projects.

Although green bonds bring more benefits than conventional bonds, this is not the primary consideration for an investor investing. Research [4] explains that the risk preference characteristics of an investor will change due to two opposite things, gains/losses. Correspondingly, there is empirical evidence from a study showing that investors' risk preferences are positively correlated with market performance. Betterment further supports this claim, as its clients tend to increase portfolio risk after periods of strong market performance [5].

The purpose of green bonds as a financial instrument that is expected to contribute to the financing of various green projects and investments is also stated by [6] that green bonds or climate bonds are bonds that try to increase efficiency and support environmental initiatives. These bonds raise funds to improve the energy efficiency of existing structures (brownfield projects), build new sustainable infrastructure (greenfield projects), develop green transportation, sustainable agriculture, and other similar initiatives. This market segment has overgrown and has allowed investors to profit and make environmental contributions.

Green bonds in Indonesia are not much different from what is applied by the World Bank. POJK No.60/POJK.04/2017 explains that activities that green bonds can finance are only Environmental Business Activities. In addition, the proceeds from the Public Offering of Green Bonds must at least 70% be used to finance environmental business activities. The business activity selection mechanism will be carried out by environmental experts who refer to individuals, groups, or institutions with the competence or expertise to assess, verify, or test only environmental business activities [7].

When referring to the performance of green bonds, a study [2] shows that the results of green bonds have a positive return announcement, an increase in long-term value and operating performance, an increase in environmental performance, an increase in green innovation, and an increase in long-term ownership for investors. It shows that green bonds bring more benefits than conventional bonds, but this is not the primary consideration for an investor in investing in green bonds. Based on research [4] explains that the risk preference characteristics of an investor will change due to two opposite things, namely gains/losses. Therefore, it is essential to know if green bonds have a high return or not compared to conventional bonds, as this factor has a crucial role in deciding what instrument to invest in.

In economics, risk preference more often refers to the tendency to engage in behaviors or activities that involve higher variance in returns, regardless of whether these represent gains or losses, and is often studied in monetary payoffs involving lotteries [8]. On the other hand, [9] said that risk preference comes into play when a decision has to be made under uncertainty. Risk preference refers to the attitude of the decision-maker or investor to risk, that is, her degree of tolerance to risk.

2. Methods

The present paper utilizes the time series dataset constituting the monthly observations of Indonesia's green bonds and conventional bonds prices and yield from Bloomberg Terminal for 2018 - 2021. This study is an adaptation of previous research [4], which believe that gain/loss are the variables that influence investors' risk preference where investors are willing to take risks after getting previous profits. Therefore, this research used to represent gain/loss that can be analyzed from green and conventional bonds yield.

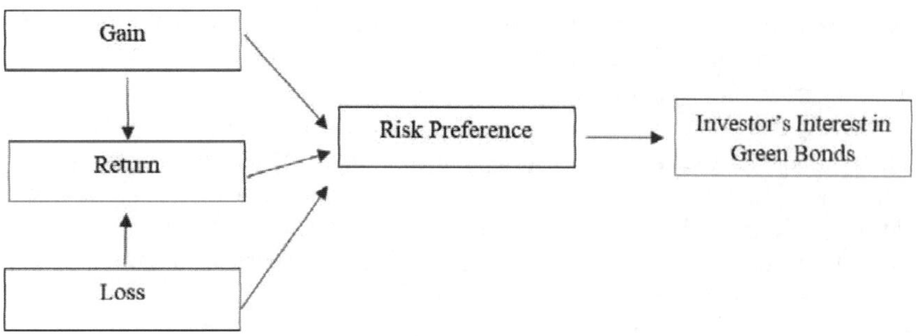

Figure 1. Conceptual framework

Before comparing the yield of green bonds and conventional bonds, we would like to see the performance of green bonds and conventional bonds in Indonesia, which can be seen in the figure below.

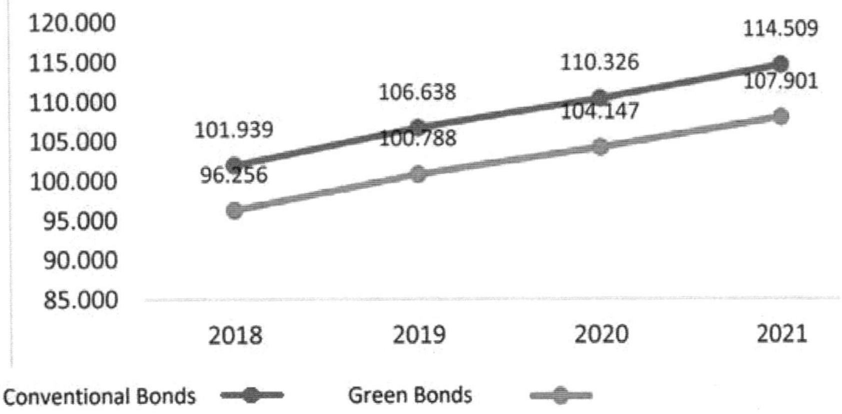

Figure 2. Indonesian green bonds and conventional bonds' performance

In the picture above, it can be seen that since it was issued in 2018, the price of green bonds has slowly increased every year and has a pattern of having an increase of approximately four thousand each year except from 2020 to 2021, the increase is slightly decreasing, namely from IDR 104.147 to IDR 107.901. The slowdown in growth in green bonds during this period was possibly influenced by the entry of the Covid-19 pandemic into Indonesia, namely in 2020, and continuing to spread until 2021. This is supported by data from the Ministry of Finance [10], which explains that Indonesia is experiencing an economic contraction due to Covid–19, minus 2.07 percent.

Similar to green bonds, the movement of conventional bonds shows an increase every year but has a higher number when compared to green bonds. In 2018, conventional bonds had an average price of Rp101,939, and in the following year, it increased to Rp106.638. However, there is a difference in that conventional bonds do not have a pattern of gains, unlike green bonds. From 2020 to 2021, there is a less significant increase. It is predicted that this is similar to what is experienced by green bonds, which are affected by the pandemic.

Overall, the price movement of conventional bonds is still higher than green bonds. This is because conventional bonds have already been in the investment market, so investors are more comfortable making business investments in conventional bonds. In addition, based on the researcher's questionnaire to bond investors, most of them have an investment goal to increase income, which means that profit is still the main factor in addition to impacting the environment. Indonesia's green bonds are not yet following the "green" value. It indirectly illustrates the low awareness of local investors about the environment, so more Indonesian green bonds are issued in foreign markets [1].

2.1 *Independent Sample T-Test*

The t-test is carried out by comparing the difference between the two average values with the standard error of the difference between the two samples. This test method was chosen in this study because it wanted to compare the average yield of green bonds with conventional bonds where the two had no relationship, so an independent sample t-test was used. Using the T-test method, researchers can answer one of the research objectives related to whether or not there is a significant difference in yields of green bonds with conventional bonds.

3. Result and Discussion

Table 1. Normality test[a]

Yield	Kolmogorov – Smirnov			Shapiro - Wilk		
	Statistic	**df**	**Sig.**	**Statistic**	**df**	**Sig.**
	0.086	28	0.200*	0.954	28	0.243

[a] Sources: SPSS – processed by researchers (2022)

The results of the normality test show that the data has a normal distribution with the value of Sig. of both showed > 0.05 with a gain of 0.200 and 0.243.

Table 2. Descriptive analysis

	Type	**N**	**Mean**	**Std. Deviation**	**Std. Error Mean**
Yield	Green Bonds	12	5,3054	2,48902	0,71852
	Conventional Bonds	16	4,2007	2,25456	0,56364

[a] Sources: SPSS – processed by researchers (2022)

Table 2 shows that green bonds' average yield is more significant than conventional bonds' average yield, which is 5.3054, while conventional bonds are 4.2007. So, statistically descriptive, it can be concluded that there is a difference between the average yield of green bonds and the average yield of conventional bonds. However, from this comparison, it cannot be concluded whether the two have significant differences. For this reason, hypothesis testing will be carried out in the next stage using an independent sample t-test.

3.1 *Independent sample T-test*

Table 3. The Result of Independent Sample T-Test[a]

	Levene's Testor Equality of Variances		Mest for Equality of Means						95% Confidence Interval of the Difference	
	F	Sig	T	df	Significance One-sided p	Significance Two-sided p	Mean DIfference	Std.Error Difference	Lower	Upper
Equal variances assumed	0.261	0.614	1.228	26	0.115	0.231	1.10473	0.89994	-0.74513	2.95459
Equal variances not assumed			1.210	22.465	0.119	0.239	1.10473	0.91321	-0.78689	2.99635

[a] Sources: SPSS – processed by researchers (2022)

The Processing of Secondary Data used in this research yields from Indonesia's green bonds and Indonesia's conventional bonds. As a basis for decision-making in the independent sample t-test, the researcher has determined the following hypothesis:

H0: Yields of green bonds and conventional bonds are not significantly different

$$\mu GB = \mu KO \quad (1)$$

H1: Yields of green bonds and conventional bonds are significantly different

$$\mu GB \neq \mu KO \quad (2)$$

Referring to Table 3, several values need to be considered. The first is Levene's Test for Equality Variances, where this value will determine whether the variance of the two data is equal (equal variance assumed) or not equal (equal variances not assumed). The value will be said to be the same if the significance of the probability (Sig.) > 0.05 and declared unequal if Sig. < 0.05. The results of Levene's Test for Equality Variances show that the value of Sig. of 0.614, so the data has the same variance. The values in the equal variance assumed line or the first line in Table 3 would be used. Furthermore, based on the assumed equal variance line, it is known to have a Sig value. (two-sided) is 0.231 or > 0.05, then, as the basis for decision-making in the independent sample t-test, it shows that H0 is accepted and H1 is rejected. Thus, it can be concluded that there is an insignificant difference in yields of green bonds and conventional bonds.

In the table above, it is also known that the mean difference is 1.10473. This value shows the difference between green bonds' average yield and conventional bonds' yield, or 5.3054 – 4.2007 = 1.10473. At the same time, the difference interval is - 0.74513, the lower limit (lower) to 2.95459, and the upper limit (upper) is stated in the 95% confidence interval of the difference section. In other words, with a 95% confidence level, the difference between the average yield values of the two bonds is between -0.74513 to 2.95459.

The independent sample t-test results align with the existing data, where the data used does not show a significant price difference between green and conventional bonds. This becomes influential because the measurement is carried out, and there is a yield to maturity of each bond in which there is an element of price in the calculation. Considering the average price of conventional bonds during 2018-2021, the value is higher than green bonds. For example, in 2018, the average price of conventional bonds was IDR 101,939, while the average price of green bonds in the same year was IDR 96,256. For this reason, it is necessary to look at the return of each bond because, in return, there is a coupon or annual interest income. In this element, there is a difference. Namely, the interest held by green bonds is higher than conventional bonds, so indirectly, the return from green bonds is higher, which ultimately affects the yield to maturity so that it is slightly higher than conventional bonds.

Based on the research results, it was found that the yield of green bonds and yields of conventional bonds did not have a significant difference. Referring to the yield, green bonds have a slightly higher yield compared to conventional bonds. However, if we look at the performance of the two bonds in terms of price movements, conventional bonds are still higher than green bonds. Most likely, this is influenced by the popularity of conventional bonds, which have been issued earlier than green bonds. Meanwhile, green bonds have a slightly higher yield, possibly because some Indonesian green bonds use the United States Dollar while conventional bonds use Rupiah.

The same results can be seen from previous research [11], which compared bonds in several countries, including Indonesia, showing that Indonesian green bonds decreased after the first date of issuance. Furthermore, conventional bonds with the same maturity were included for the same comparison, but the yield was worse than green bonds because they declined more sharply. He further explained that this is likely because Indonesian conventional bonds are denominated in Rupiah, not in United States Dollars like Indonesian green bonds.

Based on the research results, it can be said that green bonds have many benefits for both the environment and profit. By knowing this, it is hoped that investors will be more interested in investing in green bonds to help the process of environmental improvement, especially on climate change issues. According to the International Energy Agency (IEA), using existing or new technologies, requires US$46 trillion to halve global emissions by 2050. Of this amount, the debt securities market, including bank loans and bonds, are expected to be able to contribute as much as US$6 trillion [7].

4. Conclusion

Indonesian green bonds and conventional bonds have increased performance every year. The yield of green bonds and conventional bonds did not significantly differ. It is because some Indonesian green bonds use the United States Dollar, while conventional bonds use Rupiah. Further research is suggested to conduct a deeper analysis of investors' risk preferences for green bonds with different

IOP Conf. Series: Earth and Environmental Science 1111 (2022) 012069 doi:10.1088/1755-1315/1111/1/012069

variables and a more significant number of samples.

References

[1] Redaksi Pagar Bisnis 2019 Tantangan Indonesia dalam Mewujudkan Green Finance *https://pagarbisnis.com/tantangan-indonesia-dalam-mewujudkan-green-finance/*

[2] Kanamura T 2020 Are green bonds environmentally friendly and good performing assets? *Energy Econ.* **88** 104767

[3] Kleimeier S and Viehs M 2021 Pricing carbon risk: Investor preferences or risk mitigation? *Econ. Lett.* **205** 109936

[4] Wen F, He Z, Gong X and Liu A 2014 Investors' risk preference characteristics based on different reference point *Discret. Dyn. Nat. Soc.* **2014**

[5] Alsabah H, Capponi A, Ruiz Lacedelli O and Stern M 2021 Robo-Advising: Learning Investors' Risk Preferences via Portfolio Choices *J. Financ. Econom.* **19** 369–92

[6] Baker H K, Nofsinger J R and Spieler A C 2020 Being a Lender not a Borrower: The Bond Market *Savvy Investor's Guid. to Build. Wealth Through Tradit. Investments* 59–87

[7] OJK 2016 Laporan Kajian Pengembangan Green Bond Di Indonesia *Jakarta Otoritas Jasa ...* i–iii, 1–162

[8] Mata R, Frey R, Richter D, Schupp J and Hertwig R 2018 Risk preference: A view from psychology *J. Econ. Perspect.* **32** 155–72

[9] Ananth B and Shah A 2013 Preference Theory *Financial Engineering for Low-Income Households* (SAGE Publishing India)

[10] Kementerian Keuangan 2021 Pertumbuhan Ekonomi Indonesia Tahun 2020 Diatas Rata – rata Negara di Asia Tenggara

[11] Siswantoro D 2018 Performance of Indonesian green sukuk (islamic bond): A sovereign bond comparison analysis, climate change concerns? *IOP Conference Series: Earth and Environmental Science* vol 200 p 012056

Greener Communication: To Make Better Understand the Interaction between Aging and the Environment

Irwansyah[1*], **Niken Febrina Ernungstya**[2]
[1] Department of Communication, Faculty of Social and Political Science, Universitas Indonesia, Depok, 16422, Indonesia, ORCID ID: 0000-0002-5047-1746
[2] Postgraduate Program of Communication, Institut Komunikasi dan Bisnis LSPR, Jakarta, 10220, Indonesia, ORCID ID: 0000-0002-9618-8794
[1*]irwansyah09@ui.ac.id, [2]niken.f@lspr.edu

Abstract. As a part of a marginalized and vulnerable group, global older adult and environmental change bring together two key policy challenges that must be addressed to ensure a sustainable future. In terms of the communication perspective of sustainable development goals, communication performance should address environmental factors for the effectiveness of successful aging. The success of the communication strategies and attitudes scale in two previous studies was applied in the current study. This study applied a survey to assisted-specific living communities with the 71 items of communication scale for older adults. The result showed that older adults' communication strategies and attitudes scale has high validity and reliability. Several items of communication strategies and attitudes were modified based on the local context of a specific region in Indonesia. This study concluded that an Indonesian version of the greener communication self-assessment scale of older adults (Indonesian-Greener CSOA) could be developed for more local contexts in specific regions of Indonesia, including the integration with the model that applied a greener communication perspective. The existing greener communication perspective could make a better understanding of the interaction between global aging and the environment to prevent and minimize disproportionate negative impacts on older people.

1. Introduction

"Leave no one behind" commitment is an opportunity to be concerned about aging. The 2030 agenda of sustainable development goals (SDG) recognizes the particular situation of older persons [1]. As a part of a marginalized and vulnerable group [2], global older adult and environmental change bring together two key policy challenges which need to be addressed to ensure a safe, secure, equitable, and sustainable future [3]. The concern of the communication perspective is to participate in the SDGs agenda. A previous study recommends that communication performance should address environmental factors for the effectiveness of successful aging [4]. The importance of communication attitudes and strategies would be beneficial to recognize and adjust to the new ecosystem for older adults [5]. Moreover, the previous study about communication attitudes and strategies should be applied to people in more institutionalized environments such as assisted living communities or long-term care facilities [5].

A previous study successfully developed communication strategies and attitudes of a specific country, such as Korea [6], from the origin country [5]. Another study found the environmental footprint of the aging population by promoting greener communication strategies and attitudes, including the behavior of individual lifestyle choices [3]. Communication strategies refer to actions such as the convenience of the communication process, improvement of communication, and modification of the communication process[6]. The dimension communication strategy could be divided into three types of strategies: (1) facilitation strategy, (2) repair strategy, and (3) preliminary strategy. The facilitation strategy included adaptation strategy, structural strategy, instruction strategy, and anticipation strategy. The repair strategy included writing requirements and spelling requirements. Finally, preliminary strategies included retention strategy (starting a conversation), retention strategy (taking your turn), and maladjustment strategy[6]. The communication attitude scale consisted of one's attitude, family's attitude, and the attitude of others. The explanation of communication strategy and attitude types was considerable in identifying the ecosystem for the elderly who entered an aging society [5].

Unfortunately, two previous studies [5,6] were still of considerable clinical value in examining communication strategies and attitudes of the elderly who has entered an aging society. The current study aimed to re-examine the communication strategies and attitudes based on the experience and the perception of older adults in specific assisted living communities. This study also aimed to identify the environmental communication aspect of the older adults who enter the new ecosystem in the local context of Indonesia.

2. Methods

Previous studies applied the scale of communication strategies and attitudes for older adults as a part of the quantitative approach with the survey method [5,6]. The population of this study was an assisted and specific living community [7] in Semarang, Surabaya, and Yogyakarta. The data collection was conducted from June to July 2022. This study applied total sampling with a high response rate (87%). However, the current study analyzed 34 respondents for statistical purposes after the cleaning process. A previous study confirmed that a minimum of 30 observations is sufficient to conduct significant statistics[8].

The communication scale for older adults (CSOA) variable as a single measurement scale consisted of two dimensions. The first dimension is the communication strategies, consisting of 41 items such as "You are talking with someone about an environment you do not know well. You do not understand. You ask her to repeat". The second dimension is the communication attitude, consisting of 30 items such as "I feel embarrassed when I do not understand our environmental situation ." This study used the suggested response format on a 5- point scale of (1) never, (2) rarely, (3) often (sometimes), (4) mostly always, and (5) always [6]. The adaptation and proposed version of CSOA or Indonesian-Greener CSOA (I-Greener CSOA) were translated into Bahasa Indonesia from the origin CSOA. It is similar to the previous study of Korean-CSOA that was translated into Korean from the origin CSOA. The translation was reviewed by five experts, similar to the previous study [9].

Statistical processing was performed using IBM SPSS (Statistic Packages for Social Science, IBM, Inc., USA) Version 26.0 [10]. To confirm the proposed CSOA construct validity, the correlation between the total score and the sub-items of communication strategy and attitude was checked, and Pearson's Correlation Coefficient was used to check the correlation between items [5,6]. To verify the reliability of the proposed CSOA, the Cronbach Alpha Coefficient was used to check the internal consistency [5,6].

3. Results and Discussion

3.1. The validity of the communication scale for older adult

The construct of the Indonesia-Greener CSOA communication scale for an older adult was valid. It could be shown that the result of Pearson's correlation coefficient was high. The validity result was divided into two dimensions: Communication strategy and communication attitude dimension. The first or the communication strategy dimension showed three categories: Preliminary strategy, facilitation strategy, and repair strategy. The correlation between the total average score of the communication strategy and the preliminary strategy score showed a high correlation ($r = .788$, $p <$

.001). Then, the correlation between the total mean score of the communication strategy and the facilitation strategy score showed a high correlation (r = .747, p < .001). Finally, the total average score of the communication strategy and the score of the modified strategy also showed a high positive correlation (r =.726, p <.001) (Table 1).

Table 1. Construct validity of communication strategy dimension

	Total	Preliminary Strategy	Facilitation Strategy	Repair Strategy
Total	One			
Preliminary Strategy	.788**	One		
Facilitation Strategy	.747**	.717**	One	
Repair Strategy	.726**	.702**	.658**	One

** p <.001.

Pearson's correlation between dimensions of the communication strategy scale was confirmed (Table 1). The correlation between the preliminary and facilitation strategies was highly positive (r = .717, p <.001). A high correlation was also found between preliminary strategy and repair strategy (r = .702, p <.001) and between facilitation strategy and repair strategy (r = .658, p <.001).

The second or the communication attitude dimension showed three categories: Self-attitude, family attitude, and other people's attitudes. The correlation between the total average score of the communication attitude and the self-attitude score showed a high correlation (r = .844, p < .001). Then, the correlation between the total mean score of the communication attitude and the family attitude score showed a high correlation (r = .813, p < .001). Finally, the total average score of the communication attitude and the score of the other attitude also showed a high positive correlation (r =.802, p <.001) (Table 2).

Table 2. Construct validity of communication attitude dimension

	Total	Self-attitude	Family attitude	Other people attitude
Total	One			
Self-attitude	.844**	One		
Family attitude	.813**	.720**	One	
Other people attitude	.802**	.719**	.702**	One

** p <.001.

Pearson's correlation between dimensions of the communication attitude scale was confirmed (Table 2). The correlation between self-attitude and family attitude was highly positive (r = .720, p <.001). A high correlation was also found between self-attitude and another attitude (r = .702, p <.001) and between family attitude and another attitude (r = .658, p <.001).

3.2. The reliability of the communication scale for older adult

A total result of 34 respondents was confirmed and was shown high consistency in the I-CSOA communication strategy and communication attitude scale. The result of the total mean of the communication strategy and the internal consistency of 15 preliminary strategy items was high (α=.821). The high consistency was also found in the result of the total mean of the communication strategy and the internal consistency of 10 items of the facilitation strategy (α=.816) and the result of the total mean of the communication strategy and the internal consistency of 16 items of the repair strategy (α=.802) (Table 3).

Table 3. Internal consistency of communication strategies dimension

Communication strategies categories	Cronbach's Alpha	Item
Preliminary strategy	.821	15
Facilitation strategy	.816	10
Repair strategy	.802	16
Total	.802	41

The confirming result of the total mean of the communication attitude and the internal consistency of 19 items of the self-attitude was high consistency (α=.834). The high consistency was also found in the result of the total mean of the communication attitude and the internal consistency of seven items of the family attitude (α=.816) and the result of the total mean of the communication attitude and the internal consistency of 5 items of other people's attitude (α=.802) (Table 4).

Table 4. Internal consistency of communication attitude dimension

Communication attitude categories	Cronbach's Alpha	Item
Self-attitude	.834	19
Family attitude	.831	7
Other people attitude	.810	5
Total		31

3.3. The proposed Indonesia-Greener Communication Scale for Older Adult

This study proposed the context of the greener communication scale for older adults in Indonesia better to understand the interaction between aging and the environment. The study adopted the Communication Scale for Older Adults (CSOA) developed [5]. The study also conducted the translation and modification of Korean-CSOA [6]. The adaptation, modification, and translation of CSOA in this study were called as Indonesian Greener Communication Scale for Older Adults (Indonesia-CSOA).

The original CSOA and Korean-CSOA were applied to older adults with aural problems. However, this study surveyed the observed total sampling in the assisted living elderly community in a specific environment. The observed sampling could understand and respond to the instrument concerning older adults with hearing loss. This study was similar to the previous study [6], which used the instrument for the regular elderly group. Interestingly, the observed sampling of the current study gave several notes for the next revision of the instrument for non-aural rehabilitation.

The Indonesian-Greener CSOA was shown to have high validity and reliability in examining communication strategies and attitudes of the elderly in a specific environment in Indonesia. The tool could be a significant and valuable test for the elderly who has entered an aging society. This study confirmed the previous study [6] that communication strategies can be used to improve environmental aspects, especially communication in everyday life. In addition to communication attitudes, the tool could be important for evaluating and coping with older adults who may have aural problems. Similar to a previous study [6], the communication attitudes showed that the perception of older adults of aural difficulty affects the feeling and attitudes of individuals, families, and other people. As a result, the Indonesian-Greener CSOA could be recognized as a tool to accommodate the 2030 agenda of sustainable development goals (SDG) on "Leave No One Behind" [1]. A previous study found that in developing countries, a lack of basic infrastructures such as clean water and sanitation, health, and social care combined with poverty and malnutrition make older adults vulnerable to environmental threats[11]. This study also confirmed that older adults feel unsafe to environmental threats without successful communication strategies and attitudes [12].

The I-Greener CSOA, which consisted of communication strategies and attitudes dimensions, was an essential aspect of the development of older adults from an environmental perspective[5] in Indonesia. A report from an international consortium [3] was similar to the result of the current study that the need to raise awareness of the effects of a changing environment on older people across the world, including from a greener communication perspective [13]. Older people are particularly vulnerable to environmental change. Therefore there is a need for communication strategies and attitudes to adjust their living in the new ecosystem and environment of the aging era [14]. The existence of a greener communication perspective could make a better understanding of the interaction between global aging and the environment to prevent and minimize disproportionate negative impacts on older people [12].

4. Conclusion

This study successfully adapted, modified, and translated the original CSOA dan Korean-CSOA scale into Indonesian-Greener CSOA. The success of Indonesian-Greener CSOA was confirmed based on a high correlation of validity test and high consistency of reliability test. The Indonesian-Greener CSOA also confirmed the dimension of communication strategies and communication attitudes. At the same time, the current study also successfully confirmed three categories of communication strategies (preliminary, facilitation, and repair strategies) and communication attitudes (attitude of individual, family, and other people attitude). The Indonesian-Greener CSOA could be a part of tools to recognize "Leave No One Behind" for elderly people on the 2030 agenda of SDG, especially for the changing of the new environmental ecosystem. The communication strategies and attitudes were several aspects of aligning the new environmental change, especially for vulnerable and marginalized older adults. The greener communication perspective was a possible interaction between aging and the environment.

Future studies should revise the instrument for ordinary older adults in the local context based on the notes of respondents. Due to the small sample size of older adults, the result of the current study could not be generalized or interpreted into a larger population. The following study should increase the size of the sample and the size of the font on the instrument, including fewer items from observed variables.

Acknowledgments
This research was supported by PDKN DIKTI Grant No. NKB: 909/UN2.RST/HKP.05.00/2022.

References

[1] UNDESA 2016 Leaving No One Behind — SDG Indicators *United Nations Stat. Div. Dev. Data Outreach Branc*
[2] Migita D and Banerjee D 2020 ‘An Invisible Human Rights Crisis’: The Marginalization of Older Adults During the COVID-19 Pandemic – An Advocacy Review *Psychiatry Res.* **292** 113369
[3] Haq G, Snell C, Gutman G, and Brown D 2013 Global Aging and Environmental Change: Attitudes, Risks and Opportunities *Manag. Environ. Qual.* **24**
[4] Demorest M E and Erdman S A 1987 Development of the Communication Profile for the Hearing Impaired *J. Speech Hear. Disord.* **52** 129–43
[5] Kaplan H, Bally S, Brandt F, Busacco D and Pray J 1997 Communication Scale for Older

Adults (CSOA). *J. Am. Acad. Audiol.* **8** 203–17
[6] Kim E Y and Kim N R 2021 Validity and Reliability of the Korean Version English Version *J. Korean Acad. Nurs. Adm.* **27** 282–90
[7] Reviani and Wulandari R D 2020 An Evaluation of the Implementation of the Elderly Health Program in Indonesia: A Case Study *J. Heal. Transl. Med.* **23** 177–81
[8] Sharma A 2020 Is n = 30 really enough? A popular inductive fallacy among data analysts *Tower. Data Sci.*
[9] Memon M A, Ting H, Cheah J-H, Thurasamy R, Chuah F and Cham T H 2020 Sample Size for Survey Research: Review and Recommendations *J. Appl. Struct. Equ. Model.* **4** i–xx
[10] IBM 2019 *IBM SPSS Categories 26* (Armonk, New York: IBM)
[11] Ezbakhe F, Giné-Garriga R and Pérez-Foguet A 2019 Leaving No One Behind: Evaluating Access to Water, Sanitation and Hygiene for Vulnerable and Marginalized Groups *Sci. Total Environ.* **683** 537–46
[12] Haq G 2013 New study highlights impact of environmental change on older people - SEI *Stock. Environ. Inst.*
[13] Gasiorek J and Fowler C 2019 Effects of "Environmental Chatter" about Age and Aging on Middle-Aged and Older Adults' Communication and Perceptions of Aging Efficacy *West. J. Commun.* **0** 1–25
[14] Butler R N and Pfitzer E A 1987 *Aging in Today's Environment* (Washington D.C: National Academy Press)

Environment, Social, Governance, and Performance in Far East Asia for the Period of 2018-2021

H A B Sarwono[1*]
[1] Department of Management, Faculty of Economics and Business, Universitas Indonesia, Depok, 16424, Indonesia, 0000-0002-2847-0552
[1]*haryo.ariobimo@ui.ac.id

Abstract. This study investigates how banking performance in the Far East Asia region is impacted by environmental sustainability through sustainability reporting using ESG Index as a proxy. For the 2018–2021 period, 89 Far East Asia stock exchange-listed companies that routinely provide ESG scores are the sample for this study. The environment sustainability dimension in this study uses the ESG index as a proxy. The research analysis starts with descriptive statistical analysis to summarize the sample data. Then regression was performed using the Fixed Effect Model method for all research models to test the variables simultaneously. This study discovered a detrimental effect of ESG as a proxy of environmental sustainability on banking performance. However, they also discovered a beneficial relationship between one ESG dimension—corporate governance—on banking performance.

1. Introduction

There is a need for a second report that can provide information on the company's sustainability, which includes ESG (Environment, Social, Governance), as delivering financial reports alone is no longer sufficient to satisfy shareholders. Governments and businesses throughout the world now share the broad goal of sustainability. Companies are now compelled to focus not just on profitability but also on issues like corporate governance, social responsibility, and the environment due to the growth of environmental concerns in the global population. Companies are urged to share how their operations have impacted and helped to address environmental, social, and corporate governance challenges. The implementation of ESG in Asia is still a bit late compared to the US and Europe. According to study results from Willis Towers Watson, asset owners in Asia are still catching up, and as many as 31% of organizations have ESG policies. Only 11% of investor money is invested in sustainable assets or goods in the Asian region due to the low ESG policy activity. Sustainable inflows of money and corporate ESG disclosures have increased quickly, even though overall ESG implementation in the Asian area is still lagging in terms of both number and magnitude of investments.

The relevance of publishing sustainability reports to the public is still not well understood by businesses, even in more specialized industries like banking. The social dimension is more concerned than the environmental and corporate governance dimensions of the sustainability report's three pillars of environmental, social, and corporate governance (ESG) [1]. Banks' direct environmental impact is measured by the consumption of utilities they use, such as the use of office equipment, waste produced in cities, and air pollution. Banks do not directly harm the environment in the same way mining companies conduct business activities and profit from using natural resources. The bank's indirect environmental impact is also a result of its services, some of which potentially harm the environment [2]. By participating in social activities and distributing funds for the welfare of the local community, banks typically contribute more to the social component. Because banks typically have superior corporate governance than businesses in other industries, banks also benefit from ESG. According to World Bank Treasury Data, reserve management is gradually using ESG. Only 17% of central banks have made ESG a part of their investment strategies.

Additionally, 9% of central banks have integrated ESG considerations into their investment management framework. But few institutions have included ESG issues in their investment decisions. Nearly half of central banks said they had neither; 29% said they had discussions but no implementation. Asia's economic growth and development still heavily rely on the banking sector. Sustainable development can be integrated into business strategies, operations, and systems in the financial sector. Based on the idea of fulfilling the primary goal of banking activities—creating money for shareholders—while considering environmental sustainability and social development.

Country-specific aspects of the economy, labor, and government substantially impacted the company's sustainability reporting. Investigating the link between environmental sustainability using the ESG Index and company success in country-specific characteristics is crucial. To look into the connection between sustainability reporting and business performance [3], a study considers the ESG levels in banks listed on stock exchanges in the European Union. Over ten years, from 2007 to 2016, the study used data from 235 banks. According to this study's descriptive analysis, ESG tends to be more prevalent in banks with significant financial leverage and assets. The study also discovered that banks in nations with higher GDP have more important disclosures regarding corporate governance. However, banks in low GDP and poorly governed nations have more significant social and environmental disclosures. The regression model results demonstrate that ESG has a very beneficial effect on corporate performance [3]. However, the outcomes may alter if the factors are examined separately. ROE and TQ were found to benefit from environmental disclosure. Meanwhile, the ROA, ROE, and TQ models are negatively impacted by corporate social responsibility disclosure. Then it was discovered that corporate governance transparency has a detrimental impact on financial and operational performance (ROA and ROE). The detrimental effect of ESG scores on corporate performance in developed and developing nations [4]. [3] research is replicated in this study, but it does so with a new focus on the Far East Asia region.

This study considers the company's performance as measured by ROA and ROE and the company's value as measured by Tobin's Q. To determine the relationship between ESG as a proxy of environment sustainability and bank performance. This study also considers the macroeconomic consequences of corporate governance and the worth of a nation's GDP. As an ESG indicator, the Governance Pillar and ESG indicators are used as an ESG index for the environmental sustainability dimension. Corporate governance affects a company's performance and ESG investments in a significant way because it determines the decisions taken by management for ESG policies. Good governance can result in better company sustainability as well. The study examines how listed banks in the Far East region perform concerning the sustained environment and social and economic development. This study should determine how bank sustainability affects the region of Far Eastern Asia's development sustainability. The unique aspect of this study is that it uses annual data on ESG disclosure from the banking industry in the Far East Asia region for the years 2018 through 2021.

2. Methods

The objective of this study is to objectively investigate the impact of environmental sustainability through sustainability reporting using ESG Index as a proxy for the financial results of bank firms listed on various exchanges in the Far East Asia region between 2018 and 2021. Refinitiv, an independent organization that publishes ESG disclosure score values based on GRI criteria, is where the data comes from. Descriptive statistical analysis is the first step in the research analysis process to view a sample data summary. Then, this study chooses the optimal regression technique to be applied for statistical tests. The variables were then simultaneously tested using a regression for all study models. The data used are from 89 banks listed on stock exchanges in the Far East (Taiwan, Japan, South Korea, Hong Kong, China, Indonesia, Singapore, Malaysia, Thailand, and the Philippines) during the four-year research period, 2018-2021. There is also a selection of banks publishing ESG and corporate governance scores. For the environmental sustainability dimension in this study, the ESG index is used as a proxy.

Data collection, testing, and regression were processed using Stata 14 software. The research analysis starts with descriptive statistical analysis to summarize the sample data. Then regression was performed using the Fixed Effect Model method for all research models to test the variables simultaneously. Data collection, testing, and regression were processed using Stata 14 software. This study replicates the research model made by [3].

The following are the equations used in the study;

Model I

$$ROA_{itg} = \beta_0 + \beta_3 ESG_{itg} + \beta_4 CGD_{itg} + \beta_5 TA_{itg} + \beta_6 FLEV_{itg} + \beta_7 GDP_{itg} + \beta_8 GOV_{itg} + \varepsilon_{itg} \quad (1)$$

Model II

$$ROE_{itg} = \beta_0 + \beta_3 ESG_{itg} + \beta_4 CGD_{itg} + \beta_5 TA_{itg} + \beta_6 FLEV_{itg} + \beta_7 GDP_{itg} + \beta_8 GOV_{itg} + \varepsilon_{itg} \quad (2)$$

Model III

$$TQ_{itg} = \beta_0 + \beta_3 ESG_{itg} + \beta_4 CGD_{itg} + \beta_5 TA_{itg} + \beta_6 FLEV_{itg} + \beta_7 GDP_{itg} + \beta_8 GOV_{itg} + \varepsilon_{itg} \quad (3)$$

ROA = Return on Assets
ROE = Return on Equity
TQ = Tobin's Q
ESG = ESG Index as a proxy of environment sustainability
CGD = Corporate Governance Disclosure Score
TA = Total Assets
FLEV = Financial Leverage
GDP = Gross Domestic Product
GOV = Public Governance Level

3. Results and Discussion

Table 1 is the regression result of Model I, Model II, and Model III. The regression was carried out to determine the effect of sustainability reporting as indicated by the ESG (ESG Score) and CGD (Corporate Governance Score) variables on the company's performance. This test is carried out together with four control variables, namely FLEV (Financial Leverage), TA (Total Assets), GDP (Gross Domestic Product), and GOV (Governance Level).

Table 1. Regression Results

Variable	Model 1 ROA		Model 2 ROE		Model 3 TQ	
	Coef	Prob	Coef	Prob	Coef	Prob
ESG	0.000	0.005	-0.001	0.001	-0.001	0.012
CGD	0.000	0.027**	0.000	0.034**	0.000	0.215
FLEV	0.000	0.432	-0.001	0.436	0.000	0.673
TA	0.000	0.031**	0.000	0.000***	0.000	0.114
GDP	0.018	0.000***	0.218	0.000***	0.099	0.000***

GOV	0.000	0.014**	0.002	0.043**	0.000	0.327
Cons.	0.001	0.785	0.030	0.603	0.073	0.032**
R-Squared	0.211		0.248		0.145	
Prob > F	0.000		0.000		0.000	

Based on Table 1, ESG as a proxy of environmental sustainability has a negative and significant effect on the company's operational performance (ROA) in Model 1. This hostile environment sustainability effect using ESG Index follows the research of [4], which found a negative impact of environmental sustainability using ESG Index on all company performance indicators in both developed and developing countries [4]. The high costs incurred by companies for environmental sustainability using ESG indicators worry shareholders. Meanwhile, the company's operations are required to make more efforts to reduce the impact of its business on sustainability. In Model II it can also be seen that environment sustainability using ESG Index as a proxy has a negative and significant relationship to ROE. Several previous studies show that the effect of environmental sustainability using ESG Index on financial performance is not entirely positive. This is because not all sustainability activities are not related to the company's finances, even some of them negatively affect financial performance. Based on the trade-off theory, the fulfillment of responsibilities on environmental sustainability using ESG Index can harm short-term financial performance because it requires additional investment for significant sustainability. Then the ESG Index variable in Model III has a significant but negative relationship to the company's market performance or TQ. This finding follows previous research [5], which states that for any further involvement of the company in environmental sustainability, the costs incurred tend to increase, the company's profits tend to fall and incur losses in terms of value from a shareholder perspective. The companies that prioritized sustainability did not experience improved performance, and better market assessments compared to companies that tended to be neutral and did not even contribute to environmental sustainability and green strategies.

Then in Model I, it can be seen that CGD has a positive and significant effect on ROA. This finding is in line with Saygili, Arslan, & Birkan (2022), who found that stakeholder participation which includes the rights of shareholders and the board of directors in managing the company, leads to the company's operational efficiency [6]. In Model II, which uses ROE or the company's financial performance as the dependent variable, the regression results found that CGD has a positive and significant effect on ROE. These results follow previous research, which found a positive impact of the board of directors' policies on financial-related performance in the company [7]. In Model III, it can be seen that CGD has a positive but insignificant effect on the TQ variable.

In Model I, control variables such as Total Assets have a positive and significant effect on ROA, which shows that bank assets affect bank operational performance. Meanwhile, the regression results show that financial leverage has no significant effect on ROA. The macroeconomic control variables, namely GDP and Governance Level, show a positive and significant influence on the ROA variable. In Model II, the control variable appears to have a positive and significant effect on ROE. In contrast, the financial leverage variable has an insignificant effect. According to several previous studies, such as Gaur & Kumar (2018), the effect of financial leverage cannot be seen in the short term because it can be an indicator of bank performance in the long term [8]. Then in Model III, control variables such as financial leverage, total assets, and governance level do not significantly affect TQ. Meanwhile, the GDP variable has a positive and significant effect on TQ.

4. Conclusion

The research has concluded that, in the short term, ESG has a negative and significant impact on bank operational performance (ROA), bank financial performance (ROE), and bank market performance (TQ). The findings do not support the original presumptions about this research. Based on these findings, environment sustainability using the ESG indicator is a corporate investment that cannot immediately yield a return. Investments in environmental sustainability using ESG indicators are costly, but they have long-term rewards. There is a negative correlation between the ESG Index as a proxy of environment sustainability and bank performance because banks are thought to have gained little from their sustainability initiatives. Longer-term returns will result from the sustainability investment. Organizations with a connection between environmental sustainability using ESG Index as a proxy and sustainability initiatives will improve their long-term value and performance. According to Mohammad & Wasiuzzaman (2021), the long-term nature of environmental sustainability using the ESG Index disclose a complete description of the relationship between the ESG Index as a proxy of environment sustainability and firm performance [9]. The Corporate Governance component, one of the ESG components, is crucial and can have a short-term positive impact on the organization's performance.

References

[1] Sobhani FA, Amran A, Zainuddin Y. Sustainability disclosure in annual reports and websites: A study of the banking industry in Bangladesh. J Clean Prod 2012;23:75–85. https://doi.org/10.1016/j.jclepro.2011.09.023.

[2] Igbudu N, Garanti Z, Popoola T. Enhancing bank loyalty through sustainable banking practices: The mediating effect of corporate image. Sustain 2018;10:1–11. https://doi.org/10.3390/su10114050.

[3] Buallay A. Is sustainability reporting (ESG) associated with performance? Evidence from the European banking sector. Manag Environ Qual An Int J 2019;30:98–115. https://doi.org/10.1108/MEQ-12-2017-0149.

[4] Buallay A, Fadel SM, Alajmi J, Saudagaran S. Sustainability reporting and bank performance after financial crisis: Evidence from developed and developing countries. Compet Rev 2020;31:747–70. https://doi.org/10.1108/CR-04-2019-0040.

[5] THUY PBG, DUC VH. Corporate Governance and Firm's Performance: Empirical Evidence from Vietnam. J Econ Dev 2013;218:62–77. https://doi.org/10.24311/jed/2013.218.08.

[6] Saygili E, Arslan S, Birkan AO. ESG practices and corporate financial performance: Evidence from Borsa Istanbul. Borsa Istanbul Rev 2022;22:525–33. https://doi.org/10.1016/j.bir.2021.07.001.

[7] Lin WL, Law SH, Ho JA, Sambasivan M. The causality direction of the corporate social responsibility – Corporate financial performance Nexus: Application of Panel Vector Autoregression approach. North Am J Econ Financ 2019;48:401–18. https://doi.org/10.1016/j.najef.2019.03.004.

[8] Gaur A, Kumar M. A systematic approach to conducting review studies: An assessment of content analysis in 25 years of IB research. J World Bus 2018;53:280–9. https://doi.org/10.1016/j.jwb.2017.11.003.

[9] Mohammad WMW, Wasiuzzaman S. Environmental, Social and Governance (ESG) disclosure, competitive advantage and performance of firms in Malaysia. Clean Environ Syst 2021;2:100015. https://doi.org/10.1016/j.cesys.2021.100015.

Analysis of Students' Participation in Practicing the Concept of Green Campus

DE Eprilianto[1*], MF Ma'ruf[2], GW Pradana[2], AN Hilmi[2], and UN Habibah[2]
[1]Department of Public Administration, Faculty of Social Science and Law, State University of Surabaya, Surabaya, 60231, Indonesia, 0000-0003-2917-8446
[2]Department of Public Administration, Faculty of Social Science and Law, State University of Surabaya, Surabaya, 60231, Indonesia
[1*]debyeprilianto@unesa.ac.id, [2]muhammadfarid@unesa.ac.id,
[3]galihpradana@unesa.ac.id, [4]ahmadhilmi@unesa.ac.id
[5]uminur.20036@mhs.unesa.ac.id

Abstract. Campus's future contribution to sustainable development will continue, driven by the global community. Green Campus becomes a promoted program to answer that challenges. This study aims to know how students participate in the concept of Green Campus. This study was conducted with an online survey method using Google Forms by 205 respondents who are students of the State University of Surabaya. The survey was conducted in the period time June - August 2022. Data analysis techniques are carried out with approach qualitative general by inductive. This study shows students' participation in practicing the concept of Green Campus through eight indicators measurement. Obtained average participation of 80%. The highest percentage was found in the maintenance of eco-friendly infrastructures, and the lowest was found in research or innovation in the environment and renewable energy field. It shows that students participate well in realizing the Green Campus at the State University of Surabaya.

1. Introduction

The global community often echoes discussions related to environmental issues. The crisis during the Covid-19 pandemic has also contributed to increasing public awareness of nature and the environment [1]. This awareness is encouraged to thrive in higher education. It is proved by the existence of *The Higher Education Sustainability Initiative* (HESI) Global Forum, which has been a cooperation between the United Nations (UN) and the college community since 2012 [2]. In the future, the role of college will keep going encouraged and highlighted in sustainable development.

Green Campus became one of the alternative programs that not only focused on providing a green area on campus but also created a supportive ecosystem for the learning process. It affects stakeholders on campus, where their satisfaction and quality of life on campus who apply to Green Campus are higher than those on campus who do not run the program [3]. The creation ecosystem on this Green Campus could be working as the practice of eco-friendly. Moreover, Green Campus became vital because it combines the practice of eco-friendly with education to promote sustainable practices on campus.

Green Campus is the answer to the need for a green environment healthy and beautiful campus. Besides that, the campus is part of the urban ecosystem, so the existence of a Green Campus can balance the utilization of land cities that impact temperature and air. Also, Green Campus can encourage the implementation of science and technology in the field environment.

As a planned program that includes the whole campus environment, Green Campus is necessary supported by the academic community, mainly by the students. Furthermore, students' environmental caring attitude also faced obstacles, such as difficulties understanding the theory and the balance of field practice and theory not yet optimal [4]. In order to increase participation, it required promotion through specific approaches to increase bottom-up students' involvement. Besides, it is essential to reduce the gap between ideal sustainable goals with what can be understood and practiced by students [5]. So, this research was conducted to know how students participate in practicing the Green Campus concept at the State University of Surabaya.

2. Methods

The researcher collects data with a survey online method using Google Forms. As for the survey held in the period time June-August 2022. This study used a simple random sampling method and involved 205 State University of Surabaya students from various departments. All participants were ready to answer several questions about their understanding and experience of the Green Campus program. The study was conducted with permanent guard confidentiality of data from participants.

The survey consists of 5 entries for respondents' data and 13 entries for the answers to questions related to research. The questions on the survey have evaluation validity by experts. It was conducted to measure suitability questions with existing indicators. Those indicators are sorting waste culture, plastic waste minimization, the usage of the pedestrian in the campus area, water and electricity minimization, paper waste minimization, participation in tree planting activities, maintenance of eco-friendly infrastructures, and research or innovation in the field of environment and renewable energy. The survey can be done in 8-15 minutes. The responses that have entered analyzed use approach qualitative inductive. The data is processed using processing software numbers for then grouped based on existing indicators.

3. Results and Discussion

Green Campus or Eco-Campus at the State University of Surabaya has been implemented since 2015. There are nine Green Campus programs such as waste paper management, office-friendly environment, water and waste management program, electricity management program, manufacturing internal transportation that is eco-friendly, integrated waste management program, reforestation and forest city program, revitalization of the Eco-Campus based master plan, and optimizing alternative energy. Green Campus is not only implemented at the university level but also the faculty. The distribution role assists the adaptation, goal attainment, integration, and latency process that can increase students' competencies in sustainable practices [6].

This research shows students' participation in practicing the concept of Green Campus at the State University of Surabaya, Indonesia.

Table 1. Respondents' Participation in Practicing the Concept of Green Campus

Variable	Percentage (%)	Category
Sorting waste culture	86	Very Good
Plastic waste minimization	84	Good
The usage of the pedestrian in the campus area	72	Good
Water and electricity minimization	83	Good
Paper waste minimization	88	Very Good
Participation in tree planting activities	70	Enough
Maintenance of eco-friendly infrastructures	89	Very Good
Research or innovation in the field of environment and renewable energy	68	Enough
Mean	**80**	Good

Table 1 shows that the sorting waste culture is excellent, with a percentage of 86%. The trash can at the State University of Surabaya has been customized based on the type, i.e., organic waste and non-organic waste. The number of trash cans perceived by respondents needs to be added to increase accessibility. The culture of sorting trash waste needs to follow up with a cycle repeated by internal parties or by cooperating with a third party [7]. In *reuse* and *recycle* ting, the State University of Surabaya has formed a waste bank and hydroponic area, as well as activities recycling plastic waste to be product [8].

Respondents' participation in plastic waste minimization shows a good result with a percentage of 84%. The usage of plastic for food amount a little compared to drinks because drinks tend to be packed with plastic even though the State University of Surabaya canteen has provided food and drink with non-plastic packaging, i.e., plates and glasses, so that it could reduce plastic waste. Even though sometimes students buy a drink with plastic packaging from outside, many students still bring tumblers from home. This awareness needs support from the campus, for example, by providing water refills at some points [7]. The student has not yet gotten that facility.

The usage of pedestrians in the campus area by students belonging to good though is on the verge limit the lower category with a percentage of 72%. Pedestrians on campus are one of the mobilization facilities that should be upgraded in quality. Good pedestrians should notice segment pedestrians, conflict pedestrians, the accessibility for disabilities, vegetation, street furniture, buffers, maintenance, and aesthetics [9]. At the State University of Surabaya, pedestrians are designed considering those aspects. However, there are still inequality aspects in some paths that do not review more deeply in this study.

Water and electricity minimization get 83% or included in the excellent category. Respondents use related facilities such as AC, lamps, and water wisely. It is also supported by the time regulations on rooms which could push students to utilize the garden or their organization's room for their activities. Decreasing energy consumption naturally significantly impacts minimizing carbon footsteps and reaching sustainable developments. Moreover, the energy audit could be one of the alternatives for controlling electricity and water usage and could help to find out the best way to save energy without disturbing any activities [10].

Waste paper minimization shows a percentage of 88% and includes an outstanding category. College institutions contribute to the high number of waste paper. If the recycling process does not follow, it will create new problems. The usage of online media and soft files is one of the alternatives for reducing waste paper generation [11]. The State University of Surabaya makes efforts on waste paper reduction, such as through online text-based learning activities and online service administrations. Online lectures during the pandemic also contribute to accelerating that transition.

Participation in tree planting activity was categorized as enough with a percentage of 70%. Creating a comfortable and beautiful campus environment could be a medium to promote and develop the campus' construction culture. Setting up the trees is different from other environment settings because it emphasizes culture and heritage and becomes part of the campus' construction culture [12]. The planting tree activity occurred in the State University of Surabaya, mainly on a particular day, for example, on earth day [13]. Students' participation is not yet maximum because some representatives usually do it.

Respondents' maintenance of eco-friendly infrastructures gets 89% or includes the outstanding category. Green Campus facilities and infrastructure are encouraged to consider energy consumption. Although using low consumption energy, it has to show high performance [14]. The State University of Surabaya has utilized 80% of LED lights on the campus, and 76.34% of sinks are using *the eco plant.* Besides that, a green area called 'Laboratorium Kampus Merdeka' became education, recreation, and conservation of some flora and fauna [8]. Utilizing *eco*-friendly infrastructure and students' support could chance to speed up the transition of others by going to campus friendly environment.

Research or innovation in the environment and renewable energy include in enough category with a percentage of 68%. Individual understanding of the environment and renewable energy at the State University of Surabaya yet ensures the enhancement of research or innovation. It needs collaboration between the academic community so that the method, content of conventional education, and research could tend to a more sustainable approach, for example, in the syllabus, projects research, publication, students' activities, and investment [15].

Based on the measurement of all variables, it is known that the average student participation is good. The highest score was found in maintaining eco-friendly infrastructures (89%). This can be influenced by awareness to use infrastructure wisely. On the other hand, research or innovation in the environment and renewable energy is the variable with the lowest score (68%). Students must be encouraged by the system and stakeholders to produce environmental studies related to their knowledge.

4. Conclusion

Students' participation in practicing the concept of Green Campus at the State University of Surabaya achieves an average score of 80 % and is in a suitable category. The highest score was found in the maintenance of eco-friendly infrastructures (89%), and the lowest was found in research or innovation in the environment and renewable energy (68%). This study describes students' participation in general, so it has limitations in measuring the depth of each variable. The researcher suggests that further research can provide detailed results from these variables or provide other alternatives to contribute to the development of the research or innovation in the environment and renewable energy field.

References

[1] Rousseau S and Deschacht N 2020 Public Awareness of Nature and the Environment During the COVID-19 Crisis *Environ. Resour. Econ.* **76** 1149–59
[2] Tiyarattanachai R and Hollmann N M 2016 Green Campus initiative and its impacts on quality of life of stakeholders in Green and Non-Green Campus universities *Springerplus* **5** 1–17
[3] Darmadi S and Haryanto R 2017 Implementation of Environmental Education to Support Sustainability of Green Campus Program in Universitas Riau 978–9
[4] Pereira Ribeiro J M, Hoeckesfeld L, Dal Magro C B, Favretto J, Barichello R, Lenzi F C, Secchi L M de L C R and S O de A G 2021 Green Campus Initiatives as sustainable development dissemination at higher education institutions: Students' perceptions *Clean* **3** 312
[5] Triadi S N A, Akmalah E and Ainun S 2018 Pengembangan Indikator Waste (Sampah) pada Penerapan Konsep Green Campus di Itenas *J. Online Inst. Teknol. Nas.* **4** 122–32
[6] UNESA UNESA Paparkan 'Kampus Hijau Berkelanjutan' di Hadapan Asesor UI Greenmetric 1
[7] Paramastri S and Pamungkas T 2018 Walkability pada Jalur Pedestrian di Area Kampus Universitas Negeri Malang *J. Mhs. Jur. Arsit. Univesitas Brawijaya* **6** 1–9
[8] Getu B N and Attia H A 2016 Electricity audit and reduction of consumption: Campus case study *Int. J. Appl. Eng. Res.* **11** 4423–7
[9] Agrawal M, Joshi R R H and K A 2021 Waste Paper Collection for Recycling : Overview and their Reuse Potential 2021
[10] Wang, W., Yi H 2017 The Strategy of Planting Arrangement in College Campus Culture Construction 76 484 – 7 **76** 484–97
[11] Fakultas Bahasa dan Seni 2019 Laporan Kegiatan Peringatan Hari Bumi 1–20
[12] Abdou A A A E G I O and M A 2018 Reducing Energy Consumption Strategies in University Buildings in Egypt Acad. Res. Community Publ. **2** 212
[13] Najafian S M and Karamidehkordi E 2018 Challenges of sustainability efforts of universities regarding the sustainable development goals: A case study in the University of Zanjan, Iran *E3S Web Conf.* **48** 0–4
[14] Wang Z, Lin T and Chen W 2020 Occurrence and removal of microplastics in an advanced drinking water treatment plant (ADWTP) *Sci. Total Environ.* **700** 134520
[15] Drempetic S, Klein C and Zwergel B 2020 The Influence of Firm Size on the ESG Score: Corporate Sustainability Ratings Under Review *J. Bus. Ethics* **167** 333–60

IOP Conf. Series: Earth and Environmental Science 1111 (2022) 012073 doi:10.1088/1755-1315/1111/1/012073

Community Resilient from Anak Krakatau Eruption: Lesson Learn from The Tale of Lampung Submerged to Increase Environmental Awareness

Azhar Firdaus[1]*, Suraya Abdulwahab Afiff[2], and Herdiansyah[3]
[1] School of Environmental Science, Universitas Indonesia, Jakarta 15430, Indonesia
[2] Faculty of Social and Political Science, Universitas Indonesia, Depok 16424, Indonesia
[3] Cluster Interaction, Community Engagement and Social Environment, SIL, Universitas Indonesia
[1]*azharfirdaus@ui.ac.id, [2]suraya.afiff@ui.ac.id, [3]martoing151515@gmail.com

Abstract. Natural disasters in Indonesia have a high frequency because a whole land and sea in Indonesia be hemmed in tectonic and volcanic plates. Some natural disasters are not caused by those plates, such as landslides, floods, hurricanes, etcetera. In the research location (beginning of 2018), there was a natural disaster called the silent tsunami caused by landslides at Mount Anak Krakatau in the middle of the ocean. In large quantities, these landslides produced high waves in the coastal area in Lampung and Banten, and there was no early warning system for this silent tsunami. The article aims to study the local knowledge related to the Anak Krakatau Volcano eruption in the coastal around it. Also, the article will present the history of Anak Krakatau volcanoes and the lessons from communities within the local knowledge. For example, it is from the Tale of Lampung Submerged as secondary data for this article. This tale will be analyzed how this tale could understand the local knowledge in terms of the Mount Krakatau eruption. This local knowledge will help the communities' strategies to face natural disasters and how the recovery process is carried out to increase environmental awareness.

1. Introduction

According to Indonesian history, ten large explosions in volcanoes have impacted the community around the world. Two of them took place in Indonesia. The first was Mount Tambora, which erupted in 1815, killing over 80,000 people, and the second was Mount Krakatau, which erupted in 1883, killing 36,000 people [1]. It emphasizes that volcanic eruptions are one of Indonesia's threats to natural disasters. Several volcanoes are in the group of islands in the territories of Indonesia, including the Islands of Sumatra, Java, Maluku, and Sulawesi [2]. Volcanic areas in Indonesia are fertile agricultural areas and are always densely populated, although they cannot be separated from the threat of volcanic eruptions. According to the Centre for Volcanology and Geological Disaster Mitigation records, active volcanoes in Indonesia are classified into three types based on their eruption history [3–5], namely Type A (79), which are volcanoes that have erupted since 1600. Type B (29 fruit) erupted before 1600, and type C (21 units) are solfatara and fumarole fields. Studies have been conducted on type A, namely volcanoes that have erupted since 1600 [4,5]. There are differences in the character of the eruption, which are directly related to the potential threat of eruption hazards. The potential threat of volcanic eruptions is related to the state of the peak landscape (the shape of the crater) and the type and dynamics of the eruption.

Volcanic eruptions in Indonesia related to the caldera formation occurred on Mount Krakatau in 1883. This eruption affected the world climate and caused geological disasters in several parts of the earth [6–11], marked by a decrease in temperature. The eruption of Mount Krakatau in 1883 through more than 10 km^3 of pyroclastic material, both in the form of eruption hot clouds and eruptive ash. This eruption killed over 36,000 people because of tsunami waves caused by the collapse of the volcanic crater wall (sector failure) and the flow of hot eruption clouds into the sea [5,12,13].

The first indication of the Mount Anak Krakatau embryo's existence occurred on December 29, 1927, to January 5, 1928, as an underwater volcano at a depth of 28 meters below sea level [6,14]. This mountain grows as a volcanic island (Mount Anak Krakatau), which until now, in 202, has reached an altitude of about 230 meters above sea level. Approximately four years ago, in mid-December 2018, Mount Anak Krakatau showed an increase in eruptive activity, which resulted in a silent tsunami wave due to an underwater landslide following the massive eruption of Mount Anak Krakatau on December 22, 2018, on the coast in Pandeglang Regency, Banten Province.

Based on this historical background, it can be concluded that the coast's condition around the Mount Anak Krakatau area significantly impacts the community. The significant impacts that can be experienced for these coastal communities due to the eruption of Mount Anak Krakatau in the future are the loss of buildings and houses, loss of livelihoods, and deteriorating health conditions due to volcanic ash, as well as the psychological condition of the community.

The significant impact is stated on one of the tales in Lampung submerged [15] with line 204: "… *Tebalnya pasir pula dikata, Kira2 ada duanya hasta, Bertaburan di laut penglihatan mata, Adalah sepert pantai semata ...* (The thickness of the sandbar I should guess, Was around one meter deep, but no less, Spread across the sea, as far as the eye's reach, The ocean was transformed into a vast beach)." and line 214: "… *Ada setengah naik ke rumah, Menunggu harta sepanjang tanah, Hilang akal pikiran berekah, Dipukul air habislah punah* … (Perch on roofs, many people were found, One eye on belongings below on the ground, They did not know what to do or say, When the wave came and swept them away)."

Related to the community, Indonesia has much diversity of indigenous people. These indigenous people have many cultures related to norms and religious processes in the whole activity. There is a cultural justification for natural disasters that have been researched in the last few years. It is because there is some increased attention to how culture (norms, customs, and knowledge) contributes to natural disasters and whether to improve or reduce the impact of natural disasters. The interconnection between belief and knowledge systems of various types and how external actors become significantly involved with indigenous knowledge has experienced a resurgence in several literature studies [16–19]. This development is seen in the policy of comprehensive architecture from the humanitarian sector, who more realizes it is essential to put a people and community as an object of the impact of natural disasters [20]. By shifting the emphasis from short-term disaster management to long-term disaster risk management based on the Sendai Framework for Disaster Risk Reduction 2015-2030, there is more room for structural determinants of larger disaster-prone areas, such as cultural considerations or community wisdom [21].

The article will present the lesson learned within the local knowledge from the Tale of Lampung Submerged to resilience to natural disasters, especially with the volcano eruption in Anak Krakatau volcanoes. Lessons learned from the tale will make a harmonious relationship within the environment, so it will save the people who live there and conserve the environment.

2. Methods

Provide This research employs a focused review only from the Tale of Lampung Submerged to better understand the community resilience related to the Anak Krakatau volcano eruption. Studying the tale of Lampung submerged [15] also observes the extent of natural disasters' strength, especially with volcano eruptions. The extent means how are the scale and the impact of the natural disasters on the communities in coastal areas.

This tale was published in 1883 in *Jawi* script, with three subsequent editions issued within five years [15]. The tale appeared in four lithographed editions between 1883 and 1888, with four slightly different titles published in Singapore. The poem's first edition was forty-two pages long and titled *Syair Negeri Lampung yang Dinaiki Oleh Air dan Hujan Abu* (A poem about Lampung when engulfed by water and rain of ash). The colophon notes that this edition was published in the Muslim Year 1301 AH (November 1883 to October 1884). One copy of this small book is preserved in the National Library of the Republic of Indonesia in Jakarta and another in the Russian State Library in Moscow.

The stage of the research is divine, with three stages. The first stages are to identify the history of the eruption in the mountains based on the newsletter information. The second stages are to review the tale of Lampung submerged and complete it with the third stage, how the lesson learned could be implemented in the future in terms of increasing awareness of Mount Anak Krakatau.

3. Results and Discussion

3.1. The history of Krakatau

The volcanic island of Krakatau has a long and complex geological history. An ancient volcano estimated to be two hundred meters in height was destroyed in eruptions that left three principal remnants: the Islands Sertung, Rakata, and Rakata Kecil. Layers of lava covered Rakata from a basaltic volcano that eventually grew to eight hundred meters. Two submarine volcanos, Danan and Perbuatan, ultimately emerged and joined with Rakata to form the pre-1883 Rakata Island. Before the 1883 eruption, the most recent substantial activity from these volcanic centers was recorded in 1680-81. Fumaroles, or gas vents, we reported as far back as 1939.

Early in 1883, a marked increase in earthquake activity was recorded in the region. On May 20, 1883, volcanic eruptions began, first from the cone of Perbuatan and, within a few days, from a second, newly formed vent. In July that year, the violence of the eruptions increased, and by the middle of August, three central vents and numerous smaller ones were spewing large volumes of ash and steam. The last three days of activity, from August 27 to 28, were marked by a succession of explosive blasts, at first separated by intervals of ten minutes but later becoming continuous.

The eruption's effects were felt not just in Indonesia and its immediate neighbors; the ash entered the atmosphere and cooled the global climate for years. The Krakatau Committee of the Royal Society of London published an extensive report in 1888 titled The Eruption of Krakatau and Subsequent Phenomena. Since that time, Krakatau has been the subject of numerous reports and publications, both scholarly and literary.

In the next section, we describe how the communities in Lampung experienced the eruption as it was written in the poem of Syair Lampung Karam (The tale of Lampung Submerged).

3.2. The tale of Lampung submerged.

Syair Lampung Karam appeared in four lithographed editions between 1883 and 1888, with four slightly different titles, published in Singapore [15] by Muhammad Saleh. This tale was printed in Jawi letter with the Melayu language and Riau dialect. The tale is preserved in Lampung when the eruption of Mount Krakatau happened. The tale of Lampung Submerged can be categorized as a "journalism poem" (*syair kewartawanan*) to borrow a term used by the classical Malay scholar Sri Wulan Rujiati Mulyadi. This kind of *syair* typically contains eyewitness reports on real-life events, including events of historical importance and political developments. Nevertheless, in this poem, the author's purpose was to share the story about the natural disaster he had experienced rather than explore it. In table 1, some poems have been selected that focus on how the volcano erupted and how the community managed to survive.

Table 1. Some of the Poem in the Tale of Lampung Submerged [15]

Line of Poem	In *Jawi* script	In English
6	.... *Mula pertama asalnya itu.* *Pada bulan Rajab datanglah abu,* *Dua jari tebalnya tentu.* *Tiga hari kerasnya itu.*	..... Let us hearken to the start of it all, Rajab, the seventh month, when ash began to fall. For three long days, ash descended thick and hard. Till a layer two fingers deep covered the yard.
7	*Kheranlah orang di dalam negeri,* *Habu nan dating tidak terperi,* *Apalah sebab demikian peri,* *Rupanya putih lekat sekali.*	Everyone was astounded on that day. Why would ash be falling in such a way? What was the reason for this state of affairs, This clinging white ash suffusing the air?
10	..... *Lalu terdengar orang sekalian,* *Seperti guntur bunyinya, Tuan,* *Teduh angin tiada kedengaran,* *Berulang2 di dalam demikian.*	..... Around that time, there was heard a tremendous boom, Like thunder it was, a pronouncement of doom, Drowning out the sound of the wind and rain, The volley repeated, time and again.
11	*Di dalam negeri diamlah, Tuan,* *Angin keras tampak kelihatan,* *Lamanya itu lebih tiga bulan,* *Barulah ia serta merusakkan.*	The country fell. Still, life stopped, my good sir. A hard wind was rising. The sky was astir. For three months, it continued unabated, A prelude to the destruction that awaited.
12	*Tahun 1300 pada Nabi kita,* *Kekasih Allah, Tuhan semata,* *Pada bulan Syawal pula dikata,* *Hari 22 sudahlah nyata.*	Thirteen Hundred in the year of the Prophet, Beloved of Allah, Creator of this planet, The month was Syawal, the tenth of the year. The day, the twenty-second, that too is clear.
13	*Hari Ahad nyatalah tentu,* *Pukul empat jam di situ,* *Berbunyi guruh menderu-deru,* *Dikatakan kapal apinya itu.*	"Twas Ahad, a Sunday, a day I recall well, And still quite early, the time of the fourth bell, There came a booming sound of the constant report, Like the horn of a steamer coming into port.
14	*Gaduhlah orang di dalam negeri,* *Mengatakan datang kapalnya api,* *Lalu berjalan berperi-peri,* *Nyatalah Rakata empunya bunyi.*	A most anxious frown appeared on each face. Why would a steamer be coming to this place? When they went out to see what it could be, They found it was Krakatau, out there in the sea.
17	..... *Mangkin malam bertambah keras,* *Airnya datang terlalu deras,* *Mana yang kuat tahan berempas,* *Mana yang buruk hilanglah lalas,*	..... As the gloom grew, so did the waters' force. The tide did not follow its ordinary course, Only the strongest could stand their ground. The weak were soon lost, never to be found,
34	..... *Kepada masa Ketika itu,* *Datang gelombang bukan suatu,* *Banyaklah umat berhiru-hiru,* *Lari-lari seperti dikerjar hantu.*	..... But then there came again, in from the coast, Not just one wave, but a veritable host, Once again, an actual test of faith occurred, With people fleeing as if chased by a wraith.

Line of Poem	In *Jawi* script	In English
85	..... *Di kampung Teba hamba perikan,* *Tempatnya tinggi nyatalah, Tuan,* *Pergilah orang berkawan2,* *Membawa harta serta pakaian.*	..... Teba village, which I will now describe to thee, is far higher than the sea. So it was there that many did defect, With bundles of clothing and other effects.
86	*Ahli akal ada di situ,* *Orang mengikut katanya itu,* *Berkampung mereka banyak di situ, Laki2-perempuan ramailah tentu.*	A seer lived in Teba, it had been said, And so, people went there as the news spread. Many people came to his kampong, Both men and women, a great throng.
87	*Katanya: Hendak kemanalah pergi?* *Tempatnya kita sudahlah tinggi,* *Jikalau air sampai ke sini,* *Sudah kiamat isinya negeri.*	"There is no place to go," the seer did expound, "Where we are right now is very high ground, "if the waters should reach as far as this place, "It will be the end of the world, our very race."
222	*Rumah yang buruk habislah rubuh,* *Habis rusak ditimpa habu,* *Banyaklah mati orang di situ,* *Tiadalah tinggal barang suatu."*	Poorly built houses became piles of timber When felled by the weight of ashes and cinder. Within the houses, few people were alive, Tiny, if anything, could survive.

Based on table 1, two main points are stressed in the poem. The first is about how the process of the eruption. The second is about how the community survived the disaster. The process of the eruption starts with the ash in a range of three days, then continues to the repetitive explosion in a range of three months, after the massive explosion that affected the tsunami in around the mountain (especially with Lampung and Banten) and the ash entered the atmosphere and cooled the global climate for years.

Second, it is about how the community survived the disaster, such as to build robust infrastructure of the houses. Moreover, the communities need to reach a higher position to make a rescue from the high wave in the sea. These two central points will be discussed regarding the lesson learned to increase Environmental awareness.

3.3. The lesson teaches to increase Environmental awareness.

The lesson learned from the tale of Lampung Submerged to increase environmental awareness is concerned with two things. The first is about the construction of the building. Building construction is needed to improve based on the characteristics and aspects of the environment. For example, in the Baduy community, they build houses from wood, bamboo, palm fiber, and thatch that all have the same shape [22]. Baduy houses also believe that the house, as the focal point, represents a neutral power between the underworld and the world above.

Still related to the construction of the building, the communities must be knowledgeable about awareness of the resilience process. The knowledge is about building materials and practices that will significantly impact the cost, ability, and even possibility of homeowners rebuilding their homes [23]. The second learning is about a sense of belonging. This means the community around the volcano needs to adapt and survive the possibility of eruption and tsunami if the volcano erupts again. These processes are willing and responsible for making changes to the particular aspects of its culture that do not violate its core values. This process of sense of belonging has been implemented in Mount Merapi [24] and Mount Sinabung [25].

4. Conclusion

The article concluded that the lesson learned from the tale of Lampung Submerged is essential to face the potential eruption of the Anak Krakatau volcano. The novelty of this research is the increasing environmental awareness based on the tale or poem passed down from generation to generation. This environmental awareness includes how to identify the condition of the Anak Krakatau mountain before the big eruption happens, building a good infrastructure of the houses, and how to make a safe escape route if the eruption happens. These also will not disturb the core values of the environment itself.

Acknowledgments

This study was funded by the Ministry of Education, Culture, Research And Technology (KEMDIKBUDRISTEK) grant number [091/E5/PG.02.00.PT/2022 and 172/PKS/WRIII DRP/UI/2022] and grant number [NKB-1041/UN2.RST/HKP.05.00/2022], Research and Development (Risbang), Universitas Indonesia. We thank Sinan Vidi Lazuardi and Nova Amalia Sakina from Cluster of Interaction, Community Engagement & Social Environment, School of Environmental Science, Universitas Indonesia, who has helped with article editing and ethical clearance.

References

[1] Pratomo I and Abdurahman K 2004 Characteristics of The Indonesian Active Volcanoes and Their Hazards *Miner. Energi* **2** 56-60
[2] USGS 2022 Cascades Volcano Observatory
[3] Bemmelen R W 1949 *The Geology of Indonesia* (Martinus Nyhoff, Netherland: The Haque)
[4] Padang N Van 1951 *Catalogue of Volcanic Activity and Solfatara Fields* (International Volcanological Association)
[5] Kusumadinata K 1979 *Data Dasar Gunung Api Indonesia* (Bandung: Direktorat Vulkanologi)
[6] Simkin T and Fiske R S 1983 *Krakatau 1883: The volcanic Eruption and Its Effects* (Washington DC: Smithsonian Institution Press)
[7] Newhall C G and Dzurisin D 1988 *Historical Unrest at Large Calderas of The World* vol 1
[8] Sigurdsson H and Carey S 1989 Plinian and Co-Ignimbrite Tephra Fall From the 1815 Eruption of Tambora Volcano *Bull. Volcanol.* **51** 243-270
[9] Lipman P W 1997 Subsidence of Ash-Flow Calderas: Relation to Caldera Size and Magma-Chamber Geometry *Bull. Volcanol.* **59** 198–218
[10] Kartadinata M N, Budianto A, Wirakusumah A D and Hadisantono R D 1997 *Pyroclastic Flow Deposits Erupted by The 1815 Tambora; Distribution, Characteristic of The Deposits and Interpretation of The 1815 Eruption Mechanism*
[11] Sutawidjaja I S, Sigurdsson H, Rachmat H and Pratomo I 2005 *The Deadliest Volcanic Eruption of 1815 Tambora Volcano, Sumbawa, Indonesia* (Proc. Int. Semin. On Quart. Geol)
[12] Camus G and Vincent P M 1983 Discussion of a New Hypothesis for The Krakatau Volcanic Eruption in 1883 *J. Volcanol. Geothermal Res.* **19** 167-173
[13] Valentine G and Fisher R V 2000 *Pyroclastic Surges and Blasts In: H. Sigurdsson* (Academic Press)
[14] Alcántara-Ayala I 2002 Geomorphology, Natural Hazards, Vulnerability and Prevention of Natural Disasters in Developing Countries *Geomorphology* **47** 107-124
[15] Saleh M 2014 *Krakatau The Tale of Lampung Submerged (Syair Lampung Karam)* (Singapore: NUS Press)
[16] Ager A and Ager J 2015 Faith, Secularism, and Humanitarian Engagement: Finding the Place of Religion in the Support of Displaced Communities *Journal of Refugee Studies* (New York: Palgrave Macmillan)
[17] Bankoff G 2003 *Cultures of Disaster: Society and Natural Hazard in the Philippines* (London: Routledge)
[18] Mercer J and Kelman I 2010 Living Alongside a Volcano in Baliau, Papua New Guinea *Disaster Prev. Manag. An Int. J.* **19** 412-422

[19] Wilkinson O 2018 'Faith Can Come In, But Not Religion': Secularity and Its Effects on The Disaster Response to Typhoon Haiyan *Disasters* **42** 459-474

[20] Nations U 2015 *Sendai Framework for Disaster Risk Reduction 2015 – 2030* (Geneva: UNISDR The United Nations Office for Disaster Risk Reduction)

[21] World Humanitarian Summit 2015 *Restoring Humanity Global Voices Calling for Action: Synthesis of the Consultation Process for the World Humanitarian Summit* ed J Mahmood (New York: United Nations)

[22] Permana R C E, Nasution I P and Gunawijaya J 2011 Kearifan Lokal Tentang Mitigasi Bencana Pada Masyarakat Baduy *Makara Sos. Hum.* **15** 67-76

[23] Allen B L 2007 Environmental Justice, Local Knowledge, and After-Disaster Planning in New Orleans *Technol. Soc.* **29** 153-159

[24] Pawirodikromo W 2018 Achievements and Lesson Learned Assessment of The 2010 Merapi Eruption Disaster Management: An Initial Stage to Sustainable Volcano Disaster Risk Reduction *MATEC Web Conf.* **229**

[25] Mori M, McDermott R, Sagala S and Wulandari Y 2019 Sinabung Volcano: How Culture Shapes Community Resilience *Disaster Prev. Manag.* **28** 290-303

Determination Strategies for HDI Improvement Using SWOT Matrix Toward Sustainable Development and Good Environment

Nurul Listiyani[1], Rizka Zulfikar[2*], Achmad Jaelani[3] and Didi Susanto[4]
[1] Law Department, Faculty of Law, Islamic University of Kalimantan MAB Banjarmasin, South Kalimantan, Indonesia, 0000-0002-2572-3673
[2] Management Department, Economic Faculty, Islamic University of Kalimantan MAB Banjarmasin, South Kalimantan, Indonesia, 0000-0002-9257- 6781
[3] Animal Husbandry Department, Agricultural Faculty, Islamic University of Kalimantan MAB Banjarmasin, South Kalimantan, Indonesia, 0000- 0002-1908-5847
[4] Communication Department, Social and Politics Faculty, Islamic University of Kalimantan MAB Banjarmasin, South Kalimantan, Indonesia, ORCID ID : 0000- 0002-1908-5847
[1]nurullistiyani05@gmail.com, [2*]rizkazulfikar@gmail.com,
[3]ach_jaelaniborneo@yahoo.com, [4]didisusanto@fkip.uniska-bjm.ac.id

Abstract. The SWOT matrix is used in this study. This study aimed to determine strategies to improve the Human Development Index through the dimensions of education in Banjar Regency, South Kalimantan, Indonesia. The data used primarily comes from interviews with respondents who can represent a conclusion. The results of this study indicate that the SWOT matrix can be used to determine strategies to increase the Human Development Index in Banjar Regency is still very large, considering the program synergy carried out by the Education Office and the Ministry of Religion in the commission for education improvement with various policies including the implementation of PKBM and Equality Education in Islamic Boarding Schools. In order to increase the HDI of Banjar Regency through the education dimension, especially the indicator of the average length of schooling, and produce seven strategies that policymakers can use in Banjar Regency. The results of this study, the educational dimension is an essential factor that must be a concern in improving HDI. The education dimension is also part of the SDGs that the government strives to continue to implement in order to create human resources that have an environmental orientation.

1. Introduction

Economic development from a broad perspective is seen as a multidimensional process that includes various fundamental changes to the social structure, society's attitudes, and national institutions, and continuing to pursue the acceleration of economic growth, reducing income distribution inequality, and alleviating poverty [1]. One of the indicators of development progress is economic growth. This indicator measures a country's ability to develop.

Human development is defined as expanding more choices to the population through empowerment efforts that prioritize improving basic human capabilities to participate in all areas of development two fully. In principle, the primary purpose of this development is to increase the choice of the population where these options are not limited and can change at any time so that the population can get greater access to knowledge and education, better nutrition and health services, safe livelihoods, etcetera. The elements of human development expressly underline the goals to be achieved, namely healthy and long-lived, educated and able to enjoy life decently [2]. This means that human development aims to improve the welfare of society related to the quality of people and society. Therefore, humans are central to the development process [3].

One of the benchmarks for increasing public welfare is the *Human Development Index* (HDI) increase. HDI is an important indicator to measure success in efforts to build the quality of life of humans (society/ population). HDI can determine the ranking or level of development of a region/country. For Indonesia, HDI is a strategic data because apart from being a measure of government performance, HDI is also used as one of the allocators for determining the General Allocation Fund (DAU) [4]. A high HDI indicates high public welfare as well and can be interpreted as his government's success in developing and improving community welfare. HDI explains how residents can access development outcomes in income, health, education, etcetera. The United Nations Development Programme (UNDP) introduced the HDI in 1990, and it is published regularly in the annual Human Development Report (HDR). HDI is formed by three primary dimensions: longevity and healthy life, knowledge, and a decent standard of living. HDI is an important indicator to measure success in efforts to build the quality of life of humans (society/ population). Therefore, HDI can determine a region or country's ranking or level of development.

The current development paradigm is economic growth as measured by human development, seen by the level of quality of human life in each country. One of the benchmarks used in looking at the quality of human life is the Human Development Index (HDI) which is measured through the quality of education, health, and economic levels (purchasing power). Through the improvement of these three indicators, it is hoped that there will be an improvement in the quality of human life. This is because the heterogeneity of individuals, geographical disparities, and diverse social conditions of society cause the income level to no longer be the leading benchmark in calculating the success rate of development [5].

Placing human development as the ultimate goal of the development process is expected to create opportunities that directly contribute to efforts to expand and improve human capabilities and their quality of life, among others, through the improvement of health services, primary education, and social security [6]. As the development implementer, the government needs qualified human beings as the primary capital for development [7].

Humans, in their role, are subjects and objects of development, which means that humans, in addition to being development actors, are also development targets [8]. In this case, various facilities and infrastructure are needed to encourage the role of humans in development. Therefore, investment is needed to create productive human resources [9]. Government spending can be used as a reflection of policies taken by the government in an area. Government spending is used to finance critical public sectors, including investment in the education and health sectors, which are expected to improve human resources quality [4].

One form of successful development can be seen in an area's level of human development. Therefore, the government's seriousness is needed in dealing with the problem of improving human development [10]. One of the proofs of the seriousness of the government is through government spending, which is a reflection of concrete evidence of the government's role in regulating the economy. The government spending sector that is quite important and affects human development is expenditure on education and health [11].

Previously study showed that education factors, especially the average indicator of school length, were the dominant factor for determining strategies for increasing HDI in an area [12] and in the province of South Kalimantan, which is the location and focus of research, it showed that this indicator is an indicator that is still not of concern to local governments and so it should be sought that prioritizes increasing education sector [13]. HDI has a strong correlation with the environmental conditions of an area, where the higher the HDI indicator will show, the more advanced the area and the macro-environmental conditions will be better, and the higher the HDI, the average greenhouse emissions per capita will be more significant [14].

SWOT analysis is a straightforward tool that can help build and develop a strategy to achieve much better results[15,16]. SWOT is a term and abbreviation for Strength, Weakness, Opportunities, and Threats. SWOT is used as a benchmark for business people to look at the plan carefully, whether the plan can have a good impact in the short or long term. In simple terms, SWOT analysis means approaching several factors related to Strengths, Weaknesses, Opportunities, and Threats. If interpreted in more detail, SWOT analysis is a technique carried out for planning a strategy that can be used to evaluate [17,18]. In order to determine strategies that can improve the Human Development Index in Banjar Regency, the SWOT matrix is used as a strategy determination analysis tool.

2. Methods

This research was conducted in the province of South Kalimantan, starting from June 2021 – July 20, for one year. This research is descriptive quality research described in the form of S-O (Strength-Opportunity), W-O (weakness – Opportunity), S-T (Strength-Threats), and W-T (Weakness- Threats) strategies. The data used are primary data derived from the respondents' interviews that can represent concluding.

2.1 Qualitative Design

The design in this study is qualitative research with a descriptive approach because the data obtained qualitatively by the researcher is intended as secondary data that strengthens the findings of the quantitative data results so that what is obtained from quantitative research is strengthened and deepened by qualification analysis.

2.2 Validity of Qualitative Data

The secondary data assessment on a descriptive qualitative approach still pays attention to the technique of data validity through data triangulation, with source triangulation, namely multiplying data from various parties, which are considered able to explain the influence of education indicators as leverage for increasing the human development index (HDI) of Banjar Regency.

The qualitative analysis used in this study using the opinions of [10] suggests that analysis data (*data analysis*) consists of three interrelated sub-processes. Data reduction, data presentation, and conclusion/verification. Our Qualitative Instrumental Grid is shown in table 1.

Table 1. Research indicators and sub-indicators

Indicators	Sub Indicators
Factors Affecting HDI	a. Old School Expectations b. Expenditure Figures
HDI Improvement Strategy	a. Equitable Distribution of Education b. Quality and Competitiveness Improvement c. Strengthening Governance

The results of the qualitative analysis will then be plotted in the SWOT Analysis Matrix, which aims to formulate the results of the study by considering the Strengths, Weaknesses, Opportunities, and Threats presented in the form of a matrix (S-O, S-T, W-O, W-T *strategy*) which will then be used as a policy recommendation. The overview of the SWOT matrix is illustrated in table 2.

Table 2. SWOT Matrix Components

	Strength – S Strength	**Weakness – *W*** Debilitation
Opportunity – *O* Chance	**S-O Strategy** Using power to seize opportunities	**W-O Strategy** Overcoming weaknesses by seizing opportunities
Threat – *T* Threat	**S-T Strategy** Use force to avoid threats	**W-T Strategy** Minimize weaknesses and avoid threats

3. Results and Discussion

The average length of schooling (RLS) in Banjar Regency is still increasing; it is suspected that the most influential on RLS are Islamic Boarding Schools, while many Islamic boarding schools have been open to policies on education equity from 38 numbers. There are 26 Islamic boarding schools in Banjar

Regency that provide equal education, the rest are still in the process of being submitted, and others still need socialization.

From 2012 to 2022, the number of participants for equality education recorded by the Ministry of Religion of Banjar Regency was 18,210. There are around 7,438 students at the Ula, Wustha, and Ulya levels. Divided into 915 educational activities at the Ula level, 4,549 students at the Wustha level, and 1974 students and students at the Ulya level.

The percentage of students who take equivalence education is still not optimal when compared to the number of students, as many as 18,210 students compared to the number of equivalence students of 7,438 students, which is about 41% of students who have attended equality education. School Operational Fund Assistance (BOS) is often delayed, so it is feared that it will disrupt the fun of the learning process inequality education in Islamic boarding schools.

The School Operational Funds (BOS) received by Salafiyah Boarding Schools in remote areas, such as the Rantau Balai Kecamtan Aranio, have not been sufficient for the operational activities of the cottage, let alone coupled with the school's equity program activities. There has been continuous socialization and education from the Education Office through PKBM education and equality education conducted by the Ministry of Religion and socialization carried out by the "Pondok" so that people are aware of the importance of equality education. The SWOT analysis result was:

3.1 Strength
The resulting strength points are (a) appointment of P3K Teachers and contract workers following the needs of the district, (b) the Banjar School Santri Program gets two diplomas or known as (sajadah), and (c) incentives for equality of education teaching staff, (d) there are already 26 Islamic boarding schools that carry out equality education, (e) equality education is already included in the category of recipients of School Operational Assistance (BOS).

3.2 Weakness
The resulting weakness points are (a) the geographical condition of the Banjar district is extensive so that the distribution of schools is uneven in various sub-districts, *and* (b) there has not been an inclusion of students who go to school in the calculation of PBS, Banjar Regency, (c)There are still some sub-districts that do not have high schools (SMA), (d) Only 7 of the 26 cottages received financial assistance for the payment of teaching staff, (e) there is no regulation on the requirement that a public official must have formal education in Banjar district, (f) teachers who provide equality education are still paid far from what was expected, namely between IDR 200,000 to IDR 300,000 and paid within every six months, (g) teachers in some Salafiyah huts with an equalization system are still not following the field of only cottage graduates and with information that they have studied general lessons only and most of them are equivalent high school graduates.

3.3 Opportunity
The resulting opportunity points are (a) Pondok is increasingly open to local government policies with the implementation of PBM education and Equality Education, and (b) the number of recipients of school operational assistance (BOS) in equalization schools in Islamic boarding schools is increasing, (c) the number of huts that provide equality education in Banjar Regency will continue to increase, because some cottages still complement the requirements in the Ministry of Religious Affairs of Banjar Regency, (d) the law on Islamic boarding schools paves the way for huts to increase independence as an effort to take part not only in the field of religion but also in other public fields, (e) the cooperation between the local government and the community through religious leaders has been well established, (f) adding and providing unique locations for equalization schools in every sub-district that is prone to dropping out of school, for example, sub-districts that have been isolated.

3.4 Threats
The resulting threats points are (a) delays in the distribution of school operational assistance and incentives in Islamic boarding schools are often felt by cottage managers in implementing equality education, and (b) perceptions of the school community of equality change the educational traditions of Islamic boarding school, (c) people still feel that general education and equality are less critical than religious education in the cottage, (d) there are already 26 cottages that carry out equalization education,

but about only seven cottages have received the assistance of equalization teachers, (e) some Islamic boarding schools still rely on independent funding in implementing equality education, (f) there are still regions that until now have difficulty using the internet network, making it difficult to take the national equalization examination system online. Based on the results of the SWOT analysis above, a strategy can be formulated using a SWOT matrix, a combination of all components in the analysis shown in table 3.

Table 3. Strategy analysis results

		Strength – S		**Weakness – *W***
Opportunity – *O*		**S-O Strategy**		**W-O Strategy**
	1.	Increased education budget for incentives for equality education teaching staff.	1.	Policy/rulemaking on the conversion of subjects from trainees to formal schools.
	2.	Increased School Operational Fund (BOS) Assistance	2.	Data collection of the latest educational status of Islamic boarding school students
	3.	Determination of 12-Year Compulsory Education Program		
Threat – *T*		**S-T Strategy**		**W-T Strategy**
	4.	Integrated policymaking between the Education Office and the Ministry of Religious Affairs.	3.	Implementation of equalization activities that can be carried out with an online system.
			4.	Increased assistance for transitional scholarships between school levels, especially to underprivileged students

4. Conclusion

This research is a qualitative descriptive study that is described in the form of S-O (Strength-Opportunity), W-O (Weakness – Opportunity), S-T (Strength- Threatens), and W-T (Weakness-Threatens) strategies. Various obstacles and disturbances can be overcome by researchers with the help of various parties so that the implementation of the research cycle can run smoothly. The results of the study carried out resulted in several conclusions, including the following : (1) The opportunity to increase HDI through efforts to increase the average length of schooling in Banjar Regency is still huge, considering the synergy of programs carried out by the Education Office and the Ministry of Religion in the commission to improve education with various policies including the implementation of PKBM and Equality Education in Islamic Boarding Schools; (2) In order to increase the HDI of Banjar district through the education dimension, especially the average length of schooling indicator, the recommendations that can be given are (1) Increased education budget for incentives for equality education teaching staff and School Operational Fund (BOS) Assistance; (2) Determination of 12-Year Compulsory Education Regulation; (3) Policy/rulemaking on the conversion of subjects from boarding schools to formal schools by Designing a policy for converting subjects from entrepreneurs to formal schools; (4) Data collection of the latest educational status of Islamic boarding school students with digital data collection; (5) The integrated policy between the Education Office and the Ministry of Religious Affairs; (6) Implementation of equalization activities that can be carried out with an online system by Making guidelines for the implementation of online equalization activities; and (7) Increased assistance for transitional scholarships between school levels, especially to underprivileged students by Submission of a proposal to increase the education budget in the Regional Budget of Banjar Regency for transition between school levels and underprivileged students.

This study's limitation was using qualitative data, which is sourced from the results of observations and interviews, so that further research can be carried out using a mixed method that combines quantitative data and qualitative data.

References

[1] Todaro M P and Smith S C 2006 *Pembangunan Ekonomi* (Jakarta: Penerbit Erlangga)
[2] Haq M 1995 *Reflections on Human Development: How the Focus of Development Economics Shifted from National Income Accounting to People-centred Policies, Told by One of the Chief Architects of the New Paradigm* (Oxford University Press)
[3] UNDP 2000 *Human Development Report 2000* (New York)
[4] Alam H V, Baruwadi M H, Mopangga H, Tantawi R and Yulia Akib F H 2021 Human Capital Quality Development Strategy in Efforts To Improve Economic Growth in Boalemo District, Gorontalo Province . *Ilkogr. Online* **20** 296–312
[5] McGillivary Mark and White Howard 1993 Measuring Development ? The UNDP 's Human Development Index *J. Int. Dev.* **5** 183–92
[6] Rescue G 2021 Wealth-adjusted Human Development Index *J. Clean. Prod.* **318** 128587
[7] Türk U, Östh J, Toger M and Kourtit K 2021 Using individualised HDI measures for predicting educational performance of young students—A Swedish case study *Sustain.* **13** 1–13
[8] Khan N H, Ju Y and Hassan S T 2019 Investigating the determinants of human development index in Pakistan: an empirical analysis *Environ. Sci. Pollut. Res.* **26** 19294–304
[9] Indri Arrafi Juliannisa and Siswantini T 2019 Improving Human Development Index (HDI) by Illiteracy Eradication (Case Study in Cibadak Village) *J. Samudra Ekon. dan Bisnis* **11** 29–42
[10] Alam H V and Baruwadi M H 2020 The Role of Local Governments in Increasing the Human Development Index as a Barometer in Achieving SDGs in Boalemo District *Conf. Ser.* 817–25
[11] Desmaryani S 2021 Strategy To Improve Human Development Index In Education Author *Multidisciplinary Multidimential J.* **1**
[12] Baqtiar Arifin and Ahmadi Murjani 2017 The Impact Of Local Government Spending On Education, Health, And Poverty On The Human Development Index (Study On The Regencies/Cities In South Kalimantan) *J. Transform. Adm.* **7** 1352–68
[13] Listiyani N, Zulfikar R, Jaelani A and Susanto D The Years Of Schooling Contribution As Factors That Most Considered In Increasing Human Development Index 1599–606
[14] Ramani A 2014 Hubungan Indeks Pembangunana Manusia dengan Indikator Penyakit, Lingkungan dan Gizi Masyarakat *Ikesma* **10** 13–21
[15] Benzaghta M A, Elwalda A, Mousa M, Erkan I and Rahman M 2021 SWOT analysis applications: An integrative literature review *J. Glob. Bus. Insights* **6** 55–73
[16] Delu N, Weibin W, Xiyong H, Zhiyuan F and Ting M 2019 Development status and SWOT analysis of tree nuts industry in Yunnan province . 650201
[17] Quezada L E, Reinao E A, Palominos P I and Oddershede A M 2019 Measuring performance using SWOT analysis and balanced scorecard *Procedia Manuf.* **39** 786–93
[18] Bordoloi R 2021 Contribution Of Kkhsou To Human Resource Development : A Swot Analysis *Online J. Sci. Technol.* **11** 95–108

Performance of Green Environmental Governance at State University of Surabaya, Indonesia

Suci Megawati[1*], Raden Roro Nanik Setyowati[2], Revienda Anita Fitrie[1], Trenda Aktiva Oktariyanda[1], Jauhar Wahyuni[3], Meki Wastiti[1], Tamianingsih[1]

[1] Department of Public Administration, Faculty of Social Sciences and Law, Universitas Negeri Surabaya, Surabaya, Indonesia 60231. ORCID ID: 0000-0002-1798-6879

[2] Department of Pancasila and Citizenship Education, Faculty of Social Sciences and Law, Universitas Negeri Surabaya, Surabaya, Indonesia 60231

[3] Communication Department, Faculty of Social Sciences and Law, Universitas Negeri Surabaya, Surabaya, Indonesia 60231

[1*]sucimegawati@unesa.ac.id

Abstract. Sustainable development of the environment is one of the things that must be fulfilled, especially in the university environment, including at the State University of Surabaya. Through the eco-campus program, the State University of Surabaya (Unesa) participates in green environmental management that supports sustainable development. The purpose of this study was to describe the performance of green environmental governance at the State University of Surabaya. The research method uses a literature/library study approach. Data analysis methods with data minimization, data visualization, data analysis, data validation, and concluding. The results of the study stated that Unesa has six aspects that are of concern in the sustainability of environmental development: policies and infrastructure, energy and climate change, waste management, water management, transportation, and education. From the six aspects, Unesa has carried out activities that support energy saving and environmental awareness. The total final score for the assessment of the six aspects is 65.75% which is good but still needs to be improved to make it even better. In conclusion, Unesa already has activities and programs supporting sustainable development, but it still needs improvement to improve the environment.

1. Introduction

Environmental problems are the main problems of this century that must receive full attention. Scientists continue to warm to the notion that the Earth is entering (or has already entered) an epoch characterized by human-induced changes to Earth's surface condition, the "Anthropocene." On the other hand, more and more attention is being paid to the ecological consequences of uninterrupted population growth. Moreover, other consequences are associated with economic activities and human consumption patterns in an ever-industrializing world [1]. Ecological intelligence is humans' ability to adapt to the environment they are in to respond to various conditions around them [2]. Therefore, sensitivity, awareness, understanding, critical thinking, and environmental ethics are needed in the context of environmental competence [3]. Ecological intelligence is developed through education. Environmental awareness competence is needed to indicate students' awareness of a healthy environment [3]. Environmental problems require cooperation from all elements of society, including the academic community as an academic who is expected to provide solutions or solutions related to environmental problems. A growing interest in environmental issues has led to a research stream dedicated to university sustainability [4]. In line with environmental problems that demand solutions and solutions, there are various movements, programs, and policies on dealing with environmental problems, including programs being developed by universities, namely green campuses or eco campuses.

In particular, the eco-campus represents an opportunity to guide the implementation of innovative environmental measures in areas such as public transit, heating, design, and construction [5]. The eco campus program believes the campus should be clean, shady, and healthy. As a result of the increasing awareness of sustainability and environmental concerns, universities are now a part of environmental sustainability not only through research but also through improving their campus infrastructure into a more environment-friendly setting, as well as updating their curricula to cover courses on the environment and sustainability [6]. As a campus that received an eco campus award from the Mayor of Surabaya in 2016, the State University of Surabaya (Unesa) continues to be committed to realizing a sustainable and comprehensive eco campus. The Unesa eco campus has programs already running to support the go green-based Unesa campus. Among them are establishing a waste bank, saving electricity and water, constructing a small reservoir in the Lidah Wetan campus area as a rainwater reservoir, and forming an eco-campus team and its divisions. Furthermore, to support the eco-campus program, Unesa also received assistance in the form of plants and trees from the Surabaya City Environment Agency [6].

Unesa's eco-campus program has seen how campuses can effectively and efficiently utilize available resources and infrastructure, such as managing transportation, using electricity, processing waste, and implementing lifestyles. Appropriate measurements and assessments are needed to maintain the sustainability of the eco-campus program at Unesa to find out what things need to be improved and improved. This study will describe the performance of green environmental governance at the State University of Surabaya, Indonesia.

2. Methods

This research used a literature or library study approach [7]. The data source will be collected from relevant information on the eco campus program at the State University of Surabaya through scientific books, journals, research reports, regulations, and encyclopedias, both printed and electronic. The data analysis method for this research is analytical at the time of data collection, capturing the essence of the research aims through the collected sources. Then re-analyze the collected data as raw data that need to be identified with each other. Analytical activities include data minimization, visualization, validation, and conclusion.

3. Results and Discussion

As a university that cares about sustainable development, Unesa has an eco-campus program with several aspects of concern. These aspects include (1) policies and infrastructure, (2) energy and climate change, (3) waste management, (4) water management, (5) transportation, and (6) education. This aspect measures the achievement index of sustainable development at Unesa. The results of the measurement of each aspect are listed in Table 1.

Unesa is still ranked 313 out of 956 universities in the world related to eco-green campuses as a commitment to overcome environmental problems and impacts [8]. Dari penilaian yang dilakukan oleh UI Green Matric, untuk kebijakan dan sarana prasarasna State University of Surabaya mencapai poin 875 yakni 58,33% dari maximum point 1500. Pada segi energi dan perubahan iklim, State University of Surabaya mencapai poin 1450, yakni 69.05% dari 2100. Pada segi waste management, State University of Surabaya mencapai poin 900, yakni 50% dari maximum point 1800. Pada segi water management, State University of Surabaya mencapai poin 800, yakni 80% dari maximum point 1000. Pada segi transportation, State University of Surabaya mencapai poin 1250, yakni 69.44% dari maximum point 1800. Pada segi education, State University of Surabaya mencapai poin 1300, yakni 72.22% dari maximum point 1800. Sehingga State University of Surabaya secara keseluruhan mencapai poin 6575, yakni 65.75% dari maximum point 10000.

Table 1. Unesa's green environmental governance performance in 2021 [9]

Category	Point	% Achievements	MaximumPoint
Policy and Infrastructure	875	58.33%	1500
Energy and Climate Change	1450	69.05%	2100
Waste Management	900	50.00%	1800
Water Management	800	80.00%	1000
Transportation	1250	69.44%	1800
Education	1300	72.22%	1800
Total Score	6575	65.75%	10000

3.1. Policy and Infrastructure

Unesa campus management provides infrastructure that supports the implementation of the eco campus in the university environment. Facilities and infrastructure can be understood as all the supporting tools needed in carrying out a process and activity. Eco campus facilities and infrastructure provided by Unesa management include the availability of trash bins. The available trash cans are almost evenly distributed throughout the building and have been provided in 2 types, namely organic and inorganic waste bins. In addition to the availability of trash cans, Unesa's eco campus facilities can also be found in its green open space, which provides a gazebo for students to gather and relax while doing assignments or just discussing.

Eco campus infrastructure in Unesa includes green open fields, mango gardens, open gardens in each faculty, and a recreation park named the Independent Learning Laboratory at Unesa Lidah Wetan. Besides functioning as a green open space, the park also functions as a comfortable place for interaction and expression for faculty members. Even though it has provided facilities and infrastructure that support the implementation of the eco campus, the ideal achievement score index is still at 875 per 1500, or 58.33%. This can be seen based on findings in the field, which show that there are still things that need to be improved.

Figure 1. Unesa Lab Merdeka Belajar Facilities [10].

3.2. Energy and Climate Change

Climate change is driven by the dominant use of natural energy, such as fossils, which causes greenhouse gas emissions. Global warming and climate change's impact on various sectors naturally prove the exploration of unsynchronized development and un-environment-friendly resources [11]. Renewable energy and climate change are interrelated. Unesa has been committed to saving and using renewable energy, which is known to come from the sun, water, and wind.

Solar panels are one of Unesa's efforts to save energy in line with the existing climate change and require renewable energy. The use of solar panels in Unesa reaches 10% of the total energy used, both in offices and for street lighting [12]. In addition, this figure was also achieved based on the campus response to climate change in terms of other things such as reforestation, water recycling, paperless policies, no plastic policies when shopping at university convenience stores, as well as policies on saving the use of Air Conditioners and LED lights in the campus environment. The index for measuring green environmental governance based on energy and climate change at Unesa reached 1450 out of 2100 points, or 69.05%. Even though it is more than 50%, there is still a need for improvement in the application of efficiency and energy, for example, by holding the implementation of equipment with procurement policies tailored to sustainable development goals.

Figure 2. Solar Panels in the Unesa Environment, research documentation (2022)

3.3. Waste Management

Waste management is a mandatory thing that universities must do to support sustainable development [13]. Unesa has innovatively managed organic waste into compost with the Unesa Compost House. The Unesa Compost House was established in 2011 and consistently processes organic waste consisting of leaves, food waste, and perishable grass using aerobic methods.

In addition to compost processing, Unesa's efforts and commitment to reducing the use of plastic and paper are carried out as follows: a). Enforce an E-office or paperless policy by utilizing the website and campus applications for several administrative arrangements; b). Collection of student assignments and reports in the form of soft files, c). Innovation to turn plastic waste into valuable items such as recycling banner waste into tiktokers bags with various functions.

The green environmental management measurement index based on waste management aspects at Unesa reached 900 out of 1800, or as much as 50%. This indicates that things still need to be improved, such as the paperless policy, which has not been implemented thoroughly at the faculty and department levels. Several files and document management still require direct interaction and hardcopy, mainly in document management administration.

3.4. Water Management

Unesa also implements water conservation, water recycling, water efficiency programs, and drinking water production as part of a sustainable program related to water issues. Unesa supports a zero delta-Q policy for each drainage to reduce surface loss by providing several ponds equipped with door facilities and pump houses. This is necessary to maintain the surface and surface layer of the drainage system to optimize its function.

Water quality from existing ponds is maintained by keeping fish and regular water checks. The campus forest can objectively be seen as an effort to maintain green open space and is an infiltration area to increase the capacity of groundwater as an effort to conserve water within the campus area. This effort has conserved at least 50% to 75% of water.

Approximately 50% of the water within the University has been recycled and used to water plants in the forest area of the campus and around the campus. The recycled water is also used to water the sports field in the campus area. In terms of water efficiency, 50% of all buildings in Unesa have installed tools for efficient water use, including tap water and flushing in toilets. In addition, Unesa's drinking water production is carried out by providing packaged drinks and gallons of water treatment in the campus environment. The Unesa water treatment index score was satisfactory at 80.00%.

Figure 3. Unesa Drinking Water Factory, research documentation (2022)

3.5. Transportation
Campus transportation is needed to support the high mobility of lecturers and students. Along with the high mobility, the campus transportation mode must be environmentally friendly to support the reduction of emissions from the gas produced. Some of the steps taken by Unesa in reducing gas emissions are as follows: a). Provide campus bicycle facilities near the parking area, b). Providing bus facilities with large capacity, c). Providing minibus facilities, d). Provides campus bus and minibus services equipped with bus schedules, and e). Providing integrated bus stops with two to four parking spaces with sidewalks [12].

The Unesa Lidah Wetan bus stop is also integrated with Trans Semanggi Suroboyo, which has the Unesa-ITS route so that it can be used by campus residents and the public who want to visit from or to Unesa. The green environmental governance measurement index based on the transportation aspect at Unesa is at 69.44%, 1250 out of 1800 points. There are still students who do not understand how to access and get transportation services at Unesa. Therefore there is still a need for more optimal socialization.

3.6. Education
Unesa is committed to supporting sustainable development in the realm of environmental governance. In terms of education, Unesa offers 62 courses with themes such as education, transportation, environment, buildings, and the natural environment. The courses are spread across nine faculties and include 1,278 courses students can take. Some courses are mandatory in specific courses. Courses that support sustainability can be accessed through http://siakadu.unesa.ac.id/prodi_[12]. In the future, local wisdom of environmental management can be used as input into effective, efficient, fair, and effective forestry formulation policies to tackle deforestation [14]. In line with that, the character of caring for the environment at the University becomes an important thing to develop.

To support the sustainability of research and education, the funds dedicated to supporting sustainable research reached 18500.3 US Dollars. The average total scientific publications on environmental, economic, and social management from 2019 to 2021 is nine publications. The green environmental governance measurement index based on the education aspect at Unesa reached 1300 out of 1800, 72.22% [12].

4. Conclusion
As a university that supports sustainable development programs in environmental management, Unesa has at least six aspects that require attention and are used as measurements of environmental governance, including: (1) policies and infrastructure; (2) energy and climate change; (3) waste management; (4) water management; (5) transportation; and (6) education. The total score in Unesa's green environmental governance performance measurement index in 2021 shows 65.75%, which Unesa already has activities

and programs supporting sustainable development. Even though it is at more than 50%, it is still necessary to improve environmental governance by providing more massive socialization to campus residents so that all parties can carry out their respective roles to realize sustainable development in the field of a better environment. Green environmental governance is still possible to continue to be researched. This study's limitation is examining green environmental governance, which only focuses on the State University of Surabaya. This paper opens opportunities to be continued, such as discussing the success of the campus eco concept nationally and internationally. In addition, it can also be developed by comparing success with relying on developing concepts.

References

[1] Mohammadalizadehkorde M and Weaver R 2018 Universities as Models of Sustainable Energy-consuming Communities? Review of Selected Literature *Sustain.* **10** 1–17
[2] Goleman D 2010 *Ecological Intelligence. Kecerdasan Ekologis: Mengungkap Rahasia di Balik Produk-Produk yang Kita Beli* (Jakarta: PT Gramedia Pustaka Utama)
[3] Megawati S, Niswah F, Mahdiannur M A and Segara N B 2022 A-G-I-L Scheme as Social System to Build Characters Caring for the Environment Through the Adiwiyata Program *IOP Conf. Ser. Earth Environ. Sci.* **950**
[4] Statistik B P 2019 *Lingkungan hidup indonesia 2019* (Jakarta)
[5] Finlay J and Massey J 2012 Eco-campus: Applying the Ecocity Model to Develop Green University and College Campuses *Int. J. Sustain. High. Educ.* **13** 150–65
[6] Atici K B, Yasayacak G, Yildiz Y and Ulucan A 2021 Green University and Academic Performance: An Empirical Study on UI GreenMetric and World University Rankings *J. Clean. Prod.* **291**
[7] Snyder H 2019 Literature Review as A Research Methodology: An Overview And Guidelines *J. Bus. Res.* **104** 333–9
[8] Anon UI Green Metric 2021 Overall Rankings 2021
[9] Redaksi Unesa Jadi Contoh Eco Campus di Surabaya
[10] Anon Official Unesa 2021 Profil Lab Merdeka Belajar Unesa
[11] Anis M, Afiff A Z, Kiswanto G, Suwartha N, and Sari R F 2018 Managing University Landscape and Infrastructure Towards Green and Sustainable Campus *E3S Web Conf.* **48** 1–6
[12] Surabaya S U of 2021 *2021 Sustainability Report*
[13] Zhang N, Williams I D, Kemp S and Smith N F 2011 Greening Academia: Developing Sustainable Waste Management at Higher Education Institutions *Waste Manag.* **31** 1606–16
[14] Megawati S and Mahdiannur M A 2021 Implementation of Forest Conservation Policies based on Local Wisdom of the Ammatoa Kajang Indigenous Community *IOP Conf. Ser. Earth Environ. Sci.* **940**

Stakeholder Roles on Tourism Management, Environment Protection, and Conservation in Post Covid-19

D Handayani[1*] and S V Lazuardi[2]

[1]Faculty of Economics and Business, Universitas Indonesia, Depok, 16424, Indonesia, 0000-0002-1021-9677
[2]Department of Sociology, Faculty of Social and Political Sciences, Universitas Indonesia, Depok, 16424, Indonesia, 0000-0002-7770-8459
[1*]dwiniarianto@yahoo.com, [2]sinan.laz1999@gmail.com

Abstract. The tourism sector has a strong attraction to visitors. The increasing number of visitors encourages the creation of sustainable tourism. Sustainability is strongly related to stakeholders, so it can not be excluded. This study aims to analyze stakeholder roles in environmental management in the tourism sector. This study uses a literature review as the method. This study shows that stakeholders with various roles supported the tourism sector. Visitors are the critical actor in environmental protection, and the community was another actor with support from the government to the tourism business actors. Various stakeholder interests and priorities were not obstacles to the management. To unite the goals and vision, it needs commitment and cooperation to reach environmental protection.

1. Introduction

Tourism has an essential role in economic growth in each territory. Tourism has various impacts by providing job opportunities, welfare increment, and helping build infrastructure [1]. This impact makes tourism mentioned as a transformation agent to a resilient economy [2]. However, another perception shows that tourism is one of many factors contributing to environmental degradation [3]. Previous research shows waste issues, pollution, declining ecosystem, and biodiversity loss [1,4]. The tourism sector continues to grow until this day. However, overtourism [5,6] is still a problematic issue and endangering the sustainability of tourism. Also, anthropogenic activities have been pressuring the tourism sector. This issue implies that the tourism sector needs to elaborate on the environmental aspects as a solution. The negative impacts caused by tourism should be well managed to reach sustainable tourism.

Sustainable tourism is still not practical because many managements are only for a short-term goal [7]. Also, not only management aspects but lack of stakeholder involvement in tourism management contributes to ineffective sustainable tourism. The problem in reaching sustainable tourism can be seen in Gili Trawangan that the Gili EcoTrust dealing with sustainable development issues must involve many stakeholders [8]. However, it is not an easy task to reach sustainable tourism. There is some previous research on conservation and tourism [1,9], tourism and stakeholder [10,11], and sustainable tourism [5,7,12]. It needs a long-term goal by incorporating dynamics in environmental aspects and the dynamic interaction between actor and environment.

This study aims to analyze the role of stakeholders in environmental protection by explaining the role of stakeholders, what stakeholders will contribute to environmental protection, and the motivation of stakeholders. This study is a process of building communication and cooperation by benefitting stakeholder roles in environmental protection to reduce environmental damage on reaching sustainable tourism in environmental management. In reaching sustainable tourism that protects the ecosystem, it should not only focus on the environmental impacts, but it needs to facilitate economic growth and social welfare [13].

2. Method

This study is qualitative research by using a literature review from journal articles. The keywords used in the literature review were "stakeholder AND environmental protection" and "stakeholder AND sustainable tourism." The literature will be analyzed qualitatively.

The data that was collected ranged from 2015-2022. The articles were collected from journal databases such as Web of Science and Scopus. The collected articles are read and selected according to the topic of this study and focus only on tourism and environmental sciences. Also, the article was selected according to the appropriate substantive, so the articles that were selected will be the article that is appropriate to the study's topic and answer the aim of this study.

3. Result and Discussion

3.1. Various Stakeholders in Tourism Sector

The tourism sector has complicated networking because it relies not only on the industry but also on society, visitors, and local people, so it can not stand alone. Also, various stakeholder involvement contributes to the problem. However, there are some benefits to the environment in the tourism sector. Previous research denies the negative impact on the environment caused by tourism [5]. The tourism sector's development can also be an alternative to environmental protection aside from economic and social benefits. This can be happening because the awareness of environmental conditions caused various efforts to protect the environment, even though anthropogenic behavior can not be excluded from resource utilization. The development of tourism also encourages the interest in more sustainable management.

Various stakeholder also facilitates the development of tourism [4]. Every stakeholder should be responsible for pro-environmental behavior to reach sustainable tourism [3]. According to the table below (Table 1), it can be concluded that the stakeholders that were involved in the tourism sector are visitors, business actors, local people/communities, non-governmental organizations, and the government.

Table 1. Stakeholder roles in environmental protection

Stakeholder	Roles	Reference
Visitors	Willingness to pay Pro-environmental behavior: no littering and not destructing the environment	[14]
Community	Environment protection	[15]
Business Actors	Nature Resource Management Environmental Conservation Controlling Environmental Damage	[16]
NGO	Mentoring and Empowerment	[17]
Government	Policy creator Facilitator	[1,4]

3.1.1 The Visitors

The visitors are not only deemed a consumer, but it is one of stakeholders [4]. The visitors are one of the parties that affect tourism conditions, but sometimes their role is abandoned in planning and implementation in developing tourism. The increase in visitors was developed by giving knowledge because a lack of visitor knowledge makes visitors not have environmental awareness.

The smart visitor is becoming a new concept to introduce in reaching sustainable tourism. Visitors can be doing pro-environmental behavior by reducing their footprint [3]. Visitors can participate in the management by participating in environmental protection activities. Participation in sustainable tourism is crucial because visitors strongly tend to contribute to the management [14]. The involvement of visitors in management can encourage other parties to empower. Other literature mentioned that visitors could give feedback as information [7]. The feedback from visitors can be used as a reference for understanding the visitor's attitude and preference for sustainable management.

3.1.2 Community

The community was transforming from beneficiaries to one stakeholder in the tourism sector. However, involving the community participating in the management can be used community-based management concept [18]. Involving the community to participate in management can make the community feel rewarded aside from encouraging pro-environmental behavior [19]. Also, the involvement of communities can encourage communities as environment protection actors [20]. Involving the communities also can empower adat people by giving the capacity to the management, increasing the welfare and the environmental sustainability. This involvement has become necessary because tourism has reduced community degradation [21]. Local communities have local wisdom, and their management practices are more environmentally friendly because they come from their ancestors [15].

However, environmental management by communities has barriers [7]. Lack of knowledge of the environment makes the behavior not environmentally friendly. This issue can be solved by community empowerment. It needs community empowerment through education, involving the communities to develop the tourism sector's planning, implementation, and decision-making process [7].

The development of tourism based on the community can be reached by developing ecotourism [12,15]. This concept can empower communities to participate and protect environmental sustainability. Also, it has possibilities for job opportunities that can increase development and social welfare. This consent can be one of the alternatives for managing sustainable tourism because it benefits on environment and tourism sustainability. This concept has been supported by previous research on how the community protects its environment by its voluntary act of observing and collecting data on the environment [17].

3.1.3 Business Actors

Industry in the development aspect can implement pro-environment behavior by managing resources, protecting the environment, and observing tourism. Industry actors can contribute to sustainable tourism by providing waste management infrastructure in tourist places, encouraging visitors to pro-environmental behavior, reducing waste, increasing recycling, and using environmentally friendly products [16].

3.1.4. Government

To reach sustainable tourism, the government can be contributed by providing and enforcing the regulation on tourism development [4]. This effort can make the industry actors protect the environment from their economic activities. The government can facilitate empowerment, increasing pro-environmental behavior, giving training, and raising awareness of communities.

3.2. *Opportunities and Barriers to Reaching Sustainable Tourism.*

Environmental behavior is complex because it is interdisciplinary by integrating social and natural sciences [9]. The behavior on environmental responsibility was according to the attitude to reduce the anthropogenic impacts [3]. This behavior is also according to pro-environment behavior. To reach sustainable tourism, investment is essential from the community or individual [5]. It needs to know about how individual's perspective on sustainability issues. However, there are many factors on pro-environment behavior that stakeholders did. Stakeholder perception affects pro-environment behavior from socio-economic factors such as education, income, and socio-cultural factors [10]. Cultural factors can be seen in the individual's perception of the environment. The knowledge can be used as a reference for the attitude towards the environment. Also, cultural knowledge can be obtained from experience, especially in their close relation with tourist places can give more cultural knowledge. This cultural knowledge can be used for understanding environmental issues. Other research supports this notion by stating that stakeholder participation in environmental protection was affected by close relations with tourist places [9,14,17]. Also, environmental behavior was affected by environmental issues [14]. According to the table below (Table 2.), there are barriers and solutions to reaching sustainable tourism.

Table 2. Barriers and Solutions to Reaching Sustainable Tourism

Barriers	Solution	Reference
Duality value of Pro-environment behavior	Integrating socio-economic and environmental aspects to managing the tourism	[9]
	Waste management can be one example of environmental protection and adding more revenue.	[3,9]
Different interests and motives by stakeholders.	Knowing each stakeholder's needs and value	[5]
	Increasing collaboration among each stakeholder to reach a common goal	[17]
Inequalities between stakeholders	Empowerment	[11,18]
Lack of Synergy between stakeholders	Emphasize communication and coordination between stakeholders.	[22]

Pro-environmental behavior has become a duality value [9]. This can happen because stakeholders have a dilemma between tourism development and environmental protection. The sustainability concept has been integrating socio-economic and environmental aspects. This integration of these aspects has many benefits; Economic aspects are the primary stimulus for communities to involve in environmental protection so people can be more ecocentric [9]. This can happen through community participation in wildlife conservation to attract more visitors and increase revenue. In the literature, the effort in waste management in tourism can give the managers more revenue [3,9]. It can be implied that economic and environmental aspects have interactions that can be dependent on each other. This interaction can explain that environment-friendly activities can increase economic revenue and vice versa. This can happen because stakeholders depend on the economic benefit and the incentives.

However, multi-stakeholder approaches have many challenges. Various ideas need mutual agreement on the implementation because there are structural difficulties in management and operational aspects [12]. Every stakeholder has a different orientation toward sustainability, so it needs mutual agreement [23]. The difference between ideas can be problematic in facing conflicts [17]. Making a shared vision can become the biggest obstacle to reaching sustainability. To create a shared vision, there are some strategies for knowing the needs and values of each stakeholder [5]. The collaboration between stakeholders can give various ideas on problem-solving. This can happen through collaboration and learning between stakeholders by facilitating ideas and supporting each other in reaching common goals [17].

It should be noted that every stakeholder has rights in planning, implementation, and decision making, has an equal position and should be active in the participation. To make stakeholders participate, it needs to be empowered to make the stakeholder equal, especially in the communities. The inequality between stakeholders implies the not optimum tourism management because some stakeholders are not well represented. Empowerment can be a solution because it involves local stakeholders and can increase the capacity of the communities [11,18]. It can happen because empowering means giving chances and priority to community interests. However, the barrier to empowerment is a lack of continuity [24].

Stakeholders have dynamic behavior [13]. This means that their priorities for development constantly change, especially in unstable environmental conditions. Also, it makes the understanding of the environmental condition crucial. It needs coordinated and sustainable observation by stakeholders to reach effective management. Such wildlife conservation needs environmental sustainability, but the visitors must be limited by observing the environment. Appropriate regulation is also needed to protect the tourism sectors and the environment. Tourism development was recommended for more qualitative development, not only focusing on quantity so it can protect the ecology [5]. Inclusive tourism is just one solution to reaching sustainable tourism. There are other variables, such as stakeholders' participation and commitment to the limits on visitors not reducing the revenue.

Reaching sustainable tourism is not about identifying stakeholder and their roles. It is about how stakeholders can work together to understand the roles and motives to reach sustainable tourism as their common goal [25]. This solution must emphasize communication and teamwork for long-term goals [22]. The relationship between stakeholders can be maintained by involving the representative of each stakeholder in the discussion. However, the lack of synergy between stakeholders became a barrier [12]. This can happen because there is unequal justice in the relationship between stakeholders [11]. Tourism development to reach sustainability should coordinate ecology, socio-cultural, and economic aspects [7]. This effort is supported by arguments that sustainable goals can be reached when there are transparency and communication between stakeholder [4]. Collaboration between stakeholders was proven to help develop sustainable tourism and protect biodiversity because the barriers to reaching sustainable tourism were connected.

4. Conclusion

This study concluded that tourism sectors need involvement from various stakeholders. All stakeholders have different interests, goals, and roles. It needs commitment and teamwork to reach sustainable tourism and protect the environment. Communication and equality are the main factors in reaching sustainability. However, there are many challenges to reaching sustainability. To reach the synergies of each stakeholder, it needs to solve the duality, different motives, and inequality issues. This study is limited to identifying tourism stakeholders' roles, opportunities, barriers, and solutions to reaching sustainability in general. Future research can focus on implementing a multi-stakeholder approach, so the community and visitors can reach optimum environmental protection.

Acknowledgment

This study was funded by the Ministry of Education, Culture, Research And Technology (KEMDIKBUDRISTEK) grant number [021/E5/PG.02.00.PT/2022] and grant number [NKB-774/UN2.RST/HKP.05.00/2022], Research and Development (Risbang), Universitas Indonesia.

References

[1] Hanafiah M H, Jamaluddin M R and Kunjuraman V 2021 Qualitative assessment of stakeholders and visitors perceptions towards coastal tourism development at Teluk Kemang, port Dickson, Malaysia *J. Outdoor Recreat. Tour.* **35** 100389

[2] Athletic I 2020 Transforming the (tourism) world for good and (re)generating the potential 'new normal' *Tour. Geogr.* **22** 467–75

[3] Fenestra R M, Premananto G C, Sedera R M H, Abbas A and Laila N 2022 Environmentally responsible behavior and Knowledge-Belief-Norm in the tourism context: The moderating role of types of destinations *Int. J. Geoheritage Park.* **10** 273–88

[4] Roxas F M Y, Rivera J P R and Gutierrez E L M 2020 Mapping stakeholders' roles in governing sustainable tourism destinations *J. Hosp. Tour. Manag.* **45** 387–98

[5] Burbano D V., Valdivieso J C, Izurieta J C, Meredith T C and Ferri D Q 2022 "Rethink and reset" tourism in the Galapagos Islands: Stakeholders' views on the sustainability of tourism

development *Ann. Tour. Res. Empir. Insights* **3** 100057
[6] Cahyadi H S and Newsome D 2021 The post COVID-19 tourism dilemma for geoparks in Indonesia *Int. J. Geoheritage Park.* **9** 199–211
[7] Zhang X, Zhong L and Yu H 2022 Sustainability assessment of tourism in protected areas: A relational perspective *Glob. Ecol. Conserv.* **35**
[8] Partelow S and Nelson K 2020 Social networks, collective action and the evolution of governance for sustainable tourism on the Gili Islands, Indonesia *Mar. Policy* **112** 103220
[9] Imran S, Alam K and Beaumont N 2014 Environmental orientations and environmental behaviour: Perceptions of protected area tourism stakeholders *Tour. Manag.* **40** 290–9
[10] Bekele K, EW Linders T, Eschen R, Shiferaw H, Haji J, Legesse B, Choge S, Eckert S, Rima Mbaabu P and Schaffner U 2022 How well do local stakeholders' perceptions of environmental impacts of an invasive alien plant species relate to ecological data? *Ecol. Indic.* **137**
[11] Shafieisabet N and Haratifard S 2020 The empowerment of local tourism stakeholders and their perceived environmental effects for participation in sustainable development of tourism *J. Hosp. Tour. Manag.* **45** 486–98
[12] Su M M, Wall G and Ma Z 2014 Assessing Ecotourism from a Multi-stakeholder Perspective: Xingkai Lake National Nature Reserve, China *Environ. Manage.* **54** 1190–207
[13] Bu Y, Wang E and Yu Y 2021 Analysis on asymptotic stabilization of eco-compensation program for forest ecotourism stakeholders *Environ. Sci. Pollut. Res.* **28** 29304–20
[14] Chen J S 2015 Tourism stakeholders attitudes toward sustainable development: A case in the Arctic *J. Retail. Consum. Serv.* **22** 225–30
[15] Prasetyo N, Carr A and Filep S 2020 Indigenous Knowledge in Marine Ecotourism Development: The Case of Sasi Laut, Misool, Indonesia *Tour. Plan. Dev.* **17** 46–61
[16] Qian X and Schneider I E 2016 Waste minimization practices among tourism businesses : A multi-year comparison ☆ *TMP* **19** 19–23
[17] Webster S E, Donovan E C, Chudoba E, Miller Hesed C D, Paolisso M and Dennison W C 2022 Identifying and harmonizing the priorities of stakeholders in the Chesapeake Bay environmental monitoring community *Curr. Res. Environ. Sustain.* **4** 100155
[18] Dodds R, Ali A and Galaski K 2018 Mobilizing knowledge: determining key elements for success and pitfalls in developing community-based tourism *Curr. Issues Tour.* **21** 1547–68
[19] Mendoza-Moreno J, Cruz-Coria E and González-Cruz T F 2021 Socio-technical innovation in community-based tourism organizations: A proposal for local development *Technol. Forecast. Soc. Change* **171**
[20] Izurieta G, Torres A, Patiño J, Vasco C, Vasseur L, Reyes H and Torres B 2021 Exploring community and key stakeholders' perception of scientific tourism as a strategy to achieve SDGs in the Ecuadorian Amazon *Tour. Manag. Perspect.* **39**
[21] Jørgensen M T, Hansen A V, Sørensen F, Fuglsang L, Sundbo J and Jensen J F 2021 Collective tourism social entrepreneurship: A means for community mobilization and social transformation *Ann. Tour. Res.* **88** 103171
[22] Obradović S, Stojanović V, Kovačić S, Jovanovic T, Pantelić M and Vujičić M 2021 Assessment of residents' attitudes toward sustainable tourism development - A case study of Bačko Podunavlje Biosphere Reserve, Serbia *J. Outdoor Recreat. Tour.* **35**
[23] Sørensen F and Grindsted T S 2021 Sustainability approaches and nature tourism development *Ann. Tour. Res.* **91**
[24] López-Guzmán T, Borges O and Castillo-Canalejo A M 2011 Community-based tourism in Cape Verde - a case study *Tour. Hosp. Manag.* **17** 35–44
[25] Boluk K A and Rasoolimanesh S M 2022 Introduction to the special issue on "Deepening our understandings of the roles and responsibilities of the tourism industry towards the United Nations Sustainable Development Goals (SDGs)" *Tour. Manag. Perspect.* **41** 100944

Preventing Fuel Station Accidents: The Importance of Community Involvement

Aryo Wibowo[1], Fatma Lestari[2*], Robiana Modjo[3]
[1]Faculty of Public Health, Universitas Indonesia, Depok, Jawa Barat 16424, Indonesia,
[2]Faculty of Public Health, Universitas Indonesia, Depok, Jawa Barat 16424, Indonesia 0000-0002-1389-4360,
[3]Faculty of Public Health, Universitas Indonesia, Depok, Jawa Barat 16424, Indonesia 0000-0002-4435-1473
[1]aryo.wibowo@ui.ac.id, [2*]fatma@ui.ac.id, [3]bian@ui.ac.id

Abstract. From 2016-2018 there were 120 fuel station accidents in Indonesia. Fuel station accidents may harm not only workers but also the public. Major accidents in fuel stations have caused multiple fatalities and injuries. Various research regarding fuel stations has been done, but research on the role of the community in preventing fuel station accidents is limited. Therefore, studying the cause and effect of fuel station accidents is essential. The study aims to collect accident information and determine the impact, cause, and possible prevention programs for fuel station accidents. Study methods use data collection and classification, followed by qualitative analysis. Data were collected from the top 10 national news portals and tabulated in a structured manner. Twenty-nine fuel station accidents were identified and evaluated. These data were then categorized and analyzed to obtain the dominant effects of accidents. Accident causes were also investigated to provide an understanding and propose accident prevention programs adopted from previous research in various fields. It is concluded that fuel station accidents in Indonesia are dominated by vehicle fires, which involve consumer vehicles. The causes of accidents vary, but most of them were concerning the technical condition of vehicles, and a few were unsafe operations. To prevent future accidents, it is essential to ensure compliance with regulations and standards of facilities and equipment. Furthermore, community involvement is also crucial. Therefore, more efforts are needed to increase public awareness of fuel station hazards and how the community can be involved in preventing future accidents.

1. Introduction

Being a hazardous public space, accidents that occur at fuel stations might cause harm to people. Not only workers but also members of the community who happen to be in the vicinity of the accident site. It can be those buying fuel for their vehicles, stopping by to purchase goods at the convenience store, or being at a fuel station for other businesses. Ghana's 2015 fuel station fire caused up to 152 fatalities [1]. Another explosion occurred at Hinch, Haiti, in March 2016, killing seven people and 30 others severely injured [2]. During 2016-2018, an average of 4.150 fuel station fires were reported [3]. Data from the National Fire Protection Association (NFPA) shows that 56% of accidents in fuel stations were dominated by vehicle fire [3]. Indonesia's ministry of energy and mineral resources reported 120 accidents between 2016-2018 [4]. These accidents caused losses to the business and harm to the public. In Indonesia, several fire accidents in the past have killed at least six people who were fuel station customers [5-8]. These facts show the importance of studying and analyzing fuel station accidents, especially in Indonesia.

Various research regarding fuel stations has been done and can be classified into five major topics. The first topic discusses the health effects on workers. Mulia et al. [9] evaluated the safe concentration limits for benzene in a fuel station environment. A study by Pasiga et al. [10] reviewed lead exposure in Indonesian fuel station officers. An assessment of occupational hazards and health problems of petrol stations has been conducted by Johnson & Umoren [11]. A study on risk assessments for fuel stations has been done by Ahmed et al. [12], which identified hazards related to petrol fuel station operation and maintenance. A previous study from the same author created an assessment framework for identifying fuel station safety and risk [13]. Studies on hazard identification and risk assessment on fuel stations have also been done [13-15]. Environmental impacts have also been studied, as shown by the work of Andreev et al. [16], which attempted to recover fuel vapors from motor vehicles at fuel stations. Analysis of fire & explosion at fuel stations by Setyawan [17] and Huang & Ma [2], who assessed explosions during the gas station refilling process.

Among the studies carried out by other researchers, a scant amount has been found regarding the analysis of fuel station accidents. Therefore, it is necessary to look at previous accidents and systematically analyze them to answer the following research questions: (1) What kind of accidents are dominant at fuel stations in Indonesia? (2) What are the causes of the accidents? Furthermore, (3) How can accidents in fuel stations be prevented?

The study discussed in this article aims to collect accident data and determine the impact, cause, and possible prevention programs of fuel station accidents. Identifying the dominant effect and causes of the accident is essential, so a more suitable prevention program can be developed and employed. A successful accident prevention program could reduce the frequency of fuel station accidents and might save people's lives.

2. Methods

The method used in this research is deductive reasoning with qualitative analysis. As an initial study, the scope of this study is limited to one year, i.e., from September 2021 to August 2022. Area-wise, fuel station accidents across Indonesia were considered. Data for this study is collected using the document analysis technique [18]. Data collection was done from August 10-16, 2022. Finally, data were analyzed using the classification method [19] with a modified set of classifications.

This research is divided into three key steps. This study's first step was gathering and recording fuel station accidents on the web. The top ten news portals were selected, then searched using the keyword *"kebakaran SPBU"* or "fuel-station fire". News related to fuel station accidents was recorded, and duplicate accident events were discarded. Notes were taken for each event, including the date and time of the accident, location, brief description, direct cause of the accident, related factors contributing to the accident, the number of casualties (death and injury), and asset loss. The number of casualties was also split into those who were workers and those who were from the community. For this study, the community is defined as people not related to the operation of fuel stations whose activities at that moment are connected to fuel stations (e.g., fuel consumers, vehicle drivers, etcetera). This first step would provide enough information to identify the dominant types of accidents in fuel stations and those affected.

The following essential step is to analyze the cause of the accident. Direct causes of accidents were obtained from statements of witnesses and authorities written in the news. They were then categorized accordingly to facilitate analysis. Any other information that could be contributing factors was recorded as related factors. Most of the information posted by news portals was not actual investigation reports. Therefore, the cause identified for each accident might not be accurate. To overcome this shortcoming, further analysis is done by cross-referencing one event to another news source, if any. This second step would provide explanations to answer research question number two.

The final step of this study is to explore potential prevention measures. Various approaches were examined from the literature. The three most appropriate ones were carefully selected and explained, with details on how they can be applied to prevent fuel station accidents.

3. Results and Discussion

3.1. Results

Ten news portals were searched and initially returned 295 news after filtering out unrelated articles and removing duplicate events. The data interpretation for each event was made by recording the date, location, and the number of deaths/injuries. The direct causes were classified according to the information available in the news text. Asset loss was also grouped into cars, motorcycles, or buildings. Data interpretation results are shown in Table 1.

Table 1. Study Result: Fuel station accidents in Indonesia between September 2021 and August 2022.

Date	Location	Direct Cause	Death	Injured	Asset Loss
14-Sep-21	Deli Serdang, Sumatera Utara	Short circuit in the vehicle	-	2	1 car
15-Sep-21	Nganjuk, Jawa Timur	Short circuit in the vehicle	-	-	1 car
13-Oct-21	Buton, Sulawesi Tenggara	Unsafe action	-	-	-
15-Nov-21	Klaten, Jawa Tengah	Short circuit in structure	-	-	1 building & 1 motorcycle
10-Feb-22	Pacitan, Jawa Timur	Filling jerry can	-	-	1 car
04-Mar-22	Mojokerto, Jawa Timur	Smoking	-	1	1 building
16-Mar-22	Kendari, Sulawesi Tenggara	Filling jerry can	-	-	9 stores
29-Mar-22	Banyuwangi, Jawa Timur	Short circuit in the vehicle	-	1	1 car
31-Mar-22	Depok, Jawa Barat	Engine running	-	-	1 motorcycle
15-Apr-22	Banyuwangi, Jawa Timur	Tank modification	-	1	1 car
26-Apr-22	Aceh Besar, Aceh	Tank modification	-	1	1 car
28-Apr-22	Semarang, Jawa Tengah	Short circuit in the vehicle	-	2	1 car
01-May-22	Sukabumi, Jawa Barat	Engine running	-	1	1 car
10-May-22	Bandung Barat, Jawa Barat	Gas leak in vehicle	-	-	7 motorcycles
20-May-22	Jakarta Timur, DKI Jakarta	Unknown	-	-	1 motorcycle
25-May-22	Banda Aceh, Aceh	Wrong filling	-	-	1 car
31-May-22	Konawe, Sulawesi Tenggara	Filling jerry can	-	1	1 car
31-May-22	Bojonegoro, Jawa Timur	Short circuit in the vehicle	-	1	1 car
01-Jun-22	Mandailing Natal, Sumatera Utara	Short circuit in the vehicle	-	-	1 car
01-Jun-22	Kendari, Sulawesi Tenggara	Short circuit in the vehicle	-	-	1 car
07-Jun-22	Medan, Sumatera Utara	Short circuit in vehicle	-	-	1 car
17-Jun-22	Sidoarjo, Jawa Timur	Short circuit in structure	2	1	1 building
23-Jun-22	Pekalongan, Jawa Tengah	Overfilling	-	-	1 motorcycle
16-Jul-22	Polewali Mandar, Sulawesi Barat	Short circuit in the vehicle	-	1	1 car

19-Jul-22	Pacitan, Jawa Timur	Short circuit in the vehicle	-	-	1 car
02-Aug-22	Jombang, Jawa Timur	Short circuit in the vehicle	-	-	1 car
04-Aug-22	Luwu, Sulawesi Selatan	Filling jerry can	-	2	1 car
04-Aug-22	Palopo, Sulawesi Selatan	Tank Modification	-	-	1 motorcycle
10-Aug-22	Lebak, Banten	Technical fault in installation	-	1	-

3.2. Discussion

3.2.1. Type of Accidents

All accidents identified involved fire. Most of them (82.75%) involved vehicle fires, which fuel station consumers own. This figure is more significant than data on gas station fires in the United States of America [3]. The total number of injured was 16 people, almost all of them members of the community, with only one being a fuel station worker. There were two fatalities, both family members of the owner of an unofficial fuel retailer. Three fire accidents involved fire in a building structure, which also occurred on unofficial fuel retailers.

This unofficial retailer usually sells gasoline fuel in the kiosk on side roads [20]. Gasoline is stored in jerry cans or small drums. Traditional hand pumps and mini fuel dispensers were used to sell gasoline [21]. In Indonesia, these retailers were called *"Pertamini,"* trying to resemble small stations of Pertamina, the nation's state-owned oil & gas company, which operates most of the official fuel stations in Indonesia. *Pertaminis* are illegal and pose a safety threat to the community. First, it operates without a permit from the government. Hence there is no guarantee that the facilities and equipment used are safe. Second, individuals are managed and operated without proper training or safety precautions [22]. Third, usually, there is no safety apparatus available. Nevertheless, many of these retailers were available everywhere throughout the country because people find it practical to fill up fuel in *Pertaminis*.

The second group affected party was consumers buying fuel using the vehicle with either modified fuel tanks or jerry cans. There were four cases of jerry can filling and two cases of the vehicle with modified fuel tanks, involving one car and the other a motorcycle. Filling gasoline into jerry cans is dangerous, especially using non-approved fuel containers. The National Institute of Occupational Safety and Health had issued a hazard warning on filling portable gas cans on pickup trucks. It warns that fire might spontaneously ignite when filling portable fuel containers on a vehicle [23]. The Health and Safety Executive of the United Kingdom and the Occupational Safety and Health Administration (OSHA) require portable fuel storage containers [24,25]. In Indonesia, however, regulations on this matter have yet to be issued.

The last affected group is people buying fuel at the fuel station. Some of them were fortunate to walk out of their vehicle without any injury, leaving their vehicle severely burnt. Others suffer from first-degree to second-degree burns, usually from flash fires or not being fast enough to evacuate a burning vehicle.

3.2.2. Accident Causes

In most cases, fire initially started from within the vehicle, be it a car or a motorcycle. Several eyewitnesses reported that they saw smoke escaping from inside the car. Smoke in a vehicle is typically the effect of burning cables from a short circuit. Short vehicle circuits often happen in old cars or cars without proper maintenance. If a short circuit occurs in a car within fuel station premises, the car might burst into flames faster due to higher fuel vapor concentrations.

The second cause of the fire was filling modified fuel tanks and jerry cans, which are the dangers described in the previous section. Fueling a vehicle while keeping the engine running also causes fire. Turning the engine off when refueling a vehicle is mandatory, but sometimes people contravene the procedures and accidents. One fire case was caused by someone smoking while transferring fuel, which is not a good practice.

There were a few instances where the cause of fire came from technical aspects of the fuel station itself. For example, there were two cases of short circuits in the fuel station structure and one case of fire that started at an underground storage tank facility.

From the previous explanation, the causes of fire can be divided into two groups. First, fires were caused by fuel station-controlled facilities and equipment, and the second was fires that were caused by consumer vehicles. In this case, the latter has more significant contributions as the cause of fire at fuel stations.

3.2.3. Accident Prevention
To prevent accidents in the future, it is essential to ensure compliance with all the regulations and standards. Inspection and maintenance are crucial to assure the safe operation of fuel stations [26]. Leak detection devices could also be installed to prevent fire development [27]. Moreover, organizational and legal measures are also vital to provide safe fuel stations [28]. This includes strict enforcement of government regulations. Well-trained operators and fuel station managers are also essential because they are the frontline in delivering safety precautions to customers.

3.2.4. Community Involvement
Community involvement is critical in addressing the accident contribution factor from consumer vehicles. In other fields, such as road traffic safety, efforts have been made to engage the community in building safety awareness. These efforts can be replicated to increase the tacit knowledge about barriers to accident prevention and behavior change [29]. Developing a compelling campaign message and methods for disseminating them is the next critical part of community involvement in accident prevention [30]. The use of social media and radio are also considered effective in delivering safety campaigns and persuasion to the public [31].

4. Conclusion
A considerable amount of information has been acquired by collecting data on fuel station accidents during a one-year period. Identification of casualties and accident impacts shows that a vehicle fire dominates fuel station accidents in Indonesia. Accident classification gives us knowledge of aspects that cause these accidents. Many factors could contribute to fire at fuel stations, but in these cases, fire is mainly caused by consumer vehicle technical failures, such as short circuits on wires. Other conditions might as well lead to fire, such as tank modifications, filling to jerry cans, and unofficial fuel kiosks.

Fuel station facilities and equipment must comply with regulations and standards to prevent accidents. However, most factors causing accidents in fuel stations are related to consumers who buy fuel. This is an external factor from the perspective of fuel station business owners. Therefore, future accident prevention programs must involve the community.

This study has its limitation as data was collected from news websites, and the research period was limited to only one year. Consequently, the cause of accidents is not as accurate as accident investigation reports, and other accident cases have not been identified. The authors suggest that further research should be done for a more extended period and use more accurate data, such as accident or investigation reports.

Acknowledgments
This study was funded by the Ministry of Education, Culture, Research And Technology (KEMDIKBUDRISTEK) grant number [091/E5/PG.02.00.PT/2022 and 172./PKS/WRIII-DRP/UI/2022] and grant number [NKB-949/UN2.RST/HKP.05.00/2022], Research and Development (Risbang), Universitas Indonesia.

References

[1] Asumadu-Sarkodie S Owusu P A and Rufangura P, 2015 Impact analysis of flood in Accra, Ghana *Pelagia Res. Libr.* **6**, October p. 53–78.

[2] Ma G and Huang Y, 2019 Safety assessment of explosions during gas stations refilling process *J. Loss Prev. Process Ind.* **60** p. 133–144.

[3] Ahrens M, 2020, Service or Gas Station Fires.

[4] 2018 *Peraturan Menteri Energi dan Sumber daya Mineral No. 18 Tahun 2018* Jakarta: Kementerian Energi dan Sumber Daya Mineral.

[5] Fatir M D, 07-Aug-2021, Korban meninggal akibat kebakaran di SPBU Pangkep bertambah satu orang, *Antara News*. [Online]. Available: https://www.antaranews.com/berita/2314146/korban-meninggal-akibat-kebakaran-di-spbu-pangkep-bertambah-satu-orang. [Accessed: 11-Feb-2022].

[6] Wahyono D, 2018, Korban Kebakaran di SPBU Pangkalpinang Meninggal Dunia. [Online]. Available: https://news.detik.com/berita/d-4169472/korban-kebakaran-di-spbu-pangkalpinang-meninggal-dunia. [Accessed: 12-Dec-2020].

[7] Hidayat F, 2018, Kebakaran di SPBU, Sopir Bus Tewas, *wartaekonomi.co.id*. [Online]. Available: https://www.wartaekonomi.co.id/read202425/kebakaran-spbu-sopir-bus-tewas. [Accessed: 01-Jun-2021].

[8] Nainggolan M, 14-Feb-2021, Mobil Terbakar di SPBU Batam, Dua Balita Meninggal Dunia, *IDN Times*. [Online]. Available: https://www.idntimes.com/news/indonesia/margaretha-nainggolan/mobil-terbakar-di-spbu-batam-dua-balita-meninggal-dunia/full. [Accessed: 01-Jun-2021].

[9] Mulia S A Rahman Z F Sugiharta A M B Susanti L and Tualeka A R, 2020 Evaluation of benzene threshold value in benzene exposed work environment: Case study at ciputat gas station *Indian J. Forensic Med. Toxicol.* **14**, 1 p. 413–418.

[10] Pasiga B D Samad R Pratiwi R and Akbar F H, 2019 Identification of lead exposure through saliva and the occurrence of gingival pigmentation at fuel station Indonesian officers *Pesqui. Bras. Odontopediatria Clin. Integr.* **19**, 1.

[11] Tanaka T Azuma T Evans J A Cronin P M Johnson D M and Cleaver R P, 2007 Experimental study on hydrogen explosions in a full-scale hydrogen filling station model *Int. J. Hydrogen Energy* **32**, 13 p. 2162–2170.

[12] Ahmed M M Kutty S R M Khamidi M F and Dominic P D D, 2013 Hazards related to petrol fuel stations during operation and maintenance *Malaysian Constr. Res. J.* **12**, 1 p. 99–116.

[13] Ahmed M M Kutty S R M Shariff A M and Khamidi M F, 2011 Petrol Fuel Station safety and risk assessment framework in *2011 National Postgraduate Conference - Energy and Sustainability: Exploring the Innovative Minds, NPC 2011*.

[14] Nakayama J Sakamoto J Kasai N and Shibutani T, 2016 Preliminary hazard identification for qualitative risk assessment on a hybrid gasoline-hydrogen fueling station with an on-site hydrogen production system using organic chemical hydride *Int. J. Hydrogen Energy* **41**, 18 p. 7518–7525.

[15] Varta O and Krocova S, 2016 The location of LPG filling stations and potential risks of incidents *Commun. - Sci. Lett. Univ. Zilina* **18**, 1 p. 85–88.

[16] Andreev V V Dmitriyev S M Duncev A V Obidina K A Tarasova N P and Utyatnikov A E, 2020 Built-in low-temperature system for capturing and recovering motor fuel vapours at gas stations *Neft. Khozyaystvo - Oil Ind.* **2020**, 4 p. 101–104.

[17] Setyawan H, 2020 The determinant of fire disaster mitigation (fire practices study in gas station operator Surakarta, Indonesia) in *IOP Conference Series: Earth and Environmental Science* **423**, 1.

[18] Gross J M S, 2018, The SAGE Encyclopedia of Educational Research, Measurement, and Evaluation. SAGE Publications, Inc., Thousand Oaks,.

[19] Sakamoto J Sato R Nakayama J and Kasai N, 2016 Leakage-type-based analysis of accidents

involving hydrogen fueling stations in Japan and USA *Int. J. Hydrogen Energy* **41**, 46 p. 21564–21570.

[20] Robert Hutauruk M Ghozali I Sutarmo Y Mushofa A Astri Yulidar M and Yanuarta W, 2019 Application Of Marketing Mix: Study On Two-Wheeled Vehicle Users In Deciding To Buy Fuel On Roadside Unofficial Kiosks (At Samarinda Indonesia) *Int. J. Sci. Technol. Res.* **8**, 12 p. 1275–1279.

[21] Stefanno Reinard Sulaiman, 22-Jan-2019, Consumers find Pertamini convenient, appear safe from crackdown, *The Jakarta Post*. [Online]. Available: https://www.thejakartapost.com/news/2019/01/22/consumers-find-pertamini-convenient-appear-safe-from-crackdown.html.

[22] Nisak K Kurniawan B and Widjasena B, 2019 Faktor-Faktor yang Berhubungan dengan Kesiapan Mitigasi Kebakaran pada Pom Bensin Mini (Studi di Kecamatan Banyumanik dan Tembalang Kota Semarang) *Kesehat. Masy.* **7**, 4 p. 2356–3346.

[23] Flesch J P *et al.*, 1998, Fire Hazard from Filling Portable Gas Cans in Pickup Trucks and Cars.

[24] 2009 *PART 1926-SAFETY AND HEALTH REGULATIONS FOR CONSTRUCTION* OSHA, p. 198–213.

[25] 2014, Portable Petrol Storage Containers, *Health and Safety Executive*. Health and Safety Executive.

[26] de Araújo Sousa L E Stopiglia F S de Mello Filho L V F Guimarães D H P and de Moraes Gomes Rosa M T, 2021, Prototype Proposal for Control and Inspection of Gas Stations, *Smart Innovation, Systems and Technologies*, **202**. Center of Science and Technology, Mackenzie Presbyterian University, Brazil Avenue, Campinas, SP 1220, 13073-148, Brazil, p. 357–366.

[27] Yurismono H Sumarsono D A Indradjaja I M and Nugroho S, 2022 Development of lower flammability limits testing apparatus for gasoline vapor – Air mixture using an internal vaporizer at ambient conditions *Transp. Res. Interdiscip. Perspect.* **14**, April p. 100613.

[28] Kolosov A Yaremenko S Garmonov K and Sklyarov K, 2020 Influence of gas stations on the ecology of the urban environment in *E3S Web of Conferences* **164**.

[29] Austin E K and Green K N, 2019, The Central Role of Community Participation in Traffic Safety Culture, in *Traffic Safety Culture*, N. J. Ward, B. Watson, and K. Fleming-Vogl, Eds. (Emerald Publishing Limited), p. 129–143.

[30] Hamann C J Daly E Schwab-Reese L Askelson N and Peek-Asa C, Dec. 2021 Community engagement in the development and implementation of a rural road safety campaign: Steps and lessons learned *J. Transp. Heal.* **23** p. 101282.

[31] Earlyanti N I Putro B East W and Police J R, 2020, The Role of Radio in Digital Era and Community Participation in Maintenance Safety and Traffic Safety in East Java.

The Effect of Age, Knowledge, and Community Attitudes Against COVID-19 Prevention Behaviour in Balikpapan City, East Kalimantan

Desy Arianti[1]* and Ratna Purwaningsih[2]
[1]Industrial Engineering, Engineering, Universitas Diponegoro, Semarang, 50275, Indonesia, 0000-0003-1870-0980
[2]Industrial Engineering, Engineering, Universitas Diponegoro, Semarang, 50275, Indonesia, 0000-0002-3016-4730
[1]*desyarianti97@gmail.com, [2]ratna.purwaningsih@ft.undip.ac.id

Abstract. The SARS-CoV-2 pandemic, also known as COVID-19, has shaken the world, claimed thousands of lives, and impacted economies worldwide. So far in Balikpapan City, East Kalimantan, the situation is still very threatened and facing the reality of preparing to coexist with COVID-19. This study aimed to determine the effect of age, knowledge, and attitudes on COVID-19 prevention behaviour. Collecting data through an interview with the Head of the P2P Division at the Balikpapan City Health Office and distributing online questionnaires using the google form platform to the community of Balikpapan City. Data were processed using SPSS software with a binary logistic regression method. The results showed that 36.7% or 22 people had poor knowledge of COVID-19, while 63.3% or 38 had good knowledge of COVID-19. In addition, 16.7% or ten respondents have a terrible attitude when dealing with COVID-19, while 83.3% or 50 others have a good attitude when dealing with COVID-19. Partially, age and attitude have no significant effect on COVID-19 prevention behaviour.

1. Introduction

In December 2019, a new viral disease that had never been identified in humans was discovered in Wuhan, China. The outbreak of the SARS-COV-2 virus pandemic, also known as COVID-19, has shaken the world, claimed thousands of lives, and impacted economies worldwide. The virus spread rapidly in more than 160 countries in 3 months resulting in more than 400,000 deaths and more than 8.9 million people affected globally [1]. Most people infected with the coronavirus have mild to moderate symptoms and recover without specific treatment. However, some people experience severe pain and require medical attention. Globally, as of 18:21 CET, November 26, 2021, there were 259,502,031 confirmed cases of COVID-19, including 5,183,003 deaths, reported to the World Health Organization (WHO). As of November 24, 2021, a total of 7,702,859,718 doses of the vaccine have been administered [2]. The Indonesian Ministry of Health reported the number of COVID-19 (cumulative) cases in Indonesia on November 26, 2021, namely 4,255,268 positive cases and 4,103,379 cases recovered 143,796 cases died [3]. Based on information from the Head of the P2P Division, the Balikpapan City Health Office, on October 24, 2021, there were 38,221 confirmed cases, 36,283 cases recovered, and 1,874 cases died [4].

Various prevention recommendations aimed at reducing the spread of the COVID-19 virus include maintaining distance, using face masks, always maintaining hand and personal hygiene, limiting activities through regional quarantine and temporary closure of places of business, and prohibiting people from gathering or gathering for religious, work purposes, education, and recreation and the COVID-19 vaccination plan as one of the main preventive measures in controlling COVID-19 [5]. According to the research [6], the highest chance of contracting the coronavirus is only 3% if we wear masks. On the other hand, if you do not wear a mask, there is a 17% chance of transmission. [7] Emphasized that there is no genuinely protective intervention in preventing COVID-19; it needs to be combined with social distancing, face masks, and other interventions. This is the best way to prevent COVID-19. So far in Balikpapan City, East Kalimantan, the situation is still very threatened and facing the reality of preparing to coexist with COVID-19.

This study aimed to determine the effect of age, knowledge, and attitudes on COVID-19 prevention behaviour. Education is needed to increase knowledge and ways to prevent COVID-19. Cooperation between the government and community efforts and social control is needed to increase knowledge and understanding of implementing health protocols [8]. According to research from [9], a positive attitude also leads to appropriate actions to prevent the spread of COVID-19. Advanced age is also associated with mortality rates; those with congenital diseases may be at higher risk for more severe diseases. Based on existing data, the age of patients infected with COVID-19 ranges from 30-89 years [10]. [11] Further studies proved that anxiety, optimism, fear of death, and social isolation significantly mediate the effects of age on protective behaviour. Furthermore, the middle-aged group has an optimistic effect on protective behaviour, while the younger and older groups affect social isolation, which reduces protective behaviour. Therefore, it is suspected that people respond to mitigating behaviour against COVID-19 differently based on age.

According to research from [12] that attitudes, knowledge, education level, and socio-cultural environment have a significant relationship with physical distancing policy compliance. Research results differ from [13] did not find a significant relationship between good knowledge and positive attitudes, and good behaviour in preventing COVID-19. Research by [14] Knowledge and attitudes influence a person's behaviour in preventing the transmission of COVID-19. A good attitude toward preventing COVID-19 is essential in the practice of preventive actions taken by employees in the Airmadidi Bottled Drinking Water Industry in North Sulawesi [15]. There is a relationship between the level of knowledge and attitudes towards COVID-19 prevention practices in the internet user community in Semarang City. Improved health education is needed to improve literacy and prevention practices for COVID-19 [16].

2. Methods

This study uses an analytical, quantitative method approach with a cross-sectional design. The research was conducted at the Balikpapan City Health Office, East Kalimantan. The step for this research is the data collecting, data processing by regression analysis, continue by analysis and conclude the result of data processing:

2.1. Data collecting

The primary data collection technique by interviewing the Head of the P2P Division at the Balikpapan City Health Office and Balikpapan resident. First questionnaire was designed to obtain information regarding indicators of community behaviours that affect the prevention of COVID-19 and other supporting data for the background of research writing. The second questionnaire is distributed online to determine the perceptions or habits of a population based on respond from resident of Balikpapan used the technique of purposive sampling. Based on the purposive sampling technique, the authors determined that the number of respondents used in the study was 60 respondents.

Secondary data were obtained from literature review to obtain the hystorical data from the official website of the World Health Organization (WHO), the Balikpapan City Health Office's official website of the Indonesian Ministry of Health, journals, and books. The result of literature review is the model to measure Knowledge of respondent using six questions about COVID-19 symptoms, the severity of the vulnerability, and transmission and prevention. Attitudes were investigated using six questions about respondents' beliefs about responding to COVID-19 and their attitudes towards COVID-19 prevention and response. The behaviour surveyed with six questions is the respondent's behaviour to prevent the spread of COVID-19 and behaviour from adapting to new habits. The dependent variable in this study is COVID-19 prevention behaviour, while the independent variables are age, knowledge, and attitudes. Data processing was carried out using SPSS software using a binary logistic regression method to determine the simultaneous effect of the X variable on the Y variable. Starting with validation and reliability tests, Omnibus and R squared tests, Hosmer and Lemeshow tests, and hypothesis tests.

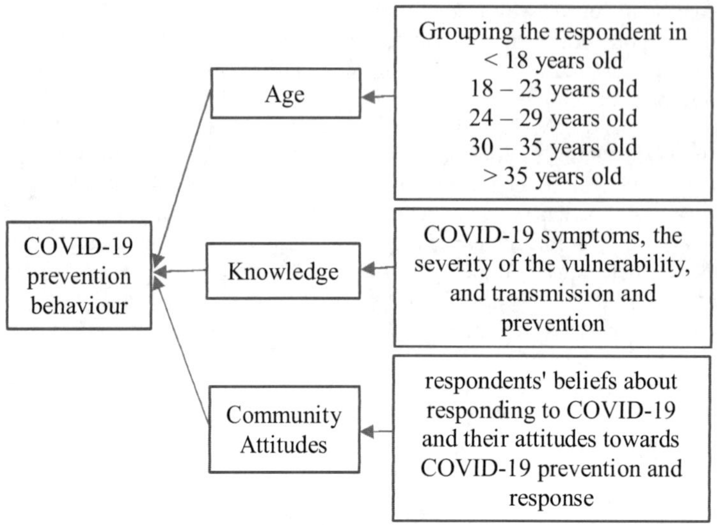

Figure 1. Flow chart.

2.2. Data processing using logistic regression model

The data collection then processed using the binary logistic regression analysis is a logistic regression between the response variable (y) and the predictor variable (x), where the y variable produces two categories, namely 0 and 1 [17]. So that the variable (y) follows the Bernoulli distribution with the probability function as follows.

$$(f(y) = \pi^y(1-\pi)^{1-y};\ y\ =\ 0{,}1 \qquad \textbf{(1)}$$

Where, if y = 0 then $f(y) = 1 - \pi$ and if y = 1 then $f(y) = \pi$. The logistic regression function can be written as follows.

$$f(z) = \frac{1}{1+e^{-z}}\ equivalent\ f(z)\ = \frac{e^z}{1+e^z} \qquad \textbf{(2)}$$

With

$$(z = \beta_0 +\ \beta_1 x_1 + \ldots + \beta_p x_p) \qquad \textbf{(3)}$$

If the value of z is between $-\infty$ and ∞ f(z) is between 0 and 1 for any given z value. Then, the logistic model describes the probability or risk of an object. The logistic regression model is as follows.

$$\pi_{(x)} = \frac{e^{\beta_0+\ \beta_1 x_1+\ldots+\beta_p x_p}}{1+e^{\beta_0+\ \beta_1 x_1+\ldots+\beta_p x_p}} \qquad \textbf{(4)}$$

Where p = number of predictor variables

$$g_{(x)} = \ln\left(\frac{\pi_{(x)}}{1-\pi_{(x)}}\right) = \beta_0 + \beta_1 x_1 + \ldots + \beta_p x_p \quad \textbf{(5)}$$

2.3. Research hypotheses and model

There are four hypotheses in this study:

H_1: Age has a significant effect on preventive behaviour

H_2: Knowledge has a significant effect on preventive behaviour

H_3: Attitude has a significant effect on preventive behaviour

H_4: Age, knowledge, and attitude simultaneously have a significant effect on preventive behaviour.

The proof of the hypothesis was carried out using logistic regression, a variable is said to have an effect between one independent variable on the dependent variable if it has a significance value of $t < 0.05$. Conversely, the variable is said to have no significant effect if the significance value $t > 0.05$ between one independent variable on the dependent variable.

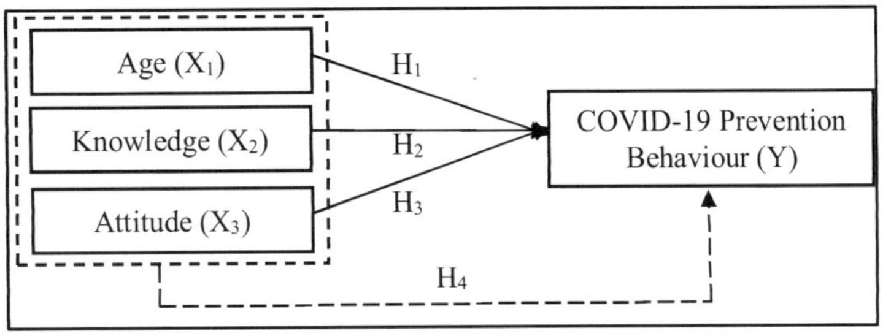

Figure 2. Relationship between research variables

3. Results and Discussion

The study was conducted in the community of Balikpapan City with a total of 60 respondents. Research respondents were dominated by the age group of 18-23 years, consisting of 21 or 35% men and 39 or 65% women. Then, various fields of work from the respondents such as 28.3% students, 1.7% Civil Servants, 35% private employees, 10% entrepreneurs, and 25% others.

Table 1. Age distribution of the people in Balikpapan city who filled out the questionnaire.

		Age			
		Frequency	Per cent	Valid Percent	Cumulative Percent
Valid	< 18 years old	4	6.7	6.7	6.7
	18 – 23 years old	26	43.3	43.3	50.0
	24 – 29 years old	17	28.3	28.3	78.3
	30 – 35 years old	4	6.7	6.7	85.0
	> 35 years old	9	15.0	15.0	100.0
	Total	60	100.0	100.0	

This study shows that behavioural responses to COVID-19 vary by age. However, adolescent and adult age has no significant effect on COVID-19 prevention behaviour. The results showed that 36.7% or 22 people had poor knowledge of COVID-19, while 63.3% or 38 people had good knowledge of COVID-19, meaning that people know this virus is hazardous. These results also show that most of the people of Balikpapan city have good knowledge skills about COVID-19. Respondent's attitudes towards the COVID-19 pandemic were varied. There are 16.7% or ten respondents who have a terrible attitude when dealing with COVID-19, while 83.3% or 50 others have a good attitude when dealing with COVID-19.

IOP Conf. Series: Earth and Environmental Science 1111 (2022) 012078 doi:10.1088/1755-1315/1111/1/012078

Not only that, in dealing with COVID-19, respondents have different behaviours. It is known that 23.3% or 14 respondents behaved poorly in preventing COVID-19. So based on this percentage, it can be seen that most people have a good attitude about the dangerous coronavirus. People's behaviour shows how they respond to COVID-19. While 76.7% or 46 others had good behaviour in preventing COVID-19, it can be concluded that this study is dominated by people who behave well toward COVID-19 prevention behaviour. This study's results align with studies [12] and [14], which state that knowledge significantly influences COVID-19 prevention behaviour. This research is also reinforced by [16], which states that knowledge strongly correlates with COVID-19 prevention behaviour. Therefore, understanding and knowledge of the dangers of COVID-19 can increase public awareness to carry out COVID-19 prevention behaviours. Even so, these results contradict research [13], where no significant relationship was found between knowledge of COVID-19 prevention measures. This result also contradicts research [15] which explains that there is no significant relationship and influence between the level of knowledge on COVID-19 prevention measures. This is because there are still people with good knowledge but do not want to take preventive measures against COVID-19.

3.1. Validity and reliability test

The validity test results in the correlation table show that the six questionnaire items have a calculated r-value greater than the r-table value of 0.244, so it can be concluded that the questionnaire items used are valid and have proven validity. The reliability test results of each variable are knowledge 0.766, attitude 0.781, and behaviour with a value of 0.729. [18] The instrument is said to be reliable if the Cronbach's Alpha value is more significant than 0.06. Based on the results above, it can be said that the questionnaire items used are reliable or consistent.

Table 2. Validity test result.

Validity Test				
Variables	Questionnaire Items	r value	r table	Description
Knowledge	Item 1	0.704	0.244	Valid
	Item 2	0.650	0.244	Valid
	Item 3	0.498	0.244	Valid
	Item 4	0.736	0.244	Valid
	Item 5	0.732	0.244	Valid
	Item 6	0.771	0.244	Valid
Attitude	Item 1	0.696	0.244	Valid
	Item 2	0.397	0.244	Valid
	Item 3	0.795	0.244	Valid
	Item 4	0.702	0.244	Valid
	Item 5	0.738	0.244	Valid
	Item 6	0.769	0.244	Valid
Behaviour	Item 1	0.591	0.244	Valid
	Item 2	0.454	0.244	Valid
	Item 3	0.699	0.244	Valid
	Item 4	0.723	0.244	Valid
	Item 5	0.725	0.244	Valid
	Item 6	0.723	0.244	Valid

3.2. Omnibus and r squared test

When using a significance level of 0.05, the value is 0.067 > 0.05, meaning there is no significant effect between the variables of age, knowledge, and attitude towards COVID-19 prevention behaviour.

Table 3. Omnibus test results and R squared.

Omnibus Test of Model Coefficients				
		Chi-square	df	Sig.
Step 1	Step	7.150	3	.067
	Block	7.150	3	.067
	Model	7.150	3	.067

However, if using a significance level of 0.10, the value is 0.067 < 0.10, so it can be said that the variables of age, knowledge, and attitude significantly affect COVID-19 prevention behaviour. Research to know the effect or relationship between variables like this generally uses a significant level of 0.05 (5%).

3.3. Hosmer and Lemeshow test

Based on the above test, it is known that the significance value obtained is 0.081 > 0.05. These results indicate that the study meets the model's feasibility test.

Table 4. Hosmer and Lemeshow test results.

Hosmer and Lemeshow Test			
Step	Chi-square	df	Sig.
1	19.259	8	.081

3.4. Hypothesis test

The test results above show that the variables of age and attitude have a significance value of 0.127 and 0.168. This value is more significant than 0.05, meaning that partially age and attitude have no significant effect on COVID-19 prevention behaviour. Another result is shown by the Knowledge variable, which has a significance of 0.044. Again, the value is smaller than 0.05.

Table 5. Hypothesis test results.

Variables in the Equation									
								95% C.I.for EXP (B)	
		B	S.E.	Wald	df	Sig.	Exp(B)	Lower	Upper
Step 1[a]	Age	.520	.341	2.333	1	.127	1.682	863	3.280
	Knowledge	1.279	.634	4.071	1	.044	3.594	1.037	12.456
	Attitude	-1.587	1.151	1.899	1	.168	.205	.021	1.954
	Constant	.765	2.416	.100	1	.751	2.150		

[a] Variable (s) entered in step 1: Age, Knowledge, Attitude

Referring to the logistic regression analysis results, it is known that the knowledge variable significantly influences COVID-19 prevention behaviour. Indicates that the higher the public's knowledge, the more aware they will be of taking preventive measures against COVID-19. Even so, age and attitude variables did not affect COVID-19 prevention behaviour significantly. These results indicate that age and people's attitudes are not the main factors influencing COVID-19 prevention behaviour. Maturity or not, as well as whether or not the community's attitude is good, has nothing to do with the COVID-19 prevention behaviour.

4. Conclusion

This study's results concluded that age and attitude partially had no significant effect on COVID-19 prevention behaviour, while partially, knowledge significantly influenced COVID-19 prevention behaviour. This research also has a limitation, such as the sample used in this study was only conducted in Balikpapan city, so the results obtained might make a difference if carried out in other areas in Indonesia. In addition, the data collection method only used questionnaire data, and the variables used in this study do not yet represent all the behavioural factors that affect the prevention of COVID-19.

Further research suggestions are necessary to add other variables to discover more about people's perceptions of COVID-19 prevention behaviour. Based on this study's results, respondents have good knowledge about COVID-19 but do not want to take COVID-19 precautions. This research is also expected to assist the Balikpapan city government in making decisions based on research results that provide information about the influence of age, knowledge, and public attitudes towards COVID-19 preventive behaviour so that the government can take further action.

Acknowledgements
We want to thank the Government of the Balikpapan City Health Office, the Head of the P2P Division of the Balikpapan City Health Office, and the community of Balikpapan City, who have supported and assisted in providing data and information to complete this research.

References
[1] Bodapati S, Bandarupally H and Trupthi M 2020 COVID-19 Time Series Forecasting of Daily Cases, Deaths Caused and Recovered Cases using Long ShortTerm Memory Networks 2020 IEEE 5th International Conference on Computing Communication and Automation (ICCCA) p 525
[2] World Health Organization 2021 WHO Coronavirus (COVID-19) Dashboard Covid19.who.int Online: https://covid19.who.int/
[3] Kementerian Kesehatan RI 2021 Infeksi Emerging Kementerian Kesehatan RI Infeksiemerging.kemkes.go.id Online: https://infeksiemerging.kemkes.go.id/dashboard/covid-19
[4] Balikpapan Satuan Tugas COVID-19 2021 Data Pemantauan COVID-19 Kota Balikpapan Online: http://covid19.balikpapan.go.id/
[5] Kementerian Perencanaan Pembangunan Nasional 2021 *Studi Pembelajaran Penanganan COVID-19 Indonesia* (DKI Jakarta: ISBN 978-623-96020-3-1)
[6] Chu D, Duda S, Solo K, Yaacoub S, and Schunemann H 2020 Physical Distancing, Face Masks, and Eye Protection to Prevent Person-to-Person Transmission of SARS-CoV-2 and COVID-19: A Systematic Review and Meta-Analysis Journal of Vascular Surgery 72 1500
[7] MacIntyre C and Wang Q 2020 Physical distancing, face masks, and eye protection for prevention of COVID-19 The Lancet 395 1950-1951
[8] Sofianto A, Zuhri M, Febrian L and Ambarwati O 2022 Penerapan Protokol Kesehatan Covid-19 Pada Ruang Publik Di Jawa Tengah Jurnal Kesehatan Masyarakat (Undip) 10 20-30
[9] Peng Y, Pei C, Zheng Y, Wang J, Zhang K, Zheng Z and Zhu P 2020 A cross-sectional survey of knowledge, attitude and practice associated with COVID-19 among undergraduate students in China BMC Public Health 20
[10] Handayani D, Rendra Hadi D, Isbaniah F, Burhan E and Agustin H 2020 Penyakit Virus Corona 2019 Jurnal Respirologi Indonesia 40 121
[11] Pasion R, Paiva T, Fernandes C and Barbosa F 2020 The AGE Effect on Protective Behaviors During the COVID-19 Outbreak: Sociodemographic, Perceptions and Psychological Accounts Frontiers in Psychology 11
[12] Bawiling N, Toar J and Lamhot P 2021 Determinan Kepatuhan Masyarakat Kabupaten Minahasa Terhadap Kebijakan Physical Distancing Dalam Pencegahan Covid-19 Jurnal Kesehatan Masyarakat UNIMA 2 1-7
[13] Gunawan S, Sinsin I and Zani A 2021 Hubungan Antara Pengetahuan dan Sikap dengan Perilaku Pencegahan COVID-19 Pada Peserta Seminar Online STIKes Raflesia 7 April 2020 Perilaku dan Promosi Kesehatan: Indonesian Journal of Health Promotion and Behavior 3 47
[14] Linawati H, Helmina S, Intan V, Oktavia W, Rahmah H and Nisa H 2021 Pengetahuan, Sikap, dan Perilaku Pencegahan COVID-19 Mahasiswa Media Penelitian dan Pengembangan Kesehatan 31 125-132
[15] Soeratinoyo D, Doda D and Warouw F 2021 Hubungan antara Pengetahuan dan Sikap dengan Tindakan Pencegahan Penyebaran COVID-19 pada Perusahaan Produsen Air Minum Dalam Kemasan Jurnal Biomedik:JBM 13 317

[16] Sutiningsih D 2021 Hubungan Tingkat pengetahuan dan Sikap dengan Praktik Pencegahan COVID-19 pada Masyarakat Pengguna Internet di Kota Semarang Jurnal Epidemiologi Kesehatan Komunitas 6 300-306.
[17] Hosmer D, Lemeshow S and Sturdivant R 2013 Applied logistic regression (Hoboken, New Jersey: Wiley)
[18] Ghozali I 2006 Aplikasi analisis multivariate dengan program SPSS (Semarang: Badan Penerbit Universitas Diponegoro)

Leadership as the Main Driving Factor of Regional Sustainable Development Cooperation: A Case Study of ASEAN Tuna Ecolabelling (ATEL)

Andre Notohamijoyo[1], Martani Huseini[2], Hendra Sugandhi[3], Elishabeth Srihayu Harsanti[4], Adhi Setya Wiyata[5], and Mustaidz Billah[5]

[1]Coordinating Ministry of Human Development and Cultural Affairs Republic of Indonesia, Jakarta, 10110, Indonesia, 0000-0002-3350-3590

[2] Faculty of Administrative Science, University of Indonesia, Kampus Baru UI Depok, Indonesia

[3] Secretary General of Indonesian Tuna Association 2014-2020

[4] Organization for Agriculture and Food, National Research and Innovation Agency, Cibinong, West Java, Indonesia

[5] Ministry of Marine Affairs and Fisheries, Jakarta, 10110, Indonesia

[1*]andre.notohamijoyo@kemenkopmk.go.id, [2]martani0703@yahoo.com, [3]hendrasugandhi1164@gmail.com, [4]esharsanti@gmail.com, [5]adhisetya@dkp.or.id, [6]mustaidz@gmail.com

Abstract. ASEAN Tuna Ecolabelling (ATEL) has agreed on a scheme of tuna ecolabelling after the approval at the Ministerial of ASEAN Ministry of Agriculture and Forestry (AMAF) meeting in 2018. It makes ATEL the first regional seafood eco-label scheme in the world. It becomes an achievement of ASEAN in supporting Sustainable Development Goals (SDGs), especially for tuna fisheries management. Unfortunately, after four years, its scheme has not yet been completed and takes no support from ASEAN countries, including Indonesia as the initiator. Lack of leadership in ASEAN is the leading cause of the delay in the implementation of ATEL in its region. Leadership is the primary key to encouraging sustainable development, especially during the Covid-19 pandemic. The study aims to analyze the lack of leadership as the main factor in the low effectiveness of regional cooperation in ASEAN. This research is expected to be a solution for completing the ATEL implementation process in that region.

1. Introduction

Fish is a commodity whose highest economic value is in a live or fresh state. Grouper and lobster, for example, have the highest economic value when sold alive. Meanwhile, if it is included in grade A or the highest quality and food safety rating, tuna has the highest value. When tuna is caught and directly consumed in the form of sushi and sashimi, the tuna will be much higher in value than in processed forms. This makes the management of fisheries resources require specific handling.

On the other hand, tuna management faces a severe threat in the form of sustainability. The World Bank and FAO, in their 2009 report entitled "Sunken Billion: The Economic Justification of Fisheries Reform," stated that since 2006 it was stated that 75 percent of global fishery resources have been under threat of depletion or stock decline due to excessive fishing practices and damaging the environment [1]. This condition worries many parties about the scarcity of fishery resources to meet the consumption needs of the global community. The 1992 Earth Summit became the starting point for environmental observers and activists to push for a sustainable production and consumption control mechanism through environmental labeling or eco-labeling as one of the items of Agenda 21 [2].

Based on the FAO Code of Conduct of Responsible Fisheries in 1995 [3], FAO issued the FAO Guidelines on Ecolabelling for Fish and Fishery Products from Marine Capture Fisheries in 2009 as a reference for fisheries eco-label certification worldwide [4]. It aims to encourage sustainable fisheries management by applying a fisheries eco-label scheme. Various definitions of eco-label have emerged, among others, as stated by [5]: Eco-labelling is the granting of product labels (usually applied voluntarily) by a private or public organization to inform consumers about the environmental impact of a product. Eco-label initially did not touch food products, including fisheries. In its development, fisheries eco-label were also developed as follows: Eco-labels are derived from certification processes and are a market-based approach that attempts to influence consumer behavior toward fisheries products generated through sustainable practices [6].

Tuna (*Thunnus* sp) is a type of fish with the highest value and the most popular in the world. So that many countries and international organizations have an interest in the management of these fish. Tuna is a fish that has a wide range in the sea and swims across various countries, so its management requires cross-country cooperation. If for palm oil and wood products, the form of certification is national, then for tuna, it is more effective if the form of certification is cross-country.

Several countries have implemented various certification systems for tuna traded nationally and internationally. The exciting thing is that the certification system is still minimally regional. ASEAN has started consolidation to start the regional system. ASEAN countries are starting to realize the complexity of problems in tuna fisheries, such as illegal and destructive fishing, stock declines, environmental pollution, ecolabelling provisions to export market requirements, and tariff and non-tariff barriers. This problem is difficult to solve alone, so it requires comprehensive cooperation between ASEAN countries. In 2010 ASEAN formed the ASEAN Tuna Working Group (ATWG) as a forum for tuna fisheries cooperation within that region. The organization is an entry point for cooperation in the sustainable management of tuna fisheries and building joint alliances in dealing with regional and international issues. Indonesia was appointed as the lead country in that forum.

Since 2012, Indonesia has proposed the ATEL concept at the tuna working group level. ATEL is an environmentally friendly tuna certification initiative based on the principle of sustainability and socially responsible practices. The purpose of ATEL is to encourage the sustainable management of tuna fisheries, inclusive and equitable. The main commodity of global capture fisheries with high economic value becomes the livelihood of many parties, including small-scale fisheries. The high economic value of tuna causes each country to compete to use it [7]. The idea of ATEL appears from a mutual awareness of the various groups that fishermen and crews must be thoroughly involved in the sustainable management of fisheries and the need for cooperation among countries in the region in tackling IUUF practice. If ATEL is implemented, the entire tuna products traded intra-ASEAN is a product that is captured in an environmentally friendly, sustainable, and free of labor abuse and consider justice for fishermen.

Institutions that assess these aspects are independently accredited institutions designated by ASEAN member states. Business people who apply for ATEL certification must be authorized by the official representatives of ASEAN Focal Point. During the initial initiation by Indonesia, ATEL received many rejections from ASEAN countries. Indonesian negotiators' negotiation ability and comprehensive understanding of sustainable fisheries management have caused various ASEAN countries to accept the concept slowly. After various rigorous negotiation processes, ASEAN countries agreed to the ATEL concept.

ATEL agreed on a scheme of tuna ecolabelling in South East Asia in the ASEAN Ministry of Agriculture and Forestry (AMAF) meeting in Hanoi, Vietnam, 12-13 October 2018. The approval at the Ministerial level makes ATEL the world's first regional seafood eco-label scheme. It has also become an achievement of ASEAN in supporting Sustainable Development Goals (SDGs). It is a concrete manifestation of the implementation of SDGs, especially the 14th Goal, Life Below Water [8].

Although it has been ratified, all ASEAN countries still have homework related to implementing standards and integration among member countries. This is the key to the successful implementation of ATEL in the Southeast Asia region. Support from all ASEAN countries is needed to encourage the integration process. The success of ATEL will create a new brand for tuna products from Southeast Asia. Unfortunately, after three years of running, the follow-up from ATEL has not been completed. Indonesia's initiatives as the lead sector in the ATWG have stagnated.

Several things must be considered in the three years. The first thing is the leadership change in the Ministry of Marine Affairs and Fisheries of the Republic of Indonesia. The year 2019 was the transition of President Joko Widodo's leadership from the first and second periods. During the transition period, there was a leadership change at the Ministry level as assistant to the President. There was a change in the position of the Minister of Marine Affairs and Fisheries, who had different views and policies. The second thing is the Covid-19 virus pandemic, which has caught the attention of all countries worldwide, including the Southeast Asian region. The Covid-19 pandemic has caused the concentration of each country to be prioritized on severe handling of the spread of the virus and its impact on society. The third thing is the mindset of officials at the Ministry of Marine Affairs and Fisheries. They support the certification system initiated by non-governmental organizations and multinational companies. This is reflected in the support from the Ministry of Marine Affairs and Fisheries for that certification system.

Based on the above mentioned situation, leadership is the main factor driving the implementation of ATEL in the ASEAN region. The leadership of one country influences the running of a program in the Southeast Asia region. ASEAN member countries need a stronger push so that cooperation programs in a sector can run well. The principle of the ASEAN Economic Community (AEC), which has been echoed for a long time and was proclaimed effective starting January 1, 2015, has proven to be not running optimally due to the absence of leadership from one of the countries in the region. Likewise, with the implementation of ATEL in the ASEAN region. Leadership is needed in pushing the ATEL platform so that it can build concrete cooperation in the region.

2. Methods

This study aims to analyze the leadership descriptively in Southeast Asia in implementing sustainable development regionally. In a previous study on ATEL [9] in 2018, ATEL had just been approved as a tuna fisheries eco-label scheme in Southeast Asia. The study discusses the ATEL negotiation process in ASEAN until it is finally approved. The implementation of ATEL is a case study on leadership in the Southeast Asian region, especially ASEAN. The next step is the study of literature on eco-label implementation in various countries.

The literature study in this study aims to compare the application of eco-label schemes in various regions. The comparison was conducted to explore further the role of leadership in implementing the eco-label scheme. After the literature study, the study of leadership issues in the implementation of ATEL was discussed descriptively and strengthened by the conditions that occur in Indonesia, the ASEAN region, and globally. This stage aims to further examine the role of leadership and commitment from ASEAN countries, especially in implementing ATEL.

3. Results and Discussion

3.1 The journey of ATEL

The ATEL concept originated from Indonesia's involvement in cooperation in the agriculture and forestry sector in ASEAN. In ASEAN, the fisheries sector is part of agriculture and forestry. The ASEAN Minister of Agriculture and Forestry (AMAF) Meeting is the highest form. The AMAF meeting is held annually and reviews the overall development of cooperation in the agriculture and forestry sectors, including fisheries in ASEAN.

One of the cooperation platforms in the sector is the Memorandum of Understanding (MoU) of the ASEAN Cooperation on Agricultural and Forestry Product Promotion Scheme [10]. The MoU encourages cooperation in developing agricultural and forestry commodities among ASEAN countries. The objectives of the MoU are: strengthening the bargaining position with ASEAN countries in trade, expanding product coverage through diversification, intensifying processing (downstream) and value-added activities, improving product quality and safety, and building a stronger foundation for member countries' economic cooperation. Furthermore, preserve resources to ensure sustainable production.

Collaborated commodities include cocoa, rubber, crude palm oil, pepper, tea, carrageenan, and tuna. Indonesia was mandated to become the top country for tuna, tea, and pepper commodities. The Ministry of Marine Affairs and Fisheries was appointed as the National Focal Point (NFP) for tuna commodities. The ASEAN Tuna Working Group (ATWG) was formed as a first step. Establishing the ASEAN Tuna Working Group (ATWG) in 2011 is an excellent start to encouraging intra-trade in the region. As one community, it is appropriate that ASEAN economic and trade cooperation is comprehensive and valuable for people in the region. As an implementation of the Strategic Plan of Action 2011-2014, this WG has produced the ASEAN Tuna Ecolabelling [11]. The initiative is the answer to the challenges in the tuna fishery. ATEL is the most critical part of AWG. ATEL is an Indonesian initiative in the form of a scheme to regulate the intra-ASEAN eco-friendly tuna trade. Threats to the sustainability of tuna fisheries resources are an essential issue in the international tuna trade.

Data from The State of World of Fisheries and Aquaculture FAO in 2020 shows that the Southeast Asia region is the highest area for marine fish catches at 20 percent of the total world catch [12]. Four of the world's top ten largest fish producers are from ASEAN, Indonesia, Thailand, Viet Nam, and the Philippines [13]. Southeast Asia is also the world's primary producer of tuna and the world's leading exporter of canned tuna, with a value of USD 7 billion [14]. The increase in global fish consumption has spurred hunting for fishery commodities. This hunting encourages the practice of overfishing, overcapacity, destructive fishing, and illegal, unreported, and unregulated fishing (IUUF). Overfishing is a condition where the catch in an area has exceeded the environmental carrying capacity of fishery resources [15].

All countries, including ASEAN, face sustainability problems caused by IUUF practices. The ATEL scheme is expected to encourage ASEAN countries to implement sustainable tuna management in the region. In addition, ATEL is expected to answer various challenges in managing the tuna trade in ASEAN, such as growing market requirements, logistics and transportation, IUUF, inefficient fishing practices, equal access to small-scale fisheries, and lack of active collaboration among countries in the region. The need for a fishery eco-label certificate is still difficult to fulfill for small-scale fishermen [16]. There needs to be an adjustment from the certificate agency to this. The fast-growing world population in recent decades led to each country's increasing burden in fulfilling their people's food needs. Food security is also closely related to poverty alleviation, and then if this problem can be resolved with comprehensive action, the problem of poverty can be overcome.

3.2 Implementation of eco-label schemes in various countries

The implementation of area-based eco-label schemes is minimal, and their implementation is not yet optimal. However, several initiatives need further study. On fisheries, eco-label initiatives in several countries such as Japan, Iceland, Alaska, Canada, and the United States are one of the references [17]. How fishery eco-labels need to be applied follows the characteristics of each country. Eco-labels that are transnational lead to the monopoly of an organization. The emergence of national eco-labels is the answer from various countries to developing transnational eco-labels such as the Marine Stewardship Council (MSC). This study strengthens the previous research [18] on applying national eco-label certificates in Norway and Iceland.

Strengthening evidence of how the national fisheries eco-label scheme has become a demand from stakeholders in Indonesia [9]. The national fisheries eco-label scheme can be the backbone of ATEL implementation in Southeast Asia. The development of the European Ecolabel or EU Flower in Spain is one of the pieces of evidence for applying eco-labels in a region [19]. Future application of ATEL needs to consider the application of the EU Ecolabel as comparison material. Implementing the EU Ecolabel faces various obstacles, such as competition with other eco-labels and public perceptions. Research shows that several factors strengthen the implementation of regional eco-labels, including public management, communication strategies, criteria for sustainable public procurement, and regional income per capita to international trade incentives.

The application of fisheries eco-labels in the ASEAN region also supports the need for the application of fisheries eco-labels in the Southeast Asian region [20]. ASEAN countries are generally oriented towards the international legal framework of sustainable fish management. The fisheries eco-label is more of a supporter of the implementation of sustainable fish management that refers to the Responsible Fisheries Code of Ethics and the Food and Agriculture Organization (FAO) Guidelines for Ecolabelling of Fish and Fishery Products from Marine/Inland Fisheries. Fishery certification requires a process of improvement to the concept of adaptive management and a precautionary approach. In addition, fisheries monitoring, control and supervision, and synchronization of fisheries regulations are needed to support the certification process. The research was carried out before ATEL was approved at the ASEAN Ministerial level [20].

ATEL shows that the certificate is needed to strengthen regional cooperation and encourage integrated tuna management in the Southeast Asian region [21]. This research supports the ATEL integration process that must be carried out between various eco-label schemes implemented in the ASEAN region. The leadership of one or more countries in ASEAN is the key to how the integration process is carried out. The research shows that most ASEAN countries have implemented eco-label schemes in their respective countries, both from government and non-government initiatives. This shows that ASEAN countries know the need for environmental conservation through this scheme. Implementing ATEL can be faster if the integration process between the eco-label schemes can be realized. However, until now, this process has not been discussed at all. As the initiator of ATEL, Indonesia is not active in encouraging this process.

This makes the implementation of ATEL difficult. There has been no significant development from the ASEAN Ministers of Agriculture and Forestry agreement on the ATEL scheme from 2018 until 2022. The Indonesian Government, in this case, the Ministry of Marine Affairs and Fisheries, even tends to ignore ATEL. Indonesia has pushed for the ATEL initiative since 2012 and has gone through a relatively long process for six years. Indonesia's leadership in ATEL seems to have disappeared after the change in leadership at the Ministry of Marine Affairs and Fisheries in 2019 after the second period of the Government of President Joko Widodo.

There has been a shift in priorities from the Ministry of Marine Affairs and Fisheries toward developing the marine and fisheries sector. This looks like an anomaly from Indonesia as the initiator of ATEL, which is no longer enthusiastic about encouraging its implementation. Unfortunately, Indonesia's inactive condition did not receive input or correction from the leaders of other ASEAN countries. As a result, the implementation of ATEL is suspended without any input or correction from the Indonesian Government.

The Government's role in building public trust is still high. Stakeholders in the fisheries sector, especially tuna species, are still waiting for the Government's attitude regarding the implementation of ATEL. This is where the evidence is that leadership plays a vital role in the entire policy process in ASEAN. A comprehensive cooperation policy is difficult to realize without strong leadership and initiative from a member country. Government leadership is urgently needed to support the application of eco-labels. Consumers in the US, UK, Denmark, and Sweden prove that eco-labeling can be successful with substantial support or full government support at all stages [22]. The level of consumer confidence in eco-label products increases with the more intense involvement of the Government.

Although currently, in various regions of the world, the role of Government is increasingly being reduced to various governance platforms without Government, on the other hand, the role of Government is very much needed for various implementations of environmentally friendly platforms [23]. His research stated that the greater the management without the Government would reduce the possibility of the system operating effectively. The role of the Government must be decisive so that non-government actors have incentives to cooperate and the Government does not worry that their position will be replaced.

On the other hand, the form of non-government management is referred to [24] in his research as Private Informational Governance, or PIG for short. PIG has become a trend in various European countries when direct regulatory mechanisms from the Government are considered ineffective in overcoming environmental damage and increasing public awareness of the importance of protecting the environment. Although PIG is considered an effective alternative to direct government regulation, PIG needs to address gaps in practice. In his research, this gap was stated by [25], which identified six gaps in the application of private informational governance (PIG), namely geographical, informational, communication, compliance, power, and legitimacy. This gap is difficult to overcome because the eco-label certification body does not want to understand and is one-way. As a result, the certification process stalled, and the stakeholders gradually left the certification process.

3.3. The next step of ATEL
The ATEL scheme needs to be encouraged by the leaders of countries in Southeast Asia or ASEAN. The ATEL platform requires integrated support between countries so that its implementation can run smoothly [26]. The ATEL scheme is perfect and touches all levels of stakeholders in the ASEAN fisheries sector, including small fishermen. It is hoped that big countries in ASEAN, such as Indonesia, can play a more vital role.

As ATEL's initiator, Indonesia must fight for its implementation in ASEAN again. The Indonesian Government needs to show leadership in ASEAN and take sides with small fishermen and the sustainability of tuna fish resources. ATEL's initiation in 2012 is an excellent example of Indonesia's attitude toward sustainable management of fishery resources. This is also in line with Indonesia's commitment to implementing the Sustainable Development Goals. Indonesia also reminded all ASEAN countries to be committed to sustainable tuna fisheries management and combating the practice of IUU fishing.

Leadership is a problem that must be resolved immediately in ASEAN. In the last 20 years, not a single ASEAN leader has dared to take the initiative to cooperate. This is in line with [27] that the global political situation is in trouble due to the absence of leadership on various issues or G-zero (G-0). Likewise, with regional cooperation in the Southeast Asia region as it is today. Unfortunately, as the initiator of ATEL, Indonesia encouraged the implementation of the other international fisheries eco-label, an initiative of International NGOs and Multinational Companies [28]. In this case, the Indonesian Government of Marine Affairs and Fisheries promotes that eco-label [29]. This is counterproductive to the mission of implementation of ATEL. ASEAN countries are currently walking alone in implementing the tuna fishery eco-label scheme. Indonesia's move caused other ASEAN countries to no longer be active in supporting ATEL. This is the main problem with implementing ATEL in ASEAN.

4. Conclusion
Leadership is a critical factor for the success of regional cooperation on sustainable development in a region, including ASEAN. ATEL certification is a clear example of how the follow-up management in the region is constrained by the leadership of each country in the Southeast Asian region. Leadership is the primary driver of cooperation in the region. Without strong leadership, a region will find it challenging to develop cooperation, especially on strategic issues such as sustainable development. Strong leadership is needed to drive sustainable development cooperation in a region. Achieving the Sustainable Development Goals is challenging if regional cooperation in the region cannot be realized.

References
[1] The World Bank and FAO 2008 *The Economic Justification for Fisheries Reform*
[2] United Nations 1992 *UN Conference on Environment and Development* vol Agenda 21
[3] FAO 1995 *Code of Conduct for Responsible Fisheries* (Rome)
[4] FAO 2009 *Guidelines on Ecolabelling For Fish and Fishery Products From Marine Capture Fisheries* (Rome: FAO Rome)
[5] Vitalis V 2002 Round Table on Sustainable Development: Private Voluntary Eco-labels: Trade

Distorting, Discriminatory, and Environmentally Disappointing *Organ. Econ. Cooperation Dev.* 1–18

[6] Potts T and Haward M 2007 International trade, eco-labelling, and sustainable fisheries - Recent issues, concepts and practices *Environ. Dev. Sustain.* **9** 91–106

[7] Notohamijoyo A, Huseini M and Fauzi S 2018 ASEAN tuna ecolabelling (ATEL): The challenge and opportunity of the first seafood regional ecolabelling in the world *E3S Web Conf.* **74**

[8] Department of Economic and Social Affairs 2021 The 17 Goals *United Nations*

[9] Notohamijoyo A 2016 *The Implementation of Sustainable Development Principles on Seafood Ecolabel Certificate (Case study on the failure of the implementation of Marine Stewardship Council certification in Indonesia)* (Universitas Indonesia)

[10] Melrose J, Perroy R and Careas S 2015 Memorandum of Understanding on ASEAN Co-operation in Agriculture and Forest Products Promotion Scheme *State. Agric. L. Use Baseline 2015* **1**

[11] ASEAN 2016 *Strategic Plan of Action on ASEAN Cooperation on Fisheries*

[12] FAO 2020 *The State of World Fisheries and Aquaculture 2020. Sustainability in action* (Rome)

[13] The Association of Southeast Asian Nations 2021 *Investing in ASEAN 2021 - 2022*

[14] Hasnan L 2019 Southeast Asia's lucrative tuna industry *Asean Post*

[15] Fauzi A 2011 *Ekonomi Perikanan* (Jakarta: Kompas Gramedia)

[16] Wakamatsu M and Wakamatsu H 2017 The certification of small-scale fisheries *Mar. Policy* **77** 97–103

[17] Foley P and Havice E 2016 The rise of territorial eco-certifications: New politics of transnational sustainability governance in the fishery sector *Geoforum* **69** 24–33

[18] Kvalvik I, Noestvold B H and Young J A 2014 National or supranational fisheries sustainability certification schemes? A critical analysis of Norwegian and Icelandic responses *Mar. Policy* **46** 137–42

[19] Prieto-Sandoval V, Mejía-Villa A, Ormazabal M and Jaca C 2020 Challenges for ecolabeling growth: lessons from the EU Ecolabel in Spain *Int. J. Life Cycle Assess.* **25** 856–67

[20] Lieng S, Yagi N and Ishihara H 2018 Global ecolabelling certification standards and ASEAN fisheries: Can fisheries legislations in ASEAN countries support the fisheries certification? *Sustain.* **10** 1–17

[21] Notohamijoyo A, Huseini M, Koestoer R H and Fauzi S 2020 The Integration of the National Ecolabel in Southeast Asia to Support Asian Tuna Ecolabelling (Atel) 651–6

[22] Sønderskov K M and Daugbjerg C 2015 Eco-Labelling , the State and Consumer Confidence Eco-Labelling , the State and Consumer Confidence Kim Mannemar Sønderskov * and Carsten Daugbjerg ** ** Research School of Social Sciences , Australian National University and

[23] Börzel T A 2010 *Governance with / out Government False Promises or Flawed Premises ?*

[24] Pristupa A O, Lamers M and Amelung B 2015 Private informational governance in Post-Soviet waters: Implications of the Marine Stewardship Council certification in the Russian Barents Sea region *Fish. Res.* **182** 128–35

[25] Boström M, Jönsson A M, Lockie S, Mol A P J and Oosterveer P 2015 Sustainable and responsible supply chain governance: Challenges and opportunities *J. Clean. Prod.* **107** 1–7

[26] ASEAN Tuna Working Group 2018 *Policy Paper on the Establishment of ASEAN Regional Ecolabelling Scheme* (Viet Nam)

[27] Reykers Y 2013 Ian Bremmer (2012). Every Nation for Itself: Winners and Losers in a G-Zero World. New York: Portfolio/Penguin, ISBN 978-0670921041, 229 pp., £ 14.99. *Hague J. Dipl.* **8** 343–4

[28] Biro Humas dan Kerjasama Luar Negeri 2021 Protecting Global Market, Minister Trenggono Asks for MSC Tuna Certificate to be Maintained *https://kkp.go.id/artikel/27621-jaga-pasar-global-menteri-trenggono-minta-sertifikat-msc-tuna-dipertahankan*

[29] Humas Ditjen PDSPKP 2021 Dahsyat! 71% Konsumen Seafood Inginkan Produk Perikanan Berkelanjutan dan Ramah Lingkungan *https://kkp.go.id/artikel/36154-dahsyat-71-konsumen-seafood-inginkan-produk-perikanan-berkelanjutan-dan-ramah-lingkungan*

IOP Conf. Series: Earth and Environmental Science 1111 (2022) 012080 doi:10.1088/1755-1315/1111/1/012080

Economic aspects of Green Energy Development in The Context of Maintaining Strategic Sustainability and Environmental Conservation

Andrey Zaytsev[1*], Nikolay Dmitriev[2] and Tonny Sebbaggala[3]
[1] Graduate School of Industrial Economics, Peter the Great St. Petersburg Polytechnic University, St. Petersburg, 195251, Russian Federation, 0000-0002-4372- 4207
[2] Graduate School of Industrial Economics, Peter the Great St. Petersburg Polytechnic University, St. Petersburg, 195251, Russian Federation, 0000-0003-0282- 1163
[3] Graduate School of Industrial Economics, Peter the Great St. Petersburg Polytechnic University, St. Petersburg, 195251, Russian Federation
[1*]andrey_z7@mail.ru, [2]dmitriev_nd@spbstu.ru, [3]tonymayambala.tm@gmail.com

Abstract. The energy needs of human civilization are increasing every year, causing the search for opportunities to meet them while observing the principles of sustainable development. The work aims to analyze the economic aspects of green energy as a fundamental component of a sustainable development strategy. To achieve the goal, key areas for the development of renewable energy were identified. The authors characterized the advantages of introducing innovations in the energy industry, which will allow: to reduce the negative anthropogenic impact on the environment; allow more intensive use of resource potential; prevent the exhaustion of the natural resource base; and contribute to the development of sectors of the national economy that are not related to the exploitation of carbon raw materials, etc. As a result, it was revealed that the measures for the formation of green energy are currently insufficient; from the point of view of the authors of the article, facilitating the creation and development of a technological base in the field of green energy will not only provide economic benefits in the strategic perspective and keep up with global trends but will also have a positive impact on sustainability and stability in the environment and society.

1. Introduction

Since 1992, the outlook for climate change has inspired a broad movement dedicated to a sustainable future. A severe transformation of society has led to a change in economic thinking and the search for tools necessary for economic thinking of an environmental problem [1]. One of the critical areas for solving this problem is energy management, which must fully comply with energy efficiency and conservation principles. At the same time, back in 1984, the emphasis was on the industry, the energy consumption of which should not violate the energy security of the country and regions [2]. Despite the recognized importance of strategic sustainability, since that time, the energy requirements of human civilization have increased.

In the 21st century, industrial production is gaining unprecedented momentum and consumes over 90% of the world's energy. There are acute problems with the quality of the energy resources used and their impact on strategic sustainability. One of the potential ways to solve the problem is the transition to a green economy, which is based on the use of renewable energy sources as part of the energy-producing system of the national economy. However, the development of green energy is not always economically feasible [3]. From investors' perspective, raising capital for production projects in political uncertainty is not attractive enough. Increased risk and low return on investment in the short term delay green projects [4].

Sustainable development is directly related to the implementation of economic and social transformations. These processes aim to meet human needs while ensuring a sufficient quality of life for people in the long term [5]. Maintaining strategic sustainability is possible only with the rational use of resources, ensuring optimal use with a low amount of untreated waste. Green energy reduces waste, which is the first step toward sustainable development. Thus, the relevance of research in building a study of the effectiveness of sustainable business is increasing [6].

The improvement of existing and the search for new renewable energy sources attracts public attention, as they allow for the formation of "public" efficiency. Green energy provides a range of opportunities to replace fossil fuels, reduce dependence on carbon raw materials, transform the industry, reduce the adverse anthropogenic impact on the environment, and reduce emissions of harmful substances into the atmosphere [3,7,8]. The resolution of environmental problems in the context of maintaining strategic sustainability is in the interests of society. It is one of the primary conditions for the transition to the VIth technological mode. The convergence of digitalization into everyday reality makes it possible to significantly accelerate the processes of integrating information technologies into various business areas, including the energy sector [9].

The economic effect confirms the significance of the work in technologically advanced countries from the development of renewable energy sources. Currently, over 3% of the world's electricity is generated by solar and wind installations, which are traditional sources of green energy. Entrepreneurs in technologically advanced countries actively invest in green areas, relying on government benefits and support. At the same time, the capacity of green energy is doubling approximately every four years. So, from 2003-2013. the use of green energy on a global scale has increased 4 times; in Europe, the increase was 500% [1,5,10].

If we turn to the positive economic aspect for Russia, the development of green energy at the regional level will reduce long-distance fuel transportation, which is not always profitable on the territorial scale of the country. Also, despite the significant reserves of traditional energy resources, the environmental situation and the population's quality of life are not at a high enough level. The share of green energy in the energy balance of Russia as of the beginning of 2020 (pre-pandemic values) is approximately 2-4%. These figures cannot be called high, which also actualizes the solution to the problem in the field of economic feasibility of developing renewable energy in the country [3,11].

It should be noted that modern scientific and technological progress is directly related to the economic component. Ensuring economic growth in the modern model of the world order is associated with a focus on sustainable development. Maintaining strategic sustainability is impossible without environmental stability, the achievement of which, in turn, is impossible without the active dissemination and implementation of green technologies. The socio-economic essence of the development of green energy is associated with the qualitative improvement and renewal of the scientific and technological base and, therefore, is interconnected with the economic component. The result of the development of green energy is to improve the quality of life and the environmental situation. Thus, the importance of this problem is recognized at the global level, but the economic aspects are still not fully disclosed. It is proposed within the framework of this work to study the problem, which will allow us to substantiate the range of measures to implement the transition to a green economy.

This work aims to analyze the economic aspects of the development of green energy as a primary component of a sustainable development strategy. To achieve this goal, the authors analyzed the essence of green energy, identified its key areas and economic aspects, and presented the economic consequences of developing green energy from a strategic perspective. The features and problematic provisions in developing green energy in Russia were also considered, given the high potential of traditional energy in the country.

2. Methods

The methodological basis of the work is analysis, synthesis, and theoretical generalization because the authors propose to analyze and describe the existing trends in the development of green energy from an economic point of view. The study is theoretical. To conduct the study, the article analyzed the works and articles of the authors in the field of the use of renewable energy sources, the importance of developing green technologies, and building a green economy. The authors also touched upon studies on sustainable development, entrepreneurship, and innovative technologies associated with green energy development. Further, the information was synthesized and summarized, which made it possible to draw the author's conclusions on the problem under consideration.

Despite the sufficient scientific groundwork on this problem, there is still ambiguity in allocating its economic feasibility. The volume of research on this issue makes it possible to form an expanded base for substantiating the economic significance of the development of green energy. Also, the scope of the study is not limited to the environmental and economic benefits of the spread of green energy. For example, in the actual work of [12], the conceptual provisions of the "blue" economy were considered, thanks to the transition to which it is possible to ensure the protection of the global ecosystem when creating new jobs through the construction of innovative industries, including by modernizing the energy industry along the green path.

3. Results and Discussion

An environmentally responsible economy seeks to socially construct a political economy that successfully copes with the environmental problems facing human civilization. The active development of green energy makes it possible to use inexhaustible natural sources, primarily renewable ones. The primary green sources include wind, sun, the heat of the earth, and water [10]. Getting green energy does not cause damage to the environment of human life, and its sources are endless. To maintain strategic sustainability, the most important is integrating renewable and alternative systems into existing infrastructure. Studies of the synchronization of environmental and development needs with other social problems are based on the need to justify the economic feasibility of their development [13]. Technologically advanced countries are pursuing green solid energy policies beneficial in phasing out carbon-based commodities. These countries include Germany, Japan, Scandinavian countries, and others. Nevertheless, a complete rejection of traditional energy is impossible and irrational [14].

The development of science leads to the creation of technologies that allow cleaning the earth and air from industrial waste. This waste poses a primary threat to maintaining strategic sustainability since it is the industry that consumes the main volumes of energy. The thesis is that the extensive growth of industrial production affects the standard of living of the population throughout the planet. On the one hand, the satisfaction with the continuously growing needs of the population is growing; on the other hand, serious externalities are being provoked that harm the environment, biological diversity, and public health and do not destabilize many social processes [7,15].

Ensuring sustainable development is recognized internationally. For example, the UN report [16,17] noted that to ensure strategic sustainability and prevent large-scale natural disasters, it is necessary to keep global warming at the level of 1.5 ºC compared to the pre-industrial era. The average annual temperature has risen by 1 ºC due to the negative anthropological impact on the ecosphere. To prevent environmental degradation, UN experts justify the reduction of CO2 emissions to zero by 2050. Achieving these values in current conditions causes profound skepticism. However, one key step to achieving these targets is moving away from traditional fossil fuel energy. Reducing dependence on traditional fossil fuels encourages states to intensify programs for the development of green energy.

If we turn to long-term forecasts, the most exciting scenario for transforming the world energy balance is presented in Table 1 [11] (Note: MMTOE – million tons of oil equivalent). It should be noted that even if the forecast values are reached (the share of green energy will reach 24- 40%), the quantitative use of traditional energy sources will also increase. Given the energy challenges emerging as early as 2022 due to the restructuring of markets, one should be even more skeptical about the development of green energy at the current activity level. This situation threatens the stability of the ecosystem and the maintenance of the quality of life of future generations [3].

Table 1. World Energy Balance: Strategic Forecast

Energy source	2000		2030		2050	
	MMTOE	%	MMTOE.	%	MMTOE	%
Oil	3,7	40	4,3	24	4,0	16
Gas	2,1	22	5,4	30	8,3	33
Coal	2,2	24	4,5	25	6,2	25
Nuclear power	0,6	6,5	2,0	11	3,5	14
Green energy	1,3	14	3,8	21	6-10	24-40
Total	9,3	100	18,0	100	27-31	100

The energy crisis of the 2010s has affected the need to revise energy strategies. Likely, the crisis of 2022 will also impact the prospects for sustainable development. Now, scenarios for the development of human civilization and the trajectory of maintaining strategic stability are being determined. Let's turn to the prospects for world energy development. The study of the Institute of Energy Strategy [18] is of interest, in which three potential scenarios for energy development are indicated: inertial, stagnation, and innovative.

It is proposed to express the perspectives of each of the potential scenarios:

• Inertial scenario. According to this scenario, average economic growth rates are projected to remain at the beginning of the 21st century. First, high economic growth rates will be observed among developing countries, leading to a significant increase in the consumption of traditional energy sources. This scenario assumes the onset of an ecological catastrophe and socio-ecological decline due to the deterioration in the quality of life of a significant part of the world's population.

• Stagnation scenario. Under this scenario, the growth of information technology in society is projected to accelerate, and the influence of complex political, social, legal, and economic mechanisms to curb the growth of consumption of traditional energy sources and the gradual phase-out of carbon fuels will increase. This scenario will not lead to a global environmental catastrophe. Still, it will harm economic growth, especially in developing countries, leading to increased macroeconomic imbalances and disproportions in planetary development.

• Innovative scenario. According to this scenario, a scientific and technological breakthrough is predicted. As a result of technological development, there will be a significant reduction in the consumption of "dirty" fuel, and the use of renewable energy sources will become economically viable. This scenario is favorable since it will allow for maintaining strategic stability. Still, from an economic point of view, it is costly since its onset is possible only if a considerable amount of investment is attracted.

Technological problems currently complicate the prospects for the development of green energy. For example, solar and wind stations operate only when the sun shines, and the wind blows. The development of batteries that can store energy for a long time is currently in infancy [7,8]. Consequently, obtaining a "green" rent can be achieved only with the progressive growth of technology. However, economic benefits and social effects remain no less attractive [19].

It is necessary to ensure supranational coordination to implement an innovative scenario and a breakthrough in the field of green technologies. A prime example of such coordination is China's Belt and Road Initiative (BRI). As part of this megaproject, many investment projects are being implemented that affect the development of green energy. Thanks to coordinating actions between countries and effective technology transfer, the initiative's participants are already receiving economic benefits from green projects [14]. The economic benefit is the driving element in a market economy, and without the entrepreneurial sector, states cannot solve the global problems facing humanity.

Innovative progress depends on the complex efforts of various market players, and entrepreneurial activity allows for launching innovation development processes in the most significant areas. The business contributes to the creation of supply and demand in the market and influences the formation of trends, including in the field of energy consumption. In many ways, these processes are interconnected with the market value, which the business is focused on when implementing its innovative activity [20,21]. It is also noted that an environmental organizational identity can positively influence the effectiveness of green innovation indirectly through environmental commitment and environmental and organizational legitimacy. Many industries that use "dirty" energy sources are implementing green innovations, increasing their efficiency and market value. For example, in metallurgy, innovations are being introduced that help reduce the negative impact on the environment and increase the efficiency of the intensive use of resources [22,23].

Investments in green energy in 2017 exceeded $340 billion. More than $50 billion is spent annually on energy-efficient technologies to save the generated energy. It is predicted that by 2030 the level of investment in green energy will exceed 1.9 trillion. $. It should be taken into account that their payback period exceeds 5 years. Therefore, the social responsibility of investors increases, and there is a potential for approaching the innovative scenario of energy development [11].

The transformation of the ecological and economic space in many territories leads to the developing and further diffusion of green innovations as a necessary condition for developing a green economy. In the scientific community, environmental indicators are developed, and energy consumption and generation's impact on territories' ecological and economic development is assessed [24,25]. In practice, to actively promote green energy, the concept of "green tariffs" was developed (2006 - 45 countries, 2019 - 125 countries), the essence of which is to maintain the economic potential of renewable energy sources by increasing tariffs for traditional electricity and receiving compensation for negative consequences for the environment from the use of "dirty" fuel [3].

If we consider the Russian practice of developing green energy, unlike many countries that are dependent on energy resources (for example, the countries of the European Union), Russia has a rich resource potential and a high level of energy security. However, even for Russia, there is the economic feasibility of developing green energy. Even though Russia is a key market player in the supply of traditional energy resources, the development of green energy is in the interest of ensuring the country's economic security. For example, it can carry out internal energy diversification while maintaining export potential in the external market. In the medium term, prices for traditional energy sources will remain high, making it possible to receive additional rental income, which, if used rationally, can be used for innovative development and a gradual reduction in dependence on the sale of natural resources.

Since 2013, the main imperatives for supporting green energy have been formed in Russian practice. Financing mechanisms are aimed at meeting domestic energy needs through implementing projects to develop alternative energy. In turn, the financing itself is local and does not contribute to increasing innovation activity at the required level. To increase interest in this area among entrepreneurs-innovators, it is required to form a detailed support roadmap at different levels and with different structures. Current trends assume the development of generating capacity based on renewable energy in Russia to about 5.5 GW by 2024, which will be 2-3% of the country's total energy mix. For comparison: hydrocarbons - 50%, coal - 15%, atomic energy - 14%. Without an increase in the activity of private investment, the development of green energy will remain an underdeveloped area.

From the standpoint of energy security, the underdevelopment of green energy cannot threaten Russia, since the energy balance is in excess. At the same time, there are constraining elements that impede the development of green energy and, as a result, impede the maintenance of strategic sustainability. Curbing the development of green energy and the investment component is associated with another economic problem - a technological one. A large proportion of the technological components required for the construction of renewable energy facilities are imported, the purchase of which is impossible due to sanctions pressure and requirements for the localization of components and technologies [11]. There is an urgent need for scientific and technological support. At the same time, considering this direction at the government level will allow the developing of many related areas and preparing highly qualified personnel, the uniqueness of which will allow you to receive scientific rent.

The territorial expanses of Russia represent a considerable natural, technical, and economic potential. Unfortunately, most of it remains unused. For energy purposes, there is an opportunity to implement projects in the thermal processing of municipal solid waste, the objects of which are constantly growing. So, now, about 1.5% of municipal solid waste is processed for energy purposes, and a significant part of it is disposed of without further processing into any good. In many regions of Russia, there are favorable conditions for the spread of wind energy. Investments in the region of $ 0.5 billion allow the creation of 2 GW of wind generation, the maintenance of which will be cheaper than using carbon fuel. This direction is especially relevant for remote territories, where the delivery of traditional minerals is costly from a logistical point of view.

If you pay attention to solar energy, then in the southern regions of Russia, there is great potential for using solar panels due to the favorable climate. At the same time, these territories contain significant agricultural lands and agro-industrial enterprises, which can provide a significant share of their energy consumption using solar energy and significantly expand their production by saving on the purchase of more expensive traditional energy. The listed areas are not the only ones; other areas can bring social and economic benefits. It should be noted that green energy will reduce social tensions from pollution and waste, which are increasingly manifested in connection with "garbage" problems in the regions. Although hydrocarbons cover the country's needs, focusing only on their use from a strategic perspective will lead to missed opportunities. It will not allow increasing the level of manufacturability of individual industries. It should be recognized that in the long term, the processes of development of green energy in the world are irreversible, and it is necessary to find benefits from following the global trend, including economic benefits.

The main task of the Russian economy is to achieve economic growth by increasing the competitiveness of enterprises based on the generation of new knowledge, introducing innovative technologies in business processes, and implementing technologically complex projects. Innovative energy projects will inevitably create new jobs, primarily for highly qualified personnel. Thus, Russia has the potential to develop green energy without dependence on energy imports and form a high-tech industry for producing components for green energy with a high export potential [3,26].

It is proposed to note that the innovative areas of green energy correspond to the key provisions of the "blue" economy, according to which another can replace any resource if it is necessary for production needs; there is no waste in nature; any by-product is a source for a new product. In the context of maintaining strategic sustainability, the high importance of integrating environmental and social aspects of life in civilizational progress is determined, and green energy is becoming one of the steps in the transition of humankind to a new level of development [12; 15]. Implementing energy and climate policy following sustainable development goals will affect many areas of the economy. It will affect the functioning of all industries, including construction, transport, agriculture, services, etc. [5,27].

Clean energy is based on recombinant innovations that can have practical applications in many industries. The innovative potential of alternative energy is also an essential factor for human development since it allows us to talk about the transition to circular development models [28,29]. Trends in the global economic space are associated with a focus on sustainable development. One of the key directions in achieving sustainable development is the creation of favorable conditions for the organization of circular production. Linear models of the economy remain in the past because they do not correspond to the principles of sustainability and cannot be viable in the socio-economic systems of the future. The circular economy should solve many of the global problems of humankind. For economic progress, green energy and the transition to circular models should become the primary dominants, supporting the rebuilding of business structures, reorienting them towards the so-called sustainable business [30,31].

The last decade's trends show that more and more enterprises, regardless of type and size, are realizing the inevitability of making an active contribution to society, both socially and environmentally, while maintaining their financial profitability and ensuring economic efficiency from their activities. There has been a realization that it is no longer possible to waste limited resources and ignore the expectations of both internal and external stakeholders. To survive in an era of change, you must adapt and rebuild your business models [32]. The circular economy and sustainable development are being increasingly popularized yearly in academia, industry, and politics. The conceptual landscape is changing in creating economic conditions for development, and green energy is the foundation of the future of human civilization [33]. The circular economy and circular business models represent the latest attempt to conceptualize the integration of economic activity and environmental well-being in a sustainable manner. As a result, the circular economy has been operationalized in business and politics, and green energy has become a generally recognized trend, which brings particular economic preferences [34]. Thus, based on the theoretical analysis, it becomes evident that the development of green energy is conditioned by maintaining socio-ecological stability and obtaining significant economic efficiency, and in the long term, obtaining a "green" rent [35,36].

Interest in science and practice in such issues as climate change, energy security, and alternative energy sources is at an all-time high. The consequences of finding solutions now fall on the shoulders of future generations. There are no simple solutions to solve global energy problems, so costs and benefits must be considered when evaluating all energy alternatives. Green energy has advantages and disadvantages, but its development is economically feasible. In turn, the adoption of managerial decisions on production and energy use should be built based on sound economic analysis. It examines market forces and political goals, including economic prosperity, environmental protection, and other factors affecting the well-being of society [37]. Figure 1 proposes a step-by-step process for developing green energy and the positive effects of their achievement [3]. It becomes evident that the economic need for the development of green energy is due to several positive parameters since its development will reduce the negative anthropogenic impact on the environment; allow more intensive use of resource potential; prevent the exhaustion of the natural resource base; will contribute to the development of sectors of the national economy that are not related to the exploitation of carbon raw materials, etc

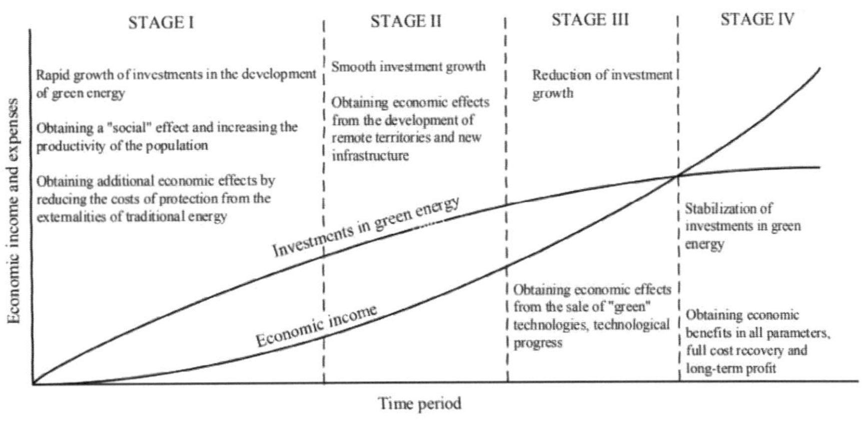

Figure 1. Economic efficiency from the development of green energy

4. Conclusion

As part of the article, it was noted that developing green energy is one of the promising areas for maintaining strategic sustainability. However, the economic aspects of activating green technologies are not always unambiguous. The issue of green energy raised in this article confirms the economic feasibility of its development. As a result, it was revealed that the measures for forming green energy are currently insufficient, but in the future, its development will provide positive effects, including economic effects.

The authors listed the energy development scenarios. The innovative development scenario is the most favorable for maintaining strategic sustainability, which requires the implementation of complex scientific and technological projects. Under such a scenario, the potential for economic growth will be preserved without a threat to the livelihoods of future generations.

From the article's authors' point of view, facilitating the creation and development of a technological base in the field of green energy will provide economic benefits from a strategic perspective and keep up with global trends. Still, it will also positively impact sustainability and stability in the environment and society.

The economic benefits of green energy development in the short term are associated with "social" effects, allowing you to receive market value or increase labor productivity; in the long term, the development of green energy will make it possible to receive "green" rent and form strategic directions for the development of related industries. The departure from traditional energy is associated with technological progress since, without developing the most promising technologies, one can end up in a position of backwardness and significantly slow down economic growth.

The theoretical analysis carried out can be used in future studies to substantiate specific measures aimed at implementing an innovative growth scenario and increasing the role of green energy in ensuring sustainable development.

References

[1] Bermejo R 2004 *Handbook for a sustainable economy* (New York: Springer Science)
[2] Fahey L 2017 *Energy management in industrial firms* (London: Routledge)
[3] Dmitriev N, Zaitsev A and Zhiltsov S 2020 Economic Aspects of The Need to Develop Green Energy *Business. Educ. Law* **53** 63–60
[4] Detemple J and Kitapbayev Y 2020 The value of green energy under regulation uncertainty *Energy Econ.* **89**
[5] Bhattacharyya S C 2019 *Energy economics: Concepts, issues, markets and governance* (New York: Springer)
[6] Scott T 2013 *The Sustainable Business: A Practitioner's Guide to Achieving Long-Term Profitability and Competitiveness* (London: Routledge)
[7] Bhowmik C, Bhowmik S and Ray A 2022 Green energy sources selection for sustainable planning: a case study *IEEE Trans. Eng. Manag* **69** 1322–34
[8] Bhowmik C, Baruah A, Bhowmik S and Ray A 2018 Green energy sources selection for sustainable energy planning using multi-criteria decision-making approach *IOP Conf. Ser.: Mater. Sci. and Engin.* **377**
[9] Vlasova N V, Kuznetsov D V, Mehdiev S Z, Timofeeva E S and Chistyakov M S 2021 Information Technologies in the Context of Forming the Synergy of Post-industrial Consciousness and Digital Economy *Mod. Glob. Econ. Sys.: Evol. Devel.t vs. Revol. leap*
[10] Soderbaum P 2013 *Ecological economics: Political economics for social and environmental development* (London: Routledge)
[11] The Energy Research Institute of the Russian Academy of Sciences 2013 *Global and Russian energy outlook up to 2040* (Moscow: ERI RAS, ACRF)
[12] Gunter P 2010 *The Blue Economy: 10 Years, 100 Innovations, 100 Million Jobs* (Boulder: Paradigm Publications)
[13] Azzopardi B 2017 *Sustainable development in energy systems* (Cham: Springer)
[14] Chernysheva N A, Perskaya V V, Petrov A M and Bakulina A A 2019 Green Energy for Belt and Road Initiative : Economic Aspects Today and in the Future *Inter. J. Ener. Econ. and Pol.* **9** 178–85
[15] Rodionov D G, Konnikov E A and Konnikova O A 2018 Approaches to ensuring the sustainability of industrial enterprises of different technological levels 277–82 *J. Soc. Sci. Res.*

277–82
[16] IPCC 2018 *Global Warming of 1.5 oC. The Intergovernmental Panel on Climate Change*
[17] SR15 2018 *Special Report on Global Warming of 1.5 °C*
[18] Belogoryev, A M, Bushuev, V V, Gromov, A I, Kurichev, N K, Mastepanov, A M and Troitsky A 2011 *Trends and scenarios of global energy development in the first half of the XXI century* (Moscow: Institute of Energy Strategy)
[19] Common M and Stagl S 2005 *Ecological economics: An introduction* (Cambridge: Cambridge University Press)
[20] Zaytsev A, Dmitriev N and Asaturova Y 2020 Developing innovative activity management tools as a way to increase the market capitalization of an industrial enterprise *Proc. Eur. Conf. Innov. Entrep.*
[21] Zaytsev A, Rodionov D, Dmitriev N and Faisullin R 2020 Building a Model for Managing the Market Value of an Industrial Enterprise Based on Regulating its Innovation Activity *Acad. Strateg. Manag. J.* **19** 1–13
[22] Savchenkov S A, Bazhin V Y and Volkova O 2020 Tendencies of innovation development of the russian iron and steel industry on the base of patent analytics for the largest national metallurgical companies *CIS Iron Steel* **20** 76–82
[23] Chang C and Chen Y 2013 Green organizational identity and green innovation *Manag. Decis* **51** 1056–70
[24] Shabunina T V, Shchelkina S P and Rodionov D G 2017 An innovative approach to the transformation of eco-economic space of a region based on the green economy principles *Acad. Strateg. Manag. J.* **16** 176–85
[25] Kichigin, O E 2017 Fossil fuel production impact on regional eco-economic development *Inter. J. Ecol. Econ. Stat.* **38** 12–22
[26] Kalinina O, Kapustina I, Barykin S, Balchik E and Sedyakina A 2019 Development of a combined approach of innovative and traditional scenarios of the company's strategy *Proc. 33rd IBIMA Conf.*
[27] Doukas H, Flamos A and Lieu J 2018 *Understanding risks and uncertainties in energy and climate policy: Multidisciplinary methods and tools for a low carbon society* (Cham: Springer)
[28] Zaytsev A, Dmitriev N, Rodionov D and Magradze T 2021 Assessment of the Innovative Potential of Alternative Energy in the Context of the Transition to the Circular Economy *Inter. J. Tech.* **12** 1328–38
[29] Guan J and Yan Y 2016 Technological proximity and recombinative innovation in the alternative energy field *Res. Pol.* **45** 1460–73
[30] Lewandowski M 2016 Designing the business models for circular economy-towards the conceptual framework *Sustainability* **8** 1–28
[31] Dmitriev N and Zaytsev A 2021 Effectiveness of Lean Business Model in Circular Manufacturing *Proc. 20th Eur. Conf. Res. Methodol. Bus. Manag. Stud.*
[32] Durst S, Hammoda B, Nguyen H D and Asl M M 2021 Sustainable business models and small- and medium-sized enterprises – A literature revie *Sust.Devel. Engin. Econ.* **1** 54–79
[33] Geissdoerfer M, Savaget P, Bocken N and Hultink E 2017 The Circular Economy – A new sustainability paradigm? *J. Clea. Prod.* **143** 757–68
[34] Murray A, Skene K, Haynes K, Murray A, Skene K and Haynes K 2015 The Circular Economy : An Interdisciplinary Exploration of the Concept and Application in a Global Context *J. Bus. Ethics* **3** 369–80
[35] Zaytsev A 2020 Rental income structure in economy as a basis for sustainable agrarian relations in the agro-industrial complex *Sustainability*, **18**
[36] Dmitriev, N., Zaytsev, A. and Gorovoy A. 2020 Development of the methodology for calculating the intellectual rent of an industrial enterprise *Proc. 17th Inter. Conf. Intel. Cap., Know. Manag. Organ. Lear.*
[37] Schwarz P M 2017 *Energy Economics* (London: Taylor & Francis Inc)

Environmental Self-Identity and Pro-Environmental Behavior in Climate Change Issue: Mediation Effect of Belief in Global Warming and Guilty Feeling

Muhammad Abdan Shadiqi[1]*, Ratna Djuwita[2], Silvia Kristanti Tri Febriana[1], Lulu Septiannisa[1], Muhammad Wildi[1], Yuli Rahmawati[1]
[1]Study Program of Psychology, Faculty of Medicine, Universitas Lambung Mangkurat, Banjarbaru, Indonesia, [1]ORCID ID 0000-0002-5478-6788
[2]Faculty of Psychology, Universitas Indonesia, Depok, Indonesia
[1]*abdan.shadiqi@ulm.ac.id , [2]juwita@ui.ac.id, [3]s.kristanti@ulm.ac.id, [4]lulusep2001@gmail.com, [5]muhammadwildi999@gmail.com, [6]yryuli01@gmail.com

Abstract. There are several arguments about the role of identity factors behind someone's pro-environmental behavior (PEB) encountering climate change. The authors assumed other factors might explain the correlation between identity factors and PEB. This study aims to examine the mediating effect of belief in global warming and guilty feeling on the correlation between environmental self-identity and pro-environmental behavior related to climate change issues. We conducted a cross-sectional survey involving 202 university students selected using the purposive sampling method. Using multiple mediators of model 4 from PROCESS Macro Hayes, the result shows that environmental self-identity is the most decisive variable in predicting PEB. Belief in global warming is also a significant PEB predictor, while the guilty feeling is insignificant. The mediation model analysis showed that belief in global warming partly mediated the correlation between environmental self-identity and PEB, and the guilty feeling was not a significant mediator. This finding explains that self-identity as an environmentalist is directly related to PEB, or it could also be mediated through the belief in global warming. This study implies that encouraging students to be pro-environmentalist and developing the belief in current climate change is essential to increase students' PEB.

1. Introduction

For these past few years, climate change has triggered various natural disasters. The Intergovernmental Panel on Climate Change (IPCC) report in February 2022 shows biological evidence of earth damage caused by human behavior that leads to climate change [1]. Survey data of the You Gov-Cambridge Globalism Project in 2019 found that Indonesia had the highest proportion of people who did not believe the role of humans in causing climate change [2]. Convincing the deniers that global warming is real is essential to promote pro-environmental action [3]. In psychology, one of the fascinating topics explaining human behavior regarding climate change is Pro-Environmental Behavior (PEB) [4]. PEB is considered an action to attempt as hard as possible to reduce harm or provide a benefit to the environment [5]. There are internal and external factors affecting PEB [6]. Among them is the self-identity factor in the environmental self-identity concept [7]. Research shows that environmental self-identity is one of the strongest PEB predictors [8,9]. However, this factor is insignificant to other kinds of PEB [8]. For instance, anything related to one-off home energy conservation, traveling behavior, and environmental-political action [8].

1.1. Environmental Identity in Climate Change Issue

Dan P. McAdams, a personality psychologist, explained that self-identity is how a person sees himself/herself, including all self-aspects [10]. A meta-analysis study concluded that individual (self) identity was more substantial than social identity-PEB associations [9]. Another meta-analysis study on climate change issues found that environmental self-identity plays a significant role in climate-friendly action [11]. Numerous scientific findings found a strong correlation between environmental self-identity and PEB [7,8,12], where the identity factor plays a vital role in forming climate change beliefs [13]. A previous study also found a connection between environmental self-identity and climate change beliefs [14]. In this study, we use the term 'belief in global warming to explain climate change belief. We select guilty feeling as an emotional factor associated with environmental self-identity. This variable selection aligns with research associating environmental self-identity with a global and environmental guilty feeling [15]. We assume that people with a courageous environmental self-identity tend to behave pro-environmentally, believe in facts about climate change, and feel guilty for causing climate change. Therefore, we suggest it hypothesizes as follows: H1: Environmental self-identity has a positive effect on PEB; H2: Environmental self-identity has a positive effect on belief in global warming; and H3: Environmental self-identity has a positive effect on guilty feelings.

1.2. Belief, Emotion, and PEB

The belief in global warming has become one of the defining factors of environmental concern and increasing ecocentrism [16]. To people who are not concerned about climate change issues, changing their beliefs (from unconcerned to concerned) may threaten their identity [17]. We assume that those with an environmentalist identity will find it easier to behave pro-environmentally (without feeling threatened) and be supported by a belief that global warming is undoubtedly happening. Findings of the meta-analysis study concluded two significant emotions regarding PEB: pride and guilt. However, compared to pride, the guilty feeling is more consistent in explaining the varieties of PEB [18]. Truelove et al. study on the significant mediation effect of guilty feeling and PEB became our base to determine guilty feeling as a mediator variable [15]. In this research, we add novelty of findings to present internal factors of PEB from 3 classification factors at once: self, emotion, and belief.
H4: Belief in global warming has a positive effect on PEB; H5: Guilty feeling has a positive effect on PEB; H6: Belief in global warming mediates correlation between environmental self-identity and PEB; and H7: Guilty feeling mediates correlation between environmental self-identity and PEB.

1.3. Current Study

We chose this province of South Kalimantan, Indonesia, because of data from the Central Agency on Statistics (BPS) in 2015, which placed South Kalimantan in the third highest maximum temperature in Indonesia or the first place on the Kalimantan island [19]. We collected a student sample because, in 2018-2019, there was the #savemeratus student action. The Meratus Mountains have an essential role in climate change issues [20]. This research will provide novelty findings on alternative variables from identity-PEB associations, especially in the countries not included in WEIRD (Western, Educated, Industrialized, Rich, and Developed) [21].

2. Method

This study uses a quantitative approach. We conducted a cross-sectional survey technique using measuring instruments in an online questionnaire. We investigated PEB as a critical variable (Y), environmental self-identity as a predictor (X), and the mediator variables (M) were belief in global warming and guilty feeling. This study has received ethical clearance from the Faculty of Medicine in Universitas Lambung Mangkurat [227/KEPK-FK ULM/EC/VIII/2022].

2.1. *Participant*

We have collected the data from 221 participants, yet screening results showed 19 people are identified as data outliers (according to Mahalanobis, Cook's, and Leverage). According to this matter, only 202 participants made it to the final analysis. All samples are university students aged 18—25 (Mean=21.34, SD=1.28). The majority of samples are 158 women (78.2%) and 193 Moslems (95.5%).

2.2. *Measurements*

We used *surveymonkey.com* and distributed it online. We used an attention check question to measure participants' sincerity in responding to the questionnaire. We also used a valid and reliable measurement in Bahasa Indonesia, starting with giving information about climate change to explain the research context. Pro-environmental behavior scale [22] was used to determine the frequency of conducting activities to support environmental improvement (α=.858, eight items, 1=never – 5=very often, e.g., "recycled newspaper"). We determined self-image regarding the identification of someone who behaves pro-environmentally using an environmental self-identity scale [23] (α=.810, three items, e.g., "I am the type of person who acts environmentally-friendly"). The guilt scale, our self-developed questionnaire, was used to measure guilty feelings as human beings for current environmental conditions (α=.860, three items, e.g., "I feel guilty about the way humanity treats the environment"). Additionally, we measured belief on global warming scale [16] to determine someone's belief of the fact that climate change or global warming is happening (α=.756, four items, e.g., "I noticed several signs of global warming"). These three measurements used responding options of 1=strongly disagree – 7=strongly agree.

2.3. *Statistical Analysis*

We performed mediation regression analysis with Model 4 of PROCESS Macro Hayes [24] to test whether the predictor variable of the PEB can be through mediation variables (guilt and belief in global warming). To complete the description of the results, we analyzed the bivariate correlation among variables using the Pearson correlation. The entire analysis uses the assistance of SPSS.

3. Result and Discussion

Before analysis, we scored the participant answers to the questionnaire and calculated the mean score. Before conducting the correlation test and mediation regression model, we found that all data satisfied assumption tests (normality, linearity, multicollinearity, and heteroscedastic test).

3.1. *Result*

The result of bivariate correlation (table 1) shows that environmental self-identity (r=.442, p<.01), belief in global warming (r=.366, p<.01), and guilty feelings (r=.377, p<.01) are significantly correlated to PEB.

Table 1. Bivariate correlation between variables

	Scale	*α*	Mean	SD	1	2	3	4
1. Environmental self-identity	1-7	.810	5.384	.869	1	.440**	.545**	.442**
2. Belief in Global Warming	1-7	.756	6.003	.740		1	.523**	.366**
3. Guilty Feeling	1-7	.860	5.540	.965			1	.377**
4. PEB	1-5	.858	1.744	.799				1

*Note: ** p < .01*

We tested a multiple-mediation regression model using PROCESS Macro Hayes with bootstrapping 5,000 (figure 1). Result of the regression coefficient from direct effect shows that environmental self-identity significantly predicts positively on PEB (β=.300, t=3.969, p<.001, H1 is accepted). A similar finding also is found in one of the mediator variables, belief in global warming, which significantly predicts positively PEB (β=.168, t=2.264, p<.01, H4 is accepted). Nevertheless, the guilty feeling does not significantly predict PEB (β=.125, t=1.566, p=.119, H5 is rejected).

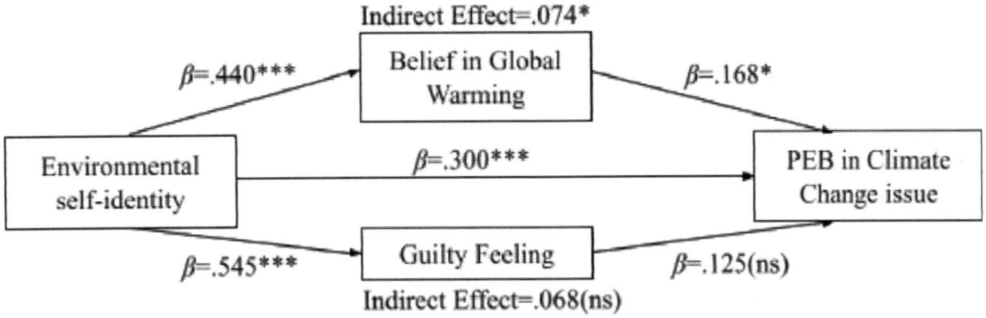

Figure 1. The result of the Mediation Model Regression Analysis
*Note: *p<.05, **p<.01, ***p<.001, ns=no significant*

Another result finds that environmental self-identity significantly predicts belief positively in global warming (β=.440, t=6.930, p<.001, H2 is accepted) and guilty feelings (β=.545, t=9.204, p<.001, H3 is accepted). In hypothesis 6, we find a significant mediation effect of belief in global warming on the correlation between environmental self-identity and PEB β=.074, Boot of CI [.017, .140], p<.05, H6 is accepted). This regression model has a determination coefficient value of R^2=.241 (F(3, 198)=21.013, p<.001). In summary, the predictor and mediators effectively contribute 24,1% to PEB.

3.2. Discussion

Based on the statistical analysis, environmental self-identity is the most vital factor in predicting pro-environmental behavior related to climate change issues. This finding is in line with several other findings [7,8,12]. People who strongly consider themselves environmentalists tend to behave pro-environmentally because they want to show who they are [25]. One of the approaches to establishing prior knowledge is to directly experience the effects caused by climate change [26]. This approach can also form an environmental self-identity in someone [26].

This research shows that belief in global warming may correlate positively to PEB. This is in line with some previous studies [27,28]. The belief in global warming also mediates the correlation between self-identity and PEB. The influence of belief in global warming on PEB can be related to global warming issue literacy through media use [29]. The flow of information drives most of the world's population to agree with climate change. Hence it is necessary to take a real solution, one of which is to establish PEB. The power of belief in global warming as a mediator can be seen through the perspective of the research sample's criteria, which all originated from South Kalimantan in Indonesia. Local warming takes a vital role in forming a belief in climate change [30], as data from Central Agency on Statistics (BPS) placed South Kalimantan in third place for the highest maximum temperature in Indonesia in 2015 [19], and ten provinces with highest maximum temperature in Indonesia in 2019 [31]. Additionally, human belief in climate change is supported by the direct effect they can experience, such as temperature increase.

There are exciting findings discovered in this study, which is also supported by the second study by Truelove et al. [15]. People with high environmental self-identity tend to feel guilty about humans' behavior toward the climate change issue. However, the guilty feeling does not predict PEB significantly either as a regression model's direct effect or mediation effect. This finding contradicted the previous study [18,32]. A meta-analysis study found that guilty feeling is not significantly correlated to PEB in a sample of university students [32]. Due to external attribution or influence from the descriptive norm from a group of students, university students have less guilty feelings and responsibility [33]. In this study, all samples are university students with a social norm that rules how they think, behave, and feel. Rather than being responsible, we believe that for students, the social norm that prevailed in their group establishes a perspective in which human guilty feeling creates responsibility for all humanity.

3.3. Limitation and Future Research Recommendations
This research is only restricted to a correlational approach. In the future, we suggest expanding the research sample not to be limited to one province only. Longitudinal research may investigate the mechanism of alteration of human belief and emotion regarding climate change. Researchers in the future can also consider experimenting with manipulating belief, emotion, or strengthened self-identity to find influence (cause and effect) on PEB.

4. Conclusion
The main finding of this research shows that people who identify as environmental-friendly will tend to conduct PEB to reduce the effect of climate change. Belief in global warming is also essential in driving people to act PEB, mediating the Identity-PEB associations, or directly predicting PEB. On the other hand, the emotional factor, a guilty feeling, is not a significant determining factor for someone to behave pro-environmentally. This finding encourages improving pro-environmental behavior by establishing self-identity and individual belief in their positive roles in climate change issues.

Acknowledgments
We want to appreciate and thank the Ministry of Education, Culture, Research, and Technology of the Republic of Indonesia for supporting our research by providing 2020 BOPTN of Higher Education Excellence Basic Research (PDUPT) for this research [grant number: 218/E5/PG.02.00.PT/2022]. We appreciate technical research support from the Social and Environmental Behavior Laboratory of Universitas Lambung Mangkurat.

References
[1] Intergovernmental Panel on Climate Change 2022 *Climate Change 2022: Impacts, Adaptation and Vulnerability*
[2] Renaldi A 2019 Indonesia Is Home to the Most Climate Change Deniers in the World *VICE*
[3] Bain PG, Hornsey M J, Bongiorno R and Jeffries C 2012 Promoting pro-environmental action in climate change deniers *Nat. Clim. Chang.* **2** 600–3
[4] Swim J, Clayton S, Doherty T, Gifford R, Howard G, Reser J, Stern P and Weber E 2009 *Psychology and Global Climate Change: Addressing a Multi-faceted Phenomenon and Set of Challenges*
[5] Steg L and Vlek C 2009 Encouraging Pro-Environmental Behaviour: An Integrative Review and Research Agenda *J. Environ. Psychol.* **29** 309–17
[6] Clark C F, Kotchen M J and Moore M R 2003 Internal and external influences on pro-environmental behavior: Participation in a green electricity program *J. Environ. Psychol.* **23** 237–46
[7] Van der Werff E, Steg L and Keizer K 2013 It Is a Moral Issue: The Relationship Between Environmental Self-Identity, Obligation-Based Intrinsic Motivation and Pro-Environmental Behaviour *Glob. Environ. Chang.* **23** 1258–65
[8] Whitmarsh L and O'Neill S 2010 Green Identity, Green Living? The Role of Pro-Environmental Self-Identity in Determining Consistency Across Diverse Pro-Environmental Behaviours *J. Environ. Psychol.* **30** 305–14
[9] Udall A M, de Groot J I M, De Jong S B and Shankar A 2021 How I See Me—A Meta-Analysis Investigating the Association Between Identities and Pro-environmental Behaviour *Front. Psychol.* **12**
[10] Mcadams D P 1995 What Do We Know When We Know a Person? *J. Pers.* **63** 365–396

[11] Vesely S, Masson T, Chokrai P, Becker A M, Fritsche I, Klöckner C A, Tiberio L, Carrus G and Panno A 2021 Climate Change Action as a Project of Identity: Eight Meta-Analyses *Glob. Environ. Chang.* **70** 102322
[12] Lot F, Quiamzade A, Falomir-Pichastor J M and Gollwitzer P M 2019 When Does Self-Identity Predict Intention to Act Green? A Self-Completion Account Relying on Past Behaviour and Majority-Minority Support for Pro-Environmental Values *J. Environ. Psychol.* **61** 79–92
[13] Pechar E and Mayer F 2015 Identity and Climate Change *Present. Bridg. David. Spaces Scholarsh. Pract. Environ. Commun.* 1–21
[14] Perera C R, Kalantari H and Johnson L W 2022 Climate Change Beliefs, Personal Environmental Norms and Environmentally Conscious Behaviour Intention *Sustain.* **14** 1–15
[15] Truelove H B, Carrico A R, Yeung K L and Wolff J M 2021 Identity and Guilt as Mediators of Pro-environmental Spillover *Front. Psychol.* **12**
[16] Heath Y and Gifford R 2006 Free-Market Ideology and Environmental Degradation: The Case of Belief in Global Climate Change *Environ. Behav.* **38** 48–71
[17] Garrett R K 2017 Strategies for Countering False Information and Beliefs about Climate Change *Oxford Res. Encycl. Clim. Sci.* 1–30
[18] Hurst K F and Sintov N D 2022 Guilt Consistently Motivates Pro-Environmental Outcomes While Pride Depends on Context *J. Environ. Psychol.* **80** 101776
[19] Badan Pusat Statistik (BPS) 2017 *Suhu Minimum, Rata-Rata, dan Maksimum di Stasiun Pengamatan BMKG (oC), 2011-2015*
[20] Banjarhits.id 2019 Aktivis Lingkungan Aksi #SaveMeratus di Barabai *kumparanNEWS*
[21] Henrich J, Heine S J and Norenzayan A 2010 The weirdest people in the world? *Behav. Brain Sci.* **33** 61–135
[22] Milfont T L and Duckitt J 2004 The Structure of Environmental Attitudes: A First- and Second-Order Confirmatory Factor Analysis *J. Environ. Psychol.* **24** 289–303
[23] van der Werff E, Steg L and Rupert A 2021 My Company is Green, So Am I: The Relationship Between Perceived Environmental Responsibility of Organisations and Government, Environmental Self-Identity, and Pro-Environmental Behaviours *Energy Effic.* **14** 1–21
[24] Hayes A F 2018 *Introduction to Mediation, Moderation, and Conditional Process Analysis: A Regression-Based Approach* (New York: Guilford Press)
[25] Gatersleben B, Murtagh N and Abrahamse W 2014 Values, Identity and Pro-Environmental Behaviour *Contemp. Soc. Sci.* **9** 374–92
[26] Young J C, Carsten Conner L D and Pettit E 2020 'You Really See It': Environmental Identity Shifts Through Interacting With A Climate Change-Impacted Glacier Landscape *Int. J. Sci. Educ.* **42** 3049–70
[27] Taylor E C, Livingston L A, Callan M J, Hanel P H P and Shah P 2021 Do Autistic Traits Predict Pro-Environmental Attitudes and Behaviors, and Climate Change Belief? *J. Environ. Psychol.* **76** 101648
[28] Duchi L, Lombardi D, Paas F and Loyens S M M 2020 How a Growth Mindset Can Change The Climate: The Power of Implicit Beliefs in Influencing People's View and Action *J. Environ. Psychol.* **70**
[29] Huang H 2016 Media Use, Environmental Beliefs, Self-Efficacy, and Pro-Environmental Behavior *J. Bus. Res.* **69** 2206–12
[30] Sugerman E R, Li Y and Johnson E J 2021 Local Warming is Real: A Meta-Analysis of The Effect of Recent Temperature on Climate Change Beliefs *Curr. Opin. Behav. Sci.* **42** 121–6
[31] Badan Pusat Statistik (BPS) 2020 *Statistik Lingkungan Hidup Indonesia 2020* (Badan Pusat Statistik (BPS))
[32] Shipley N J and van Riper C J 2022 Pride and Guilt Predict Pro-Environmental Behavior: A Meta-Analysis of Correlational and Experimental Evidence *J. Environ. Psychol.* **79** 101753

IOP Conf. Series: Earth and Environmental Science 1111 (2022) 012081 doi:10.1088/1755-1315/1111/1/012081

[33] Missing-Olson M J, Fielding K S and Iyer A 2016 Experiences of Pride, Not Guilt, Predict Pro-Environmental Behavior When Pro-Environmental Descriptive Norms Are More Positive *J. Environ. Psychol.* **45** 145–53

IOP Conf. Series: Earth and Environmental Science 1111 (2022) 012082 doi:10.1088/1755-1315/1111/1/012082

Improving Urban Resilience During COVID-19 Pandemic by Implementing Smart City Initiatives: A Case of Tangerang City, Indonesia

Ratih Dyah Kusumastuti[1], Nurmala[1], Juliana Rouli[1], Ledi Trialdi[2], and Rahayu Safitri[1]

[1] Department of Management, Faculty of Economics and Business, Universitas Indonesia

[4] Department of Economics, Faculty of Economics and Business, Universitas Indonesia

[1*]ratih.dyah@ui.ac.id, [2]nurmala85@ui.ac.id, [3]juliana.r@ui.ac.id, [4]ledi.trialdi@ui.ac.id, [5]rahayu.safitrii99@gmail.com

Abstract. Many cities in Indonesia are prone to disasters, and the Coronavirus disease (COVID- 19) pandemic taught a lesson on how important for a city to be resilient to disaster. Previous studies indicated that smart city initiatives could improve a city's resilience toward disasters. This paper sheds light on how implementing innovative initiatives can improve a city's resilience toward disasters. We review the existing literature on smart city and resilient city concepts and conduct a focus group discussion with the representative of the local government of Tangerang city, Banten province of Indonesia, to understand how the city improves its resilience towards disaster by implementing smart city initiatives. The results indicate that city has developed various software applications and community engagement programs to break the chain of transmission. The results also suggest that city leadership and citizen engagement play an essential role in developing a smart and resilient city.

1. Introduction

Indonesia is in the ring of fires that exposes most of the country's area to natural disasters [1]. In addition to natural disasters, the ongoing COVID-19 pandemic has put many cities in crisis due to an overwhelming number of cases in the early period of the pandemic and its negative economic impacts on the cities. The circumstances have made cities find ways to cope with the situation, such as smart city initiatives. Different cities around the world initially implement the smart city concept to alleviate urban problems by utilizing information and communication technology (ICT) and other modern technology as the main tools to enhance the city's "smartness," engaging the people that use the technology and improving the city system and service [2].

A smart city can be defined as "a city performing in a forward-looking way in economy, people, governance, mobility, environment, and living, built on the smart combination of endowments and activities of self-decisive, independent, and aware citizens" [3]. The dynamics of the smart city concept cannot be separated from urban problems and the goals to be achieved by applying intelligent city strategies. In general, the concept of the smart city deals with incorporating new ways and technologies to find solutions to the challenges of maintaining the quality of life of their citizens and sustaining cities' resources [4,5]. As identified in studies [6,7], cities often have insufficient knowledge about the needs of their citizens, and they do not know how to encourage their citizens' productive and positive behavior. Citizen engagement in policy decision-making is sometimes also inadequate [7].

Moreover, cities may need to communicate continuously and disseminate information because the city is constantly changing [8]. By deploying and optimizing the ICT system, smart cities may better understand their citizens' needs [6,9,10]. Engagement of the citizens may also be improved as cities can smartly provide information, motivation, education, and encouragement for them to act responsibly and proactively [10]. On top of that, cities' decision-making and quality of life, as well as interactions between citizens, government, and other city stakeholders, may be improved [9–12]. The optimal use of ICT is expected to contribute to thriving and more sustainable [6,9].

Resilience is a system's capacity to absorb disturbance and re-organize while changing to retain the same function, structure, identity, and feedback [13]. When the resilience concept is adopted in the context of natural disasters, the concept is not only about functionality but also relates to social, political, and economic structures that may have increased exposure and constrained capacity to cope with the crisis [13,14]. The concept of resilience also has been discussed in the context of cities. [15] define the resiliency of a city as an ability to activate protective qualities and processes at the individual, community, institutional, and systems levels to engage with hazards or stressors and cooperate in order to maintain or recover functionality and prosper while adapting to a new equilibrium and minimizing the accumulation of pre-existing or additional risks and vulnerabilities. In short, urban resilience means the capacity of the urban system to preserve its development and fundamental functions during emergencies [16].

It has been intensively discussed that the initiatives of smart cities will help cities maintain their level of resilience [17]. There are multiple ways in which smart-city initiatives can help cities maintain their resiliency. Smart city initiatives help cities increase their citizens' quality of living through multiple initiatives, including smart governance, smart living, smart environment, smart transportation, intelligent data, etcetera [18–20]. By having more sustainable and agile city management, a city will have a higher resilience toward any distribution, including natural disasters [18–20]. Information and technology are suggested as significant contributors to the dynamic of a smart city [20]. Information and technology can help cities manage disaster response phases more efficiently [20–22]. For example, integrated data on the city's geographical history will help citizens intensify natural disaster potentials in their area, supporting mitigation efforts [22]. Accessible data and integrated coordination also will allow cities to manage the emergency response phase when a disaster hits cities [23–25]. After an emergency response situation, a smart city can help cities to recover faster, related to multiple city aspects, including economic, social, and health services, by providing data, access, platforms, and social interactions that will boost city development [2,26,27]. Cities cannot often motivate their citizens through complicated situations [28], and smart-city initiatives can boost the motivation and engagement of citizens and motivate citizens toward productive and positive behavior, which will contribute to a more efficient reconstruction phase of disaster management and adapt to new situation equilibrium [15].

Even though some papers have already discussed how improving the "smartness" of a city can improve its resilience, more evidence is still needed. Hence, this paper aims to explain how implementing smart city initiatives can improve the resilience of Tangerang city, Banten Province of Indonesia, especially during pandemics [29]. The city is one of Jakarta's satellite cities (Indonesia's current capital). Satellite cities are those located at a distance away and become feeders to metropolises and are characterized by problems, including a lack of local employment prospects and limited economic development. This paper contributes by explaining factors that play a role in the successful implementation of the initiatives in Tangerang city and how they improve the city's resilience during the pandemic.

2. Methods

The study's main objective is to explain how implementing smart city initiatives can improve the resilience of Tangerang city towards disaster, especially during the COVID-19 pandemic. We choose

Tangerang city as the case because, at the end of 2021, the city received an intelligent city award from the Ministry of Communication and Information in the smart living category through the SIGACOR application, which has succeeded in helping the city of Tangerang to deal with the COVID-19 pandemic. We use a qualitative approach by reviewing existing literature to find how intelligent city initiatives improve the city's resilience, analyze the secondary data about Tangerang city, and conduct a focus group discussion (FGD) with seven representatives from several divisions of the local government of Tangerang city (see Table 1), to gain insights on how Tangerang city improves its resilience through smart city initiatives. The research is conducted in April-August 2022, and the FGD is conducted face-to-face by following the health protocol on 14 July 2022. The focus group discussion lasts about two hours, and the result is transcribed and analyzed.

Table 1. List of FGD participants

FGD Participant	Positions in the local government of Tangerang city
1.	Head of Department of Communication and Informatics
2.	Secretary of Regional Disaster Management Agency
3.	Secretary of Department of Public Works and Residential Housing
4.	Staff at Information and Communication Technology Governance
5.	Planner
6.	Sub-Coordinator of City's Disaster Information System
7.	Head of Division of E-Gov

3. Results and Discussion

3.1 Description of Tangerang City

Tangerang city is located in the eastern part of the Banten Province, with an area of 164.6 km^2, a population of 1,895,486 inhabitants in 2020, and a density of 11,516 inhabitants/km^2 [30]. The composition of this city's male and female population is quite balanced. Of the total population, 50.60% are Male, and 49.40% are Female [28]. Based on the age group, the population of Tangerang City is dominated by the Millennials Generation (27.51% of the total population), while the age group being the minority is the Pre-Boomer Generation (0.9% of the population) [30]. A total of 59.49% of the population in Tangerang City is categorized as a productive labor force or being economically active, and 97.37% of the labor force has a job [30]. The type of work done mainly by the residents of Tangerang City is working for others (employees). Tangerang city's gross regional domestic product (GRDP) in 2020 was IDR 143,840 trillion, and the Manufacturing sector is the highest contributor with 34.20% of the GRDP [30].

3.2 Smart city initiatives in Tangerang City

The movement toward 100 Smart Cities is a government program to guide districts/cities to prepare smart city master plans to maximize the use of technology to improve public services and the potential of their respective regions [31]. This program is carried out in stages, starting from 2017 to 2019. At the initial stage, the government selected 24 regencies/cities out of 514 regencies/cities in Indonesia to implement the smart city concept. Budget availability is one of the criteria used, and Tangerang City is one of the 24 regencies/cities.

Tangerang city started to implement smart city initiatives in 2018 by launching the mayor's regulation about the masterplan of the smart city for Tangerang 2017-2027, and the city formed the smart city council for this purpose. Tangerang city adopts the smart city concept as advised by the Ministry of Communication and Informatics. The smart city dimensions are slightly modified from [32]: smart economy, smart society, smart governance, smart branding, smart environment, and smart living (see Figure 1).

Currently, Tangerang city has a super app, namely Tangerang Live, which integrates 49 public service applications in 16 service categories that mobile devices can access by the citizens. The municipal government has conducted socialization in all sub-districts to ensure that residents are aware of and use these super apps optimally. Table 2 shows some of the applications owned by Tangerang city. In addition, the smart city initiatives are also carried out in the form of processing waste using electric power in collaboration with the private sector and monitored by several ministries, as well as sorting organic and inorganic wastes. The city government also provides sensors to monitor the air quality of Tangerang city.

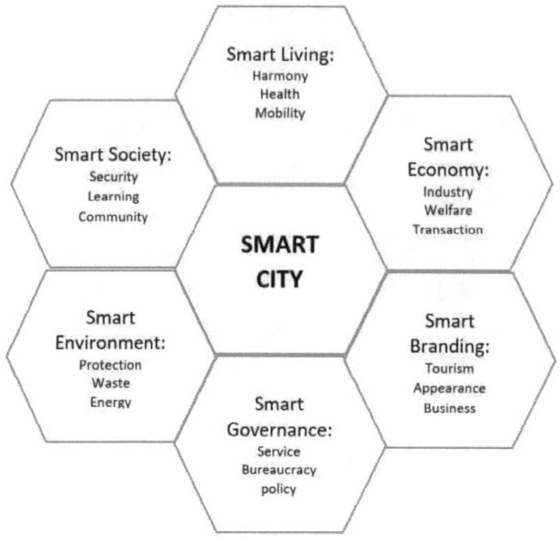

Figure 1: The smart city dimensions
Source: Tangerang Smart City internal documents (2022)

Table 2. A sample of smart city applications in Tangerang city

No.	Applications	Descriptions
1.	SIRONA	Application to monitor residents during the Covid-19 isolation period.
2.	AVATAR	Application to speed up the implementation of the Covid-19 vaccination.
3.	LAKSA	Application to submit complaints and reports and provide input to the government.
4.	SITA	Application to verify several program activities carried out by the city government.
5.	Tangerang Cakap	An application that provides various information related to improving HR competencies, how to become an entrepreneur, how to make a job application, how to apply for a job, interviews, to job training.
6.	Perak Geist	Application to respond to complaints about damage to transportation infrastructure (roads)
7.	Tangerang Belajar	An application that contains a collection of learning videos made by teachers in the city of Tangerang can be accessed by students. This application is handy during a pandemic, and learning is done online.
8.	SIGNUP	An application used to monitor the speed of handling emergencies from each related agency based on reports from residents.
9.	Aman Bersama	Application to manage the number of visitors in public spaces, such as shopping centers, hotels, and restaurants before the Peduli Lindungi application.
10.	SEGAR	Applications that contain prices for essential ingredients in the market.

Source: Tangerang Smart City internal documents (2022)

The FGD results indicate that, even though Tangerang city started to implement the smart city initiatives in 2018, at the beginning of the pandemic, the city was overwhelmed with a fast-increasing number of positive cases that affected the quality of living and its economic condition. Their capacity to cope with the circumstances is lower than warranted by the situation. Hence, just like other cities in Indonesia, Tangerang city was struggling to maintain the city's essential functions in the early period of the pandemic.

3.3 Improving the resilience of Tangerang City

The smart city programs are developed based on the master plan and the needs of the organizations. The city holds to the concept that a smart city is much more than just having many software applications. The FGD participants state that a smart city means finding solutions to urban problems more innovatively, which is similar to [33] and [5].

One of the smart city initiatives is that Tangerang city has a super application (app) for the public, called Tangerang Live, and a super app for the public officer, called E-Gov. Tangerang Live integrates various public service apps (such as free ambulance, public complaints, and emergency button) that mobile devices can access. Until May 2022, the number of users is almost 432 thousand citizens or 22.6% of the population. E-Gov is an app that integrates public officer data, including online attendance checks, monitoring of the officers' activities and performance, and attendance checks on certain events. Tangerang city also has other applications that support the government program, such as SIDATA to verify the implementation of government programs, Tangerang Bisa to register the beneficiaries of grants for micro, small and medium enterprises (MSMEs), and LAKSA, an application to receive public feedback, report, and complaints. In order to ensure that the smart city applications can be used optimally by the citizens, whenever an app is launched, the city will also launch a video manual through YouTube, which link is disseminated to the citizens through social media and other platforms. The above explanation indicates that the deployment of ICT in Tangerang city enables the city to operate more efficiently and effectively, as well as enables the city to engage and disseminate information to the citizens [8] and, more importantly, better understand what the citizen needs [6,8,9].

The FGD results also find the importance of citizen engagement in a smart city. The government of Tangerang city forms Community information groups (Kelompok Informasi Masyarakat, KIM) to help disseminate information about public services and to gather information on what kind of public services the citizens need. KIMs' activities also include training and mentoring to increase the ICT literacy of the community. Currently, there are 84 KIMs in 104 urban villages (kelurahan). In addition, Tangerang city has Tangerang Live Room, which is a command center that handles public reports or complaints through LAKSA, emergency call 112, and emergency call 119 (for health matters). The city can monitor the department or division that handles the reports or complaints in this command room. In 2020 out of 14,381 complaints, 99,27% are solved. The FGD participant also mentions that public reports/feedbacks through LAKSA and social media help the city to improve/develop better applications that can better serve the citizens. The above explanation shows that the city can provide better public services if it can communicate continuously with the citizens [8]. The smart city platform provides different means to communicate with the citizens through the smart city website, app, and call center, which make it easier for the citizens to give their opinions, report, or complaints. The explanation also indicates that the city services' co-creation behavior is only possible if the citizens are active and ICT literate (or smart people), similar to [10].

The FGD results also reveal that during the COVID-19 pandemic, the Department of Health of Tangerang City initially used an MS Excel spreadsheet to gather the data on positive cases. However, when the number of positive cases increases significantly, they work together with the ICT department to develop an app for SIRONA. So, when a patient has a test for COVID at one of the central government registered labs, the data will be linked to SIRONA. The community health facilities will verify the data. If it is verified that the patient is indeed a citizen of Tangerang city, and if the symptoms only require self-isolation, they will provide medications and monitor the patient for 14 days. The Tangerang city government also has a program called as Corona-Prepared Villages (Kampung Siaga Corona). With an app called SIGACOR, this program involves the community to monitor village residents who self-isolate due to COVID-19 and provide food for the affected families during the isolation period. The program also includes socialization/education on COVID-19 and the health protocol, routine temperature checks, and disinfecting of the area to break the chains of transmission. The city also has an app called AVATAR, which the citizens can use to register and get COVID- 19 vaccinations from the government at appointed health facilities. These apps are also supported by a program called Cageur Jasa from the Department of Health of Tangerang City, which visits the citizens' homes and provides health services. The main objective is to encourage hygiene in homes and reduce transmitted diseases, including COVID-19. The above explanation shows how the smart city initiatives help improve Tangerang city's resilience, which follows [15]. By combining a smart city platform, citizen engagement through Corona-Prepared Villages and SIGACOR, and home visits program (Cageur Jasa), Tangerang city can cope with the COVID-19 pandemic and recover quicker, hence, improving the city's resilience.

The FGD results also reveal the critical factors determining the successful implementation of the smart city concept in Tangerang city: city leadership and the program's sustainability. It requires a vision and commitment from the leaders to develop the intelligent city master plan and develop programs to implement the master plan. Another concern is the sustainability of the smart city initiatives, as the mayor holds the position for five years and can only be elected twice. The smart city initiatives require funding that should be proposed by the mayor and approved by the city parliament. The current mayor of Tangerang city is already on the second term until 2023. Hence, to ensure that the smart city initiatives will not be terminated in the next term, a set of regulations have been prepared to safeguard the smart city initiatives in the city. The above explanations indicate that a smart and resilient city needs good leadership to guide the city in the right direction and ensure the sustainability of the programs.

4. Conclusions

The still ongoing COVID-19 pandemic requires a city to be smart and resilient. This paper presents insights on how smart city initiatives can improve the city's resilience to the pandemic. Using the case of Tangerang city, Banten Province of Indonesia, we show that by deploying ICT in a smart city platform, combined with a good quality of community engagement, Tangerang city can cope with the pandemic and recover quickly. However, the successful implementation of smart city initiatives requires city leadership with a vision to guide the city in the right direction and ensure the sustainability of the programs. This research has limitations. In this paper, we only use one case on how smart city initiatives can convert a city into a smart and resilient city. More cases are needed so that the results can be compared and more reliable. Future studies can address this limitation.

Acknowledgments

The authors would like to thank the Government of Tangerang City and the FGD participants. The Indonesian Ministry of Education, Culture, Research and Technology funds this work under Grant NKB-777/UN2.RST/HKP.05.00/2022.

References

[1] Pratikto W A and Suntoyo 2015 Partnership in Building Community Resilience on Disaster in the Region of Coral Triangle, Indonesian Case *Procedia Earth Planet. Sci.* **14** 1–8
[2] Zhu S, Li D and Feng H 2019 Is smart city resilient? Evidence from China *Sustain. Cities Soc.* **50** 101636
[3] Giffer R 2007 Smart cities Ranking of European medium-sized cities *Res. Inst. Housing, Urban Mobil. Serv.* **16** 1–24
[4] Gutiérrez V, Amaxilatis D, Mylonas G and Muñoz L 2018 Empowering Citizens Toward the Co-Creation of Sustainable Cities *IEEE Internet Things J.* **5** 668–76

[5] Parasher Y, Kaur G and Tomar P 2020 *Green Smart Environment for Smart Cities* (Springer Singapore)
[6] Gao Z, Wang S and Gu J 2020 Public participation in smart-city governance: A qualitative content analysis of public comments in urban China *Sustain.* **12** 1–20
[7] González-Zamar M D, Abad-Segura E, Vázquez-Cano E and López-Meneses E 2020 IoT technology applications-based smart cities: Research analysis *Electron.* **9** 1–36
[8] Zubizarreta I, Seravalli A and Arrizabalaga S 2016 Smart City Concept: What It Is and What It Should Be *J. Urban Plan. Dev.* **142** 04015005
[9] Choque J, Diez L, Medela A and Muñoz L 2019 Experimentation management in the co-created smart-city: Incentivization and citizen engagement *Sensors (Switzerland)* **19** 1–17
[10] Toyota T, Habibzadeh H, Ekenna C, Nussbaum B and Lozano J 2019 Smart city in crisis: Technology and policy concerns *Sustain. Cities Soc.* **50** 101566
[11] Pardini K, Rodrigues J J P C, Diallo O, Das A K, de Albuquerque V H C and Kozlov S A 2020 A smart waste management solution geared towards citizens *Sensors (Switzerland)* **20** 1–15
[12] Yeh H 2017 The effects of successful ICT-based smart city services: From citizens' perspectives *Gov. Inf. Q.* **34** 556–65
[13] Walker B, Holling C S, Carpenter S R and Kinzig A 2004 Resilience, adaptability and transformability in social-ecological systems *Ecol. Soc.* **9**
[14] Paton D and Johnston D 2001 Disasters and communities: Vulnerability, resilience and preparedness *Disaster Prev. Manag. An Int. J.* **10** 270–7
[15] Patel R and Nosal L 2016 *Defining the Resilient City*
[16] Ribeiro P J G and Pena Jardim Gonçalves L A 2019 Urban resilience: A conceptual framework *Sustain. Cities Soc.* **50**
[17] Arafah Y and Winarso H 2017 Redefining smart city concept with resilience approach *IOP Conf. Ser. Earth Environ. Sci.* **70** 0–13
[18] Romano M, Díaz P and Aedo I 2016 Emergency management and smart cities: Civic engagement through gamification *Lect. Notes Bus. Inf. Process.* **265** 3–14
[19] Alsabi E, Alrwais L, Alosimai R, Alrwais R and Dahanayake 2016 *A Benchmark for Smart Cities*
[20] Criado J I and Gil-Garcia J R 2019 Creating public value through smart technologies and strategies: From digital services to artificial intelligence and beyond *Int. J. Public Sect. Manag.* **32** 438–50
[21] Schumann G J P, Brakenridge G R, Kettner A J, Kashif R and Niebuhr E 2018 Assisting flood disaster response with earth observation data and products: A critical assessment *Remote Sens.* **10** 1–19
[22] Silva T, Wuwongse V and Sharma H N 2013 Disaster mitigation and preparedness using linked open data *J. Ambient Intell. Humaniz. Comput.* **4** 591–602
[23] Bernasconi A, Canakoglu A, Masseroli M, Pinoli P and Ceri S 2021 A review on viral data sources and search systems for perspective mitigation of COVID-19 *Brief. Bioinform.* **22** 664–75
[24] Ofli F, Meier P, Imran M, Castillo C, Tuia D, Rey N, Briant J, Millet P, Reinhard F, Parkan M and Joost S 2016 Combining human computing and machine learning to make sense of big (Aerial) data for disaster response *Big Data* **4** 47–59
[25] Yagci Sokat K, Zhou R, Dolinskaya I S, Smilowitz K and Chan J 2016 Capturing Real-Time Data in Disaster Response Logistics *J. Oper. Supply Chain Manag.* **9** 23
[26] Suri N, Zielinski Z, Tortonesi M, Fuchs C, Pradhan M, Wrona K, Furtak J, Vasilache D B, Street M, Pellegrini V, Benincasa G, Morelli A, Stefanelli C, Casini E and Dyk M 2018 Exploiting smart city IoT for disaster recovery operations *IEEE World Forum Internet Things, WF-IoT 2018 - Proc.* **2018-January** 458–63
[27] Kusumastuti R D, Viverita, Husodo Z A, Suardi L and Danarsari D N 2014 Developing a resilience index towards natural disasters in Indonesia *Int. J. Disaster Risk Reduct.* **10** 327–40

IOP Conf. Series: Earth and Environmental Science 1111 (2022) 012082 doi:10.1088/1755-1315/1111/1/012082

[28] Cardullo P and Kitchin R 2019 Being a 'citizen' in the smart city: up and down the scaffold of smart citizen participation in Dublin, Ireland *GeoJournal* **84** 1–13

[29] Merrilees B, Miller D and Herington C 2013 City branding: A facilitating framework for stressed satellite cities *J. Bus. Res.* **66** 37–44

[30] Badan Pusat Statistik 2021 Tangerang city regional statistics 2021 *https://tangerangkota.bps.go.id/*

[31] Rizkinaswara L 2018 *Gerakan menuju 100 smart city* https://aptika.kominfo.go.id/2018/11/gerakan-menuju-100-smart-city/

[32] Giffer R 2015 Smart city concepts: Chances and risks of energy efficient urban development *Commun. Comput. Inf. Sci.* **579** 3–16

[33] Gutiérrez V, Theodoridis E, Mylonas G, Shi F, Adeel U, Diez L, Amaxilatis D, Choque J, Camprodom G, McCann J and Muñoz L 2016 Co-creating the cities of the future *Sensors (Switzerland)* **16** 1–27

The Effectiveness of Community Participation in Urban Water Supply: A Narrative Review

Debora E. Purba[1]* and Agung Minto Wahyu[2]
[1]Faculty of Psychology, Universitas Indonesia, 16424, West Java - Indonesia, ORCID ID: 0000-0002-0610-3209
[2]Faculty of Psychology, Universitas Indonesia, 16424, West Java - Indonesia, ORCID ID: 0000-0001-8898-4021
[1]*eflina@ui.ac.id, [2]agung.minto@ui.ac.id

Abstract. Debates on the effectiveness of top-down versus bottom-up approaches in water management in urban areas are still ongoing. This study examines the effectiveness of community participation as a bottom-up approach. The narrative review is employed to describe 16 cases of community participation. Three themes emerge from the case studies: the success story and failure of community participation from outside of Indonesia, lessons learned and best practices of community participation from Indonesia, and self-supply as an alternative water provision. Then a discussion of the findings based on the themes found in the literature was carried out. The study found that community participation would be successful if the participatory elements were satisfied, including sufficient capacity and knowledge of the community involved to ensure meaningful participation, involving local government in the education of how to manage and conserve water supply, and involving women in project sustainability. Hence, a pure community participation approach, particularly in disadvantaged urban areas, cannot be achieved. Therefore, a holistic approach combining a top-down and a bottom-up approach is recommended. Self-supply water provision emerged as another method that has largely been ignored by the government and academia but serves as a common practice, particularly in developing countries.

1. Introduction

Indonesia's total human population growth rate was predicted to reach 1.17% by the end of 2022 [1]. The high population growth will lead to increased scarcity of water [2] and pressure on water resources, especially freshwater [3]. The pressure occurs not only in the quantity and quality of water resources but also in household water consumption, especially in urban areas.[4] studied 14 metropolitan areas in Indonesia. The study's results found that water consumption in all metropolitan areas has exceeded the minimum standard of the water consumption needs of 60 liters per person per day, i.e., 169.98 l/p/d. Meanwhile, Indonesian Constitution (Undang-Undang Dasar, UUD) 1945 Chapter XIV Article 33 (3) states that "The land, the waters and the natural resources within shall be under the powers of the State and shall be used to the greatest benefit of the people." For this reason, the government must act for water sustainability in urban areas.

In the study of environmental management strategy, including water management, there have been discussions about the effectiveness of two approaches, namely the top-down and the bottom-up approaches. The top-down approach relied on a government-focused plan for water management, often led by expert knowledge [5]. The top-down approach is often seen as potentially exclusive and alienating local people [6] and seems to be primarily used in developed countries [7], especially among the Uni-European countries.

On the other hand, the bottom-up approach emphasizes community participation in water management [8], which leads to community engagement in the program implementation. Community participation (CP), also known as public participation, community-driven development, or citizen participation, is comparable to democratic governance, which is well grounded in many cultures in developing countries in Africa and Asia [9]. Participation involves empowering the poor to influence the local decision-making process.

The government's limitation emphasizes the need for involving communities in urban water management to provide sustainable water for people. In Indonesia, the average coverage of clean water service in 2018 was 65.28% [10], pipeline networks provide less than 30% of the service, and more than 70% of households rely on self-sufficient and community-based water provision [11].

Involving residents in water management aims to improve water quality, save water use, and increase community knowledge about water management [12]. Community participation is thus necessary as an approach to water sustainability. But there are still doubts about whether community participation is an effective way of urban water management [13]. Some studies found community participation is an effective approach [14–16], and some others found the approach difficult to implement [17–19]. Thus, it is interesting to uncover what makes community participation successful. Collaboration among stakeholders in urban issues is seen as ideal and desirable, but problems and issues often emerge in practice. One of the problems is the unclear purpose of participation. Often, the higher-ups involve citizens to give public comments on the projects but do not say how the comments would help the project. Citizens must be given knowledge, power, and authority to influence decisions [20]. Power imbalance among stakeholders in water management (e.g., local government, private companies, and residents) poses significant barriers in participatory projects when the elites and the experts dominate the implementation [21]. [20] research in Australia suggested for community participation to be effective, the government should establish an independent agency with power and authority to set the terms of reference for community participation, guarantee participation and input from residents, and set up forums for discussions [20]. Therefore, this narrative review aims to review the effectiveness of community participation as a form of bottom-up approach in urban water management from several cases of community participation in Indonesia and abroad. Contributions are made to the literature by showing features of a community participation approach that may enable community participation to work best from the cases employing it and by suggesting a better approach that fits the Indonesian urban communities.

2. Methods

2.1. Study selection

A narrative review is employed to answer the research question posed in this paper. A narrative review summarizes and interprets selected empirical studies to understand a topic and provide an authoritative argument to the reader [22,23]. A narrative review is suitable if the aim is to get an initial impression of the research area without going too deep into the study [24]. Following the method, analysis was conducted on minor but representative studies of community participation in the urban context.

2.2. Search strategy and network

Initially, key review papers are scanned on community participation in water management and searched further in databases subscribed by Universitas Indonesia from 2012 to 2022. Keywords used to search papers included community participation, public participation, community engagement, and urban water management (an asterisk was used to include different forms of the same word). A total of 268 papers were identified and further filtered by inclusion criteria: (1) only full-text journal articles and conference proceedings, (2) scholarly and peer-reviewed, and (3) in the areas of social sciences, sociology & social history, psychology, and water management. Finally, 16 papers were selected to be included in the content analysis.

3. Results and Discussion

The current study aims to examine the effectiveness of community participation in urban water management by reviewing case studies related to the topic. Of 16 papers selected for further review, most are case studies from developing countries, such as Mexico, Tanzania, South Africa, Kenya, Pakistan, and Indonesia [17–19,25–28]. This is in line with the argumentation that community participation is mainly used in developing countries. Several themes are found around the topic that will be further explored below.

3.1. *The success story and the failure of community participation from outside of Indonesia*

From cases involved in the review, we found that community participation is employed in developing countries, i.e., Some cases showed the success stories of the approach [25] while others reported failure [18,26]. Community participation is well grounded in African countries. Ananga et al. summarized the effect of community participation on water supply projects from six studies in different countries in Asia and Africa. They found that community participation contributed to the success of water production [25]. A survey was conducted in Kenya to examine the effects of community participation on beneficiary satisfaction. The survey's results found that provision of labor, willingness to contribute money, meeting attendance, and willingness to intervene in cases of vandalism contribute to beneficiary satisfaction.

Meaningful participation is an important indicator that determines the project's success. Boakye & Akpor studied the community in Msunduzi River, South Africa, where the community could not give meaningful participation in the project because of unmet expectations, the lack of interaction between the "community representations" and the community itself, and the lack of knowledge on water management [17]. Thus, the capacity and skills of the community in water management need to be developed for community participation to succeed. Lack of community capacity and knowledge also play roles as failure factors of the project success in Iringa District, Tanzania [18]. Donors initiated a water project in Iringa District, but the community was not involved in all project phases. Lack of communication, knowledge to engage in the project maintenance, and women participation led to the project failure. Moreover, the government posited itself as the project owner, leading to the community's low commitment to protecting the facilities [18].

Outside of Africa, Garciadiego studied the Magdalena River restoration project in Mexico City and found that an imbalance in power relations determines the project's success [26]. The academia and the government initiated the restoration, and it involved some urban neighbors with little or no relation to the river, while people mostly affected by the project were ignored. The affected people were a group of disadvantaged people in informal settlement areas. The lesson learned from the project was to constantly communicate with the community to avoid rejection and obstruction of the project [26].

3.2. *Lessons learned and best practices of community participation from Indonesia.*

In Indonesia, community participation as a method of decentralized water supply system (DWSS) for areas that are not covered by pipeline water systems is prevalent, i.e., the disadvantaged urban areas (for example, [11,29,30]). There are various types of DWSS: buying water from traveling vendors, water terminals, rainwater harvesting, public wells, and collaborative networks. Besides DWSS, self-sufficient water provision in groundwater boreholes is common in Indonesia. Household access to groundwater in Indonesian cities is almost without a permit, which eventually harms water sustainability [27]. It is also expected that no one monitors or controls the use of public water sources, and such would lead to the *tragedy of the commons* [11,31]. [32] found that the existence of power relations between formal and informal leaders and strong social ties within the community determine the willingness to engage in community-based water management and, thus, the system's sustainability. On the other hand, villagers from Wonogiri Village in Central Java, Indonesia, employed a community participation approach by forming a nonprofitable organization to provide a self-sufficient water supply [30]. Meanwhile, Roekmi et al. assessed the sustainability of five community-based water supply (CBWS) projects in Cikarang, a peri-urban area of Jakarta metropolitan, Indonesia, using Ostrom's eight design principles [27].

The study found that all five CBWS did not comply with safe water standards, but the community is strong enough to support itself. However, if the community fails to show the increase in Ostrom's eight sub-principles in the long run, all CBWS will fail to support the aim of SDG 6 and sustain. The local government should support CBWS committees in management and maintaining the safe water standard. Based on the case studies, pure community participation may not apply in developing countries without sufficient water conservation and sustainability education. Hence, because DWSS is primarily available in disadvantaged communities with lower education levels, community participation may lead to potential failure unless the government takes action in educating them about water.

3.3. Self-supply as an alternative method of water provision

Interestingly, some authors consider self-supply water provision or self-sufficient water management at the household level as an essential method [28,33,34]. Self-supply water provision is an alternative method of water management that may serve as a primary or sole source of clean water or as a complement or substitute of piped water system [34]. Raised the use of water self-supply, particularly groundwater, and the government's duty to raise awareness of the sustainability of groundwater resources [35]. The practice of groundwater supply by wells and boreholes is common in South Asia and Southeast Asia, while rainwater harvesting is dominant in the Pacific [34]. Water self-sufficiency in a regional area of New South Wales, Australia, is practiced mainly by collecting rainwater. The method poses some challenges, namely the possibility of running out of water and developing the infrastructure and equipment needed. To adapt to the self-supply method, people need to have higher self-regulation and restrict water use every day. Despite the challenges, most residents were happy and would do so again in the future [28].

Cahyadi et al. studied self-supplied clean water as an alternative approach for areas not covered by either the pipeline water system or community-based water supply in Indonesia [33]. The study contended that the self-supply system is lesser-known water governance in Indonesia that tends to be excluded from the discussion of water supply and yet practiced by many. Almost 90% of DKI Jakarta households used self-supplied water, improved and unimproved [33]. Thus, ignoring the self-supply method would likely lead to the *tragedy of the commons* [31]. Self-supply may serve as an opportunity for additional sources of funding. Self-supply also reflects that households are 'masters of their destiny rather than passive recipients of top-down approach [34]. However, self-supply is challenging to monitor and regulate and may lead to over-abstraction of groundwater [34]. It is difficult for policymakers to estimate household investment in the water supply. The government policy and regulation for self-supply, particularly in urban areas, are deemed essential to ensure the sustainability of water provision. The government should monitor the total water consumption, understand the trends of water abstraction, measure the safety and water quality and quantity aspects, and examine the disadvantaged groups being left behind [35]. Thus, future research is needed to examine the costs and benefits of self-supply and evaluate policy interventions to maximize the benefits and minimize the risks.

3.4. Recommendation for the sustainable water supply

Based on the evidence on the strengths and weaknesses of community participation, these past two decades, scientists in water management studies tried to employ a more holistic and integrated approach that combines both approaches [7,36]. To employ an inclusive development approach that relies on both a top-down approach in which the government and its agencies try to provide a centralized water supply system for all and the bottom-up approach in which the community actively participates in the development and maintenance of water supply and the government sets up policies and regulations for self-supplied water provision [37].

A study comparing the community participation approach employed by the San Francisco Bay Area in California, USA, and the top-down approach employed by the Catalan coastal region in Spain shed light on the need to balance top-down and bottom-up approaches [7]. The findings suggest that some features of the top-down approach can be important for adaptive capacity concerning the potential of the lack of leadership and coordination for the long run and the lack of future financing. Meanwhile, the bottom-up approach offers the potential for new learning and strong ownership among those involved in the project. Thus, combining both approaches may raise potential benefits for future projects.

The holistic approach has been used in Indonesia in the form of PAMSIMAS (Program Penyediaan Air Minum dan Sanitasi Berbasis Masyarakat – Community-Based Water Supply and Sanitation Program), used primarily in the rural area. PAMSIMAS is a government program to develop a water supply system in rural areas to ensure inclusive water provision for all people, initiated in 2007. Data from May 2021 showed that the program's beneficiaries exceed 20 million people in 32 thousand villages in the whole country [38]. The village community submits a proposal to the PAMSIMAS facilitator, who assesses the feasibility. Village people are then involved in the program from the construction phase, and the infrastructure maintenance is organized by a board committee [39]. The policy of the PAMSIMAS program is to have at least 40% women as board members. However, only one-fifth of all PAMSIMAS projects met the criteria. Moreover, women's participation was found to have no significant impact on the functionality of the PAMSIMAS but rather on the program's sustainability [39].

4. Conclusion

Community participation has been employed mainly in developing countries to ensure that all people are provided with water. Community participation needs every party to be involved in the project and get meaningful participation in all phases of the project. Thus, good interaction and a sense of community are required for community participation to be successful. Successful community participation also involves developing the capacity and knowledge of the community to enable them to participate fully in the project phases. Another feature of successful community participation is the role of informal leaders in motivating the community to participate. Although the project is owned and maintained by the community, the local government should be involved in educating people to ensure water conservation and monitor water use. Lastly, involving women in the project determines the sustainability of the project. Therefore, future research should thoroughly examine the government's role in community participation and project sustainability. Another alternative method many practices, but little has been done to manage its use, is the self-supply water provision. Future research should examine the method's benefits and weaknesses to ensure the water supply's sustainability for all.

Acknowledgment

This study was funded by the Ministry of Education, Culture, Research And Technology (KEMDIKBUDRISTEK) grant number [021/E5/PG.02.00.PT/2022] and grant number [NKB-831/UN2.RST/HKP.05.00/2022], Research and Development (Risbang), Universitas Indonesia.

References

[1] BPS 2022 *Laju Pertumbuhan Penduduk (Persen), 2020-2022* (Jakarta)

[2] Levy B S and Sidel V W 2011 Water rights and water fights: Preventing and Resolving Conflicts Before They Boil Over *Am. J. Public Health* **101** 778–80

[3] Vörösmarty C J, Green P, Salisbury J and Lammers R B 2000 Global Water Resources: Vulnerability from Climate Change and Population Growth *Science (80-.).* **289** 284–8

[4] Maryati S, Rahmani N I and Rahajeng A S 2018 Keberlanjutan Sistem Penyediaan Air Minum Berbasis Komunitas (Studi kasus: Hippam Mandiri Arjowinangun, Kota Malang) *J. Wil. dan Lingkungan.* **6** 131

[5] Agrawal A and Gibson C C 2001 The Role of Community in natural resource conservation *Communities and The Environment: Ethnicity, Gender, and The State in Community-based Conservation* (New Brunswick: Rutgers University Press) pp 1–31

[6] Rolston A, Jennings E and Linnane S 2017 Water matters: An assessment of opinion on water

management and community engagement in the Republic of Ireland and the United Kingdom ed S A Loiselle *PLoS One* **12** e0174957

[7] Serra-Llobet A, Conrad E and Schaefer K 2016 Governing for Integrated Water and Flood risk Management: Comparing Top-down and Bottom-up Approaches in Spain and California *Water* **8** 445

[8] Carr A 2002 *Grass roots & green tape: Principles and Practices of Environmental Stewardship.* (Sydney: Federation Press)

[9] Ananga E O, Naiga R, Agong' S G, Njoh A J and Vickers H P 2021 Examining The Contribution of Community Participation in Water Resource Production and Management: Perspectives from Developing Countries *SN Soc. Sci.* **1** 37

[10] BPS 2018 *Persentase Rumah Tangga Yang Memiliki Akses Terhadap Layanan Sumber Air Minum Layak Dan Berkelanjutan (40% Bawah), Menurut Provinsi (Persen), 2017-2019* (Jakarta)

[11] Maryati S, Firman T and Humaira A N S 2022 A Sustainability Assessment of Decentralized Water Supply Systems in Bandung City, Indonesia *Util. Policy* **76** 101373

[12] Seelen L M S, Flaim G, Jennings E and De Senerpont Domis L N 2019 Saving Water for The Future: Public Awareness of Water Usage and Water Quality *J. Environ. Manage.* **242** 246–57

[13] Druschke C G and Seltzer C E 2012 Failures of Engagement: Lessons Learned From a Citizen Science Pilot Study *Appl. Environ. Educ. Commun.* **11** 178–88

[14] Allan J V, Fielding K S, Kenway S J and Head B W 2022 Community Perspectives on Sustainable Urban Water Security *Urban Water J.* **19** 325–35

[15] Birkinshaw M, Grieser A and Tan J 2021 How Does Community-Managed Infrastructure Scale Ip From Rural to Urban? An Example of Co-Production in Community Water Projects in Northern Pakistan *Environ. Urban.* **33** 496–518

[16] Mashazi T P, Morole M S and Modley L S 2019 Evaluating Public Perceptions, Attitudes and Participation in Water Resource <management: The Case of An Urban township in South Africa *Water Pract. Technol.* **14** 726–31

[17] Boakye M K and Akpor O B 2012 Community Participation in Water Resources Management in South Africa *Int. J. Environ. Sci. Dev.* **3** 511–6

[18] Chumbula J J and Massawe F A 2018 Community Participation in Water Resource Projects Management in Iringa District Council, Tanzania. *Tengeru Community Dev. J.* **5** 80–90

[19] Harera A, Putri G L and Foster T 2021 Self-supply Service Level Assessment as Drinking Water Source in Bekasi City (Case study: Jatiluhur, Sumur Batu, and Jatirangga Villages) ed A Harakan and K Jermsittiparsert *E3S Web Conf.* **277** 04004

[20] Mahjabeen Z, Shrestha K K and Dee J A 2009 Rethinking Community Participation in Urban Planning: The Role of Disadvantaged Groups in Sydney Metropolitan Strategy *Australas. J. Reg. Stud.* **15** 45–63

[21] Fung A and Wright E O 2003 *Deepening democracy: Institutional innovations in empowered participatory governance (The real Utopias Project)* (London: Verso Books)

[22] CampbellCollaboration 2001 Campbell Collaboration Guidelines

[23] Kirkevold M 1997 Integrative nursing research — an important strategy to further the development of nursing science and nursing practice *J. Adv. Nurs.* **25** 977–84

[24] Bryman A 2012 *Social Research Methods* (Oxford: Oxford University Press)

[25] Ananga E O, Agong' S G, Acheampong M, Njoh A J and Hayombe P 2020 Examining the effect of community participation on beneficiary satisfaction with the work of water management committee in Urban Community-based Operated Water Schemes *Sustain. Water Resour. Manag.* **6** 49

[26] Garciadiego F G A 2022 Community participation in Mexico City's water management. Learning from the failure of the Magdalena River restoration project *Urban Water J.* 1–14

[27] Roekmi R A K, Baskaran K and Chua L H 2018 Community-based water supplies in Cikarang, Indonesia: are they sustainable? *Nat. Resour. Forum* **42** 108–22

[28] Wilkinson C and Gibbs L 2021 Beyond mains water: A survey of households that are self-sufficient for water in Regional Australia *Australas. J. Environ. Manag.* **28** 149–68

[29] Ari I R D, Ogi K, Matsushima K and Kobayashi K 2013 Community Participation on Water Management; Case Singosari District, Malang Regency, Indonesia *Procedia Environ. Sci.* **17** 805–13

[30] Pratiwi E, Juhadi J, Trihatmoko E, Sartohadi J, Fauzanna R and Mahmud A 2019 Community participation on water resources management in the Drought Prone Area (A case study from Wonogiri Village, Central Java, Indonesia) *1st International Conference on Environment and Sustainability Issues* (Semarang: EAI)

[31] Hardin G 1968 The Tragedy of the Commons *Sci. New Ser.* **162** 1243–8

[32] Ari I R D, Ogi K, Matsushima K and Kobayashi K 2013 Community Participation on Water Management; Case Singosari District, Malang Regency, Indonesia *Procedia Environ. Sci.* **17** 805–13

[33] Cahyadi R, Kusumaningrum D and Prasetyoputra P 2022 Self-supplied water in Indonesia: recent spatial and socio-demographic conditions and its future development *IOP Conf. Ser. Earth Environ. Sci.* **1062** 012038

[34] Foster T, Priadi C, Kotra K K, Odagiri M, Rand E C and Willetts J 2021 Self-supplied drinking water in low- and middle-income countries in the Asia-Pacific *Clean Water* **4** 37

[35] Grönwall J and Danert K 2020 Regarding groundwater and drinking water access through a human rights lens: Self-supply as a norm *Water* **12** 419

[36] Rollason E, Sinha P and Bracken L J 2022 Interbasin water transfer in a changing world: A new conceptual model *Prog. Phys. Geogr. Earth Environ.* **46** 371–97

[37] Purba D E, Arrania N, Syamila M, Pranaya D and Kusumawardhani D E 2021 Inclusive development approach to urban water services in Jakarta *IOP Conf. Ser. Earth Environ. Sci.* **940** 012022

[38] PAMSIMAS 2021 *Sebaran Program PAMSIMAS* (Jakarta)

[39] Djono T P Al and Daniel D 2022 The effect of community contribution on the functionality of rural water supply programs in Indonesia *Groundw. Sustain. Dev.* **19** 100822

Nyepi and the Efforts to Save the Environment

Ni Kadek Surpi[1*]
[1] Brahma Widya, Faculty, Hindu State University of I Gusti Bagus Sugriwa Denpasar, Denpasar, 80236 , Indonesia, ORCID ID 0000-0002-1121-179X
[1*]dosen.surpiaryadharma@gmail.com

Abstract. During the Nyepi celebration, within twenty-four hours, the island of Bali is wholly rested from various activities and business. This study analyzes how the implementation of the Nyepi Holy Day celebration has a broad impact on efforts to save the environment, which is a world issue. Data was obtained through observation, interviews, and literature studies examining Balinese environmental data. Through thematic analysis and interpretation, the Nyepi celebration significantly impacts efforts to protect nature and the environment in Bali. It is closely related to four Nyepi prohibitions, known as Catur Brata Penyepian: *Amati Geni*, *Amati Lelanguan, Amati Lelungan*, and *Amati Karya*. With these four prohibitions, the island of Bali seems to be a dead island for 24 hours, without noise and electricity even at night. All transportation activities, including airports and seaports, have stopped operating. The data shows that there has been saving in electricity use, a reduction in carbon dioxide gas emissions, fuel saving, and an increase in air quality due to the cessation of all human transportation activities. The Nyepi celebration in Bali and carried out by Hindus in Indonesia has had a tremendous impact on the earth.

1. Introduction

The celebration of the Holy Day of Silence is a special occasion on the island of Bali, Indonesia. This island, which is famous as a world tourism destination, has a day of rest for the earth for 24 hours a year. This island has become a complete standstill without any activities for a day and a night on the implementation of Nyepi Day. Nyepi Day is a celebration of the Saka new year which falls on *sasih kadasa* (the tenth month) based on the Saka year calendar. In contrast to the holiday, which seems lively and crowded, Nyepi, as the origin of the word Sepi, silent, is carried out precisely with *Catur Brata Penyepian*, namely *Amati Geni* - not lighting the fire, *Amati Karya*, which means not doing activities or working, *Amati Lelungan* which means no traveling, and *Amati Lelanguan*, a prohibition for pleasure, not doing self-entertainment or commotion. Hindus in Indonesia generally carry out this Nyepi day, but specifically on the island of Bali, it is very different because it can be done simultaneously, and Bali seems to be a dead island, quiet, silent, and completely dark at night. Transportation was almost completely paralyzed for 24 hours. I Gusti Ngurah Rai International Airport is also closed for 24 hours. The same applies to the four seaports: Benoa, Padang Bai, Gilimanuk, and Celukan Bawang Harbour. Public facilities that operate only hospitals and police stations. Television broadcasts, both national and local, are turned off for 24 hours. The island will be dark at night, so there will be a fantastic night view.

The Hindu community in Indonesia in welcoming the Nyepi day was previously coupled with ritual ceremonies, which included; *melasti* (purification ceremony to a spring or sea with a procession), and *tawur kasanga* (Balinese Hindu ceremony that aims to purify and balance nature), which coincides with the day of *tilem kasanga* (new moon in the ninth month) based on the Saka year calendar, and the next day celebrating Nyepi day (Saka new year) followed by the day of *Ngembak Geni* on the next day which marks the end of the prohibitions on Nyepi [1]. So for 24 hours, Bali is free from activities, and all the streets will be empty.

In addition to providing a religious meaning, the Nyepi celebration is closely related to the efforts to save the environment [2]. The cessation of activities for 24 hours makes energy saving and Bali's nature seem to breathe with silence. The Nyepi celebration in Bali has inspired *World Silent Day* because it is considered to provide many benefits to the earth [3].

Many figures have appreciated this Hindu religious celebration because it impacts saving the earth. Minister of Environment and Forestry Siti Nurbaya attended an art-ecological performance entitled "Nuwur Kukuwung Ranu" held at Segara Danu Temple, Batur, Regency. Bangli, on Saturday, May 14th, 2022, stated that Bali could also be an example in the actualization of *sustainable development* and the implementation of the *Paris Agreement*. Because for four years until 2019, the Ministry of Environment and Forestry recorded that Bali Nyepi Day reduced carbon emissions by an average of around 12-14 thousand tons of CO_2 equivalent in a day. Thus, the Nyepi celebration is one support form for the program to save the environment.

Hinduism does have a concept of life that respects the environment. Hindu theo-ecology has been valued as a concept that has great respect for Mother Earth [4]. Hindus in Bali also have the noble concept of Tri Hita Karana, namely three things of harmony that leads to happiness: the relationship between humans and God, humans with nature, and fellow humans [5]. Nature is essential, and humans have a sacred obligation to protect nature and the environment. In addition, the Balinese Hindu community also has the concept of Sad-Kerthih, namely, six efforts to maintain the universe's balance [6]. This article discusses how the Nyepi celebration has a good impact on the earth and as an effort to save the environment. Nyepi is, at the same time, implementing various world thoughts about protecting the natural environment from various threats due to various human activities that cause damage to the natural environment. Various studies have been conducted regarding Nyepi activities, such as the research by [7] entitled The Function and Meaning of Nyepi Ritual in Bali, stating that Nyepi Day as a series of ceremonies functions as an effort of self-control (*Catur Brata Penyepian*). It is said that the Nyepi celebration contains a noble and profound purpose or function and meaning because the whole series of Nyepi rituals is a spiritual dialogue [7]. Overall, the ceremony's implementation system on Nyepi Day meets the needs of human life biologically, socially, and psychologically and contributes to achieving the ideal hopes and goals of a harmonious and balanced Hindu life. [3] their research entitled Relation between Carbon Monoxide (Co) Concentration and Air Temperature Towards Anthropogenic Intervention (Case Study Nyepi Day 2015 In The Bali Province) concluded that during Nyepi Day or when there is no anthropogenic intervention, the relationship between CO concentration and air temperature tends to be consistent in the form of a positive relationship, while on regular days the relationship between CO and air temperature tends to be inconsistent (random and unclear). It is suspected that anthropogenic factors in various human activities (agriculture, fisheries, trade, transportation, education, and tourism in Bali Province) interfere with the relationship between CO concentration and air temperature [3]. [8] conducted a study of a religious holiday called Nyepi, which restricted human activity for one day, recording acoustic noise levels for one week in shallow water in a minor traffic area west of Bali under a flight from Ngurah Rai airport[8]. There was a significant reduction in anthropogenic noise levels during the cessation of all activities for one day. [9] stated that local activities, namely Nyepi, have adopted global thoughts about brotherhood and saving the environment[9]. Finally, [10] wrote that the religious tradition of restraint had anticipated emergencies such as the covid 19 pandemic. The pandemic and the tradition of religious restraint must be understood as a synergistic collaboration between nature and culture, warnings about productivity deterioration, and a call for more justice and a balanced yet realistic lifestyle. That is how the 'new standard should be[10]. All of these studies contributed to building the framework for this research. This study combines religious and philosophical thinking regarding implementing a religious tradition called Nyepi and its relevance to efforts to save the environment, which has become a global issue.

2. Methods

This article discusses the Nyepi celebration and the efforts to save the earth. The research was conducted in the Province of Bali, Indonesia, which is located between the islands of Java and West Nusa Tenggara (NTB) and is directly adjacent to the Bali Sea, Indian Ocean, Bali Strait, and Lombok Strait. In the history of religion in Indonesia, Bali is the last bastion of Hindu civilization, the island with a predominantly Hindu population. The research focuses on environmental attitudes in implementing Nyepi and its relation to the sustainability of Bali's nature. Data were obtained through observation, interviews, documents, and literature studies and collected in an ongoing multi-site. Data analysis using thematic analysis, namely data categorization based on key themes. Thematic analysis is also referred to as interpretative thematic analysis. Researchers process data, interpretations, and theoretical concepts to produce a satisfying narrative about the implementation of Nyepi Day to support saving the environment. Data is traced to various sources. The impact of Nyepi celebrations is widely publicized, both in the form of news and official data from related agencies. The data obtained were analyzed using descriptive interpretative methods to formulate accurate conclusions. Data on electrical energy savings are issued by the Bali State Electricity Company (PLN) regarding fuel savings by Pertamina, emission reductions by the Environment Agency, and various other relevant agencies. Data were taken in the last five years, namely Nyepi in 2022, 2021, 2020, 2019, and 2018. The data is collected for further processing and analysis.

3. Results and Discussion

The celebration of the Holy Day of Silence in Bali is a special occasion every year. For 24 hours, the entire island is stopped from all activity and noise. The Hindu community in Bali carries out a religious obligation known as *tapa brata penyepian*, namely an effort to silence themselves and perform asceticism as a continuation of the ancient tradition of Hindu Dharma. In order to support the implementation of asceticism or self-restraint, four things must be carried out. Namely, Amati Geni, prohibited from lighting fires/lamps, implies self-control. Second, Amati Karya is prohibited from doing physical activities or working. The following is Amati Lelungan, prohibited from traveling outside the house. Lastly, Amati Lelanguan is prohibited from doing self-entertainment/recreation and not making noise[11]. These four prohibitions aim to support the spiritual activities of Hindus.

Meanwhile, non-Hindu people participate by not doing prohibited activities such as not starting a fire, not working, not traveling, making noise, and not turning on electricity for 24 hours. Research (2) shows that one of the impacts of the Nyepi Day celebration is the reduction of greenhouse gas emissions resulting from human activities such as office, transportation, and industrial activities. On this holy day, the environment becomes free from anthropogenic intervention. This reduction in greenhouse gas emissions causes an increase in air quality to be healthier. It was concluded that during Nyepi Day or when there was no anthropogenic intervention, the relation between CO concentration and air temperature tended to be consistent in the form of a positive relationship, while on days outside of Nyepi Day, the relation between CO and air temperature tended to be inconsistent (random and unclear). It is suspected that anthropogenic factors in the form of various human activities, such as transportation, tourism, agriculture, fisheries, trade, and education, interfere with the relationship between CO concentration and air temperature. This celebration will reduce greenhouse gas emissions by 33%.

Based on data from the Institute for Essential Service Reform (IESR) published in 2020, during the Nyepi celebration in 2019, the island of Bali reduced its emissions to 5,462.74 tons of Co2 in one day. The implementation of Catur Brata Penyepian will directly reduce the consumption of electricity and fuel for motorized vehicles. The electricity usage the day before Nyepi slightly increases and decreases drastically on Nyepi day. On regular days the electricity usage is 21,121 MWh. Meanwhile, during Nyepi in 2019, the usage was 13,427 MWh. On average, during Nyepi, Bali saves electricity by 60%. This is equivalent to saving 4 billion or 290 megawatts[12]. Data from PT PLN Distribution states that the peak load of electricity consumption in Bali usually is 850 megawatts (MW). The need for electrical energy in Bali Province can be seen from its peak load, which reaches 684 MW, of which 559 MW or

81% is supplied by fuel generators. The average fuel consumption to generate electricity reaches 2.5 million liters per day.

The needs in the transportation sector and other industries in Bali are still entirely dependent on fuel, with a total consumption of 5.2 million liters per day. This fuel is needed to serve the needs of 360 thousand cars, 2.4 million motorcycles, industry, sea transportation, and fishermen. The most significant savings occur in using fuel for transportation because practically no vehicle operates for 24 hours. On Nyepi day, the Government can save fuel subsidies for premium fuel and diesel by IDR 12 billion. The country saved around IDR 52 billion in foreign exchange on the same day.

Meanwhile, the characteristics of LPG usage show an increase of around 5% a few days before Nyepi but a decrease of 50% during Nyepi. So that the total savings of subsidized LPG in 3 kg and 12 kg gas holders reached 575 tons and 92 tons, respectively. With a 3 kg LPG subsidy of Rp 2,415/kg, during Nyepi, the LPG subsidy savings was Rp 1.4 billion. Meanwhile, the foreign exchange that can be saved is around IDR 5 billion. Another positive impact is the reduction of pollution from carbon dioxide (CO2) gas emissions into the air. Burning fuel per liter will produce 2.7 kg CO2 gas, while burning LPG per 1 kilogram will produce 3 kg CO2 gas. During Nyepi day, there was a decrease of 23.9 thousand tons of carbon dioxide gas emissions. Data from the Head of the Environmental Agency for the Province of Bali for the Implementation of Nyepi Day, which falls on Wednesday (9/3/2016), shows a decrease in CO2 gas that can reach 20,000 tons of PPM (Part Per Million). In addition, the implementation of Nyepi is also said to reduce the effect of methane gas as a trigger for the impact of greenhouse gases[3]. For the reduction of 20,000 tons, including reducing electricity, reducing the impact of transportation, reducing waste, reducing trash, and reducing fuel and firewood.

The Meteorology, Climatology, and Geophysics Agency (BMKG) carried out emission measurements during Nyepi and compared them with the daily situation. This year's Nyepi falls on March 3rd, 2022. From 06:00 to March 4th at 06:00 WITA, residents are prohibited from leaving their houses. At night the lights are turned off (except for essential public facilities), including airport flight activities. During Nyepi, particulate pollutants in the air are expected to decrease. Therefore, it is necessary to support air quality observation data to calculate the percentage of reduction in emissions during Nyepi. Dr. Danang Eko Nuryanto, Central BMKG Researcher, said, by obtaining observational data, it is hoped that it can be used as evidence for the community and the world regarding the efforts to reduce emissions in Indonesia for climate change.

The BMKG Research and Development Center (Puslitbang) will measure air quality on Nyepi Day in 2022. This activity is also in collaboration with the MKG Center Region III Denpasar, UPT BMKG in Bali Province, the Environment Service (DLH) of West Java Province, Denpasar City DLH, and the Bali and Nusa Tenggara Ecoregion Development Control Centre (P3E) KLHK. The measurement locations were chosen, which represent urban and suburban areas. The urban area is represented by observations at the location of the Regional Office III Denpasar Bali. In contrast, the suburban area is represented by equipment placement in the Jembrana and Karangasem regencies. This measurement uses equipment to measure total dust particles (TS) using HAZ-DUST EPAM-5000 and Met One while measuring carbon monoxide (CO) and carbon dioxide (CO2) using Multigas Sensync. The measurement results showed a significant decrease in the concentration of dust particulates which varied at each location on Nyepi Day 2022 compared to other days. It was concluded that Nyepi, as a religious ritual, was proven to be able to make a real contribution to improving air quality [13].

Compared with the measurement of the concentration of dust particulates by the Bali DLHK for the data period 2020 – February 2022, the decrease is around 47.07%. Daily human activities significantly affect the concentration of pollutant gases and particulates in the air [14]. On Nyepi Day, when all human activities are reduced, the air concentration becomes better. This is indicated by a decrease in dust particles (TSP) concentration in the air. Urban areas filled with daily activities have the most significant improvement in air quality compared to suburban areas. The area with the highest concentration of carbon monoxide in Bali is Denpasar City, the capital city of Bali Province. According to Danang, the recommendation related to ecology from this calculation is to emphasize the importance of mitigating global warming and climate change. These data are measurable evidence that human activities increase the concentration of greenhouse gases. The initial results of measurements generally show a significant decrease in the concentration of pollutant gases and particulate dust, which varies at each location during Nyepi 2022 compared to regular days.

Denpasar Earth Hour Community Coordinator Andri Purba stated that Hari Raya is a genuinely positive tradition to save the earth and energy. Because, in one day, there was a reduction of 20 thousand tons of carbon dioxide (CO_2) emissions. The commemoration of Nyepi Day, marked by the cessation of all activities on the island of Bali, has proven to save electrical energy. PT PLN recorded savings in diesel fuel for power plants of IDR 3 billion during Nyepi[14]. The cessation of community activities on Nyepi in Bali made two power plants in Bali also stop operating. The two power plants that have stopped operating are the Pemaron Gas Steam Power Plant (PLTGU) which usually produces 80 MW of electricity, and the Gilimanuk Gas Power Plant (PLTG), which usually produces 130 MW of electricity. The rest of the two power plants saved 500,000 liters of diesel fuel, or the equivalent of IDR 3 billion.

World Silent Day, celebrated on March 21st every year, is inspired by Nyepi Day in Bali. Nyepi celebration is considered to have given many benefits to the earth. World Silent Day was initiated by Triple-C, Collaboration for Climate Change, a combination of Balinese NGOs, during the global warming conference (UNFCCC) in Bali in 2007[14]. This movement is expected to reduce activities that harm the earth, such as global warming or climate change.

The Provincial Government of Bali also announced that Nyepi Day in Saka 1944, which falls on March 3rd, 2022, will turn off cellular and IPTV data services. Head of the Bali Provincial Government's Communication, Information and Statistics Agency, Gede Pramana, in a press release in Denpasar, on Sunday (27/2/2022), stated that providers serving cellular and IPTV data in Bali Province and surrounding areas will turn off cellular data and IPTV on Thursday, March 3rd, 2022 at 06.00 WITA to Friday, March 4th, 2022 at 06.00 WITA. This has been reinforced by the Decree of the Minister of Communication and Information of the Republic of Indonesia Number 2 of 2022 and the Letter of Appeal for the Bali Province Religious Harmony Forum in 2022. Thus, Nyepi Day has a real impact related to saving the environment. Based on the data collected, there was an electricity saving of around 60%, namely 290 Megawatts, which if converted into IDR is 4 billion, Reduction of Carbon dioxide Gas Emissions, Saving on fuel by IDR 3 Billion, and improving air quality due to the cessation of all human transportation activities. Bali will look darker on Nyepi. The stars are seen more clearly because there are no lights, and the sky is cleaner. The data analysis shows evidence that the implementation of Nyepi in Bali contributes to efforts to save the environment. Collectively the Balinese community and the island of Bali have reduced their *carbon footprint* by stopping all activities 24 hours a year.

Nyepi Day celebration contains a deep and noble meaning: a balanced and harmonious life because it provides tranquillity and peace for human life and nature. Such a calm atmosphere is challenging, especially in urban communities that are always busy with packed routines.

4. Conclusion

Nyepi, a Hindu religious holiday, has a real impact on saving the environment. On the day of Nyepi, the Balinese people will remain silent doing asceticism with *Catur Brata Penyepian*, the cessation of all transportation activities, including airports and seaports. In addition, no lights on and the cessation of human activities. Based on the data collected, there is an electricity savings of around 60%. In one day, the Nyepi celebration in Bali can save electricity usage as much as 60 percent, namely 290 Megawatts, which if converted is around IDR 4 billion. In addition, there is a reduction in carbon dioxide gas emissions, reaching 20,000 tons of PPM. The cessation of all transportation activities also results in fuel savings of IDR 3 billion and an increase in air quality due to the cessation of all human transportation activities. Improved air quality and reduction of pollutant particles. Bali will look darker on Nyepi. The stars are seen more clearly because there are no lights, and the sky is cleaner.

The limitation of this research lies in the research process and the available data. Data on the impact of Nyepi are incomplete at all years in the study range. This means that not all agencies issue data on the impact of Nyepi every year, so this research has a weakness in data analysis efforts every year. The

IOP Conf. Series: Earth and Environmental Science 1111 (2022) 012084 doi:10.1088/1755-1315/1111/1/012084

resulting data is global data. However, it can provide an overview of how the implementation of Nyepi has a good impact on efforts to save the environment. The recommendation put forward is that in the future, more in-depth research involving different disciplines is needed to elaborate more profoundly on Nyepi's impact on the environment.

References

[1] Gorda A A . N S R 2017 Implementation of Local Wisdom Values of Smerti Veda (Hindu Law Compendium) in Hindu Banking Law *Int. J. African Asian Stud.* **40** 127–38

[2] Hobart M 2019 How Balinese Argue *J. Kaji. Bali (Journal Bali Stud.* **9**

[3] Aprilina K, Badriah I U and Aldrian E 2016 Hubungan Antara Konsentrasi Karbon Monoksida (CO) Dan Suhu Udara Terhadap Intervensi Anthropogenik *J. Meteorol. Dan Geofis.* **17** 53–60

[4] Gosling D L 2020 Ecology and Hindu Tradition *Religion and Ecology in India and Southeast Asia*

[5] Surpi N K 2021 *Implementasi Danu Kerthi dalam Upaya Menjaga Eksistensi Tiga Danau di Hulu Bali* (Denpasar)

[6] Wiana I K 2018 "Sad Kertih": Sastra Agama, Filosofi, dan Aktualisasinya *J. Bali Membangun Bali* **1** 169–80

[7] Suwena I W 2017 Fungsi dan Makna Ritual Nyepi 1–30

[8] Williams R, Erbe C, Dewantama I M I and Hendrawan I G 2018 Effect on Ocean Noise: Nyepi, a Balinese Day of Silence. *Oceanography* **31** 16–8

[9] Yupardhi W S 2018 Perayaan Nyepi Umat Hindu Bali Bertindak Lokal dan Berpikir Universal *Widya Duta J. Ilm. Ilmu Agama dan Ilmu Sos. Budaya* **13** 19

[10] Wijaya Y 2021 Sabbath, Nyepi, and Pandemic: The Relevance of Religious Traditions of Self-Restraint for Living with the 'New Normal' *Stud. Christ. Ethics* **34** 529–43

[11] Japar M, Komarudin, Fadhillah D N and Syarifa S 2019 Mapping The Balinese Social Tolerance Model (Bhinneka Tunggal Ika for Strengthening Unity in Diversity) *Opcion* **35**

[12] IESR 2020 Total Energi Listrik di Pulau Bali *Inst. Essensial Serv. Reform*

[13] Findly E B, Chapple C K and Tucker M E 2002 Hinduism and Ecology: The Intersection of Earth, Sky, and Water *J. Am. Orient. Soc.* **122**

[14] Badriyah I U 2014 Indikasi Berhentinya Urban Heat Island (Suhu) Di Bali Saat Nyepi *J. Meteorol. dan Geofis.* **15** 167–76

IOP Conf. Series: Earth and Environmental Science 1111 (2022) 012085 doi:10.1088/1755-1315/1111/1/012085

Challenges and Chances of Sustainable Construction in Indonesia: Policy Insights

MA Hapsari[1]*, WH Putri[2], and Sarju[3]
[1] Law department, Faculty of law, Universitas Janabadra, Yogyakarta, 55231, Indonesia, 0000-0002-1525-9491
[2] Accounting department, Faculty of economics, Universitas Janabadra, Yogyakarta, 55231, Indonesia, 0000-0002-1552-3993
[3] Civil engineering department, Faculty of engineering, Universitas Janabadra, Yogyakarta, 55231, Indonesia, 0000-0001-7775-460X
[1]*murtiayu@janabadra.ac.id , [2]wikaharisa@janabadra.ac.id ,
[3]sarjukesumo@janabadra.ac.id

Abstract. The construction industry in Indonesia is the fourth-largest contributor to the country's GDP; however, the industry accounts for one-third of total CO_2 emissions. As 71% of Indonesians are expected to live in urban by 2030, building sector emissions are also expected to grow at a 4.5% annual rate. The adoption of sustainable construction is required to balance the needs of the community while minimizing the impact on the environment. In 2017, Indonesia integrated its national development planning with the Sustainable Development Goals 2030. Moreover, infrastructure and policies to support the sustainable construction ecosystem began to be implemented. The research aims to assess whether the policies for sustainable construction have been sufficiently all-encompassing and holistic to be implemented appropriately. The analysis was conducted with the descriptive-analysis method. This study concluded that relevant ministries already supported the policies but struggles with execution arose due to underskilled human resources, ineffective stakeholder collaboration, and regional inequality and disparities. The Indonesian government should take advantage of the post-pandemic momentum and potential for high investment values to carry out extensive socialization and fortify the foundation of policies.

1. Introduction

The philosophical key point of sustainable construction is based on the World Commission on Environment and Development's definition of sustainable development, which is "development that meets the needs of the present without compromising the ability of future generations to meet their own needs"[1]—the fundamental idea is respecting the life of the present as well as the future. The strengthening of the discourse on sustainable construction appears because construction industries, along with power generation and automobile usage, are one of the three most significant contributors to greenhouse gas emissions that threaten the climate[2,3]. This concept was developed as an effort to lay standards, techniques, and means to support the management of greenhouse gas emission control in the construction industry [4]

The political will and caliber of supportive policy formation significantly impact how well sustainable construction policies, or policies in general, are executed. One of the significant challenges communities encounter when incorporating the policies is the complication of metric devising. It is necessary to define and explain terms like efficiency and sustainability to set attainable, measurable, and trackable goals [5]. However, compared to developed countries, the barriers to implementing sustainable development policies in developing countries are complex and multifaceted; problems range from a lack of financial incentives to over-exploitation of natural resources or simply because the government was incapable of instituting an efficient framework [6,7]. Additionally, in general, the public sectors in developing countries have minimal capacity to initiate changes [7]. In correspondence, due to issues with political and economic instability, investors frequently move their investments out of developing countries, which prevents the country from having enough financial support to function correctly [8]. Studies argue that adequate policy support and well-organized governmental mechanisms are essential for the aspired ecosystem's growth [9]. Therefore, preferential policies such as government subsidies, incentive programs, taxation, and other forms of technical assistance must be well-supported by policies. [10].

In 2004, the developing nation of Indonesia began to envision and incorporate sustainable development programs. Unfortunately, between the 2009-2014 period, there were no policies to support green development, such as funds and institutional resources for the research and development of affordable renewable energy. As a result, the government failed to make significant progress in managing and reducing carbon emissions [11]. Although it should be noted that having policies in place is not an assurance to present and performing maximum carrying capacity to achieve the desired goals; studies contend that policy failure occurs due to a lack of clear objectives, stakeholder perception differences; and decision-makers proclivity to pursue short-term goals [12]. As a case point, a previous study conducted by Sahid, Sumiyati, and Purisari in 2021 provided insights that Indonesian green policies are still very lacking both in proportion and reliability [13], further based on a literature review study by Agnes and Koestoer also conclude the existing regulations and policies on Indonesia green building require continuous improvement [14].

This study aims to provide insight into Indonesian sustainable construction policies and provide some context for whether these policies have delivered a solid foundation for establishing an effective ecosystem for sustainable construction. The study also considers variables outside the policy-making, such as socio-economic variables. Furthermore, this study will outline the government's steps to set a firm foundation for long-term policy development.

2. Methods

The study uses the descriptive policy analysis method, which focuses on evaluating specific policy categories and how they are implemented [15]. This study was conducted for a duration of 6 months. The authors gathered the data based on existing sustainable construction policies, and after that, content analysis was conducted on available and relevantly curated data. Subsequently, a conclusion was drawn based on the content analysis and the social-economic climate in Indonesia to provide appropriate solutions for the issues that were discovered. Variables such as law/legislation making, the policy substances, and socio-economic are considered while conducting the study. Further, the authors only include policies and documents related to sustainable construction that have been released as of mid-2022 in their study.

3. Results and Discussion

As one of the basic human needs, buildings or shelters are becoming more in demand as the world's population grows. Due to the unfortunate fact that the construction industry accounts for one-third of total CO2 emissions, energy consumption will continually increase due to population and building construction growth [16]. Sustainable construction is a concept that aims to address these problems by creating a balance between environmental, social, and economic concerns to ensure the industry's viability for future generations. The idea of sustainable construction is an all-encompassing and holistic approach to the industry. It is ranged from the designing process to demolition of the building, construction management aspects, as well as a matter of materials selection. Based on the International Council for Research and Innovation in Building and Construction (CIB) agenda further explained that sustainable construction is achievable through the following principles:

a. Reduction of construction resources
b. Resource reuse
c. Use of recycled materials
d. Protection of the natural environment
e. Exclude the use of toxic materials.

f. Guaranteed low costs for maintenance of construction sites.
g. Emphasis on quality[9,17].

These principles apply to all stages and cycles of building construction, from planning to demolition[18]. Furthermore, the resources required for construction are also frontrunners in managing construction processes based on overall quality management [9].

The embodiment of sustainable construction is essential to provide and support an enviable natural and social environment with the ultimate goal of enhancing communities' quality of life [19]. Thus, Raynsford argues that it is essential for the government to take part in the construction industry, in this case, as a decision-maker, to ensure communities' sustainability which is the core of all future development. Based on this reasoning, the authors argue that a comprehensive policy back-up from the government is mandatory to execute a strategic-sustainability program. Díaz-López further argues that governments need to take legal, financial, technical, and social measures to promote sustainability in building construction by using a comprehensive and integrated approach to balance the economic, environmental, and social aspects.[10].

The construction industry in Indonesia is the fourth-largest contributor to the country's gross domestic product (GDP), with a 10% average growth per year. Emissions from the industry are expected to grow at 4.5% per year, while around 71% of the population will live in urban areas by 2030 [20]. The fact that Indonesia is still categorized as a developing country gives some layers of complexity in the context of sustainable construction. Several researchers concluded that implementing sustainable construction practices is obstructed by various factors, including financial constraints, the incompetence or ignorance of the actors involved, and ineffective communication among key stakeholders. In general, it is possible to conclude that a significant obstacle to sustainable development is a less holistic approach [21–24]. Similar problems are also found in other low-income communities in developing countries [25,26]. Furthermore, the mainstream discourse on the topic of sustainable development, in general, emphasizes the favorability of economic interests over environmental issues; in the sense that economic interests have more influence in decision making, resulting in a lack of infrastructure and basis to promote sustainable construction in general [6,7]. To ensure a productive and healthy living environment, environmental protection, effective use of limited resources, and consideration for future generations, the government must support adequate management of the construction object throughout its life cycle [9].

It will be possible to determine the political will, commitment, and capacity of the Indonesian government in carrying out the policies by conducting an analysis of sustainable construction-related policies and other relevant environmental policies in Indonesia. In 2017, the Indonesian government integrated the national development programs with Sustainable Development Goals (SDGs) 2030 by constructing a "Sustainable Development Action Plan" (Rencana Aksi Pembangunan Berkelanjutan) in regulation frameworks, followed by standardized guideline programs in 2020. These indications mark the beginning of progress of sustainable construction in the right direction [27,28]. Moreover, some of the development program's plans have been precisely defined with more technical regulations issued directly by the relevant State Ministries. Furthermore, several regions in Indonesia also have issued regional sustainable construction regulations focusing on the development of Bangunan Gedung Hijau's "green building."[29]. In line with this commitment, the government has stated its objectives to achieve sustainable infrastructure development during its 2022 G20 Presidency as chairman [30]. Specific legislation directly relating to sustainable construction in Indonesia is shown in Table 1

Table 1. Sustainable construction-related policy in Indonesia

No	Legal instrument	Summary
1	Act No 28 of 2002 on Building	Building should be functional and in accordance with the environment
2	Government Regulation No 16 of 2021	Categories of Green Building Buildings, Principles of Green Building Buildings, Implementation stage (from planning to demolition), BGH certification mechanisms, incentive provisions

3	Public Works and Housing Ministerial Regulation No 21 of 2021 on Green Buildings Assessment (replaced Public Works and House Ministerial Regulation No 05/PRT/M/2015	Standardized measurement of green building, green building's principles and exertion, assessment mechanism, clauses on mandatory/voluntary green building
4	Environmental Ministerial Regulation No 8 of 2010 on Environmental Building Standards and Certification	Clauses on building eco-labelling, eco-label builing certification mechanism, clauses on environmental-friendly materials usage.
5	Environmental Ministerial Regulation No P.5/MNLHK/SETJEN/KUM.1/2/2019	Environment-friendly houses certification mechanism, requirements of environment friendly houses, eco-friendly materials labelling

The Indonesian Green Taxonomy mechanism by the Financial Services Authority (Otoritas Jasa Keuangan) serves as a manual for green economic activity. It contains a list of classifications of economic activities that support environmental protection and mitigation and adaptation to climate change. This mechanism is one of the institutions the government has built to improve the overall ecosystem for sustainable construction. The Green Taxonomy of Indonesia serves as the foundation for financial incentive policies to further promote sustainable construction [31]. Financial incentives are also provided in the form of the tax deduction for green-certified buildings [32]. Additionally, the Ministry of Energy and Mineral Resources has added eco-friendly mining regulations to the ecosystem; it is anticipated that building materials were sourced ethically [13].

Moreover, some regions have formed regional regulations to support sustainable construction, such as Jakarta, Bandung, Semarang, and Gorontalo. However, in comparison to the rest of Indonesia, this figure is negligible. Following that, regional sustainable construction practices face challenges and shortages, such as insufficient financial incentives, certification bodies using disparate standards, and perception gaps between the actors [33–35]. A 2019 study conducted in Lampung, one of the major cities in Indonesia, revealed that aspects associated with employees' partial skill and expertise, in addition to a communication gap among project stakeholders, have the highest value as a barrier to the implementation of sustainable construction [23].

Comprehensive sustainable construction-related arrangements are regarded as new regulations even though Indonesia has been working to establish a sustainable development ecosystem since 2004. Green building regulations are still in the socialization stage and the beginning of the transition period. In contrast, the integration of goals between the Indonesia Development Plan and the SDGs only started in 2017. Moreover, In 2019, the COVID-19 pandemic, which has redirected the government's attention to the health sector and economic recovery, must impede the development of sustainable construction regulations and foundations in Indonesia [36]. The pandemic has also caused some initiatives that support sustainable construction, such as improving public-private sector collaboration and increasing the value of the green construction chain, to be put on hold [36].

However, fortunately, the pandemic raised public awareness of the importance of green buildings and a healthy environment in general. [37], Several studies have shown that the "pandemic factor" can be used as a basis to promote the use of sustainable green buildings [38,39]. Further, from the financial perspective, IFC analysis of smart investment potential revealed that the investment potential in the field of green building construction in Indonesia is currently at $209 billion [40]. Based on investment opportunities and post-pandemic momentum supporting the sustainable construction ecosystem, it is expected that the government should immediately take advantage of the opportunity to maximize capitalization and socialization into other regions to strengthen the legal basis of sustainable construction throughout the nation. Other strategies include providing a strong foundation in the sustainable construction ecosystem, such as constructing a long-term macroeconomic policy to support sustainable construction firms and further pioneering strategic partnerships with the government and private sector's target stakeholders.

4. Conclusion

Indonesia has been planning for sustainable development since 2004, but the regulatory structure and infrastructure have only recently begun to be built. Following that, Indonesia made an effort toward sustainable development by integrating its national development plan with the SDGs 2030, and several ministries are involved in supporting sustainable construction policies. Several cities and regions in Indonesia have followed suit by establishing regional regulations, particularly on green buildings. The authors argue that the government should implement large-scale socialization in local governments across the country and financial incentives to improve both the quality and quantity of sustainable construction in Indonesia. Furthermore, the government can leverage potentially high-value green construction investment and positive post-pandemic momentum to strengthen the legal infrastructure of sustainable construction policy based in the long run and string strategic partnerships to connect each actor and stakeholders. Further research into the impact of Indonesia's current policy on sustainable construction post-pandemic may be required soon.

References

[1] EUR-Lex Sustainable development *EUR-Lex*
[2] Det Udomsap A and Hallinger P 2020 A bibliometric review of research on sustainable construction, 1994–2018 *J. Clean. Prod.* **254**
[3] IPCC 2019 2019 Refinement to the 2006 IPCC guidelines for national greenhouse gas inventories *IPCC*
[4] MATHUR V S, FAROUQ M M and LABARAN Y H 2021 The carbon footprint of construction industry: A review of direct and indirect emission *J. Sustain. Constr. Mater. Technol.* **6**
[5] Rahangmetan R 2021 Sustainable Development in Developing Countries *Cose Eur.*
[6] Ogunmakinde A . E, Sher W . D and Maund K 2016 Obstacles To Sustainable Construction in Developing Countries *Australas. Univ. Build. Educ. Assoc. Conf. 2016, Pullman Cairns Int. Cairns, Queensland, Aust. Vol. 40*
[7] Howes M, Wortley L, Potts R, Dedekorkut-Howes A, Serrao-Neumann S, Davidson J, Smith T and Nunn P 2017 Environmental sustainability: A case of policy implementation failure? *Sustain.* **9**
[8] Zainul Abidin Nazirah N 2010 Investigating the awareness and application of sustainable construction concept by Malaysian developers *Habitat Int.* **34**
[9] Marichova A 2020 Role of the Government for Development of Sustainable Construction *Ovidius Univ. Ann. Constanta - Ser. Civ. Eng.* **22**
[10] Díaz-López C, Navarro-Galera A, Zamorano M and Buendía-Carrillo D 2021 Identifying public policies to promote sustainable building: A proposal for governmental drivers based on stakeholder perceptions *Sustain.* **13**
[11] Wiryomartono B 2015 ‘Green building’ and sustainable development policy in Indonesia since 2004 *Int. J. Sustain. Build. Technol. Urban Dev.* **6**
[12] Hudson B, Hunter D and Peckham S 2019 Policy failure and the policy-implementation gap: can policy support programs help? *Policy Des. Pract.* **2**
[13] Sahid S, Sumiyati Y and Purisari R 2021 Strengthening green building policies in Indonesia *IOP Conference Series: Earth and Environmental Science* vol 780
[14] Agnes M and Koestoer R H 2021 A Review on Sustainable Construction Regulations in Asian Countries: Savvy Insights for Indonesia *J. Ilmu Lingkung.* **19**
[15] Patton C V., Sawicki D S and Clark J J 2015 *Basic methods of policy analysis and planning*
[16] Neill P 2020 Construction industry accounts for 38% of CO2 emissions *Environ. J.*
[17] Ganiyu B O 2016 *Strategy to enhance sustainability in affordable housing construction in South Africatle* (Cape Peninsula University of Technology)
[18] Ding G K C 2013 Life cycle assessment (LCA) of sustainable building materials: An overview *Eco-Efficient Construction and Building Materials: Life Cycle Assessment (LCA), Eco-Labelling and Case Studies*
[19] Raynsford N 2000 Sustainable construction: The Government’s role *Proc. Inst. Civ. Eng. Civ. Eng.* **138**
[20] Indonesia M R 2022 The Booming Growth of the Construction Industry in Indonesia *Mark. Res. Indones.*
[21] Ross N, Bowen P A and Lincoln D 2010 Sustainable housing for low-income communities:

Lessons for south africa in local and other developing world cases *Constr. Manag. Econ.* **28**
[22] Häkkinen T and Belloni K 2011 Barriers and drivers for sustainable building *Build. Res. Inf.* **39**
[23] Susanti B, Filestre S F H and Juliantina I 2019 The Analysis of Barriers for Implementation of Sustainable Construction in Indonesia *IOP Conference Series: Earth and Environmental Science* vol 396
[24] Heni F and Ajayi S 2022 Barriers to sustainable practices in the Indonesian construction industry *J. Environ. Plan. Manag.*
[25] Babalola A and Harinarain N 2022 Policy barriers to sustainable construction practice in the Nigerian construction industry: an exploratory factor analysis *J. Eng. Des. Technol.*
[26] Ahmed A M, Sayed W, Asran A and Nosier I 2021 Identifying barriers to the implementation and development of sustainable construction *Int. J. Constr. Manag.*
[27] Kementerian PPN 2020 *Pedoman Teknis Penyusunan Rencana Aksi - Edisi II Tujuan Pembangunan Berkelanjutan/ Sustainable Development Goals (TPB/SDGs)*
[28] Bappenas K P 2020 Pedoman Teknis Penyusunan Rencana Aksi Tujuan Pembangunan Berkelanjutan (Tpb)/ Sustainable Development GoalS (SDGs) *Bappenas* **53**
[29] AsiaToday 2019 Applying Green Building, DKI Jakarta and Bandung Save Energy IDR 2.58 Trillion *Asiatoday*
[30] RI B K F K K 2022 Indonesia's G20 Presidency Encourages Collaboration to Accelerate PostPandemic High-Quality and Sustainable Infrastructure Development *Kementeri. Keuang. Republik Indones.*
[31] Keuangan O J 2022 Taksonomi Hijau Indonesia *Otoritas Jasa Keuang.*
[32] Adhetiya Safitra D 2022 Insentif Pajak Atas Bangunan Hijau: Sebuah Studi Komparasi *J. Pajak dan Keuang. Negara* **3** 331–42
[33] Pahnael emy R N, Anton Soekiman and Mia Wimala 2020 Penerapan Kebijakan Insentif Green Building Di Kota Bandung *J. Infrastruktur* **6** 1–13
[34] Nurman T R 2021 Analisa Penerapan Gedung Bangunan Hijau pada Tahap Pelaksanaan Konstruksi *Syntax Idea* **3**
[35] Widiati I R 2019 Tinjauan studi analisis komparatif bangunan hijau (green building) dengan metode asesmen sebagai upaya mitigasi untuk pembangunan konstruksi yang berkelanjutan *Pros. Konf. Nas. Pascasarj. Tek. Sipil X 2019*
[36] Indonesia K P U dan P R R 2021 *Buku I Konstruksi Indonesia 2021: Era Baru Konstruksi Berkarya Menuju Indonesia Maju* (Jakarta: Kementerian PUPR)
[37] Widyartanti J E 2021 Green Building pada Desain Rumah Pasca Pandemi, Ciptakan Hunian Ramah Penghuni dan Bumi *IDEA Onine*
[38] Wardani F K 2021 Green Building in the midst of Pandemic *Smart City* **1**
[39] Leony M and Sari S R 2022 Arsitektur hijau mendukung adaptasi perilaku di masa pandemi Coronavirus disease (COVID-19) di Alun-Alun Kota Batam *Din. Lingkung. Indones.* **9**
[40] Corporation I F 2016 Climate Investment Opportunities in Emerging Markets An IFC Analysis *IFC World Bank Gr.*

Affective Dimensions in Community Responses to the Free COVID-19 Vaccination Program and Its Relationship with Public Health in Sidoarjo Regency

Trenda A. Oktariyanda[1]*, Meirinawati[2], Eva H. Fanida[2], Fitrotun Niswah[2], and Ngukup C. Achmadja[2]
[1]Departement of Public Administration, Faculty of Social and Law, Universitas Negeri Surabaya, Jalan Ketintang, Surabaya 60231, Indonesia, ORCID ID: 0000-0002-7099-427X
[2]Departement of Public Administration, Faculty of Social and Law, Universitas Negeri Surabaya, Jalan Ketintang, Surabaya 60231, Indonesia
[1]*trendaoktariyanda@unesa.ac.id, [2]meirinawati@unesa.ac.id, [3]evafanida@unesa.ac.id, [4]fitrotunniwah@unesa.ac.id, [5]ngukup.18098@mhs.unesa.ac.id

Abstract. The increasingly sloping development of COVID-19 cases cannot be separated from the COVID-19 Free Vaccination Program. The implementation of the vaccination program is oriented towards mass immunity against COVID-19. This is the link between the response and the public health aspect. This study discusses the community's response to the affective dimension (awareness and expectation) about public health in 100 community respondents in Sidoarjo Regency. This research uses a quantitative method with a descriptive type. Several 100 samples were obtained from the formula for measuring samples developed by Slovin. The results showed that most of the respondents on the indicators of awareness and hope chose the "enough" answer. The rest answered, "poor." Interestingly, none of the respondents in the two indicators answered "good." Of course, this is a note for policymakers (government), especially in the free vaccine program, that must be evaluated and further improved to improve the effective response of the community.

1. Introduction

Before going into the core background, let us take a moment to explore the journey of the COVID-19 pandemic case that hit the country of Indonesia, starting from the beginning of its entry until the critical moment in 2021 when it became the peak of cases. The COVID-19 pandemic, which entered Indonesia in the first quarter of 2020, showed an increasing trend of severe cases. It was recorded that in March 2021, there were 129,844 positive people, 1,290 cases were declared cured, 790 people died, and 39,550. The highest record of adding daily positive COVID-19 cases was 54,000 on July 14, 2021 [1]. The Indonesian government has taken various steps to deal with the COVID-19 pandemic. National data in August 2022 showed positive as many as 52,009 people, and the number of cases declared cured as many as 6,097,334 people [2]. The distribution from East Java Province is 939 positive people, the number of cases declared cured 560,978 people, and the number of cases died is 31,079 people [3]. Specifically, the distribution in Sidoarjo was 71 positive, 45,442 people recovered, 1,019 people died, and 0 suspects.

One of the paths taken by the Government of Indonesia in dealing with COVID-19 is through

Presidential Regulation 99 of 2020 concerning Vaccine Procurement and Vaccination Implementation in the Context of Eradicating the Corona Virus Disease 2019 (COVID-19) Pandemic.

Based on the Presidential Decree, the government, through the Minister of Foreign Affairs in 2020, will conduct international diplomacy to gain access to the Covid-19 vaccine and budget support for multilateral cooperation and other necessary support. The result of this policy is that more than 300 million COVID-19 vaccines will be injected from various manufacturers and countries in early 2022 [4]. The success of the procurement of vaccines and national COVID-19 vaccinations will not run smoothly without the Minister of Health Regulation Number 84 of 2020 concerning the Implementation of Vaccinations in the Context of Combating the 2019 Corona Virus Disease (COVID-19) Pandemic. It can also be said that the government's efforts are also oriented towards achieving mass immunity from COVID-19 in particular and fulfilling optimal public health in general. Public health is the science and art of maintaining, protecting, and improving public health through community organizing efforts [5]. This aspect of public health is urgent because it also concerns long-term life expectancy and community productivity.

Departing from the above background, the researcher is interested in how the community's behavioral or affective response is related to vaccine regulation. Does the community fully support the government program or vice versa? The diversity or diversity of the majority or minority of the community's affective response is one of the vital pieces of information that cannot be separated from seeing how the government is performing, especially in dealing with COVID-19 and improving the health status of the community. Based on the description above, this research studies the vaccination response. The research location is focused on Sidoarjo Regency, with the letter from the Ministry of Health that the priority of the first phase of the Sinovac vaccine is Surabaya, Gresik, and Sidoarjo. Therefore, researchers are interested in taking the research title "Affective Dimensions in Community Responses to the Free COVID-19 Vaccination Program and Its Relation to Public Health in Sidoarjo Regency". The purpose of this study was to capture and analyze the public's response to the government's vaccination program, especially from the affective dimension and its relationship to public health in Sidoarjo Regency, Indonesia.

2. Methods

This study uses a quantitative research approach with a descriptive type. Research [6] states that quantitative research is research whose data type is in the form of numbers or qualitative information that is numbered. The selection of the quantitative approach in this study was due to the adjustment of the research objectives, namely to measure and analyze the results of community responses which were included in the affective dimension of the free vaccination program in suppressing the spread of COVID-19 in Sidoarjo Regency and its relationship to public health. The sample is 100 people of Sidoarjo Regency based on calculations with Slovin's formula [7]. Questionnaires were distributed online to 100 respondents in Sidoarjo Regency. The sample is a resident of Sidoarjo Regency who is of the appropriate age to be given the vaccine, namely the age of 17 years to 65 years and over.

3. Results and Discussion

3.1 Characteristics of Respondents

In this scientific article based on empirical research, the authors classify the characteristics of respondents based on age, education level, type of work, and vaccination status. Regarding the age range of respondents, as many as 93 percent of respondents are aged 18 to 64 years. Then followed by 3 percent of respondents aged 12 to 17 years and 4 percent of respondents aged 65 years and over. Hurlock revealed that the older the level of maturity and strength of a person would be more mature they are in thinking and working [8]. Thus, the more mature the respondent is, the more accurate the information that can be provided in this study will be. Research [9] states that education is the ability to absorb knowledge. The higher the level of education, the easier it is for someone to absorb knowledge. In this study, 39 percent of respondents had undergraduate education, then 36 percent of respondents had high school education/equivalent, and 16 percent had postgraduate education.

Furthermore, at the last three education levels, 6 percent have graduated with a diploma, 2 percent have just graduated from junior high school/equivalent, and only 1 percent have not finished school. This shows that respondents can understand the questions asked by the research team in collecting research data. So that the data obtained is accurate. The respondents' occupations were 40 students, 19 private employees, 18 civil servants, and 23 percent working in other sectors/not working. In addition, the last respondent's characteristics are related to whether the respondent received the COVID-19 vaccination or not. When the team conducted data collection, as many as 82 percent of the respondents had received the first or second dose of vaccine injection. Meanwhile, the remaining 18 percent have not received any vaccine injections.

3.2 Response Level

The response is a psychological term used to describe responses to stimuli received by the five senses. The things that support and underlie the size of the response are attitudes, perceptions, and participation. A person's attitude precedes the response in the process because attitude is a person's tendency or willingness to behave when faced with a specific stimulus. In general, it can be said that three factors influence a person's response, namely: 1) The person concerned sees. The person concerned who sees and tries to give an interpretation of what he sees, which is influenced by his attitude, awareness, and expectations; 2) Response targets. In the form of people, things, or events. The characteristics of the sign usually influence the reaction of the viewer. In other words, the movement, sound, size, action, and other characteristics of the response target also determine how people perceive it; 3) Situational factors. The response can be seen contextually, meaning that it gets attention in situations where the answer appears. The problem is a factor that plays a role in forming a person's response. The response is essentially a behavioral response or attitude that becomes reverse behavior, which is also a process of organizing the stimulus in which the proximal stimulus is arranged in such a way that there is a phenomenal representation of the proximal stimulus. A person's attitude precedes the response in the process because attitude is a person's tendency or willingness to behave when faced with a specific stimulus. So, attitude determines whether someone responds or does not respond to something.

3.3 Affective Dimension

The affective dimension relates to the tendency to act, operate, think and feel in the face of objects, ideas, situations, and values-driven by the existence of goals, values, and motivational needs. The following are the results of research on the affective dimension.

Table 1. Affective Dimension Response Category.

Response Category	Overall Score	Total	
		f	%
Good		0	0
Enough	2.984	85	85
Bad		15	15

Table 1 shows that the overall affective dimension of the community's response to the free vaccination program is still quite adequate. Where most respondents gave a sufficient rating of 85 percent, this shows that the free vaccination program is still considered "only" sufficient by the community, even though 15 percent of the community has a bad or unfavorable view.

3.3.1. Indicators of Awareness.

The awareness indicator in this affective dimension is the first consideration because the awareness wrapped in a pattern or way of thinking is one of the fundamental factors in a person's behavior. [9] defines consciousness as the result of how a group of people thinks, and each thought is separate from another [10]. The following are the results of research on awareness indicators.

Table 2. Awareness Indicator Response Category

Response Category	Overall Score	Total	
		f	%
Good		0	0
Enough	2.426	87	87
Bad		13	13

Based on table 2, it is known that the overall awareness indicator is quite adequate, where there is 87 percent of respondents choose sufficient and reasonable. This shows that the respondents consider the free vaccination program "merely" sufficient. However, a small proportion (13 percent) of respondents still do not yet have or meet these knowledge indicators. One of the interesting findings here is that there are no respondents who rate well. This reflects the government's performance, especially in the free vaccine program, which must be evaluated and further improved so that the community's affection response improves.

3.3.2 Indicators of Expectations

According to the theory of SOR (Stimulus-Organism-Response), the size of the influence of the message stimulus on the response depends on the amount of attention given to being understood and accepted by the recipient of the message. Furthermore, [11] revealed that a person's attention is influenced by psychological factors such as willingness, desire, motivation, hope, etcetera, and biological and socio-cultural factors. The following are the results of research on the expectation indicator.

Table 3. Expectation Indicator Response Category.

Response Category	Overall Score	Total	
		f	%
Good		0	0
Enough	558	84	84
Bad		16	16

Based on table 3, it is found that the overall expectation indicator is in the excellent category. That is, as many as 84 percent of respondents chose the answer enough. This data shows that respondents' experience with the government's free COVID-19 vaccination program is still at a sufficient level. The remaining 16 respondents considered the free vaccination program in the wrong category. This needs more attention from the government, considering that this vaccination program has entered its fourth dose and may continue in the future.

3.4 The Relationship with Public Health Factors

The concept of public health is one of the essential foundations to see how the government's free vaccination program results and objectives are achieved from the community's perspective, which can also be seen from the community's response (affective dimension). According to [12], public health is the science and art of preventing disease, prolonging life, promoting physical and mental health and efficiency through organized community efforts to improve environmental sanitation, infection control in the community, education of individuals about personal hygiene, and organization of medical services. Moreover, for early diagnosis, disease prevention, and social development, nursing will support everyone in society to have a substantial standard of living to maintain their health. The scope of public health includes two main disciplines, namely biomedical sciences and social sciences, in line with the development of public health sciences, among others; Biology, medicine, chemistry, physics, environment, social, anthropology, education, etcetera. The disciplines supporting public health are biology, medicine, chemistry, physics, environment, social, anthropology, education, etcetera. The disciplines that support public health are as follows; 1) Epidemiology; 2) Biostatistics/health statistics; 3) Environmental health; 4) Health education/behavioral science; 5) Public Health Administration; 6) Community nutrition; 7) Occupational health.

According to [13], four factors influence public health status, namely: 1) Behavior, 2) Environment, 3) Health Care Services, and 4) Heredity. From the four factors above, it turns out that the influence of behavior is quite significant, followed by environmental factors, health care services, and heredity. The four factors above are closely related and influence each other. Healthy behavior will support increasing health status. This can be seen from the number of behavior and lifestyle-based diseases.

3.4.1 Behavior Factor
Lifestyle is the first factor that affects public health because health and an unhealthy environment are highly dependent on human behavior. In addition, it is also influenced by habits, customs, beliefs, education, socioeconomic and other behaviors inherent in him. For example: in a society undergoing a transition from a traditional society to modern society, there will be changes in lifestyle that have an impact on health status. Specific examples: some people still think that vaccine injections are not halal, or there is also an assumption that vaccine liquids harm the body and are only projects for the benefit of the government

3.4.2 Environment Factor
This environment includes the physical environment (both natural and artificial), such as waste, water, air, and water. Housing and socio-cultural (economy, education, employment, and others). In the physical environment, health will be affected by the quality of environmental sanitation where humans are located and influenced by the spread of diseases in the vicinity (e.g., COVID-19, which has spread everywhere). This is because many diseases originate from poor quality of environmental sanitation. For example, clean water availability in an area will affect health status because water is a basic human need, and humans constantly interact with water in their daily life. At the same time, the social environment is related to the economic condition of society. The poorer the individual/community, the more difficult it is to access good health.

3.4.3 Health Care Services Factor
Health services are the third factor that affects the degree of public health because the existence of health facilities is very decisive in health recovery services, prevention of disease, treatment, and nursing, as well as groups and communities that require health services. The availability of facilities is very influential by the location, whether it can be reached by the community or not, health workers who provide services, information and community motivation to do so visiting facilities to obtain services, as well as whether the health service program itself is following the needs of the community itself.

3.4.4 Heredity Factor
Hereditary/genetic factors are also very influential on health status. This is because several diseases are passed down through genetics or factors that already exist in humans that are brought from birth, for example: from hereditary diseases, including diabetes mellitus, bronchial asthma, epilepsy, mental retardation, hypertension, and color blindness.

From the description above, the relationship between public health factors, especially people's behavior with the affective dimension of the community's response to the government's vaccination program becomes evident and straightforward. This happens because the behavioral factor here is a manifestation or a tangible manifestation of the awareness and expectations of the community, which are contained in the affective dimension of the community's response itself. It becomes illogical when people can be moved if there is no form of mental, physical, or psychological awareness. A link between public health behavioral factors and the affective dimension of community response also strengthens its influence in observing and assessing government vaccination programs from the community's perspective.

IOP Conf. Series: Earth and Environmental Science 1111 (2022) 012086 doi:10.1088/1755-1315/1111/1/012086

4. Conclusion

Based on the data analysis of the research results, it is known that the majority of the community, in terms of their affective dimension, considers the government vaccination program to be "merely" sufficient. One of the interesting findings here is that from the two indicators (awareness and expectation), there are no respondents who rate well. This reflects the government's performance, especially in the free vaccine program, which must be evaluated and further improved to improve the community's affective response. Therefore, this community response reflects "at a glance" the Sidoarjo Regency Government and the central government, which incidentally still bears quite a lot of homework. Hence, the government needs to work together to provide more socialization, education, and counseling to the community about the benefits of free vaccination programs. This is expected to dispel doubts and disinformation regarding the free vaccination program.

References

[1] Indonesia C N and B C 2021 RI Cetak Rekor Terus, Kasus Baru Covid-19 Tembus 54.000 *CNMB Indones.*
[2] Force C-19 H T 2022 Peta Sebaran COVID-19
[3] Government E J P 2022 Peta Sebaran COVID-19 Jatim
[4] Rokom 2022 300 Juta Dosis Vaksin COVID-19 Telah Disuntikkan Kepada Masyarakat
[5] Simorangkir O P 1987 *Kesadaran, Pikiran dan Tanggungjawab* (Jakarta: Penerbit Yagrat)
[6] Sugiyono 2012 *Statistika untuk Penelitian* (Bandung: Alfabeta)
[7] Ariola M M 2006 *Principles and Methods of Research* (Sampaloc: Rex Bookstore Inc)
[8] Hurlock E B 1978 *Perkembangan anak; Jilid 1* (Jakarta: Erlangga)
[9] Koentjaraningrat S 1997 *Metode-Metode Penelitian Masyarakat* (Jakarta: PT.Gramedia)
[10] Jamanti R 2014 Pengaruh Berita Banjir Di Koran Kaltim *J. Ilmu Komun.* **2** 17–33
[11] Mulyana E, Mujidin M and Bashori K 2015 Peran Motivasi Belajar, Self-Efficacy, dan Dukungan Sosial Keluarga Terhadap Self-Regulated Learning pada Siswa *PSIKOPEDAGOGIA* **4** 165–73
[12] Leavell H . and Clark E G 1965 *Preventive Medicine for Doctor in his Community* (New York: McGraw-Hill Book Company)
[13] Blum H L 1974 *Planning for Health, Development and Aplication of Social Changes Theory* (New York: Human Sciences Press)

The Green Economy and Decentralization of Natural Resources Management (DNRM) Policy in Indonesia under the International Law Framework: *Quo Vadis*?

Dian Parluhutan[1]*, Satya Arinanto[2] and Velentina Napitupulu[3]
[1]Faculty of Law, Pelita Harapan University, Karawaci, 1558, Indonesia, ORCID ID: 0000-0001-7244-7918
[2]Faculty of Law, University of Indonesia, Depok, 16424, Indonesia, ORCID ID: 0000-0002-2148-9840
[3]Faculty of Law, University of Indonesia, Depok, 16424, Indonesia, ORCID ID; 0000-0002-8772-142X
[1*]dian.parluhutan@uph.edu, [2]satya_arinanto@yahoo.com,
[3]velentina.napitupulu@yahoo.com

Abstract. The Green Economy, as the multilateral paradigm for national developments, emerges to counter an obsolete mindset of development. A green economy fosters a responsible long-term developmental attitude, emphasizing environmentally sustainable economic progress to foster low carbon and socially inclusive development. There must be a catalyzation of investment and innovation that will underpin sustained growth and give rise to new economic opportunities. Indonesia is obliged to implement green economy principles and mandates. However, on the national level, there is no specific regulation concerning implementing a Green economy. Since 1999, Indonesia has been implementing decentralization (regional autonomy). The natural resources management policy has been under the authority of regional governments. However, the decentralization has resulted only in the central government's willingness to redistribute responsibilities but not willing to share decision-making authorities with the regional governments. This phenomenon could jeopardize the green economy implementation. Hitherto, there are a few kinds of literature discussing the implementation of a green economy and the rule of international law in Indonesia. Accordingly, this paper attempts to analyze implementing a green economy and decentralizing natural resources management policy in Indonesia. The doctrinal and comparative law methods will be employed in this research.

1. Introduction

The Green economy emerged as a multilateral paradigm for national development and has been prompted by a new mindset to challenge an obsolete development mindset, namely "grow first, clean up later." According to the United Nations Environment Programme (UNEP), a green economy is a low-carbon, resource-efficient, and socially inclusive development approach. Hence, growth in employment and income is driven by public and private investments into economic activities supporting that notion. Put differently, foreign direct investments are directed towards infrastructure development and assets that allow reduced carbon emissions and pollution, including resource efficiency [1,2].

Accordingly, in implementing a green economy, the authorities must involve sustainable consumption and production and resource efficiency for sustainable development. Sustainable consumption and production aimed to consolidate production processes in reducing waste generation and resource consumption. Resource efficiency aims to reduce the resources needed, emissions, and waste generated per service unit [1–4].

Historically, the holistic concept of the green economy is based upon the Convention on Biological Diversity (CBD), as one of the key agreements adopted during the Earth Summit in Rio de Janeiro in 1992. Initially, the concept of a green economy stemmed from the UNED in October 2008 to tackle the opportunity towards greening the world. In February 2012, the World Bank and the UNEP, OECD, and GGGI initiated a new international knowledge-sharing platform in the Mexico-Green Growth Knowledge Platform (GGKP). Subsequently, in 2012, the Mexican Presidency of G-20 introduced 'inclusive green growth' as the cross-cutting priority in the G-20 Presidency. Furthermore, in 2013, the Global Symposium on REDD+("Reducing Emissions from Deforestation and Forest Degradation Plus") in Green Economy was convened between the 19th and 20th of June. The goal of this Symposium was to provide a more informed platform for exchanging information for connecting REDD+ planning and investments with green economy endeavours. Afterward, between July 18th and 20th, 2013, the 8th Annual Green Economics Institute Conference was held at Oxford University [1,2,5,6].

In the meantime, in Indonesia, the concept of a green economy was initiated based upon President Yudhoyono's remark before the international community, as follow said that they will need to work together, to adapt our lifestyle and our national policies toward a green global economy based on the principle of joint, but differentiated responsibilities and respective capabilities. In this spirit, we in Indonesia have adopted a 4 - track development strategy, which is pro-growth, pro-job, pro-poor, and pro-environment [1,5].

Reasonably, Indonesia's green economy is vital to boost national development. Empirical evidence shows that the incorporation of the green economy in the Indonesian developmental plan would bring beneficial impacts, for example, improvement of income per capita (GDP) and greenhouse gas emission (CO2 emission) in the long term [1,2,5].

2. Methods

The implementation of the green economy in Indonesia and international practices such as in the UNEP and the United Nations Economic and Social Commission for Asia and the Pacific (UNESCAP) Member Countries are discussed in this study. This study focuses on the differences in laws between countries, particularly the definition and scope of green economy policy and the necessary apparatus to undertake the green economy policy devised by the central and local governments. The normative legal method is used in this study. The rules and regulations concerning the green economy policy and its implementing regulations, such as the Law on Forestry and corresponding regulations in Indonesia, given international practices, are then compared. The research is normative and descriptive in nature. Primary research materials, secondary research materials, and other legal normative research materials are included. The primary materials relevant to each country's rules are the focus of this study. Secondary research materials include research materials, scientific articles, and other items about the green economy and forestry regulations. Other materials that explore regulations are referred to as tertiary research materials [7].

3. Result and Discussion

3.1. Green Economy as the Paradigm for Multilateral and National Developmental Policies

The emergence of the green economy as a multilateral paradigm for national developments had been inspired by a motivation to challenge an obsolete developmental paradigm: "grow first, clean up later ."Hence, the Green economy fosters a responsible long-term developmental attitude by emphasizing environmentally sustainable economic progress to foster low carbon and socially inclusive developments. Historically, the idea of a green economy is derived from the Sustainable Development

Objective, developed by the United Nations Conference on Environment and Development (UNCED). Initially, the green economy was introduced during the 5th Ministerial Conference on Environment and Development (MCED 2005) on March 2005, attended by the Member Countries of UNESCAP (United Nations Economic and Social Commission for Asia and the Pacific) [2,5].

According to the Organization of Economic Cooperation and Developments (OECD), the concept of a Green economy refers to the following notion. The green economy fosters economic growth and development while ensuring that the natural assets continue to provide the resources and environmental services on which our well-being relies. To do this, it must catalyze investment and innovation, underpin sustained growth and give rise to new economic opportunities. If governments pursue policies designed to promote green growth, they need indicators that can raise awareness, measure progress and identify potential opportunities and risks. According to *Hepburn, Pfeiffer, and Teytelboym*, the notion of a Green economy has become widely adopted as a core objective of international institutions and national governments.

In the hitherto practices, there are two typologies of Green Growth: so-called 'Strong Green Growth' and 'weak green growth .'This differentiation could be soundly explained as follows Strong green growth' involves no trade-off between growth and natural capital protection, even in the short term. Under 'weak green growth,' natural capital protection may reduce growth rates below business-as-usual growth in the short term, but growth will be higher in the long term. The potent variant of 'green growth' particularly pleases politicians and policymakers; the absence of a short-run trade-off between environmental protection and economic growth enables proponents to claim to have it all. Whereas the Green economy is not intended merely to serve as a developmental idea but as the developmental praxis of states, the *Organization for Economic Cooperation and Developments* has introduced the 'measurement framework' to evaluate the implementation and identify relevant factors for the legislators and other stakeholders. This 'measurement framework' was designed by considering interactions between the economy, natural (environmental) resources, and policy or legislative measures, which is illustrated as follows [2,5,6].

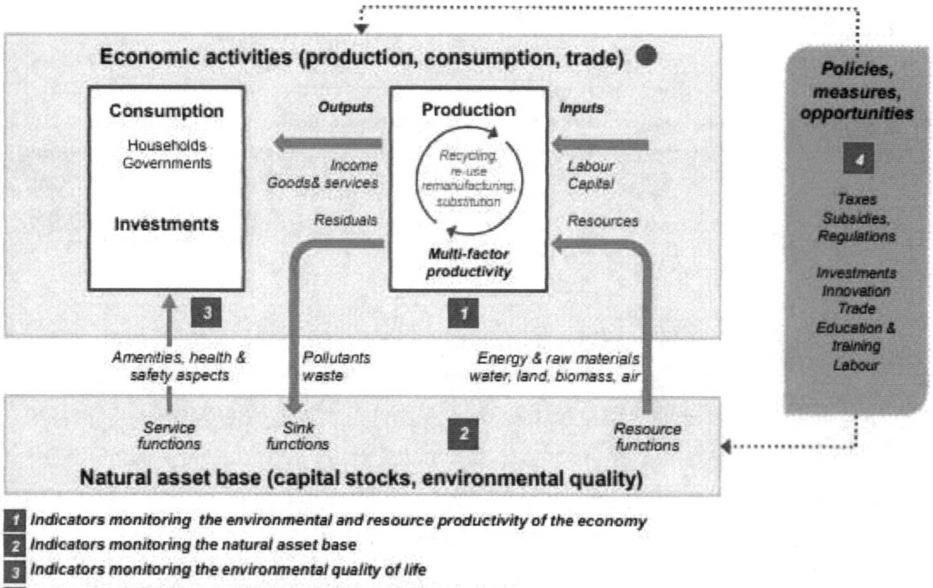

Figure 1. Measurement Framework for Implementation of Green Economy in a country

Equally important, the *Organization for Economic Cooperation and Developments* envisages as well, several key parameters to assess whether it has been successfully implemented, which can be portrayed as follows [2,5,6]:

Table 1. Headline Indicators for Environmental and resource productivity

Environmental and resource productivity	
Carbon & Energy productivity	1. CO2 productivity
Resource productivity	2. Non-energy material productivity
Multifactor productivity	3. Environmentally adjusted multifactor productivity
Natural asset base	
Renewable & non-renewable stocks	4. Natural resource index
Biodiversity & ecosystems	5. Changes in land cover
Environmental quality of life	
Environmental health & risks	6. Population exposure to air pollution (PM2.5)
Economic opportunities and Policy responses	
Technology & innovation	7. Technology innovation that promotes environmental sustainability
Environmental goods and services	8. Air pollution (population exposure to PM 2.5)
Regulation and management approaches	Regulation pro for environmental and energy sustainability at the central and municipal levels.

Implementing the Green economy is critical in the Indonesian long-term developmental plan. Arguably, the incorporation of a Green economy in the Indonesian developmental plan will bring beneficial impacts in terms of improvements in (1) Gross Domestic Product (GDP); (2) Green House Gas (GHG) Emissions; (3) Energy use; (4) Energy productivity, as could be seen in the following figure.

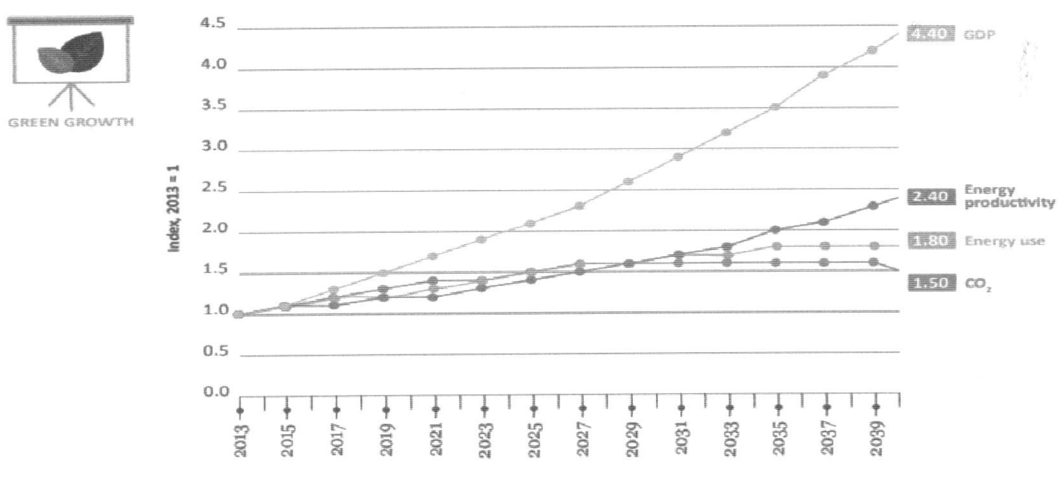

Figure 2. Positive Impacts due to the Implementation of Green Economy in Indonesia

3.2. Green Economy as the Subset of the United Nations Sustainable Development Goals

Arguably, the Green economy as a developmental strategy is not a new concept. Indeed Green economy originally crystallized from a broad and multi-faceted concept of 'Sustainable Development Goals (SDGs) [17,18,20,21]. Hence, it is argued that the Green economy is a subset of, not a replacement, of the sustainable development objectives, which encompasses the social equity pillar therein.

3.3. International Legal Character of the Green Economy and UN Sustainable Development Goals (SDGs)

From the International law perspective, the UN General Assembly Resolution on Sustainable Development Goals (commonly known as Global Agenda 2030) is considered a Law-making treaty that codifies existing customary law of UN Member States. Furthermore, the UN Resolution on SDGs 2030 must be understood as "autorité juridique plus élevée," in which the UN Member States have bounded themselves to the contents of the UN Resolution on SDGs 2030. Further, on the secondary level of analysis, the UN Resolution on SDGs 2030 can be seen as a manifestation of the International customary law because it embodies both UN Member States' practice which is obligatory (*opinion Juris sive necessitates*) and similar continuous practices of UN Member States (*longa consulted*) [2,3,5,6,8].

On the other hand, according to *Huck, Kurkin, and Spießhofer*, the UN Resolution on Sustainable Development Goals 2030 can be seen as a 'soft law.' The 'soft-law' is very important within interpretations of international treaties and for the gradual transformation of an international treaty's contents. However, in recent praxis of international law, the principles in SDG 2030 are developing into the 'soft-law' but with 'hard-law' sanctions [3,8].

3.4. International Law Analysis of Green Economy in Indonesia

The green economy in the context of sustainable development is not new because Indonesia has considered the importance of sustainable development since 1972. The term "sustainable development" is officially used in TAP MPR No.IV/MPR/1999 concerning the Broad Outline of State Policy (Garis-Garis Besar Haluan Negara, GBHN). Then, the term "environmentally sustainable development" is employed in Law No.23 of 1997 concerning Environmental Management. The popular term used is "pembangunan yang berwawasan lingkungan" (eco-development) [5,6,8,9]. In 1992, Indonesia played an active role at the United Nations Conference on Environment and Development (UNCED), held in Rio de Janeiro, Brazil. In 1997, Indonesia issued a National Agenda 21 containing references to incorporate the principles of sustainable development into national development planning. Then, the National Agenda 21 was followed by Sectoral Agenda 21 issued in 2000, which included the mining, energy, housing, tourism, and forestry sectors. Some local governments have also begun to formulate Local Agenda 21, which could provide guidance on development planning at the sub-district level and be a reference for various parties to draw up action plans.

Indonesia then ratified many international environmental agreements/conventions, namely the Convention on International Trade of Endangered Species (CITES), Basel Convention on Hazardous Waste, Vienna Convention on the Protection of the Ozone Layer and Montreal Protocol, United Nations Convention on Biological Diversity (UNCBD) and Cartagena Protocol on Biosafety, and United Nations Framework Convention on Climate Change (UNFCCC) and Kyoto Protokol, and United Nations Convention to Combat Desertification (UNCCD) [1,2,5,6].

From 1992 to 1999, Indonesia issued regulations supporting the implementation of sustainable development. These include but are not limited to Law No. 23 of 1997 concerning Environmental Management, Law No. 4 of 1982 concerning Basic Provisions for Environmental Management, and Law No. 41 of 1999 concerning Forestry. Then, considering the importance of a green economy, in the Working Meeting (*Rapat Kerja*), which was held at Tampak Siring Palace, Bali, on 19-21 April 2010, the President of the Republic of Indonesia, Susilo Bambang Yudhoyono, emphasized ten directives to encourage national economic development for the period 2010-2014. The directives, which were known as the "Deklarasi Tampak Siring" (Declaration of Tampak Siring), were based on the spirit of sustainable development and the philosophy of "triple bottom line." Deklarasi tersebut berorientasi pada pembangunan berkelanjutan atau menggunakan pola pembangunan yang pro-growth, pro-poor, pro-job, and pro-environment. Furthermore, the declaration was oriented towards sustainable development or employed development patterns that were pro-growth, pro-poor, pro-job, and pro-environment. Furthermore, the declaration was in line with Indonesia's commitment to the world to reduce carbon emissions by 26% (14% of the land use and forestry sector, 6% of the energy sector, and 6% of the waste sector), so there is no other choice so that Indonesia must be more serious work on the green economy [10,11].

The implementation of a directed and comprehensive green economy in Indonesia must be supported by Government policies to ensure the success of its implementation. During the United Indonesia Cabinet II, the National Medium-Term Development Plan (*Rencana Pembangunan Jangka Menengah Nasional*, RPJMN) for 2010-2014 was used as the basis for the sustainable development of the Indonesian economy. For example, the application of green budgeting to organizing financial management of government budgets and implementing green procurement in public policy.

According to Prof. Otto Sumarwoto, there are several reasons for Indonesia to implement Green Economy significantly and systematically: (1) Because there have been conflicts of interest between economic utilization and environmental preservation, and various natural resource management problems, such as: depletion of natural resources for economic interests; reduced ability of environment to neutralize pollutants; forest deforestation; unsustainable and non-environmentally friendly exploitation of mineral and marine resources; (2)The implementation of pro-growth, pro-job, pro-poor, pro-environment development strategies has not been optimal, and the integration between mainly pro-environment strategy into growth strategy, poverty reduction and expansion of employment opportunities has not yet occurred; (3) Indonesia can no longer rely on growth from the primary natural resource sector, but has to step on the secondary sector which has high added value to sustain ever-increasing growth rate along with the increasing community needs; (4) Green economy needs to be built to meet consumer needs in the production sector, not only primary sector, but also secondary and tertiary sectors as well as accomodating graduates and educated workforce (and no longer want to work in the primary sector).

Considering the four reasons mentioned above, Indonesia must step on significant sustainable development and develop a green economy that relies on resource efficiency and a more environmentally friendly economic structure. Law No. 32 of 2009 concerning Environmental Protection and Management has been applied in various forms of licensing and regulations. For example, sanctions have been imposed on individuals performing business activities without environmental permits.

The government's commitment to reducing greenhouse gas emissions is implemented with the issuance of Presidential Regulation No. 61 of 2011 concerning the National Action Plan for Reducing Greenhouse Gas Emissions. To implement the Reducing Greenhouse Gas Emissions, the Ministry of National Development Planning/National Development Planning Agency formed the Climate Change Coordination Team, which consists of ministries/institutions related to the implementation of Reducing Greenhouse Gas Emissions. To implement the commitment at the regional level, the program was prepared by the provincial government in accordance with Presidential Regulation No. 61 of 2011 concerning the National Action Plan for Reducing Greenhouse Gas Emissions.

To this day, there has been no legislation in agriculture whose regulations can regulate all agricultural activities in Indonesia to meet the criteria of environmentally friendly agriculture. Concerning Green Economy in the agriculture sector, Law No. 41 of 2009 on the Protection of Sustainable Food Crops Farmland has been an excellent first step. Article 33, paragraph (1) of this law stipulates that the utilization of agricultural land designed as Sustainable Food Crops Farmland shall ensure land and water conservation. The following paragraph states that the government and regional governments are responsible for implementing land and water conservation, including protection of land and water resources, preservation of land and water resources; land and water quality management; and pollution control. Article 34 of this Law stipulates that any person with land titles determined as Sustainable Food Crops Farmland must utilize the land according to its designation and prevent irrigation damage. Owners of these agricultural land titles must maintain and increase soil fertility; prevent land damage, and preserve the environment. Specifically, the principle of sustainability in oil palm plantations began to be implemented with the implementation of Indonesian Sustainable Palm Oil (ISPO) established through Regulation of the Minister of Agriculture No.19/Permentan/0T.140/3/2011 concerning Guidelines on Indonesian Sustainable Palm Oil (ISPO). Specifically, the principle of sustainability in oil palm plantations began to be implemented with the implementation of Indonesian Sustainable Palm Oil (ISPO) established through Regulation of the Minister of Agriculture No.19/Permentan/0T.140/3/2011 concerning Guidelines on Indonesian Sustainable Palm Oil (ISPO).

The implementation of ISPO aims to ensure that the management of palm oil plantations meets the Green Economy principles. Under the such provision, any oil palm plantation shall possess ISPO certification. This means that the implementation of ISPO is mandatory. With the implementation of ISPO, Indonesia has standard provisions for sustainable palm oil plantations. This step is essential for adopting sustainable production and the beginning of Indonesia's oil palm diplomacy on the world market. In accordance with the concepts and policies of the rapidly developing green economy in Indonesia, the agricultural system is an organic farming system. Indonesia started to implement go-organic in May 2010. According to the Indonesian Organic Alliance (AOI), the total area of organic agricultural land in Indonesia in 2010 increased 10 percent from the previous year. In 2009, Indonesia's total area of organic agricultural land was 231.687,11 ha. Organic farming has been developed by the community, NGO's and government and attracted more consumers. However, the development of organic agriculture in Indonesia has been prolonged. This is due to many obstacles, including market constraints, and many small farmers think the organic certification process is complicated.

In addition, developing a green economy needs to be used as an opportunity to prioritize high-value local varieties. This is necessary considering that, currently, the availability of seeds is dominated by imported seeds. A green economy needs to be implemented to improve Indonesia's agricultural identity, implement sustainable agriculture, help overcome farmer poverty and realize food sovereignty. Among the challenges faced in expanding the implementation of a green economy in the agriculture sector, the decentralization issue is one of the toughest challenges. Decentralization impedes institutional efforts of farmers, for example, regarding farmer independence. The decentralized system has eliminated the production movement nationally as in the past. Politically, farmers have been weakened because farmer organizations do not entirely belong to the farmers. It has not been proven that farmer organizations can prioritize issues and outcomes benefiting smallholder farmers.

A decentralized system also weakens the role of agricultural extension workers, which is currently more necessary than during the centralized system. With a decentralized system, agriculture is only an optional assignment. The main priority is common infrastructure and not improving farmers' welfare, although infrastructure may directly impact mobility and reduce transportation costs.

The need to increase production by using high-yield varieties has developed the seed industry to continue to seek new seeds. At the same time, with the decentralized Center/Technical Implementation Unit (UPT), the seeding business that used to be the backbone of the distribution of superior seeds has closed down and is not functioning. Then today, the growth of multinational/global industries to continue to grow and take over markets requires the availability of superior seeds. As a result, the seed breeding system by seed breeders is marginalized. They can serve small-scale production, but the widespread implementation of a green economy requires revitalizing the national seed system.

The green economy is vital to be implemented in the forestry sector, given the importance of forests being managed for other purposes. Indonesia's forests have experienced degradation and deforestation. The role of forests in contributing to climate change is listed in Forestry Law No.41/1999 and its amendment, namely Law No.19/2004, including carbon sequestration efforts constituting utilization of forest environmental services. Technically, carbon sequestration has been regulated in the Forestry Minister Regulation No.P.36/Menhut-II/2009 concerning Procedure for Licensing of Commercial Utilization of Carbon Sequestration and Storage of Carbon in Production and Protected Forests issued as implementing regulation of Article 33 paragraph (1), Government Regulation No. 6 /2007 in conjunction with Government Regulation No. 3/2008 concerning Forest Management and Making Forest Management Plan, and Forest Utilization, stated in one of the clauses that: *"one form of environmental services utilization in production and the protected forest is sequestration and storage of carbon."*

Sustainable Forest Management is an integral part of the forestry sector in the green economy. Forest sustainability will only be achieved if forest management is carried out correctly. According to Law No. 41 of 1999, article 10 paragraph, forest administration (management) shall include (1) forestry planning; (2) forest management; (3) research and development, education and training, as well as counseling in forestry and supervision. Sustainable forest management is developed through various efforts, including: (a) strengthening of forest management unit; (b) implementation of timber certification system; (c) reforestation of forest area, restoration of degraded forests, and expansion of community forest; (d) development of plantation forest.

To realize sustainable forest management, the government divided forest management areas in Indonesia into several Forest Management Units (FMU). The establishment of FMU is clearly stated in Government Regulation No. 44/2004 concerning Forestry Planning and Government Regulation No. 3/2008 concerning Forest Management and Making Forest Management Plan and Forest Utilization. Two (2) essential elements of FMU establishment are: (i) the existence of areas or areas managed by FMU; and (ii) organization. The Minister of Forestry determines FMU. This site-level organization is responsible for forest management implementation. In addition, FMU shall be managed by professional human resources in the forestry sector. The determination of FMU areas which is the authority of the Minister of Forestry can be evaluated for the sake of efficiency and effectiveness and whether there is a change in spatial planning. The Forestry Minister Regulation No.P.42/Menhut-II/2009 concerning General Patterns, Criteria, and Standards for Integrated River Basin (DAS) Management has also determined the utilization of forest hydrological services.

The challenges to realizing the potential for sustainable/green economic development are: (1) arrangement of forest area boundaries that remain low, with numerous "open access"; (2) stretches of forest that shall be physically monitored, especially to implement extensive sustainable forest management; (3) incomplete instruments for managing forests sustainably, for example, maps with sufficient scales, arrangement of authority limit; (4) number and capacity of human resources to manage and supervise as well as carry out law enforcement over the area management to comply with applicable regulations and following sustainable principles.

The limited rate of boundary arrangement has resulted in 43.9 million ha of production forests status as ‘open access because they have not been granted concessions rights to certain management units so that the production forests are vulnerable to encroachment and illegal logging. The plan for establishing an FMU is to establish a monitoring "unit" as a measure of adequate control range. The establishment of FMU is also carried out to share supervisory authority with the regions. Without forest management at the site level through FMU, forest destruction will be brutal to stop, and forest rehabilitation and critical land will be hard to monitor for its progress and accountability. However, in terms of the government budget and sufficient human resources capacity, technical capabilities only generated 120 out of 600 FMUs needed until 2014. Breakthrough steps and firm commitments are needed to realize an adequate number of FMUs and implement sustainable forest management through the FMU.

Mining policies in Indonesia encouraging low carbon are carried out starting from the initial licensing process, process, and post-mining. According to Law No. 4/2009 concerning Mining Management, mining in Indonesia shall be managed based on the principles of benefit, justice, and equitable distribution; participation, transparency, and accountability; favoring the nation's interest; sustainability. Moreover, care for the environment. The mining business activities shall: (i) be effective, efficient, and competitive; (ii) support and promote the national capacity to be more capable of competing nationally, regionally, and internationally; and (iii) guarantee raw materials and sources of energy for domestic need.

Regarding post-mining of green mining efforts, Law 4/2009 in Article 99 regulates that: (i) any holder of IUP and IUPK shall be required to submit reclamation and post-mining plans on the application for IUP Production Operational; and (ii) the reclamation and post-mining activities shall be executed following the designation of the post-mining area. Post-mining management regulation is also strengthened by Government Regulation No. 78/2010 concerning Reclamation and Post-Mining, with environmental principles embodied in the policies so that post-mining shall at a minimum cover: (i) protection of the quality of surface water, groundwater, seawater, and land and the air; (ii) protection of bio-variety; (iii) guarantee on the stability and safety of closing rock piles, tailing pools, former mine areas, and other artificial structures; (iv) utilization of former mine areas following their allotment; and (v) observing local social and cultural values; and protection of the quantity of groundwater. To make more significant savings, specifically in early 2011, the President issued Presidential Instruction No. 13 of 2011 on Energy and Water Saving to all government institutions both at the central and regional levels, including BMN/D (State-Owned Goods/Regional Owned Assets) to carry out measures and innovations in saving energy and water in the surrounding of their respective institutions.

The essential instructions include: (i) 20% electrical saving and 10% water saving, calculated from the average consumption of electricity and water in their respective surroundings within 6 (six) months before the issuance of Presidential Instruction; and (ii) 10% subsidized fuel saving, through the regulation of the use of subsidized fuel for vehicles within respective institution settings, and within BUMN and BUMD settings, carried out as long as non-subsidized fuel is available in their respective regions. Implementing a Green Economy in Indonesia's energy and mining sectors faces various problems. One of the problems is that decentralization has resulted in overlapping land claims and energy investment processes, and intransparent and uncertain licensing has resulted in the slow implementation of the investment.

4. Conclusion

The implementation of a directed and comprehensive green economy in Indonesia must be supported by Government policies to ensure the success of its implementation. During the United Indonesia Cabinet II, the National Medium-Term Development Plan (Rencana Pembangunan Jangka Menengah Nasional, RPJMN) for 2010-2014 was used as the basis for the sustainable development of the Indonesian economy. For example, the application of green budgeting to organizing financial management of government budgets and implementing green procurement in public policy. According to Prof. Otto Sumarwoto, there are several reasons for Indonesia to implement a Green Economy significantly and systematically: (1) Because there have been conflicts of interest between economic utilization and environmental preservation, and various natural resource management problems, such as depletion of natural resources for economic interests; reduced ability of the environment to neutralize pollutants; forest deforestation; unsustainable and non-environmentally friendly exploitation of mineral and marine resources; (2) The implementation of pro-growth, pro-job, pro-poor, pro-environment development strategies has not been optimal, and the integration between mainly pro-environment strategy into growth strategy, poverty reduction and expansion of employment opportunities has not yet occurred. (3) Indonesia can no longer rely on growth from the primary natural resource sector but has to step on the secondary sector, which has high added value, to sustain the ever-increasing growth rate along with the increasing community needs; (4) Green economy needs to be built to meet consumer needs in the production sector, not only the primary sector but also secondary and tertiary sectors as well as accommodating graduates and educated workforce (and no longer want to work in the primary sector).

Considering the four reasons mentioned above, Indonesia must step on significant sustainable development and develop a green economy that relies on resource efficiency and a more environmentally friendly economic structure. For example, Law No. 32 of 2009 concerning Environmental Protection and Management has been applied in various forms of licensing and regulations to consolidate the green economy implementation in Indonesia. For the implementation of the commitment at the provincial and municipal levels, the program had been prepared by the provincial government by Presidential Regulation No. 61 of 2011 concerning the National Action Plan for Reducing Greenhouse Gas Emissions.

Acknowledgments
The authors wish to acknowledge financial support from the Directorate of Research and Development (RISBANG) of the University of Indonesia (UI) and the Faculty of the Law University of Pelita Harapan (UPH) for providing resources to undertake the research and thus finish the manuscript, without whom this research work would not be finished without the supports of the University of Indonesia.

References
[1] APEC Leaders 2010 *The APEC Leaders' Growth Strategy*
[2] ESCAP 2005 Seoul Initiative on Green Growth 4 1–9
[3] Butt S 2014 The Position of International Law within the Indonesian Legal System *Emory Int. Law Rev.* 28 1
[4] Gautama S 2006 *Business Law in Indonesia* (Bandung: Citra Aditya Bakti)
[5] OECD Ministerial Council Meeting 2011 *Towards Green Growth*
[6] OECD Green Growth Strategy 2014 *Green Growth Indicators 2014*
[7] Soekanto S and Mamudji S 2018 *Penelitian Hukum Normatif Suatu Tinjauan Singkat* (Bandung: Rajawali Press)
[8] Huck W and Kurkin C 2018 Die UN-Sustainable Development Goals (SDGs) im transnationalen Mehrebenensystem *ZaöRV* 78 375–424
[9] World A and Conference B 2000 *Annual World Bank in Development Latin America and the Caribbean*
[10] Permatasari A 2011 *Analisis Wacana Media: Peta Kecenderungan Aktor Dalam Isu Moratorium* (Seri Kajian HuMa)
[11] Purnomo A 2011 *Menjaga Hutan Kita: Pro-Kontra Kebijakan Moratorium Hutan dan Gambut* (KPG)

IOP Conf. Series: Earth and Environmental Science 1111 (2022) 012088 doi:10.1088/1755-1315/1111/1/012088

Environmental Governance in Response to Pollution Threats from Dry Bulk Cargo Spill Accidents at Sea

Silmina Sabila[1]*, Suyud Warno Utomo[2] and Haruki Agustina[2]
[1] Environmental Science, School of Environmental Science, University of Indonesia, Jakarta, 10430, Indonesia, 0000-0002-3573-3111
[2] Environmental Science, School of Environmental Science, University of Indonesia, Jakarta, 10430, Indonesia
[1]*silminasabila@gmail.com, [2]suyud.si@ui.ac.id, [3]harukiagustina@gmail.com

Abstract. Dry bulk cargo involves in shipping accidents often overlooked as pollution threats. Some of them are classified as hazardous noxious substances (HNS), and the impacts are precarious due to their various type, properties, and behavior in the marine environment. Thus, it is pertinent to understand how current governance responds to this matter. First, environmental governance in legislation frameworks provided by global authorities is reviewed. Then, how do some regions and selected nations internalize this legislation with their legislation and institutional arrangements? Indeed, this includes comparative analysis to provide suggestions for addressing challenges faced by Indonesia. This paper aims to give an overview of environmental governance in global, regional, and selected nations in response to pollution threats from dry bulk cargo spills in accidents at sea. Fragmented coordination, lack of implementation capacity, and lack of regional coordination are identified. Indonesia should build a national contingency plan and enhance ASEAN coordination regarding this matter to ameliorate its environmental sustainability.

1. Introduction

Marine transportation is the backbone of world trade, with more than 80% of product distribution depending on it [1]. The increase of marine transportation activity along with economic growth brings concerns to environmental sustainability: ships as marine transportation give negative externalities in the form of marine pollution [2]. Ship accidents may become a source of pollution to the marine environment that could lead to domino effects with fatal consequences like oil spills [3], which is well known for their devastating impacts on the marine environment in both the short and long term [4]. Consequently, there is already proper management for oil spill combat methods, technologies for countermeasures [5,6], financial estimation of its degradation to the marine environment [7], frameworks, and future strategy [8].

However, there is another risk imperiling the marine environment from ship accidents, and it lacks attention compared to oil spills [9,10]. Dry bulk cargoes that are carried with ships could enter and give exposure directly to the marine environment. This type of cargo is also associated with its considerable amount and homogeneity; some are coal, iron ore, nickel ore, and cement [11]. Furthermore, it is prone to sinking because of its low dynamic stability [12]. Thus, if the ships are involved in an accident, this could pose a threat instantly at its location. It could be worse if that cargo is classified as *hazardous and noxious substances* (HNS). HNS is defined as any substances other than oil that, if introduced into the marine environment, are likely to create hazards to human health, harm living resources and other marine life, damage amenities, and interfere with other legitimate uses of the sea [13]. Main dry bulk cargo poses inherent threats to the marine environment: coal spills could reduce the growth and reproduction rate of fish, potentially impact the respiratory of coral fish and seagrass, insofar lethal effects on coral [14,15]. Due to various types of HNS and the nature of the marine environment, the actual impacts of pollution are still unclear [16]. In addition, cargo spill not only brings pollution threats but also cause damage to properties, economic loss, or even loss of life.

To address these problems, we need to see how environmental governance pertaining this matter. Indeed, this paper aims to overview environmental governance in response to dry bulk cargo spill accidents on a global scale, regionals, and nations. Furthermore, it is also considered to address challenges faced by Indonesia as one of the major exporters of dry bulk cargo coal in the global market share and developing country by offering suggestions to enhance the sustainability of its marine environment.

2. Methods

This study provides an overview of environmental governance on different scales: global, regional, and selected nations in response to dry bulk cargo spills. Environmental governance in this matter refers to how authorities provide legislation frameworks regarding preparedness, response, recovery, or compensation of dry bulk cargo spills in an accident and how institutional arrangements work. Data sources in this study are from available information found on the authority's official website, reports, article journals, and books. First, legislation frameworks provided by global authorities are reviewed. Then, how do some regions and selected nations work with their representative or institutional arrangements to internalize this framework to their legislation? Subsequently, comparative analysis to address challenges faced by the government of Indonesia pertains to the response to dry bulk cargo spill accidents at sea.

3. Results and Discussion

Bulk carrier is one of the significant ship types involved in accidents globally in 2010 – 2019, and the severity of these is affected by some factors [17]. The impact of this accident comes in different degrees and the extent to the environment (marine organisms, water quality), economic (property damage or cargo loss), and social (wounded, loss of life, human health, and activities). For instance, the inherent hazard of dry bulk cargo such as nickel ore, iron ore, and coal is prone to liquefy, resulting in loss of stability and accident [18]; coal also emits methane and potentially creates flame and explosion. In addition, these cargoes may contain variable amounts of trace metals associated with toxic elements. Depending on each type's biodegradation and bioaccumulation characteristics, they may accumulate in marine biota or sediment. It is noted that the real impact of this spill accident on marine biota and its environment is precarious because of lack of information following up after an accident. It not only harms marine biota, but it could also endanger human health, activities, and loss of life. Based on the limited information available, the total loss for bulk carrier casualty amounts to an average of 230,000 death weight tonnage per year and an average of nine lives lost annually from 2012 – 2019 [19].

3.1. Global Governance

International Maritime Organization (IMO), under the auspices of the United Nations, is responsible for setting international standards in shipping safety, security, and protection of marine environments. Some of their works include conventions: The International Convention of Prevention Pollution from Ships (MARPOL), Safety of Life at Sea (SOLAS); protocols: The Protocol on Preparedness, Response and CooperationCooperation to pollution Incidents by Hazardous and Noxious Substances (OPRC-HNS Protocols); codes: The International Maritime Solid Bulk (IMSBC) Codes. In short, some instruments that pertain to dry bulk cargo spill accidents are listed in **Table 1.**

Table 1. Instruments by global governance pertain to dry bulk cargo spill accidents.

Instrument	Description
UNCLOS	Law governing the world's ocean, all uses, and its resources.
SOLAS	Provisions for maritime and ship safety.
MARPOL Convention	Pollution prevention from ships in six annexes.
OPRC-HNS Protocol	Preparedness and response system for dealing with pollution incidents involving HNS.
IMSBC Code	Standard information and procedure of safe stowage, shipment, and dangers associated with solid bulk cargoes.
Hazard Evaluation Procedure	Criteria for evaluating hazards of chemicals to the marine environment from ships.

The United Nations Convention on the Law of the Sea (UNCLOS) is the foundation for the legislative framework of the sea around the world, including the management of the marine environment and its natural resources. HNS Convention addresses the compensation system for HNS incidents, but due to an insufficient number of participants, this does not impose. Then IMO developed and adopted practical guidance 2010 HNS Protocol that is OPRC – HNS Protocol. It covers preparedness and response to chemical pollution incidents. As of February 2022, 41 countries adopted this protocol, representing 52,57% of world tonnage [20]. While many countries do not participate in this legislation, each country may use its policy regarding this matter because protocols act as additional legislation. Each contracting parties of IMO conventions obligate to impose these conventions by facilitating through national regulation and providing technical guidance. In addition, each nation needs to impose sanctions if there is a violation [21].

IMSBC Codes is the extended mandatory provision under SOLAS for solid bulk cargoes but excludes grain bulk cargo because it is regulated under International Grain Code [22]. IMSBC Codes provide information on the dangers associated with cargoes and how to deal with the safety of stowage. This code categorizes cargo into three groups: Group A for cargo which may liquefy, Group B for cargo with chemical hazards, and Group C for cargo that is neither in Group A nor B. Even though IMSBC Codes contain provisions for bulk cargoes in solid form, classifications to determine substances as harmful to the marine environment lie in IMDG Codes.

As part of the UN, IMO also works with the Group of Experts on the Scientific Aspects of Marine Environmental Protection (GESAMP) with other UN agencies to provide a scientific-based approach to marine environmental policy. One of their works is the hazard evaluation procedure of chemical substances carried by ships with consideration of potential effects on the marine environment, such as interference with other uses of the sea, damage to living resources, human health, and reduction of amenities. Even though it was designed to help assist MARPOL implementation, it can also provide information on environmental impact assessments on a short and long-term basis. In addition, it assists authorities in analyzing responses during maritime emergencies by determining potential chemical-released [23].

It is worth noting that IMO has no rights over the national jurisdiction. All regulatory frameworks are voluntary, adopted by state parties, and only mandatory for contracting state parties. Capacities building responsibility for the implementation also lies in contracting parties, and they can also build their capabilities through bilateral or multilateral coordination.

3.2. Regional Governance
Maritime emergencies result in cargo spills, especially HNS, which could spread across a national boundary. Therefore some countries in the same area coordinate to enhance their capabilities and capacities. Governance in some regions in response to pollution threats from HNS spills is selected.

Atlantic Region Coastal Pollution Response (ARCOPOL) is a project that aims to improve maritime coastal pollution preparedness, response, and mitigation in Atlantic regions from oil and HNS spill. This project involves countries of the Atlantic Area: Ireland, the UK, Portugal, and Spain, with representatives from respective local and regional agencies and authorities, research centers, and international organizations. Five main scopes addressed contingency planning, HNS, modeling and decision support tools, environmental monitoring, and training and awareness. It is worth noting that ARCOPOL started in 2009 and developed with ARCOPOLplus and ARCOPOLplatform until 2015. Outcomes of these projects are accessible to their respective members on their websites [24,25].

The Western Mediterranean Region Marine Oil and HNS Pollution Cooperation (West MOPoCo) is a project to support namely Western Mediterranean Region – Algeria, France, Italy, Malta, Morocco, Spain, and Tunisia in collaboration with Monaco for strengthening cooperationcooperation in preparedness and response to oil and HNS marine pollution. This inter-regional effort consists of a representative of established authority from the region: Mediterranean Sea - Regional Marine Pollution Emergency Response Centre for the Mediterranean Sea (REMPEC); Greater North Sea and its approaches - Bonn Agreement; Baltic Sea - Helsinki Commission (HELCOM) resulting with comprehensive information from regulations, hazard classification, preparedness, response, post-spill management, until case studies pertaining HNS spill in forms of marine HNS response manual [26]. Moreover, the objective of this manual is to provide operational guidance for first responders and decision-makers during maritime incidents at sea or in ports involving oil and HNS. It also systematically considers the framework of international regulations by incorporating it into preparedness steps. This document is easily accessible through the website, so it benefits the purpose of sharing knowledge and may act as a benchmark for other regions.

On the other hand, The Association of Southeast Asian Nations (ASEAN), which consists of Brunei Darussalam, Cambodia, Indonesia, Lao DPR, Malaysia, Myanmar, Philippines, Singapore, Thailand, and Vietnam, still does not have a regional contingency plan in response to HNS spill, yet. However, they already have ASEAN regional contingency plan for oil spills [27] and a regional action plan for combating marine debris [28]. Moreover, only Malaysia and Singapore have ratified OPRC-HNS Protocol [29]. Indeed proves the lack of attention in ASEAN regarding this matter.

3.3. National Governance
The weak implementation of international regulatory frameworks in marine environmental governance comes from the national level [30]. Here we consider an overview of Australia's response to this matter as a benchmark, noting that Australia is a significant exporter of dry bulk commodities such as iron ore [11] and has already built a well-established response. In contrast, Indonesia is considered a significant coal exporter.

3.3.1 Australia
Australia already ratified OPRC-HNS Protocol, and it forced the government to have the capability in response to the emergency of an HNS spill at sea [31]. As a result, the government released a national plan for maritime environmental emergencies. This document includes a legal and administrative basis for national plan governance and management, prevention, preparation, response, and recovery actions needed in marine pollution incidents, including HNS spills [32]. Furthermore, lists of each international legislation, bilateral, and multilateral agreements that partake in this national plan are stated concisely and systematically, such as OPRC-HNS Protocol, MARPOL, and Pacific Islands Regional Marine Spill Contingency Plan (PACPLAN). It also visually includes a visual governance structure that gives clear coordination pictures between stakeholders, their roles, and responsibilities. The Australian Maritime Safety Authority (AMSA) manages the national plan and acts as representative of the Australian government in IMO.

In short, access to information provided by the government regarding this matter is easily accessible. A national contingency plan is comprehensive, integrated, and well-documented. Moreover, governance also releases a series of national emergency management handbooks. The annual review also states government compliance with international obligations and programs' effectiveness.

3.3.2 Indonesia

Indonesia has already ratified SOLAS, MARPOL 73/78, and IMSBC Code. However, there is no comprehensive national contingency plan for maritime emergencies regarding HNS spills. Consequently, we need to see its national legislation that covers response to this matter.

Fundamental of maritime governance in Indonesia is regulated under Law No. 32 of 2014 on Marine [33]. Meanwhile, national regulations regarding the response to dry bulk cargo (HNS) spills in accidents are regulated under the Ministry of Environment and Forestry (Regulation No. P.74/MENLHK/SETJEN/KUM.1/10/2019) and Ministry of Transportation (Regulation No. 58 of 2013 with amendments No. 39 of 2021). On the other hand, insurance on the safety of the marine environment from shipping is regulated under Government Law Number 31 of 2021, even though Indonesia has not yet ratified the HNS Convention and OPRC-HNS Protocol. This legislative framework is maybe sufficient for the government in this matter. However, implementation of all these regulations and post-monitoring following the accidents remain unclear because there is a lack of adequate information available from reports or scientific journals.

From organization and institutional arrangements perspectives based on Ikrami and Bernard's review of Indonesia's maritime governance, 21 state organs are related to maritime and ocean laws: 18 executive agencies, two judicial organs, and a *quasi*-judicial body [34]. Moreover, highlights rule pertains to the protection of the marine environment is regulated under some law, which is Government Regulation No. 19 of 1999 (the Control of Marine Pollution and Degradation), Law No.27 of 2007 as amended by Law No.1 of 2014 manages coastal areas and small islands, Law No.32 of 2009 (Protection and Management of the Environment), and Law No. 17 of 2008 on Shipping also contains protection of the marine environment. This hinted that there are issues of overlapping functions and powers in enforcing ocean and maritime laws between Indonesian government agencies; it is also discussed in the governance of marine protected areas (MPAs) and marine mammals [35,36]. Boundaries between their functions are not stated concisely. As stated by Browman and Stergiou, in many countries, there is a strong power division between some sector institutions and ministries; it reflects in the difference in coordination among sub-sector [37]. A lack of coordination between government agencies regarding response to spill accidents or marine governance, in general, could hinder the purpose of marine environmental protection.

In addition, the government faces problems with management capacity to implement regulations that have already been ratified because there are no technical or procedural rules yet [38]. National capacity and competency to enforce established policies must be reviewed to ensure that their policies follow international law [39]. Overall, the comparison highlights governance between Australia and Indonesia listed in **Table 2.**

Table 2. Comparison of environmental governance in response to dry bulk cargo HNS spill.

	Australia	Indonesia
National Instruments	SOLAS, MARPOL, OPRC-HNS Protocol, IMSBC Codes, national contingency plan	SOLAS, MARPOL, IMSBC Codes.
Response coordination	Integrated	Fragmented
Regional coordination	PACPLAN	-

IOP Conf. Series: Earth and Environmental Science 1111 (2022) 012088 doi:10.1088/1755-1315/1111/1/012088

4. Conclusion

This paper has presented an overview of environmental governance in response to pollution threats from dry bulk cargo spills in maritime accidents in global, regional, and nations. Moreover, consider addressing Indonesia's challenges in this issue to ensure sustainability in the marine environment. First, Indonesia needs to consider ratifying HNS Convention [40] and OPRC-HNS Protocol because one of the methods to manage marine pollution is through ratifying international legal instruments [41]. Then government must build national capacity by providing technical and procedural rules. After that, official authorities must create a national contingency plan to respond to HNS spills and maritime accidents. It should also consider incorporating and adjusting international regulations into national legislation frameworks concisely and systematically in the national plan. Building a national contingency plan will hopefully solve fragmented coordination between government agencies. Furthermore, the lack of national capacity in implementing international legislation could be addressed by strengthening coordination in the ASEAN region to enhance capabilities[42].

This study only covers matters regarding response to dry bulk cargo spills in maritime accidents, especially those that classify as HNS, and does not cover aspects of a specific type of dry bulk cargo or HNS. Further research on the effects of HNS spill in the short or long term on marine species and the quality of the marine environment is highly suggested. Post monitoring after spill accidents based on risk prioritization also needs to be considered. Then, an evaluation of policy implementation would give a real insight into optimizing the policy's purpose.

References

[1] UNCTAD 2019 *Review of Maritime Transport 2019* (United Nations Publications)

[2] Jägerbrand A K, Brutemark A, Barthel Svedén J, and Gren I M 2019 A review on the environmental impacts of shipping on aquatic and nearshore ecosystems *Sci. Total Environ.* **695**

[3] Wang L, Huang R, Shi W and Zhang C 2021 Domino effect in marine accidents: Evidence from temporal association rules *Transp. Policy* **103** 236–44

[4] Han Y, Nambi I M and Prabhakar Clement T 2018 Environmental impacts of the Chennai oil spill accident – A case study *Sci. Total Environ.* **626** 795–806

[5] Yang J, Wan J, Ma Y, Zhang J, and Hu Y 2020 Characterization analysis and identification of common marine oil spill types using hyperspectral remote sensing *Int. J. Remote Sens.* **41** 7163–85

[6] Tayeb A M, Farouq R, Mohamed O A and Tony M A 2020 Oil spill clean-up using combined sorbents: a comparative investigation and design aspects *Int. J. Environ. Anal. Chem.* **100** 311–23

[7] Chen J, Di Z, Shi J, Shu Y, Wan Z, Song L, and Zhang W 2020 Marine oil spill pollution causes and governance: A case study of Sanchi tanker collision and explosion *J. Clean. Prod.* **273** 122978

[8] Chen J, Zhang W, Wan Z, Li S, Huang T and Fei Y 2019 Oil spills from global tankers: Status review and future governance *J. Clean. Prod.* **227** 20–32

[9] Neuparth T, Moreira S, Santos M M and Reis-Henriques M A 2011 Hazardous and Noxious Substances (HNS) in the marine environment: Prioritizing HNS that pose major risk in a European context *Mar. Pollut. Bull.* **62** 21–8

[10] Galieriková A, Dávid A, Materna M and Mako P 2021 Study of Maritime Accidents with Hazardous Substances Involved : Comparison of HNS and Oil Behaviours in Marine Environment *Transportation Research Procedia* vol 55 (Elsevier BV) pp 1050–64

[11] Chen S, Meersman H, Voorde E Van de and Frouws K 2014 *Modelling and Forecasting in Dry Bulk Shipping* (New York, NY: Routledge)

[12] Chen J, Bian W, Wan Z, Yang Z, Zheng H and Wang P 2019 Identifying factors influencing

total-loss marine accidents in the world: Analysis and evaluation based on ship types and sea regions *Ocean Eng.* **191**
[13] IMO 2019 Protocol on Preparedness, Response and CooperationCooperation to pollution Incidents by Hazardous and Noxious Substances, 2000 (OPRC-HNS Protocol)
[14] Berry K L E, Hoogenboom M O, Brinkman D L, Burns K A and Negri A P 2017 Effects of coal contamination on early life history processes of a reef-building coral, Acropora tenuis *Mar. Pollut. Bull.* **114** 505–14
[15] Berry K L E, Hess S, Clark T D, Wenger A S, Hoogenboom M O and Negri A P 2021 Effects of suspended coal particles on gill structure and oxygen consumption rates in a coral reef fish *Mar. Pollut. Bull.* **169** 112459
[16] Grote M, Mazurek N, Gräbsch C, Zeilinger J, Le Floch S, Wahrendorf D S and Höfer T 2016 Dry bulk cargo shipping — An overlooked threat to the marine environment? *Mar. Pollut. Bull.* **110** 511–9
[17] Wang H, Liu Z, Wang X, Graham T and Wang J 2021 An analysis of factors affecting the severity of marine accidents *Reliab. Eng. Syst. Saf.* **210**
[18] Nwigwe T and Kiyokazu M 2022 Assessment of Factors Contributing to the Risks of Cargo Liquefication Accident in Bulk Carriers (A Qualitative Approach) *TransNav* **16** 149–52
[19] INTERCARGO 2022 *Bulk Carrier Casualty Report* (London)
[20] Markus T 2018 Handbook on Marine Environment Protection *Handbook on Marine Environment Protection* ed M Salomon and T Markus (Cham: Springer International Publishing) pp 545–62
[21] Zis T P V and Cullinane K 2020 The desulphurisation of shipping: Past, present and the future under a global cap *Transp. Res. Part D Transp. Environ.* **82** 102316
[22] IMO 2019 International Maritime Solid Bulk Cargoes (IMSBC) Code
[23] GESAMP 2019 *GESAMP Hazard evaluation procedure for chemicals carried by ships* (London)
[24] ARCOPOL 2015 ARCOPOL
[25] Perez P, Fernandez M L, Ayensa G, Montero P, Navarro L, Rodriguez I, Jimenez M D, Montilla P, Silva R, Miguel Y de, Santiago A, Gonzales C J, Reyes E, Fernandes R, Santos M M, Oliveira H, Tuohy T, Shanahan P, Kibble A, Harold P, Gioia R, Kirby M F, Sexto L, Novas A, Muttin F, Campbell R, Caumont P, Perisse D, Berry A, Cronin M, Louchart P and John D 2015 ARCOPOLplatform. Protecting the Atlantic European coast from maritime pollution
[26] Alcaro L, Brandt J, Giraud W, Mannozzi M and Nicolas-Kopec A 2021 *Marine HNS Response Manual Multi-Regional Bonn Agreement, HELCOM, REMPEC*
[27] IMO 2018 ASEAN Regional Oil Spill Contingency Plan adopted
[28] ASEAN 2021 ASEAN Regional Action Plan for Combating Marine Debris in the ASEAN Member States (2021 – 2025)
[29] IMO 2022 Status of Conventions
[30] Grip K 2017 International marine environmental governance: A review *Ambio* **46** 413–27
[31] Government A 2020 Hazardous and Noxious substances (HNS) training to help ships with chemical cargo accidents at sea
[32] Australian Maritime Safety Authority 2020 *National Plan for Maritime Environmental Emergencies - 2020 edition*
[33] Ardyantara D M 2019 The Harmonization of Maritime Law (UU No. 32 of 2014) for Indonesia's Sovereignty in Marine Resource Management Facing UN Policy about Area beyond National Jurisdiction *Proceedings of the 2nd International Conference on Indonesian Legal Studies (ICILS 2019)* (Paris, France: Atlantis Press)
[34] Ikrami H and Bernard L 2019 Indonesia's Maritime Governance *Korean J. Int. Comp. Law* **6** 134–71
[35] Dirhamsyah 2016 Setbacks in the development of marine protected areas in Indonesia *Aust. J. Marit. Ocean Aff.* **8** 87–100

[36] Sahri A, Liza P, Mustika K, Yoga H and Murk A J 2020 A critical review of marine mammal governance and protection in Indonesia *Mar. Policy* **117** 103893
[37] Browman H I and Stergiou K I 2004 Perspectives on ecosystem-based approaches to the management of marine resources *Mar. Ecol. Prog. Ser.*
[38] Purwendah E K 2019 Precautionary Principle Through the Obligations of Marine Insurance for Oil Tankers in the Indonesian Legal System *Proc. 1st Int. Conf. Islam. Dev. Stud.*
[39] Ikrami H and Bernard L 2018 Indonesia's Maritime Governance *Korean J. Int. Comp. Law* **6** 134–71
[40] Chika S and Imania N 2020 Evaluation Ratification HNS Convention in Indonesia *Int. J. Innov. Sci. Res. Technol.* **5** 726–32
[41] Yuni N 2020 Marine Pollution Ditinjau dari Perbandingan Praktik Negara Terhadap Instrumen Hukum Internasional *SIGn J. Huk.* **2** 55–71
[42] Hung P Van, Kim K S and Lee M 2019 Cooperative response to marine hazardous and noxious substances and oil spill incidents in the ASEAN region *Aust. J. Marit. Ocean Aff.* **11** 61–72

AUTHOR INDEX